Lecture Notes in Physics

For information about Vols. 1–151, please contact your bookseller or Springer-Verlag.

Vol. 152: Physics of Narrow Gap Semiconductors. Proceedings, 1981. Edited by E. Gornik, H. Heinrich and L. Palmetshofer. XIII, 485 pages. 1982.

Vol. 153: Mathematical Problems in Theoretical Physics. Proceedings, 1981. Edited by R. Schrader, R. Seiler, and D.A. Uhlenbrock. XII, 429 pages. 1982.

Vol. 154: Macroscopic Properties of Disordered Media. Proceedings, 1981. Edited by R. Burridge, S. Childress, and G. Papanicolaou. VII, 307 pages. 1982.

Vol. 155: Quantum Optics. Proceedings, 1981. Edited by C.A. Engelbrecht. VIII, 329 pages. 1982.

Vol. 156: Resonances in Heavy Ion Reactions. Proceedings, 1981. Edited by K.A. Eberhard. XII, 448 pages. 1982.

Vol. 157: P. Niyogi, Integral Equation Method in Transonic Flow. XI, 189 pages. 1982.

Vol. 158: Dynamics of Nuclear Fission and Related Collective Phenomena. Proceedings, 1981. Edited by P. David, T. Mayer-Kuckuk, and A. van der Woude. X, 462 pages. 1982.

Vol. 159: E. Seiler, Gauge Theories as a Problem of Constructive Quantum Field Theory and Statistical Mechanics. V, 192 pages. 1982.

Vol. 160: Unified Theories of Elementary Particles. Critical Assessment and Prospects. Proceedings, 1981. Edited by P. Breitenlohner and H.P. Dürr. VI, 217 pages. 1982.

Vol. 161: Interacting Bosons in Nuclei. Proceedings, 1981. Edited by J.S. Dehesa, J.M.G. Gomez, and J. Ros. V, 209 pages. 1982.

Vol. 162: Relativistic Action at a Distance: Classical and Quantum Aspects. Proceedings, 1981. Edited by J. Llosa. X, 263 pages. 1982.

Vol. 163: J.S. Darrozes, C. Francois, Mécanique des Fluides Incompressibles. XIX, 459 pages. 1982.

Vol. 164: Stability of Thermodynamic Systems. Proceedings, 1981. Edited by J. Casas-Vázquez and G. Lebon. VII, 321 pages. 1982.

Vol. 165: N. Mukunda, H. van Dam, L.C. Biedenharn, Relativistic Models of Extended Hadrons Obeying a Mass-Spin Trajectory Constraint. Edited by A. Böhm and J.D. Dollard. VI, 163 pages. 1982.

Vol. 166: Computer Simulation of Solids. Edited by C.R.A. Catlow and W.C. Mackrodt. XII, 320 pages. 1982.

Vol. 167: G. Fieck, Symmetry of Polycentric Systems. VI, 137 pages, 1982.

Vol. 168: Heavy-Ion Collisions. Proceedings, 1982. Edited by G. Madurga and M. Lozano. VI, 429 pages. 1982.

Vol. 169: K. Sundermeyer, Constrained Dynamics. IV, 318 pages. 1982.

Vol. 170: Eighth International Conference on Numerical Methods in Fluid Dynamics. Proceedings, 1982. Edited by E. Krause. X, 569 pages. 1982.

Vol. 171: Time-Dependent Hartree-Fock and Beyond. Proceedings, 1982. Edited by K. Goeke and P.-G. Reinhard. VIII, 426 pages. 1982.

Vol. 172: Ionic Liquids, Molten Salts and Polyelectrolytes. Proceedings, 1982. Edited by K.-H. Bennemann, F. Brouers, and D. Quitmann. VII, 253 pages. 1982.

Vol. 173: Stochastic Processes in Quantum Theory and Statistical Physics. Proceedings, 1981. Edited by S. Albeverio, Ph. Combe, and M. Sirugue-Collin. VIII, 337 pages. 1982.

Vol. 174: A. Kadić, D.G.B. Edelen, A Gauge Theory of Dislocations and Disclinations. VII, 290 pages. 1983.

Vol. 175: Defect Complexes in Semiconductor Structures. Proceedings, 1982. Edited by J. Giber, F. Beleznay, J.C. Szép, and J. László. VI, 308 pages. 1983.

Vol. 176: Gauge Theory and Gravitation. Proceedings, 1982. Edited by K. Kikkawa, N. Nakanishi, and H. Nariai. X, 316 pages. 1983.

Vol. 177: Application of High Magnetic Fields in Semiconductor Physics. Proceedings, 1982. Edited by G. Landwehr. XII, 552 pages. 1983.

Vol. 178: Detectors in Heavy-Ion Reactions. Proceedings, 1982. Edited by W. von Oertzen. VIII, 258 pages. 1983.

Vol. 179: Dynamical Systems and Chaos. Proceedings, 1982. Edited by L. Garrido. XIV, 298 pages. 1983.

Vol. 180: Group Theoretical Methods in Physics. Proceedings, 1982. Edited by M. Serdaroğlu and E. Ínönü. XI, 569 pages. 1983.

Vol. 181: Gauge Theories of the Eighties. Proceedings, 1982. Edited by R. Raitio and J. Lindfors. V, 644 pages. 1983.

Vol. 182: Laser Physics. Proceedings, 1983. Edited by J.D. Harvey and D.F. Walls. V, 263 pages. 1983.

Vol. 183: J.D. Gunton, M. Droz, Introduction to the Theory of Metastable and Unstable States. VI, 140 pages. 1983.

Vol. 184: Stochastic Processes – Formalism and Applications. Proceedings, 1982. Edited by G.S. Agarwal and S. Dattagupta. VI, 324 pages. 1983.

Vol. 185: H.N. Shirer, R. Wells, Mathematical Structure of the Singularities at the Transitions between Steady States in Hydrodynamic Systems. XI, 276 pages. 1983.

Vol. 186: Critical Phenomena. Proceedings, 1982. Edited by F.J.W. Hahne. VII, 353 pages. 1983.

Vol. 187: Density Functional Theory. Edited by J. Keller and J.L. Gázquez. V, 301 pages. 1983.

Vol. 188: A.P. Balachandran, G. Marmo, B.-S. Skagerstam, A. Stern, Gauge Symmetries and Fibre Bundles. IV, 140 pages. 1983.

Vol. 189: Nonlinear Phenomena. Proceedings, 1982. Edited by K.B. Wolf. XII, 453 pages. 1983.

Vol. 190: K. Kraus, States, Effects, and Operations. Edited by A. Böhm, J.W. Dollard and W.H. Wootters. IX, 151 pages. 1983.

Vol. 191: Photon Photon Collisions. Proceedings, 1983. Edited by Ch. Berger. V, 417 pages. 1983.

Vol. 192: Heidelberg Colloquium on Spin Glasses. Proceedings, 1983. Edited by J.L. van Hemmen and I. Morgenstern. VII, 356 pages. 1983.

Vol. 193: Cool Stars, Stellar Systems, and the Sun. Proceedings, 1983. Edited by S.L. Balliunas and L. Hartmann. VII, 364 pages. 1984.

Vol. 194: P. Pascual, R. Tarrach, QCD: Renormalization for the Practitioner. V, 277 pages. 1984.

Lecture Notes in Physics

218

Ninth International Conference on Numerical Methods in Fluid Dynamics

Edited by Soubbaramayer and J.P. Boujot

Springer-Verlag
Berlin Heidelberg GmbH 1985

Editors

Soubbaramayer
C.E.N.-Saclay, Departement de Physicochimie
F-91191 Gif-sur-Yvette, France

J. P. Boujot
C.I.S.I.
B.P. 24, F-91190 Gif-sur-Yvette, France

ISBN 978-3-540-13917-1 ISBN 978-3-540-39144-9 (eBook)
DOI 10.1007/978-3-540-39144-9

Originally published by Springer-Verlag Berlin Heidelberg New York Tokyo in 1985

2153/3140-543210

Editors' Preface

This volume contains the papers presented at the Ninth International Conference on Numerical Methods in Fluid Dynamics, held at the Centre d'Etudes Nucléaires de Saclay in France, June 25-29, 1984. The papers presented were selected from abstracts submitted from all over the world, by three papers selection committees, one in the USA, another in the USSR and the third in Europe. The papers selection committees were headed by M. HOLT (USA), the late N.N. YANENKO (USSR) and R. TEMAM (Europe).

The book includes the welcome talk by M. COMMELIN, the inaugural lecture by J.L. LIONS, the invited lectures by R. DAUTRAY, S.C.R. DENNIS, C.A.J. FLETCHER, D. GOTTLIEB and S.G. RUBIN, plus contributed papers arranged in alphabetical order of the first author's name.

The conference was attended by over 270 scientists. In addition to the strong representation from France, the participation of delegations from the USA, West Germany, United Kingdom, Netherlands, China, Israel, Belgium, Sweden, Italy, Switzerland, USSR, Czechoslovakia, Canada, Brazil, Australia, Japan, Taiwan, Algeria and Sudan, showed the continuously increasing interest in this conference throughout the world.

The editors served as the general conference co-chairmen. We are indebted to our many colleagues who helped with the details of the meeting, but especially to Jacqueline BLOCH, who coordinated all of the local arrangements and to Martine MOITIE, the conference secretary.

Financial support for the conference was provided by

Commissariat à l'Energie Atomique
Compagnie Internationale de Services en Informatique
Direction des Recherches et Etudes Techniques
Centre National d'Etudes Spatiales
Electricité de France
Framatome - Novatome

We are indebted to Dr. W. BEIGLBÖCK and C. PENDL for valuable assistance in preparing these proceedings.

September 1984.

SOUBBARAMAYER JP.BOUJOT

(Editors)

IC9NMFD - ALLOCUTION D'ACCUEIL DE MONSIEUR COMMELIN

Directeur-Adjoint du Centre d'Etudes Nucléaires de Saclay

Messieurs les Présidents,
Mesdames, Messieurs,

Je suis heureux d'accueillir au Centre d'Etudes Nucléaires de Saclay les participants à la 9ème Conférence Internationale sur la Mécanique des Fluides. Il me semble que le choix de ce centre de recherche qui a maintenant plus de trente ans d'existence, pour la tenue de vos travaux n'est pas entièrement fortuit.

Le recours aux méthodes numériques pour résoudre certains types de problèmes est classique en Mécanique des Fluides. Ceci est dû à la complexité même des équations générales de la Mécanique des Fluides qui ne se prêtent pas toujours à une solution analytique exacte. Cette méthode d'approche numérique s'est intensifiée depuis quelques années, facilitée ou motivée par deux phénomènes importants :

1) Développement des techniques de l'analyse numérique, associé d'ailleurs à une progression technologique considérable des grands ordinateurs scientifiques IBM, CDC, CRAY, etc...

Or, vous trouvez précisément dans les laboratoires qui constituent ce centre de près de 8 000 personnes, un réseau informatique très élaboré, et notamment à la CISI, ces gros ordinateurs.

2) Des utilisateurs appartenant à des branches nouvelles (scientifiques, techniques ou industrielles) très différentes se trouvent confrontés à des problèmes de la même classe "Fluide" et pour lesquels ils ont besoin de solutions concrètes. Je citerai parmi "les utilisateurs", bien sûr l'aviation et l'espace, mais aussi l'océanographie, la météorologie, la marine, l'armée de l'air, le génie nucléaire, donc des horizons très différents.

Comme vous pourrez le constater en visitant nos installations, ce sont des thèmes de recherche sur lesquels différentes équipes du Centre de Saclay sont aujourd'hui orientées.

La forte participation active de nos ingénieurs et chercheurs venant non seulement de Saclay mais aussi de Limeil, Bruyères-le-Châtel, Grenoble, Cadarache, etc ... est d'ailleurs la preuve de l'intérêt que présente pour la poursuite des programmes du CEA le développement de l'analyse numérique.

Il me reste maintenant à vous inviter à visiter quelques unes des installations du Centre dans l'après-midi de mercredi que les organisateurs du congrès ont réservé à cette intention.

Je vous souhaite de bons et fructueux travaux en formant le voeu que dans les annales de vos assises le congrès de Saclay 1984 ait gardé l'image d'un plein succès.

CONTENTS

INAUGURAL TALK

INVITED LECTURES

CONTRIBUTED PAPERS

REMARKS ON APPROXIMATION SCHEMES

J.J. LIONS
Collège de France.
PARIS

INTRODUCTION.

In this short paper, we want to give some indications on two classes of methods in the Numerical Analysis of "Large Systems".

The first method comes from homogenization theory ; in Homogenization theory one deals with a (very) complicated material, (a composite material, a perforated material, a porous media etc...) and one wants to "replace" this complicated material by a simpler one -the homogenized material- In order to obtain, in a constructive manner, the coefficients (the so-called effective coefficients) of the homogenized material, we use asymptotic expansion techniques. In Section 1 we briefly indicate how these techniques may be thought of as related to classical "splitting up" or "fractional steps" methods.

The second class of methods comes from optimal control theory. It seems to us that using (simple) ideas form control theory of distributed systems one can obtain useful algorithms as it has already been demonstrated in some works (we refer to R. GLOWINSKI and J. PERIAUX [1] and to the Bibliography there in). Some indications on these ideas are given in Section 2.

1. SPLITTING UP AND HOMOGENIZATION.

Let us consider, in a formal manner, the evolution equation

$$\frac{\partial u}{\partial t} + A(u) = f \tag{1.1}$$

where A is an unbounded operator acting on functions (or vector functions) $u(x,t)$; $x \in \Omega, \Omega$ being an open set in $\mathbb{R}^n$ ($n = 1,2,3$ in most -but not all- of the applications), t = time > 0. The solution $u = u(x,t)$ of (1.1) is subject to boundary conditions on $\partial\Omega = \Gamma$, $t > 0$, and to initial conditions at $t = 0$; we shall assume for simplicity that

$$u(x,0) = 0 \quad \text{in } \Omega , \tag{1.2}$$

and we shall not make precise the boundary conditions.

It often happens, in a large variety of applications, that A -that we shall assume to be linear, but this is by no means essential- appears "naturally" in the form

$$(1.3) \qquad A = A_1 + \dots + A_q$$

where the A_j's are "simpler" operators.

We are then interested in finding approximation's schmemes for the equation

$$(1.4) \qquad \frac{\partial u}{\partial t} + (A_1 + \dots + A_q)u = f$$

subject to (1.2) and to appropriate boundary conditions. □

Let us now introduce -and this is admittedly artificial at this stage- functions

$$(1.5) \qquad m_1(\tau), m_2(\tau), \dots, m_q(\tau)$$

which are periodic in τ, with period 1, piecewise continuous and such that

$$(1.6) \qquad m_j(\tau) \geq 0, \int_0^1 m_j(\tau)d\tau = 1, \quad j = 1,\dots,q.$$

We then consider the equation

$$(1.7) \qquad \frac{\partial u_\varepsilon}{\partial t} + \sum_{j=1}^{q} m_j(t/\varepsilon)\, A_j u_\varepsilon = f,$$

where u_ε is subject to the same appropriate boundary conditions than u (it is in fact slightly more complicated, since the m_j's can be zero on some intervals), and u_ε is also subject to

$$(1.8) \qquad u_\varepsilon(x,0) = 0 \quad \text{in } \Omega.$$ □

<u>We claim that u_ε is an approximation of u</u>. Let us convince ourselves of this fact, in the following fashion. We look for u_ε in the form (ansatz)

$$(1.9) \qquad u_\varepsilon = u^0 + \varepsilon\, u^1 + \dots$$

where

$$(1.10)\quad \left|\begin{array}{l} u^j = u^j(x,t,\tau) \quad \text{is defined for} \\ x \in \Omega,\ t > 0,\ \tau \in \mathbb{R},\ u^j \text{ is periodic in } \tau \\ \text{with period 1} \end{array}\right.$$

and where in the expansion (1.9) we replace τ by t/ε.

Remark 1.1. The ansatz given by (1.9) corresponds to multi-time asymptotic expansions. We are here thinking in terms of periodic structures ; cf. A. BENSOUSSAN, J.L. LIONS, G. PAPANICOLAOU [1]. □

We notice that, if $\tau = t/\varepsilon$,

$$(1.11)\quad \frac{\partial}{\partial t}(u^j(x,t,\tau)) = \varepsilon^{-1}\frac{\partial u^j}{\partial \tau} + \frac{\partial u^j}{\partial t}$$

where we replace τ by t/ε at the end of the computation. Therefore using (1.11) and (1.9) in (1.7), and by identifying terms in the ε-expansion, we find

$$(1.12)\quad \frac{\partial u^0}{\partial \tau} = 0$$

$$(1.13)\quad \frac{\partial u^1}{\partial \tau} + \frac{\partial u^0}{\partial t} + \Sigma\, m_j(\tau)\, A_j u^0 = f$$

$$(1.14)\quad \frac{\partial u^2}{\partial \tau} + \frac{\partial u^1}{\partial t} + \Sigma\, m_j(\tau)\, A_j u^1 = 0.$$

But (1.12) means that $u^0 = u^0(x,t)$ does not depend on τ; equation (1.13) admits a solution u^1 periodic (with period 1) in τ iff

$$(1.15)\quad \int_0^1 \left(\frac{\partial u^0}{\partial t} + \Sigma\, m_j(\tau)\, A_j u^0\right)d\tau = \int_0^1 f\, d\tau = f$$

Using (1.6), (1.15) reduces to

$$(1.16)\quad \frac{\partial u^0}{\partial t} + \sum_{j=1}^{q} A_j u^0 = f$$

where

$$(1.17)\quad u^0(x,0) = 0,$$

i.e. $\mathbf{u}^0 = u$. This explains why u_ε is indeed an "approximation" of u (the above procedure is formal, but it can be justified, under suitable hypothesis on the A_j's). □

Let us now indicate how all this is related to splitting up (or fractional steps) methods.

We choose m_j on [0,1] by

(1.18) $$m_j(\tau) = q \text{ for } \frac{j-1}{q} < \tau \le \frac{j}{q},\ 0 \text{ outside}$$

and we extend m_j to $\mathbb{R}$ in a periodic manner.

Let us take the implicit approximation scheme for (1.7) which is given by

(1.19) $$\left|\begin{array}{l} \dfrac{u_\varepsilon(\varepsilon^{k/q}) - u_\varepsilon(\varepsilon\,\frac{(k-1)}{q})}{(\frac{\varepsilon}{q})} + \\[2ex] + \displaystyle\sum_{j=1}^{q} m_j(\tfrac{k}{q})\, A_j\, u_\varepsilon(\varepsilon k/q) = f(\tfrac{k\varepsilon}{q}), \\[2ex] k = 1,\dots,q, \end{array}\right.$$

where $u_\varepsilon(0) \doteq 0$.

Let us set

$$\varepsilon = \Delta t,$$

$$u_\varepsilon(\tfrac{\varepsilon k}{q}) = u^{k/q}\ .$$

We observe that $m_j(k/q) = 0$ if $j \neq k$, $= q$ if $j = k$. Therefore (1.19) reduces to

(1.20) $$\left|\begin{array}{l} \dfrac{u^{k/q} - u^{(k-1)/q}}{\Delta t} + A_k\, u^{k/q} = \dfrac{1}{q}\, f\,(k\,\frac{\Delta t}{q}) \\[2ex] k = 1,\dots,q, \\[2ex] u^0 = 0. \end{array}\right.$$

We then proceed and define in this way $u^{1+k/q},\dots,u^{n+k/q}$.

This is one of the classical fractional step methods.

Remark 1.2. This approach allows a lot of flexibility. Indeed :

(i) we can take for the m_j's continuous smooth functions. Then the schemes are more complicated but the approximation is smoother ;

(ii) we can take for the m_j's random functions. □

Remark 1.3. As we indicated in Remark 1.1, the ansatz (1.9), (1.10) is reminiscent of the ansatz used in homogenization theory (for composite materials) ; cf. A.BENSOUSSAN, J.L. LIONS, G. PAPANICOLAOU [1], E. SANCHEZ-PALENCIA [1]. Similar techniques are also useful in perforated materials and in porous media (assuming a periodic structure) ; cf. J.L. LIONS [1] and the Bibliography therein. We also wish to point out the interesting work connecting ideas coming form homogenization theory to more classical approaches in turbulence theory ; we refer to O. PIRONNEAU [1] and to the Bibliography therein. □

2. ALGORITHMS AND OPTIMAL CONTROL.

Let us consider the problem of finding u, solution of

(2.1) $- \Delta u = F(u)$ in Ω, Ω open set of $\mathbb{R}^n$,

(2.2) $u = 0$ on $\partial\Omega = \Gamma$

where F is a non linear function from, say, $\mathbb{R} \to \mathbb{R}$.

Let us assume firstly that (2.1)(2.2) uniquely defines a solution u.

There are of course many methods giving approximation schemes for computing u.

We want to give here some indications on the possibilities of applying methods coming from the Optimal Control of distributed systems. □

Let v be a control variable (function) ; we shall make more precise below the function spaces where we consider v. Let $y = y(v)$ be the solution of the Dirichlet's problem

(2.3) $- \Delta y = F(v)$ in Ω,

(2.4) $y = 0$ on $\partial\Omega$.

Equation (2.3), subject to the boundary condition (2.4), admits a unique solution $y = y(v)$. We think of $y(v)$ as the state of the system.

We then introduce a cost function

(2.5) $J(v) = ||y(v) - v||_X$

where X is a Banach function space to be chosen, and $|| \ ||_X$ denotes the norm in the Banach space X.

A key problem is how to choose X. Let us admit, for the time being, that X is given.

We then consider the problem of optimal control

(2.6) $\inf . J(v), \quad v \in X.$

This problem admits a unique solution u (the optimal control), given by the solution of (2.1) (2.2), and of course

(2.7) $J(u) = 0.$

The question of finding approximation schemes for u, solution of (2.1)(2.2), is then "reduced" to the problem of approximating the optimal control, solution of (2.6). □

Remark 2.1. For the optimal control of distributed systems we refer to J.L. LIONS [2]. □

Remark 2.2. The (formal) approach given above is of course completely general, and applies to "all" possible systems of non linear partial differential equations. For applications, for instance, to Stokes problem, we refer to R. GLOWINSKI and O. PIRONNEAU [1] and the Bibliography therein. □

Remark 2.3. For obvious reasons, one will choose for X a Hilbert space.

But the choice of the Hilbert space is by no means unique and, as we already saw it is a crucial point.

We can for instance take

(2.8) $X = L^2(\Omega)$

Hence (we replace $\| \ \|_X$ by $\| \ \|_X^2$ in (2.5))

$$(2.9) \qquad J(v) = \int_\Omega (y(v) - v)^2 dx$$

A "smoother" cost function is given by taking

$$(2.10) \qquad X = H_0^1(\Omega) \text{ (}^1\text{)}$$

and

$$(2.11) \qquad J(v) = \int_\Omega |\nabla(y(v) - v)|^2 dx.$$

cf. J. CEA and G. GEYMONAT for algorithms based on (2.9) and R. GLOWINSKI and J.PERIAUX for algorithmes based on (2.11). □

Let us indicate now how these ideas can be, under suitable modifications, be applied to <u>multi state systems</u>. Let us return to (2.1)(2.2) but now <u>without</u> the assumption that it admits a <u>unique</u> solution. For instance let us consider

$$(2.12) \qquad -\Delta u = u^3 + f, \quad \text{in } \Omega \subset \mathbb{R}^3,$$

u subject to (2.2), where f is given in, say, $L^2(\Omega)$. It has been shown by several authors that (2.12) admits in general an <u>infinite number of solutions</u>. The question is then to find either some branches of solutions or a solution which is "as close as possible" from a given function (chosen for physical reasons).

We use here ideas coming from the theory of <u>Singular</u> Distributed Systems, as in J.L. LIONS [3]. We take

$$(2.13) \qquad v \in L^6(\Omega)$$

and we define <u>the state</u> y(v) of the system by

$$(2.14) \qquad \left| \begin{array}{l} -\Delta y = v^3 + f, \\ \\ y = 0 \quad \text{on } \partial\Omega \end{array} \right.$$

which admits a unique solution $y = y(v)$. We introduce next the <u>cost function</u>

---

(1) $H_0^1(\Omega)$ denotes the (Sobolev) space of functions ϕ such that $\frac{\partial \phi}{\partial x_i} \in L^2(\Omega)$ and $\phi=0$ on Γ. One can take $\|\phi\|_X = (\int_\Omega |\nabla\phi|^2 dx)^{1/2}$.

$$(2.15) \qquad J(v) = \| y(v) - \zeta \|^6_{L^6(\Omega)} + k \, \| y(v) - v \|^2_{L^2(\Omega)} , \quad k > 0$$

where ζ is given, and we look for

$$(2.16) \qquad \inf. \, J(v), \quad v \in L^6(\Omega).$$

Remark 2.4. One has here to use the space $L^6(\Omega)$ in the first part of the cost function in order the problem to make sense.

The parameter k is a penalty term ; if k is "large" then the term $k \, \|y(v)-v\|^2_{L^2(\Omega)}$ in (2.15) "obliges" y(v) to be "close" to v, hence y(v) to be "close" to a solution of (2.12).

Of course it is always preferable not to introduce "large" parameters in the computations. A way out of this difficulty is to use **augmented Lagrangian methods**. cf. M. FORTIN and R. GLOWINSKI [1] and J.L. LIONS [3]. □

Remark 2.5. Techniques arising from optimal control theory are also used in Meteorology. We refer to F.X. LE DIMET [1] and to the Bibliography therein. □

BIBLIOGRAPHY

A. BENSOUSSAN, J.L. LIONS, G. PAPANICOLAOU [1] Asymptotic Analysis for Periodic Structures. North Holland. Pub. 1978.

J. CEA et G. GEYMONAT [1] Une méthode de linéarisation via l'optimisation. Institut Naz. di Alta Mat., Symp. Math. 10, Bologna (1972), p. 431-451.

M. FORTIN and R. GLOWINSKI [1] Augmented Lagrangian methods : applications to the numerical solution of boundary value problems. North Holland. 1983.

R. GLOWINSKI et J. PERIAUX [1] Finite Element, Least squares and domain decomposition methods for the numerical solution of non linear problems in fluid dynamics. In Simulation Numérique en Mécanique des Fluides, INRIA, 1984, p.45-157.

R. GLOWINSKI and O. PIRONNEAU [1] On numerical methods for the Stokes problem, in Simulation Numérique en Mécanique des Fluides, INRIA, 1984, p. 159-188.

F.X. LE DIMET [1] A general formalism for variational Analysis in Meteorology. Tellus. 1984.

J.L. LIONS [1] Some methods in the Mathematical Analysis of Systems and their control. Science Press, Beijing 1981. Gordon Breach. 1981.

[2] Sur le contrôle Optimal des systèmes gouvernés par des équations aux dérivées partielles. Paris, Dunod. Gauthier Villars 1968 (English Translation by S.K. Mitter, Springer, 1971).

[3] Contrôle des systèmes distribués singuliers. Paris, Gauthier Villars. 1983.

O. PIRONNEAU [1] Simulation numérique de la Turbulence par homogénéisation des petites structures. Cours INRIA "Simulation Numérique en Mécanique des Fluides", 1984. p. 415-454.

E. SANCHEZ - PALENCIA [1] Non homogeneous Media and Vibration Theory. Lecture Notes in Physics. Springer Verlag, 127. 1980.

TOPICS IN THE NUMERICAL SIMULATION OF HIGH TEMPERATURE FLOWS

R. Chéret (1), R. Dautray, J.C. Desgraz, B. Mercier, G. Meurant, J. Ovadia, B. Sitt (2)

Abstract

We review some numerical methods used in the field of multifluid flows, radiation hydrodynamics, detonation and instability of related flows.

Introduction

In the fields of inertial confinement fusion, astrophysics, detonation, or other high energy phenomena, one has to deal with multifluid flows involving high temperatures, high speeds and strong shocks initiated e.g. by chemical reactions or even by thermonuclear reactions.

The complexity of such flows is a real challenge for computer simulation.
In the present paper, we review some of the methods we use in the field of computer simulation of multifluid flows, radiation hydrodynamics and detonation.

The outline of this paper is as follows.

Section 1 is devoted to the simulation of multifluid flows : we first review Lagrangian methods which have been successfully applied in the past. Then we describe our experience with newer adaptive mesh methods, originally designed to increase the accuracy of Lagrangian methods.

Finally, we recall some facts about Eulerian methods, with emphasis on the EAD scheme [1] which has been recently extended to the elasto-plastic case [2].

In section 2 we turn to high temperature flows described by the equations of radiation hydrodynamics. We show how one can ensure conservation of energy while solving the radiative transfer equation via the Monte Carlo method.

In section 3 devoted to detonation, we review some models introduced to describe the initiation of detonation in heterogeneous explosives.

Finally, in section 4, we say a few words about instability of these flows.

(1) Centre d'Etudes de Vaujours, B.P. 7, 93270 SEVRAN
(2) Centre d'Etudes de Limeil-Valenton, B.P. 27, 94190 VILLENEUVE-ST-GEORGES

1. MULTIFLUID FLOWS

a. Lagrangian methods

The most standard way of solving the system of conservation laws of mass, momentum, energy, completed with an equation of state, in the case of multifluid flows is to use Lagrangian methods.

The main feature of these methods is that the mesh follows material motion, so that the interface between two materials is fixed with respect to the mesh.

The procedure to update the mesh is straightforward when the degrees of freedom for the velocity field are chosen at the vertices of the cells, assumed to be quadrilaterals in 2-D. The solution of the momentum equation at time t_n gives the new velocity field from the old one, so that the coordinates of the vertices are updated according to formula

$$\frac{x^{n+1} - x^n}{\Delta t} = v^{n+1/2} \tag{1}$$

From the variation of volume of each cell, one is able to compute the new densities ρ^{n+1} from the old ones. Note that the density ρ is naturally cell centered. So are the other thermodynamical quantities p (pressure) and ε (internal energy), which are obtained by solving simultaneously, (and in an implicit way) the energy equation and the equation of state.

The source of one of the main disadvantages of Lagrangian methods is the need for an artificial viscosity. Velocities and internal energies are not centered at the same place in the mesh. Thus, we solve the internal energy equation which however is not in conservative form.

An artificial viscosity is then needed to take into account the entropy jump across a shock [4].

Another disadvantage of Lagrangian methods is of course the limitation due to distorsion of the mesh. In case of large shear strains, the method may actually fail because of twisted cells.

However Lagrangian methods have been widely used in the past, and have still a bright future for complex flows, when many pieces of information about the material have to be carried with the flow. In a Lagrangian method those pieces of information will always be attached to a given cell. Another advantage, we would like to point out, is that Lagrangian methods have a good resolution in regions of high compression.

b. Adaptive mesh methods

The purpose of adaptive mesh methods is to increase the accuracy of Lagrangian methods by using a mesh which has better approximation properties than the Lagrangian mesh.

Such a mesh should be sufficiently regular, since distorted meshes lead to a loss of accuracy, and also refined in the zones of strong gradient of the flow.

When a shock propagates in the material this might give a mesh which is much finer than the Lagrangian mesh in the neighborhood of the shock. The method can roughly be divided into 3 steps

(i) Lagrangian phase
(ii) construction of an appropriate mesh
(iii) remapping phase.

The Lagrangian phase is the same as the one described above. It starts from the physical quantities $v^{n-1/2}$, ρ^n, p^n, ε^n defined on an "old" mesh whose coordinates are called x^n, and leads to some new physical quantities $\tilde{v}^{n+1/2}$, $\tilde{\rho}^{n+1}$, $\tilde{p}^{n+1}$, $\tilde{\varepsilon}^{n+1}$ defined on a Lagrangian mesh such that

$$\tilde{x}^{n+1} = x^n + \Delta t\, \tilde{v}^{n+1/2}$$

as in formula (1).

Step ii) consists then of constructing an appropriate mesh, which is sufficiently regular but refined in some zones (e.g. the zones where the pressure gradient is large). The coordinates of the vertices of the adaptive mesh are denoted by x^{n+1}. Finally step (iii) starts from the physical quantities $\tilde{v}^{n+1/2}$, $\tilde{\rho}^{n+1}$, $\tilde{p}^{n+1}$, $\tilde{\varepsilon}^{n+1}$ defined on the Lagrangian mesh ($\tilde{x}^{n+1}$) and leads to the corresponding quantities $v^{n+1/2}$, ρ^{n+1}, p^{n+1}, ε^{n+1} on the adaptive mesh (x^{n+1}).

Note that if we choose $x^{n+1} = x^n$ for all n, then we obtain an Eulerian method.

Many methods have been proposed for step ii) in the proceedings of this conference. We also refer the reader to Brackbill-Salzmann [5] who proposed a method based on non linear optimization.

Let F denote the mapping from a fixed mesh to the current mesh, they minimize some functional

$$I \equiv I_R + \lambda_0 I_0 + \lambda_W I_W$$

where I_R is a regularity term :

$$I_R = \int_{\mathcal{D}} \left(|\mathrm{grad}\, \xi|^2 + |\mathrm{grad}\, \eta|^2 \right) dx\, dy$$

where $\mathcal{D}$ is the physical domain, x,y the coordinates of F, and ξ, η the coordinates on the fixed mesh ;

$$I_0 = \int_{\mathcal{D}} (\mathrm{grad}\, \xi \cdot \mathrm{grad}\, \eta)^2\, dx\, dy$$

is an orthogonality term, and

$$I_W = \int_{\mathcal{D}} W(x, y)\, J\, dx\, dy$$

is an adaptation term, with J denoting the Jacobian of F.

Finally λ_0 and λ_W are given positive constants. Since I is to be minimized, product WJ should never be too large. If W is large in some zone, then J should be small, which means a refinement of the mesh in that zone.

Practically, the weight function W may be chosen equal to $|\mathrm{grad}\, p|/p$ or to $|\mathrm{grad}\, \rho|/\rho$.

The remapping phase (step (iii) above) should not be underestimated. Combining conservative form and accuracy is indeed a difficult task. As an example, let us consider the remapping of density $\tilde{\rho}^{n+1}$ which is assumed piecewise constant on the Lagrangian mesh ($\tilde{x}^{n+1}$). To get a piecewise constant ρ^{n+1} on the new mesh x^{n+1}, a natural idea is to choose for ρ^{n+1} on a given cell the average of $\tilde{\rho}^{n+1}$ on this cell.

This process is obviously conservative, however it happens to be too much diffusive.

This is also true for the other physical quantities to be remapped. In particular, internal energy and momentum are conserved, but kinetic energy is eventually underestimated, leading to dissipation of energy.

To increase the accuracy of the remapping phase, Dukowicz [6] suggests a method which appears as an extension in 2-D of Van Leer's method [7]. Other methods have been proposed by Zalesak [8] and Bailey [9] in 2-D and by Boris-Book [10] and Woodward-Collela [11] in 1-D.

The idea of these methods lies in the fact that, for instance, $\tilde{\rho}^{n+1}$ being cell centered, it is possible to construct, from its values at the cell centers, a better approximation to the exact solution than the piecewise constant function used above. Van Leer and Dukowicz, as an example, use discontinuous piecewise linear functions. On the other hand, Woodward and Collela use continuous piecewise parabolic functions. The accuracy of such methods is second order with respect to the cell size Δx ; however to avoid the well known oscillations of 2nd order scheme, one has to be careful and accept to be only 1st order in some zones (see [7],[10],[11]).

We compare in Figure 1, the effect on a square density profile of 200 successive remappings with a Courant number of .4, which means that the square density profile is shifted $.4\Delta x$ further at each cycle.

In Figure 2, we compare a Lagrangian method to three adaptive mesh methods, on a shock tube problem defined by Sod [12].

We show the internal energy profiles at a given time to the exact solution. The first adaptive mesh is actually Eulerian, since the mesh is fixed. The second one corresponds to $W = (\text{grad } p/p)^2$ and $\lambda_W = 320$. The third one corresponds to $W = (\text{grad } \rho / \rho)^2$ and $\lambda_W = 185$. Note the good results obtained in the last two cases.

c. Eulerian methods

As we have said above, Lagrangian methods have difficulties to handle great deformations. The study of phenomena such as impacts blasts or jets cannot be performed with Lagrangian methods. Adaptive mesh methods are potentially the best to solve such problems. However tracking material interfaces on an arbitrary mesh is a difficult problem which we have not yet properly solved. This is why we have developed 2-D Eulerian multifluid methods.

Generally these schemes are first order accurate in space and time, then shocks are spread over three or four computational cells.

We have developped the E.A.D. (Eulerian with Anti Diffusion) algorithm [1] which computes 2D multifluid flows with second order accuracy in time and space. Hydrodynamic or elasto-plastic material behavior can be handled. This algorihm is able to compute flows involving strong shocks, rarefaction waves, detonations, free surfaces, material interfaces, and great deformations.

The main features of the EAD scheme are :

- The introduction of the "Flux Corrected Transport" technique [10] for cells which are filled by several materials.

- The "fictitious fluid" method to compute free surfaces, even for great deformations [13].

- The numerical procedure for elasto-plastic materials.

Main features of the discretisation

- We consider an orthogonal mesh, in 2-D plane or cylindrical geometry.

- Each eulerian cell can be filled by one or several materials. Each material is defined by an index α, volume V_α, density ρ_α, velocity $\vec{v}_\alpha = (u_\alpha, v_\alpha)$, total energy E_α, inside the cell at time t_0.

- The material interface tracking is computed with the SLIC method [14].

- The discretisation in time uses an alternating direction procedure, which leads to a decomposition in two half cycles, the first one in the x-direction, and the second one, in the y-direction.

- Each half-cycle is decomposed into two steps : we shall consider the first half-cycle in the x-direction only.

The first step is a lagrangian one

We compute velocity $u^{1/2}$, pressure $p^{1/2}$ at time $t_0 + \frac{\Delta t}{2}$.

The second step is decomposed into four stages :

. stage 1-we project all values on an intermediate mesh defined so that it divides each cell of the initial mesh into two equal volumes.

. stage 2-Lagrangian phase : we solve the conservative equations using the intermediate mesh, which is moving with velocity $u^{1/2}$ defined in the first step.

Stresses in the momentum and in the total energy equations are taken into account by solving the equations :

$$\frac{\partial \rho u}{\partial t} = \frac{\partial}{\partial x}\sigma_{xx} + \frac{\partial}{\partial y}\sigma_{xy} - \rho\, g_x$$

$$\frac{\partial \rho E}{\partial t} = \frac{\partial}{\partial x}(\sigma_{xx}u) + \frac{\partial}{\partial y}(\sigma_{xy}u) - \rho\, u\, g_x$$

where σ is the stress tensor and g is the gravitational acceleration.

Stress tensor σ is to be updated in the following way. Let

$$\tilde{\sigma} \equiv R\left[\sigma(t_0) + \Delta t\,(2\mu D + \lambda(\mathrm{div}\,\vec{v})\,\delta)\right] R^{-1}$$

where $R = \delta + \Omega\cdot\Delta t$ denotes the rotation tensor ; Ω (resp.D) denotes the skew symmetric (resp. symmetric) part of tensor $\nabla\vec{v}$, and finally λ, μ denote some elasticity coefficients.

The deviatoric part $\tilde{s} = \tilde{\sigma} - 1/3\ \mathrm{tr}(\tilde{\sigma})\delta$ of $\tilde{\sigma}$ is then computed.

In the elastic case $\mathrm{tr}(\tilde{s}^2) \leq \frac{2}{3} Y_0^2$ then $\sigma(t_0+\Delta t) = \tilde{\sigma}$.

In the plastic case, on the other hand, we let

$$\sigma\,(t_0+\Delta t) = -p\delta + s$$

where p is computed from the equation of state and $s = \theta\,\tilde{s}$ where θ is chosen such that $tr(s^2) = 2/3\, Y_0^2$.

- stage 3-We project all the physical quantities on the initial Eulerian mesh.
- stage 4-To obtain a second order accuracy, we add an antidiffusion term, like in the F.C.T. method.

Numerical results

To illustrate the current capabilities of the code three calculations are presented :

1. The same shock tube problem as before to see the accuracy of the method in the hydrodynamic case.
2. A shock propagation in a piece of aluminium to show the accuracy in the elasto-plastic case, extracted from [2].
3. An impact problem to show the capability of the method for great deformations, also extracted from [2].

2. RADIATION HYDRODYNAMICS

In most high temperature flows one gets high, though non relativistic, speeds : v/c, where c denotes the speed of light, is typically smaller than 1%.

However some relativistic terms in the radiation hydrodynamic equations have to be kept, as we shall see, in order to conserve energy.

Assuming the specific radiative intensity I to be given in the comoving frame (rather than in the laboratory frame), the equations of radiation hydrodynamics can be written in 1-D (see Buchler [15]).

(2) $\frac{D\rho}{Dt} + \rho\frac{\partial v}{\partial z} = 0$

(3) $\rho\frac{Dv}{Dt} + \frac{\partial}{\partial z}(p + P_R) = 0,$

(4) $\rho\left[\frac{D\varepsilon}{Dt} + p\frac{D}{Dt}\left(\frac{1}{\rho}\right)\right] = 2\pi\int_0^{\infty} d\nu \int_{-1}^{+1} (\chi I - S)\, d\mu,$

(5) $\rho\frac{D}{Dt}\left(\frac{I}{\rho}\right) + c\mu\frac{\partial I}{\partial z} + \frac{\partial}{\partial \mu}(aI) - \frac{\partial}{\partial \nu}(\nu g I) + (g + c\chi)I = cS,$

where

$$P_R = \frac{2\pi}{c}\int_0^{\infty} d\nu \int_{-1}^{+1}\mu^2\, I\, d\mu,$$

$$a = \mu(\mu^2-1)\frac{\partial v}{\partial z},$$

$$g = \mu^2\frac{\partial v}{\partial z};$$

finally, χ denotes the opacity of the material, and S the emission.

Note that in (5) g is of order v/c compared to $c\chi$.

However, we suspect that gI might be of the same order of magnitude as $c(S-\chi I)$; in any case we shall prove that neglecting g would lead to improper energy balance.

In fact, let

$$E_R = \frac{2\pi}{c} \int_0^\infty d\nu \int_{-1}^{+1} I \, d\mu$$

$$F_R = 2\pi \int_0^\infty d\nu \int_{-1}^{+1} I \, d\mu$$

By integration with respect to μ, ν, (5) gives the following radiation energy balance :

$$(6) \quad \rho \frac{D}{Dt}\left(\frac{E_R}{\rho}\right) + \frac{\partial F_R}{\partial z} + P_R \frac{\partial v}{\partial z} = 2\pi \int_0^\infty d\nu \int_{-1}^{+1} (S - \chi I) \, d\mu$$

On the other hand, a combination of (3) and (4) gives the material energy balance, where $E \equiv \rho\,(\varepsilon + v^2/2)$:

$$(7) \quad \rho \frac{D}{Dt}\left(\frac{E}{\rho}\right) + \frac{\partial}{\partial z}(pv) + v \frac{\partial P_R}{\partial z} = 2\pi \int_0^\infty d\nu \int_{-1}^{+1} (\chi I - S) \, d\mu \, .$$

Finally (6) and (7) give

$$\rho \frac{D}{Dt}\left(\frac{E+E_R}{\rho}\right) + \frac{\partial F_R}{\partial z} + \frac{\partial}{\partial z}(pv) + \frac{\partial}{\partial z}(P_R \, v) = 0$$

which shows conservation of energy.

Had we neglected g in (5), the $P_R \frac{\partial v}{\partial z}$ term, representing the work of the radiative pressure P_R, would miss in (6), and we would not get conservation of energy.

As far as numerical simulation is concerned this remark is very important. In fact, in the transfer equation g takes into account the frequency shift due to Doppler's effect.

When the Monte Carlo method is used for solving the transfer equation (5) (see [16], the weight m(t) of the Monte Carlo photons should satisfy

$$(8) \quad \frac{dm}{dt} + (g + c\chi)\, m = 0$$

along the characteristics, in order to include Doppler's effect. On the other hand, the energy loss of a photon in a given cell should be divided into two parts : the energy really absorbed by material and the work of the radiative pressure.

A new Monte Carlo method including these modifications has been programmed (see [17]). In some extreme cases it appears to conserve energy much better than the standard Monte Carlo method (with g = 0), see Figure 6.

3. DETONATION

Besides autonomous detonation, the modelization of which has been made by Chapman and Jouguet (see [18]), one would like to study phenomena like transition from shock to detonation, or extinction of detonation due e.g. to boundary effects.

Then, one cannot ignore what happens in the reaction zone which depends on the intimate structure of the explosive material. This is particularly true in the heterogeneous case. In fact, people have shown in this case that a strong shock may first activate some "hot spots" which are responsible for the initiation of the reaction in the material.

To study such phenomena, we have at our disposal three different models

a. Wilkins' model (1964)
b. The "Forest Fire" model [19]
c. The "Krakatoa" model [20]

We first recall some facts about Von Neumann's theory (see e.g. [21]). Let us define the reaction rate m to be zero for the solid phase and m=1 for the detonation products in the (p, V) plane (where $V \equiv 1/\rho$), we have a Hugoniot curve for m=0, and another one for m=1 which is usually called Crussard's curve (see Fig. 7). Starting from the initial point (p_0, V_0), the thermodynamical state (p, V) of the explosive in a detonation running at velocity D, will be located on a straight line, called Rayleigh line, the slope of which is proportional to D. If the chemical reactions are not instantaneous, pressure p should jump from p_0 to some value p_A such that (p_A, V_A) be on the Hugoniot (m=0), and then, as the chemical reactions take place, decrease to some value p_B such that (p_B, V_B) be on the Hugoniot (m=1) but on the same Rayleigh line.

We recall that in the particular case where the Rayleigh line is tangent to the Hugoniot (m=1), we have a CJ detonation.

a. Wilkins' method

Let $p = g(V,\varepsilon)$ denote the equation of state of detonation products, a coefficient $f \in [0,1]$ is defined such that $f = (V_0-V)/(V_0-V_{CJ})$ if $V \geqslant V_{CJ}$ and f=1 otherwise. Then, in the standard Lagrangian equations, one uses $p=f.g(V,\varepsilon)$ as an equation of state instead of $p=g(V,\varepsilon)$.

This method is very simple, since one needs an equation of state only for m=1. However it is valid only for CJ detonations. Also the transient phase is not correct and pressure p cannot be greater than p_B, unlike what is predicted by Von Neumann's theory.

b. "Forest Fire" model

The "Forest Fire" model is a Z.N.D. model (Zeldovitch - Von Neumann - Doering, see [21]), i.e. a model which adds to the usual mass, momentum and energy equations, an equation for the reaction rate of the following type

(9) $$\frac{dm}{dt} = \varphi(m, p)$$

Function φ is determined from experimental data on build-up distances in a corner of explosive initiated by shocks of variable strength.

Also it requires knowledge of an equation of state for solid phase, and some thermodynamical assumptions in the reaction zone $0 < m < 1$, where both equations of state have to be mixed.

We refer the reader to [22] for an example of computation with this method implemented in a 2-D Eulerian code.

c. Krakatoa model

It is also a ZND model, but where function φ involved in the reaction rate equation (9) is assumed from theoretical considerations to have the following form :

$$\varphi(m,p) = A \exp\left(\frac{I}{I_0}\right) p^{\alpha}(1-m)\left[\mathrm{Log}(1-m)\right]^{2/3}$$

where I is the strength of the shock, and A, I_0, α denote some parameters to be determined from experimental data.

The same kind of thermodynamical assumptions is needed as in the previous model. We refer the reader to [20] for some numerical results in 1-D, which show a correct behaviour of the pressure profile.

4. INSTABILITY OF HIGH SPEED FLOWS

There are many kinds of unstable behaviour in high-speed compressible flow, particularly of the convective and of the Rayleigh-Taylor type 23 . The latter, which occurs when the acceleration is directed from a lighter to a heavier fluid, is of fundamental importance in multifluid flow.

The analysis of such instabilities can be carried out following the separation into three natural phases : linear, non-linear, turbulent.

The linear phase has been the subject of a large amount of work. A Lagrangian approach has been proposed and used by L. Brun and B. Sitt in 1976 (see [23]), and later by other authors. In the case of a laser imploded spherical target, interesting quantitative result have been obtained recently [24].

The non linear phase is usually approached through spectral methods [25], and we only refer to some work in progress at Limeil about the Rayleigh-Bénard instability in a compressible fluid.

Finally, the turbulent phase requires the derivation of some specific closure models, like the one recently proposed by Gauthier [3]. This model involves an equation for the turbulent kinetic energy, usually denoted by K. It has been applied successfully to compute the diffusion of a turbulent mixing layer observed at the air-helium interface in a shock tube experiment [26].

ACKNOWLEDGEMENTS

We would like to thank J.P. Chabard, C. Coste, F. Galaup, B. Meltz and M. Patron who provided us the numerical results.

REFERENCES

[1] C. Coste, B. Meltz, J. Ovadia, Computing methods in applied sciences and Engineering, North Holland (1982), p. 369.

[2] J.P. Chabard, C. Coste, to be published.

[3] A.Froger, S. Gauthier, to be published.

[4] Richtmyer, Morton, Difference methods for initial value problems, J. Wiley

[5] J.U. Brackbill, J.S. Saltzmann, J. Comp. Phys. 46,3 (1982), pp.342-368.

[6] J.K. Dukowicz, "An improved accuracy general remapping algorithm", to appear.

[7] B. Van Leer, J. Comp. Phys. 23 (1977), pp. 276-299.

[8] Zalesak, J. Comp. Phys. 31 (1979), pp. 335-362.

[9] D. Bailey, to appear.

[10] J.P. Boris, D.L. Book, J. Comp. Phys. 11 (1973), pp. 38-69
cf also Vol 18 (1975), pp. 248-283 and Vol 20 (1976), pp. 397-431.

[11] P. Woodward, P. Collela, J. Comp. Phys. 54 (1984), pp. 174-201.

[12] G. Sod, J. Comp. Phys. 27 (1978), pp. 1-31.

[13] N. Legrand, J. Ovadia, Méthode Numériques dans les sciences de l'ingénieur, Dunod (1979), p. 347

[14] W.F. Noh, P. Woodward, 5th International Conference on Numerical Methods in Fluids Dynamics (1976)

[15] J.R. Buchler, JQSRT 30 (1983), pp. 395-408.

[16] B. Mercier, to be published.

[17] G. Meurant ,M. Patron, J. Tassart, to be published.

[18] R. Courant, K.O. Friedrichs, Supersonic flow and shock waves, Springer, 1951.

[19] C.L. Mader, C.A. Forest, Los Alamos Scientific Laboratory report 6259 (1976)

[20] G. Dammame, M. Missonnier, 7th Symposium on Detonation, Annapolis (1981), p. 641.

[21] W. Ficket, W.C. Davis, Detonation, University of California press, (1979).

[22] P. Donguy, N. Legrand, 7th Symposium on Detonation, Annapolis (1981), p. 695.

[23] L. Brun et al. in Laser interaction and related plasma phenomena, Plenum Press, vol 4 (1977), p. 1059 ; L. Brun, B. Sitt, CEA Report R5012 (1979).

[24] J.M. Dufour, D. Galmiche, B. Sitt, in Laser Interaction and related plasma phenomena, Plenum Press, Vol 6 (1984), p. 709.

[25] D. Gottlieb, S.A. Orszag, Numerical Analysis of Spectral methods, SIAM (1977).

[26] Andronov et al., J.E.T.P. 44 (1976), p. 424.

Fig.1
Advection of a square density profile;
(a) exact solution
(b) first order remapping
(c) Van Leer's remapping
(d) P.P.M. remapping

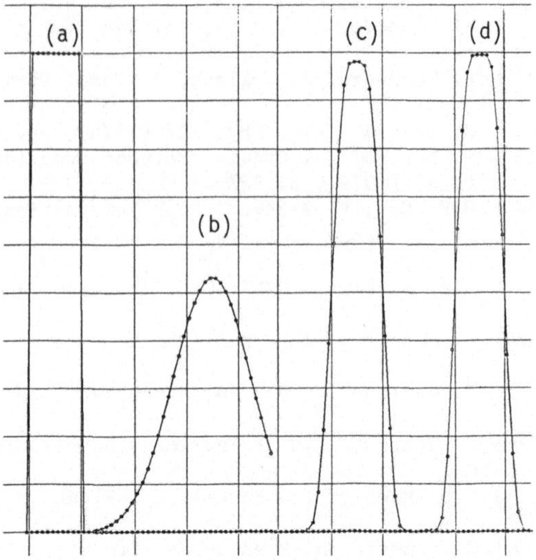

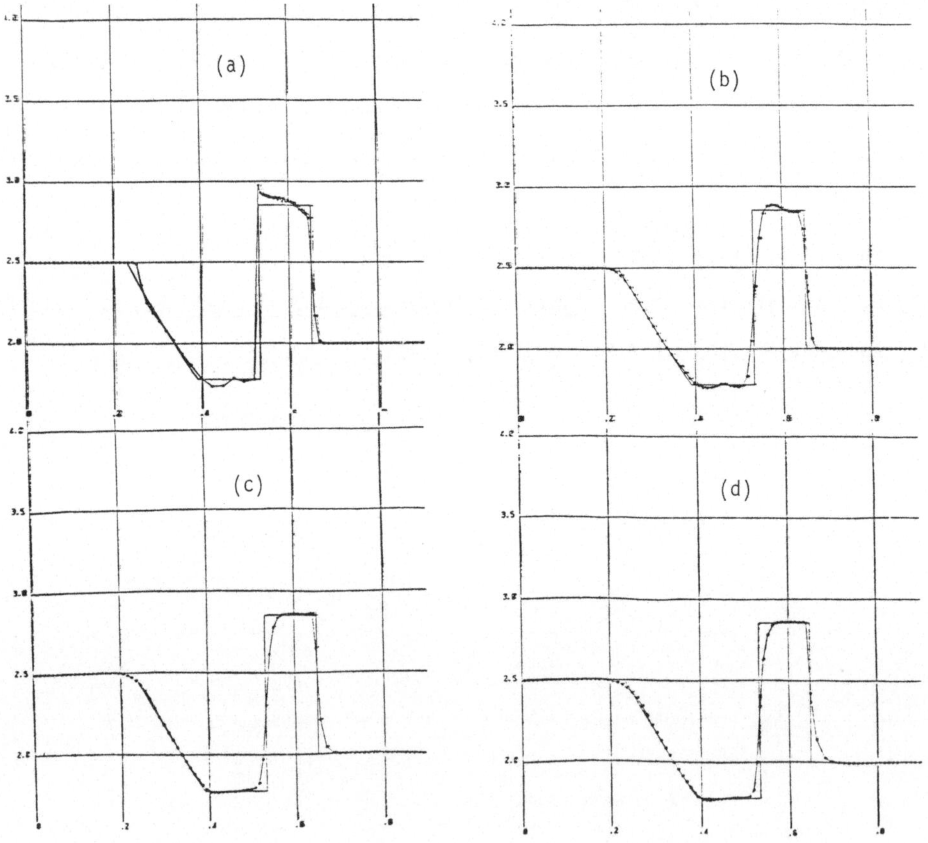

Fig.2. Sod's shock tube problem. Internal energy profile.
(a) Lagrangian method .
(b) W=0 (Eulerian method)
(c) $W = (\text{grad } p / p)^2$, $\lambda_W = 320$.
(d) $W = (\text{grad } \rho / \rho)^2$, $\lambda_W = 185$.

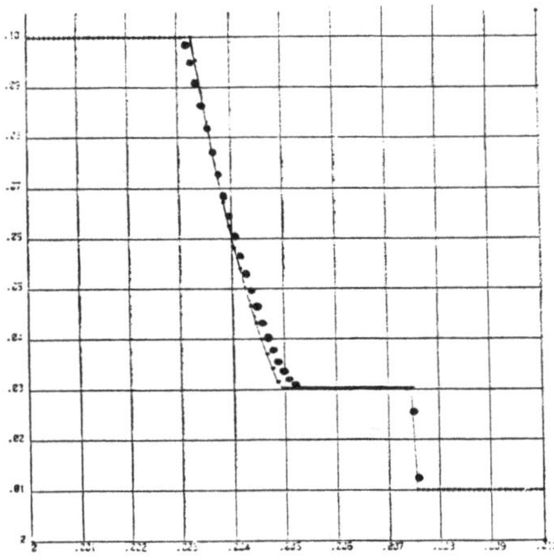

Pressure profile.

Fig.3 -Sod's shock tube problem.

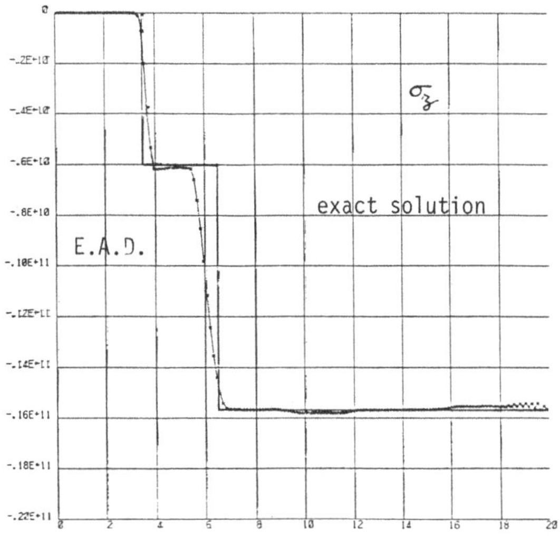

Fig.4 - Elasto-plastic shock. Uniaxial strain

Fig.5
Impact problem.

(a) t=0.

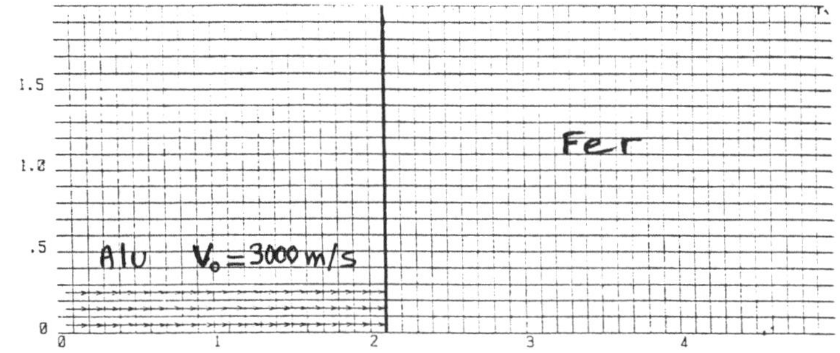

(b) t=21.5 s.

vmax=102.7 m/s

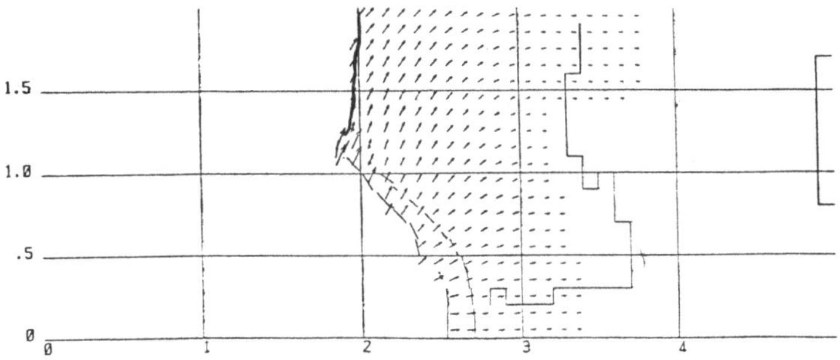

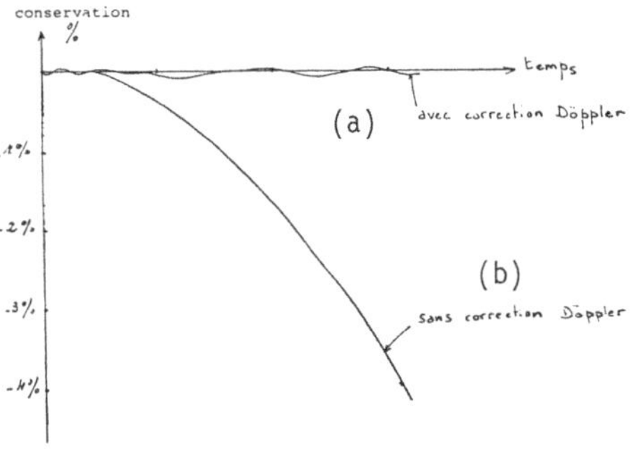

Fig.6- Energy conservation

(a) with Döppler's correction

(b) without.

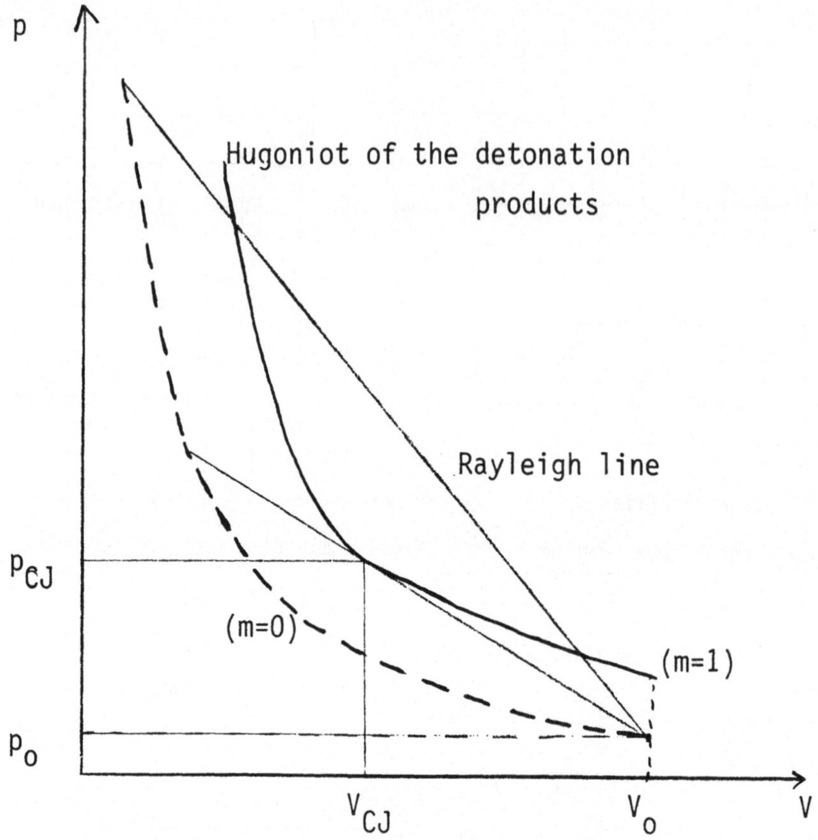

Fig.7. Hugoniot diagram

COMPACT EXPLICIT FINITE-DIFFERENCE APPROXIMATIONS TO THE NAVIER-STOKES EQUATIONS

S.C.R. DENNIS
DEPARTMENT OF APPLIED MATHEMATICS
UNIVERSITY OF WESTERN ONTARIO
LONDON, ONTARIO, CANADA

ABSTRACT

A review is given of some methods of obtaining explicit compact finite-difference formulae which approximate operators of the type occurring in the Navier-Stokes equations governing the motion of incompressible fluids. In their original form the coefficients which multiply the dependent variable in the formulae contain exponentials, but these can be removed by suitable expansions giving formulae with generally satisfactory computational properties. The results are developed first for operators in one space dimension and can then at once be extended to more space dimensions and time by suitable combination techniques. Approximations in which the truncation error can be either of order h^2 of h^4 in the spatial grid size h are considered.

INTRODUCTION

The object of the present paper is to review some explicit finite-difference approximations to differential operators typical of those which occur in problems associated with the Navier-Stokes equations for incompressible fluid motion. In particular, one is interested in higher-order compact approximations to such operators, where the description compact is here defined as involving the minimum number of adjacent grid points. Most of the explicit approximations discussed in detail are derived from a basic method proposed by Dennis (1960). However, some of them have been developed more recently (Dennis and Hudson, 1978, 1979, 1980, 1984) particularly those dealing with higher-order approximations. The higher-order methods will be discussed in some detail; very accurate results can be obtained using them.

In the present paper we consider an operator L such that

$$L\phi = \phi'' - u\phi' = r \ , \tag{1}$$

where the prime denotes differentiation with respect to a space variable, say x. There has been much recent interest in approximations, particularly those of high order, to such an operator. It models the Navier-Stokes type of operator. The function u is a velocity component and is usually multiplied by a Reynolds number parameter which can be large, here omitted for convenience. For an ordinary differential equation (1) where $\phi = \phi(x)$, then $r = r(x)$; but generally r is a function of more than one variable and then further defining equations are necessary. For example, the vorticity transport equation for the unsteady vorticity $\phi(x, y, t)$ can be written

$$\partial\phi/\partial t = \nabla^2\phi - (\underset{\sim}{v} \cdot \underset{\sim}{\nabla})\phi \ , \tag{2}$$

where the Reynolds number factor multiplying the velocity vector $\underset{\sim}{v} =$ (u, v) has again been omitted. Equation (2) can be expressed as the three equations

$$\partial^2\phi/\partial x^2 - u\partial\phi/\partial x = r = s - \partial^2\phi/\partial y^2 + v\partial\phi/\partial y \ , \tag{3}$$

$$\partial\phi/\partial t = s \ . \tag{4}$$

In the general case all functions depend upon x, y and t. For steady-state flow, $s \equiv 0$ and (3) then defines the two equations sufficient to determine $\phi(x, y)$. Similarly if $\phi = \phi(x, t)$, $r \equiv s$, then (3) and (4) give the two equations

$$\partial^2\phi/\partial x^2 - u\partial\phi/\partial x = r = \partial\phi/\partial t \ . \tag{5}$$

It is easy to approximate these sets of equations once basic approximations to (1) have been obtained.

The best known finite-difference approximation to (1) is the explicit central-difference formula

$$(1 - \tfrac{1}{2}hu_o)\phi_1 + (1 + \tfrac{1}{2}hu_o)\phi_3 - 2\phi_o - h^2r_o = 0 \tag{6}$$

in the simplified notation in which the subscripts 0, 1, 3 are associated with values at x_o, $x_o + h$, $x_o - h$ in the space variable x, and h is the grid size. The truncation error in approximating the differential operator in (1) using this approximation is $O(h^2)$; thus it is often termed h^2 accurate. It is also known to have limitations in that the associated matrix for determining approximations to ϕ ceases to be diagonally dominant if $|hu_o| > 2$ at one or more grid points in the solution domain. This can lead to unsatisfactory features in the solution and also can cause convergence problems and often lead to divergence of iterative methods of solution. Upwind methods give associated matrices which are diagonally dominant and therefore solution procedures are generally more satisfactory, but they are only of O(h) accuracy and computed results using them are often quite inaccurate.

There does exist one scheme which is both h^2 accurate and has an associated matrix which is diagonally dominant, namely the method of Allen and Southwell (1955). This method assumes that in equation (1), for example, we can approximate u and r by their values u_o and r_o at the grid point 0. The resulting ordinary differential equation is solved exactly and the solution fitted to the values ϕ_o, ϕ_1, ϕ_3. This gives the tridiagonal approximation

$$E\phi_1 + E^{-1}\phi_3 - (E + E^{-1})\phi_o + h(E - E^{-1})r_o/u_o = 0 , \qquad (7)$$

where

$$E = \exp(-\tfrac{1}{2}hu_o) \sim 1 - \tfrac{1}{2}hu_o + h^2u_o^2/8 \quad . \qquad (8)$$

Actually, Allen and Southwell applied the method only to the steady-state two-dimensional equations (3) when $s \equiv 0$ but Allen (1962) applied it to other problems, including cases of (1) treated as an ordinary differential equation. The method was investigated by Dennis (1960) who showed that it was of the same h^2 order of accuracy as the central-difference approximation. This is readily shown from (7) by substituting the expansion (8) for E and the corresponding one for E^{-1} and noting that the grouping of terms

$$\alpha h^2u_o^2 (\phi_1 + \phi_3 - 2\phi_o) \approx \alpha h^4u_o^2 (\partial^2\phi/\partial x^2)_o = O(h^4) , \qquad (9)$$

where $\alpha = 1/8$ in the present case. The matrix associated with the tri-diagonal system (7) is clearly diagonally dominant and this was demonstrated by Dennis (1973).

The method of Dennis (1960) was more general than that of Allen and Southwell. It lead to an approximation to (1) involving exponential coefficients but the associated matrix is not in general diagonally dominant. However, the method is capable of yielding compact approximations of h^4 accuracy. For example, in the case of the steady vorticity transport equation given by (3) with $s \equiv 0$, we can obtain an explicit h^4 approximation involving the nine-point star centred on a grid point (x_o, y_o). We use the term explicit in this paper to mean that only the variable ϕ and known functions appear in the final approximation; in particular the derivative of ϕ in (1) does not appear as one of the dependent variables in the final matrix to be inverted. This distinguishes it from implicit, or Hermitian, methods in which the derivative is a dependent variable.

There has been considerable interest recently in compact h^4 approximations of implicit type. For an equation of type (1) they employ both

ϕ and $F = \phi'$ as dependent variables. Then (1) can be written as the two equations

$$\phi' = F \quad ; \quad \phi'' = uF + r = G \quad . \tag{10}$$

A typical procedure would now be to use accurate three-point formulae to approximate each of (10), for example

$$\phi_1 - \phi_3 = (h/3)(F_1 + 4F_o + F_3) \tag{11}$$

for the first equation and

$$\phi_1 - 2\phi_o + \phi_3 = (h^2/12)(G_1 + 10G_o + G_3) \quad . \tag{12}$$

for the second. This last equation is generally ascribed to Numerov (1927); a brief account of its derivation is given by Hartree (1958). Thus with some defining equation for r, equations (11) and (12) give a tridiagonal system for the determination of ϕ and F. If $u \equiv 0$ in (1), only the set of equations typified by (12) is necessary. The analogue corresponding to this for the pair of equations (3) when $s \equiv 0$ is Laplace's equation, for which it is known that a compact h^4 accurate method is available in the form of a nine-point approximation. This type of approximation can be extended to the more general case $u, v \neq 0$ using the methods of Dennis (1960).

A large number of investigations of various compact implicit methods have been made. Many of them have recently been reviewed by Hirsh (1983). In some of the earlier methods [Krause (1971), Krause et al. (1973), Peters (1975)] the ancillary variables are eliminated completely to give finally an explicit relationship in terms of the function itself. There are numerous versions and applications of the implicit methods, some involving the use of splines. They cover the period from the work of Hirsh (1975) through to the work of Ciment, Leventhal and Weinberg (1978) on operator compact implicit (OCI) methods and then through ensuing investigations to recent work such as that of Leventhal (1982) on OCI methods of exponential type. They will not be reviewed here since we are largely concerned with explicit methods and sufficient reference can be found in the review of Hirsh (1983).

One common feature of the compact implicit methods is the utilization of the derivative $F = \phi'$ as a dependent variable. However, for an equation such as (1) the variables ϕ and F will not both be known at boundary points; for example, if two-point conditions are given for ϕ there will be none for F. This difficulty is overcome by expansion methods at the boundaries; such procedures are not necessary for explicit methods in which only ϕ appears. The present explicit methods are based on the method of Dennis (1960); the first derivative does not appear because it is removed by a prior local transformation before the equation is put into finite differences. For (1) it is defined in $x_o - h \leq x \leq x_o + h$ by

$$\chi = \phi g = \phi \exp\{-\tfrac{1}{2} \int_{x_o}^{x} u(\xi) d\xi\} \quad . \tag{13}$$

It follows the classical method used to remove the first derivative and it is the equation for χ which is approximated by finite differences rather than (1), thereby avoiding the presence of the first derivative. We shall now review this procedure and note some explicit approximations of h^2 accuracy which can be obtained. Some approximations of h^4 accuracy are then considered, including some recent results derived by Dennis and Hudson (1984) by expanding the exponentials appearing in the finite-difference formulae in powers of their arguments. Some numerical illustrations are given but the main details are given by Dennis and Hudson (1984).

BASIC METHOD AND APPROXIMATIONS

We can start with the one-dimensional equation (1) and then utilize the results in cases involving more space dimensions and time. If we make the substitution (13) in (1) locally in $x_o - h \leq x \leq x_o + h$, the local equation for χ is

$$\chi'' + f\chi = rg \tag{14}$$

where

$$f = \tfrac{1}{2}u' - \tfrac{1}{4}u^2 \quad . \tag{15}$$

We can now express (14) in finite differences of χ using any standard polynomial procedure; then we eliminate χ using (13). In this way, an h^2 approximation to (1) is

$$g_1\phi_1 + g_3\phi_3 - (2 - h^2 f_o)\phi_o - h^2 r_o = 0 \quad . \tag{16}$$

In order to evaluate the coefficients g_1 and g_3 in (16) it is necessary to evaluate the integral in (13) when $x = x_o - h$ and $x = x_o + h$. It was pointed out by Dennis (1960) that this can be done using any suitable quadrature formula and some examples were given. In particular, it is obvious that if the variable u in the integral in (13) is approximated by the leading term u_o of the Taylor expansion of u about $x = x_o$, the coefficients g_1 and g_3 in (16) approximate to the coefficients E and E^{-1} appearing in the Allen and Southwell approximation (7). In this way Dennis (1960) was able to relate (16) to Allen and Southwell's method. Both are h^2 accurate approximations with clear interconnections. There seems to be some similarity between these ideas and those of El-Mistakwy and Werle (1978) and subsequently utilized in the exponential OCI method of Leventhal (1982) although their conclusions regarding the accuracy of Allen and Southwell's method may not be quite the same as ours.

The approximation (16) was taken up again by Dennis and Hudson (1978) who considered a power series expansion method for the exponential coefficients g_1 and g_3 and obtained an h^2 accurate expanded form. The expanded form is

$$(1 - \tfrac{1}{2}hu_o + h^2u_o^2/8)\phi_1 + (1 + \tfrac{1}{2}u_o + h^2u_o^2/8)\phi_3$$
$$- (2 + \tfrac{1}{4}h^2u_o^2)\,\phi_o - h^2r_o = 0 \quad . \tag{17}$$

The same expanded form can be obtained from the Allen and Southwell formula (7) by using the expansion (8) and the corresponding one for E^{-1}. We note that the grouping of terms (9) which are dropped in the central-difference approximation are retained in (17). The reason for this is to retain the diagonal dominance of the tridiagonal matrix associated with (17). Lindroos (1981) has pointed out that the grouping of terms (9) can be included in (17) with any value of $\alpha \geq 1/16$ and that the matrix remains diagonally dominant, as it is for $\alpha = 1/8$. Of course one could say that inclusion of the terms (9) adds an artificial diffusion term in any case. However, it will be seen when h^4-accurate approximations are considered that the terms (9) form part of them and that the best value of α is probably $\alpha = 1/12$.

Actually, the investigation of Dennis and Hudson (1978) was concerned with the steady-state vorticity transport equation defined by (3) with $s \equiv 0$. It gave the corresponding two-dimensional form of (17) obtained from the similar two-dimensional form of (16) given by Dennis (1960). This was obtained by applying the same techniques used for (1) in the x direction to the equation $\partial^2\phi/\partial y^2 - v\partial\phi/\partial y + r = 0$ in the y direction and then eliminating r_o from (16) using this equation. This type of technique had already been used by Allen and Southwell (1955). It leads to an approximation over a five-point star centred at the point (x_o, y_o) which can generally be written in the form

$$a_1\phi_1 + a_2\phi_2 + a_3\phi_3 + a_4\phi_4 - a_o\phi_o = 0 \quad , \tag{18}$$

where the Southwell notation (see Smith, 1965, p.142) is used and the a_n are to be identified in any particular case. For example, for the two-dimensional approximation corresponding to (17) we find

$$a_1 = 1 - \tfrac{1}{2}h\,u_o + h^2u_o^2/8\,, \quad a_2 = 1 - \tfrac{1}{2}h\,v_o + h^2\,v_o^2/8\,,$$
$$a_3 = 1 + \tfrac{1}{2}h\,u_o + h^2u_o^2/8\,, \quad a_4 = 1 + \tfrac{1}{2}h\,v_o + h^2\,v_o^2/8\,, \tag{19}$$
$$a_o = 4 + \tfrac{1}{4}h^2(u_o^2 + v_o^2).$$

This approximation is quite satisfactory even when $h\,u_o$, $h\,v_o$ are not small. For example, it was used to compute solutions for Reynolds numbers up to 2000 for steady flow in a stepped channel by Dennis and Smith (1980) and for steady flow in a branching channel by Bramley and Dennis (1982). Solutions for steady flow in a curved tube of circular cross-section were given for Dean numbers to 5000 by Dennis (1980) and for steady flow external to a rotating sphere for Reynolds numbers up to 5000 by Dennis, Ingham and Singh (1981), both by adaptations of this method. However, it is only of $O(h^2)$ accuracy so we now seek compact approximations of higher accuracy.

APPROXIMATIONS OF HIGHER ACCURACY

The first derivative of χ is absent from (14) and thus we can use the Numerov method to obtain an explicit tridiagonal approximation in terms of χ alone which we can then replace by $g\phi$. If we note again that $g_o = 1$ we obtain the result

$$(1+h^2f_1/12)g_1\phi_1 + (1+h^2f_3/12)g_3\phi_3 - (2-5h^2f_o/6)\phi_o$$
$$-h^2(g_1r_1 + 10r_o + g_3r_3)/12 = 0\ . \tag{20}$$

This result was in fact given by Dennis (1960) for the equation (1) considered as an ordinary differential equation, where it was described as an extension of Numerov's method to differential equation in which the first derivative is not absent. A numerical example of this h^4 method was given which clearly demonstrated its superiority over existing h^2 approximations. However, the extension to two-dimensional problems of the type (3) with $s \equiv 0$ was not given until the paper by Dennis and Hudson (1979). In that investigation the approximation (16) together with a higher-order difference correction were taken to make up the full approximation (20). Thus the matrix corresponding to the h^2 approximation, but with the addition of a right-hand side corresponding to the h^4 corrections, was repeatedly inverted in an iterative process until a final h^4 accurate solution was obtained. This type of deferred-correction method is not new; it was used by Fox (1948) and probably earlier. The coefficients g_1 and g_3 in (20) involve exponentials, of course, and a somewhat different formulation, again involving exponential coefficients, was given by Dennis and Hudson (1980). However, the main point was that both of these methods were, in effect, compact in that they only involved for the equation (3), with $s \equiv 0$, values of ϕ at the nine-point star centred on (x_o, y_o).

Dennis and Hudson (1984) have considered a procedure in which the exponential coefficients g_1 and g_3 in (20) are expanded in powers of their arguments. The process is carried far enough to retain h^4 accuracy and the final result is a tridiagonal relationship which we can write

$$c_1\phi_1 + c_3\phi_3 - c_o\phi_o - h^2r_o + C_o = 0 \quad . \tag{21}$$

After an amount of algebraic reductions we can finally express the coefficients as

$$\begin{aligned} c_1 &= b_1^2 + \frac{1}{48}h^2u_o^2 - \frac{1}{24}h^3u_o'' \quad ; \\ c_3 &= b_3^2 + \frac{1}{48}h^2u_o^2 + \frac{1}{24}h^3u_o'' \quad ; \\ c_o &= b_1^2 + b_3^2 + \frac{1}{24}h^2u_o^2 \quad ; \\ C_o &= -\frac{1}{12}h^2\{(1 - \tfrac{1}{2}hu_o)r_1 + (1 + \tfrac{1}{2}hu_o)r_3 - 2r_o\} \end{aligned} \tag{22}$$

where

$$\begin{aligned} b_1 &= 1 - \tfrac{1}{4}hu_o - \frac{1}{12}h^2u_o' \quad , \\ b_3 &= 1 + \tfrac{1}{4}hu_o - \frac{1}{12}h^2u_o' \quad . \end{aligned} \tag{23}$$

It may be noted that although the expression for C_o in (22) contains a term in r_o it is considered as separate from the term in r_o in (21) because the whole of C_o represents a higher-order correction term.

With the definitions (22), the matrix associated with the tridiagonal relationship in ϕ on the left side of (21) is diagonally dominant if

$$h^3|u_o''| \leq 24b^2 + \tfrac{1}{2}h^2u_o^2 \tag{24}$$

at every grid point, where b is the smaller of the absolute values of the coefficients b_1 and b_3. Such a condition should be considerably easier to satisfy than the condition $h|u_o| \leq 2$ which is necessary at every grid point to ensure that the matrix associated with the tridiagonal operation on ϕ in the finite-difference equations (6) is diagonally dominant. It has also been verified that in calculating the derivatives in (22) and (23) one can still retain the necessary level of accuracy for the h^4 method by using three-point central-difference formulae. Thus all the operations can be carried out over three adjacent grid points.

If (24) is not satisfied at any grid point we can still retain diagonal dominance by using an upwind scheme to deal with only the terms in (21) which involve u_o''. From (22) these terms amount to

$$(h^3/24)u_o''(\phi_3 - \phi_1) \approx -(h^4/12)u_o''\phi_o' \qquad (25)$$

on the left side of (21). We can write the term $\phi_3 - \phi_1$ as

$$2(\phi_3 - \phi_o) - D_o \text{ if } u_o'' > 0; \; 2(\phi_o - \phi_1) + D_o \text{ if } u_o'' < 0$$

where

$$D_o = \phi_1 + \phi_3 - 2\phi_o \quad . \qquad (26)$$

If we neglect D_o and group the remaining terms with similar terms in (21) we obtain a diagonally-dominant matrix in ϕ but an h^3 accurate method rather than h^4. On the other hand we can retain D_o as a deferred correction following the manner that was used by Dennis and Chang (1969) and since by other workers to upgrade the first-order upwind scheme to central-difference accuracy. In the present case we in effect have a higher-order upwind scheme as a first approximation, with a deferred correction to achieve h^4 accuracy. This method is given by Dennis and Hudson (1984). It may also be noted in general that if u is constant with respect to x in (1), the coefficients of the terms in u_o^2 in c_o, c_1 and c_3 in (22) correspond to setting $\alpha = 1/12$ in (9).

PROBLEMS INVOLVING SEVERAL SPACE VARIABLES AND TIME

From the basic approximations of h^4 accuracy to the operator in (1) one can readily obtain results for operators involving several space variables and time. For example, the pair of equations (5) can be approximated in time using any suitable method on the understanding, of course, that an additional truncation error in the time variable is involved. Thus the Crank-Nicolson method applied to the equation $\partial\phi/\partial t = r$ at the spatial grid point $x = x_o$ gives the equation

$$\phi_o(t+k) - \tfrac{1}{2}kr_o(t+k) = \phi_o(t) + \tfrac{1}{2}kr_o(t) \qquad (27)$$

with an $O(k^3)$ error on the right-hand side, where k is the time step. Then with neglect of this error term we can obtain h^2-accurate implicit procedures by expressing the terms involving r_o at the two time levels in (27) explicitly in terms of the variable ϕ using any of (6), (16) or the expanded form (17). The right-hand side is known from the previous time step and the left-hand side defines a matrix inversion to determine $\phi_o(t+k)$, assuming that boundary conditions for ϕ are known.

An h^4-accurate method can likewise be obtained by substitution of r_o at the two time levels into (27) from (21). The result is now implicit because the term C_o in (22) depends upon r. However, we may express C_o in terms of values of ϕ using the equation $\partial\phi/\partial t = r$ and expressing the derivative as a backward difference in time at each spatial grid point. For example, one can evaluate $C_o(t+k)$ using the approximations

$$r_n(t+k) \approx [\phi_n(t+k) - \phi_n(t)]/k, \quad (n = 0, 1, 3) \quad . \tag{28}$$

Then the left-hand side of (27) again, after transferring some terms to the right-hand side, gives rise to a tridiagonal matrix whose inversion determines the vector components $\phi_o(t+k)$. We shall not discuss the truncation errors necessarily introduced in the time variable nor the conditions under which the matrix to be inverted is diagonally dominant. However, it may be noted that $C_o(t+k)$ is itself a term of order h^4 and one way of performing the matrix inversion for $\phi_o(t+k)$ is to use an iterative method in which $C_o(t+k)$ is treated as a deferred correction which is repeatedly evaluated and added to the right-hand side of (27). Such an iterative scheme can be arranged to employ an associated matrix which is certainly diagonally dominant if (24) is satisfied and, if not, the problem can be formulated using the correction D_o, adding an appropriate multiple of the deferred correction $D_o(t+k)$ to the right-hand side of (27) in addition to $C_o(t+k)$.

In order to formulate an h^4 compact method for the steady-state vorticity equation defined by (3) with $s \equiv 0$, we write down the corresponding approximation to (21) when the equation $\partial^2\phi/\partial y^2 - v\partial\phi/\partial y + r = 0$ is approximated to h^4 accuracy in the y direction and then eliminate the term $h^2 r_o$ by addition of this equation to (21). This gives a finite difference equation involving ϕ_o, ϕ_1, ϕ_2, ϕ_3, ϕ_4 together with the correction C_o in (22) and a similar correction C_o^* involving r_o, r_2 and r_4. These latter three quantities may now be eliminated from C_o^* using (6) or (17), with corresponding approximations in the x direction for r_2 and r_4. It is not necessary to include any higher-order terms in (6) or (17) and the corresponding equation for r_2 and r_4 in order to preserve h^4 accuracy in the final result. Similarly, the quantities r_o, r_1, r_3 are eliminated from C_o using approximations similar to either (6) or (17) in the y direction obtained from the equation $\partial^2\phi/\partial y^2 - v\partial\phi/\partial y + r = 0$ along with corresponding approximations for r_1 and r_3. In this way the equation (2) with $\partial\phi/\partial t \equiv 0$ can be represented by a nine-point compact formula of the form

$$\sum_{n=1}^{8} d_n\phi_n - d_o\phi_o + B_o = 0 , \qquad (29)$$

where the Southwell notation (Smith, 1962, p. 142) has again been used, e.g. the subscript 5 refers to the point $(x_o + h, y_o + h)$. Here B_o could include the effect of a forcing term if such a term depending upon x and y were added to the left-hand side of (2); or it could depend solely on a linear combination of the ϕ_n if it were a deferred correction term and no forcing term were present.

Dennis and Hudson (1979, 1980) gave deferred-correction methods, in effect equivalent to (29), to obtain h^4-accurate approximations to (2) with $\partial\phi/\partial t \equiv 0$. In these the central-difference approximation (6) and its equivalent in the y direction were used to evaluate the higher-order terms. The coefficients d_n in (29) involved the exponential function. Numerical examples were given in cases in which a forcing term was either present or absent on the left-hand side of (2). Dennis and Hudson (1984) use approximations of the form (17) to evaluate the higher-order terms and arrive at formulae of the type (29) in which the exponential coefficients have been removed from d_n by suitable expansion procedures. The term B_o in (29) is generally a higher-order deferred correction (assuming no forcing term present) which is chosen so that the matrix associated with the remaining terms shall be diagonally dominant, viz.

$$\sum_{n=1}^{8} |d_n| \leq d_o \qquad (30)$$

at all grid points, assuming $d_o > 0$.

The most desirable form of (29) as an approximation to (2) when $\partial\phi/\partial t \equiv 0$ and no forcing term is present is one in which $B_o \equiv 0$ but with (30) satisfied. In considering such an approximation it may be noted that the expression which multiplies the factor $-h^2/12$ in the definition of C_o in (22) may be replaced by the expression

$$(1 - \tfrac{1}{2}hu_o + \beta h^2u_o^2)r_1 + (1 + \tfrac{1}{2}hu_o + \beta h^2u_o^2)r_3 - 2(1 + \beta h^2u_o^2)r_o \qquad (31)$$

with no change in the overall truncation error in approximating (2) since the additional terms which appear in (31) are $O(h^4)$ on aggregate. If $\beta \geq 1/16$ the coefficients of the terms in r in (31) are all positive and it is found that some satisfactory approximations of the form (29) can be found. A full discussion of this is given by Dennis and Hudson (1984). When $u = v = 0$ in (2), with $\partial\phi/\partial t \equiv 0$, the approximation (29) reduces to the standard nine-point approximation to Laplace's equation. This approximation was discussed by van de Vooren and Vliegenthart (1967), who considered convergence rates of standard iterative procedures of solutions of Laplace's equation using both the five-point and nine-point approximations to the Laplace equation. It was found, particularly in numerical experiments, that rates of convergence were superior for the nine-point approximation with considerably enhanced accuracy. The basic structure of the nine-point formula (29) is more or less a generalization of the corresponding approximation to Laplace's equation. Some future investigation of convergence rates might be worthwhile.

NUMERICAL ILLUSTRATIONS

We shall only give brief details here of a few typical results which have been described in greater detail by Dennis and Hudson (1984). Consider in the first place the problem in one space dimension and time defined by

$$\partial\phi/\partial t = \partial^2\phi/\partial x^2 + 2x\partial\phi/\partial x \tag{32}$$

with boundary conditions for $\phi(x, t)$ given by

$$\phi(0, t) = 1, \quad \phi(\infty, t) = 0, \quad t > 0 \tag{33}$$

and with the initial condition

$$\phi(x, 0) = 0, \quad \text{for all } x \geq 0 \ . \tag{34}$$

The equation (32) can be integrated in time by the Crank-Nicolson formula (27) with $r_o(t)$ and $r_o(t+k)$ defined by any suitable expression. As $t \to \infty$, $\partial\phi/\partial t \to 0$ and a steady state is achieved. The steady-state solution for ϕ is given by

$$\phi(x, \infty) = 1 - \mathrm{erf}(x) \ . \tag{35}$$

The expressions for $r_o(t)$, $r_o(t+k)$ depend upon how we approximate the operator on the right side of (32). Several methods were used including the standard upwind method of $O(h)$ accuracy, the formulae (6) and (17) which are both of $O(h^2)$ accuracy, and finally the h^4-accurate method (21).

Solutions were obtained for various spatial grid sizes and time steps. We present here only some steady-state results since the main interest is in the spatial accuracy. A comparison is made of the various methods for a fixed grid size $h = 0.2$ in table 1, using the condition $\phi(5, t) = 0$ as an approximation to the condition $\phi(\infty, t) = 0$. It is clear from these results that the use of the upwind approximation is grossly inaccurate even in this simple example. Also in this example, the use of the Dennis and Hudson (1978) method expressed by (17) is clearly superior to central-difference approximation; similar results have been found in many other examples. Finally, the clear superiority of the h^4 method is demonstrated by the last two columns.

x	Upwind	Eq.(6)	Eq.(17)	Eq.(21)	1-erf(x)
0.2	0.7915	0.7754	0.7771	0.777296	0.777297
1.0	0.1962	0.1524	0.1577	0.157307	0.157299
1.8	0.0238	0.0094	0.0113	0.010919	0.010909
2.0	0.0127	0.0038	0.0050	0.004685	0.004678

Table 1. Comparisons of Steady-State Solution of Eq. (32)

Dennis and Hudson (1979, 1980) have already published some illustrative results of solutions of the two-dimensional steady-state Navier-Stokes equations using explicit h^4-accurate methods of the present type. They correspond to using a nine-point formula of type (29), but with exponential coefficients d_n; the h^4 correction was added as a deferred correction. Dennis and Hudson (1984) have reconsidered one of these examples (Dennis and Hudson 1979, pp 47 - 51) using an explicit nine-point formula (29) with expanded forms of the exponential coefficients. The results are found to be even an improvement on those previously found; the problem has a simple exact solution and the new results differed from it nowhere by more than two units in the fifth decimal place.

In summary, we have reviewed in this paper some approximations of both h^2 and h^4 accuracy which are capable of giving explicit representations of the Navier-Stokes equation in a compact form, even for the h^4-accurate formulae. The basic operator considered is of the form (1). It may be noted that in practical problems the variable u in (1) is often multiplied by a Reynolds number parameter which may be large. This has been omitted for convenience, but must always be thought of as present. Thus when it is demonstrated that (7) is h^2-accurate, we are considering the asymptotic behaviour as $h \to 0$, even though u may be large. The question of the behaviour when u itself (because of the Reynolds number factor) becomes large is a separate one.

REFERENCES

Allen, D.N. De G. and Southwell, R.V. 1955 Quart. J. Mech. Appl. Math. 8, 129.

Allen, D.N. De G. 1962 Quart. J. Mech. Appl. Math. 15, 11.

Bramley, J.S. and Dennis, S.C.R. 1982 Lecture Notes in Physics 170, 155.

Ciment, M., Leventhal, S.H. and Weinberg, B.C. 1978 J. Comp. Phys. 28, 135.

Dennis, S.C.R. 1960 Quart. J. Mech. Appl. Math. 13, 487.

Dennis, S.C.R. and Chang, G.-Z 1969 Phys. Fluids Supp. II, 12, II-88.

Dennis, S.C.R. 1973 Lecture Notes in Physics 19, 120.
Dennis, S.C.R. and Hudson, J.D. 1978 Proceedings of the First International Conference on Numerical Methods in Laminar and Turbulent Flow, Swansea, United Kingdom: Pentech Press, London, p. 69.
Dennis, S.C.R. and Hudson, J.D. 1979 J. Inst. Math. Applics. 23, 43.
Dennis, S.C.R. and Hudson, J.D. 1980 J. Inst. Math. Applics. 26, 369.
Dennis, S.C.R. 1980 J. Fluid Mech. 99, 449.
Dennis, S.C.R. and Smith, F.T. 1980 Proc. Roy. Soc. Lond. A 372, 393.
Dennis, S.C.R., Ingham, D.B. and Singh, S.N. 1981 Quart. J. Mech. Appl. Math. 34, 361.
Dennis, S.C.R. and Hudson, J.D. 1984 to be published.
El-Mistakwy, T.M. and Werle, M.J. 1978 AIAA J. 16, 749.
Fox, L. 1948 Proc. Roy. Soc. Lond. A 190, 31.
Hartree, D.R. 1958 Numerical Analysis 2nd Ed., Clarendon Press, Oxford, p. 142.
Hirsh, R.S. 1975 J. Comp. Phys. 19, 90.
Hirsh, R.S. 1983 Higher order approximations in fluid mechanics - compact to spectral. Von Karman Institute for Fluid Dynamics Lecture Series 1983-04. Computational Fluid Dynamics, March 7 - 11, 1983.
Krause, E. 1971 Mehrstellenverfahren zur integration der grenschichtgleichungen. DLR Mitt 71 - 13, 109.
Krause, E., Hirschel, E.H. and Kordulla, W. 1973 Fourth-order "mehrstellen" integration for three-dimensional turbulent boundary layers. AIAA Computational Fluid Dynamics Conference, Palm Springs, July, 1973.
Leventhal, S.H. 1982 J. Comp. Phys. 46, 138.
Lindroos, M. 1981 Lecture Notes in Physics 141, 272.
Numerov, B.V. 1927 Astron. Nachricht 230, 359.
Peters, N. 1976 Lecture Notes in Physics 59, 313.
Smith, G.D. 1962. The numerical solution of partial differential equations. Oxford University Press.
van de Vooren, A.I. and Vliegenthart, A.C. 1967 J. Engng Math. 1, 187.

TIME-SPLITTING AND THE FINITE ELEMENT METHOD

C.A.J. Fletcher
University of Sydney, Sydney, NSW 2006, Australia

1. INTRODUCTION

Time-splitting in conjunction with the finite element method will be discussed as an efficient means of solving implicit equations to obtain the steady-state solution to viscous flow problems via a pseudo-transient algorithm. However, with minor modifications, the algorithms described below are equally suitable for transient problems. Here the expression, "time-splitting", is used in the sense of Gourlay (1977). That is, a perturbation is added to the implicit terms to permit a product splitting of the time-dependent terms. This process is also called approximate factorisation or tensor product construction.

Although time-splitting (or approximate factorisation) has been used extensively with the finite difference method (e.g. Briley and McDonald, 1977; Beam and Warming, 1978), its adaptation to the finite element method has been more recent, initially as an ADI implementation (Fletcher, 1981, 1982), subsequently as a genuine splitting (Fletcher and Srinivas, 1983) in the sense of Gourlay (1977). The extension of the time-split finite element method to distorted computational domains modelled in generalised coordinates (Srinivas and Fletcher, 1984b) has necessitated heavy reliance on the *group* finite element formulation (Fletcher, 1983).

Except for very low Reynolds numbers, flow problems are dominated by the convective behaviour. For incompressible flow the convective terms in the momentum equations contain quadratic nonlinearities; for compressible flow they contain cubic nonlinearities. The conventional finite element method handles convective nonlinearities in a rather inefficient manner.

This can be illustrated by considering the two-dimensional x-momentum equation governing compressible viscous flow,

$$\partial(\rho u)/\partial t + \partial(\rho u^2)/\partial x + \partial(\rho uv)/\partial y + \partial p/\partial x = \{\text{viscous terms}\}. \qquad (1)$$

The conventional finite element method introduces a separate trial solution for each dependent variable. For example, for linear rectangular elements,

$$\rho = \sum_{j=1}^{4} \Phi_j(x,y)\bar{\rho}_j \qquad (2)$$

where $\Phi_j(x,y)$ is a bilinear interpolating function and $\bar{\rho}_j$ are the nodal values of ρ.

Application of the Galerkin finite element method produces a large number of products of nodal values associated with the discretised form of the convective terms (Fletcher, 1984). The subsequent manipulation of this large number of terms (in evaluating the equation residual, for example) implies an uneconomic method.

If three-dimensional flows are considered or if higher-order elements (quadratic or cubic) are introduced, this problem of increased connectivity associated with the nonlinear convective terms is seriously aggravated. Connectivity, here, is understood to mean the number of nodal groups appearing in the algebraic expressions after application of the finite element method.

The problem of increased connectivity is substantially alleviated by introducing a single trial solution for each *group* of dependent variables appearing in eq.(1). For example, using rectangular elements.

$$\rho uv = \sum_{j=1}^{4} \Phi_j(x,y)\,(\overline{\rho uv})_j \quad . \tag{3}$$

After application of the Galerkin finite element method far fewer nodal groups occur in the discretised equations, than when the conventional finite element is applied.

In time-split formulations the evaluation of the equation residual at each time-step is usually a major contribution to the overall execution time. Consequently, an operation count comparison for the evaluation of the equation residual formed by the conventional and group finite element formulations, respectively, will provide a comparison of the relative economy of the alternative formulations. Such a comparison is shown in Table 1.

With reference to Table 1 the two and three dimensional Burgers' equations have the same structure for the convective terms as do the incompressible Navier-Stokes equations. The residual operation counts are the number of equivalent additions (one multiplication or division equals three additions or subtractions) to evaluate the steady-state equation residuals. The operation counts and connectivities are based on the use of linear trial function in rectangular (two dimensions) and brick (three dimensions) elements.

The results shown in Table 1 indicate that the connectivity of the conventional finite element treatment of the convective terms grows rapidly with an increase in the order of the nonlinearity or in the dimension of the problem. The group finite element formulation demonstrates a small increase in connectivity with dimension but none with the order of the nonlinearity.

Table 1. Comparison of Conventional and Group Finite Element Methods

Equation System	Convective non-linearity	Conventional F.E.M. Connectivity (convective non-linearity)	Conventional F.E.M. Residual operation count	Group F.E.M. Connectivity (convective non-linearity)	Group F.E.M. Residual operation count	Conventional R.O.C. / Group R.O.C.
2-D Burgers' equations	quadratic	49	828	9	206	4
3-D Burgers' equations	quadratic	343	12603	27	1309	9
2-D viscous comp. flow	cubic	225	6772	9	404	16.8
3-D viscous comp. flow	cubic	3375	217065	27	2349	92.4

The increase in the connectivity is reflected in the size of the residual operation count. In particular the ratio of the residual operation counts shown in the last column corresponds to the ratio of the execution times for the two formulations. It is clear that the conventional finite element method does not handle the convective nonlinearity in an economical manner.

Computational solutions to the two-dimensional Burgers' equations (Fletcher, 1983b) indicate a ratio of execution times of about two and a half, which is consistent with the ratios shown in Table 1, when the execution time to construct and solve the tridiagonal systems of equations are included. Steady-state solutions to the two-dimensional Burgers' equations indicate that the group finite element formulation produces solution that are slightly more accurate than the conventional finite element method. Thus, it is apparent that the group finite element formulation is more efficient, computationally, than the conventional finite element method.

2. MASS OPERATORS

The development of a consistent time-splitting algorithm is facilitated by the explicit extraction of *directional mass and difference operators*. This can be illustrated by considering the two-dimensional vorticity transport equation, in conservation form,

$$\partial\zeta/\partial t + \partial(u\zeta)/\partial x + \partial(v\zeta)/\partial y - \{1/Re\}(\partial^2\zeta/\partial x^2 + \partial^2\zeta/\partial y^2) = 0 , \quad (4)$$

where ζ is the vorticity. Application of the Galerkin group finite element formulation with linear rectangular elements produces the following system of ordinary differential equations,

$$M_x \otimes M_y \dot{\zeta} + M_y \otimes L_x u\zeta + M_x \otimes L_y v\zeta - \{1/Re\}(M_y \otimes L_{xx} + M_x \otimes L_{yy})\zeta = 0, \quad (5)$$

where $\dot{\zeta} \equiv d\zeta/dt$ and $\otimes$ is the tensor (or outer) product. The directional mass (M) and difference (L) operators appearing in eq. (5) are defined as follows,

$$M_x \equiv \{1/6,\ (1+r_x)/3,\ r_x/6\}\ ,\quad M_y^t \equiv \{r_y/6,\ (1+r_y)/3,\ 1/6\} \quad (6)$$

$$L_x \equiv \{-1,\ 0,\ 1\}/2\Delta x,\quad L_y^t \equiv \{1,\ 0,\ -1\}/2\Delta y \quad (7)$$

and $L_{xx} \equiv \{1, -(1+1/r_x),\ 1/r_x\}/\Delta x^2;\ L_{yy}^t \equiv \{1/r_y, -(1+1/r_y), 1\}/\Delta y^2$, (8)

where the grid ratio parameter, r_x and r_y, are defined in Fig. 1.

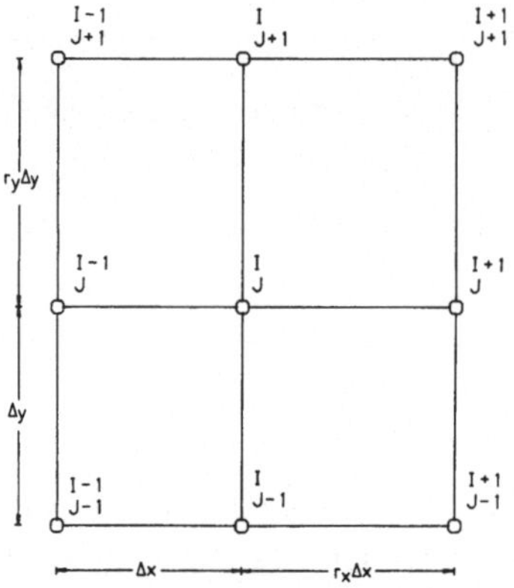

Fig. 1 Nonuniform rectangular grid

The directional difference operators are recognisable as being equivalent to three-point finite difference formulae, at least on a uniform grid. The integral nature of the Galerkin formulation is responsible for the appearance of the mass operators, M_x and M_y. The mass operators can be interpreted as providing transformations from three-point finite difference formulae to nine-point and twenty-seven point finite element formulae in two and three dimensions, respectively.

The greater accuracy associated with the finite element method comes from the mass operators. A Taylor expansion indicates that the finite element scheme typically has smaller dissipative and dispersive errors than an equivalent finite difference scheme. On a uniform grid the finite element discretisation of first derivatives, e.g. $\partial(u\zeta)/\partial x \Rightarrow M_y \otimes L_x u\zeta$, is *fourth-order accurate* at nodal points.

The mass operators also have important smoothing properties. Recent numerical experiments with the flow over a backward-facing step

(Fletcher and Srinivas, 1984) indicate that, with the mass operators retained, stable smooth solutions can be obtained at cell Reynolds Numbers of about 80. When the mass operators are removed a converged (steady-state) solution is not obtained. Although the mass operators effectively generate nine-point formulae in two dimensions the overall execution time is only about 18% greater than for an equivalent finite-difference vorticity stream function formulation when special ordering techniques (Fletcher and Srinivas, 1984) are used to evaluate the equation residuals.

3. SPLITTING ALGORITHM

Attempts to march the system of ordinary differential equations, (5), to obtain the steady-state solution in the minimum number of steps or to obtain accurate transient solutions requires the introduction of a splitting algorithm to achieve acceptable economy.

An efficient marching algorithm can be constructed by introducing a three-level finite difference representation for $d\zeta/dt$ in eq.(5). The result is

$$M_x \otimes M_y\{\alpha\Delta\zeta^{n+1}/\Delta t + (1-\alpha)\Delta\zeta^n/\Delta t\} = \beta RHS^{n+1} + (1-\beta)RHS^n \tag{9}$$

where $RHS = (1/Re)\{M_y \otimes L_{xx} + M_x \otimes L_{yy}\}\zeta - M_y \otimes L_x u\zeta - M_x \otimes L_y v\zeta$ (10)

and $\Delta\zeta^{n+1} = \zeta^{n+1} - \zeta^n$ and $\Delta\zeta^n = \zeta^n - \zeta^{n-1}$.

In eq.(9) α and β weight the time levels n and n+1 at which $\Delta\zeta$ and RHS are evaluated. Appropriate values of α and β will be indicated below. In eq.(9) the solution is known up to time-level n, and is sought at time-level n+1. This is achieved most effectively by constructing a linear system of equations for $\Delta\zeta$. This requires expanding RHS^{n+1} about RHS^n as a Taylor series, i.e.

$$RHS^{n+1} = RHS^n + \left\{\frac{\partial(RHS)}{\partial\zeta} + \frac{\partial(RHS)}{\partial u}\frac{\partial u}{\partial t} + \frac{\partial}{\partial v}(RHS)\frac{\partial v}{\partial t}\right\}\Delta t\ldots \tag{11}$$

Truncating eq.(11) at the point shown introduces an error of $O(\Delta t^2)$. Substituting into eq.(9) produces the result,

$$\alpha(M_x \otimes M_y - \Delta t\{\beta/\alpha\}\partial(RHS)/\partial\zeta)\Delta\zeta^{n+1} = \Delta t\ RHS^{n,\beta} - (1-\alpha)M_x \otimes M_y\Delta\zeta^n. \tag{12}$$

In $RHS^{n,\beta}$, u and v are evaluated at $t^{(n)} + \beta\Delta t$ by extrapolation; all other terms are evaluated at $t^{(n)}$. By extrapolating (explicitly) to obtain u and v, the system of equations, (11), is tridiagonal in $\Delta\zeta$ rather than block-tridiagonal in $\Delta\zeta$, Δu and Δv.

Expanding $\partial(RHS)/\partial\zeta$ in eq.(12) and adding the following additional term to the left-hand side of eq.(12),

$$\Delta t^2\ \{\beta^2/\alpha\}\ (\{1/Re\}L_{xx}-L_x u) \otimes (\{1/Re\}\ L_{yy}-L_y v)\Delta\zeta^{n+1},$$

allows the following product construction,

$$\alpha[M_x-\Delta t\{\beta/\alpha\}(\{1/Re\}L_{xx}-L_x u)] \otimes [M_y-\Delta t\{\beta/\alpha\}(\{1/Re\}L_{yy}-L_y v)]\Delta\zeta^{n+1}$$
$$= \Delta t\ RHS^{n,\beta} - (1-\alpha)M_x \otimes M_y\Delta\zeta^n \ . \tag{13}$$

Equation (13) is consistent with eq.(12) to $0(\Delta t^2)$, and permits the mass operator structure shown in eq.(12) to be preserved. It is well-known that the retention of the mass matrix ($M_x \otimes M_y$) multiplying time-dependent terms (e.g. $\partial\zeta/\partial t$ in eq.(5)) produces smaller dispersion errors (Baker, 1983) than if these terms are lumped. Therefore, the present consistent splitting, although developed to obtain steady-state solutions, is expected to provide accurate solutions to transient problems. Previous finite element splittings have used a lumped form (Fletcher, 1982). That is, the mass operators M_x and M_y on the left-hand side of eq.(13) are replaced by $\{0,1,0\}$.

The splitting shown in eq.(13) indicates that each implicit factor contains operators associated with a single direction. Therefore, eq. (13) can be implemented as an efficient two-stage algorithm as

$$[M_x-\Delta t\{\beta/\alpha\}\ (\{1/Re\}L_{xx} - L_x u)]\Delta\zeta^* = \{\Delta t/\alpha\}RHS^{n,\beta}-(1/\alpha-1)M_x \otimes M_y\Delta\zeta^n \tag{14}$$

and $$[M_y - \Delta t\{\beta/\alpha\}(\{1/Re\}L_{yy} - L_y v)]\Delta\zeta^{n+1} = \Delta\zeta^* . \tag{15}$$

Equation (14) indicates that subsystems of equations, associated with each gridline in the x-direction, can be solved independently of the other gridlines. Each subsystem is tridiagonal if linear elements are used and alternating tridiagonal and pentadiagonal if quadratic elements are used. For this second case the directional mass and difference operators eqs.(6) to (8), would have three and five components in an alternating pattern. For both cases efficient algorithms are available (Fletcher, 1984, pp. 300-301) to solve the subsystems.

During the second stage eq.(15) is solved for subsystems of equations associated with each gridline in the y-direction. As with eq.(14) each subsystem is tridiagonal if linear elements are used and is alternating tridiagonal and pentadiagonal if quadratic elements are used.

The structure of eqs(14) and (15) for the present formulation may be compared with the equivalent equations for a finite difference formulation. The solution of the subsystems of equations is essentially the same for both. The major difference is that the evaluation of $RHS^{n,\beta}$ and $M_x \otimes M_y\Delta\zeta$ are less economical for a finite element formulation. As indicated in Section 2 the overall execution time is only about 18% greater in two dimensions if special ordering techniques are exploited (Fletcher and Srinivas, 1984).

A number of different choices of α and β in eq.(9) are possible while retaining a second-order temporal accuracy. The choice $\alpha = 1.0$, $\beta = 0.5$ gives rise to the Crank-Nicolson scheme for which only two levels of data need be stored. However, when used to obtain steady-state solutions this scheme demonstrates relatively slow convergence as the (rms) magnitude of the equation residual is reduced below about 10^{-3}. The three-level fully implicit method, $\alpha = 1.5$, $\beta = 1.0$, is more robust and gives faster convergence to the steady state (Fletcher and Srinivas, 1983).

4. GENERALISED COORDINATES

For problems in irregular domains e.g. the flow around an isolated aerofoil, it is necessary to employ a distorted grid in the physical plane. Traditionally the isoparametric formulation has been used with the finite element method. However, the isoparametric formulation requires a computationally expensive evaluation of the algebraic coefficients. The formulation also introduces errors if the elements become distorted (Strang and Fix, 1973).

We prefer to follow a different path. First the equations are recast in generalised coordinates and then the group finite element formulation is applied *in the transform plane*. When a uniform grid is used in the transform plane for first derivatives a truncation error analysis indicates that the method is fourth-order accurate spatially. By avoiding complicated numerical integrations the method is very economical.

To facilitate a comparison with the group formulation in the physical plane, the vorticity transport equation, (4), will be transformed into generalised coordinates, $\xi = \xi(x,y)$ and $\eta = \eta(x,y)$. Equation (4) becomes

$$\frac{\partial \zeta^*}{\partial t} + \frac{\partial F}{\partial \xi} + \frac{\partial G}{\partial \eta} - \left\{\frac{\partial^2 R}{\partial \zeta^2} + \frac{\partial^2 S}{\partial \xi \partial \eta} + \frac{\partial^2 T}{\partial \eta^2}\right\} = 0 \qquad (16)$$

where $\zeta^* = \zeta/J$, $F = [U_c + \{1/Re\}(\xi_{xx} + \xi_{yy})]\zeta^*$

$$G = [V_c + \{1/Re\}(\eta_{xx} + \eta_{yy})]\zeta^* \;, \quad R = (\xi_x^2 + \xi_y^2)\zeta^*/Re \qquad (17)$$

$$S = 2(\xi_x\eta_x + \xi_y\eta_y)\zeta^*/Re \;, \quad T = (\eta_x^2 + \eta_y^2)\zeta^*/Re.$$

In the above expressions, U_c and V_c are the contravariant velocities and given by

$$U_c = \xi_x u + \xi_y v \quad \text{and} \quad V_c = \eta_x u + \eta_y v. \qquad (18)$$

In equations (17) the transformation jacobian is given by $J = 1/(x_\xi y_\eta - x_\eta y_\xi)$ and $x_\xi \equiv \partial x/\partial \xi$ etc. The various terms ξ_x etc. are evaluated from $\xi_x = Jy_\eta$, $\eta_x = -Jy_\xi$, $\xi_y = -Jx_\eta$ and $\eta_y = Jx_\xi$,

and the grid transformation parameters are obtained from

$$x_\xi = [x_{i+1,j} - x_{i-1,j}]/[(1+r_\xi)\Delta\xi] \ , \ x_\eta = [x_{i,j+1} - x_{i,j-1}]/[(1+r_\eta)\Delta\eta]. \tag{19}$$

Similar expressions are obtained for $x_{\xi\xi}$ and y_ξ etc. Equation (19) can be interpreted as a one-dimensional lumped Galerkin finite-element formulation with linear Lagrange elements (Srinivas and Fletcher, 1984b).

The structure of eq. (16) is similar to that of eq. (4), except for the appearance of the term $\partial^2 S/\partial\xi\partial\eta$. The group finite element formulation is applied *directly* to the terms ζ^*, F, G, R, S and T in eq. (16). That is trial solutions of the following form are introduced,

$$F = \sum_{j=1}^{4} \phi_j(\xi,\eta)\bar{F}_j \tag{20}$$

where $\bar{F}_j$ represents a nodal value of F.
Application of the Galerkin finite element method with linear rectangular elements produces the following system of ordinary differential equations,

$$M_\xi \otimes M_\eta d\zeta^*/dt + M_\eta \otimes L_\xi F + M_\xi \otimes L_\eta G - \{M_\eta \otimes L_{\xi\xi} R + L_\xi \otimes L_\eta S$$
$$+ M_\xi \otimes L_{\eta\eta} T\} = 0 \ , \tag{21}$$

where $M_\xi \equiv \{1/6,\ (1+r_\xi)/3,\ r_\xi/6\}$ and $L_{\xi\xi} \equiv \{1,\ (1+1/r_\xi),\ 1/r_\xi\}$. (22)

The grid growth parameters r_ξ and r_η play the same role in the transform (ξ,η) space as do r_x and r_y in Fig. 1. Equations (6) to (8) can be used to deduce the form of M_η, L_η etc.

Attempts to construct a split marching scheme from eq. (21) equivalent to eqs. (14) and (15) are restricted by the need to treat $L_\xi \otimes L_\eta S$ explicitly. The result is a two-stage algorithm,

$$[M_\xi - \Delta t\{\beta/\alpha\}(L_{\xi\xi}\partial R/\partial\zeta^* - L_\xi \partial F/\partial\zeta^*)](\Delta\zeta^*)^i = \{\Delta t/\alpha\}RHS^A$$
$$- (1/\alpha - 1)M_\xi \otimes M_\eta (\Delta\zeta^*)^n \tag{23}$$

and $[M_\eta - \Delta t\{\beta/\alpha\}(L_{\eta\eta}\partial T/\partial\zeta^* - L_\eta \partial G/\partial\zeta^*)](\Delta\zeta^*)^{n+1} = (\Delta\zeta^*)^i$. (24)

In eq. (23) $RHS^A = RHS^{n,\beta} + \beta L_\xi \otimes L_\eta \partial S/\partial\zeta^* (\Delta\zeta^*)^n$, where $RHS^{n,\beta}$ is equivalent to $RHS^{n,\beta}$ in eq. (14). During the first stage eq. (23) provides a tridiagonal subsystem of equations along each ξ-line. During the second stage eq. (24) provides a tridiagonal subsystem of equations along each η-gridline. As before the tridiagonal subsystems can be solved efficiently.

5. COMPRESSIBLE NAVIER-STOKES EQUATIONS

The governing equations for two-dimensional compressible viscous flow can be written in vector conservation form as

$$\partial \underset{\sim}{q}/\partial t + \partial \underset{\sim}{F}/\partial x + \partial \underset{\sim}{G}/\partial y - \{\partial^2 \underset{\sim}{R}/\partial x^2 + \partial^2 \underset{\sim}{S}/\partial x \partial y + \partial^2 \underset{\sim}{T}/\partial y^2\} , \tag{25}$$

where $\underset{\sim}{q}^t \equiv \{\rho, \rho u, \rho v\}$

$\underset{\sim}{F}^t \equiv \{\rho u, p+\rho u^2 - \sigma_x, \rho uv - \tau_{xy}\}$

$\underset{\sim}{G}^t \equiv \{\rho v, \rho uv - \tau_{xy}, P + \rho v^2 - \sigma_y\}$

$\underset{\sim}{R}^t \equiv \{\theta\rho, 4u/3, v\}/Re$, $\underset{\sim}{S}^t \equiv \{0, v/3, u/3\}/Re$, $\underset{\sim}{T}^t \equiv \{\theta\rho, u, 4v/3\}/Re$

and $\sigma_x = \{2\varepsilon/3\}(2\partial u/\partial x - \partial v/\partial y)$, $\sigma_y = \{2\varepsilon/3\}(2\partial v/\partial y - \partial u/\partial x)$, $\tau_{xy} = \varepsilon(\partial u/\partial y + \partial v/\partial x)$ where ε is the eddy viscosity and σ_x, σ_y and τ_{xy} are Reynolds stresses. The energy equation is not included in eq. (25) since only subsonic and transonic Mach numbers are of interest (Fletcher, 1982).

In generalised coordinates eq. (25) becomes

$$\partial \underset{\sim}{q}^*/\partial t + \partial \underset{\sim}{F}^*/\partial \xi + \partial \underset{\sim}{G}^*/\partial \eta - \{\partial^2 \underset{\sim}{R}^*/\partial \xi^2 + \partial \underset{\sim}{S}^*/\partial \xi \partial \eta + \partial^2 \underset{\sim}{T}^*/\partial \eta^2\}. \tag{26}$$

Equation (26) is similar to eq. (16) except that F^* etc are three-component vectors rather than scalars. The term F^* is given by

$$\underset{\sim}{F}^* \equiv \frac{1}{J}\begin{bmatrix} \rho U_c \\ \xi_x P + \rho u U_c + \left\{\frac{4}{3}\xi_{xx} + \xi_{yy}\right\}\frac{u}{Re} + \left\{\frac{1}{3}\xi_{xy}\right\}\frac{v}{Re} - \xi_x\sigma_x - \xi_y\tau_{xy} \\ \xi_y P + \rho v V_c + \left\{\xi_{xx} + \frac{4}{3}\xi_{yy}\right\}\frac{v}{Re} + \left\{\frac{1}{3}\xi_{xy}\right\}\frac{u}{Re} - \xi_x\tau_{xy} - \xi_y\sigma_y \end{bmatrix} \tag{27}$$

where the contravariant velocities, U_c and V_c, are given by eq. (18). Equivalent expressions to eq. (27) for the other terms are given by Srinivas and Fletcher (1984b).

Application of the Galerkin group finite element formulation with linear elements to eq. (26) produces a system of ordinary differential equations equivalent to eq. (21). That is

$$M_\xi \otimes M_\eta d\underset{\sim}{q}^*/dt + M_\eta \otimes L_\xi \underset{\sim}{F}^* + M_\xi \otimes L_\eta \underset{\sim}{G}^* - \{M_\eta \otimes L_{\xi\xi}\underset{\sim}{R}^* + L_\xi \otimes L_\eta \underset{\sim}{S}^*$$
$$+ M_\xi \otimes L_{\eta\eta}\underset{\sim}{T}^*\} = 0 . \tag{28}$$

Equation (28) provides a three-component vector equation centered at every grid point. This may be contrasted with eq. (21) which produced a scalar equation centered at every grid point.

To obtain steady-state solutions of eq. (28) a split marching algorithm is constructed in the same way as for eq. (5) and (21). The result is a two-stage algorithm,

$$[M_\xi - \Delta t\{\beta/\alpha\}(L_{\xi\xi}\partial \underset{\sim}{R}^*/\partial \underset{\sim}{q}^* - L_\xi \partial \underset{\sim}{F}^*/\partial \underset{\sim}{q}^*)](\Delta \underset{\sim}{q}^*)^i = \{\Delta t/\alpha\}\underset{\sim}{RHS}^A$$

$$- (1/\alpha-1) M_\xi \otimes M_\eta (\Delta \underset{\sim}{q}^*)^n \qquad (29)$$

$$\text{and } [M_\eta - \Delta t\{\beta/\alpha\}(L_{\eta\eta}\partial \underset{\sim}{T}^*/\partial \underset{\sim}{q}^* - L_\eta \, \partial \underset{\sim}{G}^*/\partial \underset{\sim}{q}^*)](\Delta \underset{\sim}{q}^*)^{n+1} = (\Delta \underset{\sim}{q}^*)^i \qquad (30)$$

$$\text{where } (\underset{\sim}{RHS})^A = (\underset{\sim}{RHS})^n + \beta \, L_\xi \otimes L_\eta \, (\partial \underset{\sim}{S}^*/\partial \underset{\sim}{q}^*)(\Delta \underset{\sim}{q}^*)^n \, .$$

Equation (29) produces a system of block (3x3) tridiagonal equations associated with each grid line in the ξ direction. Equation (30) produces a system of block (3x3) tridiagonal equations associated with each gridline in the η direction. The extension of the scalar algorithm to solve block tridiagonal systems of equations is given by Isaacson and Keller (1966).

The above split marching schemes have been applied to laminar (Fletcher, 1982) and turbulent (Srinivas and Fletcher, 1984a) flows over obstacles, backward-facing steps (Fletcher and Srinivas, 1983) and aerofoil trailing-edge flows (Srinivas and Fletcher, 1984b). Here we illustrate the above formulation for the high Reynolds number ($3x10^7$) flow past an asymmetric aerofoil trailing edge (Fig. 2) at $M_\infty = 0.40$.

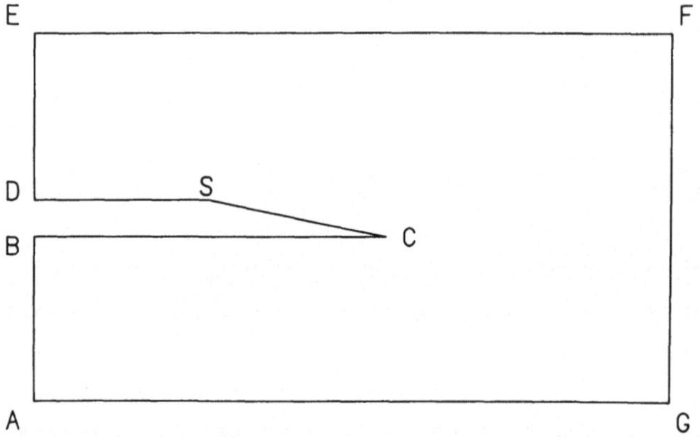

Fig. 2. Trailing-edge geometry.

The undersurface, BC, is completely flat. The top surface has a wedge angle of 12.5°. In the generalised coordinate (ξ,η) domain the finite thickness wedge collapses onto a zero-thickness line. A typical velocity distribution behind this wedge is shown in Fig. 3. These results were obtained on a 41 x 82 grid with a modified algebraic eddy viscosity turbulence model. The solutions are seen to be in good agreement with the experimental results of Cleary et al. (1980) particularly behind the upper surface. The computational results due to Cleary et al. were obtained using a finite difference method on a 60 x 100 grid with a two-equation turbulence model. The superior results produced by the present method on a coarser grid are self-evident.

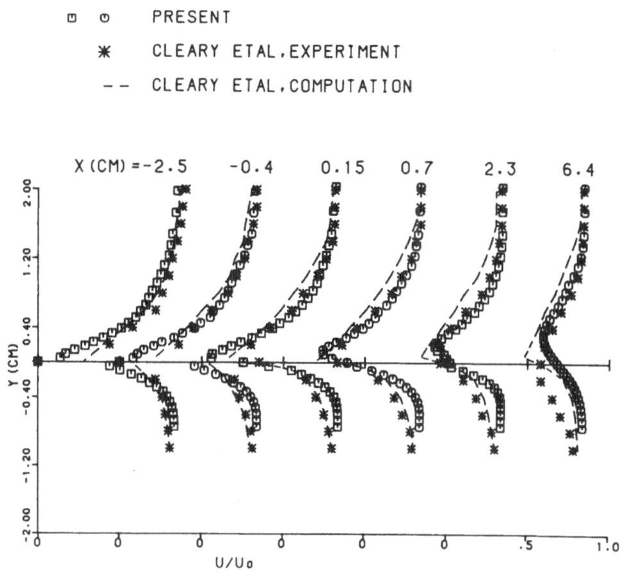

Fig. 3. Velocity distribution behind an asymmetric trailing edge.

REFERENCES

A.J. Baker (1983), *Finite Element Computational Fluid Mechanics*, McGraw-Hill, New York.

R.M. Beam and R.F. Warming (1978), *A.I.A.A. J.*, 16, 393 -402.

W.R. Briley and H. McDonald (1977), *J. Comp. Phys.*, 24, 372-397

J.W. Cleary, P.R. Viswanath, C.C. Horstman and H.L. Seegmiller (1980), AIAA Paper-80-1396.

C.A.J. Fletcher (1981), *Lecture Notes in Physics*, 141, Springer, New York, 182-187.

C.A.J. Fletcher (1982), *Comp. Meth. App. Mech. Eng.*, 30, 307-322.

C.A.J. Fletcher, (1983a), *Comp. Meth. App. Mech. Eng.*, 37, 225-243.

C.A.J. Fletcher, (1983b), *J. Comp. Phys.*, 51, 159-188.

C.A.J. Fletcher (1984), *Computational Galerkin Methods*, Springer-Verlag, New York.

C.A.J. Fletcher and K. Srinivas (1983), *Comp. Meth. App. Mech. Eng.*, 41, 297-322.

C.A.J. Fletcher and K. Srinivas (1984), "On the Role of Mass Operators in the Group Finite Element Formulation", *Comp. Meth. App. Mech. Eng.*, to appear.

A.R. Gourlay (1977) in *The State of the Art in Numerical Analysis* (ed. D. Jacobs), Academic Press, London, 757-796.

E. Isaacson and H.B. Keller (1966), *Analysis of Numerical Methods*, Wiley, New York.

K. Srinivas and C.A.J. Fletcher (1984a), *Int. J. Num. Meth. Fluids*, 4, 421- 439.

K. Srinivas and C.A.J. Fletcher (1984b), "A Three-level Generalised-Coordinate Group Finite-Element Method for Compressible Viscous Flow", submitted.

G. Strang and G.F. Fix (1973), *Analysis of the Finite Element Method*, Prentice-Hall, Englewood Cliffs, N.J.

SPECTRAL METHODS FOR COMPRESSIBLE FLOW PROBLEMS

David Gottlieb
Tel-Aviv University, Tel-Aviv, Israel and
Institute for Computer Applications in Science and Engineering
NASA Langley Research Center, Hampton, VA 23665

Introduction

In the last decade spectral methods have been used very successfully in the numerical simulations of incompressible flows. Spectral methods have also emerged as a major tool in computational meteorology. This has led many researchers to look into the possiblity of applying spectral methods to simulate compressible flows that are of interest to aeronautical engineers. The aim of this article is to give a brief review of the major developments in this field in the last few years. In particular we would like to discuss the notion of the information that is contained in the numerical result. We argue that spectral methods yield more information about the exact solution than low order methods. This information is hidden in the form of numerical oscillations when the exact solution is discontinuous or contains extreme gradients. The structure of these wiggles depends on the nature of the discontinuity and, in some cases, a very accurate solution can therefore be extracted.

2. Spectral Methods

There are basically two steps in obtaining a numerical approximation $u_N(x)$ to a solution $u(x)$ of a differential equation. First, an appropriate finite or discrete respresentation of the solution must be chosen. This may take the form of an interpolating function between the values $u(x_j)$ at some suitable points x_j or a series coefficient in the finite representation

$$u_N(x) = \sum_{k=0}^{N} a_k \phi_k(x) \tag{2.1}$$

with given expansion functions $\phi_k(x)$. The second step is to obtain

Research was supported in part by the Air Force Office of Scientific Research under Contract No. AFOSR 83-0089 and in part by the National Aeronautical and Space Administration under NASA Contract No. NAS1-17070 while the author was in residence at ICASE, NASA Langley Research Center, Hampton, VA 23665.

equations for the discrete values $u_N(x_j)$ or the coefficients a_k from the original equations. This second step involves finding an approximation for the differential operator in terms of the grid point values of u_N or, equivalently, the expansion coefficients. For example, the pseudospectral Chebyshev approximation to the equation

$$\begin{aligned} &u_t = u_x, && |x| < 1 \\ &u(x,0) = u_0(x), && u(1,t) = h(t) \end{aligned} \tag{2.2}$$

is obtained in the following manner. For a given time t we assume that $\{u_N(x_j,t)\}$ is known where $x_j = \cos \frac{\pi j}{N}$. We then interpolate these values to get

$$u_N(x,t) = \sum_{j=0}^{N} u_N(x_j,t)\, g_j(x) \tag{2.3}$$

where

$$g_j(x) = \frac{(-1)^{j+1}}{N^2} \frac{(1-x^2)T_N'(x)}{c_j(x-x_j)}, \qquad \begin{aligned} &c_0 = c_N = 2 \\ &c_j = 1, \qquad 0 < j < N. \end{aligned}$$

Note that $g_j(x_k) = \delta_{jk}$. Equivalently, since

$$g_j(x) = \frac{2}{N} \sum_{n=0}^{N} \frac{T_n(x_j)\, T_n(x)}{c_n}$$

where $T_n(x) = \cos(n \cos^{-1} x)$ is the Chebyshev polynomial of degree n, one gets

$$\begin{aligned} u_N(x,t) &= \sum_{n=0}^{N} a_n T_n(x) \\ a_n &= \frac{2}{c_n N} \sum_{j=0}^{N} u(x_j,t) \frac{\cos(\pi jn/N)}{c_j}, \qquad j = 0,\cdots,N. \end{aligned} \tag{2.4}$$

The next step is to differentiate (2.3) to get the system of ordinary differential equations

$$\frac{\partial u_N(x_k,t)}{\partial t} = \sum_{j=0}^{N} u_N(x_j,t) g_j'(x_k), \qquad j = 1,\cdots,N$$

$$\frac{\partial u_N}{\partial t}(x_0,t) = h'(t) \tag{2.5}$$

or using (2.4)

$$\frac{\partial u_N(x_k,t)}{\partial t} = \sum_{n=0}^{N} a_n T_n'(x_k) = \sum_{n=0}^{N-1} b_n T_n(x_k), \quad j = 1,\cdots,N$$

$$\frac{\partial u_N}{\partial t}(x_0,t) = h'(t) \tag{2.6}$$

where

$$b_N = 0, \qquad b_{N-1} = 2N\, a_N, \qquad b_n = \frac{1}{c_n}\left[b_{n+2} + 2(n+1)a_{n+1}\right].$$

Equations (2.5) and (2.6) are, in fact, identical. Equation (2.5) points out the possibility of applying the pseudospectral Chebyshev method by mulitplying the vector $u(x_j,t)$ by the matrix $g_j'(x_n)$ whereas the asymptotically efficient implementation of (2.6) is by using a Fast Fourier Transform.

In general, consider the system of equations

$$u_t = L(u)$$

$$u(t=0) = u_0, \tag{2.7}$$

where L is a nonlinear operator that involves only spatial derivatives. In spectral methods we define a finite dimensional subspace B_N which is the space of polynomials (or trigonometric polynomials) of degree N, and a projection operator P_N that maps the original space to B_N. An example of such a P_N is given in (2.3). In fact, given a function $f(x)$, $-1 \leq x \leq 1$, then (2.3) defines $P_N f = \sum_{j=0}^{N} f(x_j) g_j(x)$.

We then seek a solution u_N belonging to B_N such that

$$\frac{\partial u_N}{\partial t} = P_N L(u_N),$$
$$u_N(t=0) = P_N u_0. \tag{2.8}$$

For a more complete description of spectral methods we refer the reader to [3], [6].

Spectral methods are global in nature, i.e., in order to get an expression for $\frac{\partial}{\partial x} u_N$ we use all the grid points x_k, $k = 0,\cdots,N$ (see (2.5)). Together with the choice of the points x_k this explains their high order accuracy. The accuracy of spectral methods depends on the total number of points N, and the number of smooth derivatives of u. For smooth flows, great savings of computer storage and time is gained by using spectral methods since only a small number of grid points is required to get the same accuracy obtained by other methods.

3. Spectral Methods and Shock Waves

The use of any formal high order method for the numerical simulation of flows with shocks poses theoretical and practical problems. The error estimates obtained for spectral methods depend on the smoothness of the solution and it is not clear at all that any degree of accuracy can be achieved for discontinuous solutions. On the one hand, it has been proven that for linear problems, high accuracy can be maintained within spectral methods far away from the discontinuity; on the other hand, it may be thought that for nonlinear problems the overall accuracy in the presence of discontinuities is limited to first order. However, in [10] Lax has argued that more information about the solution is contained in high resolution schemes, even in the nonlinear case. In fact, Lax has shown that the ε-capacity of the set of approximate solutions is closer to the ε-capacity of the set that includes the projections of exact solutions if the numerical scheme is a high order scheme. Typically, when a spectral method is used to simulate flows with shocks it yields an oscillatory solution. The oscillations are global, that is they occur not only in the neighborhood of the shock but all over the flow field. Several methods of overcoming these oscillations were suggested. Historically, the first attempts to get nonoscillatory results concentrated on using finite difference type artificial dissipation.

Taylor, et al. [15] used the method of Boris and Book of adding diffusion and antidiffusion terms for some model problems. Sakell [12] has checked a version of the Von Neumann-Richtmyer artificial dissipation for the wedge flow problem. Cornille [2] has used a version of the Lax-Wendroff scheme with inherent dissipation. Zang and Hussaini [16] simulated slightly viscous flows and treated the viscosity term by finite differences. Two real life flows were simulated using the above ideas. Reddy [11] introduced Fourier representation in the azimuthal direction in the three-dimensional Navier-Stokes code of Pulliam and Steger. In this problem there is enough dissipation coming from the discretization in the other directions. Reddy reports substantial improvement over the finite difference code. Streett [14] simulated transonic flow around an airfoil. His code is a full potential algorithm with retarded density. His results indicate that for subsonic flows, spectral methods are superior to the finite difference codes, whereas for transonic flow they are comparable. The results obtained by these methods indicate that a highly structured flow field is well-represented along with the front of the shock. However, the shock profiles are smeared and the accuracy in the smooth part of the flow is perhaps no longer spectral.

A different approach advocated first by Hussaini, Salas and Zang [9] is to fit the shock. This approach has been used to simulate various physical problems, most of them concerned with shock wave interactions. Since they were interested in the behavior of the flow on only one side of the shock, a coordinate transformation was employed so that the shock wave became a coordinate boundary. The Rankine-Hugoniot conditions were used both to determine the flow variables immediately upstream of the shock and to determine the shock position. Since all the physical quantities on the downstream of the shock were prescribed the flow variables on the upstream side were obtained from the Rankine-Hugoniot relations. Note that the shock boundary is supersonic and therefore all the quantities must be specified and no special boundary treatment is necessary. The fluid motion was modeled by the two-dimensional Euler equation in nonconservation form. Also a spectral filtering in which the high modes were filtered every fifty time steps was employed to avoid nonlinear instability. Beautiful results were obtained for various shock interactions and for the blunt body problem.

In the third approach proposed in a forthcoming paper by Abarbanel and Gottlieb, the oscillations are being used to recover accurate information about the solution. Oscillations may arise from different sources; e.g., incorrect treatment of the boundaries in hyperbolic systems; nonlinear instabilities, etc. Usually these oscillations build up and finally cause explosive instabilities. One interesting class of numerical oscillations occur when flows with extreme gradients or local discontinuities are simulated. This type of oscillations does not cause instabilities even after many time steps. It has been observed (see [7]) that the wiggles are caused by the fact that the mesh is not fine enough to resolve the sharp gradients. In the case of a finite gradient a local refinement of the mesh often gets rid of the wiggles. For a very impressive demonstration of this fact, see [17]. Of course for a shock wave, no refinement of the mesh can remove the oscillations.

To better understand the origin of the oscillatory solution, consider the model equation

$$\begin{aligned} u_t &= u_x \\ u(x,0) &= H(x,x_\ell) \end{aligned} \tag{3.1}$$

where $H(x,x_\ell)$ is the Heaviside function

$$H(x,x_\ell) = 0 \qquad x < x_\ell$$

$$H(x,x_\ell) = 1 \qquad x \geq x_\ell$$

$$X_\ell = \cos \frac{\pi}{N} (\ell + \tfrac{1}{2}), \qquad \ell \text{ integer.}$$

When (3.1) is discretized by the pseudospectral Chebyshev method we get as the initial condition

$$u_N(x,0) = S(x,x_\ell) = \sum_{k=0}^{N} A_k T_k(x) \tag{3.1a}$$

where $T_k(x)$ is the Chebyshev polynomial of order k, and

$$A_0 = \frac{1}{N} (\ell + \tfrac{1}{2}), \qquad A_N = \frac{1}{2N} \sin \pi(\ell + \tfrac{1}{2})$$

$$A_k = \frac{1}{N} \sin \frac{k\pi}{N} (\ell + \tfrac{1}{2})/\sin \frac{k\pi}{2N}, \qquad 1 \leq k \leq N-1.$$

At the grid points, $x_j = \cos \frac{\pi j}{N}$

$$S(x_j, x_\ell) = H(x_j, x_\ell),$$

Thus, no oscillations occur. However, after the numerical solution is convected by equation (3.1), it becomes oscillatory. This is because initially it is oscillatory between the grid points (see Fig. 1). Observe that the oscillations disappear when the discontinuity is exactly in the middle between two grid points. This demonstrates the fact that the structure of the oscillations provides information about the position and magnitude of the shock.

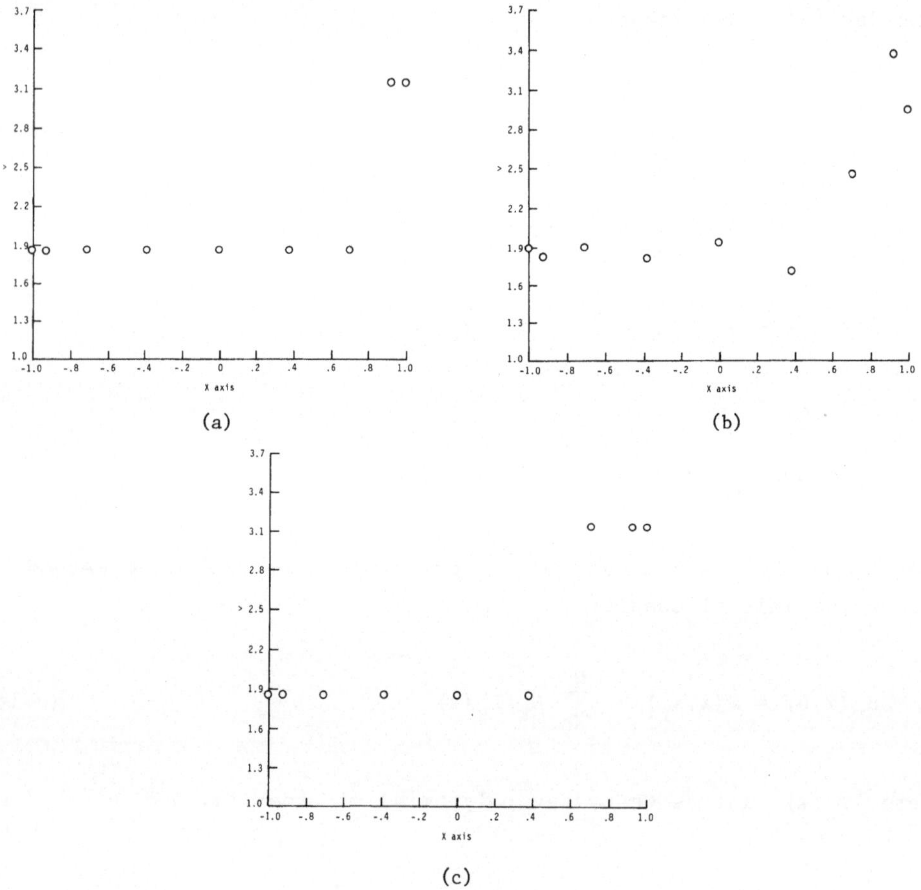

Figure 1

In general, consider (2.7) - (2.8) where now L is a linear operator and u_0 is discontinuous. From the last example it is clear that u_N does not approximate well $P_N u$ since $P_N u$ coincides with u at the grid points. We introduce an auxiliary equation

$$\frac{\partial v}{\partial t} = Lv$$
$$v(t=0) = P_N u_0. \tag{3.2}$$

For fixed N, v is a smooth function in contrast to the solution u of (2.7). We argue that u_N approximates (at the grid points) v rather than u. In fact from (2.8) and (3.2) one gets

$$\frac{\partial}{\partial t}(u_N - P_N v) = P_N LP_N(u_N - P_N v) + P_N L(P_N v - v)$$
$$(u_N - P_N v)(t=0) = 0. \tag{3.3}$$

Thus

$$u_N - P_N v = \int_0^t [\exp P_N LP_N(t - \tau)][P_N L(P_N v(\tau) - v(\tau))]d\tau.$$

The operator $\exp P_N LP_N(t - \tau)$ is bounded. This is, in essence, the notion of stability. The term

$$P_N L(P_N v - v)$$

is small because v is a smooth function. This shows that u_N approximates $P_N v$, hence at the grid points u_N approximates v.

In the last example we have demonstrated the fact the v is, in general, oscillatory. It is therefore no surprise that u_N is oscillatory. It is also clear that the structure of the oscillations may be used to extract a better approximation to u.

We will demonstrate now the possibility of extracting information from an oscillatory solution even in the nonlinear case. The physical problem is the well-known wedge flow. A plate is inserted in a uniform flow, and an oblique shock develops. The time dependent Euler equations in two-space dimensions were discretized by the pseudospectral Chebyshev method in space with a 9×9 grid and a modified Euler scheme was used for the time discretization (see [5]). Since we are interested in the steady state only, the accuracy of the time integration is of no importance. In order to be sure that a steady state is reached the code was run until all the physical quantities did not change to 11 significant figures over a span of 100 time steps. The values of the density in the steady state at the grid points together with the grid points themselves are given in Fig. 2.

	ρ									Y
	1.862	1.851	1.869	1.871	1.837	1.865	1.892	1.885	1.878	1.
	1.862	1.870	1.867	1.820	1.870	1.954	1.899	1.803	1.759	.961
	1.862	1.854	1.852	1.904	1.877	1.770	1.782	1.864	1.900	.853
	1.862	1.871	1.876	1.812	1.838	1.969	1.975	1.884	1.841	.691
	1.862	1.848	1.842	1.935	1.899	1.703	1.710	1.890	1.984	.5
	1.862	1.883	1.894	1.729	1.832	2.429	2.994	3.255	3.316	.308
	1.862	1.808	1.810	2.387	3.133	3.375	3.224	3.054	3.002	.146
	1.862	2.115	2.868	3.288	3.176	2.965	3.006	3.136	3.187	.038
	1.862	3.083	3.046	2.975	3.087	3.108	3.024	3.013	3.016	0
X	0	.038	.146	.308	.5	.691	.853	.961	1.	

Figure 2

Note that at the stations: $x_0 = 1$; $x_1 = .9619$; $x_2 = .85355$, the jump takes place between the grid points $y = .3086$ and $y = .5$, whereas the corresponding correct shock location is $y = .434$ for x_0, $y = .417$ for x_1 and $y = .370$ for x_2. Note also that the oscillatory behavior of the density is very similar to the behavior of $P_N\, v$, the solution of (3.2) at the grid points (see Fig. 1).

We therefore fit a step-function of the form $d_1 + d_2\, S(y,y_\ell)$ where $S(y,y_\ell)$ is defined in (3.1a) to the numerical results $\rho(y)$ in Fig. 2, at any station x_j, regarding d_1, d_2 and ℓ as unknowns. This yields three equations

$$
\begin{aligned}
d_1\, f_0 + d_2\, f_1 &= S_1 \\
d_1\, f_1 + d_2\, f_2 &= S_2 \qquad (3.4) \\
d_1\, f_4 + d_2\, f_3 &= S_3
\end{aligned}
$$

where $f_0 = N$; $f_1 = \sum_{j=0}^{N} S(y_j,y_\ell)\, \frac{1}{c_j}$; $\quad f_2 = \sum_{j=0}^{N} S(y_j,y_\ell)^2\, \frac{1}{c_j}$;

$f_3 = \sum_{j=0}^{N} S(y_j,y_\ell) \frac{\partial}{\partial \ell} S(y_j,y_\ell) \frac{1}{c_j}$; $f_4 = \sum_{j=0}^{N} \frac{\partial S}{\partial \ell} (y_j,y_\ell) \frac{1}{c_j}$;

$S_1 = \sum_{j=0}^{N} \rho(y_j) \frac{1}{c_j}$; $S_2 = \sum_{j=0}^{N} \rho(y_j) S(y_j,y_\ell) \frac{1}{c_j}$;

$S_3 = \sum \rho(y_j) \frac{\partial S}{\partial \ell} (y_j,y_\ell)$. Equation (3.4) yields the following nonlinear equation for the shock location y_ℓ

$$\begin{vmatrix} f_0 & f_1 & S_1 \\ f_1 & f_2 & S_2 \\ f_4 & f_3 & S_3 \end{vmatrix} = 0 \qquad (3.5)$$

Surprisingly, from (3.5) we recover the correct location of the shock at each x-station within the fourth significant digit. In this sense the information is indeed hidden in the form of oscillations.

It should be noted that in (3.4) we do not use the point values of $\rho(y)$ but rather the quantities S_1, S_2, S_3 which are equivalent to the integral of $\rho(y)$ against 1, $S(y,y_\ell)$ and $\frac{\partial}{\partial \ell} S(y,y_\ell)$. If $\rho(y)$ approximates well the first N modes of the solution $\rho_{ext}(y)$, then

$$\int_{-1}^{1} (\rho(y) - \rho_{ext}(y)) \frac{\phi(y)}{\sqrt{1-y^2}} = 0$$

where $\phi(y)$ is either 1 or $S(y,y_\ell)$ or $\frac{\partial S}{\partial \ell} (y,y_\ell)$. This may be the reason for the highly accurate values of the location of the shock obtained by (3.4).

Finally, we would like to describe another way of recovering correct point values from an oscillatory approximation. For simplicity we consider the spectral Legendre method although this idea has been generalized to other spectral methods. Our approach is motivated by the work of Mock and Lax (see [10]).

Suppose that $f(x)$ is a C^∞ function at $|x| < 1$ except for one point of discontinuity. Suppose also that $f(x)$ has the following expansion in terms of the Legendre polynomials

$$f(x) = \sum_{k=0}^{\infty} a_k P_k(x)$$

and that

$$f_N(x) = \sum_{k=0}^{N} a_k P_k(x) .$$

Even for large N, $f_N(x)$ is an oscillatory function. Let y be a point such that $f(x)$ is C^∞ in the interval $y-\varepsilon < x < y+\varepsilon$. Let

$$\psi(x) = \begin{cases} \frac{1}{\varepsilon}(1-\xi^2)^q \sum_{k=0}^{P} (2k+1)P_k(0)P_k(\xi) & |\xi| < 1 \qquad \xi = \frac{x-y}{\varepsilon} \\ 0 & |\xi| > 1 \end{cases} .$$

It is clear that

$$\int_{-1}^{1} f_N(x)\psi(x)dx = \int_{-1}^{1} f(x)\psi(x)dx + \int_{-1}^{1} (f_N - f)\psi dx .$$

The function $\psi(x)$ has the expansion

$$\psi(x) = \sum_{k=0}^{\infty} b_k P_k(x)$$

and since $\psi(x)$ has $q-1$ continuous derivative the function $\psi_N(x)$

$$\psi_N(x) = \sum_{k=0}^{N} b_k P_k(x)$$

approximates ψ with high accuracy. Moreover, since $\psi_N(x)$ is a polynomial of degree N

$$\int_{-1}^{1}(f_N - f)\psi dx = \int_{-1}^{1}(f_N - f)(\psi - \psi_N)dx < \|f - f_N\| \, \|\psi - \psi_N\|$$

$$< \kappa \frac{\|\psi^{(q-1)}\|}{N^{q-1}} .$$

The last estimate can be found in [1].

It is therefore clear that

$$\int f_N \psi \, dx = \int f\psi \, dx + E_1$$

where E_1 is small. Moreover,

$$\int_{-1}^{1} f(x)\psi(x)dx = \int_{-1}^{1} f(y+\varepsilon\xi)(1-\xi^2)^q \sum_{k=0}^{P} (2k+1)P_k(0)P_k(\xi)d\xi.$$

Let

$$g(\xi) = f(y + \varepsilon\xi)(1- \xi^2)^n .$$

$g(\xi)$ is a C^∞ function for $|\xi| < 1$ and therefore has a rapidly converging expansion of the form

$$g(\xi) = \sum_{k=0}^{\infty} c_k \, P_k(\xi).$$

Therefore

$$\int_{-1}^{1} g(\xi) \sum_{k=0}^{P} (2k+1)P_k(0)P_k(\xi)d\xi = \sum_{k=0}^{P} c_k \, P_k(0)$$

$$= q(0) - \sum_{k=p+1}^{\infty} c_k = f(y) + E_2.$$

This shows that

$$\int f_N \psi \, dx$$

approximates $f(y)$ to a high order of accuracy. This filter had been successfully used by Gottlieb and Gruberger for several problems.

In conclusion we have demonstrated that numerical solutions obtained by spectral methods contain information about the correct solution that may be extracted to yield a high order approximation in the regular sense.

References

[1] Canuto, C. and Quarteroni, A., "Approximation results for orthogonal polynomials in Sobolev spaces," Math. Comput., 38, 1982, pp. 67-86.

[2] Cornille, D., "A pseudospectral scheme for the numerical calculation of shocks," J. Comput. Phys., 47, 1982, pp. 146-159.

[3] Gottlieb, D., Hussaini, M. Y., and Orszag, S. A., Theory and Applications of Spectral Methods, Proc. of the Symposium of Spectral Methods for Partial Differential Equations, SIAM, 1984, pp. 1-55.

[4] Gottlieb, D., Lustman, L. and Orszag, S. A., "Spectral calculations of one-dimensional inviscid compressible flow," SIAM J. Sci. Statis. Comput., 2, 1981, pp. 296-310.

[5] Gottlieb, D., Lustman, L. and Streett, C., "Spectral methods for two-dimensional flows," Proc. of the Symposium on Spectral Methods for Partial Differential Equations, SIAM, 1984, pp. 79-96.

[6] Gottlieb, D. and Orszag, S. A., Numerical Analysis of Spectral Methods: Theory and Applications, CBMS Regional Conference Series in Applied Mathematics, 26, SIAM, 1977.

[7] Gresho, P. and Lee, R. L., "Don't surpress the wiggles, they're telling you something," Comput. & Fluids, 1981, pp. 223-254.

[8] Hussaini, M. Y., Kopriva, D. A., Salas, M. D., and Zang, T. A., "Spectral methods for Euler equations," AIAA-83-1942-CP, Proc. of the 6th AIAA Computational Fluid Dynamics Conference, Danvers, MA, July 13-15, 1983.

[9] Hussaini, M. Y., Salas, M. D., and Zang, T. A., "Spectral methods for inviscid, compressible flows," in Advances in Computational Transonics, W. G. Habshi, ed., Pineridge Press, Swansea, UK, 1983.

[10] Lax, P. D., "Accuracy and resolution in the computation of solutions of linear and nonlinear equations," in Recent Advances in Numerical Analysis, Proc. Symp., Mathematical Research Center, University of Wisconsin, Academic Press, 1978, pp. 107-117.

[11] Reddy, K. C., "Pseudospectral approximation in three-dimensional Navier-Stokes code," AIAA J., Vol. 21, No. 8, 1983, pp. 1208-1210.

[12] Sakell, L., "Solution to the Euler equation of motion, pseudospectral techniques," Proc. 10th IMACS World Congress System, Simulation and Scientific Computing, 1982.

[13] Salas, M. D., Zang, T. A. and Hussaini, M. Y., "Shock-fitted Euler solutions to shock-vortex interactions," Proc. of the 8th International Conference on Numerical Methods in Fluid Dynamics, Lecture Notes in Physics 170, (E. Krause, ed.), Springer-Verlag, 1982, pp. 461-467.

[14] Streett, C. L., "A spectral method for the solution of transonic potential flow about an arbitrary airfoil," AIAA-83-1949-CP, Proc. of the 6th AIAA Computational Fluid Dynamics Conference, Danvers, MA, July 13-15, 1983.

[15] Taylor, T. D., Myers, R. B., and Albert, J. H., "Pseudospectral calculations of shock waves, rarefaction waves and contact surfaces," Comput. Fluids, 9, 1981, pp. 469-473.

[16] Zang, T. A. and Hussaini, M. Y., "Mixed spectral/finite difference approximations for slightly viscous flows," Lecture Notes in Physics 141, Springer-Verlag, 1980, pp. 461-466.

[17] Zang, T. A, Kopriva, D. A. and Hussaini, M. Y., "Pseudospectral calculation of shock turbulence interactions," Proc. of the 3rd International Conference on Numerical Methods in Laminar and Turbulent Flow, (C. Taylor, ed.), Pineridge Press, 1983.

GLOBAL RELAXATION PROCEDURES FOR A REDUCED FORM
OF THE NAVIER-STOKES EQUATIONS

S.G. RUBIN
UNIVERSITY OF CINCINNATI
CINCINNATI, OHIO

INTRODUCTION

In appropriate streamline coordinates, a reduced form of the Navier-Stokes equations (RNS) can be considered for complex problems where viscous/inviscid pressure interaction is an essential element of the flow behavior. This system is solved by global (multi-sweep) relaxation methods and defines subsonic and transonic inviscid behavior, as well as viscous boundary layer, triple deck and separated flow interactions. The reduced system of equations includes the Euler or full potential subset, as well as the viscous component describing second-order boundary layer, interacting boundary layer (IBL) and single-sweep parabolized Navier-Stokes (PNS) approximations.Unlike coupled inviscid/boundary layer techniques that match the inner and outer solutions, a single RNS system is applied throughout the flow domain.

Two procedures are considered: (1) direct primitive variable (PV) relaxation for the pressure[1-9,22], and (2) a composite velocity (CV) procedure[11-14] that is formulated in the spirit of matched asymptotic expansion theory and results in a relaxation technique for a pseudo-potential function ϕ. Solutions have been obtained for incompressible, subsonic and transonic full potential and Euler equations, as well as, with laminar and turbulent models for the complete RNS system. Trailing edge, trough, boattail, nozzle and airfoil geometries have been considered.

GOVERNING EQUATIONS

The governing RNS equations are presented here for a body fitted conformal coordinate system (ξ,η) with metric $h(\xi,\eta)$. For primitive variables, we define the velocity components (u,v), the pressure p, the density ρ and the temperature T. For the composite system, we define new (U,ϕ,G) variables as follows. The velocity and modified pressure variable G components are related by

$$u = U(1+\phi_\xi)/h = Uu_e, \quad v = \phi_\eta/h, \quad G = \frac{\gamma p}{(\gamma-1)\rho} + \frac{u_e^2 + v^2}{2} - \left(\frac{\gamma}{\gamma-1}\frac{p_\infty}{\rho_\infty} + \frac{u_\infty^2}{2}\right) + \phi_t .$$

The continuity and momentum equations are of the following form:

Continuity:

$$\frac{\partial\rho}{\partial t} + (\rho h y^\varepsilon u)_\xi + (\rho h y^\varepsilon v)_\eta = 0 \tag{1a}$$

or

$$\frac{\partial\rho}{\partial t} + (\rho y^\varepsilon U\phi_\xi)_\xi + (\rho y^\varepsilon \phi_\eta)_\eta = -(\rho y^\varepsilon U)_\xi \tag{1b}$$

ξ-momentum:

$$\frac{\partial\rho u}{\partial t} + \frac{1}{hy^\varepsilon}(\rho h y^\varepsilon u^2)_\xi + \frac{1}{hy^\varepsilon}(\rho uvhy^\varepsilon)_\eta + (\rho uvh)_\eta/h - (\rho uv)_\eta - (\rho v^2 h)_\xi/h$$
$$+ (\rho v^2)_\xi = -p_\xi + \frac{1}{y^\varepsilon}\left[\frac{y^\varepsilon}{h^2}\mu(hu)_\eta\right]_\eta \tag{2a}$$

or $$(\rho u_e U)_t + \frac{\rho(U-1)}{h}\phi_{\xi t} + \frac{1}{y^\varepsilon h^2}\left[(\rho h y^\varepsilon u_e^2(U^2 - U))_\xi + (\rho h y^\varepsilon u_e v(U-1))_\eta\right]$$
$$+ \frac{\rho h_\eta}{h^2} u_e v(U-1) + \frac{\rho}{h}(U-1)u_e u_{e_\xi} = -\frac{\rho}{h} G_\xi + \frac{\rho T}{h} s_\xi + \frac{1}{y^\varepsilon}\left[\frac{y^\varepsilon}{h}\mu(U(1+\phi_\xi)_\eta\right]_\eta \qquad (2b)$$

η-momentum:

$$- p_\eta = \frac{1}{hy^\varepsilon}(\rho uvhy^\varepsilon)_\xi + (\rho uvh)_\xi/h - (\rho uv)_\xi$$
$$+ \frac{1}{hy^\varepsilon}(\rho v^2 hy^\varepsilon)_\eta - (\rho u^2 h_\eta)/h + (\rho u^2)_\eta + (\frac{\rho}{h}\phi_\eta)_t \qquad (3a)$$

or $$Ts_\eta = G_\eta + (U-1)\left[(\frac{u_e^2}{2})_\eta - \frac{h_\eta}{h} u_e^2 U\right] \qquad (3b)$$

where s is the entropy; $\varepsilon = 0,1$ for two-dimensional and axi-symmetric flow, respectively. The energy and state equations close the system. For transonic flow, the CV form of the ξ-momentum equation must be modified slightly if entropy variations in the inviscid flow are to be included. As presently formulated the inviscid region is represented by the full potential equation. The PV system, as given, allows for entropy variations and for $\mu \to 0$, the full Euler system results.

The governing equations include the elliptic viscous/inviscid pressure interaction. This manifests itself through the p_ξ term in the PV system and through the u_{e_ξ} or $\phi_{\xi\xi}$ terms appearing in the CV system. The usual three-point central difference approximation for $\phi_{\xi\xi}$ automatically introduces the upstream or elliptic influence in subsonic regions. The artificial compressibility concept is applied for supersonic regions[13,14]. For the (u,v,p) variables, upstream influence is introduced through the difference form of p_ξ. From an eigenvalue or stability analysis[1-5,10], it has been shown that $p_\xi = \omega(p_\xi)_h + (1-\omega)(p_\xi)_e$, where $0 \le \omega \le \omega_M \le 1$ and ω_M is a function of the local Mach number. The subscripts h and e denote the "hyperbolic" or marching portion of p_ξ and the elliptic or "downstream interaction" portion of p_ξ, respectively. These terms are differenced with "backward" and "forward" forms, respectively. In addition, for shock capturing, ω must be suitably adjusted to insure that the Rankine-Hugoniot condition is satisfied[8]. For both the PV and CV calculations explicit artificial viscosity is not added. However, numerical viscosity does appear if first-order upwind differencing is applied for the ξ-derivatives in the PV system, or through the artificial compressibility introduced in the CV formulation for transonic flow.

COMPOSITE VELOCITY FORMULATION

The CV form of the RNS equations (1b,2b,3b) reduces directly to the interacting boundary layer approximation when the η or normal momentum equation is completely uncoupled from the continuity and ξ-momentum equations. This implies that the term $\rho G_\xi - \rho TS_\xi$ in (2b) is negligible, even in viscous regions. Ordinary boundary layer theory is recovered if, in addition, u_e and u_{e_ξ} are prescribed. Inviscid potential flow equations are obtained for $\mu = 0$, $U = 1$.

CV BOUNDARY CONDITIONS

For geometries that are unbounded, at the inflow $\xi = \xi_0$, $U = 1$, $\phi = 0$, $H = H_\infty$, except at $\eta = 0$, where $U = 0$, $H = H_w$. H is the stagnation enthalpy, so that

$H = G + \frac{u_e^2(U^2-1)}{2}$. Since the boundary layer growth begins at $\xi = \xi_0$, the downstream solution will be significantly affected by the ratio of ξ_0/R, where R is a typical body dimension. Therefore, the inflow location can alter strong pressure interactions and regions of recirculation. On the surface $\eta = 0$, the no slip and zero injection conditions are $U = \phi_\eta = 0$, respectively; at the upper boundary $\eta \to \infty$, $U \to 1$, $\phi \to 0$, $H \to H_\infty$, $s \to 0$. At the outflow $\xi = \xi_1$, $U_{\xi\xi} = 0$ and $\phi_\xi = h-1$. The latter condition implies a standard non-interacting boundary-layer approximation; i.e., the pressure is prescribed. This is typically the condition used in time-dependent calculations. As $\xi \to \infty$, $h \to 1$ and $\phi_\xi \to 0$. Additional considerations of symmetry are imposed at $\eta = 0$ for the upstream and downstream wake regions. For inviscid calculations a connection condition is required for $U(\xi,0)$. This can be inferred from the vorticity $\Omega = [(1-U)(1+\phi_\xi)]_\eta$. At the surface $\eta = 0$, $\Omega = -(1+\phi_\xi)U_\eta$.

CV SOLUTION PROCEDURE

The governing equations have been discretized using second-order accurate central differencing for all derivatives of ϕ. Both central differencing and second-order upwinding have been applied for the ξ-convective terms in the momentum equation. For the $(\rho y^\varepsilon U)_\xi$ term in the continuity equation, it was found that three-point backward differencing near the surface, with central differencing elsewhere, provides the best simulation of viscous and inviscid regions, respectively[13,14]. Second-order accuracy of this term and the convective terms in ξ-momentum (2) is achieved via a deferred correction procedure[14,15]. Only the first-order accurate two-point backward differencing is considered implicitly; the remainder of the difference expression is included explicitly, e.g., for central-differencing $A_\xi = \left(\frac{A_{i,j} - A_{i-1,j}}{\Delta\xi}\right) + \left(\frac{A_{i+1,j} - 2A_{i,j} + A_{i-1,j}}{2\Delta\xi}\right)^{n-1}$, where n is the global iteration counter. This deferred correction procedure can easily be implemented in situations where different expressions for the correction are used in different regions. In order to realize improved stability characteristics for large Reynolds numbers and larger values of Δt, it was found advantageous to reinforce the implicit terms by appropriately modifying the deferred corrector[13,14]. This idea was used earlier for spline correctors in reference 16.

COUPLED STRONGLY IMPLICIT PROCEDURE (CSIP)

In an earlier paper[17], the CSIP procedure has been presented for the vorticity-streamfunction system. This algorithm has the distinct advantage of being implicit in both the ξ and η directions, as well as allowing for the coupling of all boundary conditions. Furthermore, the method has strong stability properties, allows for arbitrarily large Δt and is relatively insensitive to the grid aspect ratio. The discretized version of the equations can be written as: $(A + P)V^n = G + PV^{n-1}$, where P is chosen such that (A + P) can be decomposed into a lower and upper triangular form having a sparsity pattern similar to the original matrix A. This leads to a solution algorithm of the following form:

$$V^n = \begin{vmatrix} U_{ij} \\ \phi_{ij} \end{vmatrix}^n = \begin{vmatrix} GM_{1_{ij}} \\ GM_{2_{ij}} \end{vmatrix}^{n-1} + \begin{vmatrix} T_{1_{ij}} & T_{3_{ij}} \\ T_{5_{ij}} & T_{7_{ij}} \end{vmatrix}^{n-1} \begin{vmatrix} U_{i,j-1} \\ \phi_{i,j-1} \end{vmatrix}^n + \begin{vmatrix} T_{2_{ij}} & T_{4_{ij}} \\ T_{6_{ij}} & T_{8_{ij}} \end{vmatrix}^{n-1} \begin{vmatrix} U_{i-1,j} \\ \phi_{i-1,j} \end{vmatrix}^n$$

As shown in reference 17, the recurrence relationships can easily be obtained. Although the coupling accelerates the rate of convergence, it increases the storage requirement. A scheme for reducing storage is presented in reference 14.

CV SOLUTIONS

References 11-14 contain a variety of CV solutions. Several representative examples are included here. In Fig. 1 a typical error plot for the CSIP is shown for different Δt increments. Although convergence can be achieved for all values of Δt, for $\Delta t > 1$ there is no gain in convergence rate and more care must be exercised in the choice of initial conditions, especially for fine grids[14]. Therefore, all CV calculations are for $\Delta t = 1$. Somewhat smaller values may be reqiured for finer meshes during the early stages of the iteration process[13]. Figs. (2a,2b) compare turbulent RNS and inviscid solutions for boattail and airfoil geometries. A comparison with experimental data is given in Fig. 2b. For the boattail configuration, Figs. (3a,3b) depict the sensitivity of the solution to the inflow boundary location ξ_0 and to the mesh resolution. It is evident that meaningful viscous interaction solutions require very fine meshes and accurate inflow conditions. A transonic result is presented in a later section.

PRIMITIVE VARIABLE FORMULATION

The RNS equations in the form (1a,2a,3a) were first applied for hypersonic problems where the contribution of the pressure gradient p_ξ in the momentum equation is negligible[18,19]. The system is then mathematically parabolic and can be solved as an initial value problem by a single pass or PNS marching technique. For lower Mach numbers, where p_ξ must be retained, single sweep PNS marching then leads to an ill-posed initial value problem and exponentially growing departure solutions appear[19] for step sizes $\Delta\xi < (\Delta\xi)_{min}$, where $(\Delta\xi)_{min}$ is proportional to the extent of the subsonic portion of the flow[7]. For incompressible flow, $(\Delta\xi)_{min}$ is proportional to the total extent of the computational boundary η_M in the surface normal direction[1-3]. This implies that for $\eta_M \to \infty$, $(\Delta\xi)_{min} \to \infty$, and the entire p_ξ contribution is elliptic. For subsonic flows, $\omega \ll 1$ over a range of η values so that $(\Delta\xi)_{min} \gg 1$.

In order to circumvent the ill-posedness of single sweep PNS methods, a global pressure relaxation or repeated marching procedure has been proposed. This requires an appropriate "forward" or mid-point difference treatment of the $(p_\xi)_e$ contribution. Consistent ($\Delta\xi$ arbitrary), departure free ($\Delta\xi \to 0$) and rapidly convergent solutions have been obtained for viscous and inviscid flows. Strong pressure interactions and separation have been captured with the global RNS procedure.

PV DIFFERENCE EQUATIONS AND SOLUTION TECHNIQUE

The equations (1a,2a,3a) are differenced on the staggered grid of Fig. 4. The unknown pressure p_i is a distance $(1-\omega)\Delta\xi$ upwind of the velocity u_i. For incompressible flow, where $\omega = 0$, this is one mesh point, while for supersonic flow, where $\omega = 1$, the two locations coincide. The equations are shown herein for non-conservation cartesian coordinates, with $\Delta t = \infty$ and uniform meshes in ξ and η. In fact, non-uniform meshes, conservation equations and conformal coordinates are used for all of the problems considered herein.

Continuity: centered at c on Fig. 4.

$$[(\rho u)_{i,j} - (\rho u)_{i-1,j} + (\rho u)_{i,j-1} - (\rho u)_{i-1,j-1}]/2 + \frac{\Delta\xi}{\Delta\eta}[(\rho v)_{i,j} - (\rho v)_{i,j-1}] = 0 \quad (4)$$

ξ-momentum: centered at ξ

$$(\rho u)_{i,j}(u_{i,j} - u_{i-1,j}) + (\rho v)_{i,j}\frac{\Delta\xi}{2\Delta\eta}(u_{i,j+1} - u_{i,j-1}) + p^{n-1}_{i+1} - p_{i,j} + C_{ij}\Delta\xi$$

$$= \frac{\Delta\xi\mu_{i,j+\frac{1}{2}}}{\Delta\eta^2}(u_{i,j+1} - u_{i,j}) - \frac{\Delta\xi\mu_{i,j-\frac{1}{2}}}{\Delta\eta^2}(u_{i,j} - u_{i,j-1}) + S_{ij} \quad (5)$$

η-momentum: centered at η

$$\left[\frac{(\rho u)_{i,j} + (\rho u)_{i,j-1}}{4\Delta\xi}\right][v_{i,j} - v_{i-1,j} + v_{i,j-1} - v_{i-1,j-1}]$$

$$+ \left(\frac{(\rho v)_{i,j} + (\rho v)_{i,j-1}}{2\Delta\eta}\right)(v_{i,j} - v_{i,j-1}) + \frac{p_{i,j} - p_{i,j-1}}{\Delta\eta} = 0 \quad (6)$$

where S_{ij} was introduced to enhance the relaxation process for $\omega = 0$[20,3].

$$S_{ij} = T_{ij} + p_{ij} \text{ and } T_{ij} = T_{i-1,j} - (p^{n-1}_{ij} - p_{i-1,j})$$

S_{ij} vanishes at convergence. The quantity C_{ij} in (5) is defined by

$$C_{ij} = \omega\frac{(p^{n-1}_{i+1,j} - 2p_{ij} + p_{i-1,j})}{\Delta\xi} \text{ or } C_{ij} = \omega\left(\frac{p^{n-1}_{i+1,j} - 2p^{n-1}_{ij} + p_{i-1,j}}{\Delta\xi}\right) \quad (7)$$

The former expression is equivalent to the following representation of p_ξ in (2a)

$$p_\xi = \omega\frac{(p_{ij} - p_{i-1,j})}{\Delta\xi} + (1-\omega)\left(\frac{p^{n-1}_{i+1,j} - p_{ij}}{\Delta\xi}\right) = \omega(p_\xi)_h + (1-\omega)(p_\xi)_e \quad (8)$$

The latter expression in (7) increases the range of ω for which stable marching solutions are possible ($\omega \geq \omega_M$)[8]. An alternate form for (7,8) is

$$p_\xi = (p_\xi)_e - \omega\Delta\xi(p_{\xi\xi}) + O(\omega\Delta\xi^2), \text{ where } C_{ij} = -\omega\Delta\xi\, p_{\xi\xi} + O(\omega\Delta\xi^2) \quad (9)$$

Therefore, C_{ij} can be considered as a compressibility correction to the incompressible ($\omega = 0$) pressure differencing for p_ξ. For subsonic flow, the choice of ω has little effect on the overall accuracy of the calculation[8]. The influence of ω on the accuracy and convergence rate for transonic flow is presented in reference 8.

The quasi-linearized tridiagonal system (4-6) is solved for $(u,v,p)^T_{ij}$ by standard LU decomposition. The pressure is replaced by the density, stagnation enthalpy H and velocities (u,v) through the state equation and definition of H. The multi-sweep solution procedure is terminated when the change in maximum pressure and skin friction between global iterations is less than 10^{-4}.

PV BOUNDARY CONDITIONS

At the inflow $\xi = \xi_0$, $u(\xi_0,\eta) = U(\eta)$ and $v_\xi(\xi_0,\eta) = V(\eta) = \Omega(\eta) + U'(\eta)$, where $\Omega(\eta)$ is the inflow vorticity; for uniform conditions $U = 1$ and $V = 0$ (zero vorticity). A condition for the pressure is not required for incompressible flow. As seen from Fig. 4, the inflow pressure is calculated during the first marching step. For subsonic flow $(\omega p_{i-1,j})$ is specified at the inflow. At the upper surface

$\eta = \eta_M$, $u = 1$, $p = 0$; i.e., free-stream conditions are applied. This requires that η_M be sufficiently large, e.g., outside of the domain of the triple deck interaction. A boundary condition on v is not required. At the outflow $\xi = \xi_1$, only the pressure $p(\xi,\eta)$ or derivative $p_\xi(\xi_1,\eta)$ are prescribed. There are only slight differences in the solutions. At wall $\eta = 0$, for viscous flow $u(\xi,0) = v(\xi,0) = 0$ is specified. For inviscid flow $v(\xi,0) = 0$ and a zero vorticity connection condition for $u_\eta(\xi,0)$ is required. A boundary condition on the pressure is not required. Symmetry conditions, $u_\eta(\xi,0) = v(\xi,0) = 0$ are applied upstream and downstream where necessary.

CONVERGENCE AND MULTI-GRID PROCEDURE

From the global stability analysis[2-4], it has been shown, for incompressible flow, that the spectral radius λ of the linear system defining the line relaxation procedure is of the form $(\lambda-1) \sim (\Delta\xi/\eta_M)^2$. Therefore, as $\Delta\xi \to 0$, or for $\eta_M >> 1$, the rate of convergence will decrease markedly. It is interesting that the parameter $\Delta\xi/\eta_M << 1$, that leads to the departure solution for single sweep PNS methods, is the convergence factor for the global relaxation RNS procedure. It can be shown that for compressible flow the convergence will be more rapid as ω increases toward ω_M.This is particularly true for high subsonic or supersonic Mach numbers[8].

In order to improve the convergence rate, a one-dimensional (in ξ) multi-grid procedure following the full approximation scheme of reference 21 has been applied. This approach is less efficient, in terms of storage requirements, than a full two-dimensional multi-grid application; however, relaxation for p is only in the ξ-direction as the calculation is fully implicit in η. Also, a highly non-uniform mesh is required to accurately describe the boundary layer, triple deck and inviscid regions. This is not ideally suited for interpolation required by the multi-grid procedure. Of particular note, error transfer from coarse to fine grids is applied only for the pressure, even in regions of reversed flow where the velocity $u^{n-1}_{i+1,j}$ is also relaxed. These values are fixed on each grid from the previous global (n-1) iteration[3,4].

PV SOLUTIONS

PV results can be found in references (1-9,22). Some typical examples are presented here. Comparisons with interacting boundary layer solutions for a trough configuration (Fig. 5) verify the PV/RNS model for interactions on the scale of triple deck theory. Solutions for the trailing edge region of a flat plate have been obtained for laminar[2,3], turbulent[3,22] and compressible flow[8], see Fig. (6). Airfoil results are shown in Figs. (2b,7,9) for RNS and Euler equations[22].

TRANSONIC FLOW AND SHOCK CAPTURING

The direct (u,v,p) and composite (U,ϕ,G) formulations have recently been applied to transonic flow problems. Boattail, airfoil and nozzle geometries have been considered. As noted earlier the CV calculations imply a potential flow character for the outer inviscid region, so that only weak shock waves have been considered to date. The artificial compressibility concept is introduced for supersonic regions[13]. Preliminary results with a more "conservative" form of the CV equations have also been recently obtained. This system allows for entropy production in the inviscid or Euler region; however, the results are not completely satisfactory and further analysis is required. The PV equations in their present form reduce to the full Euler system in the inviscid region.

In order to test the PV formulation for flows with shock waves, the inviscid or Euler form of the equations was applied for a quasi-one-dimensional converging-diverging channel with a subsonic inflow. The global relaxation procedure described previously was applied. No artificial viscosity or compressibility was introduced; however, the ω parameter must be appropriately specified through the shock wave in order to insure that the Rankine-Hugoniot jump conditions are satisfied. Outside of the shock region, it is possible to specify any $\omega < \omega_M$ and obtain a high degree of accuracy for a fine mesh. The rate of convergence improves significantly when $\omega \sim \omega_M$ locally[8]. The spectral radius is decreased and the characteristic domain of dependence is more closely satisfied.

Figure 8a depicts the channel area distribution and the solutions with $\omega = \alpha\omega_M$ throughout the channel. The back pressure is specified and the exact solution leads to a sharp shock at $x = 0.5$. The inflow conditions are subsonic so that the inflow pressure is a free parameter and is determined, with the mass flow, from the calculation. The solution for $\Delta x = 0.02$ and $\alpha = 0$ ($\omega = 0$) leads to a shock wave slightly upstream of the correct location. For $\alpha = 0.3$, the shock wave location improves: however, for the $\alpha > 0.3$, the shock moves downstream and for $\alpha = 1$, the shock wave is located at the exit. This is, of course, incorrect and is due to the lack of ellipticity or upstream influence in supersonic regions where $\omega = 1$ or $p_x = (p_x)_h$. The shock jump conditions are not satisfied. It can be shown[8] that the shock jump conditions will be satisfied exactly for $\omega = 0$; for this ω value, $p + \rho u^2$ is conserved.

For fine meshes and $\omega = 0$, the computed shock wave approaches the exact location. For coarser meshes, a more accurate solution is obtained with $\omega = \omega_M$ ahead of the shock and $\omega = 0$ through the shock. There is little difference in the solution with $\omega = \omega_M$ or $\omega = 0$ behind the shock wave. Finally, in Fig. 8b the entropy variation in the channel and the effect of refining the mesh are shown. The solution is quite good and the entropy jump is very close to that given by the Rankine-Hugoniot conditions.

Viscous flow solutions are shown for an airfoil geometry in Fig. 9 for the CV and PV procedures. In the PV calculations, the shock waves are captured with $\omega = 0$. For supersonic regions $\omega = 1$. The results are reasonable, although the grids and mapping procedure as well as the ω distribution in the shock wave region should be modified to provide improved accuracy and reduce shock smearing. A similar comment applies to the artificial compressibility procedure for the CV calculations. Additional CV and PV transonic inviscid and viscous RNS solutions are given in references (6,8,13,14,22).

ACKNOWLEDGEMENT

The research reviewed here was supported in part by the Office of Naval Research under contract N00014-79-C-0849 and in part by the Air Force Office of Scientific Research under grant AFOSR 80-0047. The detailed version of this paper has been published elsewhere.

REFERENCES

1. Rubin, S.G. and Lin, A. (1980), "Marching with the PNS Equations, Israel Journal of Technology 18, pp. 211-222.
2. Rubin, S.G. (1982), "A Review of Marching Procedures for Parabolized Navier-Stokes Equations," Numerical and Physical Aspects of Aerodynamic Flows, Springer-Verlag, pp. 171-186.
3. Rubin, S.G. and Reddy, D.R. (1983), "Analysis of Global Pressure Relaxation for Flows with Strong Interaction and Separation," Computers and Fluids, 11, 4, pp. 281-306.
4. Rubin, S.G. and Reddy, D.R. (1983), "Global PNS Solutions for Laminar and Turbulent Flow," AIAA Paper No. 83-1911, see also University of Cincinnati Report No. AFL-84-101.
5. Khosla, P.K. and Lai, H.T. (1983), "Global PNS Solutions for Subsonic Strong Interaction Flows," Computers and Fluids, 11, 4, pp. 325-340.
6. Khosla, P.K. and Lai, H.T. (1984), "Global PNS Solutions for Transonic Strong Interaction Flows, AIAA Paper No. 84-0478.
7. Lin, A. and Rubin, S.G. (1981), "Three-dimensional Supersonic Viscous Flow Over a Cone at Incidence," AIAA J. 20, 11, pp. 1500-07.
8. Ramakrishnan, S.V. and Rubin, S.G. (1984), "Global Pressure Relaxation for Compressible Flows with Full Pressure Coupling and Shock Waves." University of Cincinnati, Report No. AFL-84-100.
9. Khosla, P.K. and Bender, E.E. (1984), "Solution of Parabolized Navier-Stokes Equations for Three-Dimensional Internal Flows," 9th ICNMFD, Saclay, France.
10. Vigneron, Y. et. al. (1978), "Calculation of Supersonic Flow Over Delta Wings with Sharp Leading Edges," AIAA Paper No. 78-1137.
11. Khosla, P.K. and Rubin, S.G. (1983), "A Composite Velocity Procedure for the Compressible Navier-Stokes Equation," AIAA J. 21, 11, pp. 1546-51.
12. Rubin, S.G. and Khosla, P.K. (1982), "A Composite Velocity Procedure for the Incompressible Navier-Stokes Equations," Lecture Notes in Physics, #170, Springer-Verlag, pp. 448-454.
13. Swanson, R.C., Rubin, S.G. and Khosla, P.K. (1983), "Calculation of Afterbody Flows with a Composite Velocity Formulation," AIAA Paper No. 83-1736.
14. Rubin, S.G., Celestina, M. and Khosla, P.K. (1984), "Second-Order Composite Velocity Solution for Large Reynolds Number Flow," AIAA Paper No. 84-0172.
15. Khosla, P.K. and Rubin, S.G. (1974), "A Diagonally Dominant Second-Order Implicit Scheme," Computers and Fluids, 2, 2, pp. 207-210.
16. Rubin, S.G. and Khosla, P.K. (1978), "A Simplified Spline Procedure," Lecture Notes in Physics #90, Springer-Verlag, pp. 468-476.
17. Rubin, S.G. and Khosla, P.K. (1979), "Navier-Stokes Calculations with Coupled Strongly Implicit Procedure," Computers and Fluids, 9, 2, pp. 163-180.
18. Rudman, S. and Rubin, S.G. (1968), "Hypersonic Viscous Flow Over Slender Bodies with Sharp Leading Edges," AIAA J. 6, 10, pp. 1883-1890.
19. Lin, T.C. and Rubin, S.G. (1973), "Viscous Flow Over a Cone at Moderate Incidence," Computers and Fluids, 1, 1, pp. 37-57.
20. Israeli, M. and Lin, A. (1982), "Numerical Solutions and Boundary Conditions for Boundary Layer Like Flows," Lecture Notes in Physics #170, Springer-Verlag, pp. 266-272.
21. Brandt, A. (1979), "Multi-Level Adaptive Computations in Fluid Dynamics," AIAA Paper No. 79-1455.
22. Reddy, D.R. and Rubin, S.G. (1984), "Subsonic/Transonic Viscous/Inviscid Relaxation Procedures for Strong Pressure Interactions," AIAA Paper No. 84-1627.

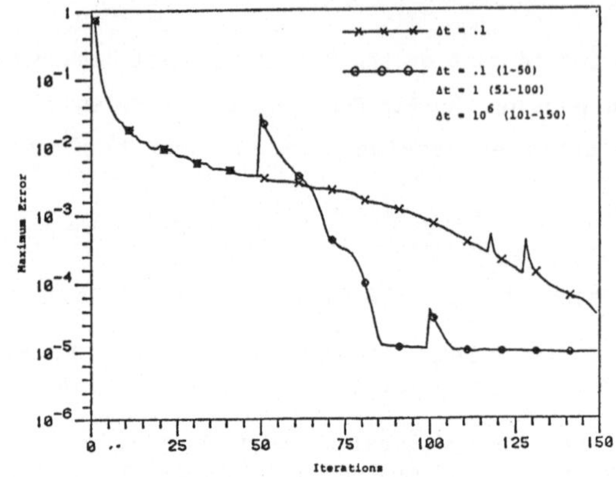

FIG. 1. MAXIMUM ERROR VS. ITERATIONS (CV).

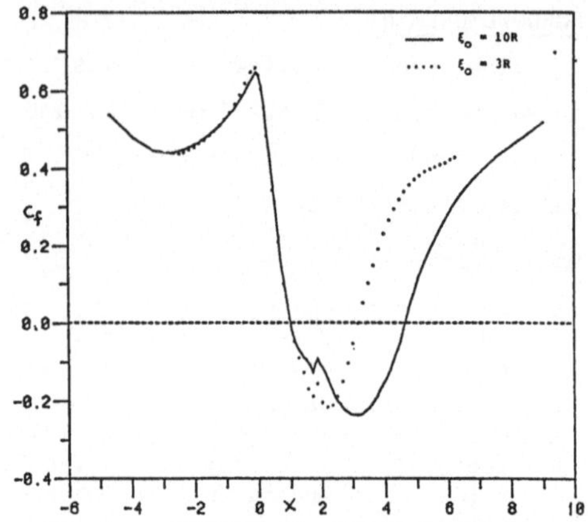

FIG. 3a. EFFECT OF INFLOW LOCATION BOATTAIL (CV).

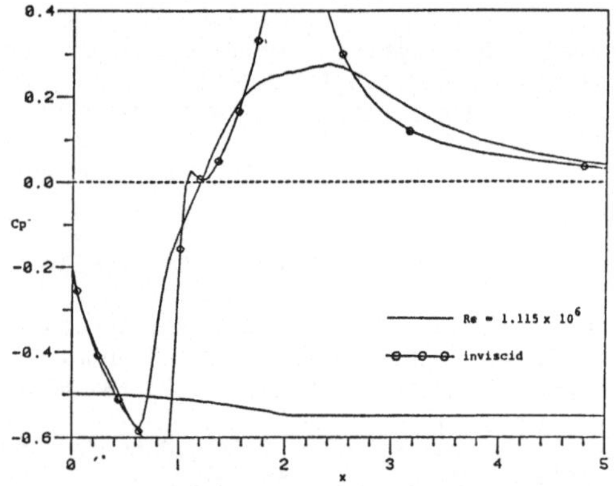

FIG. 2a. BOATTAIL, Re = 1.115 x 10^6, M_∞ = .9 (CV).

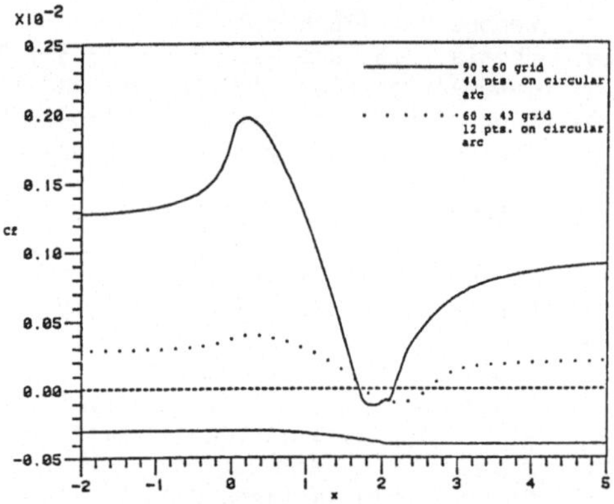

FIG. 3b. BOATTAIL, M_∞ = .8, Re = 1.115 x 10^6, (CV).

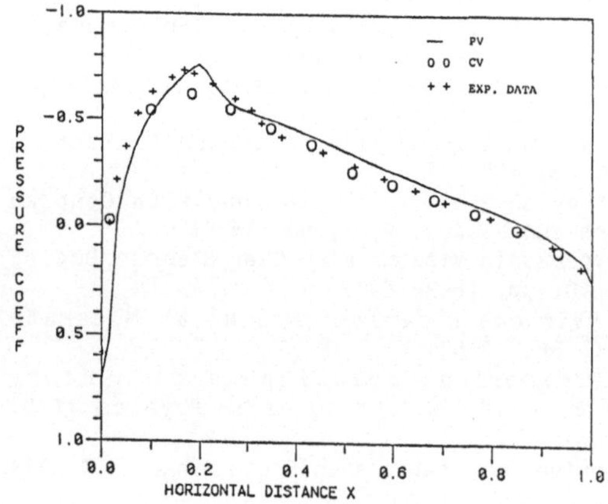

FIG. 2b. AIRFOIL, M_∞ = .75, Re = 4 x 10^6 (CV), (PV).

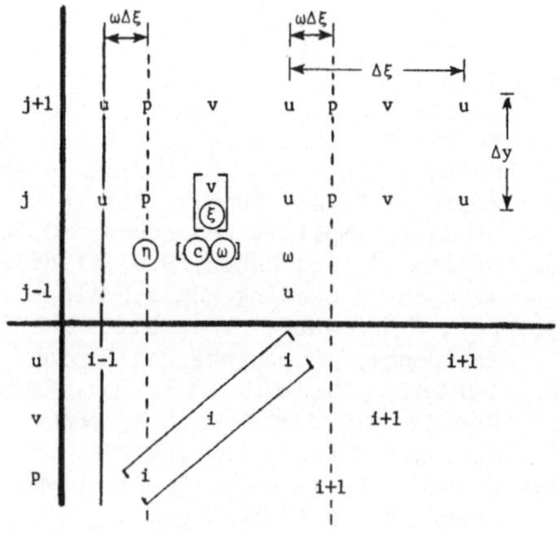

FIG. 4. STAGGERED GRID - COMPRESSIBLE FLOW (PV).

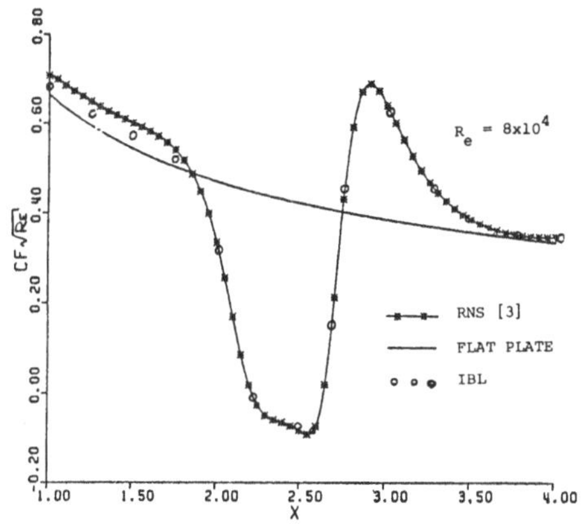

FIG. 5. TROUGH SKIN-FRICTION (PV)

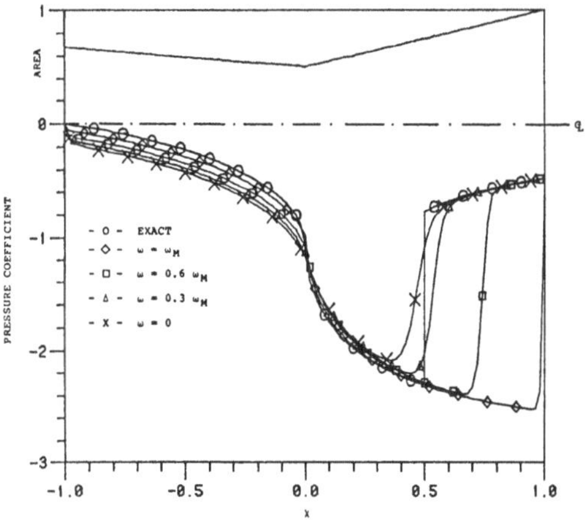

FIG. 8a. EFFECT OF OMEGA ON NOZZLE PRESSURE (PV)

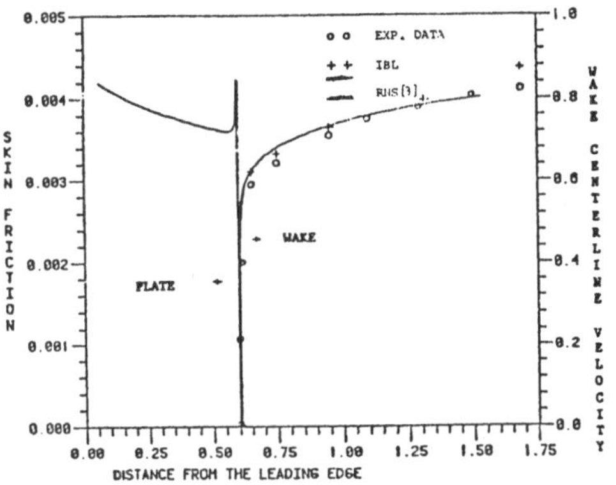

FIG. 6. FINITE FLAT PLATE (TURBULENT) RE = 650.00 (PV)

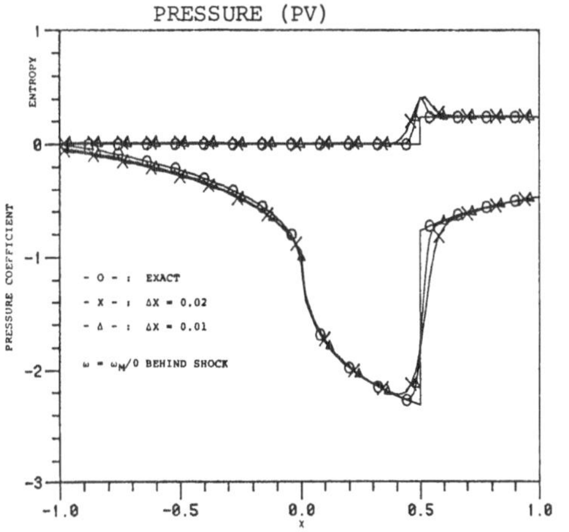

FIG. 8b. EFFECT OF X ON NOZZLE PRESSURE (PV)

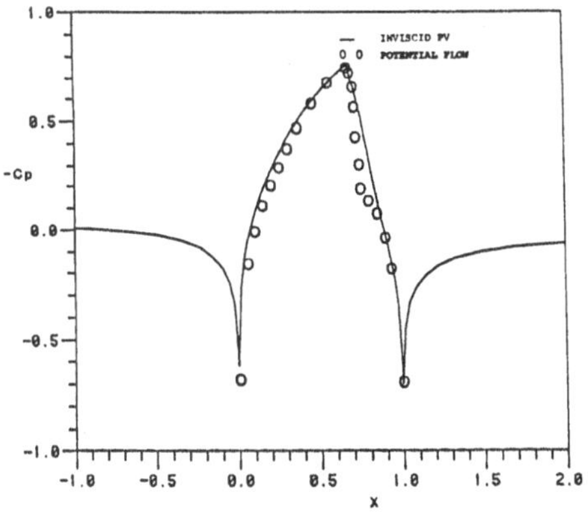

FIG. 7. 10% PARABOLIC ARC AIRFOIL, INVISCID PV, $M_\infty = 0.825$

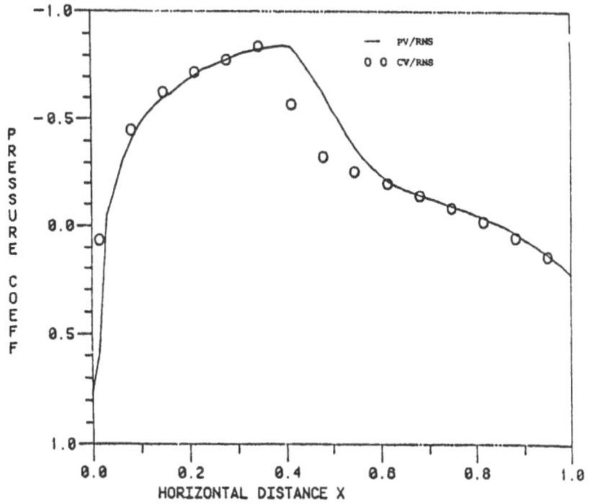

FIG. 9. NAC0012 AIRFOIL, TURBULENT, $RE = 4.1x10^6$, $M_\infty = 0.8$

MODELISATION NUMERIQUE D'ECOULEMENTS TURBULENTS INSTATIONNAIRES EN CANALISATION CYLINDRIQUE.

P. ANDRE, R. CREFF.
Laboratoire de Mécanique et d'Energétique
J. BATINA
Laboratoire d'Analyse Numérique
Université d'Orléans - 45046 Orléans Cedex France.

Abstract.

Pulsed turbulent ducted air flows and related convective heat transfer are depicted by means of a numerical model. The mixing length hypothesis joined to the eddy viscosity and diffusivity model given by CEBECI, HABIB and NA is used. Turbulent transport properties are not supposed to change with time. The equation set is solved by means of a finite difference method coupled with asymptotical developments for the different physical quantities (pressure, velocity, temperature). Assuming a fully developed dynamic regime the developing steady and unsteady thermal fluid fields are described and consequently heat fluxes at the wall for a condition of uniform temperature.

Un modèle numerique descriptif d'écoulements instationnaires périodiques et turbulents en canalisation cylindrique circulaire est présenté. Le caractère instationnaire y est traduit par une modulation périodique du gradient axial de pression générateur de l'écoulement. Le but de ce modèle est de fournir une approche nouvelle dans l'étude des transferts thermiques convectifs associés à ce type d'écoulement en supposant que le développement dynamique est déjà obtenu. Cette analyse fait suite et complète d'autres travaux expérimentaux et théoriques déjà publiés [1,2] notamment ceux relatifs à un modèle C.L.P. (Convectif-Laminaire-Pulsé) [3]

Parmi les différents modèles semi-empiriques descriptifs de la couche limite turbulente fournis dans la littérature celui de CEBECI [4] repris et modifié par HABIB et NA [5] a été retenu. Il est basé sur l'hypothèse de longueur de mélange et permet de bien caractériser le comportement dynamique du fluide non seulement au voisinage de la paroi mais aussi sur l'ensemble de la section droite de l'écoulement. Appliqué antérieurement à des écoulements stationnaires dynamiquement et thermiquement établis, il convient également au traitement d'écoulements fluides de nombre de Prandtl modéré se développant en canalisation. Ainsi, des écoulements incompressibles d'air aux pro-

propriétés physiques constantes ont été étudiés dans les applications du modèle, la condition de transfert thermique étant celle de la paroi de canalisation chauffée à température constante et uniforme. On a supposé d'autre part que, dans le domaine des fréquences étudiées, les propriétés diffusives turbulentes n'étaient pas modifiées dans le temps et qu'elles restaient donc identiques à celles de l'état stationnaire. Comme dans le modèle CLP, la résolution numérique du système d'équations obtenu a été opérée au moyen de la méthode des développements asymptotiques des grandeurs physiques de l'écoulement conjuguée à la technique des différences finies. L'intérêt de ce choix est confirmé par la souplesse du traitement qui en résulte et son applicabilité à des ordres successifs de l'instationnarité périodique de fréquences: $\omega, 2\omega \ldots$

MODELE PHYSIQUE :

Compte tenu des diverses hypothèses rappelées brièvement ci-dessus, les équations descriptives du modèle s'écrivent en coordonnées cylindriques (x,r) :

Equation du mouvement selon l'axe :

$$[AX] : \frac{\partial u}{\partial t} = -\frac{1}{\rho} \cdot \frac{\partial p}{\partial x} + \frac{1}{r}\frac{\partial}{\partial r}\left[\nu . r . \frac{\partial u}{\partial r}\right] \qquad (1)$$

Equation du mouvement selon le rayon : identiquement nulle puisque le régime dynamique est établi.

Equation de l'énergie

$$[TH] : \frac{\partial T}{\partial t} + u\frac{\partial T}{\partial x} = \frac{1}{r}\frac{\partial}{\partial r}\left[a . r . \frac{\partial T}{\partial r}\right] \qquad (2)$$

où u, p, T, ρ représentent respectivement la vitesse axiale, la pression la température et la masse volumique et où les viscosité et diffusivité ν et a sont exprimées par :

$$\nu = \nu_m + \nu_t$$

$$a = a_m + a_t$$

avec ν_m la viscosité cinématique moléculaire

a_m la diffusivité thermique moléculaire

ν_t et a_t les viscosité et diffusivité turbulentes selon HABIB et NA [5]

Les conditions aux limites s'écrivent :

- conditions d'entrée :

$x = 0 \quad u = u(r) \qquad v = 0 \; ; \; T = T_\infty, \forall r, \forall t$ avec T_∞ la température infini amont.

- conditions de symétrie sur l'axe (r=o): $\frac{\partial u}{\partial r} = \frac{\partial T}{\partial r} = 0 \quad \forall\, x, \forall\, t$

- conditions de paroi (r=R): $u = 0 \; , \; T = T_p \quad \forall\, x, \forall\, t$

La périodicité temporelle de l'écoulement est traduite par une expression du gradient axial de pression sous forme d'un développement axymptotique de fonctions sinusoïdales du temps relativement à un paramètre de perturbation ε tel que :

$$\frac{\partial p}{\partial x} = \left(\frac{\partial p}{\partial x}\right)_0 + \varepsilon\left(\frac{\partial p}{\partial x}\right)_1 e^{j\omega t} + \varepsilon^2\left(\frac{\partial p}{\partial x}\right)_2 e^{2j\omega t} + \ldots$$

avec ω la fréquence angulaire des modulations. Pour chaque rang n, on définit le gradient axial de pression en posant :

$$\left(\frac{\partial p}{\partial x}\right)_n = \tau^n\left(\frac{\partial p}{\partial x}\right)_0$$

Les inconnues u et T sont recherchées sous la forme d'un développement asymptotique de forme analogue à celui du gradient de pression. Les équations (1) et (2) sont alors découplées par un calcul d'identification aux différents ordres n. Ainsi, à titre d'illustration, l'équation $[AX]_n$ s'écrit :

$$nj\Omega \frac{r^+}{R^{+2}} \frac{u_n^+}{\tau^n} - 2\frac{r^+}{R^+} = \frac{\partial}{\partial r^+}\left[r^+ (1 + \nu_t^+) \cdot \frac{\partial(u_n^+/\tau^n)}{\partial r^+}\right]$$

L'adimensionnalisation des équations est opérée principalement à l'aide de la vitesse de frottement $u^* = \sqrt{\tau_p/\rho}$ où τ_p représente le taux de frottement à la paroi. Les nouvelles variables s'expriment ainsi :

$u^+ = u/u^*$; $r^+ = ru^*/\nu_m$; $x^+ = xu^*/\nu_m$

$t^+ = t/(R^2/\nu_m)$ avec R le rayon de canalisation

$\theta = (T_p - T)/(T_p - T_\infty)$; $\Omega = R^2\omega/\nu_m$ la fréquence réduite

Le gradient axial de pression est rapporté à la valeur connue en écoulement stationnaire établi : $-2\tau_p / R$

Pour se ramener à un domaine dont le rayon varie entre 0 et 1 en chaque section droite de l'écoulement les variables d'espaces sont réécrites ainsi :

$$x^{++} = x^+/R^+ \qquad r^{++} = r^+/R^+$$

Les expressions finales des équations $[AX]_0$ et $[AX]_1$ sont donc de la forme :

$$[AX]_0 : \quad -2R^+r^{++} = \frac{\partial}{\partial r^{++}}\left[r^{++}(1 + \nu_t^+) \cdot \frac{\partial u_0^+}{\partial r^{++}}\right]$$

$$[AX]_1 : \quad j\Omega r^{++}u_1^+ - 2R^+r^{++} = \frac{\partial}{\partial r^{++}}\left[r^{++}(1 + \nu_t^+) \cdot \frac{\partial u_1^+}{\partial r^{++}}\right]$$

Des écritures analogues sont obtenues pour les équations $[TH]_n$:

$$[TH]_1 : \quad j\Omega r^{++}\theta_1 + R^+r^{++}\left[u_0^+\frac{\partial \theta_1}{\partial x^{++}} + u_1^+\frac{\partial \theta_0}{\partial x^{++}}\right] = \frac{\partial}{\partial r^{++}}\left[r^{++}\left(\frac{1}{Pr} + a_t^+\right)\frac{\partial \theta_1}{\partial r^{++}}\right]$$

SCHEMA NUMERIQUE.

Les équations $[AX]_n$ et $[TH]_n$ sont résolues par différences finies à pas fractionnaires selon le rayon et à pas entiers selon l'axe ainsi que l'indique le schéma:

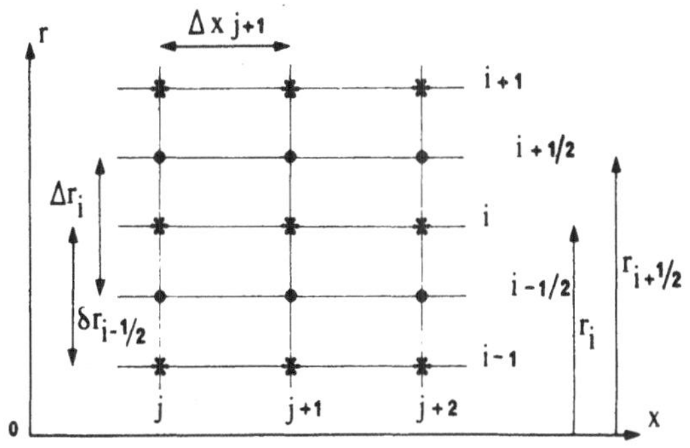

Schéma de maillage.

Pour toute fonction u ou T notée f (x,r) les différences finies s'expriment :

$$\left(\frac{\partial f}{\partial x^{++}}\right)_i^j = \frac{f_i^{j+1} - f_i^j}{\Delta x_{j+1}^{++}} + O(\Delta x_{j+1}^{++}) \text{ ; différence décentrée selon } x^{++}$$

$$\left(\frac{\partial^2 f}{\partial r^{++2}}\right)_i^{j+1} = \frac{1}{\Delta r_i^{++}}\left[\frac{f_{i+1}^{j+1} - f_i^{j+1}}{\delta r_{i+1/2}^{++}} - \frac{f_i^{j+1} - f_{i-1}^{j+1}}{\delta r_{i-1/2}^{++}}\right] + O(\Delta r_i^{++})$$

(calculée au noeud entier)

A titre d'exemple, l'équation $[AX]_o$ se discrétise comme suit :

$$-2R^+ r_i \Delta r_i = u_{i+1}\frac{r_{i+1/2}\cdot s_{i+1/2}}{\delta r_{i+1/2}} - u_i\left(\frac{r_{i+1/2}\cdot s_{i+1/2}}{\delta r_{i+1/2}} + \frac{r_{i-1/2}\cdot s_{i-1/2}}{\delta r_{i-1/2}}\right) + u_{i-1}\cdot\frac{r_{i-1/2}\cdot s_{i-1/2}}{\delta r_{i-1/2}}$$

où $s(r) = 1 + \nu_t^+ (r^{++})$ et $u_k = u_k^j \quad \forall j$

(les indices + et ++ étant par ailleurs supprimés pour alléger l'écriture).

Les vitesses u_i sont alors obtenues par simple inversion matricielle, une fois pour toutes, puisque l'écoulement est dynamiquement développé.

D'une manière similaire, les équations $[TH]_n$ ont été discrétisées et calculées à chaque pas successif j à partir de j = o où le profil de température est représenté par un front uniforme (début de la zone de paroi chauffée). La température inconnue au pas j + 1 est calculée à partir de sa valeur connue au pas j constituant ainsi un schéma progressif sans processus de type "backward-forward". Ainsi, aucune vérification de critères de convergence et de stabilité n'est requise conférant au modèle une rapidité d'exécution dans le traitement.

RESULTATS :

L'hypothèse de longueur de mélange introduite a permis de vérifier et préciser certains résultats classiques concernant les régimes dynamique et thermique stationnaires notamment les profils de vitesse axiale et les taux de transfert locaux d'un régime thermique s'établissant caractérisés par l'évolution longitudinale du nombre de Nusselt local $Nu_o(x)$. Dans le domaine instationnaire, le modèle apporte deux résultats importants :

a) le choix de l'adimensionnalisation opéré permet de mettre en évidence une identité formelle des équations de la dynamique stationnaire et instationnaire au terme d'instationnarité prés. De ce dernier, un critère sur le degré de l'instationnarité peut être dégagé sous la forme du groupement $\Omega/Re^{1,75}$ pour un écoulement de fluide donné. Au-dessous d'une valeur critique de ce nombre le comportement du fluide en mouvement peut être représenté par l'approximation des états quasi-stationnaires et dans ce cas les viscosité et diffusivité turbulentes subissent dans le temps l'effet de la modulation. Au-delà de la valeur critique $\Omega/Re^{1,75}$, cette approximation n'est plus satisfaisante et l'écoulement devient purement instationnaire ; les effets d'inertie estompent toute variation temporelle des coefficients de diffusion et le modèle s'accorde alors d'une longueur de mélange identique à celle de l'écoulement stationnaire de même nombre de Reynolds moyen.

b) outre l'existence d'un effet annulaire dans la distribution radiale des vitesses axiales instationnaires, un effet du même type est obtenu pour les températures. Le développement longitudinal de ce dernier montre que les effets instationnaires sur le transfert convectif local s'atténuent au long de l'écoulement, la rapidité d'amortissement étant manifestement liée aux effets combinés de la modulation du gradient de pression d'une part et des forces d'inertie et de frottement d'autre part.

REFERENCES.

1. P.ANDRE, R.CREFF, J.CRABOL. Int. J. Heat Mass Transfer, Vol 24, N° 7 pp 1211 - 1219 (1981)
2. R.CREFF, P.ANDRE, M.PLAN, Trans. CSME, Vol 6, N° 1, pp.27-33 (1981).
3. R.CREFF, J.BATINA, P.ANDRE, V.S.KARUNANITHI, Num. Heat Transfer, Vol 6, pp 173-188 (1983)
4. CEBECI T. J. Heat Transfer, Vol.95, pp 227-234 (1973)
5. HABIB I.S. ; NA T.Y. J. Heat Transfer, Vol 96, pp. 253-254 (1974)

ON THE USE OF RATIONAL RUNGE-KUTTA METHODS IN EULER STEADY-STATE COMPUTATIONS

F. ANGRAND*, V. BILLEY***, A. DERVIEUX**,
J.A. DESIDERI**, J. PERIAUX***, B. STOUFFLET*

The past five years, explicit time-dependent methods have regained some interest for the solution of the steady Euler Equations. This is due partly to the efficient schemes that have been recently developed, but also to the utilization of vector computers. However, the severe stability limitation on the timestep (the "CFL" condition) remains a major draw-back of these methods. For this reason several authors have directed their efforts towards increasing the maximum allowable timestep by an even modest factor. One technique to achieve this is obtained when a "post-processing" stage, itself requiring only few operations, follows each iteration of the explicit procedure. This approach is illustrated by "residual-averaging" finite-difference (FD) methods [1,2]. This post-processing stage consists in the solution of several scalar Poisson problems.

A finite-element (FE) scheme was constructed to apply this technique and here also it resulted in gains in efficiency [3]. But, if a non-structured mesh is employed a very large, completely multidimensional matrix must be generated. Then, for a size computation, an alternate method that can perhaps be less rapidly converging, but that has a more reasonable memory storage requirement may be preferred. The object of this paper is to describe one such method.

One way of solving the system of equations $B(u) = 0$ which $(u, B(u) \in \mathbb{R}^{\mathbb{N}})$ is to integrate forward in time the following ODE :

$$u_\tau = B(u) \tag{1}$$

until a steady state is achieved.

In the application considered in this paper, $B(u)$ represents a space discretization of a PDE

The time integration schemes (or "solvers") considered in this paper are the so-called "Rational Runge-Kutta" (RRK) methods introduced by Wambecq [4]. The general formulation of these methods is given by :

$$u_1 = u_0 + \Delta u \tag{4}$$

$$\Delta u = h\left(\sum_{i=1}^{\nu}\sum_{j=1}^{i} w_{ij}\, g_i\, g_j\right) / \sum_{k=1}^{\nu} b_k g_k \tag{5}$$

$$g_i = B(u_0 + h\sum_{j=1}^{i-1} a_{ij}\, g_j) \qquad (i = 1,2,\ldots,\nu) \tag{6}$$

Special definitions of the product and of the quotient of two vectors have been introduced in [4]. For an appropriate choice of the coefficients a_{ij}'s, w_{ij}'s and b_k's the method can be made ν – order accurate and perhaps more importantly A_0-stable. In this case arbitrarily large timesteps h can be used stably.

To our knowledge, Satofuka [5] was the first to use such time-integration scheme for the computation of steady flows. He solved the Navier-Stokes equations which are parabolic in nature. In this paper, we consider flows governed by the Euler equations which are hyperbolic. We have conducted a serie of numerical experiments in which the A_0- stable RRK method of order 2,3 or 4 (RRK2, RRK3, RRK4) was used for 2-D and 3-D flows represented by FE and FD approximations. To illustrate these calculations, we

* INRIA, BP 105, 78153 LE CHESNAY CEDEX, FRANCE
** INRIA, Route des Lucioles, Sophia Antipolis, 06560 VALBONNE, FRANCE
*** AMD/BA, 78 Quai Carnot, BP 300, 92214 SAINT-CLOUD, FRANCE

present Mach contours of a flow in a channel over à 4.2% thick bump at $M_\infty=.85$ obtained by the FE method on Figure 1.
Our most typical results concerning efficiency are collected in Table 1 a, b, c. There we indicate for various time-integration schemes and various spatial approximations the number of evaluations of the approximation (the function B) that was found necessary to achieve a reduction of the residual by a factor $\leqq 10^{-3}$.

Upwind-difference approximation

A first-order flux-splitting approximation was firstly used, in which the amount of numerical viscosity was such that the hyperbolic nature of the system was nearly eliminated. CFL numbers as large as 5 to 20 could be used stably. However, when comparing the efficiency of this scheme to that of the simple Euler scheme in terms of the number of flux evaluations, we observe a good improvement in the FE case (Table 1 a and Figure 3).

Central-difference approximation

Various RRK methods were applied to a central FD approximation including a small (2nd difference) artificial viscosity term.
The third-order method appeared as the most efficient solver but only marginally faster than the classical RK4 (Table 1 b).

Lax-Wendroff approximation

The RRK methods were also applied to a FE Richtmyer approximation, extending the Lerat-Peyret FD scheme [6].
Let the Richtmyer scheme writen as

$$u^{n+1} = G_R(\Delta t, \Delta u, u^n) \tag{7}$$

which in fact, defines the function G_R. There Δt is a timestep that is adjusted to produce a maximum CFL number equal to some value K. An ODE that has the same stationary solution as (7) is then

$$u_\tau = \frac{G_R(\Delta t, \Delta u, u) - u}{\Delta t} \equiv B(u)$$

in which Δt is now a fixed parameter. This ODE can then be integrated foward in time by a RRK method using a "relaxation timestep" h. In the particular case where the solver is chosen to be the forward Euler method and if $h = \Delta t$, the iteration that results is (7), but other possibilities exist ; in particular, the RRK2-Ritchmyer scheme is given by :

$$(9) \quad \left\{ \begin{aligned} g_1 &= h\, B\,(u^n) \\ g_2 &= h\, B\,(u^n + a_{21}\, g_1) \\ g_3 &= b_1\, g_1 + b_2\, g_2 \qquad (b_1 + b_2 = 1) \\ u^{n+1} &= u^n + (2\, g_{13} \cdot g_1 - g_{11} \cdot g_3) / g_{33} \end{aligned} \right.$$

in which the notation $g_{ij} = (g_i, g_j)$ is employed. This method was compared to Euler's method in a subsonic, a transonic and a supersonic flow case (Figure 2 a,b and Table 1 c). In the subsonic and transonic flow cases, the RRK2 method was observed to be more efficient than it was previously found for the central-difference scheme. For the criterion used in Table 1, an overall gain in efficiency by a factor of 2 is achieved. In the purely supersonic flow case, Euler's method converges a lot faster. Compared to RRK2 it is found to have a larger asymptotic convergence rate. However the two solvers are essentially equivalent in terms of the criterion of the Table 1.
Independently of the CFL of the RRK solver (calculated with the relaxation timestep h), the steady-state solution remains the same as that defined by the Richtmyer scheme with a CFL number K (calculated with a timestep Δt) equal to 1. This value of K is nearly optimum from the standpoint of accurary. It was also observed that after the initial phase, even larger CFL numbers such as 100 could be used stably, but this did not yield any significant improvement in convergence rate.

Conclusion

The methods presented here are attractive for the acceleration of existing codes :

- the implementation is simple,
- the possibility of vectorization is maintained,
- the accuracy of the approximation is unchanged,
- additional memory storage requirements are very small, in particular in the case of the second order method (RRK2).

A gain in efficiency of the order of 2 was obtained for almost all calculations except for the F.D first order flux-splitting approximation on a coarse grid.

At this moment, a 3-D FE simulation of a flow around a wing-body configuration in which RRK2 solver is combined with a Godunov or Lax-Wendroff FE schemes, is under investigation and will be presented in a forthcoming paper.

References

[1] A. LERAT, J. SIDES, V. DARU, An Implicit Finite-Volume Method for Solving the Euler Equations, in E. Krause (Ed.), (Proceedings of the) Eight International Conference on Numerical Methods in Fluid Dynamics (Aachen, 1982), Lecture Notes in Physics, 170, Springer, 1982, pp. 342-349.

[2] A. JAMESON, W. SCHMIDT, E. TURKEL, Steady State Solution of the Euler Equations for Transonic Flow, AIAA Paper 81-1259, 1981.

[3] F. ANGRAND, V. BOULARD, A. DERVIEUX, J. PERIAUX, G. VIJAYASUNDARAM, Triangular Finite-Element Methods for the Euler Equations, Sixth Int. Symposium on Numerical method in Engineering, Versailles, December 1983, to be published by Springer Verlag.

[4] A. WAMBECQ, Rational Runge-Kutta Methods for Solving Systems of Ordinary Differential Equations, Computing, Vol. 20, pp 333-342, 1978.

[5] N. SATOFUKA, Unconditionally Stable Explicit Method for the Numerical Solutions of the Compressible Navier-Stokes Equations, Proceedings of the 5th GAMM-Conference on Numerical Methods in Fluid Mechanics, M. PANDOLFI, R. PIVA Eds, Notes on Numerical Fluid Mechanics, Vol. 7, Vieweg and Sohn, Braunschweig/Wiesbaden, Germany, 1984.

[6] A. LERAT, R. PEYRET, Sur le choix de schémas aux différences du second ordre fournissant des profils de chocs sans oscillation, comptes rendus Acad. Sc. Paris, 277 A, 363-366 (1973).

* Cray 1-S computational costs are supported by GCCVR.

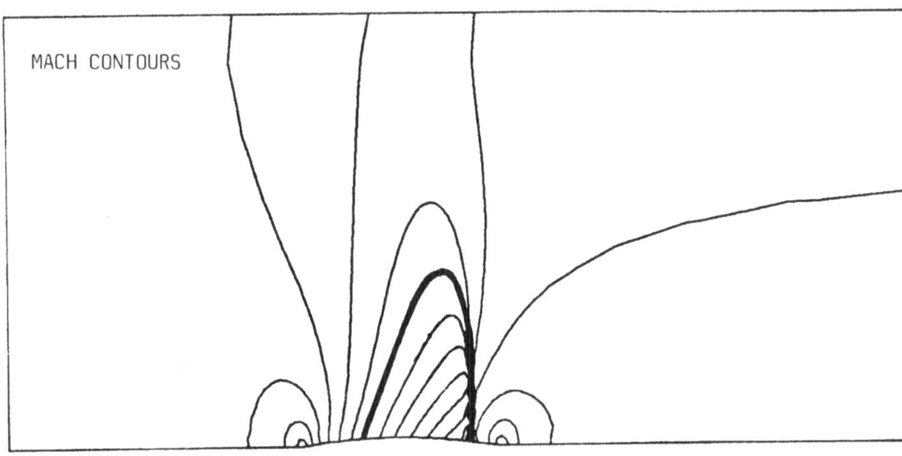

Figure 1 : Mach contours of a transonic flow in a channel. (FE)

Table 1. Number of evaluations of the spatial approximation to achieve a reduction of the residual by a factor 10^{-3}. Transonic flow regime with a shock.

Spatial approximation / Time integration Scheme	Ist-Order flux-splitting	
	Finite-difference (36x16 grid)	finite-element (72x21 grid)
Forward Euler+	680 (CFL= 0.7)*	3400 (CFL= 0.6)*
RRK2	980 (CFL = 1)* 680 (CFL = 5)*	1650 (CFL= 10)*

a) Ist-Order flux-splitting approximation

Spatial approximation / Time-integration Scheme	2nd-order finite difference approximation with 2nd-order artificial viscosity
RK4+	2000 (CFL = 2.5)*
RRK2	3000 (CFL = 2)*
RRK3	1600 (CFL = 2.25)*
RRK4	1750 (CFL = 3)*

b) 2nd-order FD spatial approximation

Spatial approximation / Time-integration Scheme	Richtmyer-type finite-element method
Forward Euler+	7000 (CFL = 0.6)*
RRK2	3400 (CFL = 5)

+ Reference case

* local time-stepping

c) 2nd-order FE Richtmyer-type spatial approximation.

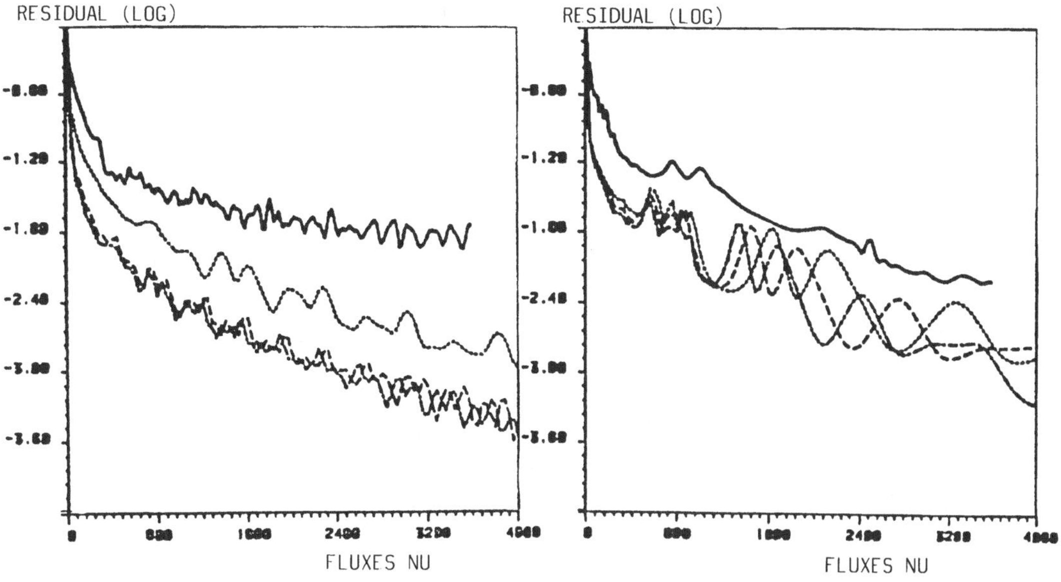

EULER $CFL_{\Delta t} = .6$
RRK2 $CFL_{\Delta t} = 1.$ $CFL_h = 1.$
RRK2 $CFL_{\Delta t} = 1.$ $CFL_h = 4.$
RRK2 $CFL_{\Delta t} = 1.$ $CFL_h = 5.$

a) Subsonic regime ($M_\infty = .50$)

EULER $CFL_{\Delta t} = .6$
RRK2 $CFL_{\Delta t} = 1.$ $CFL_h = 4.$
RRK2 $CFL_{\Delta t} = 1.$ $CFL_h = 5.$
RRK2 $CFL_{\Delta t} = 1.$ $CFL_h = 6.$

b) Transonic regime with a shock($M_\infty = .85$)

Fig. 2 : Convergence history with the FE Richtmyer scheme.

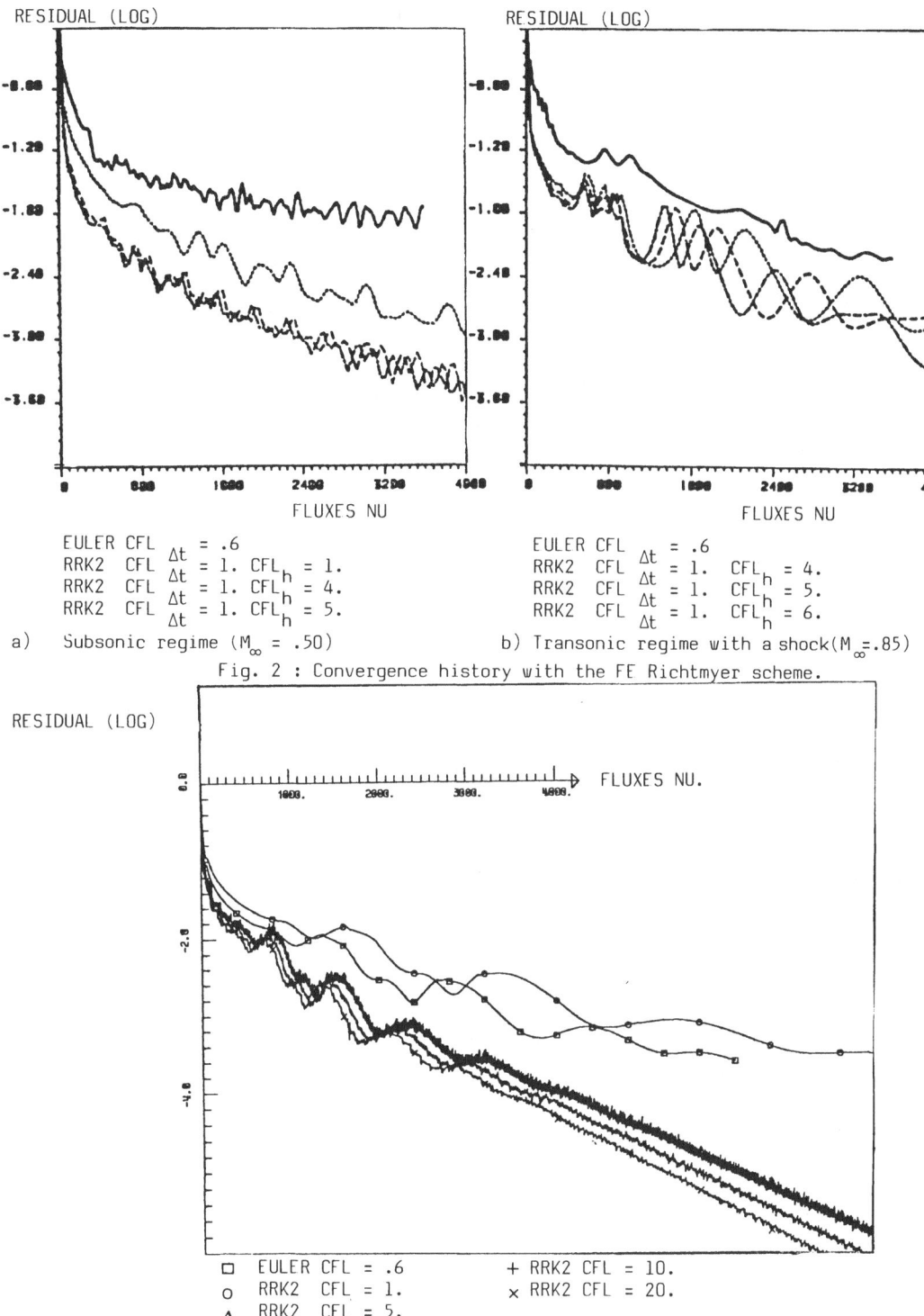

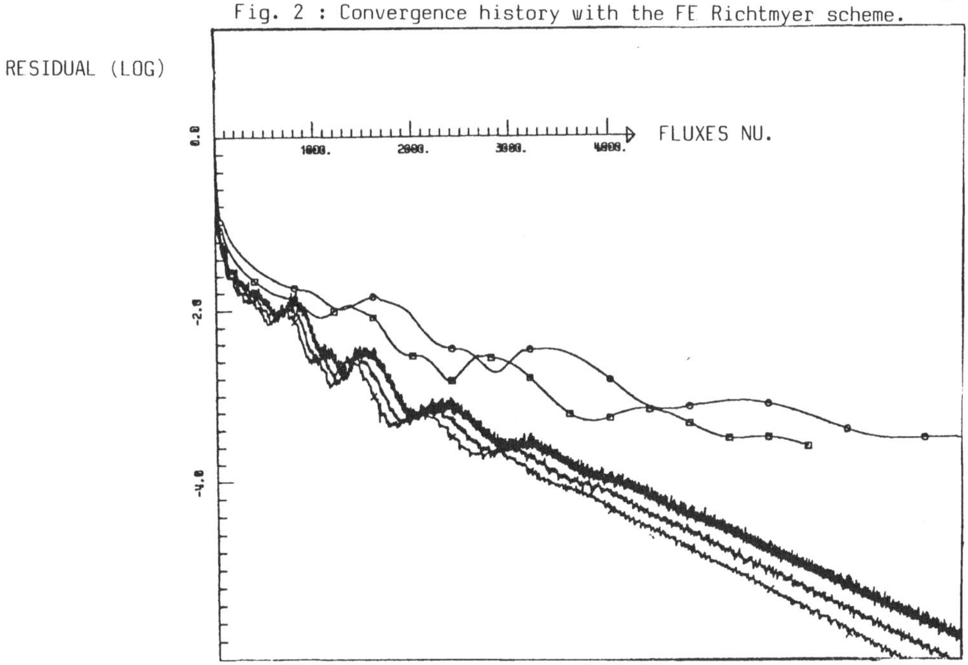

□ EULER CFL = .6
○ RRK2 CFL = 1.
Δ RRK2 CFL = 5.
+ RRK2 CFL = 10.
× RRK2 CFL = 20.

Fig. 3 : Convergence history with the FE Upwind Scheme in the transonic regime.

PROBATING A FINITE DIFFERENCE FRACTIONAL TIME STEP, NAVIER-STOKES CODE BY LARGE EDDY SIMULATION

F. Baron, D. Laurence
E.D.F. - Laboratoire National d'Hydraulique,
Chatou, 78400 France

The basic idea of large eddy simulation (LES) is to directly compute the large scale velocity field $\overline{U}$ of a turbulent field $U = \overline{U} + U'$, ($\overline{U} = G * U$), while the small scale part U' can be modeled since it is more nearly universal.

The fluctuation U' is strongly dependant on the filter G, usualy gaussian of width Δ . Ratio of filter to mesh resolution is chosen large enough to discard numerical errors, ($\Delta /h \simeq 2$).

In order to retain as much information as possible, we do not explicitely introduce such a filter, so that G is the implicit numerical filter of the code, and thus has to be defined. This choice also allows feed-back information concerning the performances of the industrial E.S.T.E.T. code to which the LES code is related [1].

I - BENCH TEST

Comte-Bellot and Corsin's experiments on grid generated turbulence is used as bench-mark. Computation of the large scale field $\overline{U}$ is carried out in physical space. From the results, the energy spectrum is computed :

$$E(k) = \int_{\|\vec{P}\| = k} \underset{\sim}{\overline{U}}(\vec{P}) \cdot \underset{\sim}{\overline{U}}(-\vec{P})\, dS \qquad \underset{\sim}{U} = \text{Fourier transform of } \overline{U}$$

Integration over the spherical shell removes random phases included in $\underset{\sim}{U}(P)$. Variations of $E(k)$ over each substeps yields information on the numerical filtering of the operators. We thus empiricaly evaluate by means of statistics the filter $G(k)$, which is equivalent to the classical amplification factor defined in computational analysis.

This procedure enables us to look for a scheme which reduces numerical diffusion and excludes artificial amplification.

II - NUMERICAL CODE

$(n + 1)_{th}$ time step is computed in three sub-steps.

. Advection of momentum :

This is resolved by a three dimensional, curvilinear characteristics method :

$$\begin{cases} d\, C_M/dt = - U^n \\ C_M(t^{n+1}) = M \end{cases}$$

then : $\tilde{\tilde{U}}^{n+1}(M) = U^n(N)$
with : $N = C_M(t^n)$

Example of 1 D Characteristic

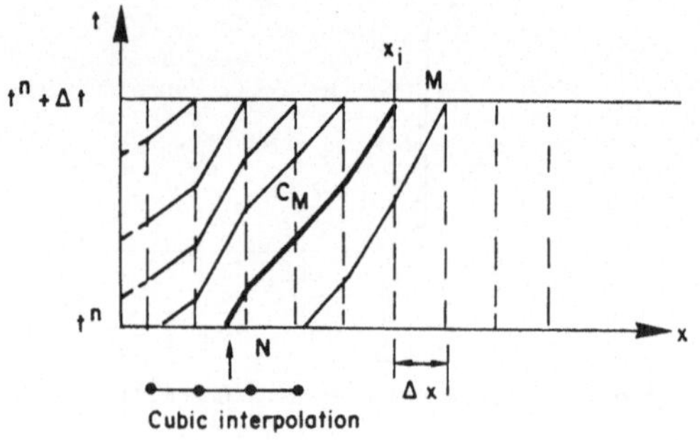

Various 3D interpolation operators are compared for translations of the turbulent field. For low CFL numbers, splitting the advection step in 3 directions allows us to use the "weak formulation" of the transport equation (advection of test functions instead of the variable U, as in finite element formulation) [2] (fig. 1).

. Diffusion of momentum (split into 3 directions) :

$$(\tilde{U}_i^{n+1} - \tilde{\tilde{U}}_i^{n+1})/\delta t = \frac{\partial}{\partial x_j}\left\{(\nu + \nu_T)\,\tilde{\tilde{D}}_{ij}\right\} \text{ (Gauss elimination method)}$$

where ν_T is SMAGORINSKY's subgrid viscosity model.

$$\nu_T = C_S^2\, h^2\, (D_{ij}\, D_{ij})^{1/2}, \quad D_{ij} = \frac{1}{2}\left(\frac{\partial U_i}{\partial x_j} + \frac{\partial U_j}{\partial x_i}\right)$$

. Continuity and pressure :

p^{n+1} is resolved by a GAUSS SEIDEL iterative method with over-relaxation under the equivalent form :

$$\frac{\partial \tilde{U}_i^{n+1}}{\partial x_i} = \frac{\delta t}{\rho}\,\frac{\partial^2 P^{n+1}}{\partial x_i\, \partial x_i}, \quad (U_i^{n+1} - \tilde{U}_i^{n+1})/\ \delta t = -\frac{1}{\rho}\frac{\partial P^{n+1}}{\partial x_i}$$

Pressure is defined on a staggered grid ("Pressure points" are at the center of "velocity cells").

Use of a 7 points Δ operator does not lead to a true divergence free velocity field. Analysis of the pressure and divergence spectra compared to exact value computed by Fourier method, shows that the error is restricted to high wave numbers, and thus can not be reduced by increasing the number of Gauss iterations. An alternative to the cumbersome 27 point compatible Δ operator (Δ 27 = Div o Grad) is found in a "Uzawa" type algorithm [3].

III - RESULTS

Initial velocity field is devised from Comte Bellot and Corrsin's experimental data [4] at station $\frac{t\, U_o}{M} = 42$.

The energy spectrum of the final computed velocity field is compared to experiment at station $\frac{t\, U_o}{M} = 98$.

The standard code, run with a small CFL number (CFL = $\sqrt{\langle U^2 \rangle}\ .\ \delta t/h = 0.1$) gives good agreement with the filtered (Gaussian, of width 2 h) experimental data (fig. 2). A faster run (CFL = 0,5) brings a high drop of energy on high wave numbers for two reasons : numerical filtering of the advection step has a non-linear dependance on the CFL number, and splitting in sub-time steps brings a drop of energy of order δt^2 in the pressure sub step (fig. 3). The maximum CFL number is actualy 5 times the r.m.s CFL, this is allowed by the unconditionaly stable caracteristics scheme.

The conservative property of the "weak formulation" allows direct comparison to unfiltered experimental data (fig. 4, 5) but is restricted to slow runs (CFL = 0,1) because only the 1 D caracteristics are available to this day.

Since we are looking forward to industrial applications of LES : higher values of CFL number must be allowed which means compensating for the underestimation of the energy transfer in the standard code. Use of the caracteristics method brings a natural separation between advecting and advected velocity fields which allows several alternative schemes :

- Energy drop in the pressure step can be reduced by centering the grad P term along the caracteristic ($\Longleftrightarrow$ in time). This mean adding $\frac{1}{2}$ grad P^n at point N to the advected velocity.

- The Energy transfer from computed large scales to computed small scales can be increased by isolating turbulent structures near the highest computed wave number k_c by double filtering : $U_{k_c} \simeq \bar{U} - \bar{\bar{U}}$ and adding this term to the advecting velocity.

- In the same manner, the diffusion step can be replaced by adding to the advecting velocity a purely random (not correlated with the computed field $\bar{U}$) velocity α with variance $< \alpha^2 > = \frac{2\nu \ T}{\delta \ t}$

Figure 6 shows results for this run with average CFL = 0.5 (maximum CFL = 2.).

CONCLUSION

This statistical procedure enables us to precisely mesure the performances of a code for all wave numbers in one single run, and is thus more thorough than, for example, the "rotating cone" test. This is not restricted to L.E.S. and some improvements were also introduced in the engineering version of the code.

Knowledge of the implicit numerical filtering is very usefull in L.E.S. for modeling the sub-grid scale stresses. The characteristic scheme allows a wider range of possibilities for S.G.S. modeling, as well as the use of high CFL numbers.

REFERENCES

[1] DEWAGENAERE, ESPOSITO, LANA , VIOLLET.
Communication to present conference.

[2] BENQUE, LABADIE, RONAT.
"Caracteristics Finite Element METHODS for Convection Dominated Flows" submited to International Journal of Numerical Methods in Fluids.

[3] GOUSSEBAILE, GREGOIRE, HAUGUEL.
"Iterative Stokes Solvers ans Splitting Techniques for Industrial Flows". Fifth International Symposium on Finite Elements and Flow Problems.

[4] COMTE-BELLOT, CORRSIN.
Simple Eulerian time correlation of full and narrow-band velocity signals in grid-generated, isotropic, turbulence.
J. Fluid Mech (1971), vol. 48.

[5] F. BARON, D. LAURENCE
"Large Eddy Simulation of a confined turbulent jet flow and homogeneous shear". Turbulent Shear Flow IV - pp. 4.7-4.12 (1983).

[6] BARON F.
"Macrosimulation tridimensionnelle d'écoulements turbulents cisaillés". Thèse de Docteur-Ingénieur - Université P. et M. Curie (6/12/82).

[7] LAURENCE D.
"Simulations numériques tridimensionnelles d'écoulements turbulents". (EDF report E41/82.11).

Figures.

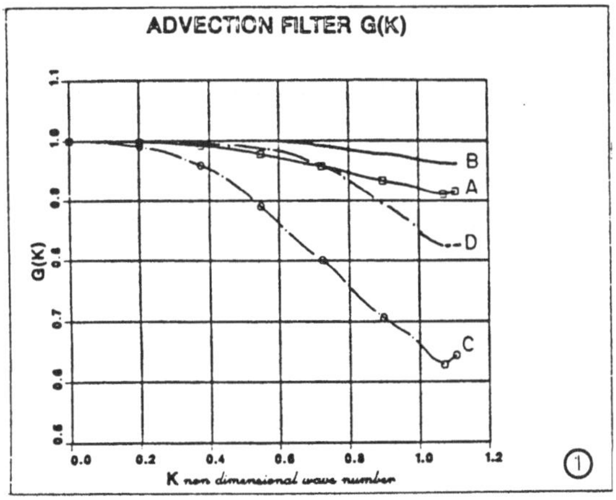

Numerical filter of the advection step

G(k) = ratio of final to initial energy spectrum while advecting by a uniform field, for CFL = 0,1.

A : G_s for standard advection scheme

B : G_w for "weak formulation"

C : $(G_s)^5$ (five steps)

D : $(G_w)^5$

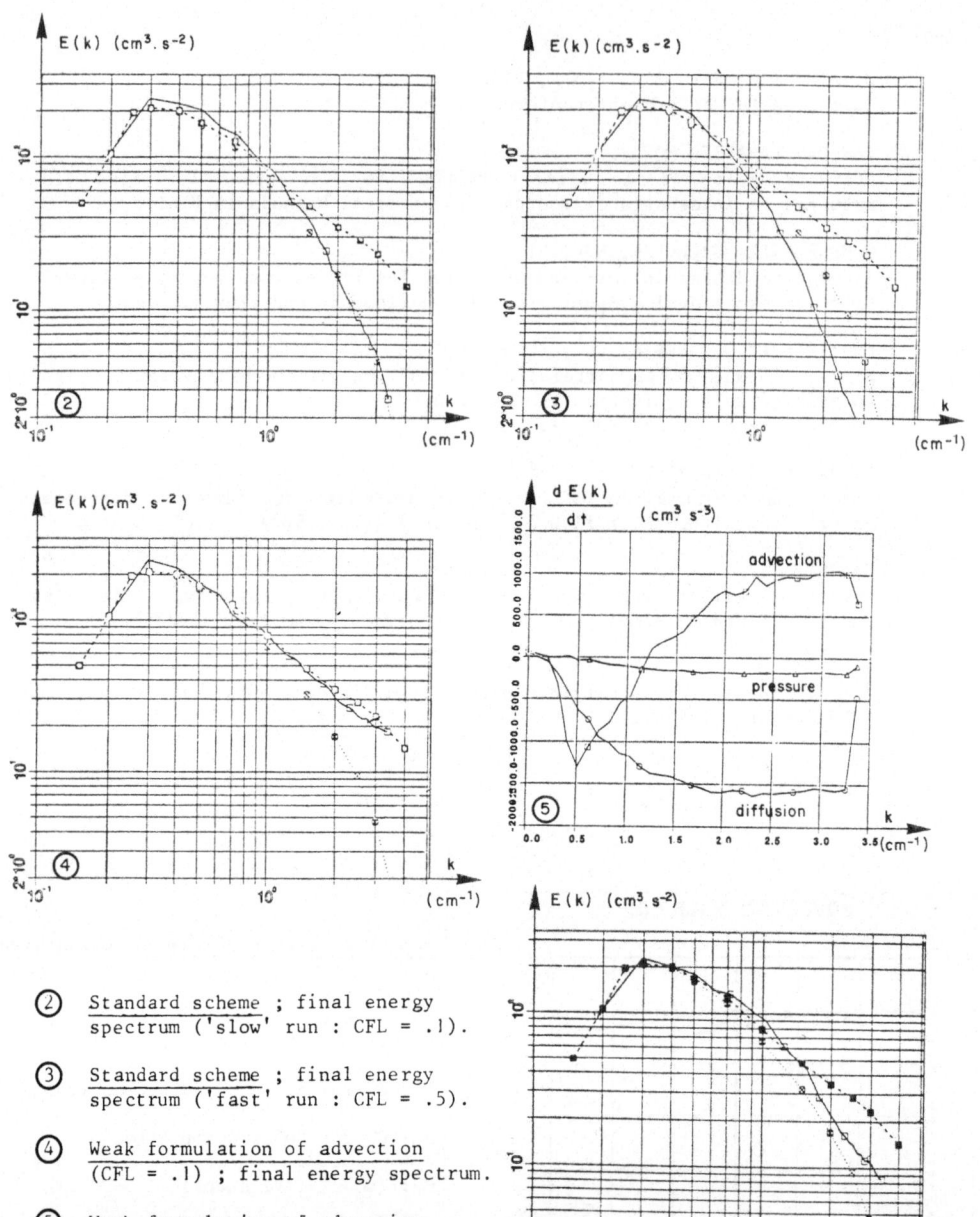

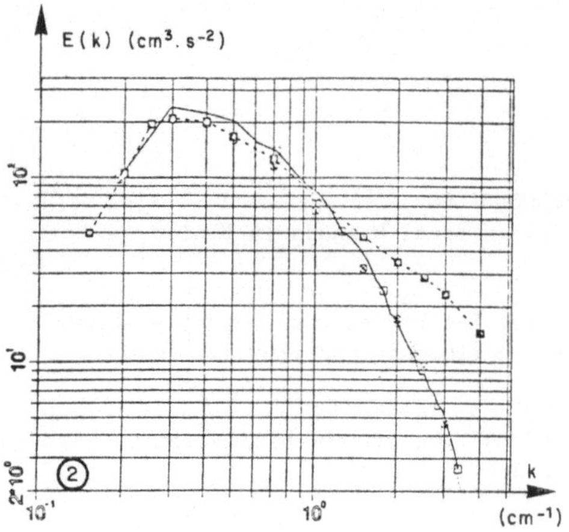

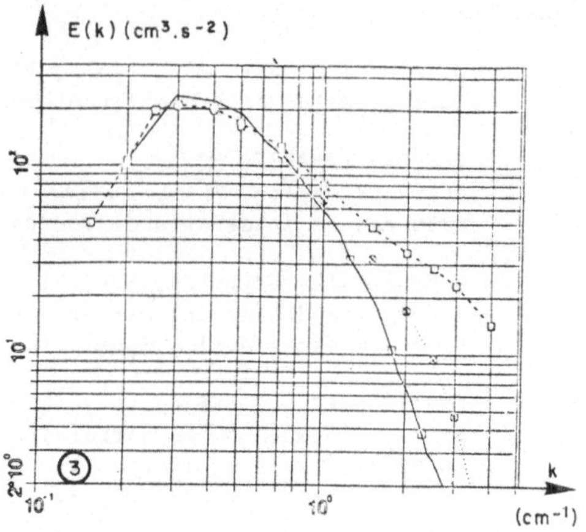

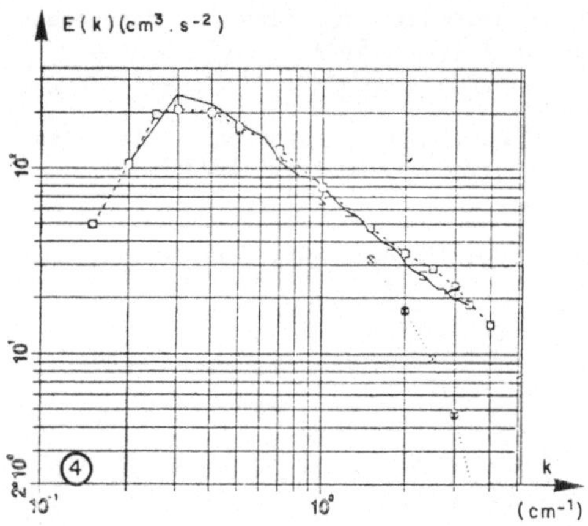

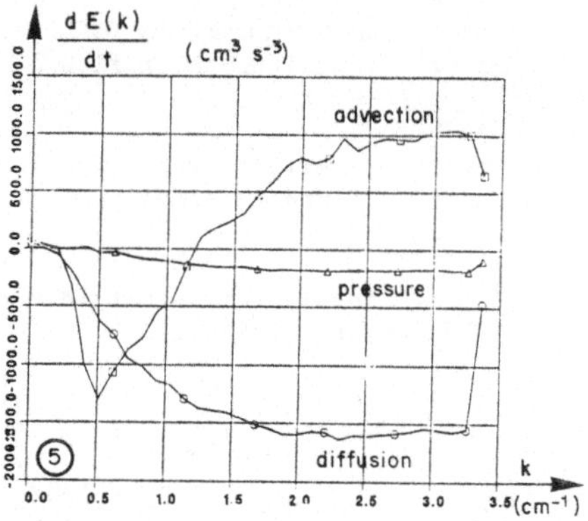

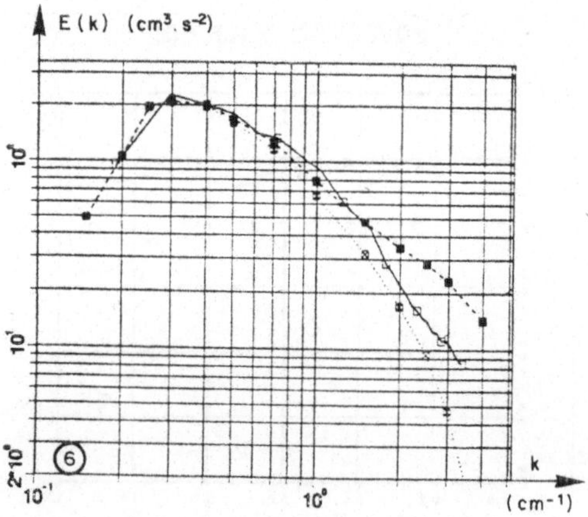

(2) Standard scheme ; final energy spectrum ('slow' run : CFL = .1).

(3) Standard scheme ; final energy spectrum ('fast' run : CFL = .5).

(4) Weak formulation of advection (CFL = .1) ; final energy spectrum.

(5) Weak formulation of advection (CFL = .1) ; energy variation per time step fraction.

(6) Modified advection scheme (centered pressure gradient ; enhanced transfer Uk_c ; stochastic diffusion α) ; final energy spectrum (CFL = .5).

[All solid lines are L.E.S. results ; dashed lines are experimental results and dotted lines are filtered experimental results].

GRP - AN ANALYTIC APPROACH TO HIGH-RESOLUTION UPWIND SCHEMES FOR COMPRESSIBLE FLUID FLOW

Matania Ben-Artzi
Dept. of Mathematics
Technion-Israel Inst. of Technology
Haifa 32000, Israel

Joseph Falcovitz
Computation Division
Rafael Ballistic Center
Haifa 31020, Israel

1. INTRODUCTION. Consider the Euler equations that model the time dependent flow of an inviscid, compressible fluid through a duct of smoothly varying cross section. Denoting by $A(r)$ the area of the cross section at the point r, these equations are,

$$A \frac{\partial}{\partial t} U + \frac{\partial}{\partial r} [AF(U)] + A \frac{\partial}{\partial r} G(U) = 0,$$
$$U = \begin{pmatrix} \rho \\ \rho u \\ \rho E \end{pmatrix}, \quad F(U) = \begin{pmatrix} \rho u \\ \rho u^2 \\ (\rho E+p)u \end{pmatrix}, \quad G(U) = \begin{pmatrix} 0 \\ p \\ 0 \end{pmatrix}, \tag{1}$$

where ρ, p, u are, respectively, density, pressure and velocity, $E=e+\frac{1}{2}u^2$ is the total specific energy (e is the specific internal energy) and an equation-of-state $p=p(e,\rho)$ is assumed.

Our approach to the time integration of (1) is based on an analytic extension of Godunov's scheme [4], and has its origin in the work of van-Leer [9]. It can be used either as a direct Eulerian method or as a Lagrangian scheme. In both frames the method yields a sequence of schemes with increasing degree of complexity. The simplest ones, denoted here as L_1, E_1, involve practically no further computational effort beyond Godunov's scheme, whereas the most complex ones, denoted by L_∞, E_∞, require a careful analysis of the structure of singularities at cell boundaries. The Lagrangian schemes are very close to the MUSCL scheme of van-Leer [9], which can be shown to be equivalent to our L_2 scheme (see [2, Appendix A]).

The following features are shared by all the schemes presented here.

(a) Upwinding, Second-order Accuracy and High Resolution of Discontinuities. The numerical solution is discretized at $t_n=n\Delta t$ as a piecewise linear function, leading to jump discontinuities at cell boundaries. These are then resolved by the main building block of the method, namely, a solution to the Generalized Riemann Problem (GRP). The equation (1) is then cast into a "quasi-conservative" difference scheme as follows.

$$U_i^{n+1}-U_i^n = -\frac{\Delta t}{\Delta V_i}[A(r_{i+\frac{1}{2}})F(U)_{i+\frac{1}{2}}^{n+\frac{1}{2}} - A(r_{i-\frac{1}{2}})F(U)_{i-\frac{1}{2}}^{n+\frac{1}{2}}] - \frac{\Delta t}{\Delta r_i}[G(U)_{i+\frac{1}{2}}^{n+\frac{1}{2}} - G(U)_{i-\frac{1}{2}}^{n+\frac{1}{2}}], \tag{2}$$

$$\Delta r_i = r_{i+\frac{1}{2}} - r_{i-\frac{1}{2}}\ , \ \Delta V_i = \int_{r_{i-\frac{1}{2}}}^{r_{i+\frac{1}{2}}} A(r)dr.$$

Observe that in (2) the jump discontinuities occur at the cell boundaries $r_{i\pm\frac{1}{2}}$ and U_i^n denotes the <u>average</u> value for the cell at time t_n. Second-order accuracy is achieved by taking,

$$U_{i+\frac{1}{2}}^{n+\frac{1}{2}} = U_{i+\frac{1}{2}}^n + \frac{\Delta t}{2}\left(\frac{\partial U}{\partial t}\right)_{i+\frac{1}{2}}^n , \tag{3}$$

and computing $F(U)_{i+\frac{1}{2}}^{n+\frac{1}{2}}$, $G(U)_{i+\frac{1}{2}}^{n+\frac{1}{2}}$ accordingly. While $U_{i+\frac{1}{2}}^n$ is obtained by solving a suitable <u>Riemann Problem</u>, the evaluation of the time derivative requires the solution to the GRP. The various schemes differ by the degree of accuracy to which the exact derivative $\left(\frac{\partial U}{\partial t}\right)_{i+\frac{1}{2}}^n$ is approximated.

b) <u>Robustness and Simplicity of Implementation</u>. Shock and contact discontinuities, as well as interactions of such singularities, are automatically taken care of. There is no need for artificial viscosity or any other dissipative mechanism, except for a simple monotonicity algorithm. Being three-point schemes, boundary conditions are easily implemented.

2. OUTLINE OF THE METHOD.

To give an idea of our solution to the GRP, namely, the exact evaluation of the time derivative in (3), consider Fig. 1 where the discontinuity is located at r=0 and variables are

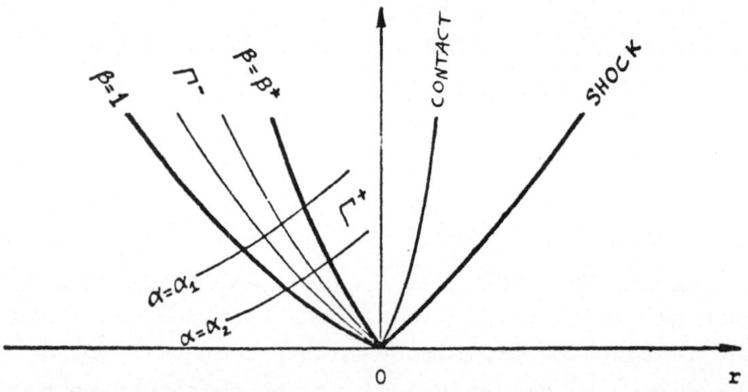

Figure 1. Wave pattern in solution to the GRP.

linearly distributed on both sides. We denote by $U_\pm$, $\left(\frac{\partial U}{\partial r}\right)_\pm$ the limiting values, as $\pm r\downarrow 0$, of the variables and their slopes, respectively. Also, we let U^* , $\left(\frac{\partial U}{\partial t}\right)^*$ be the limiting values (as $t\to 0+$) along the contact discontinuity. We have now the following theorem [3].

Theorem 1. *The derivatives* $(\frac{\partial p}{\partial t})^*$, $(\frac{\partial u}{\partial t})^*$ *are determined by a pair of linear equations,*

$$a_+(\frac{\partial u}{\partial t})^* + b_+(\frac{\partial p}{\partial t})^* = d_+ ,$$
$$a_-(\frac{\partial u}{\partial t})^* + b_-(\frac{\partial p}{\partial t})^* = d_- , \qquad (4)$$

where a_+, b_+, d_+ *(resp.* a_-, b_-, d_-*) can be determined explicitly from* U^*, U_+, $(\frac{\partial U}{\partial r})_+$ *(resp.* $U^*, U_-, (\frac{\partial U}{\partial r})_-$*).*

Clearly, once equations (4) are solved, the Lagrangian analogue of (2) is easily set up, since the fluxes depend on p,u only (note that the contact discontinuity becomes the grid-line in the Lagrangian frame). If an Eulerian scheme is desired, time derivatives along the line r=0 must be computed. Since the slope of the contact discontinuity is known, these derivatives can be obtained by using the chain rule, where Eq.(1) is used to determine the spatial derivatives at the discontinuity. This procedure is indeed the basis for the Eulerian scheme, with one important exception, namely, the Sonic Case, when the line r=0 falls within a rarefaction wave. In this case, as well as for the rigorous evaluation of a_-, b_-, d_- in (4)(using the configuration of Fig.1), we need to analyze the centered rarefaction wave. Let us use characteristic coordinates as shown in Fig.1, where (with c=speed of sound),

$$\beta = \text{limit along a fixed } \Gamma^- \text{ curve of } \frac{\rho(r,t)c(r,t)}{\rho_- c_-},$$

$$\alpha \text{ is normalized so that } t(\alpha,\beta) = -\rho_- c_- \alpha\beta^{-\frac{1}{2}} + O(\alpha^2).$$

The following theorem is basic in our method (see [3] for the proof).

Theorem 2. *Let* $a(\beta) = \frac{\partial u}{\partial \alpha}(0,\beta)$, $\beta^* \leq \beta \leq 1$. *Then* $a(\beta)$ *satisfies a differential relation of the form,*

$$a'(\beta) + H(\beta) + T(\beta) = 0,$$

where $H(\beta)$, $T(\beta)$ *reflect, respectively, the thermodynamic and geometrical non-uniformity for* $r<0$.

The second relation of (4) is obtained by expressing the directional derivative $a(\beta^*)$ in terms of $(\frac{\partial u}{\partial t})^*$, $(\frac{\partial p}{\partial t})^*$. In the sonic case, one must determine $\beta=\beta_0$, the characteristic tangent to r=0 and then use the full detail of the characteristic representation in computing the desired time derivatives

3. THE L_1, E_1 SCHEMES.

Incorporating the result of Theorem 2 in the coefficients of (4) one obtains the exact values of the time derivatives in the GRP. These can now be used in (2)-(3) to yield the L_∞, E_∞ schemes. However, the presence of sonic lines (in E_∞) and the algebraically complicated expressions are good reasons to look for simpler schemes. We note that in (2)-(3) second-order accuracy is already achieved when $(\frac{\partial U}{\partial t})^n_{i+\frac{1}{2}}$ is evaluated within an $O(\Delta t)$ error. The procedure now simplifies considerably, as is demonstrated in the next theorem.

Theorem 3. *In equations* (4) *set,*

$$a_\pm = \mp 1 \ , \ b_\pm = \rho_\pm^{-1} c_\pm^{-1} \ ,$$
$$d_\pm = - c_\pm (\frac{\partial u}{\partial r})_\pm \pm \rho_\pm^{-1} (\frac{\partial p}{\partial r})_\pm - \frac{A'(0)}{A(0)} u_\pm c_\pm . \qquad (5)$$

Then, in regions of smooth flow, $(\frac{\partial u}{\partial t})^*$, $(\frac{\partial p}{\partial t})^*$ *are obtained with an* $O(\Delta t)$ *error.*

Observe that in (5) the coefficients do not depend on U^*. The resulting scheme (2)-(3) is now only a slight modification of Godunov's scheme. The Lagrangian scheme is labeled L_1. Its Eulerian counterpart, E_1, is very simple too and does not require any special treatment of the sonic case. As a matter of fact, the coefficients in (5) are equivalent to the assumption that both waves emanating from the singularity are "acoustic" (i.e., just characteristic curves) and in particular, no shocks or rarefaction fans are present.

4. NUMERICAL EXAMPLES 1) Sod's shock-tube problem [8]. The tube extends from x=0 to x=100 and is divided into 100 equal cells. The gas is

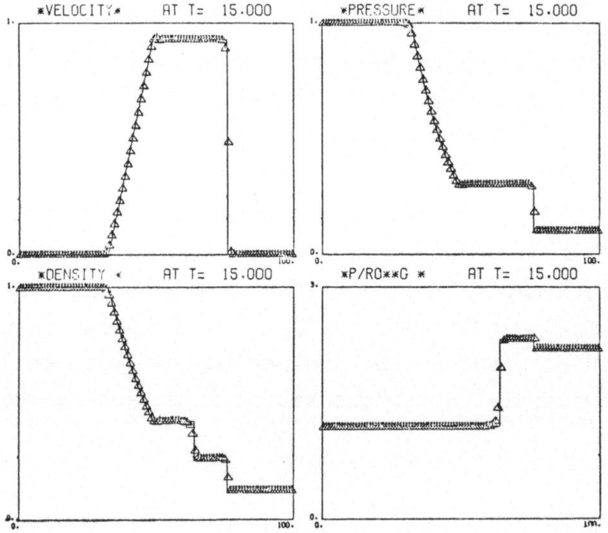

Figure 2. Sod's example, using the E_1 scheme.

initially at rest with p=ρ=1 for $0 \leq x < 50$, p=0.1, ρ=0.125 for $50 < x \leq 100$. In Fig.2 the exact solution (t=15) is plotted by a solid line, the numerical solution by dots. The E_1 scheme was used. We give pressure, velocity, density and entropy($=p/\rho^\gamma$) profiles.

2) Converging cylindrical shock. [1,5,7]. A cylindrical diaphragm of radius r=100 separates two fluids at rest. The internal one is at p=ρ=1, the external one at p=ρ=4. The interval $0 \leq r \leq 200$ is divided into 200 equal cells and the E_∞ scheme is used. In Fig.3 we give the density profiles at two times. t=56. The converging shock is about to hit the axis. t=110. After the interaction of the reflected shock and the converging contact discontinuity, a weak converging shock is produced.

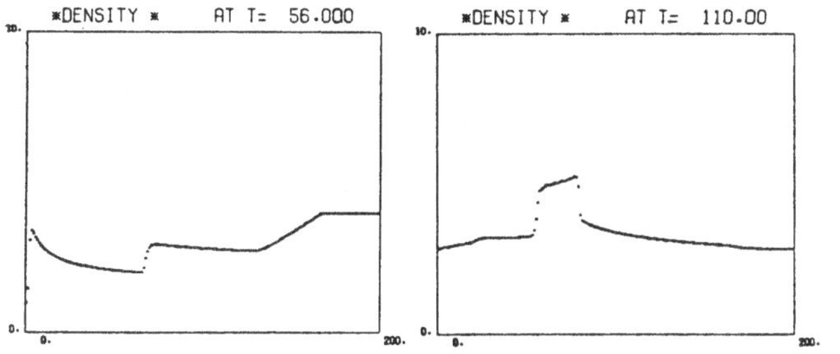

Figure 3. Converging cylindrical shock, E_∞ scheme.

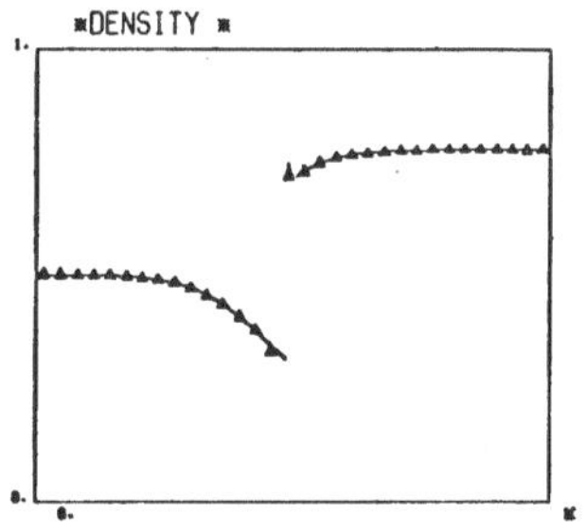

Figure 4. Density profile, quasi 1-D nozzle.

3) Steady state in a quasi 1-D nozzle [6]. In Fig.4 density profile is given, using the E_∞ scheme.

5. REFERENCES

[1] S.Abarbanel and M.Goldberg, J. Comp. Phys. 10(1972), 1-21.
[2] M.Ben-Artzi and J.Falcovitz, A second-order Godunov-type scheme for compressible fluid dynamics, J. Comp. Phys. (in press).
[3] M.Ben-Artzi and J.Falcovitz, An upwind second-order scheme for compressible duct flows, preprint.
[4] S.K.Godunov, Mat. Sbornik 47(1959), 271-295.
[5] A.Lapidus, J. Comp. Phys. 8(1971), 106-118.
[6] G.R.Shubin, A.B.Stephens and H.M.Glaz, J.Comp.Phys.39(1981),364-374.
[7] G.A.Sod, J. Fluid Mech. 83(1977), 785-794.
[8] G.A.Sod, J. Comp. Phys. 27(1978), 1-31.
[9] B.van-Leer, J. Comp. Phys. 32(1979), 101-136.

AN ADAPTIVE MULTIGRID METHOD FOR THE EULER EQUATIONS

Marsha J. Berger

Courant Institute of Mathematical Sciences
251 Mercer Street
New York University
New York, NY 10012

Antony Jameson

Princeton University
Dept. of Mechanical and Aerospace Engineering
Princeton, NJ 08544

1. INTRODUCTION

It is well known that an accurate solution to the Euler equations requires more resolution in some parts of the flow field than others. Just where these regions are, and how fine the mesh spacing must be, depends on characteristics of the solution to be computed, and changes with different flow parameters, such as Mach number, angle of attack, etc. Therefore, it cannot be known in advance of the computation. A solution technique which gets comparable accuracy over the entire flow field, thus using fewer points in smoother regions of the solution and more elsewhere, would clearly be optimal.

In this talk, we describe a method of local adaptive grid refinement for the solution of the steady Euler equations in two dimensions, which automatically selects regions requiring mesh refinement by measuring the local truncation error. Our method of refinement uses locally uniform fine rectangles which are superimposed on a global coarse grid. Possibly several nested levels of refined grids will be used until a given accuracy is attained. The fine grid patches are in the same coordinate system as the underlying coarse grid. All the data management is done in the computational plane, where, since we use rectangular grids, the data structures and bookkeeping can be very simple (see figure 1). Furthermore, the same data structures and control flow needed for adaptive grid refinement are also required for multigrid convergence acceleration. This can therefore be added with little additional cost.

Other adaptive mesh strategies have previously been proposed. Many of them are moving grid point methods, where a logically rectangular mesh is distorted to put more grid points in region where the solution error is large. Brackbill and Saltzman [3] do this by having their mesh minimize a functional which includes terms measuring the solution error, grid smoothness, and grid orthogonality. Rai and Anderson [4] attract the grid points into regions with high error by determining grid point speeds. These methods seem to work well in the examples in the literature, and they do not suffer from the difficulties with conservation that a patched grid method has at the grid interfaces. However, controlling the grid skewness in these methods is very difficult, and will be more so in three dimensions. Our method does not have this drawback. Furthermore, in a moving grid point method, it is still desirable to have a mechanism to add new points if necessary. Recent work by Murman and Usab [5] presents a similar approach to grid adaption using nested grid patches, but does not include a procedure for automatic control of the error. We add here that our adaptive method is a general purpose approach (and program) which has successfully been applied to computing other problems besides transonic flow (see e.g. [6]).

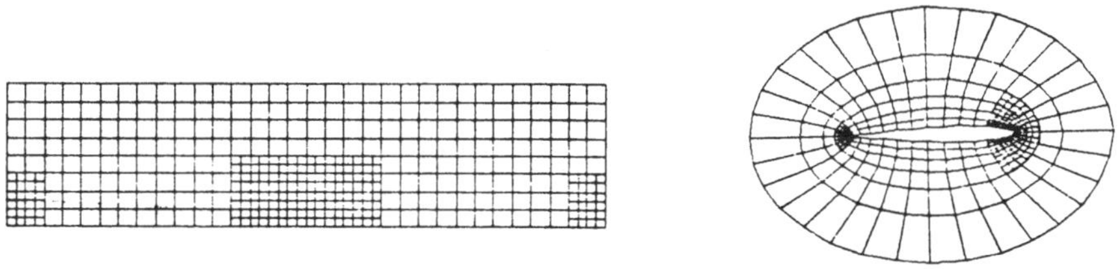

Figure 1. Fine grids at the leading and trailing edges.

The adaptive algorithm consists of four main steps. First, an automatic error estimator decides where the coarse grid accuracy is insufficient. This is based on estimates of the local truncation error, obtained by a method similar to Richardson extrapolation. In the second step, the local fine grid patches are created in the regions of high error. (This typically occurs at the leading and trailing edges of the airfoil and at a shock). Third, the solution on this multiple grid structure is integrated to steady state using the same integrator in the interior of each grid. Because of the regular grid structure, we have been able to use an existing method (FLO52) on each grid. However, special boundary conditions are needed at the interface of the fine and coarse grids to insure that the overall scheme is conservative and stable. In the separately described fourth step, the convergence of the solution is accelerated using a multigrid method with the modifications needed for this grid structure. We briefly describe each step, and show computational examples. A more detailed description of the adaptive algorithm can be found in [1], and of the underlying integrator and multigrid accelerator in [2].

2. DESCRIPTION OF THE ADAPTIVE ALGORITHM

The solution procedure begins by iterating on a single global grid, or possibly several nested grid levels, if multigrid is already being used. We wait until the residual of the solution is approximately 10^{-2} before applying the error estimator and subsequent adaptive strategy, so that the error due to lack of convergence is less than the error due to discretization. The local truncation error in the solution is estimated using an automatic procedure similar to Richardson extrapolation. In this approach, the same fine grid integrator is used for one step on a coarsened grid of mesh spacing twice the fine grid spacing. This coarsened residual is proportional to the local truncation error of the solution. Those grid points where the error estimate exceeds the specified tolerance are flagged as needing to be in a grid with finer mesh spacing. The grid generation algorithm creates rectangular grids, so that every flagged point is contained in a fine grid. We emphasize that these fine grids are not patched into one global grid, but are kept independently, each with its own solution vector. It is important that these grids are rectangles in the computational plane. This means that regardless of what method is chosen, the same integrator can be used to advance the solution of each grid, and it will vectorize on each grid. This also keeps the data structures fairly simple. The overhead in this approach is that the solution will be computed and stored on the entire coarse grid, even in those regions where it is not needed and the fine grid solution is used instead. However, we feel this is simpler than fragmenting the coarse grid and using pointers to indicate where each row and column begin.

Given the newly created fine grid structure, the solution on each grid is initialized by interpolation from the coarser grids, and the time stepping continues. For each step on the coarse grid, one (or more) steps on each fine grid are taken. When all grids have been advanced, the fine grid solution is used to update all coarse grid points which are lying underneath a fine grid. This is done by replacing the coarse grid solution value using a volume weighted interpolation from the fine grid to maintain conservation. The final step in the solution process is the specification of the boundary values for the fine grids by interpolation from coarser grids. This must be done in a way such that overall conservation in the solution is maintained at the interface between the fine and coarse grids. This is the only part of the adaptive strategy that is not independent of the chosen method of integration. Since the scheme we use to integrate the solution (FLO52) is based on a finite volume scheme, it is possible to devise a stable and conservative scheme such that the flux into the fine grid balances the flux into the coarse grid. This is described in more detail in [1].

These calculations are supported by a general purpose computer program that handles the storage and data flow management problems for multiple grids, (see [7] for details). This makes possible multiple levels of refinement with the fine grids themselves further refined in regions requiring very high resolution. These same routines support the multigrid algorithm described below. A similar approach can also be taken for calculations involving independently generated component grids.

3. DESCRIPTION OF THE MULTIGRID ALGORITHM

The method chosen to integrate the Euler equations uses a finite volume spatial discretization, augmented by dissipation terms, and a multi-stage time stepping procedure to integrate the resulting ODE's. If we write the equations as

$$\frac{d}{dt}(hw) + Qw - Dw = 0,$$

where w is the solution vector $(\rho,\rho u,\rho v,e)^T$, h the cell area, Q the terms from the Euler equations, and D the dissipative terms, the time stepping sequence looks like

$$w^{(1)} = w^{(0)} - \Delta t\alpha^{(0)}\left(\frac{Qw^{(0)}-Dw^{(0)}}{h}\right)$$

$$w^{(2)} = w^{(0)} - \Delta t\alpha^{(1)}\left(\frac{Qw^{(1)}-Dw^{(0)}}{h}\right)$$

etc. The $\alpha^{(j)}$ are chosen using criteria such as maximizing the stability region, or in the case of multigrid, maximizing the damping of the higher frequencies. Notice that the part of the right hand side in parentheses is just the residual, henceforth denoted $R_h(w)$ when on a grid with mesh spacing h.

A full multigrid acceleration strategy has been devised for use with FL052 on global grids (see [2]). In this scheme, a forcing function is defined so that on a coarser grid, with the solution obtained by conservative interpolation from the fine grid denoted by w_{2h}, the equations being solved are

$$w^{(1)}_{2h} = w^{(0)}_{2h} - \Delta t\alpha^{(0)} (Rw^{(0)}_{2h} + P_{2h})$$

$$w^{(2)}_{2h} = w^{(0)}_{2h} - \Delta t\alpha^{(1)} (Rw^{(1)}_{2h} + P_{2h})$$

where $P_{2h} = \sum R(w_h) - R(w^{(0)}{}_{2h})$. The sum is taken over the four fine cells comprising one coarse cell. This formulation insures that when the residual is zero on the fine grid, the solution on the coarse grid does not change. After one iteration on a coarse grid, the correction is interpolated back to the fine grid using bilinear interpolation.

To use an adaptive multigrid strategy, we start by iterating on a coarse grid. The Richardson error estimates are used to determine how fine a grid (global or local) is necessary to obtain a certain accuracy in the solution. Refinement based on error estimates has also been proposed in the multigrid literature. To use multigrid on fine grids that are not global means that when iterating on the global coarse grid, some cells have a forcing function from a finer grid and some do not. This is easily arranged, since the adaptive strategy outlined above already accounts for the fact that some coarse solution values are updated from a finer grid, and some are not. This updating corresponds to the restriction operation in the multigrid literature. The only new part of the algorithm that needs to be added to the adaptive code is a prolongation routine to add the correction back to the fine grid. In using multigrid on patched grids one must be careful to make sure that the solution procedures for boundary values and interfaces are compatible, so that a steady state solution is possible. For example, the far field boundary conditions for a fine grid are extrapolated for outflow boundaries using the entropy in the far field. Coarser grids however should not use this far field boundary specification, since coarse grid solution values obtained by interpolation from the fine grid give a different entropy value. Thus, using this to obtain the far field solution on the coarse grid would change a converged solution on a fine grid. We add here that although our preliminary results indicate that multigrid is effective in accelerating the solution of the Euler equations on a patched grid system, some of the assumptions of the standard multigrid optimality theory [8] are no longer valid, and it remains an open question whether a solution can be obtained in a fixed number of iterations independent of the number of grid points.

4. NUMERICAL RESULTS

Figure 2 shows a typical calculation for transonic flow over a NACA0012 airfoil with a Mach number of 0.8. Using an error tolerance of .004, the adaptive algorithm automatically generates 5 refined grids. One grid patch is at the leading edge, two are at the trailing edge (since the grid is periodic with a cut at the trailing edge), and two patches are at the top and bottom of the airfoil, surrounding the shock. Figure 3 shows the pressure coefficient computed on this grid. A total of 300 iterations were done in this calculation but only 200 of them included the five finest grids. This is approximately a factor of 6 faster than without using multigrid. The combined area of these grids was only 30% of a uniformly refined grid with 128 by 32 grid points, and so the cost of an iteration was only 1/3 the cost of an iteration of a uniformly refined grid. Faster rates of convergence have been obtained by multigrid applied to global grids without embedded patches [2], but the patched method yields savings in computational effort owing to the reduced cost per iteration.

5. REFERENCES

[1] M. Berger and A. Jameson, "Automatic Adaptive Grid Refinement for the Euler Equations", NYU Report No. DOE/ER/03077-202, October, 1983. To appear in AIAA Journal.

[2] A. Jameson, "Solution of the Euler Equations for Two Dimensional Transonic Flow by a Multigrid Method", MAE Report NO. 1613, June 1983, Applied Math. and Computation, 13 (1983), 327-356.

[3] J. Saltzman and J. Brackbill, "Applications and Generalizations of Variational Methods for Generating Adaptive Meshes", in Numerical Grid Generation, J. Thompson, ed., North-Holland, 1982.

[4] M. Rai and D. Anderson, "Grid Evolution in Time Asymptotic Problems", J. Comp. Phys. 43 (1981), 327-344.

[5] W. Usab, Jr. and E. Murman, "Embedded Mesh Solutions of the Euler Equation Using a Multiple-grid Method", AIAA Paper 83-1946-CP. Presented at the 6th AIAA Computational Fluid Dynamics Conference, Danvers, Mass. July 1983.

[6] P. Collela and H. Glaz, "Numerical Computation of Complex Shock Reflections in Gases". These proceedings.

[7] M. Berger and J. Oliger, "Adaptive Mesh Refinement for Hyperbolic Partial Differential Equations", J. Comp. Phys. 53 (1984), 484-512.

[8] W. Hackbusch, "On The Multi-Grid Method Applied to Difference Equations," Computing, 20 (1978), 291-306.

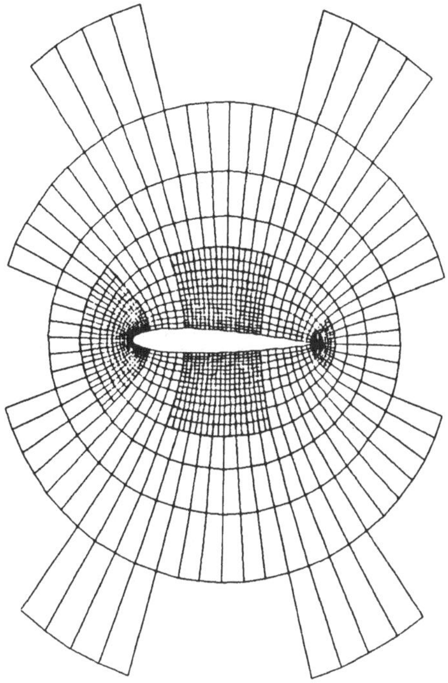

Figure 2(a)
Grid 64 x 16

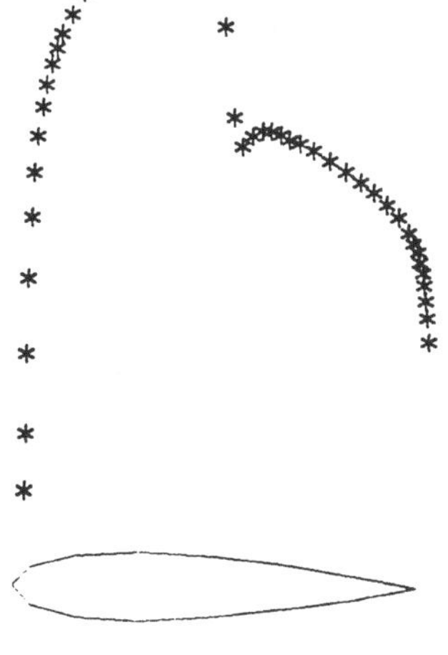

Figure 2(b)
MACH 0.800 ALPHA 0.0
CL 0.0000 CD 0.0086 CM 0.0000

DIRECT SIMULATIONS OF SPATIALLY EVOLVING COMPRESSIBLE TURBULENCE

-- TECHNIQUES AND RESULTS

J.P. Boris, E.S. Oran, J.H. Gardner, F. Grinstein+ and C.E. Oswald*
Laboratory for Computational Physics
Naval Research Laboratory
Washington, D.C. 20375, USA

INTRODUCTION

This paper describes calculations of the transition to turbulence and the behavior of coherent structures in spatially-evolving mixing layers. The focus of the new computational developments is the formulation of realistic inflow and outflow boundary conditions that can be implemented simply, accurately, and stably. These are used in FAST2D codes, which employ the Flux-Corrected Transport (FCT) continuity equation algorithm (Boris and Book, 1976) in two spatial dimensions to solve the compressible, time-dependent equations for conservation of mass, momentum and energy. Use of these boundary conditions is illustrated for four types of flows: a supersonic blast exiting from a barrel; a splitter-plate calculation to study mixing in coherent structures; flow from a cylindrical gas jet into a quiescent, ambient background; and cold flows within an idealized ramjet combustor. The last three scenarios show vorticity generation and the formation of large coherent structures. In several of these calculations the fluid speeds are high enough for sound waves and compressibility to be important.

In the calculations presented below, the grid spacing was set up initially and held fixed in time. For the Cartesian calculations used to model the splitter plate experiments, finely spaced cells were clustered around the centerline where the instability first occurs and the coherent structures form. For the cylindrical calculations used to model the gas jet the grid was finely spaced in the jet and through the region of the shear. The ramjet combustor calculations used fine zones in both directions near the inlet rim of the combustion chamber. Relatively low resolution finite-difference grids use 40-60 cells across the mixing layer and 100-150 cells in the streamwise direction. High resolution calculations use 120 × 300 cells.

*Science Applications, Inc., Mclean, Va.
+Berkeley Research Associates, Inc., Springfield, Va.

II. BOUNDARY CONDITION TESTS

Solving fluid problems with realistic outflow boundary conditions is difficult because information about the flow beyond the mesh is required to make the fluid near the boundaries behave properly. The new outflow algorithms define fluid variables at fictitious guard cells which are not actually part of the calculation, but which communicate information about the world outside the mesh to the actual boundary cells. The simplest model of outflow is to say that the momentum, energy, and density do not change, i.e., there is effectively zero gradient. This causes secular errors in long calculations since it does not ensure that the assumed outer flow relaxes to background conditions far from the interior of the mesh. Our outflow algorithm relaxes the solution just off the edge of the computational mesh toward ambient conditions (Boris, et al., 1983). The fluid inflow algorithm allows the incoming gas to respond to pressure pulses near the splitter plate or nozzle rim resulting from dynamic events such as vortex merging in the mixing layer. Nonlinear stabilizing properties of the FCT method eliminate instabilities occurring in other nonlocal methods when low order extrapolations are used (Turkel, 1980).

Boundary condition tests were conducted on the axisymmetric "barrel" shown in Figure 1 using a nested series of calculations where the outflow boundary of one grid is interior to the flowfield on a larger grid. A supersonic flow emerges into free space from the cylinder. Three grids were used: 40 × 40, 80 × 80, and 150 × 300. As can be seen from the figure, the outflow boundary condition is excellent even for the 40 × 40 case. Properties of the flow are still reproduced at times long after the fluid has effectively passed out of the small mesh.

Next we consider a problem where the speed of disturbances is much slower and fluid curl, rather than divergence, is being carried off the grid. A subsonic axisymmetric jet of neon gas is vented into quiescent air. A nested series of three calculations were performed: (1) 60 × 120 cells, step 0 to 3000, (simulating 7.6 × 26.0 cm); (2) 54 × 100 cells, steps 2000 - 3000 with interior 54 × 100 cells of #1; and restarted from step 2000 of #1; (3) 54 × 100 cells, steps 0 to 3000. Initial small perturbations at the shear interface cause the flow to become unstable. Rotating coherent structures develop which merge and flow out of the computational system. Figure 2 shows contours ranging from zero (100% air) to one (100% neon). It takes ~1000 timesteps for a fluid element entering the system at the nozzle to reach the top of the large grid if it is not slowed appreciably by interaction with the background air. Since much of the material entering at step 2000 will be exiting the grid shortly after step 3000, the calculation was continued far enough to test the boundary condition algorithm for long term fidelity and stability.

The FAST2D code has been configured in Cartesian geometry to simulate mixing and the formation of coherent structures to compare with splitter plate experiments. The computational region extended 2 cm behind the tip of the splitter plate. Calculations were performed for a system of air coflowing into air, with an initial downstream velocity ratio of 10:1. The transition from laminar to turbulent flow was triggered naturally by the pressure gradients at the shear layer near the edge of the splitter-plate. Figure 3 is early in the evolution of the flow. The lower portion shows contours which are ratios of the species from the high speed to the low speed stream. The upper portion is a mixing ratio which shows the basic asymmetric property of such flows: the area below the centerline is clearly larger than the area above during the first major roll-up but subsequent mixing seems to mitigate this imbalance.

The combustor section of a central-dump ramjet contains an entrance nozzle through which material flows into an enclosed chamber. The flow is forced to become sonic at the exit nozzle, which is a torus off the centerline. Vorticity is shed at the rearward facing step at the entrance to the chamber. The most physical results, and those which produce close agreement with experiments use an inflow boundary in which $\dot{M}$ and $\dot{E}$ are specified, and reflecting conditions are used for pressure, density, and momentum. Using an isentropic relation, the outflow boundary conditions are extrapolated to the sonic condition at an exterior interface. Figure 4 shows streamlines and vorticity contours for a typical calculation using 40 × 100 cells. Vorticity shedding off of the step is shown, as is the flow out of the exit nozzle. The calculated shedding frequency of the structure at the step is the same as the acoustic frequency characteristic of the chamber, indicating strong coupling.

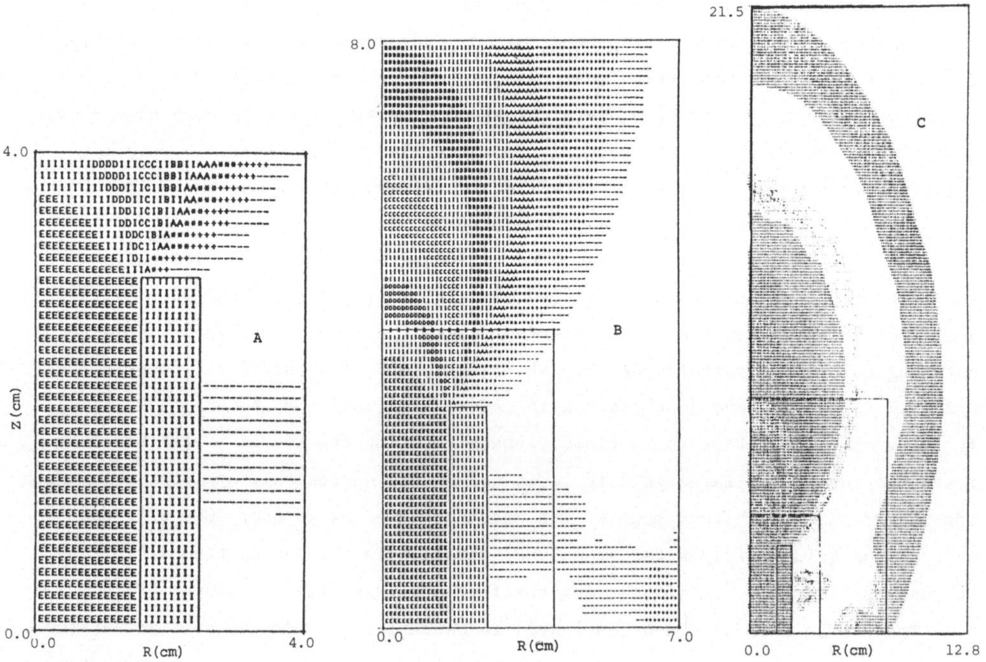

Figure 1. A nested series of calculations of a shock exiting a barrel (A) 40 x 40 cells, (B) 80 x 80 cells, and (C) 120 x 300 cells (only part of the calculation is shown).

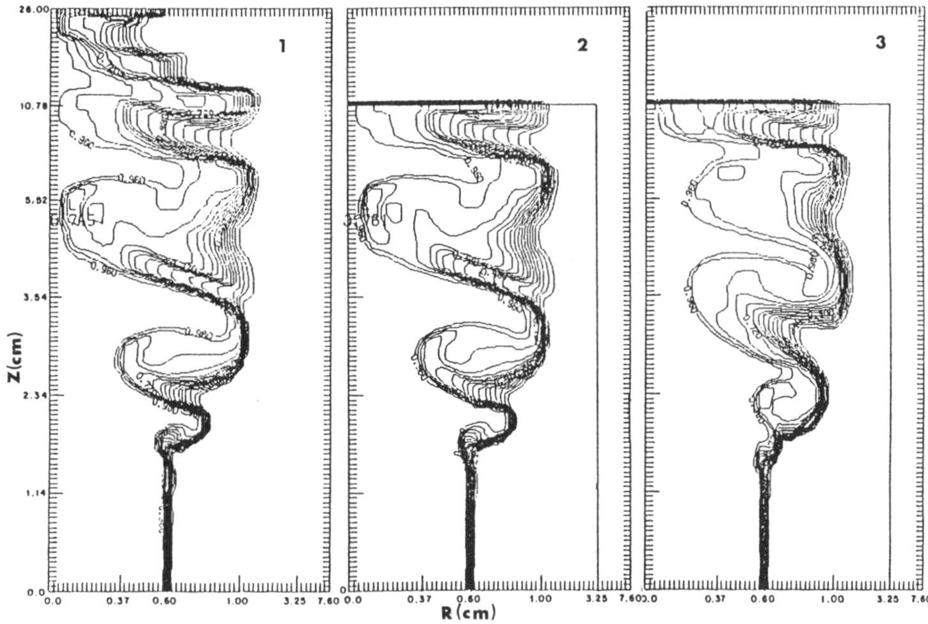

Figure 2. Comparison of contour plots of the mixing ratio of the materials in a nested series of gas jet calculations. (1) 64 x 120 cell calculation; (2) 54 x 100 cells started from step 2000 of (1); (3) 54 x 100 cells started from step 0.

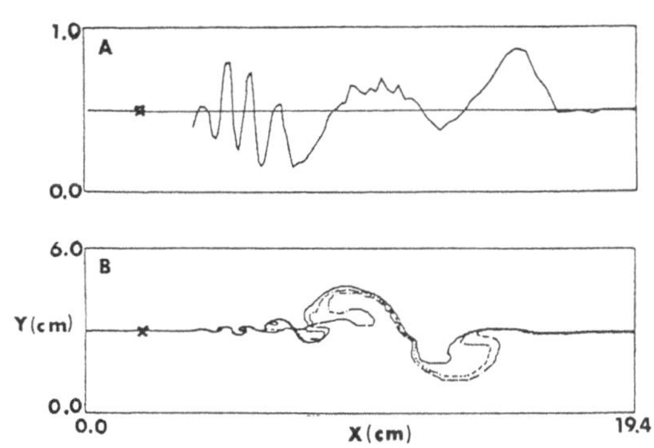

Figure 3. Cartesian calculation with splitter plate geometry. The "x" on the axis marks the tip of the splitter plate. Ratio of upper to lower flow velocities, 5:1. (A) Mixing asymmetry at a location along x-axis indicated by deviations from 0.5. (B) The number density ratio contour, indicating the amount of mixing of high speed into low speed fluid, shows the shape of the structures formed.

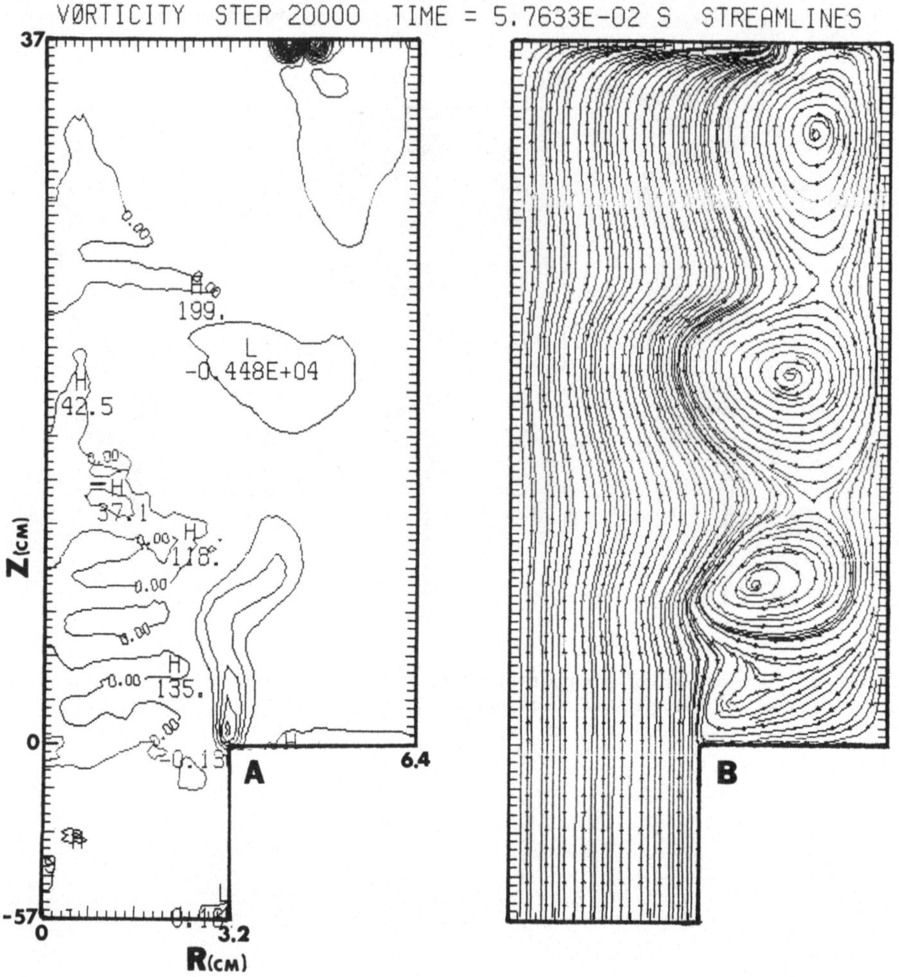

Figure 4. Calcuations of subsonic flows in an idealized ramjet combustor. Note that graphs are done per cell, so that variable spacing is reflected by the labels on the axis. (A) Contours of vorticity. (B) Streamlines.

REFERENCES

Boris J.P., and D.L. Book (1976), Solution of Continuity Equations by the Method of Flux Corrected Transport, in Methods in Computational Physics, Vol. 16, 85-129 Academic Press, New York.

Boris J.P., E.S. Oran, M.J. Fritts and C.E. Oswald (1983), Time Compressible Simularities of Shear Flows, NRL Memo Rept. 5249, Naval Research Laboratory, Washington, D.C.

Turkel, E. (1980), Numerical Methods for Large-Scale Time-Dependent Partial Differential Equations, in Computational Fluid Dynamics, pp. 128-262, ed. W. Kollman, Hemisphere Press, Wash., D.C.

FREE DECAY OF HIGH REYNOLDS NUMBER TWO DIMENSIONAL TURBULENCE

M.E. Brachet* and P.L. Sulem**

*CNRS, Observatoire de Nice, France
**School of Mathematical Sciences, Tel Aviv University Israel and CNRS, Observatoire de Nice, France

ABSTRACT

The free decay of high Reynolds two dimensional turbulence is simulated by direct numerical intergration of the Navier-Stokes equations at a resolution of 1024 x 1024 with symmetric random initial conditions. The following scenario is observed : At early times, large scale straining generates quasi-rectilinear vorticity gradient sheets with thickness decaying exponentially in time until dissipation becomes relevant. In Fourier space, the energy spectrum displays a k^{-n} - range with $n \simeq 4$, in agreement with Saffman's theory. Close to the time of maximum enstrophy dissipation, we observe a transition to an $n \simeq 3$ inertial range, consistent which the Batchelor-Kraichnan theory of enstrophy cascade. In this regime, vorticity gradients are distributed on convoluted secondary dissipative structures resulting from folding and reconnection of early time sheets.

1 . INTRODUCTION

A controversal question in high Reynolds number turbulence in two-dimensional incompressible flows is the behavior of the energy spectrum. Saffman (1971) argues that advection will bring different values of vorticity close together producing thin sheets of vorticity gradients and leading to a k^{-4} inertial energy spectrum. In contrast, the enstrophy cascade theory (Kraichnan, 1967 ; Batchelor, 1969) predicts a k^{-3} - energy spectrum with a possible logarithmic correction due to nonlocal interactions (Kraichnan, 1971). Furthermore, Kraichnan (1975) predicts that because of this non-locality, intermittency will not affect the energy spectrum. This point has been questioned by Basdevant, Legras, Sadourny and Beland (1981) who claim that intermittency will restore the predominance of local interactions and steepen the energy spectrum.

Since the first calculations of Lilly (1969), it has been recognized that very high resolutions are required to property simulate an inertial range (Herring, Orszag, Kraichnan and Fox, 1974). Preliminary calculations at $(512)^2$ - resolution presented by Orszag (1977), showed that when the large scale Reynolds number is increased from 1100 to 25000, a distinct change is observed from a k^{-4} energy spectrum to a spectrum roughly proportional to k^{-3}.

The present paper is devoted to simulations of spatially periodic solutions at $(1024)^2$ - resolution. To achieve this resolution on a 1M - word CRAY 1 computer, the "sparse mode technique" (Brachet, Meiron, Orszag, Nickel, Morf and Frisch, 1983) has been implemented : the stream function has a Fourier representation

$$\psi(x,y) = \sum_{\ell,m=0}^{N/2} a_{\ell m} \sin \ell x \sin my \qquad (1)$$

where $a_{\ell m}$ vanishes unless ℓ and m are both even or odd integers jointly. As in the Taylor-Green vortex, this representation implies flow symmetries, including reflectional invariance on the sides of an impermeable box $x = 0$ or π, $y = 0$ or π. The non-linear terms of the Navier-Stokes equation are evaluated in the form $v \times \omega$. Aliasing is suppressed by spectral truncation at a maximum wavenumber $k_m = N/3$. Time marching is done by leap-frog for the non-linear terms and Crank-Nicholson for the viscous term. Runs with deterministic initial conditions are reported in Brachet (1983). We report here on runs with gaussian random initial data with an energy spectrum $E_0(k) = c\, k \exp(-(k/k_0)^2)$. The absolute equilibrium at small k minimizes the inverse energy transfer. A sufficiently large range of scales must be excited to permit spatial averaging. We thus used $k_0 = 3.5$, $c = 0.02$ (corresponding to an energy $\Sigma = 0.132$ and an enstrophy $\Omega = 1.632$), and a viscosity $\nu = 1/30000$. With these parameters, the accuracy in the enstrophy dissipation is better than 1%. With a time-step of 0.00125, the integration up to t = 20, required five hours of CPU time on a CRAY-1 machine.

2 . HIGH REYNODS NUMBER SIMULATION

The short time behavior of the flow is observed to be dominated by the formation of vorticity gradient sheets (see Fig. 1a). A simple model of this phenomenon may be presented in terms of the stretching of vorticity gradients by velocity gradients (Weiss, 1981). This mechanism appears to dominate the early time small scale generation.

In order to extract quantitative information on the small scale generation, we have resorted to fitting the (angle averaged) energy spectrum with a function $E(k,t) = C(t)\, k^{n(t)}\, e^{-\beta(t)k}$. In this way we estimate both the magnitude β of the smallest sizeably excited scales and the inertial exponent n.

The early time sheet formation is seen to be associated with an exponential decay in time of β. This process, consistent with the above model, stops around $t \sim 1.5$ when scales are reached that are small enough for the viscosity to act. The spectral exponent n is then close to -4, in agreement with Saffman's (1971) theory (see table 1). Indeed, flow visualizations show conspicuous vorticity gradient sheets (Fig. 1a). Later, around the time $t \simeq 5$ of maximum enstrophy dissipation, a transition to a new regime characterized by an $n \approx -3.2 \pm 0.1$ inertial spectral exponent is observed* (see table 1). This developped regime persists until the end

*See Brachet and Sulem (1984) for a run at higher Reynolds number with less accuracy displaying a similar transition.

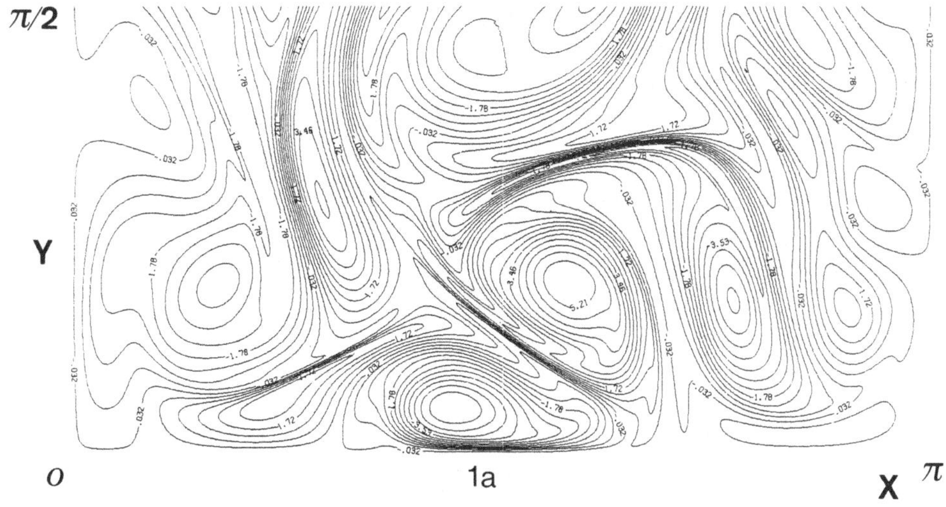

1a

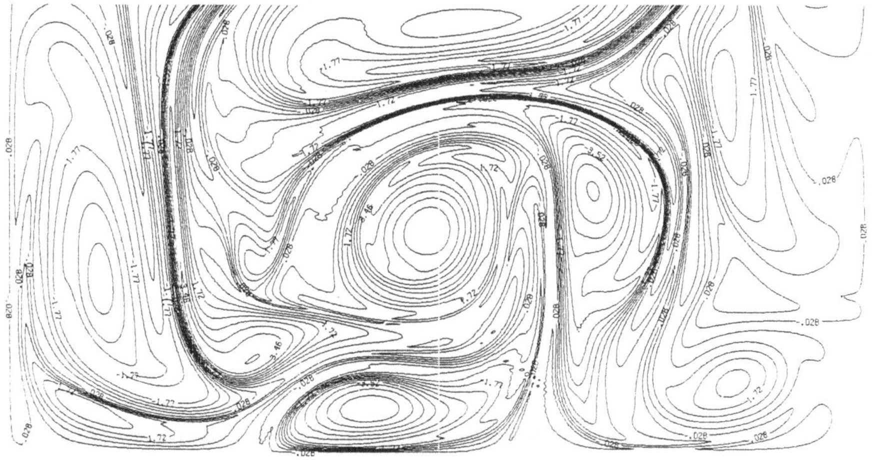

1b

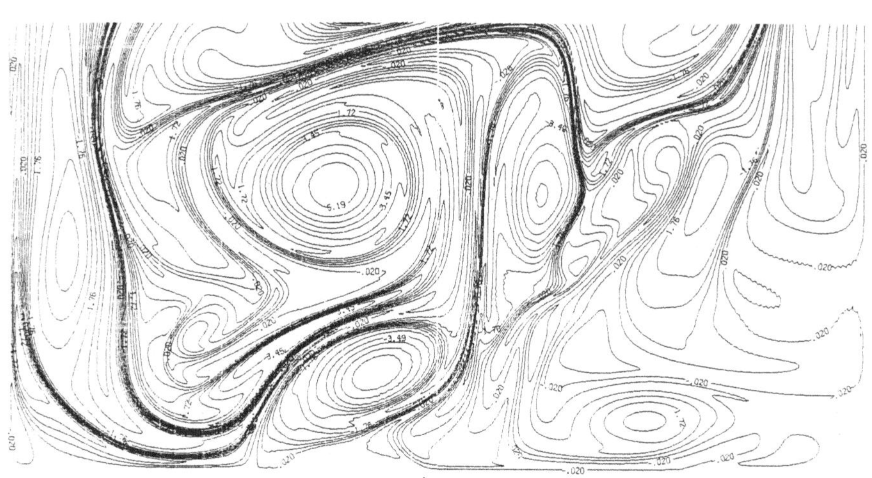

1c

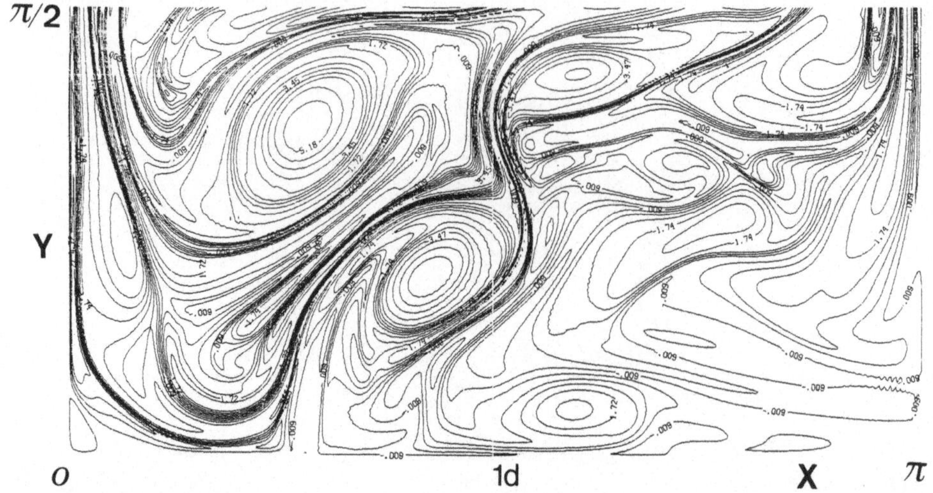

1d

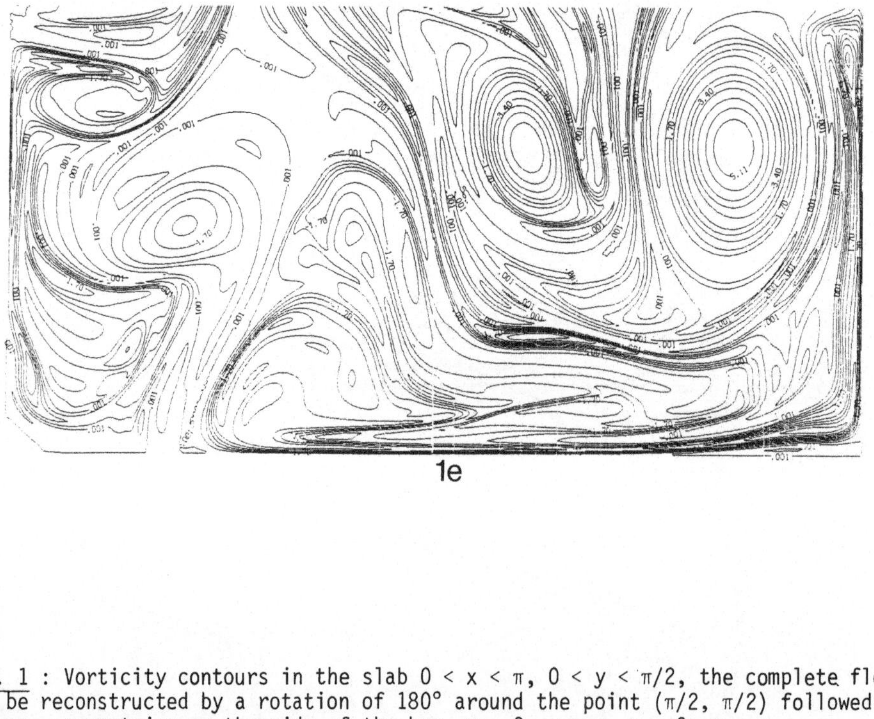

1e

Fig. 1 : Vorticity contours in the slab $0 < x < \pi$, $0 < y < \pi/2$, the complete flow can be reconstructed by a rotation of 180° around the point $(\pi/2, \pi/2)$ followed by mirror - symmetries on the side of the box $x = 0$, $x = \pi$, $y = 0$, $y = \pi$.
Fig. 1a : t = 2 , 1b : t = 4, 1c : t = 6 , 1d : t = 8 , 1e : t = 16

t	n	$\beta \times 10^2$
2	- 4.97	1.53
4	- 4.28	1.34
6	- 3.35	1.77
8	- 3.18	1.95
10	- 3.27	1.83
12	- 3.26	1.75
14	- 3.01	2.31
16	- 2.90	2.67

Table 1 : Fit of the energy spectrum $E(k) \propto k^{-n} e^{-\beta k}$ on the interval $10 < k < 341$.

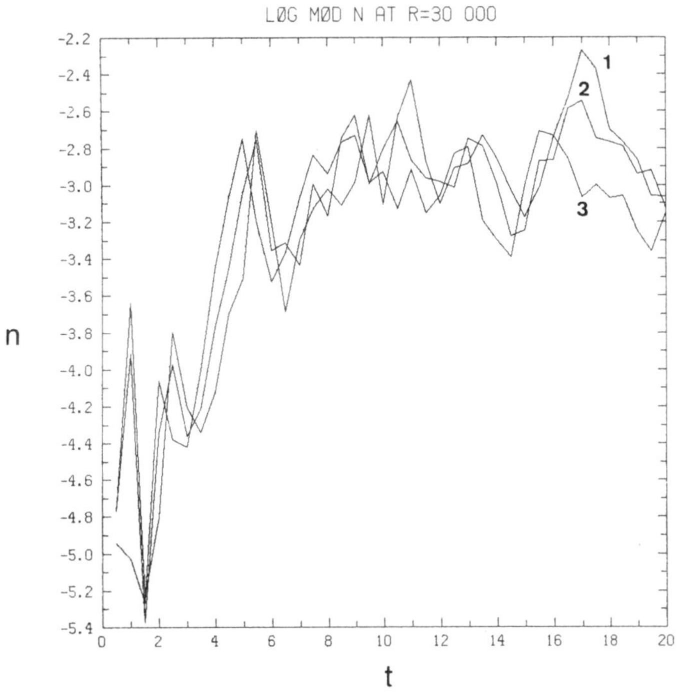

Fig. 2 : Plot versus time of n obtained by fitting E(k) with the function $A(\log(k/k_0))^{-1/3}\, k^n\, e^{-\beta k}$, on the interval $10 < k < 341$ for curve (1), $10 < k < 256$ for curve (2) and $10 < k < 170$ for curve (3).

of our run (t = 20). The n $\approx$ -3.2 spectral exponent is consistent with the enstrophy cascade theory (Kraichnan, 1967 ; Batchelor, 1969) : if we take into account Kraichnan's $Log(\frac{k}{k_o})^{-1/3}$ correction (Kraichnan, 1971), by fitting $Log(\frac{k}{k_o})^{1/3}E(k,t)$ with $Ak^{-n}\ e^{-\delta k}$, we get n $\approx$ 3.0 ± 0.1 (see Fig. 2). Flow visualizations in physical space show that vorticity gradient sheets are still present but have developed a rolled up convoluted structure (Fig. 1c-e). This leads to the formation of isolated vortices somewhat analoguous to those observed by Basdevant et al. (1981) and McWilliams (1983).

ACKNOWLEDGMENTS

We thank U. Frisch, J. Herring, R.H. Kraichnan, M. Meneguzzi and S.A. Orszag for very useful discussions and suggestions. The computations were performed on the CCVR CRAY 1. Visualizations were done with partial support of a DRET contract.

REFERENCES

Basdevant, C., Legras, B., Sadourny, R. and Beland, M. (1981), J. Atm. Sc. 38, 2305.

Batchelor, G.K. (1969), Phys. Fluids 12, 233.

Brachet, M.E. (1983), Thèse d'Etat, Université de Nice.

Brachet, M.E., Meiron, D.I., Orszag, S.A., Nickel, B.G., Morf, R.H., Frisch, U. (1983), J. Fluid. Mech. 130, 411.

Brachet,M.E. and Sulem, P.L. (1984) : in Proc. 4th Beer Sheva of MHD flows and turbulence. AIAA Progress in Astronautics and Aeronautics. To be published.

Herring, J.R., Orszag, S.A., Kraichnan, H.R. and Fox, D.G. (1974), J. Fluid Mech. 66, 417.

Kraichnan, H.R. (1967), Phys. Fluids 10, 1417.

Kraichnan, H.R. (1971), J. Fluid Mech. 47, 525.

Kraichnan, H.R. (1975), J. Fluid.Mech. 67, 155.

Lilly, D.K. (1969), Phys. Fluids Suppl. 12, II, 240.

McWilliams, J.C. (1983), The emergence of isolated, coherent vortices in turbulent flow. Preprint.

Orszag, S.A. (1977) : in Proc. 5th Int. Conf. on Numerical Methods in Fluid Mechanics p. 32. Springer Lecture Notes in Physics, 59.

Saffman, R.G. (1971), Studies in Appl. Math. 50, 377.

Weiss, J. (1981), The Dynamics of enstrophy transfer in two-dimensional hydrodynamics, La Jolla Inst. La Jolla, CA, LJI - TN - 81 - 121.

FINITE ELEMENT CALCULATION OF POTENTIAL FLOW AROUND WINGS

M. Brédif

Office National d'Etudes et de Recherches Aérospatiales (ONERA)
B.P. 72 - 92322 Châtillon - FRANCE

INTRODUCTION -

The finite element method in Computational Fluid Dynamics is generally used in connection with non structured meshes. This type of approach has been extensively studied by Glowinski et al. [1] and is able to predict potential flows around realistic aircraft configurations. On the other hand, fast algorithms have been recently developed which are based on the use of structured grids, like the ICCG method [2] and the multigrid acceleration technique [3] . The necessity to build finite element codes characterized by low computational costs has led us to develop a fast finite element method for the prediction of potential flows around airfoils using the ICCG - Multigrid algorithm on a structured mesh [4-5] .

This paper presents an extension of this approach to the three-dimensional case of flows around wings together with the development of an algebraic generation procedure of H-type meshes.

GRID GENERATION -

The grid generation technique used in the present 3-D formulation is of algebraic type, following the ideas of Cook [6] . Spanwise planes are introduced, almost equally spaced on the wing region but disposed in a geometric progression outside the wing region. Each spanwise plane is divided into six subdomains by introducing curved lines leaving the leading and trailing edges of the local profile (fig. 3). For planes outside the wing region, the profile considered is reduced to a flat plate. On the boundaries of each subdomain, nodes are defined using geometric progressions for the curvilinear abscissa. Each subdomain S is now considered as the image of the unit square $[0,1]^2$ in (ξ, η) plane by the explicit transformation $\vec{F}_S$:

$$\vec{F}_S(\xi,\eta) = (1-\eta)\vec{f}_1(\xi) + (1-\xi)\vec{f}_4(\eta) + \xi\,\vec{f}_2(\eta) + \eta\,\vec{f}_3(\xi) - (1-\eta)(1-\xi)\vec{a}_1 - (1-\xi)\,\eta\,\vec{a}_4 - \xi(1-\eta)\,\vec{a}_2 - \eta\,\xi\,\vec{a}_3$$

where $\vec{F}_S$ is the position vector in the physical plane, $\vec{a}_\ell$, $1 \leq \ell \leq 4$ designate the position vectors of the 4 corner nodes of S , and $\vec{f}_\ell$, $1 \leq \ell \leq 4$, give the parametric representations of the 4 sides of S in terms of reduced curvilinear abcissa ξ or η .

A mesh is then defined on the unit square $[0,1]^2$ in the following way (fig. 1) :

- boundary nodes on the unit square are the antecedents by $\vec{F}_S$ of the boundary nodes of S .
- interior modes are located at the intersection of straight lines joining the corresponding boundary nodes.

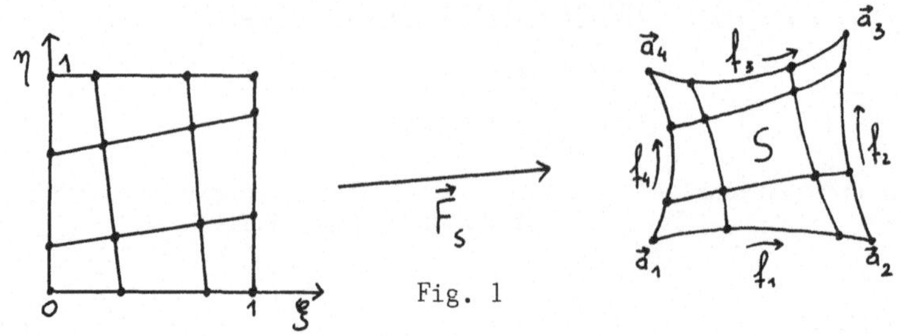

Fig. 1

The final mesh on the subdomain S is simply the image by the explicit transformation $\vec{F}_s$ of the mesh previously defined on $[0,1]^2$.

A smoothing step is added at the end of the procedure in order to reduce mesh irregularities near the boundaries between different subdomains. Another kind of modification is introduced in order to have mesh lines leaving the wing surface in an orthogonal way. If we suppose that the nodes of index J = 1 are located on the wing surface, nodes of index J = 2 are translated by a vector $\vec{\varepsilon}$ (I,K) in order to ensure orthogonality. Then all the nodes of index J greater than 2 are translated by a vector proportional to $\vec{\varepsilon}$(I,K), the coefficient of proportionality tending to zero when the nodes reach the external boundary.

The disadvantages of the H-mesh generated by our program are first the lack of regularity in the vicinity of the leading edge of the wing and second the greater number of total discretization points needed for the same number of points on the wing surface. Nevertheless, the advantages are numerous : no degenerated elements or singular lines are introduced for this type of mesh. Moreover, all the outer boundaries of the 3-D mesh are rectangular planes ; this will be an advantage for the future development of grid generation programs for complex aircraft geometries based on the multiblock approach of Lee et al. [7] .

Typical computing cost is about 2 seconds on the CRAY 1-S computer for the generation of a mesh containing more than 80.000 modes.

FINITE ELEMENT DISCRETIZATION -

The full potential equation written in conservation law form is given by :

$$\nabla\cdot(\rho\nabla\phi)=0 \tag{1}$$

together with the Bernoulli law for density :

$$\rho=\left\{1+\frac{\gamma-1}{2}M_\infty^2\left(1-|\nabla\phi|^2\right)\right\}^{\frac{1}{\gamma-1}} \tag{2}$$

All quantities are nondimensionalized by freestream values of density and velocity, and γ is the ratio of specific heats. Equation (1) is written in weak form :

$$\int_\Omega \rho\nabla\phi\nabla\theta\,dx\,dy\,dz=\int_{\partial\Omega} f\theta\,d\Gamma \quad \text{for any test function } \theta \tag{3}$$

where Ω is the physical domain, $\partial\Omega$ its boundary and f is the normal mass flux $\rho\frac{\partial\phi}{\partial n}$ to be imposed at the boundaries :

$f = 0$ on the wing surface
$f = \rho_\infty\vec{V}_\infty\cdot\vec{n}$ at infinity

A mesh is built on the domain Ω by the grid generation procedure. For each hexaedron K of the mesh, a trilinear transformation $\vec{G}_K$ is introduced which maps the unit cube $[0,1]^3$ onto the finite element K (fig. 2). The finite element discretization technique consists in approximating the exact potential function by a function ϕ_h which is assumed to be continuous and linearly dependant of each local coordinate ξ_K, η_K, ζ_K defined by $\vec{G}_K$ for every finite element K - Of course, approximate test functions θ_h are defined in the same manner. The approximate potential ϕ_h must satisfy the equality (3) :

$$\int_\Omega \rho_h \nabla\phi_h \nabla\theta_h \, dx\,dy\,dz = \int_{\partial\Omega} f\,\theta_h \, d\Gamma \tag{4}$$

for any test function θ_h

with $$\rho_h = \rho(\phi_h) = \left\{1 + \frac{\gamma-1}{2} M_\infty^2 \left(1 - |\nabla\phi_h|^2\right)\right\}^{\frac{1}{\gamma-1}}$$

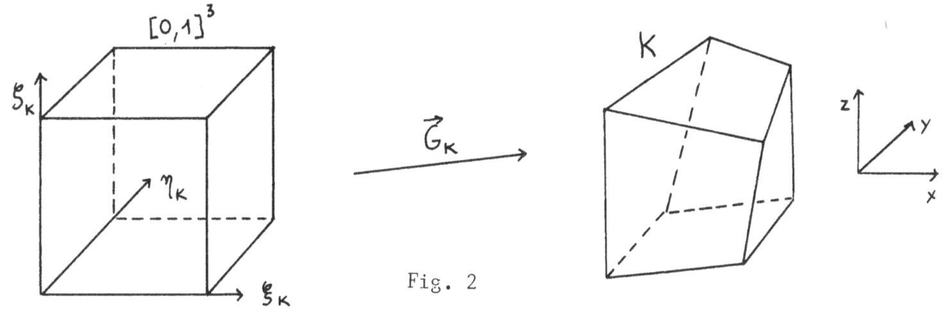

Fig. 2

In equality (4), the integral over Ω is decomposed into the sum of integrals over each element K and local coordinates are introduced :

$$\int_\Omega \rho_h \nabla\phi_h \nabla\theta_h \, dx\,dy\,dz = \sum_K \sum_{i,j=1}^{3} \int_{[0,1]^3} \rho_h \frac{g_K^{ij}}{\sqrt{|g_K|}} \partial_i\phi_h \, \partial_j\theta_h \, d\xi_K \, d\eta_K \, d\zeta_K$$

where g_K^{ij} are the contravariant components of the metric tensor associated with the local coordinates (ξ_K , η_K , ζ_K) inside each element K, and ∂_i stands for the derivation following the coordinate ξ_K (i = 1), η_K (i = 2) or ζ_K (i = 3). In the general case of arbitrary hexaedrons, the components of g_K are not polynomial functions of (ξ_K , η_K , ζ_K), and integration formulas are needed. Two integration formulas have been tested :

- The eight points Gauss formula :

$$\int_{[0,1]^3} f \, d\xi \, d\eta \, d\zeta = \frac{1}{8} \sum f\left(\frac{1}{2} \pm \frac{\sqrt{3}}{4}, \frac{1}{2} \pm \frac{\sqrt{3}}{4}, \frac{1}{2} \pm \frac{\sqrt{3}}{4}\right) \tag{5}$$

- The eight nodes formula :

$$\int_{[0,1]^3} f \, d\xi \, d\eta \, d\zeta = \frac{1}{8} \sum f\left(0 \text{ or } 1, 0 \text{ or } 1, 0 \text{ or } 1\right) \tag{6}$$

Pratically, eight evaluations of the metric tensor are needed for each element. The Gauss formula (5) is exact for polynominals up to the third degree and leads to a 27 points molecule for the discretized potential equation. The nodes formula (6) is exact only for polynominals of degree one, but leads to a reduced 19 points molecule for the discretized potential equation (4). Though less precise, the formula (6) is attractive due to a weaker amount of storage required and a smaller calculation cost per iteration.

In the supersonic part of the flow, an upwind bias is introduced by using the classical retarded density approach [8] . Upwinding formulas are the same as the formulas used for 2-D calculations [4] . A constant potential jump is introduced in each mesh line leaving the trailing edge of the wing. Two discrete equations are obtained at the double valued trailing edge point by using two separate test functions θ_h attached to this point.

SOLUTION ALGORITHM -

The basic solution algorithm is the extension in the 3-D case of the fixed point -ICCG algorithm already developed by the author [4] . The solution of the discretized non linear system resulting from equations (4) is transformed into the solution of a sequence of linear problems obtained by freezing the density. Each linear system is solved by use of the Conjugate Gradient method with Incomplete Cholesky preconditionning (the so-called ICCG method). The global matrix is in every case symmetric, so that only 10 diagonals - for the 19 points scheme - are only needed to be computed and stored for both the matrix ans its corresponding factorization.

The ICCG algorithm is very fast for the solution of the linearized system, whereas the global convergence speed is slowed down by the fixed point like linearization technique. Improvements will consist in accelerating and stabilizing the algorithm by using the multigrid technique already implemented in the two-dimensional version of the code.

COMPUTED RESULTS -

Numerical results are presented for subcritical and supercritical flows around the Onera M6 Wing. Calculations are performed using the H-type mesh presented above ; 25 spanwise planes are used ; each plane contains 97 x 34 nodes ; 128 nodes are located on the wing surface for each of the 16 sections containing the wing (fig. 3).

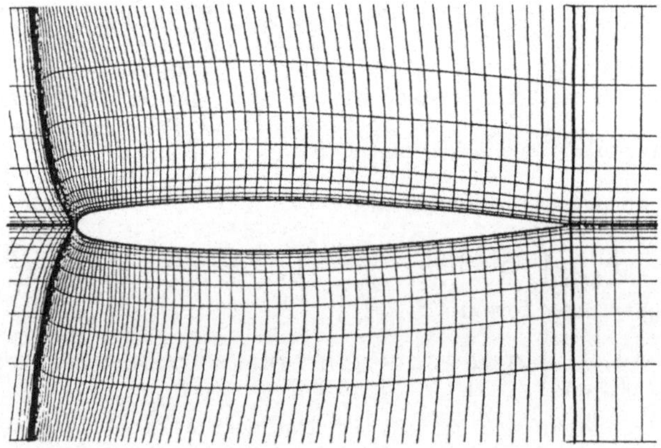

Fig. 3

The number of ICCG iterations is fixed equal to 5 for each linear system, while the total number of density updates varies between 50 and 150 in order to decrease the average residual error below 10^{-6} . Computational time varies from 5 minutes to 15 minutes on the CRAY 1-S computer, with partial vectorization.

The freestream Mach number is taken to be equal to 0.84, and angles of attack of 0° and 3,06° are considered. For both cases, comparisons are made between the 19 points finite element scheme, the 27 points finite element scheme, and a finite difference solution obtained by the ONERA non conservative code of Chattot et al. [9] . The finite difference solution and the 27 points finite element solution are in good agreement for the non lifting case (fig. 4). For the lifting case, the 27 points finite element solution exhibits shocks located more downstream than the finite difference solution, due to the conservative nature of the finite element discretization (fig. 5). The 19 points finite element solution, though less expensive, does not appear to be accurate enough in comparison with the other solutions. A comparison between theory and experiments [10] is presented for the lifting case (fig. 6). Peak suction levels and shock positions are correctly represented by the finite element solution, but the double shock structure is absent in the theoretical solution. A further improvement in the code will consist in a better treatment of the supersonic region where the scheme is presently first order accurate.

CONCLUSION -

The finite element approach has been successfully used to solve the full potential equation for flows around wings. Algebraic generation of three-dimensional meshes of H-type has been developed and the numerical results shown are satisfactory.

The next steps will concern the improvements mentioned above about discretization in the supersonic region and implementation of the multigrid technique. In parallel, more complex geometries will be studied in connection with a general multidomain approach to be defined.

REFERENCES -

[1] GLOWINSKI, "Numerical Simulation for some applied problems originating from continnum mechanics." Lecture Notes in Physics n° 195.

[2] MEIJERINK, VAN DER VORST, "An Iterative Solution method for linear Systems of which the coefficient matrix is a symmetric M-matrix." Maths of Comp., Vol. 31, Jan. 1977.

[3] BRANDT, "Multi-level adaptive solution to the boundary-value problems." Maths of Comp., Vol. 31, 1977.

[4] BREDIF, "A fast finite element method for transonic potential flows." AIAA paper 83-0507.

[5] BREDIF, "Une Méthode d'Ecoulements Finis Multigrille pour le calcul d'écoulements Potentiels Transsoniques." Proceedings of the third GAMNI Conference, March 1983, Paris.

[6] COOK, "Body oriented coordinates for generating three dimensional meshes." Int. Journal for Num. Meth. in Engineering, Vol. 8, 1974.

[7] LEE, HUANG, YU, RUBBERT, "Grid generation for general three-dimensional configurations." NASA Conference on Numerical Grid Generation Techniques, 1980.

[8] HAFEZ, MURMAN, SOUTH, "Artificial Compressibility methods for numerical solution of transonic full potential equation." AIAA paper 78-1148.

[9] CHATTOT, COULOMBEIX, TOME, "Calcul d'écoulements transsoniques autour d'ailes." La Recherche Aérospatiale 1978-4.

[10] SCHMITT, CHARPIN, "Pressure distributions on the ONERA M-6 Wing at Transonic Mach numbers." AGARD-Ar-138, 1979.

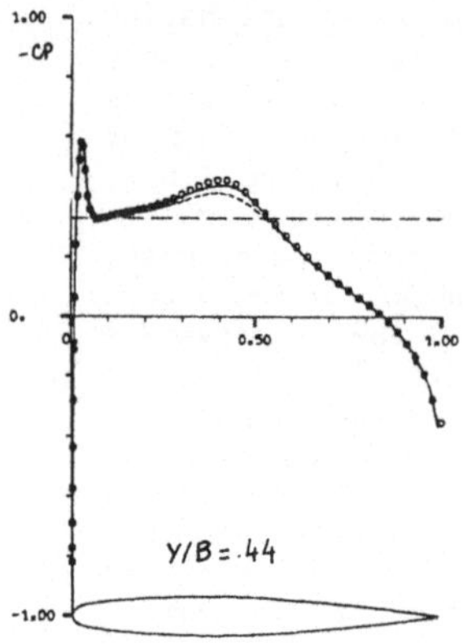

Fig. 4 M6 Wing

$M_\infty = .84 \quad \alpha = 0°$

——— 27 pts finite element result

oooo non cons. finite difference result [9]

- - - - - 19 pts finite element result

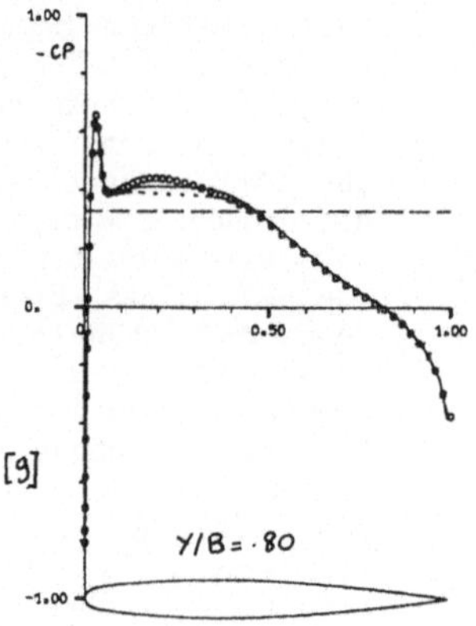

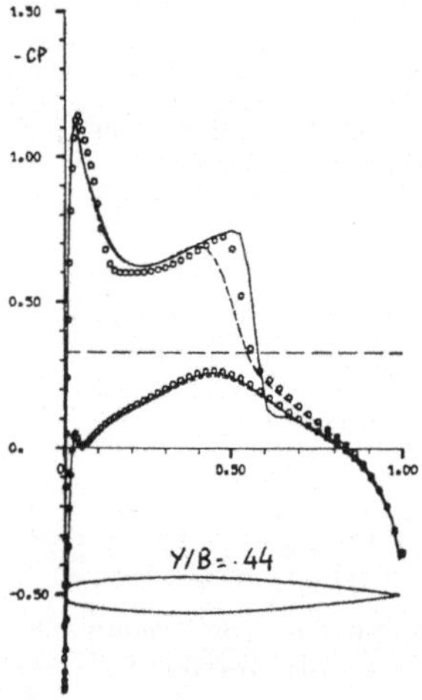

Fig. 5 M6 Wing

$M_\infty = .84 \quad \alpha = 3.06°$

——— 27 pts finite element result

oooo non cons. finite difference result [9]

- - - - - 19 pts finite element result

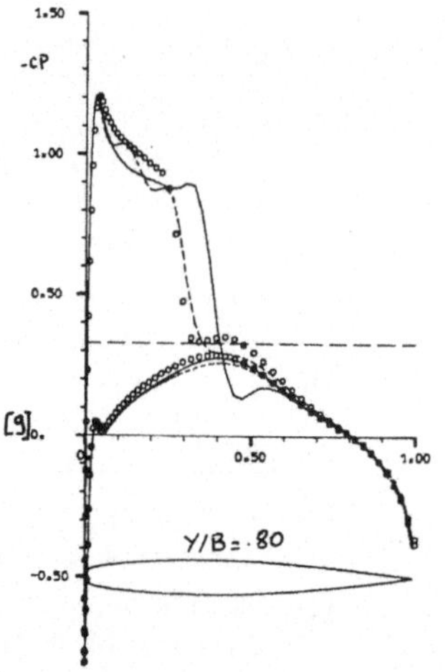

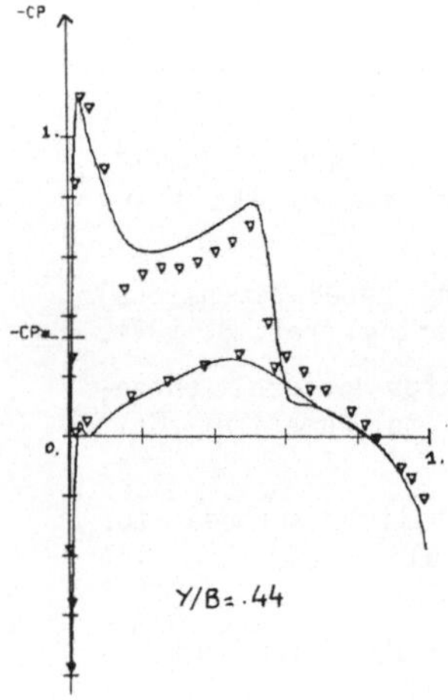

Fig. 6 M6 Wing

$M_\infty = .84 \quad \alpha = 3.06°$

——— 27 pts finite element result

▽▽▽▽ experiment [10]

$Re = 11.10^6$

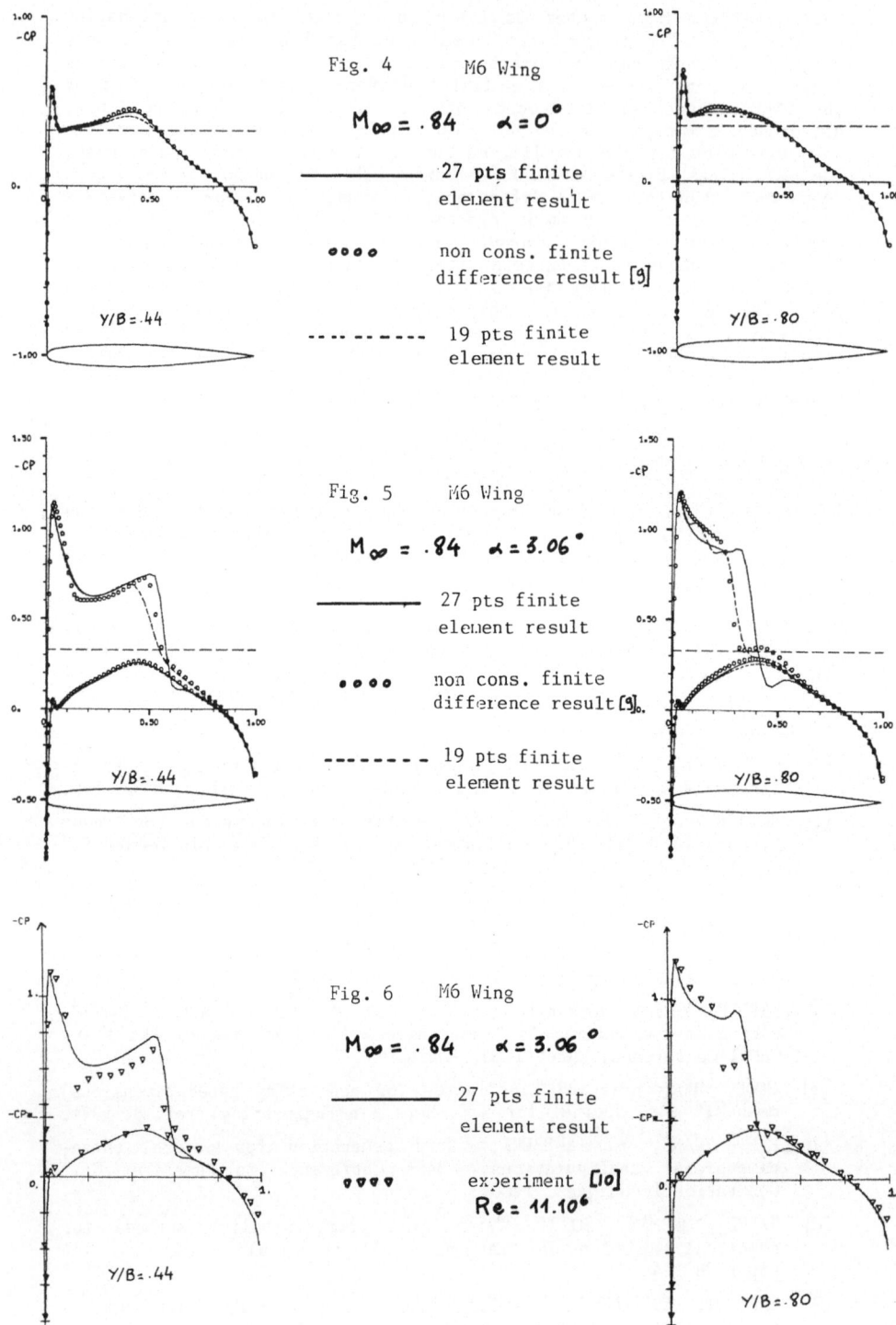

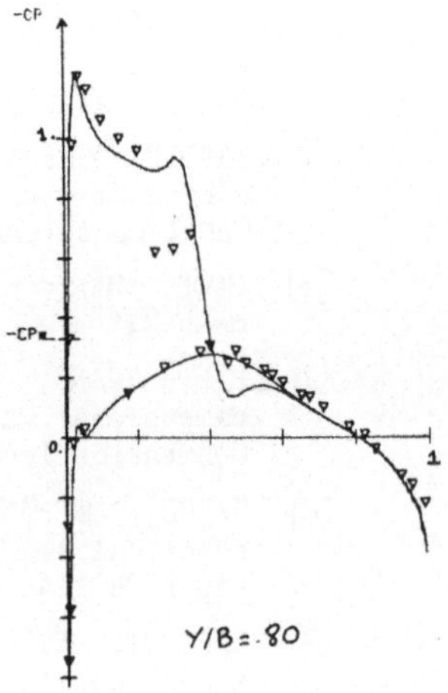

FINITE ELEMENT METHODS FOR SOLVING THE NAVIER-STOKES EQUATIONS FOR COMPRESSIBLE UNSTEADY FLOWS

M.O. BRISTEAU*, R. GLOWINSKI**, B. MANTEL***, J. PERIAUX***

Introduction

The purpose of this paper is to discuss the numerical solution of the time dependent Navier-Stokes equations modelling compressible viscous 2D or 3D flows. Most of the existing solution methods for these time dependent equations are founded on finite difference methods for both space and time discretization (see [1] - [5]) ; in this paper we use a finite difference method for time discretization and finite elements for space discretization. The general principle is to use an implicit scheme combining a generalized Stokes solver and least squares.

Governing equations

The equations to be solved numerically, written in a non-conservative form are given by the relations thereafter :

(1) $$\frac{\partial\rho}{\partial t} + \underset{\sim}{u}.\underset{\sim}{\nabla}\,\rho + \rho\,\underset{\sim}{\nabla}.\underset{\sim}{u} = 0$$

(2) $$\begin{cases} \rho\,\dfrac{\partial \underset{\sim}{u}}{\partial t} + \rho(\underset{\sim}{u}.\underset{\sim}{\nabla})\underset{\sim}{u} + (\gamma-1)\underset{\sim}{\nabla}\rho\,T = \\ \qquad \dfrac{1}{Re}[\Delta \underset{\sim}{u} + \dfrac{1}{3}\,\underset{\sim}{\nabla}(\underset{\sim}{\nabla}.\underset{\sim}{u})\,] \end{cases}$$

(3) $$\begin{cases} \rho\,\dfrac{\partial T}{\partial t} + \rho\,\underset{\sim}{u}.\underset{\sim}{\nabla}T + (\gamma-1)\rho\,T\;\;\underset{\sim}{\nabla}.\underset{\sim}{u} = \\ \dfrac{1}{Re}\,[\,\dfrac{\gamma}{Pr}\;\;\Delta T + \dfrac{4}{3}\,[\left(\dfrac{\partial u_1}{\partial x_1}\right)^2 + \left(\dfrac{\partial u_2}{\partial x_2}\right)^2 - \dfrac{\partial u_1}{\partial x_1}\;\dfrac{\partial u_2}{\partial x_2}\,] + \left(\dfrac{\partial u_1}{\partial x_2} + \dfrac{\partial u_2}{\partial x_1}\right)^2\,] \end{cases}$$

where

- ρ is the density,
- $\underset{\sim}{u}$ the velocity, $\underset{\sim}{u} = (u_1,u_2)$
- T the temperature.

To the above equations appropriate boundary conditions [6] have to be added and also initial values for the independent variables of the problem.

In many problems of interest, it can be assumed that the total enthalpy is a constant H_o and then the equation (3) can be replaced by

(4) $$\gamma T + \frac{|\underset{\sim}{u}|^2}{2} = H_o$$

where the constant H_o is defined by the boundary conditions.

---

* INRIA, B.P. 105, 78153 LE CHESNAY CEDEX, France.

** Université P. et M. Curie, LAN 189, Tour 55-65, 4 place Jussieu 75230 PARIS CEDEX 05, France and INRIA.

*** AMD/BA, 78 Quai Carnot, B.P. 300, 92214 ST CLOUD, France.

Solution methods

In order to apply the methods previously used for the incompressible problem [7] we introduce a new function σ defined by :

(5) $\sigma = \text{Log}\, \rho$.

Concerning the time approximation, for the unsteady problem, a fully implicit scheme of Gear type has been used [6], while for the constant total enthalpy model (1), (2), (4) an operator splitting method of A.D.I. type is introduced [8]. For the solution of the two formulations, the main tools are, on one hand a generalized Stokes solver [8] - [10] to solve a linear system of the following type :

$$(6) \quad \begin{cases} \alpha\,\sigma + \nabla.\underset{\sim}{u} = g \\ \alpha\,\underset{\sim}{u} - \mu\,\Delta\,\underset{\sim}{u} + \beta\,\underset{\sim}{\nabla}\,\sigma = f \end{cases}$$

with appropriate boundary conditions, on the other hand a least squares formulation [7],[11] is introduced to solve the nonlinear part of the equations written as :

$$(7) \quad \begin{cases} \alpha\,\sigma + \underset{\sim}{u}.\underset{\sim}{\nabla}\,\sigma = q \\ \alpha\,\underset{\sim}{u} - \mu\,\Delta\,\underset{\sim}{u} - \Phi(\sigma,\underset{\sim}{u},T) = \underset{\sim}{h} \\ \alpha\,T - \pi\,\Delta\,T - \chi(\sigma,\underset{\sim}{u},T) = r \end{cases}$$

The generalized Stokes solver applied to the linear part of the equations provides an accurate value of the pressure and density at the wall boundaries. The nonlinear part of the equations is identified via a preconditioned conjugate gradient algorithm.

The Stokes solver is handled either as a state equation in the first approach or as a substep in the A.D.I. method. Then in this last formulation, the state equation of the least squares substep is only a Poisson solver.

Concerning the space discretization a same kind of finite element approximation for density, velocity and temperature (continuous piecewise linear approximation) is used.

Numerical results

Several numerical experiments using the two above models (constant total enthalpy (4) or not (3)) and the associated algorithms, are presented thereafter.
Figure 1 is concerned with the numerical simulation of an unsteady flow around a NACA0012 airfoil (characteristics of the triangulation $\mathcal{C}_h$ are : Nodes : 800, Elements : 1514), using the model (1) - (2) - (3) and a fully implicit scheme of Gear type for the time discretization. Details of different aerodynamics variables (Density, Mach, Temperature lines) are plotted on Figures 1(a) - 1(b) - 1(c).
Other numerical results obtained with the ADI algorithm associated with (1) - (2) - (4) model are presented on two different geometries. The first ones are concerned with a 3-D flow around a sphere using a coarse tetraedrization (figure 2) generated by rotation of the 2-D triangulation of Figure 3 around the x-axis, with the following prescribed data $M_\infty = 0.7$, $Re = 100$, $\Delta t = 0.1$.
The Mach lines and the vorticity lines in perpendicular or parallel to the flow direction sections are plotted on Figures 4-5 after 50 time steps.
Another separated flow of more industrial interest has been simulated around and inside an idealized inlet. The triangulation used for this computation is shown on Figure 6. On Figure 7, we have compared the Mach lines of the flows at transonic range obtained with two values of the Mach number at infinity ($M_\infty = 0.7$ and $M_\infty = 0.9$).
Various 2D - 3D experiments including other boundary conditions, finer grids with combination of A.D.I. methods and least squares are under investigation.

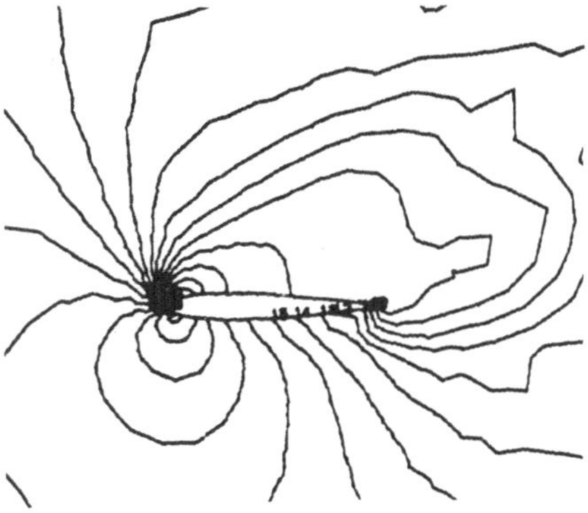

(a) Density lines

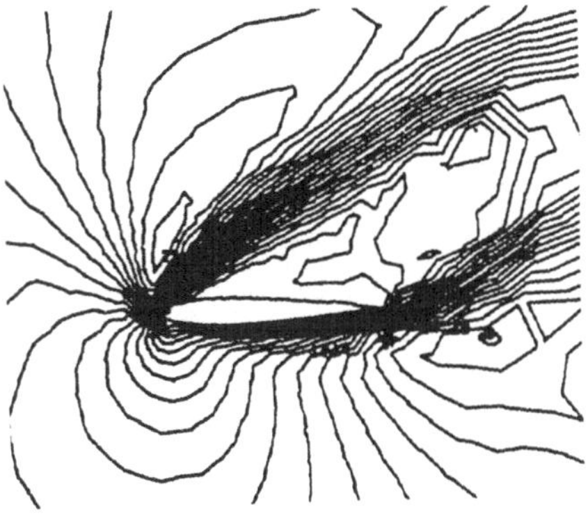

(b) Mach lines

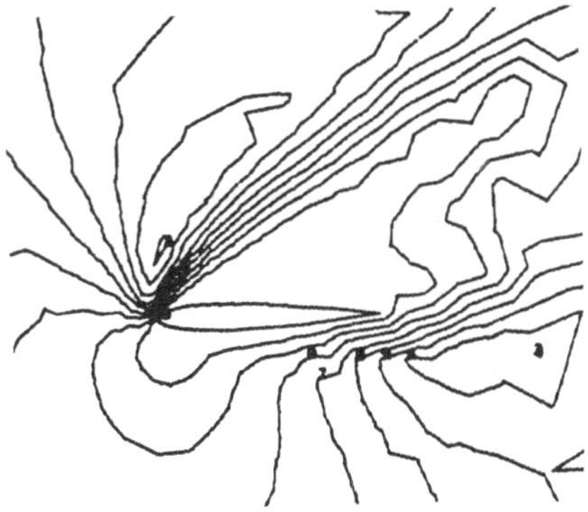

(c) Temperature lines

Figure 1
Flow around a NACA00012 airfoil
$M_\infty = 0.7$, Re = 100, $\alpha = 30°$
$\Delta t = 0.1$
Number of time steps : 50

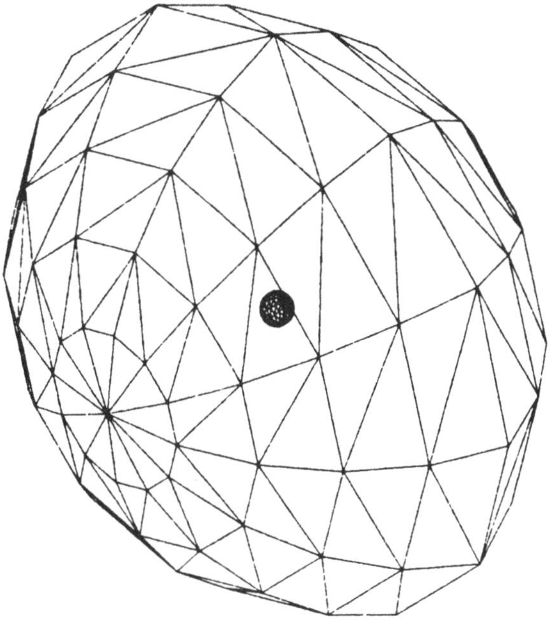

Figure 2 : 3D-Tetraedra Mesh around a sphere
Nodes : 1760 Elements : 9720

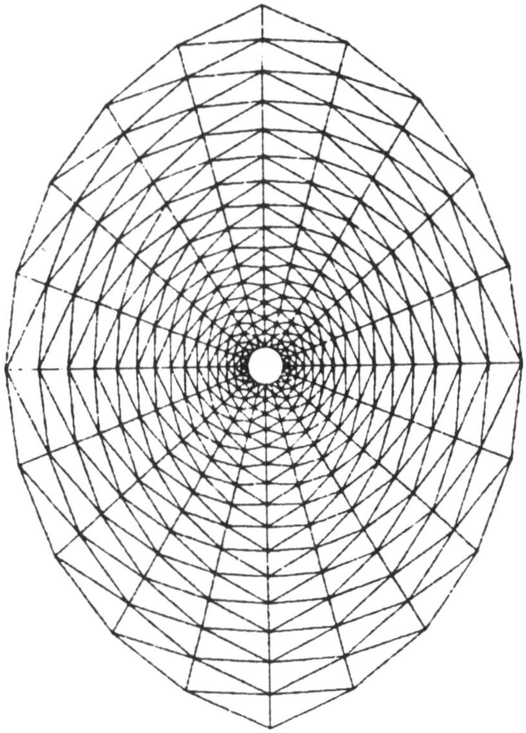

Figure 3 : 2D Triangulation

Vorticity Lines

Mach Lines

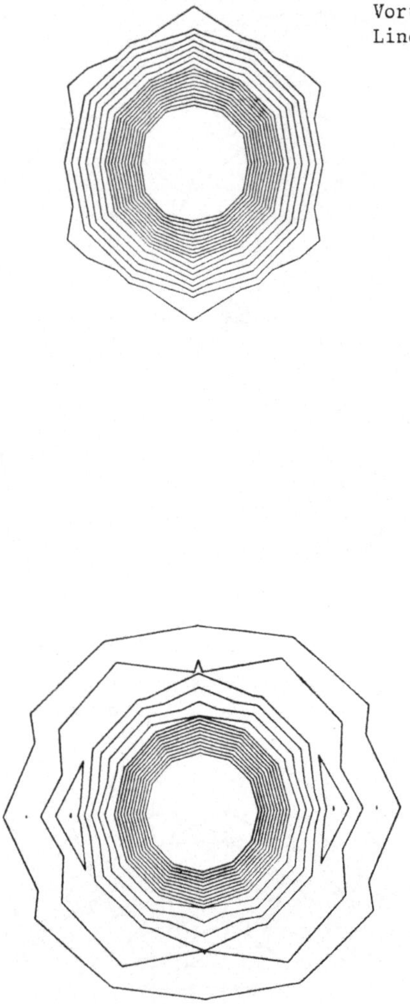

Figure 4
Perpendicular section to the flow direction.

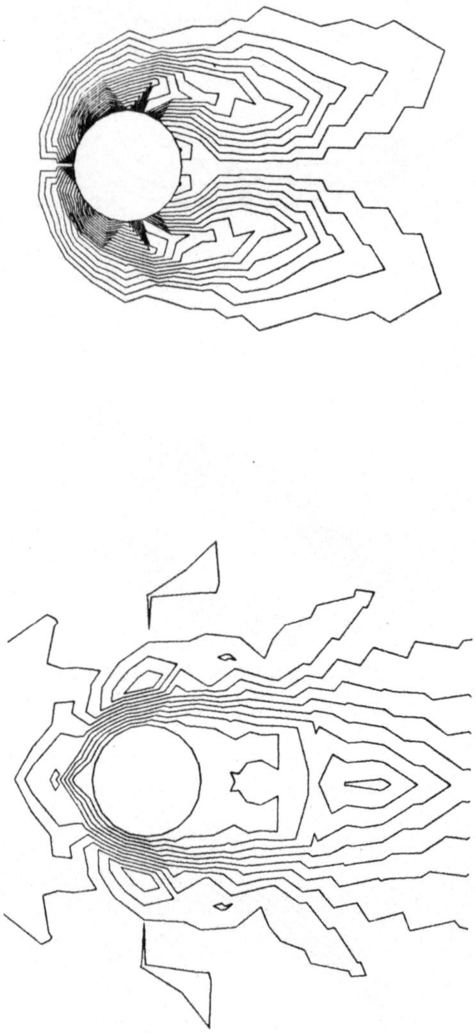

Figure 5
Parallel section to the flow direction.

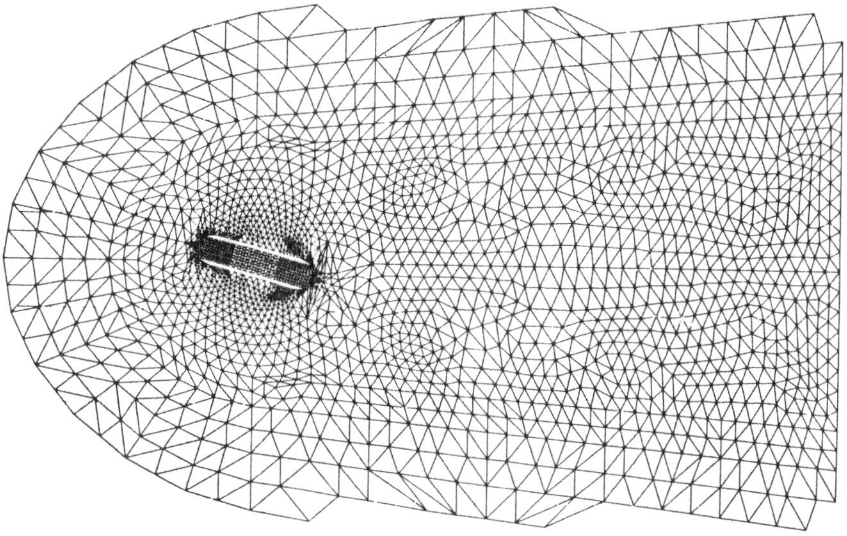

Figure 6

Triangulation around/inside an idealized inlet at angle of attach $\alpha = 15°$. Nodes = 2100, Elements = 4160.

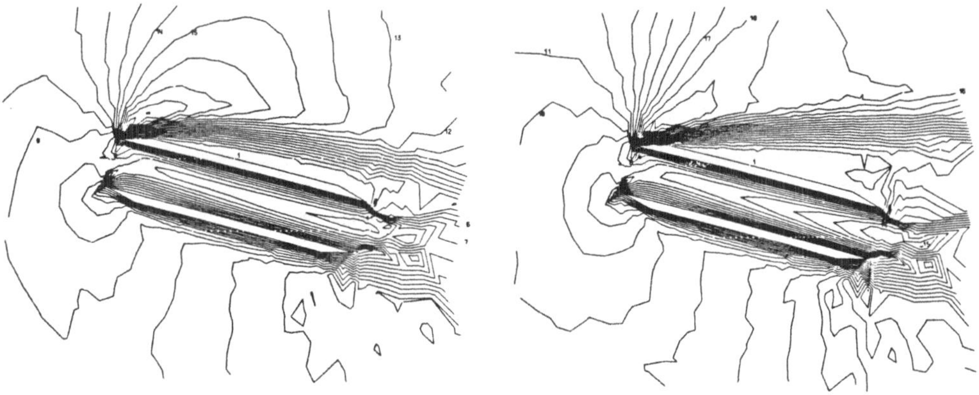

(a) $M_\infty = 0.7$ (b) $M_\infty = 0.9$

Figure 7

Mach lines near the inlet

$\alpha = 15°$, Re = 100., $\Delta t = 0.1$

Number of time steps : 100

Aknowledgements : This study was partly supported by DRET under contract n° 83/403. A part of the computations were performed on CRAY 1-S of CCVR.

References :

[1] R.W. MAC CORMACK, A Numerical Method for Solving the Equations of Compressible Viscous Flow, *AIAA J., Vol. 20,* (1982), 9, pp. 1275-1281.

[2] R.M. BEAM, R.F. WARMING, An Implicit Factored Scheme for the Compressible Navier-Stokes Equations, *AIAA J., Vol. 16,* (1978), pp. 393-402.

[3] W.R. BRILEY, H. Mc DONALD, Solution of the Multidimensional Compressible Navier-Stokes Equations by a Generalized Implicit Method, *J. Comp. Physics., Vol. 24,* (1977), pp. 372-397.

[4] R. PEYRET, H. VIVIAND, Computation of Viscous Compressible Flows Based on the Navier-Stokes Equations, *AGARD-AG-212,* 1975.

[5] R. PEYRET, T.D. TAYLOR, *Computational Methods for Fluid Flow,* Springer - New-York, 1982.

[6] M.O. BRISTEAU, R. GLOWINSKI, B. MANTEL, J. PERIAUX, P. PERRIER, Numerical Methods for the time dependent compressible Navier-Stokes equations, Computing Methods in Applied Sciences and Engineering, R. Glowinski, J.L. Lions eds., North-Holland (to appear).

[7] R. GLOWINSKI, B. MANTEL, J. PERIAUX, Numerical Solution of the Time Dependent Navier-Stokes Equations for Incompressible Viscous Fluids by Finite Element and Alternating Direction Methods, in *Numerical Methods in Aeronautical Fluid Dynamics,* P.L. Roe ed., Academic Press, London, 1982, pp. 309-336.

[8] M.O. BRISTEAU, R. GLOWINSKI, B. DIMOYAT, J. PERIAUX, P. PERRIER, O. PIRONNEAU, Finite Element Methods for the Compressible Navier-Stokes equations, in *Proceedings of the AIAA Computational Fluid Dynamics Conference, Danvers July 13-15, 1983,* AIAA paper 83-1890.

[9] J. GOUSSEBAILE, G. LABADIE, F. HECHT, L. REINHART, Finite Element Solution of the Shallow Waters Equations, by a quasi direct decomposition procedure, *Numerical Methods in Fluids* (to appear).

[10] R. GLOWINSKI, O. PIRONNEAU, On numerical methods for the Stokes problem, in *Energy Methods in Finite Element Analysis,* R. Glowinski, E.Y. Rodin, O.C. Zienkiewicz eds. Wiley, Chichester, 1979, pp. 243-264.

[11] R. GLOWINSKI, *Numerical Methods for Nonlinear Variational Problems,* Springer, New-York, 1984.

NUMERICAL SOLUTIONS OF THE EULER EQUATIONS WITH SEPARATION BY A FINITE ELEMENT METHOD

C.H. BRUNEAU*, J.J. CHATTOT**, J. LAMINIE*, R. TEMAM*

*Laboratoire d'Analyse Numérique, Bâtiment 425, Université Paris-Sud, 91405 ORSAY, E.R.A. C.N.R.S.

**M.A.T.R.A., 37 Avenue Louis Bréguet, 78140 VELIZY.

SUMMARY.

A finite element least squares method is applied to the steady Euler equations in order to capture separated flows around a cylinder. This is achieved without using artificial viscosity but only by giving a "Kutta condition" at the separation point. The sensitivity of the solution to this condition and mesh size is discussed in this paper.

INTRODUCTION.

In perfect fluid motions, separation occurs in various circumstances and is followed by vortical phenomena which are often of primary importance, for instance in term of global efforts on obstacles.

Separation along the sharp tailing edge of a wing or a rotor blade is a typical example. A vortex sheet trails the lifting surface and the accurate prediction of its location and strength requires a high degree of sophistication [1] .

More critical is the separation occuring at a sharp leading edge since the vortex sheet interacts strongly with the lifting surface, and produces non linearities on the lift characteristic [2].

Separation can also occur on smooth surfaces as in the case of the transonic flow at Mach number $M = 0.5$ past a cylinder, where the numerical solutions, computed independently by the participants of a workshop held at NASA Langley Research Center on September 1, 1981 [3] exhibit a recirculating bubble. Here separation is triggered by the stagnation pressure losses and the vorticity generated at the recompression shock.

The prediction of these flows is best handled with the Euler equations which describe the general motion of perfect fluid with vorticity. However it is not yet clear as to what kind of boundary condition can be imposed ("Kutta condition") and what accuracy can be expected in the numerical solution.

This paper addresses these last two questions by way of "numerical experiments". The tests concern separated flows past a circular cylinder at low subsonic Mach number.

I. NUMERICAL METHOD .

The steady Euler equations written in conservative form read in two dimension :

$$(1)\qquad \begin{cases} \dfrac{\partial\rho u}{\partial x} + \dfrac{\partial\rho v}{\partial y} = 0 \\[2mm] \dfrac{\partial\rho u^2+p}{\partial x} + \dfrac{\partial\rho uv}{\partial y} = 0 \\[2mm] \dfrac{\partial\rho uv}{\partial x} + \dfrac{\partial\rho v^2+p}{\partial y} = 0 \end{cases}$$

$$(2)\qquad H = \frac{\gamma}{\gamma-1}\frac{p}{\rho} + \frac{u^2+v^2}{2} \quad ; \quad (\gamma = 1.4)$$

with the following boundary conditions (Figure 1) :

On Γ_0 : the symmetry condition $(v = 0)$,

On Γ_1 : the tangency condition $(\vec{q}.\vec{n} = 0,\ \vec{q} = (u,v))$,

On Γ_2 : the free stream condition.

The method of reference [4] is used :

- Fixed point algorithm on ρ .
- Linearization by Newton's method.
- Least squares embedding.
- Use of finite element discretization for the components of the flux vector.
- Resolution by ICCG method.

Moreover the tangency condition is treated in variational form. The least squares functional is modified by addition of the following term :

$$(3)\qquad \frac{1}{2}\int_{\Gamma_1}\left[\left(\vec{\tilde{q}}_{j/q} - \vec{q}_{r_j}\right).\vec{n}\right]^2 d\Gamma$$

where $\left(\vec{\tilde{q}}_{j/q}, \tilde{\rho}_{j/q}\right)$ is the solution of the linearized equations and $\vec{q}_{r_j}$ corresponds to the previous iterate.

II. "KUTTA CONDITION".

To provoke the separation we need to impose some extra condition to the flow at a point on the cylinder. In a first attempt a zero tangential velocity component is given at the separation point $S = (1,\theta)$. A very small separation bubble occurs within a mesh row. It can be conjectured that there is not a unique solution of the Euler equations for a given separation point and that the numerical viscosity of the scheme plays a major role in the selection of the solution. By construction our scheme is centered and second order accurate and does not require the addition of artificial viscosity in subsonic regions, in contrast to unsteady methods. Thus the pressure at the separation point is imposed and yields an extra degree of freedom to control the size of the bubble.

III. NUMERICAL RESULTS.

The first tests are relevent to fully attached flow in order to check the new tangency condition (3). The results are summarized in Table 1 and show at least quadratic convergence with mesh refinement.

Three types of tests are performed in the case of separated flow to evaluate the sensitivity of the solution to the pressure at the separation point S , to the mesh size and to the location of S (θ) .

For $\theta = 129°$ and mesh of 16 x 32 points the pressure at point S , when not imposed, is at convergence $p_S = 1.05$, with a very small separation bubble. If we impose $p_S \leq 1.05$ we find a smaller bubble and even no bubble at all for $p = 0.9$, where the flow is fully attached. On the contrary if we impose $p_S > 1.05$ we find a larger bubble which increases in size with p_S (Figure 2). Even if we take $p_S > 1.35209$ (stagnation pressure of incoming flow) we get larger separation zone (Figure 3). However we note that the stagnation pressure in the separated region (point B) is always less than that of the incoming flow. It is believed that the discrepancy between the separation pressure and the stagnation pressure in the separated region comes from the dissipative effects of the scheme at the separation point where the solution exhibits large gradient.

The sensitivity to the mesh is emphasized on Figures 4 and 6 which correspond to the conditions of Figure 2 and 3 respectively with a finer 31 x 61 mesh. For a given pressure and separation location the mesh refinement yields a larger bubble. Figure 5 illustrates the vortex when the pressure of the incoming flow is imposed at point S for the finer 31 x 61 points mesh. When the separation point S is moved upstream the separated region grows as expected (Figures 6 and 7).

CONCLUSION.

The numerical results for the capture of separated flow regions with the Euler equations indicate a great sensitivity of the solution to mesh size and conditions at separation ("Kutta condition"). The proposed scheme does not need an artificial viscosity term explicitely but requires to impose both the location and the pressure of the separation point.

Preliminary tests are carried out in three dimensions to simulate separated flows past a flat plate at incidence.

REFERENCES.

[1] J.J. CHATTOT, M. BOSCHIERO and C. KOECK, Méthodes numériques de prédiction de l'aérodynamique des missiles, AGARD-CP-336, feb. 1983.

[2] C. KOECK and J.J. CHATTOT, Computation of three-dimensional Vortex flows past wings using the Euler equations and a multiple-grid scheme, Proceedings 9th Int. Conf. Meth. in Fluid Dyn., June 25-29, 1984, CEN Saclay, France, Springer-Verlag.

[3] M.D. SALAS, Recent developments in transonic Euler flow over a circular cylinder, Proceedings 10th IMACS World Congress, Aug. 8-13, 1982, Montréal, Canada, Vol. 1, p. 175-177.

[4] C.H. BRUNEAU, J.J. CHATTOT, J. LAMINIE and J. GUIU-ROUX, Finite element least square method for solving full steady Euler equations in a plane nozzle. Proceedings 8th Int. Conf. Meth. in Fluid Dyn., June 28-July 2, 1982, Aachen, Germany, Springer-Verlag.

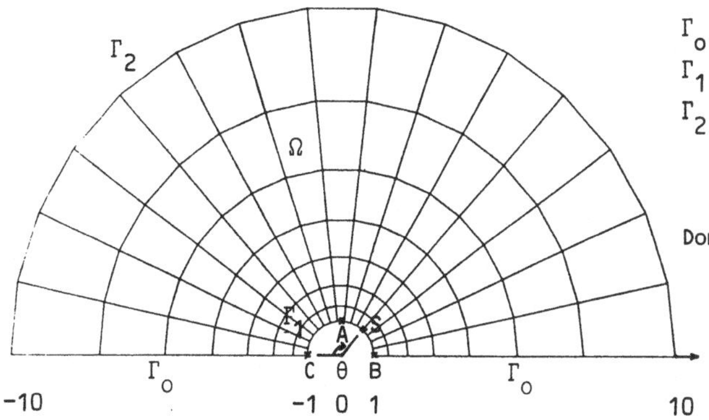

Γ_o : Symmetry condition
Γ_1 : Tangency condition
Γ_2 : Free stream condition

Domain and boundary conditions

Figure 1

Comparison of accuracy and convergence for different meshs.

Table 1

IM x JM points mesh	Velocity at the stagnation points B and C	Pressure at points B and C	Velocity at point A	Variation of entropy at point A	Value of the least-square functional
8x16	0.93×10^{-2}	1.24	1.7	11 %	1.3×10^{-1}
16x32	0.17×10^{-3}	1.32	0.85	4.5 %	2.6×10^{-3}
31x61	0.8×10^{-5}	1.35	0.93	1.7 %	4.0×10^{-4}

Stagnation pressure (1.35209) and velocity (0.425) of the incoming flow (M = 0.4).

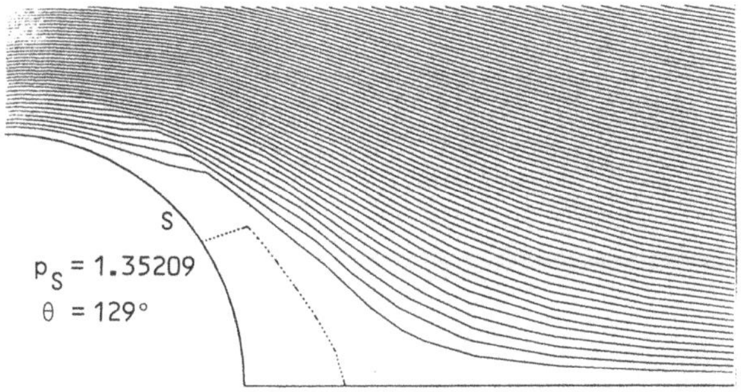

Stream lines for the 16 x 32 points mesh

Figure 2

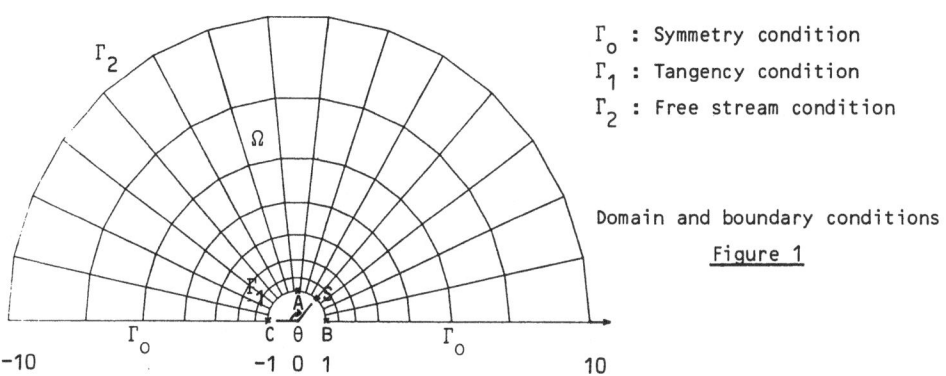

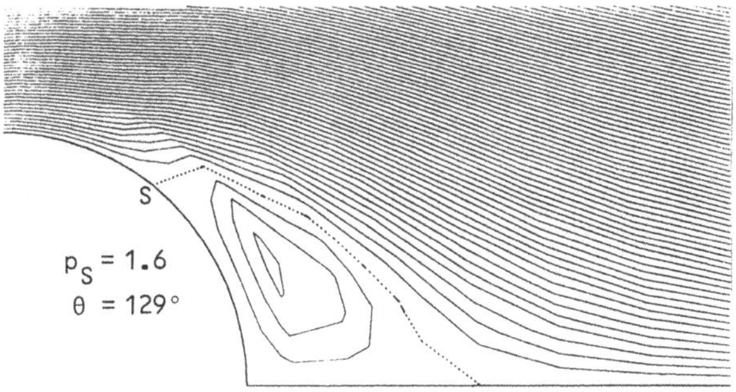

Stream lines for the 16 x 32 points mesh

Figure 3

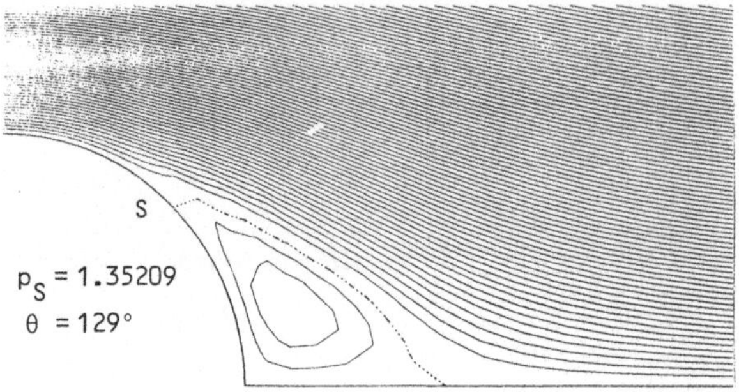

Stream lines for the
31x61 points mesh

Figure 4

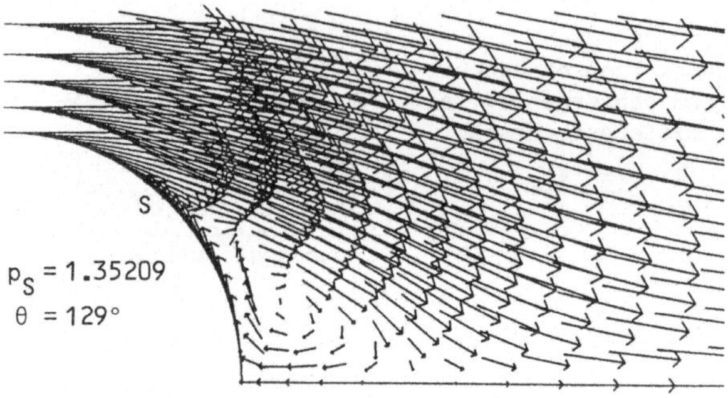

Velocity field for the
31x61 points mesh

Figure 5

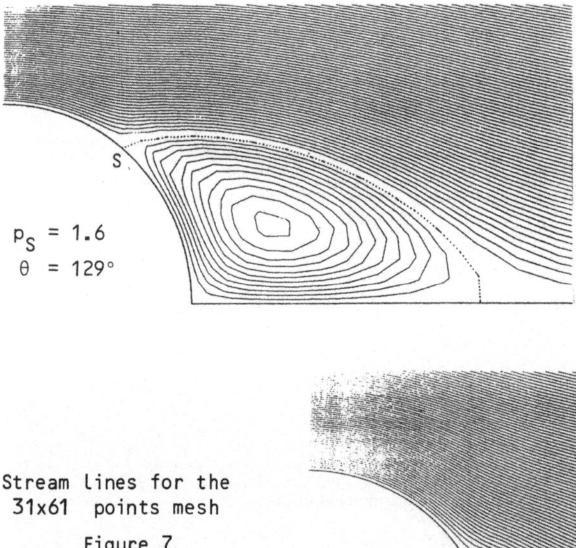

Stream lines for the
31x61 points mesh

Figure 6

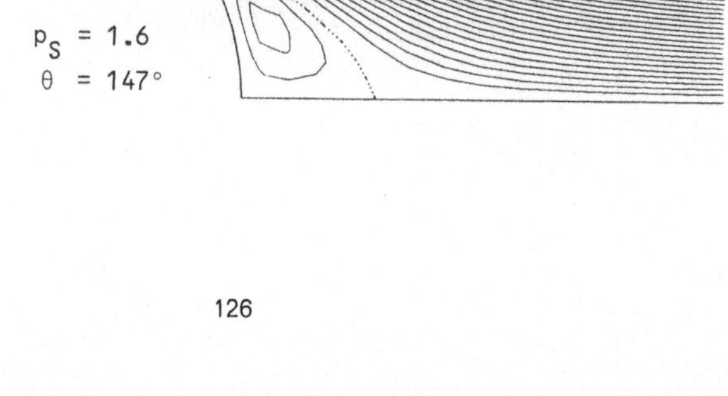

Stream lines for the
31x61 points mesh

Figure 7

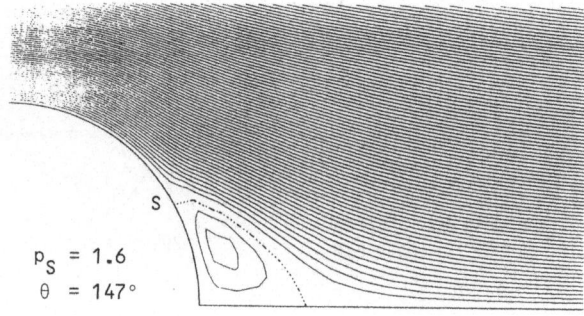

CHEBYSHEV SPECTRAL AND PSEUDOSPECTRAL SOLUTIONS OF THE NAVIER-STOKES EQUATIONS

T. Cartage, P. Demaret, M. Deville
Unité de Mécanique Appliquée
Université Catholique de Louvain
Louvain-la-Neuve, Belgium.

Summary

This paper presents numerical solutions obtained by Chebyshev approximations of thermal convection flows. In the first part, a Chebyshev Tau method with influence matrix technique for the pressure computation is applied to the simulation of convective phenomena in liquid metals. Some results are compared with those produced by finite elements. In the second part, a pseudospectral algorithm using finite element preconditioning for the momentum equations and finite difference preconditioning for the temperature equation treats thermal convection in a fluid whose viscosity is strongly temperature dependent. As finite element-pseudospectral schemes (FE-PS) are recent, some test problems are solved to assess their performance.

Chebyshev spectral method

In this section, we elaborate an algorithm for solving the Navier-Stokes equations in Chebyshev space thru the use of the Tau method (Gottlieb-Orszag, 1977). The velocity components and the pressure are the primary variables. It is well known (Deville, 1984) that splitting schemes induce an error of O(1) on the pressure gradient inside a layer $|x|-1=O((\nu\Delta t)^{1/2})$. To achieve spectral accuracy, the pressure field needs careful attention. Therefore, in order to ensure a vanishing divergence of the velocity field inside the domain and on the boundaries, the influence matrix technique proposed by Kleiser and Schumann (1980) and extended to 2-D flows by Le Queré and Alziary de Roquefort (1982) is implemented. In a preprocessing stage, Stokes problems are solved with unit pressures $\bar{p}_p$ acting on the collocation points c_i of the boundaries $\partial D(i=1,\ldots,L)$. For every pressure sollicitation, the following set of equations is solved

$$\Delta p_{p,k} = 0 \;, \quad p_{p,k} = \delta_{k,i} \;, \quad k = 1,\ldots,L \qquad (1)$$

$$\Delta \underset{\sim}{v}_k - \frac{2}{\nu \Delta t} \underset{\sim}{v}_k = \frac{2}{\nu} \nabla p_{p,k} \ , \quad \underset{\sim}{v}(\partial D) = \underset{\sim}{v}_w \quad , \qquad (2)$$

where Δ denotes the Laplacian operator, p the pressure divided by a constant density, $\underset{\sim}{v}$ the velocity field and δ_{ki} the Kronecker symbol. Eq.(2) comes from a Crank-Nicolson discretization of the time derivative, while $\underset{\sim}{v}_w$ is the velocity imposed at the walls. The influence matrix M collects in each column the divergence $\mathcal{D}_k$ of the velocity field obtained from the solutions of Eqs.(1) and (2). Every time step is decomposed as follows :

i) compute the non-linear terms by the Adams-Bashforth scheme

ii) solve a Poisson equation for the pressure with arbitrary boundary condition $\bar{p}_p$

iii) solve a Helmholtz equation to produce the new velocity field $\underset{\sim}{v}_I$

iv) compute $\mathcal{D}_I$, the divergence of $\underset{\sim}{v}_I$ at the boundary collocation points

v) solve the linear system $M\bar{p}_c = -\mathcal{D}_I$

vi) compute the correct pressure distribution at the boundary $\bar{p}_p+\bar{p}_c$ and repeat steps ii, iii which provide the velocity at the new time level.

The Poisson and Helmholtz equations are handled by a direct solver (Haldenwang et al, 1984).

The previous algorithm is applied to the Navier-Stokes and temperature equations, coupled via the Boussinesq approximation :

$$\frac{\partial \underset{\sim}{v}}{\partial t} + \underset{\sim}{v} \cdot \underset{\sim}{\nabla} \underset{\sim}{v} = -\nabla p + Pr\, \Delta \underset{\sim}{v} + Ra\, Pr(T-1)\, \underset{\sim}{e} \quad ,$$

$$\frac{\partial T}{\partial t} + \underset{\sim}{v} \cdot \underset{\sim}{\nabla} T = \Delta T \quad ,$$

where Pr is the Prandtl number, Ra the Rayleigh number and $\underset{\sim}{e}$ a unit vector with components (0,1).

The accuracy and the quality of the code were tested on the thermal square cavity problem. The benchmark solution produced by de Vahl Davis (1982) was recovered to within 1°/₀₀ for Ra between 10^3 and 10^6. Then, thermal convection in Ga-As crystal melt was investigated and spectral results were compared to those produced by transient finite element techniques (M.J. Crochet, et al., 1983). Good agreement was found between both methods. Finally, preliminary calculations of convection in molten tin (float-glass process) are carried out. Figure 1 presents the geometry and the conditions of the problem. With $Pr=5.64\ 10^{-3}$, two problems are solved at Ra=500 and 1000, respectively. Figure 2 displays the history of the nondimensional kinetic energy versus time, for two different algorithms : the first one was described hereabove while the second one is a classical splitting method (steps iv to vi are discarded; the pressure correction is neglected and in steps ii, inviscid pressure boundary conditions are imposed). At Ra=1000, it is seen that for 9x25 degrees of freedom, the pressure correction is absolutely necessary. Figure 3 presents the steady state streamlines configuration at Ra=500, while figure 4 shows the streamline pattern for Ra=1000 at real time t=5.26 hour.

Chebyshev pseudospectral solution

In Deville-Mund (1984), finite element preconditioning is considered for the solution of second-order elliptic equations. It is shown that Lagrangian bilinear or biquadratic elements are sufficient to achieve spectral accuracy within a finite number of iterations. A similar preconditioning is built for thermal convection flows at low Reynolds numbers.

Let us consider the equation

$$Lu = f \quad , \tag{3}$$

where L is a partial differential operator. The pseudospectral solution is obtained by a preconditioner represented by a sparse matrix easily invertible. Developing (3) by generalized Taylor series in the neighborhood of u^k, an approximation of u at iteration number k and assuming the existence of the Fréchet derivative of L, the truncation of the series to first-order terms gives

$$Lu^k + \frac{\partial L}{\partial u} (u-u^k) \simeq Lu = f \tag{4}$$

$$\text{or } \frac{\partial L}{\partial u} (u-u^k) = -R^k \quad . \tag{5}$$

From (5), one obtains

$$\frac{\partial L}{\partial u} \delta u^k = -R^k \quad , \tag{6}$$

with the definitions

$$R^k = Lu^k - f \quad , \tag{7}$$

$$\delta u^k = u^{k+1} - u^k \quad . \tag{8}$$

When a Chebyshev approximation to (3) is sought, Eq.(6) may be rewritten with obvious definitions as

$$\frac{\partial L_{sp}}{\partial u_N} \delta u_N^k = -R_N^k \quad , \tag{9}$$

where L_{sp} is the spectral representation of L, u_N the spectral solution, δu_N the increment of the dependent variable and R_N the residue to the p.d.e. The pseudospectral approach provides a first solution of (3) by

$$L_{ap} u^o = \bar{f} \quad , \tag{10}$$

where L_{ap} is a convenient approximation of L_{sp}(FD or FE), u^o the first guess of the spectral solution on the Chebyshev collocation grid and $\bar{f}$ the approximation of the source term at the same points. The next iterates come from the solution of (9) by the solver of the same type as in (10) and

$$u^{k+1} = u^k + \alpha \delta u^k \quad , \tag{11}$$

where α is a stabilizing underrelaxation parameter.

By this preconditioning, an iterative procedure couples a sparse algebraic solver with a residue calculation performed in spectral space, which is the driving mechanism leading the computation to machine accuracy if the solutions are smooth enough and the cutoff values sufficiently high.

To test the FE-PS method, let us solve the 2-D Stokes problem

$$-\frac{\partial p}{\partial x} + \Delta u = -4\pi^2 \sin\pi x \cos\pi y \tag{12}$$

$$-\frac{\partial p}{\partial y} + \Delta v = 0 \quad , \tag{13}$$

$$\frac{\partial u}{\partial x} + \frac{\partial v}{\partial y} = 0 \quad , \tag{14}$$

whose analytical solution is

$$u = \sin\pi x \cos\pi y \ , \quad v = -\cos\pi x \sin\pi y \ , \quad p = -2\pi\cos\pi x \cos\pi y \tag{15}$$

Problem (12-14) is subject to Dirichlet conditions in agreement with (15) on the square domain [0,1] x [0,1]. The FE code implements a Galerkin formulation of (12-14) with a 9-node Lagrangian element with biquadratic velocities and bilinear pressures. The vertices of the elements coincide with the collocation grid. Machine accuracy is obtained in a few iterations (see fig.5). Another preconditioning uses bilinear velocities and constant pressures over the elements. In this case, the checkerboard pressure mode is filtered as recommanded by Sani et al., (1981). A least square polynomial interpolation yields the pressure in spectral space. Figure 5 gives the evolution of the residue with respect to the iteration counter. The precision is limited here by the interpolation.

The last problem computes the velocity field in a rectangular cavity filled with polyisobutylene. The geometry and the conditions are described in figure 6. The temperature field comes from experimental data. The momentum equations are

$$-\underset{\sim}{\nabla} p + \underset{\sim}{\nabla} \cdot (2\mu \underline{d}) = \rho_0 g[1 - \beta(T-T_0)]\, \underset{\sim}{e} \ , \qquad (16)$$

where $\underline{d}$ is the rate of deformation tensor, g the gravity, β the expansion coefficient and ρ_0 the reference density for the reference temperature T_0. The fluid viscosity depends on the temperature according to an exponential law. Figures 7 and 8 are the steady state configurations of the streamlines and the temperature field, respectively.

Acknowledgments : T. C and P.D were supported by IRSIA.

References

M.J. Crochet, F.T. Geyling, J.J. Van Schaftingen, J. Crystal Growth, vol.65, p.166-172, 1983.

G. De Vahl Davis, Report 1982/FMT/2, School of Mech. and Ind. Eng.,1982.

M. Deville, Recent developments of spectral and pseudospectral methods in fluid dynamics, V.K.I course, 1984.

M. Deville, E. Mund, Chebyshev pseudospectral solution of second order elliptic equation with finite element preconditioning, subm.to J.C.P., 1984.

D. Gottlieb, S. Orszag, Siam monograph n°26, SIAM, Philadelphia 1977.

P. Haldenwang, G. Labrosse, S. Abboudi, M. Deville, J. Comp. Phys., vol. 54, 1984.

L. Kleiser, U. Schumann, Proc.3rd GAMM conf. on Num.Meth. in fluid Mech., p. 165-173, 1980.

P. Le Queré, T. Alziary de Roquefort, C.R. Acad.Sc.Paris, t.294, série II-p.941-944, 1982.

R.L. Sani, P.M. Gresho, R.L. Lee, D.F. Griffiths, Int. J. Num. Meth. Fluids vol.1, p.17-43, 1981.

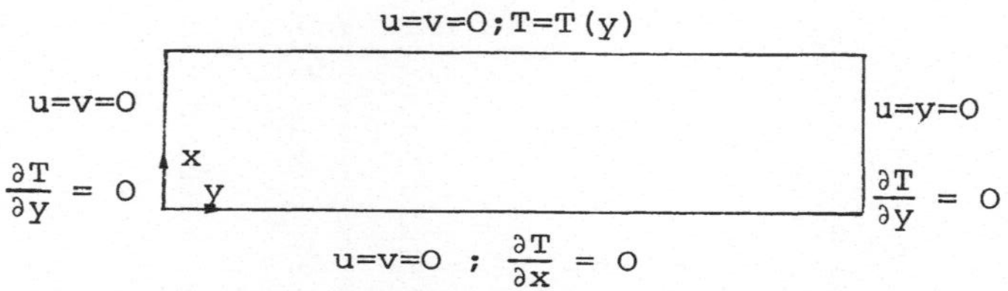

Fig.1. Geometry and boundary conditions.

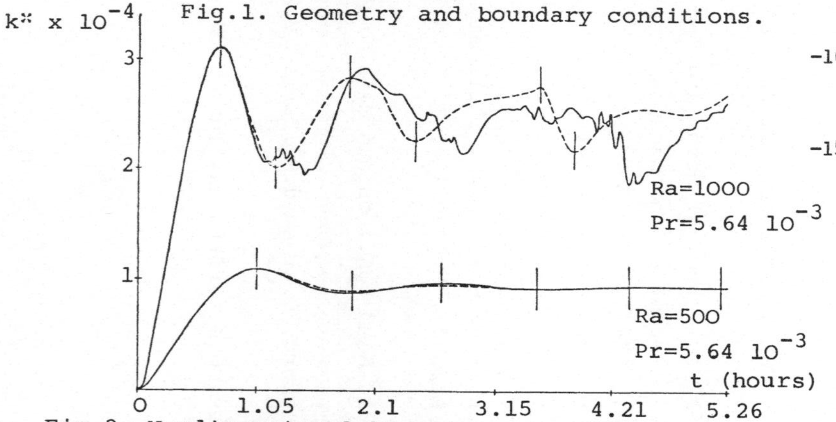

Fig.2. Nondimensional kinetic energy K*.
—— Inviscid pressure boundary condition
- - - Influence Matrix technique.

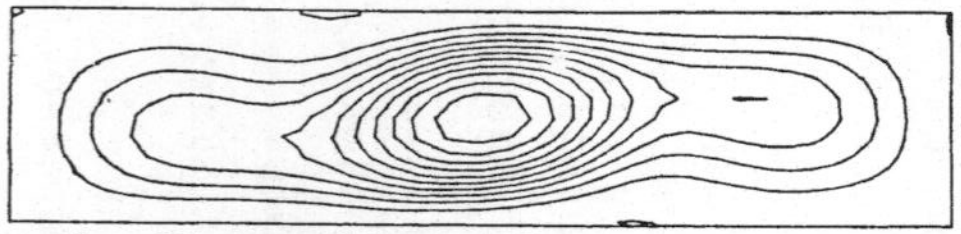

Fig.3. Streamlines for Ra=500, 0(5)55.

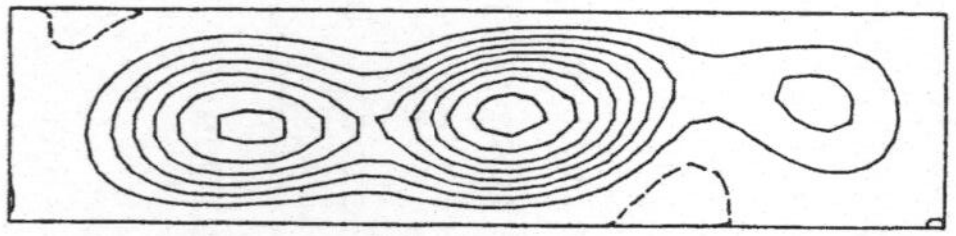

Fig.4. Streamlines for Ra=1000, 0(10)80.

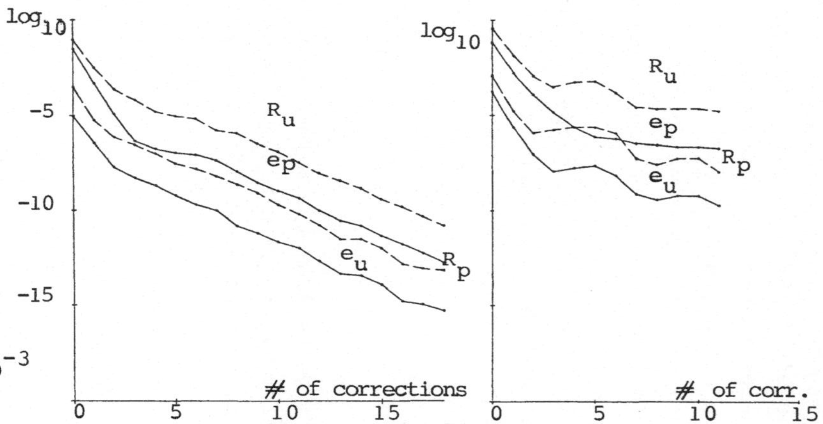

Fig.5.

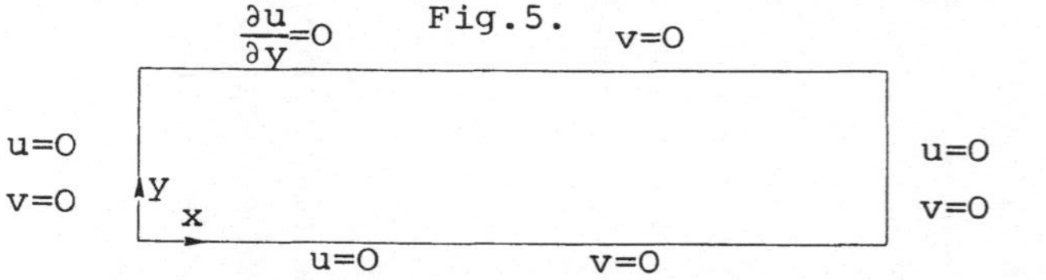

Fig.6. Geometry and boundary conditions.

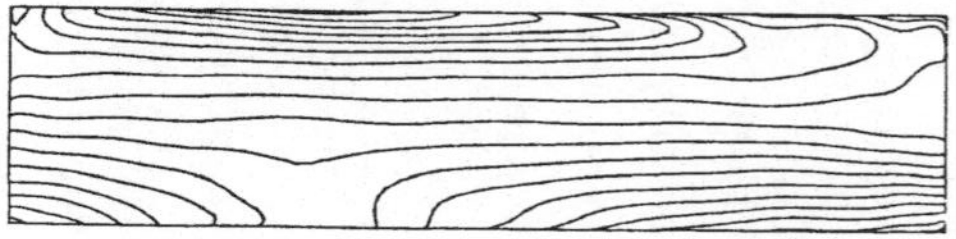

Fig.7. Isotherms 75.0(10.0) 175.0

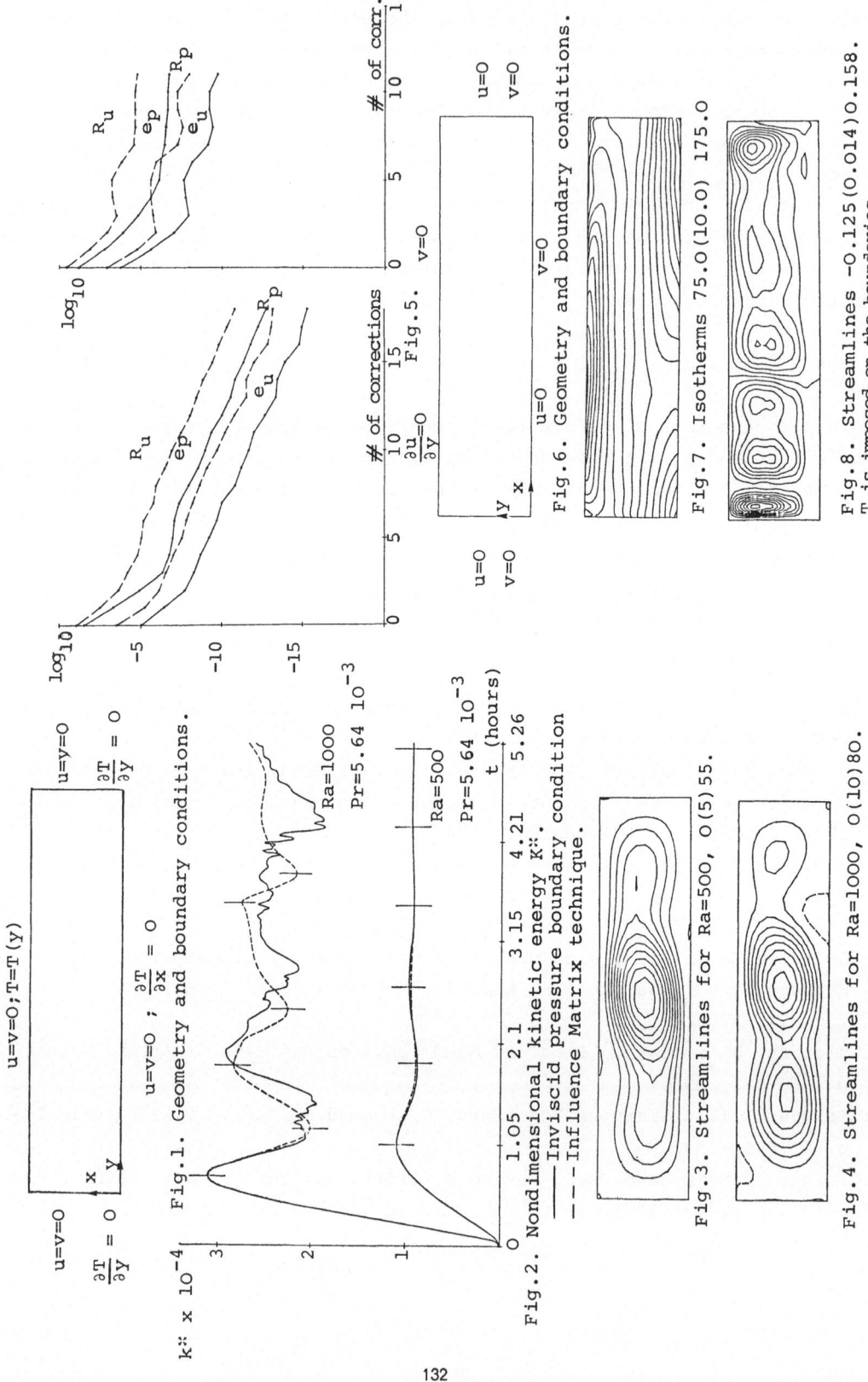

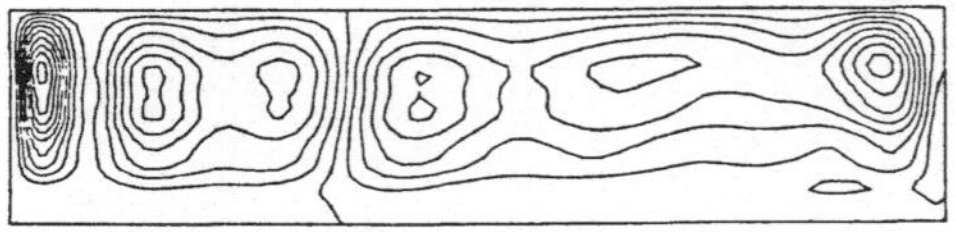

Fig.8. Streamlines -0.125(0.014)0.158.
T is imposed on the boundaries

ANALYSIS OF STRONGLY INTERACTING VISCOUS-INVISCID FLOWS INCLUDING SEPARATION

J. E. Carter, D. E. Edwards and R. L. Davis
United Technologies Research Center
East Hartford, CT 06108/USA

M. M. Hafez
Computer Dynamics
Hampton, VA 23666/USA

INTRODUCTION

Accurate prediction of the performance of various configurations for both external and internal aerodynamics depends critically on our ability to estimate the influence of viscous effects on the flow. In some instances, particularly when flow separation takes place, these effects are significant and must be determined simultaneously with an analysis of the inviscid flow. Two examples of strongly interacting viscous and inviscid flows which are addressed in this paper are that which occur when a transonic normal shock wave impinges on a turbulent boundary layer and the laminar-transitional separation bubble which often forms at the leading edge of an airfoil. In this paper, which is a condensed version of Refs. 1 and 2, a brief description and some sample results are presented for a viscous-inviscid interaction technique for the analysis of strongly interacting flows including boundary-layer separation. The analytical procedure is a generalized Interacting Boundary-Layer Theory (IBLT) method in which the inviscid and viscous flows are solved simultaneously and matched through a global semi-inverse iteration procedure. Special attention is focused in this paper on the inclusion of normal pressure gradients and the proper treatment of the streamwise convection of momentum in the reversed flow regions.

ANALYTICAL PROCEDURE

The procedure which has been used for the analysis of strongly interacting flows principally consists of a differential treatment of the defect formulation of the viscous flow equations introduced by LeBalleur[3] coupled to the inviscid stream function-vorticity method of Hafez and Lovell[4] using the global semi-inverse scheme of Carter.[5] In the defect form, the governing equations are expressed in the viscous region as the difference between the real viscous flow (RVF) and the equivalent inviscid flow (EIF) where the latter results when the inviscid flow is solved in the interior of the viscous region.[6] In strongly interacting flows the inviscid flow quantities vary across the viscous layer due to the rapid streamwise growth in the displacement thickness. The defect formulation is convenient since its use permits a smooth merging of the RVF into the EIF as the edge of the viscous layer is approached. The first-order inverse boundary-layer procedure of Carter[7] has been generalized to incorporate the defect formulation. In the present study the pressure variation across the viscous layer is treated approximately by assuming it to be equal to the inviscid pressure above the displacement thickness. Below the displacement thickness the normal pressure gradient is assumed to be zero.

In this paper the analysis of the airfoil leading edge separation bubble problem is principally focused on the effect of utilizing a windward differencing scheme to properly account for the reversed flow in the treatment of the streamwise convection of momentum. Only a first-order viscous-inviscid interaction technique has been used as even with separation the viscous layers in the airfoil transitional bubbles analyzed thus far[2,8] have remained relatively thin and hence the inviscid flow can be assumed to be invariant on this scale. However it has been observed in these calculations[8] that large reversed flow velocities, $u \simeq -.25\, u_e$, can occur in the intense vortex that forms in the separated boundary layer as the flow undergoes transition from laminar to turbulent flow. In this earlier work the FLARE[9] approximation was used in which the streamwise convection of momentum was set equal to zero in the reversed flow region. The FLARE approximation permits a simple, stable treatment of the reversed flow region in a forward marching boundary-layer analysis; however, its accuracy is unknown for large reversed flow magnitudes and hence a study has been conducted to numerically investigate the sensitivity of this interaction analysis to the treatment of the term, $u\partial u/\partial x$, in the x-momentum equation.

COMPUTED RESULTS

Transonic Shock Induced Separated Flow

Comparisons are presented between the results of the present generalized IBLT method and the data of Kooi[10] for transonic shock induced separated flow on a flat plate. Improvement in the comparison of the results with Kooi's data was found by modifying the length scale in the outer layer eddy viscosity model of Cebeci and Smith.[11] The modification to this outer model was to use the incompressible momentum thickness as the length scale instead of the incompressible displacement thickness. This length scale modification, suggested by Edwards,[12] was prompted by observations of the data reported by Simpson, et al.[13] for low speed separated flow. Figure 1 shows a comparison of the results of IBLT computations and the experimental data of Kooi for the wall pressure, displacement thickness, and skin friction for $M_\infty = 1.4$. The computed results are shown for the Cebeci-Smith (CS) model and the modified turbulence model. Figures 1(a) and 1(b) show that good agreement is obtained with the data with the present IBLT using the modified CS turbulence model. Comparison of these two figures shows that the turbulence model modification simultaneously improves the agreement for the wall pressure and displacement thickness, which substantiates the use of the displacement body concept in separated flows. The distribution of the skin friction coefficient in Fig. 1(c) shows both turbulence models agree well with the data up to the separation point, but the modified CS model substantially overpredicts the streamwise extent of the separated flow region in comparison with that deduced experimentally by Kooi. This lack of agreement for the skin friction indicates that further work is required to correspondingly improve the accuracy of the IBLT with the modified CS model in the near wall region.

Viscous-inviscid interaction calculations have also been performed for this case in which the effects of normal pressure gradients are neglected. With the pressure assumed constant across the viscous layer the present analysis reverts back to a first-order inverse boundary-layer scheme interacting with the outer inviscid flow through displacement thickness coupling. Computed results for the wall pressure, displacement thickness, and skin friction are also shown in Fig. 1 and demonstrate that the effect of including normal pressure gradients is relatively minor in this case and has less of an impact than the change in the turbulence model.

The variation of the pressure across the boundary layer from the equivalent inviscid flow (EIF) is shown in Fig. 2 for selected streamwise stations both upstream and downstream of the shock wave which is located at $x = 3.6$. The x-location of each of these normalized pressure profiles is shown in a scaled drawing inserted into Fig. 2. Since the variation in the static pressure across the viscous layer was insensitive to the turbulence model, only the IBLT results with the modified CS turbulence model are shown in Fig. 2. The arrows on each of the profiles are the location of the edge of the viscous layer. The kink in the profiles near the wall is at the displacement thickness since, as was discussed earlier, the pressure was assumed constant from δ^* to the wall. Figure 2 shows the anticipated trend in the static pressure variation in that in the immediate vicinity of the shock wave the pressure decreases with increasing y upstream of the shock wave and increases with increasing y downstream of the shock wave. The variation of the static pressure across the viscous layer reaches a maximum of about 10 percent in the immediate vicinity of the impinging shock wave; as expected, this variation vanishes upstream and downstream of this region as the flow returns to that of a conventional constant pressure boundary layer.

Airfoil Transitional Separation Bubble

Calculations with the ALESEP (Airfoil Leading Edge Separation) inviscid-viscous interaction code[2] using both the windward differencing scheme and the FLARE approximation in the reversed flow region have been made for the NACA-0010 airfoil tested experimentally by Gault[14] at 8 degrees angle of attack and a chord Reynolds number of 2×10^6. These calculations were made with a forced transition model in which the onset of transition was specified at $s/c = .0283$ and the transition length was set to .0161. Figure 3 shows the predicted distributions of pressure and skin friction, using the windward and FLARE approximations. Comparison of these results show that in general only small differences exist between the computed pressure and skin friction distributions due to the inclusion of the more accurate windward flow differencing technique. Comparison of the computed results with the experimental pressure data in Fig. 3(a) shows that the inability of the analysis to predict the constant pressure region near the peak suction pressure is not affected by the improved differencing procedure used in the reversed flow region. It is concluded from the analysis of this case which contains a maximum backflow velocity of $u/u_e = -.28$, that the FLARE approximation is quite accurate in predicting the overall results which occur when a transitional separation bubble exists near the leading edge of an airfoil. A detailed comparison of the windward and FLARE results shows only small differences which principally occur in the reversed flow region. These differences though are very interesting as the inclusion of windward differencing has revealed a new streamline pattern in the recirculating flow region. This change in bubble structure is discussed next.

It is observed in the skin friction distribution in Fig. 3(b) that the results obtained from the windward scheme in contrast with the FLARE technique shows that a small region of forward flow ($C_f > 0$) occurs in the interior of the separated flow region. Figure 4 shows a comparison of the computed streamlines in the viscous region obtained with the windward and FLARE schemes. Overall these streamline patterns are very close with the major difference being that the more accurate treatment of the reversed flow region via the windward scheme has revealed the existence of a second, counter-rotating bubble inside of the primary separation bubble. To our knowledge this is the first time such a structure has emerged from a numerical calculation of the interacting boundary-layer equations for a closed separation zone on a solid surfaces. Physically, such a structure is known to exist in separated flows as evidenced by several figures in the excellent compilation on flow visualization recently published by Van Dyke.[15] Grid refinement and convergence studies[2] have verified that the existence of the double bubble structure is insensitive to the numerical solution procedure.

CONCLUSIONS

Favorable comparisons which have been obtained with the separated flow data of Kooi ($M_\infty = 1.4$) demonstrate that the present procedure is capable of accurately resolving many of the details of transonic shock-wave, boundary-layer interaction. Second, the results show that the effect of displacement thickness interaction dominates over the effects produced by normal pressure gradients and imbedded shocks for transonic shock induced separated flows. Calculations made with an improved algebraic turbulence model demonstrate that the computed results are more sensitive to the turbulence model than to whether or not normal pressure gradients are included in the analysis. Incorporation of a windward differencing scheme in an airfoil transitional bubble analysis has demonstrated that the FLARE approximation is accurate even for large reversed flow velocities. However, use of a windward differencing scheme has revealed the existence of a new counterrotating bubble structure in the separated flow region.

ACKNOWLEDGEMENT

The authors express their gratitude for the support of this work to the following technical monitors and agencies: James D. Wilson, Air Force Office of Scientific Research (Contract F49620-81-C-0041) and Joel L. Everhart, NASA-Langley Research Center (Contract NAS1-16585).

REFERENCES

1. Carter, J. E., D. E. Edwards and M. M. Hafez: AFOSR-TR-83-1283, October 1983.
2. Davis, R. L. and J. E. Carter: NASA CR-3791, April 1984.
3. LeBalleur, J. C.: La Recherche Aerospatiale, No. 1981-3, English Edition, 1981.
4. Hafez, M. M. and D. Lovell: AIAA J., Vol. 21, No. 3, March 1983.
5. Carter, J. E.: AIAA Paper 79-1450, AIAA 4th Computational Fluid Dynamics Conference, Williamsburg, VA, July 23-24, 1979.
6. Lock, R. C. and Firmin, M. C. P.: RAE Technical Memorandum Aero 1900, 1981.
7. Carter, J. E.: NASA TP-1208, September 1978.
8. Vatsa, V. N. and J. E. Carter: AIAA Paper 83-0300, 1983.
9. Reyhner, T. A. and I. Flugge Lotz: Int. Journal of Non-Linear Mech., Vol. 3, No. 2, June 1968, pp. 173-179.
10. Kooi, J. W.: AGARD CP-168, 1975.
11. Cebeci, T. and A. M. O. Smith: Analysis of Turbulent Boundary Layers, Academic Press, 1974.
12. Edwards, D. E.: UTRC report to be published, 1984.
13. Simpson, R. L., T. T. Chew and B. V. Shivaparsad: J. Fluid Mechanics, Vol. 113, pp. 23-51, 1981.
14. Gault, D. E.: NACA TN 3505, September 1955.
15. Van Dyke, M.: An Album of Fluid Motion, Parabolic Press, Stanford, CA, 1982.

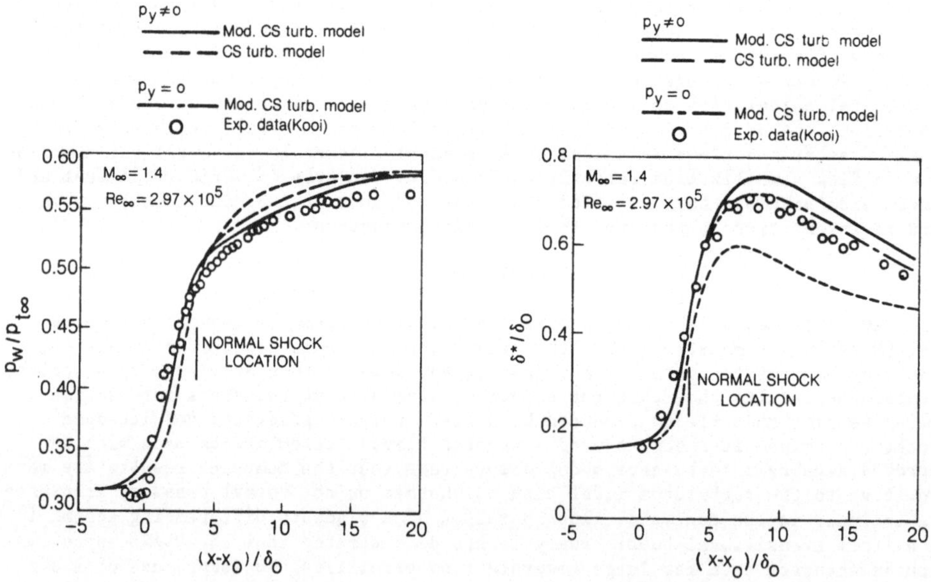

a) Wall pressure

(b) Displacement thickness

Fig. 1 Comparison of results from IBLT with experimental data for transonic normal shock-wave, boundary layer interaction

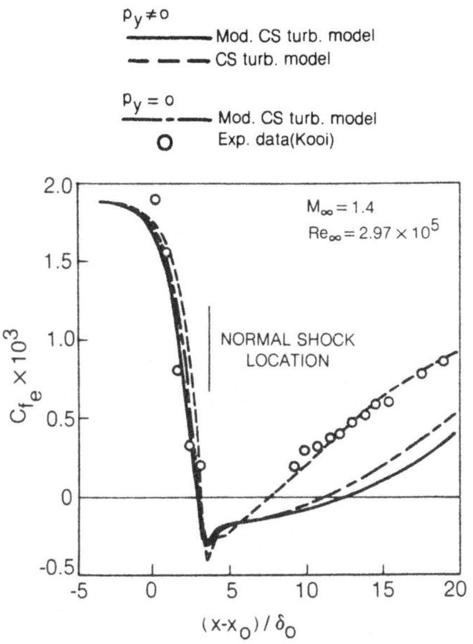

Fig. 1 (c) Skin friction

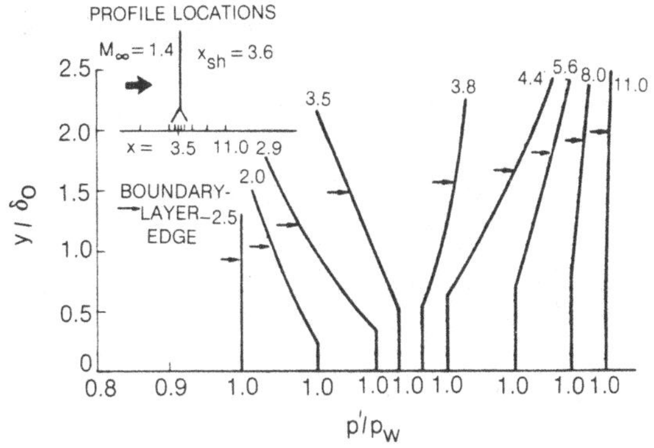

Fig. 2 Variation of pressure across boundary layer from equivalent inviscid flow (EIF) for transonic normal shock-wave, boundary-layer interaction

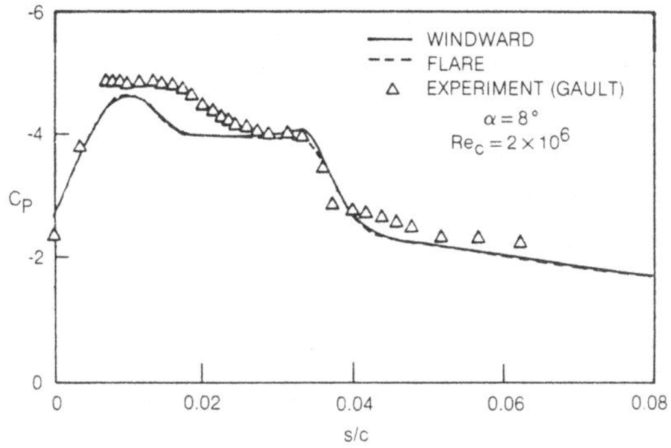

(a) Pressure distribution

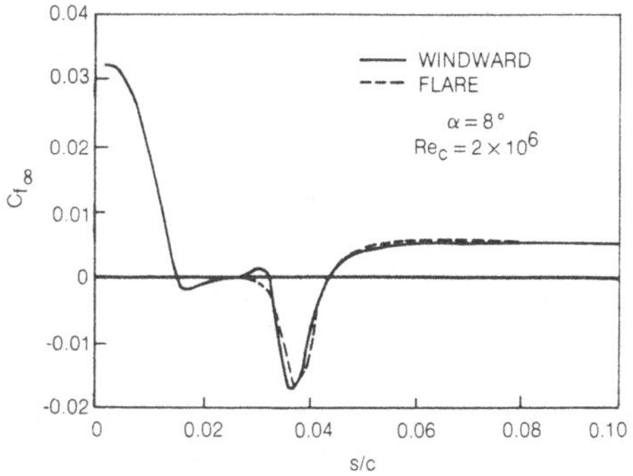

(b) Skin friction

Fig. 3 Comparison of results for windward and FLARE differencing - NACA 0010 airfoil (modified)

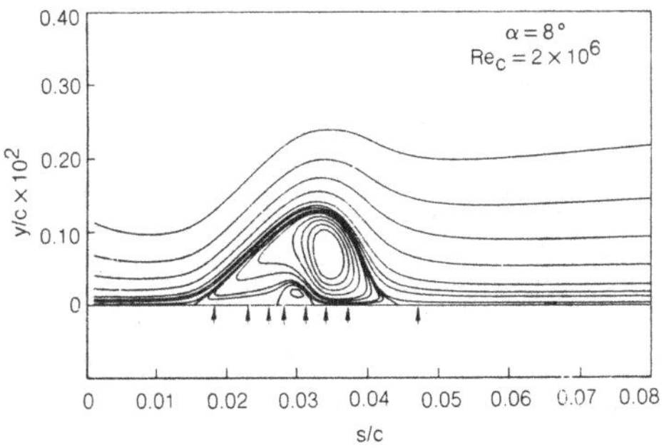

(a) Windward

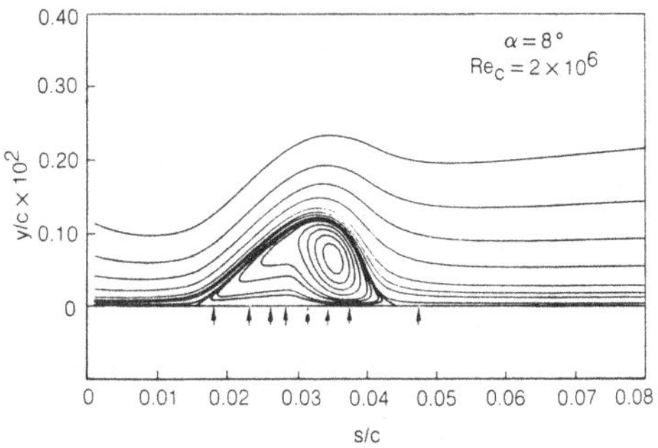

(b) FLARE approximation

Fig. 4 Transitional separation bubble streamline pattern - NACA 0010 airfoil (modified)

AN IMPROVED EULER METHOD FOR COMPUTING STEADY TRANSONIC FLOWS

D M Causon and P J Ford
Department of Aeronautical and Mechanical Engineering
University of Salford
Salford M5 4WT
England

SUMMARY

An improved Euler solver is presented for computing steady three-dimensional transonic flows around practical aircraft forebodies. The method is pseudo time-dependent, split and uses a finite-volume formulation. Shock waves are captured crisply, with minimal added smoothing, by means of an operator-switching facility which more accurately reflects the direction of propagation of signals. The method is robust, versatile, and holds promise for treating complex three-dimensional geometries economically.

INTRODUCTION

Computational aerodynamics is a revolutionary force in practical aerodynamic analysis as exemplified by the increasing number of solutions of compressible flow problems around complex configurations. With the advent of the fast, vectorised, Class 6 computers and associated increased storage capabilities, the feasibility of carrying out analyses of advanced three-dimensional aircraft configurations is becoming a reality. Pacing this advancement in computer technology is the development both of new algorithms and refinements of existing ones. The present approach is based upon the refinement of a method, which has proved to be simple to code and robust in applications to increasingly more complex and realistic configurations. In the development of this method, outlined below, the major emphasis has been on applicability and coding simplicity with the goal of providing a sound engineering tool for preliminary design and analysis. We have incorporated improvements to the method which can be proven in applications to yield clear benefits, but have avoided refinements which could result in large increases in code complexity with only slight improvements in the solutions. The resulting method shows promise for treating complex three-dimensional aircraft geometries economically. Recently [1,2] we described the method in some detail and presented a number of standard test-case solutions as evidence of the method's intrinsic validity. Here, we focus on applications of the method to some realistic aircraft forebodies in the upper transonic range of Mach numbers.

FORMULATION

Since a body-fitted mesh will be required, we cast the equations of motion in generalised co-ordinates. The Euler equations, in strong conservation form, are

$$\frac{\partial}{\partial t}\underline{U} + \frac{\partial}{\partial x^{\ell}}\,{}^{\ell}\underline{\underline{F}} = 0 \qquad \ell = 1(1)3 \tag{1}$$

where $\underline{U} = \sqrt{g}[\rho,\rho w_1,\rho w_2,\rho w_3,e]^T$, $\quad {}^{\ell}F(U) = \sqrt{g}\begin{bmatrix} \rho u^{\ell} \\ \rho w_1 u^{\ell}+p\,\dfrac{\partial x^{\ell}}{\partial z_1} \\ \rho w_2 u^{\ell}+p\,\dfrac{\partial x^{\ell}}{\partial z_2} \\ \rho w_3 u^{\ell}+p\,\dfrac{\partial x^{\ell}}{\partial z_3} \\ (e+p)u^{\ell} \end{bmatrix}$,

$\sqrt{g}$ is the Jacobian and the flow velocity $\underline{q} = u^{\ell}\underline{g}_{\ell} = w_{\ell}\underline{a}_{\ell}$, where $\underline{a}_{\ell}$ are the Cartesian unit base vectors.

Equation (2) can also be cast in integral form, which is the basis of the finite volume method

$$\underline{U}_{,t} + \iint_{S_1} {}^1\underline{\underline{F}} ds + \iint_{S_2} {}^2\underline{\underline{F}} ds + \iint_{S_3} {}^3\underline{\underline{F}} ds = 0 \quad . \tag{2}$$

The most appealing feature of the finite volume formulation becomes apparent by noting that $\partial x^{\ell}/\partial z_m = \underline{g}^{\ell}.\underline{a}_m$ and $\underline{q}.\underline{g}^{\ell} = u^{\ell}$, where upon

$$ {}^{\ell}\underline{\underline{F}}(U) = \sqrt{g}\begin{bmatrix} \rho\underline{q}.\underline{g}^{\ell} \\ \rho w_1 \underline{q}.\underline{g}^{\ell} + p\underline{g}^{\ell}.\underline{a}_1 \\ \rho w_2 \underline{q}.\underline{g}^{\ell} + p\underline{g}^{\ell}.\underline{a}_2 \\ \rho w_3 \underline{q}.\underline{g}^{\ell} + p\underline{g}^{\ell}.\underline{a}_3 \\ (e+p)\underline{q}.\underline{g}^{\ell} \end{bmatrix} = \begin{bmatrix} \rho\underline{q} \\ \rho w_1\underline{q} + p\underline{a}_1 \\ \rho w_2\underline{q} + p\underline{a}_2 \\ \rho w_3\underline{q} + p\underline{a}_3 \\ (e+p)\underline{q} \end{bmatrix} . \sqrt{g}\, \underline{g}^{\ell} = \underline{\underline{H}}(U).S_{\ell} \tag{3}$$

and we see that computations can be performed with respect to the easily constructed Cartesian flux tensor $\underline{\underline{H}}(U)$, rather than the curvilinear ${}^{\ell}\underline{\underline{F}}(U)$. Thus, we need not involve ourselves with the intricacies of the co-ordinate transformation.

DISCRETISATION

We solve equation (2) using a factored sequence of one-dimensional difference operators, where each component operator relates to its respective split differential equation

$$\underline{U}_{,t} + {}^{\ell}\underline{\underline{F}}_{,\ell} = 0 \qquad \ell = \textit{either } 1,2, \text{ or } 3 \quad . \tag{4}$$

The MacCormack difference operator $L_1(\Delta t)$ is

$$\overline{\underline{U}^{n+1}} = \underline{U}^n - \Delta t \left(\frac{\Delta_- {}^1\underline{\underline{F}}}{\Delta x^1}\right)^n \tag{5a}$$

$$\underline{U}^{n+1} = \tfrac{1}{2}\left(\underline{U}^n + \overline{\underline{U}^{n+1}} - \Delta t \left(\frac{\Delta_+ {}^1\underline{\underline{F}}}{\Delta x^1}\right)^{\overline{n+1}}\right) \tag{5b}$$

where Δ_+ and Δ_- are respectively forward and backward two-point differences. In finite volume form, the operator $L_1(\Delta t)$ becomes

$$\overline{U^{n+1}_{\sim ijk}} = U^n_{\sim ijk} - \Delta t(\underline{\underline{H}}^n_{ijk}\underline{S}_{i+\frac{1}{2}} + \underline{\underline{H}}^n_{i-1jk}\underline{S}_{i-\frac{1}{2}}) \quad , \tag{6a}$$

$$U^{n+1}_{\sim ijk} = \tfrac{1}{2}(U^n_{\sim ijk} + \overline{U^{n+1}_{\sim ijk}} - \Delta t(\overline{\underline{\underline{H}}^{n+1}_{i+1jk}}\underline{S}_{i+\frac{1}{2}} + \overline{\underline{\underline{H}}^{n+1}_{ijk}}\underline{S}_{i-\frac{1}{2}})) \quad , \tag{6b}$$

where $\underset{\sim}{U}_{ijk} = vol_{ijk}[\rho,\rho w_1,\rho w_2,\rho w_3,e]^T_{ijk}$ and $\underline{S}_{i\pm\frac{1}{2}}$ are the area vectors on opposite faces of the cell, corresponding to the surface x^1 = constant. Scheme (6) is easily coded, requiring only one level of storage and the area vectors and volumes can be evaluated from the Cartesian co-ordinates of the cell vertices [2].

The principal disadvantages of Scheme (6) are that it requires a numerical boundary condition at a supersonic exit, extra maxima and minima appear around shock waves, and additional smoothing is required around stagnation points, sonic lines and shocks. These negative features have been the motivation for much recent algorithm research and development. The present approach involves switching within a split operator between the MacCormack Scheme (6) and the upwind scheme of Beam and Warming [3], according to whether the flow is subsonic or supersonic. The upwind scheme, implemented in supersonic regions of flow, is

$$\overline{U^{n+1}_{\sim ijk}} = U^n_{\sim ijk} - \Delta t(\underline{\underline{H}}^n_{ijk}\underline{S}_{i+\frac{1}{2}} + \underline{\underline{H}}^n_{i-1jk}\underline{S}_{i-\frac{1}{2}}) \quad , \tag{7a}$$

$$\underset{\sim}{U}_{ijk}^{n+1} = \tfrac{1}{2}\left(\underset{\sim}{U}_{ijk}^{n} + \overline{\underset{\sim}{U}_{ijk}^{n+1}} - \Delta t\left(\overline{\underline{\underline{H}}_{ijk}^{n+1}} \cdot \underline{S}_{i+\frac{1}{2}} + \overline{\underline{\underline{H}}_{i-1jk}^{n+1}} \cdot \underline{S}_{i-\frac{1}{2}} + \underline{\underline{H}}_{ijk}^{n} \cdot \underline{S}_{i+\frac{1}{2}} + 2\underline{\underline{H}}_{i-1jk}^{n} \cdot \underline{S}_{i-\frac{1}{2}} + \underline{\underline{H}}_{i-2jk}^{n} \cdot \underline{S}_{i+\frac{1}{2}}\right)\right) \quad (7b)$$

where in equation (7) the local time step Δt can be twice as large as the maximum permitted in Scheme (6). Additionally, switching operators are required at the switch points to preserve numerically the strong conservation form of the equations of motion. These have been given in Reference [2].

SWITCHING CRITERIA, SMOOTHING TERMS AND BOUNDARY CONDITIONS

A simple switching criterion which works well on an H-grid is a test on the local Mach number, evaluated at the previous time level, so implemented as to ensure that the switching operators are located just within the supersonic region. On a C-grid we replace this by a test for the sign change of the eigenvalue $u^{\ell}-c$.

The smoothing term added to the right-hand side of the corrector step of the MacCormack $L_1(\Delta t)$ operator is a five-point difference replacement for the derivative $\partial^4\underline{U}/\partial x^{1^4}$. It is switched off when the upwind scheme (7) is in use and no smoothing is applied to either of splits $L_2(\Delta t)$ and $L_3(\Delta t)$.

At the inflow boundary, we hold the dependent variables at their freestream values. At the outflow boundary, a condition is needed only if the exit flow is subsonic. In such cases we place the boundary far downstream where spatial gradients are small. In the far field, we either assume solid wall conditions apply far from the body, or invoke a radiation boundary condition. At the body surface, we impose solid wall conditions and derive the required static pressure from the normal momentum balance for cells adjacent to the surface.

RESULTS

All of the meshes employed here are, essentially, of cylindrical polar, flow-conforming type and relatively coarse. Closer attention to the mesh generation methodology is expected to yield benefits. Two cases illustrate the validity of the three-dimensional code. The first relates to the axisymmetric 'aircraft forebody' at Mach 1.4 and $\alpha = 0^o$ defined analytically in Reference [1]. Figure 1 shows the axial Cp distribution and surface isobar contours. The latter are θ invariant as expected and the comparison with the earlier obtained axisymmetric results [1] is close. The second case relates to forebody No 4 from Reference [4]. The Mach number is 1.7 and $\alpha = -5^o$. Figure 2 shows axial Cp distributions as a function of θ, and isobar contours. The level of agreement with data is particularly good on this coarse mesh (40x10x8-ℓ,r,θ on a half body).

More realistic forebody shapes are considered next. Figure 3 depicts a forebody created from the axisymmetric 'forebody' by fairing-out the canopy through a surface discontinuity in θ. The canopy shock can be seen to vanish as expected over the lower half of the body. Also shown in Figure 3 are surface isobar contours. The Mach number was 1.4 and $\alpha = 0^o$. Figure 4 depicts a more complex forebody, having a 5° dipped elliptical nose-cone and relatively flat sides and underside to later accommodate canard foreplanes and chin or side intakes. The Mach number is 1.4 and $\alpha = -5^o$. The computations were performed on a coarse mesh (45x10x18-ℓ,r,θ on half body) and tests are underway using finer meshes. The examples shown in Figures 1-4 required around 400-500 seconds on a CDC 7600 computer.

CONCLUSIONS

The method presented is versatile, robust, easy to code in finite volume form, and has good shock capturing capabilities. It does not drop to first-order accuracy in the vicinity of shock waves, as do other methods, nor does it require substantial doses of added smoothing; the latter is required only within subsonic bubbles in one of the splits.

REFERENCES

[1] D M Causon and P J Ford, "Computations in external transonic flow", Proc 5th GAMM Conf on Num Meth in Fl Mech, Vieweg, 1984.

[2] D M Causon and P J Ford, "Improvements in techniques for the numerical simulation of steady transonic flows", AIAA Paper No 84-0089, 1984.

[3] R F Warming and R M Beam, "Upwind second-order difference schemes and applications in aerodynamic flows", JAIAA, Vol 14, 1976.

[4] J C Townsend, D T Howell, I K Collins, and C Hayes, "Surface data on a series of analytic forebodies at Mach numbers from 1.70 to 4.50 and combined angles of attack and sideslip", NASA TM-80062, 1979.

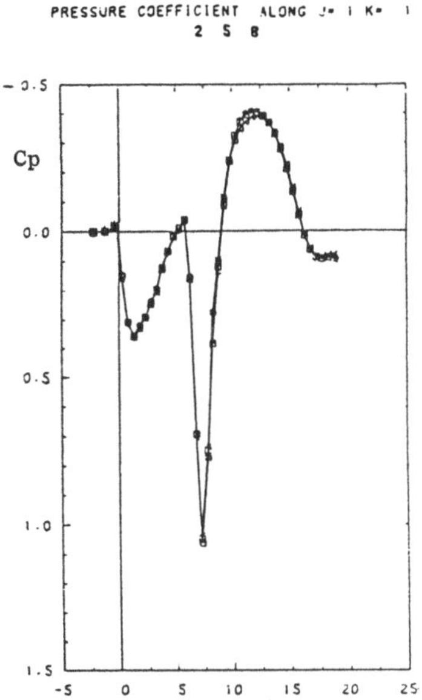

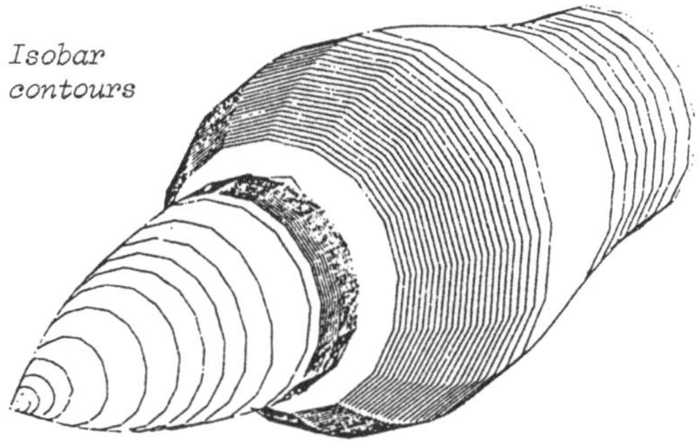

FIG 1 Axisymmetric 'aircraft forebody' at Mach 1.40 and α = 0°

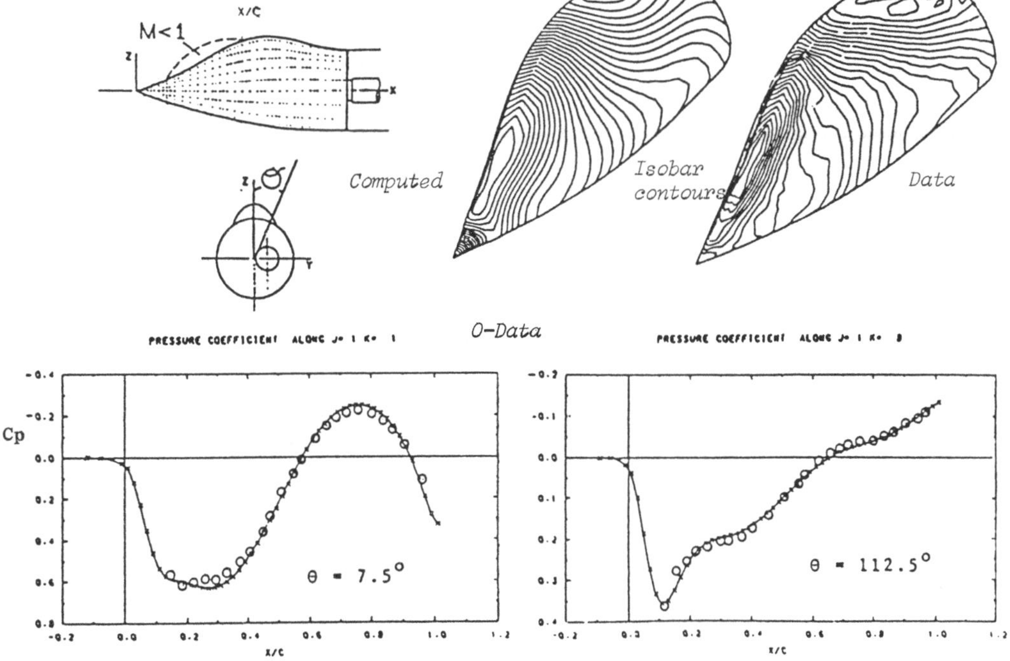

FIG 2 Forebody No 4 from NASA TM80062 at Mach 1.70 and α =-5°

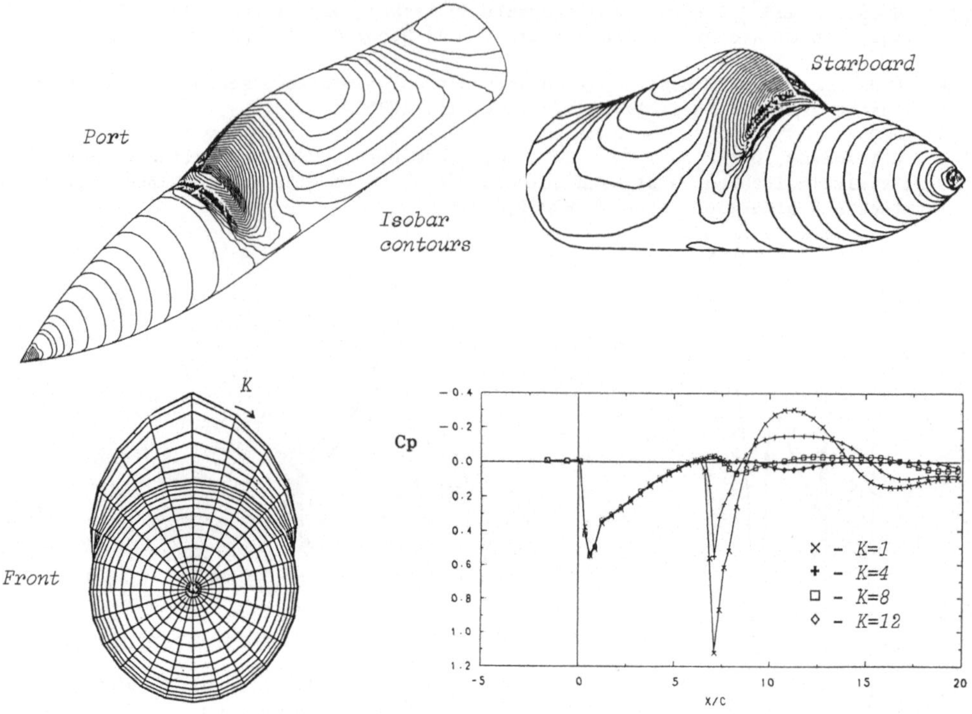

FIG 3 Hypothetical aircraft forebody at Mach 1.40 and $\alpha = 0^{o}$

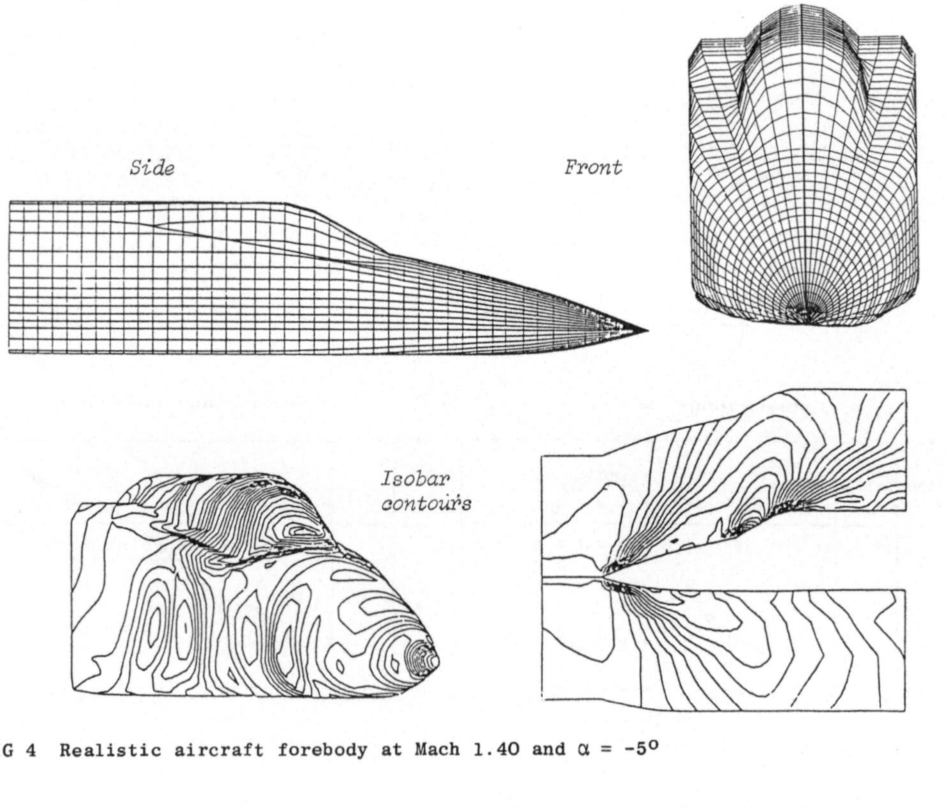

FIG 4 Realistic aircraft forebody at Mach 1.40 and $\alpha = -5^{o}$

A SEMI-DIRECT PROCEDURE USING A LOCAL RELAXATION FACTOR AND ITS APPLICATION TO AN INTERNAL FLOW PROBLEM

Sin-Chung Chang
NASA Lewis Research Center
Cleveland, Ohio, 44135, U.S.A.

1. INTRODUCTION

Generally, fast direct solvers (FDS)[1] are not directly applicable to a nonseparable elliptic partial differential equation (PDE). This limitation, however, can be circumvented by a semi-direct procedure, i.e., an iterative procedure using FDS algorithms. In this paper, we present an efficient semi-direct procedure which is easy to implement and applicable to a variety of boundary conditions. The current procedure also possesses other highly desirable properties, i.e., (1) the convergence rate does not decrease with an increase of grid cell aspect ratio, and (2) the convergence rate can be estimated using the coefficients of the PDE being solved.

Many elliptic PDE encountered in flow problems can be expressed as

$$Lu = h \tag{1.1}$$

where L is a nonseparable second-order linear elliptic operator, u the dependent variable, and h a given source term. Equation (1.1) may be solved with the iterative procedure:

$$\nabla^2 (u^{n+1} - u^n) = -\tau (Lu^n - h) \qquad (n = 0, 1, 2, \ldots) \tag{1.2}$$

Here τ is a nonzero relaxation factor and ∇^2 the Laplacian which can be directly inverted by a FDS. In the previous works[2-4] using Eq. (1.2), τ is treated as a constant and the iteration is accelerated by an optimal choice of τ. In this study, we consider τ as a spatially varying parameter. The local value of τ is chosen to optimize the local convergence rate which is evaluated using a von Neumann analysis.

2. STABILITY ANALYSIS

As an initial step, it is assumed that τ is a constant and

$$L = a\frac{\partial^2}{\partial x^2} + 2b\frac{\partial^2}{\partial x\partial y} + c\frac{\partial^2}{\partial y^2} \qquad (a > 0,\ c > 0,\ \text{and } ac-b^2 > 0) \tag{2.1}$$

where a, b, and c are constants. Using a uniform grid with grid intervals Δx and Δy, the central difference form of Eq. (1.2) at a grid point (i,j) is

$$\tilde{P}(u_{i,j}^{n+1}) - \tilde{P}(u_{i,j}^{n}) = -\tau(\tilde{L}(u_{i,j}^{n}) - h_{ij}) \tag{2.2}$$

Here h_{ij} is the source term and the operators $\tilde{P}$ and $\tilde{L}$, respectively, are defined by

$$\tilde{P}(u_{i,j}^n)=(\Delta x)^{-2}(u_{i+1,j}^n+u_{i-1,j}^n-2u_{i,j}^n)+(\Delta y)^{-2}(u_{i,j+1}^n+u_{i,j-1}^n-2u_{i,j}^n) \tag{2.3}$$

$$\begin{aligned}\tilde{L}(u_{i,j}^n)=&a(\Delta x)^{-2}(u_{i+1,j}^n+u_{i-1,j}^n-2u_{i,j}^n)+c(\Delta y)^{-2}(u_{i,j+1}^n+u_{i,j-1}^n-2u_{i,j}^n)+\\&b(2\Delta x\Delta y)^{-1}(u_{i+1,j+1}^n+u_{i-1,j-1}^n-u_{i+1,j-1}^n-u_{i-1,j+1}^n)\end{aligned} \tag{2.4}$$

The convergence rate of Eq. (2.2) will be analyzed assuming

$$e_{i,j}^n \equiv u_{i,j}^n - u_{i,j} = e^n\exp[I(ik_x\Delta x+jk_y\Delta y)], \qquad I \equiv\sqrt{-1} \tag{2.5}$$

where $u_{i,j}$ is the converged solution, e^n the error amplitude and $\vec{k} \equiv (k_x,k_y)$ the propagation vector. Assuming that $\vec{k} \neq 0$ and using the following definitions:

$$\theta_x \equiv (k_x\Delta x)/2,\ \alpha_x \equiv (\sin\theta_x)/\Delta x,\ \theta_y \equiv (k_y\Delta y)/2,\ \alpha_y \equiv (\sin\theta_y)/\Delta y \tag{2.6}$$

$$s_x \equiv \alpha_x\Big/[(\alpha_x)^2+(\alpha_y)^2]^{1/2},\ s_y \equiv \alpha_y\Big/[(\alpha_x)^2+(\alpha_y)^2]^{1/2} \tag{2.7}$$

then Eqs. (2.2) and (2.5) imply that

$$e^{n+1}/e^n = E(\tau,\vec{k}) \equiv 1-\tau G(\vec{k}) \qquad (\vec{k} \neq 0) \tag{2.8}$$

with $$G(\vec{k}) \equiv a(s_x)^2 + c(s_y)^2 + 2bs_xs_y \cos\theta_x \cos\theta_y \tag{2.9}$$

It should be noted that the von Neumann analysis (along with the condition $\vec{k} \neq 0$) can be fully justified if it is applied to a grid with periodic boundary conditions.

Let σ_{max} and σ_{min}, respectively, be the greatest and the smallest eigenvalues of the symmetric and positive-definite matrix (μ, ν = 1, 2)

$$A \equiv (\alpha_{\mu\nu}), \qquad \alpha_{11}=a,\ \alpha_{12}=\alpha_{21}=b,\ \alpha_{22}=c \tag{2.10}$$

Then it can be shown that[5]

$$\sigma_{max} \geq G\,(\vec{k}) \geq \sigma_{min} > 0 \tag{2.11}$$

with the understanding that the bounds σ_{max} and σ_{min} are sharp if all the allowable $\vec{k}(\neq 0)$ are considered. As a result, the asymptotic error amplification factor $E(\tau)$($\equiv$ the supremum of $|E(\tau, \vec{k})|$ for a given τ) will reach its minimum

$$E_0 \equiv E(\tau_0) = (\Sigma - 1)/(\Sigma + 1) < 1 \tag{2.12}$$

when $\tau = \tau_0 \equiv$ the optimal relaxation factor $= 2/(\sigma_{max} + \sigma_{min}) = 2/(a + c)$ (2.13)

Here Σ is the condition number $\equiv \sigma_{max}/\sigma_{min} > 1$.

Equations (2.1)-(2.13) can be easily generalized for N-dimensional problems. The only exception is Eq. (2.13) where the last equality sign is not valid if $N>2$.

3. LOCAL RELAXATION

The variable coefficient (VC) version of the numerical procedure presented in Section 2 is obtained by replacing the constant coefficients a, b, c, τ, τ_0, σ_{max}, and σ_{min} in Eqs. (2.2) to (2.4) and (2.13), respectively, with the grid point dependent coefficients a_{ij}, b_{ij}, c_{ij}, τ_{ij}, $\tau_{o,ij}$, $\sigma_{max,ij}$ and $\sigma_{min,ij}$. Obviously, the VC version is well defined at all interior grid points. For a grid point on a periodic or Neumann boundary, it can also be defined using the periodic condition or an extrapolation technique explained in Ref. 6.

The current procedure can be modified to solve PDE with

$$L = L'(x,y) \equiv \frac{\partial}{\partial x}\left(p(x,y)\frac{\partial}{\partial x}\right) + \frac{\partial}{\partial y}\left(q(x,y)\frac{\partial}{\partial y}\right) \tag{3.1}$$

where p and q are arbitrary positive functions of x and y. With $L = L'(x,y)$, the VC version of Eq. (2.2) can be obtained by replacing $\tilde{L}(u^n_{i,j})$ with[7]

$$\tilde{L}'(u^n_{i,j}) \equiv (\Delta x)^{-2}[p_{(i-1/2)j}\, u^n_{i-1,j} + p_{(i+1/2)j}\, u^n_{i+1,j} - (p_{(i-1/2)j}+p_{(i+1/2)j})u^n_{i,j}] +$$
$$(\Delta y)^{-2}[q_{i(j-1/2)}u^n_{i,j-1} + q_{i(j+1/2)}u^n_{i,j+1} - (q_{i(j-1/2)}+q_{i(j+1/2)})u^n_{i,j}] \tag{3.2}$$

In case that the values of p and q do not vary greatly from one grid point to its neighbors, then $\tilde{L}'(u^n_{i,j}) \doteq \tilde{L}(u^n_{i,j})$ assuming $a_{ij} = p_{ij}$, $c_{ij} = q_{ij}$ and $b_{ij} = 0$. This observation coupled with Eq. (2.13) leads us to the assumption:

$$\tau_{ij} = \tau'_{o,i,j} \equiv 2/(p_{ij}+q_{ij}) \qquad (L = L'(x,y)) \tag{3.3}$$

4. SCALING

Assuming that the definition of the operator P is broadened as

$$\tilde{P}(u^n_{i,j}) = g_x(\Delta x)^{-2}(u^n_{i+1,j}+u^n_{i-1,j}-2u^n_{i,j})+g_y(\Delta y)^{-2}(u^n_{i,j+1}+u^n_{i,j-1}-2u^n_{i,j}) \quad (4.1)$$

with g_x and g_y positive constants, then the only modifications required for Sec. 2 are to replace the coefficients α_x, α_y, a, b, and c in Eqs. (2.7), (2.9), (2.10) and (2.13), respectively, with $\sqrt{g_x}\alpha_x$, $\sqrt{g_y}\alpha_y$, a/g_x, $b/\sqrt{g_x g_y}$ and c/g_y. Since g_x and g_y are positive constants, their appearance in Eq. (4.1) does not increase the difficulty of inverting $\tilde{P}$. However, their introduction into the current iterative procedure does have an effect on the convergence rate. As a result, the current iterative procedure can be accelerated considerably by a proper choice of g_x and g_y. Obviously, this scaling technique can also be used in the solution of PDE with variable coefficients as long as the scaling coefficients g_x and g_y remain positive constants.

5. NUMERICAL EVALUATION

To facilitate this discussion, we begin with the following preliminaries:

(a) the residual norm r(n) and error norm e(n) after n iterations are defined by:

$$r(n) \equiv [\sum_{i,j} (\tilde{L}(u^n_{i,j})-h_{ij})^2]^{1/2} \text{ and } e(n) \equiv [\sum_{i,j} (u^n_{i,j}-u^\infty_{i,j})^2]^{1/2} \quad (5.1)$$

where $u^\infty_{i,j}$ is the machine accuracy solution. Moreover, we define

$$O_r(n) \equiv -\log_{10} [r(n)/r(0)] \text{ and } O_e(n) \equiv -\log_{10} [e(n)/e(0)] \quad (5.2)$$

(b) We assume that $u^0_{i,j} = 0$ at all grid points where $u_{i,j}$'s are unknowns. As a result, $e(o) = \|u^\infty\|$ and $O_e(n)$ is the number of correct digits in $u^n_{i,j}$.

(c) The coefficient $E_{o,ij}$ defined by the VC version of Eq. (2.12) may be interpreted as an asymptotic error amplification factor. As a result, the value of $O_e(n)$ (and $O_r(n)$ if $O_r(n) \sim O_e(n)$) may be predicted by the parameter

$$O_t(n) \equiv -n \log_{10} (\text{the supremum of all } E_{o,ij}) \quad (5.3)$$

where $E_{o,ij}$ is evaluated assuming $a_{ij} = p_{ij}$, $b_{ij} = 0$ and $c_{ij} = q_{ij}$ if $L = L'(x,y)$. Alternatively, the value of $O_e(n)$ or $O_r(n)$ may be predicted by $\bar{O}_t(n) \equiv$ the limit of $O_t(n)$ as the grid cell size approaches zero.

(d) The domain for all numerical problems is assumed to be $1 \geq x \geq 0$ and $1 \geq y \geq 0$. Moreover, the iterative procedure is executed using a FDE code[8] with $g_x = g_y = 1$.

The first group of PDE to be studied in this section includes:

$$(1+2x^2+2y^2)\partial^2u/\partial x^2+(1+x^2+y^2)\partial^2u/\partial y^2=1 \quad (5.4)$$

$$(1+2x^2+2y^2)\partial^2u/\partial x^2+(1+x^2+y^2)\partial^2u/\partial x\partial y+(1+x^2+y^2)\partial^2u/\partial y^2=1 \quad (5.5)$$

$$\frac{\partial}{\partial x}[\{1+(x^2+y^2)^\ell\} \frac{\partial u}{\partial x}]+\frac{\partial}{\partial y}[\{2+(x^2+y^2)^\ell\} \frac{\partial u}{\partial y}]=1, \quad \ell=2,4,6,8 \quad (5.6)$$

Ten numerical problems associated with Eqs. (5.4)-(5.6) are defined in Table 1. The parameters MX and MY, respectively, are the number of grid intervals in the x and y directions. It is assumed that u = 0 at all boundaries.

Problems 1-3 are associated with Eq. (5.4). They differ only in the grid cell size and aspect ratio. As shown in Table 1, for all three problems, the value of $O_t(20)$ or $\bar{O}_t(20)$ provides a conservative estimate of $O_r(20)$ (since $O_e(n) \sim O_r(n)$, the values of $O_e(n)$ are not listed). It is also seen that the very large aspect ratio (16:1) of the grid cell in problem 3 does not cause any substantial reduction in convergence rate. Problems 4-6 represent a parallel study for Eq. (5.5). It is evident that the appearance of a cross derivative term in Eq. (5.5) causes an increase in the conservatism of the theoretical estimate made by $O_t(n)$ or $\overline{O}_t(n)$. This is particularly true if the grid cell size or aspect ratio is large. For problems 1-6, the convergence histories are closely represented by straight lines (as an example, see Fig. 1) as would be expected from our theoretical study.

Equations (5.6) belong to the class of PDE defined by Eq. (3.1). It is evident that the variation of the coefficients p and q increases progressively as one goes from $\ell = 2$ to $\ell = 4$ and so on. For $\ell = 8$, the increase in the values of p and q from one corner (x=y=0) to another corner (x=y=1) on the unit square is of the order of 100 times. As shown in Fig. 2, the current method is still useful in this extreme case. The robustness of this algorithm is most evident in its ability to reverse the trend toward divergence during the first few iterations.

The second group of PDE to be studied includes

$$\frac{\partial}{\partial x}[\{1+(x+y)^2\}^2\frac{\partial u}{\partial x}]+\frac{\partial}{\partial y}[\{1+\sin^2(x+y)\}^2\frac{\partial u}{\partial y}]=h_1(x,y) \tag{5.7}$$

$$\frac{\partial}{\partial x}[\{1+\tfrac{1}{2}(x^4+y^4)\}^2\frac{\partial u}{\partial x}]+\frac{\partial}{\partial y}[\{1+\tfrac{1}{2}(x^4+y^4)\}^2\frac{\partial u}{\partial y}]=h_2(x,y) \tag{5.8}$$

Here $h_1(x,y)$ and $h_2(x,y)$ along with the boundary values of u are chosen such that

$$u = u_1(x,y) \equiv \sin x \sin y \text{ and } u = u_2(x,y)\equiv[x(1-x)y(1-y)]^2 \tag{5.9}$$

respectively, are the exact solutions of Eq. (5.7) and (5.8). Both Eqs. (5.7) and (5.8) are solved numerically assuming MX=MY=16. Both numerical problems are test problems used by Bank[4]. Compared with the current value of $O_e(10) = 3.47$, the values obtained by Bank for Eq. (5.7) are 3.49 without using his scaling technique and 3.87, 4.81, and 6.76 using different scaling functions. Since the operator used by Bank is a general separable operator, the computation time required for its inversion is about six times that required for the current Poisson-like operator (p. 5 of ref. 1). As a result, the current method is at least a factor of 2 more efficient even without using our scaling technique. For Eq. (5.8), compared with the current value of $O_e(10) = 8.59$, the values reported by Bank are 5.88 without using scaling technique and 14.79 if a scaling function is used. Equation (5.8) was also solved by Concus and Golub[3] and the results are comparable with ours. However their method is applicable only when $p(x,y) = q(x,y)$ as in the case of Eq. (5.8).

6. APPLICATION TO A 3-D INTERNAL FLOW PROBLEM

In a numerical study (to be published) of 3-D incompressible shear flow in a turning channel (Fig. 3), the elliptic PDE

$$\partial^2u/\partial x^2+\partial^2u/\partial y^2+\eta(x,y)\partial^2u/\partial z^2 = h, \quad (2\geq x\geq -2,\ 0.75\geq y\geq 0.65,\ 0.1\geq z\geq 0) \tag{6.1}$$

is solved in computational space. Here h is a known source term and

$$\eta(x,y) = (\cosh(\pi x) + \cos(\pi y))/(\cosh(\pi x) - \cos(\pi y)) > 0 \tag{6.2}$$

It is assumed that $u = 0$ at $x = 2$ and the normal derivative of u vanishes at other boundaries. Since Eq. (6.1) has a form specified by Eq. (1.1) and the 3-D version of Eq. (2.1), it was solved numerically using the current iterative method with a 144 x 12 x 12 uniform grid. To accelerate the convergence, the scaling coefficients of the 3-D operator P (see Eq. (4.1)) are given by

$$g_x = g_y = 1,\ g_3 = \sqrt{\eta_{max}\bullet\eta_{min}} \doteq 0.4135 \tag{6.3}$$

where η_{max} (=0.9966) and η_{min} (=0.1716), respectively, are the maximum and the minimum of η in the domain of Eq. (6.1). It should be noted that the maximum value of $\overline{O}_t(1)$ (= 0.415) is reached if the values of g_x, g_y, and g_z are chosen according to Eq. (6.3).

Equation (6.1) was numerically solved many times with different source terms h. Without exception, it was found that, after four iterations, the slope of $O_r(n)$ vs n curve is predicted by the value of $\bar{O}_t(1)$ to within ten percent.

REFERENCES

1. Hockney, R. W.: "Rapid Elliptic Solvers," Numerical Methods in Applied Fluid Dynamics edited by B. Hunt, Academic Press, 1-48 (1980).
2. D'yakonove, E. G.: Dokl. Akad, Nauk, SSSR, 138, 522-525 (1961).
3. Concus, P. and Golub, G. H.: SIAM J. Numer. Anal., 10, 1103-1120 (1973).
4. Bank, R. E.: SIAM J. Numer. Anal., 14, 950-970 (1977).
5. Chang, S. C.: "Generalizations of Two Inequalities Involving Hermitian Forms," accepted for publication in Linear Algebra and Its Applications.
6. Smith, G. D.: "Numerical Solution of Partial Differential Equations," Oxford University Press. p. 29 (1978).
7. Hageman, L. A. and Young, D. M.: "Applied Iterative Methods," Academic Press, p. 12 (1981).
8. Adams, J., Swartztrauber, P. and Sweet, R.: "FISHPAK: A Package of FORTRAN Subprograms for the Solution of Separable Elliptic Partial Differential Equations, Version 3," National Center for Atmospheric Research, Boulder, Colorado, June 1979.

Table 1. (n=20 for Probs. 1-3, n=32 for Probs. 4-10)

Prob. No.	Eq.No.	ℓ	MX	MY	$O_r(n)$	$O_t(n)$	$\bar{O}_t(n)$
1	(5.4)	NA	16	16	14.5	12.3	12.0
2	(5.4)	NA	64	64	13.8	12.1	12.0
3	(5.4)	NA	64	4	14.0	12.7	12.0
4	(5.5)	NA	16	16	14.4	9.7	9.6
5	(5.5)	NA	64	64	12.6	9.6	9.6
6	(5.5)	NA	64	4	19.0	10.0	9.6
7	(5.6)	2	16	16	17.2	15.3	15.3
8	(5.6)	4	16	16	16.8	15.3	15.3
9	(5.6)	6	16	16	13.4	15.3	15.3
10	(5.6)	8	16	16	7.6	15.3	15.3

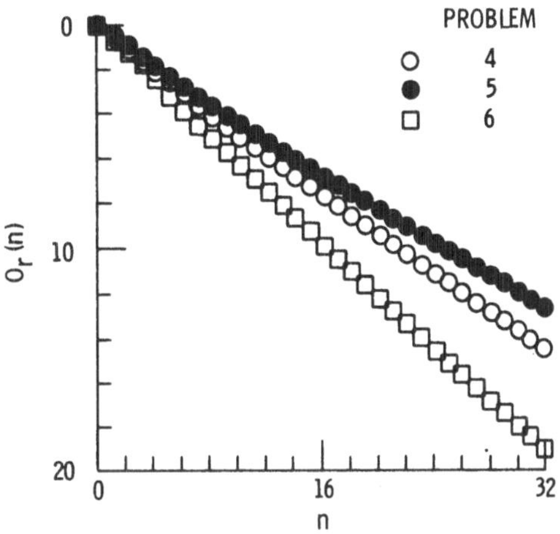

Figure 1. - Convergence histories of Problems 4-6.

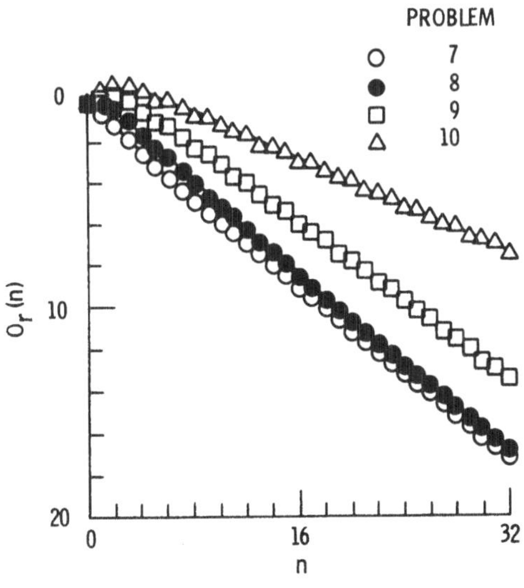

Figure 2. - Convergence histories of Problems 7-10.

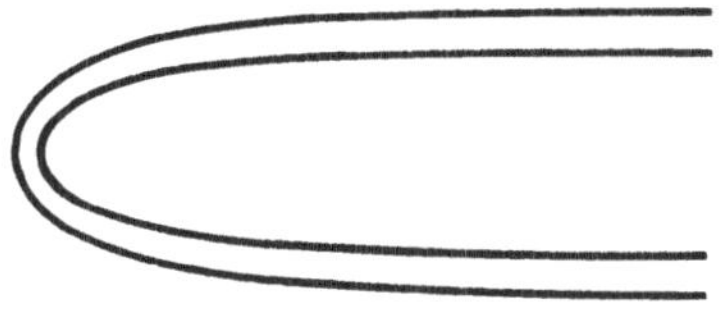

Figure 3. - A converging-diverging turning channel in physical space.

VISCOUS COMPUTATION OF A SPACE SHUTTLE FLOW FIELD

D. S. Chaussee, Y. M. Rizk, and P. G. Buning
NASA Ames Research Center
Moffett Field, CA 94035 USA

I. INTRODUCTION

Recent research efforts [1-3] have confirmed the ability of the "parabolized" Navier-Stokes (PNS) codes to predict accurately and rapidly the aerothermodynamics of the actual Space Shuttle Orbiter up to an axial station that is 50% of the overall length. This corresponds to a location where the strake ends and the 45° swept wing begins. In the past, the geometry of the Orbiter usually has been modified [4-7] such that a solution over the complete body was possible. These modifications take the form of changing the sweep angle of the wing, removing the canopy, and altering the lee of the body so that the cross section is composed of two elliptical shapes. In one instance [8], an inviscid solution was obtained for the complete Orbiter. In order to perform this calculation, a "fix" had to be used in the vicinity of the bow-shock/wing-shock interaction region. Physically, what occurs is a region of embedded subsonic flow owing to the bow-shock/wing-shock interaction which causes the resulting coalescing shock wave to be more normal to the flow for a short streamwise distance. Since the above inviscid procedure was a marching code, it failed when the Mach number in the marching direction became subsonic.

A viscous numerical procedure is described, to compute the flow over the Shuttle. Results are presented that demonstrate the capability of the method. Obtainment of these results requires the use of two computer codes. A PNS code [9,10] is used to obtain the solution up to the bow-shock/wing-shock interaction region, and an unsteady continuation code is used for the region after the shock interaction. The unsteady Navier-Stokes code [11] is also used to obtain the blunt-body starting solution. Only results from the marching code will be presented. For the flow conditions calculated, that is, $M_\infty = 7.9$, $\alpha = 25°$, $T_{wall} = 540°R$, $Re_L = 60{,}728/in.$, laminar or turbulent, the PNS code has been marched up to an $X/L = 0.7$ which is where the bow-shock/wing-shock interaction region occurs. In this work, L refers to the length of the vehicle.

II. COMPUTATIONAL TECHNIQUE

The PNS equations are obtained from the complete Navier-Stokes equations by neglecting the unsteady terms and the streamwise viscous derivative terms. The complete details of all the terms and derivations can be found in Ref. 12.

In the present formulation, ξ (the marching direction in computational space) is a function of x only (axis-normal marching planes). The governing equations are hyperbolic-parabolic in this ξ-direction if the inviscid part of the flow field is supersonic, if there is no streamwise (axial) separation, and if the pressure gradient in the viscous region near the wall is treated correctly. However, the system of equations still allows for the separation in the crossflow plane (η-ζ).

The present PNS code uses the Beam-Warming implicit algorithm to update the interior of the region and characteristic, implicit, spatially second-order-accurate boundary conditions at the outermost shock wave. An elliptic grid generator of the type developed by Steger and Sorenson [13] and further specialized to wing bodies by Rai et al. [10] is used to generate the grid for the calculations.

If the conditions in a particular region are such that the marching procedure is invalidated, the unsteady Navier-Stokes (UNS) code is used for these regions. In calculating the flow over the Space Shuttle, one such region occurs in the vicinity of the bow-shock/wing-shock interaction (a pocket of subsonic flow is encountered).

The complete details of the UNS code can be found in Ref. 11. The UNS code is extremely versatile and relatively easy to use. It uses either a Beam-Warming implicit algorithm or a hybrid scheme due to Rizk and Chaussee [11]. The outer shock wave is either fitted or captured. Usually the initial guess is furnished by the PNS code, which is modified in some manner to march through regions where it would not march before. This procedure is acceptable, since the unsteady code takes this reasonable guess and iterates in time until a steady-state solution is obtained.

The domain of this unsteady calculation encompasses the subsonic flow. The outflow boundary consists entirely of supersonic axial flow in the inviscid part of the flow field. This permits the PNS code, which is more efficient, to continue marching from this point.

III. RESULTS

Numerical results have been obtained for the following wind-tunnel conditions: $M_\infty = 7.9$, $\alpha = 25°$, $T_{wall} = 540°R$, $Re_L = 60{,}728/in.$ turbulent flow. For this calculation, the Shuttle surface coordinates were obtained from Rockwell-International Corporation. The current geometry consists of the complete Shuttle; the canopy, OMS pods, and the vertical stabilizer are included.

The three-dimensional blunt-body code originally developed by Kutler et al. [14] was used to obtain the blunt-nose solution which creates the necessary starting planes for the PNS code at $X/L = 0.0522$. This solution was then marched downstream using the elliptic grid generator to construct the grid between the body and the fitted outermost shock wave. The grid consisted of either 61 or 121 points in the meridional direction and 45 geometrically stretched radial points. An example of the grid at an $X/L = 0.66$ is shown in Fig. 1. The outermost grid line is the bow wave, which is fitted using an implicit technique.

The pressure contours in the region of the canopy are presented in Figs. 2 and 3. In Fig. 2, the contours on the lee pitch plane of symmetry between $X/L = 0.067$ and $X/L = 0.4$ are shown. In the canopy region, the coalescence of contours details the canopy shock wave followed by an expansion wave on the lee of the Shuttle. The contours for a cross section at an $X/L = 0.2$ are presented in Fig. 3. The canopy shock is once again viewed at the point on the lee where the pressure contours coalesce. The expansions which are visible on the windward are due to discontinuities in geometry.

The Mach number contours at an X/L = 0.66 are presented in Fig. 4. The main features are the wing shock and the crossflow shocks on the wing and upper body, respectively. These are denoted by the coalescence of the Mach contours.

In Fig. 5, the crossflow velocity vectors are presented at an X/L = 0.66. Two interesting features seen in this figure are the recirculation region in the wing-body juncture and the lee vortex.

The density contours in the vicinity of the wing tip at an X/L = 0.667 are shown in Fig. 6. The wing shock and the bow shock have interacted as characterized by the bulge in the outer boundary. This is due to the wing shock becoming the outermost surface, with the bow shock being captured. The bow shock appears as the coalescence of the density contours near the outer surface.

By numerically simulating oil flow on the surface of the vehicle, as in Fig. 7, many interesting features are observed. The lines of separation on both the strake-wing and the lee of the body are evident by the coalescence of the numerical oil flow. The reattachment line is visible on the Shuttle as a series of oil-flow lines diverging toward the separation lines.

The computer-generated particle paths of Fig. 8 exhibit the same trends in the flow field that are visible on the Shuttle surface via the oil flow. Specific features are the vortices on the lee which are due to the strake-wing. At this angle of attack, the vortices that are generated on the wing impact on the OMS pod.

IV. SUMMARY

A procedure has been presented for calculating the flow over vehicles that have embedded regions of subsonic flow in the inviscid part of the flow field. A PNS marching code is used to obtain the solution up to the bow-shock/wing-shock interaction region. In this interaction region, the UNS code can be employed since the region has a pocket of subsonic flow. Currently, only the results for the marching code up to an X/L = 0.667 are included. In the future, the results for the bow-shock/wing-shock region will be available.

V. REFERENCES

1. Venkatapathy, E., Rakich, J. V., and Tannehill, J. C., "Numerical Solution of Supersonic Viscous Flow over Blunt Delta Wings," AIAA Paper 82-0028, 1982.
2. Prabhu, D. K. and Tannehill, J. C., "Numerical Solution of Space Shuttle Orbiter Flow Field Including Real Gas Effects," AIAA Paper 84-1747, 1984.
3. Balakrishnan, A., "Computation of Viscous Real Gas Flow Field for the Space Shuttle Orbiter," AIAA Paper 84-1748, 1984.
4. Li, C. P., "Application of an Implicit Technique to the Shock-Layer Flow Around General Bodies," AIAA Journal, Vol. 20, 1982, p. 175.
5. Szema, K. Y., Griffith, B. J., Maus, J. R., and Best, J. T., "Laminar Viscous Flow Field Prediction of Shuttle-like Vehicle Aerodynamics," AIAA Paper 83-0211, 1983.
6. Weilmuenster, K. J., "High Angle of Attack Inviscid Flow Calculations Over a Shuttle-like Vehicle with Comparisons to Flight Data," AIAA Paper 83-1798, 1983.

7. Weilmuenster, K. J. and Hamilton, H. H., "Calculations of Inviscid Flow Over Shuttle-like Vehicles at High Angles of Attack and Comparisons with Experimental Data," NASA TP-2103, 1983.
8. Chaussee, D. S., Kutler, P., and Holtz, T., "Inviscid Supersonic/Hypersonic Body Flow Field and Aerodynamics from Shock-Capturing Technique Calculations," Journal of Spacecraft & Rockets, Vol. 13, 1976, pp. 325-331.
9. Rai, M. M. and Chaussee, D. S., "New Implicit Boundry Procedures: Theory and Applications," AIAA Paper 83-0123, 1983.
10. Rai, M. M., Chaussee, D. S., and Rizk, Y. M., "Calculation of Viscous Supersonic Flows over Finned Bodies," AIAA Paper 83-1667, 1983.
11. Rizk, Y. M. and Chaussee, D. S., "Three-Dimensional Viscous-Flow Computations Using a Directionally Hybrid Implicit-Explicit Procedure," AIAA Paper 83-1785, 1983.
12. Schiff, L. B. and Steger, J. L., "Numerical Simulation of Steady Supersonic Viscous Flow," AIAA Paper 79-0130, 1979.
13. Steger, J. L. and Sorenson, R. L., "Automatic Mesh-Point Clustering Near a Boundary in Grid Generation with Elliptic Partial Differential Equations," Journal of Computational Physics, Vol. 33, 1979, pp. 405-410.
14. Kutler, P., Pedelty, J. A., and Pulliam, T. H., "Supersonic Flow Over Three-Dimensional Ablated Nosetips using an Unsteady Implicit Numerical Procedure," AIAA Paper 80-0063, 1980.

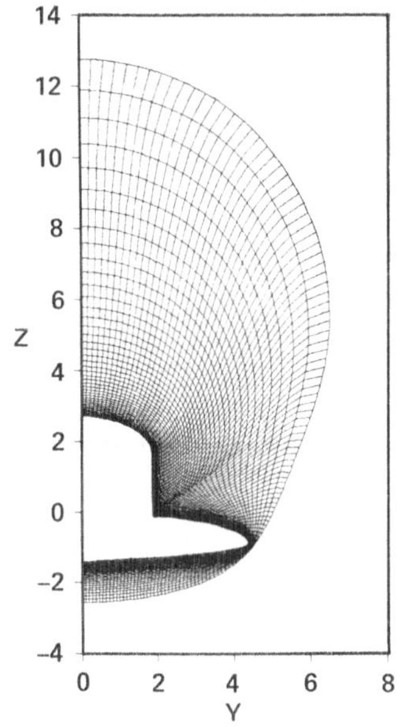

Fig. 1 Elliptic grid at X/L = 0.66.

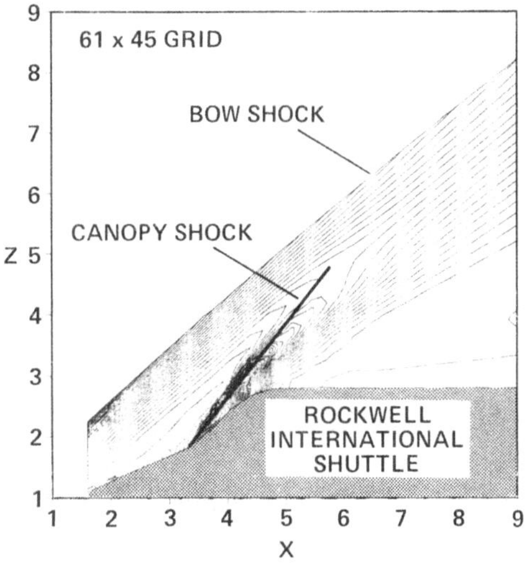

Fig. 2 Pressure contours in the lee pitch plane of symmetry.

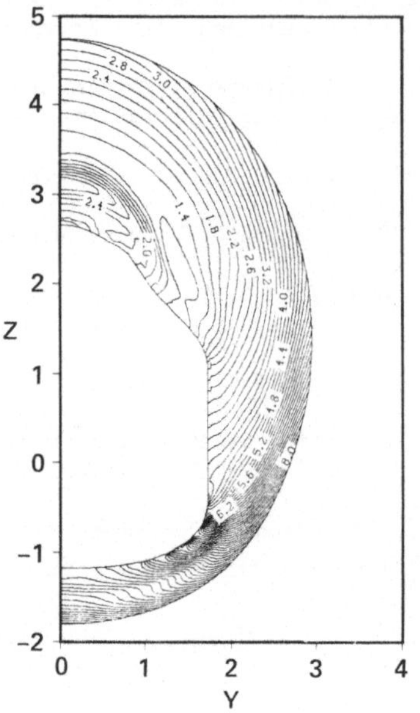

Fig. 3 Pressure contours at X/L = 0.2.

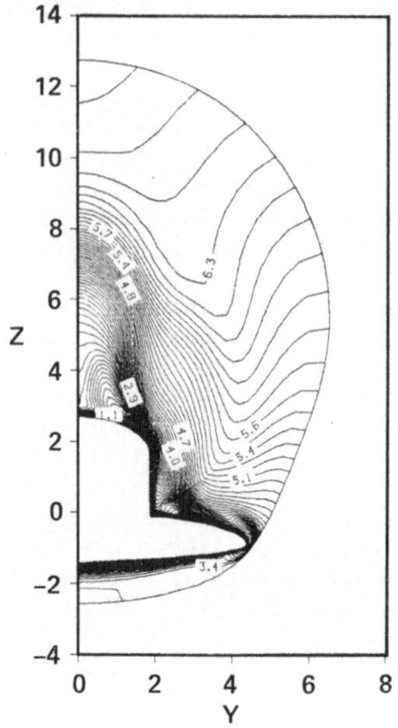

Fig. 4 Mach contours at X/L = 0.66.

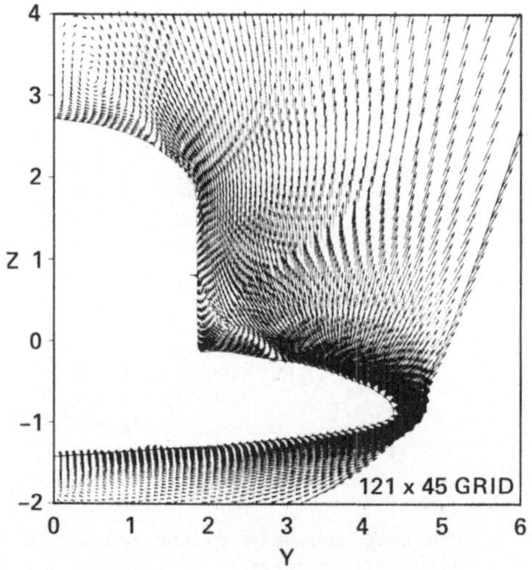

Fig. 5 Crossflow velocity vectors at X/L = 0.66.

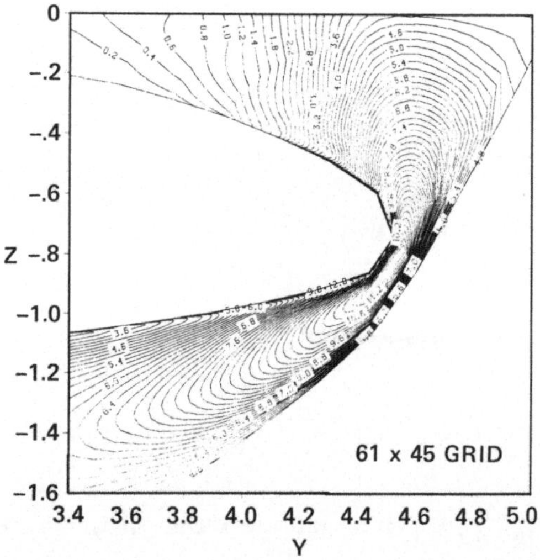

Fig. 6 Density contours near the wing tip at X/L = 0.667.

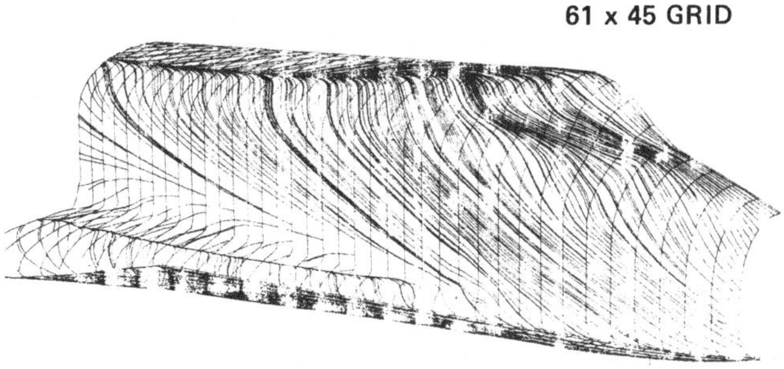

Fig. 7 Computational oil flow on the Shuttle surface up to X/L = 0.66.

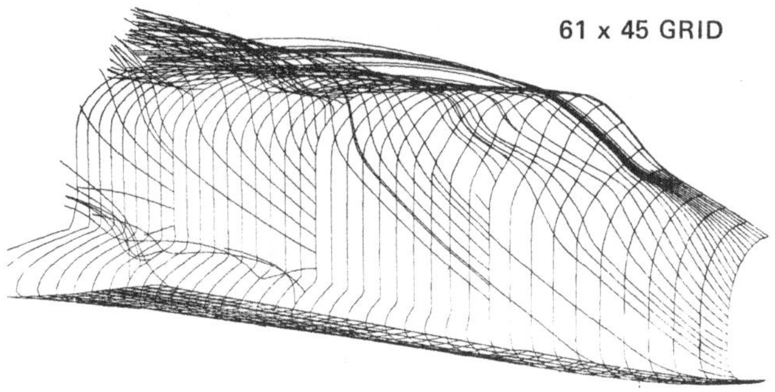

Fig. 8 Computational particle paths up to X/L = 0.66.

Numerical Calculation of Complex Shock Reflections in Gases

Phillip Colella
Lawrence Berkeley Laboratory
University of California
Berkeley, California 94720

Harland M. Glaz
Naval Surface Weapons Center
White Oak Laboratory
Silver Spring, Maryland 20910

We present here computational results using second order Godunov methods for time-dependent Eulerian gas dynamics with a general (convex) equation of state. The algorithm used in most of the calculations is described in detail in [3]. An unsplit scheme was used in conjunction with local adaptive mesh refinement for the results in Figure 3; see [1] and the references cited there for the construction of this version of the scheme.

For the well-known problem of planar shock wave diffraction by a wedge, a direct comparison of computational results and the experimental record has been completed for several different cases, [5]. Reproduced in Figure 1 are the results for a shock wave Mach number $M_s = 2.03$ and wedge angle $\theta = 27°$. The experimantal flowfield for this case of single Mach reflection is in equilibrium and viscous effects are localized at the wedge corner and the contact surface-boundary layer interaction. The infinite-fringe interferogram is shown in Fig. 1a, the numerical results with the same isopycnics (i.e., constant density lines) in Fig. 1b, and the wall density values are compared in Fig. 1c. It is seen that excellent agreement is obtained in all respects.

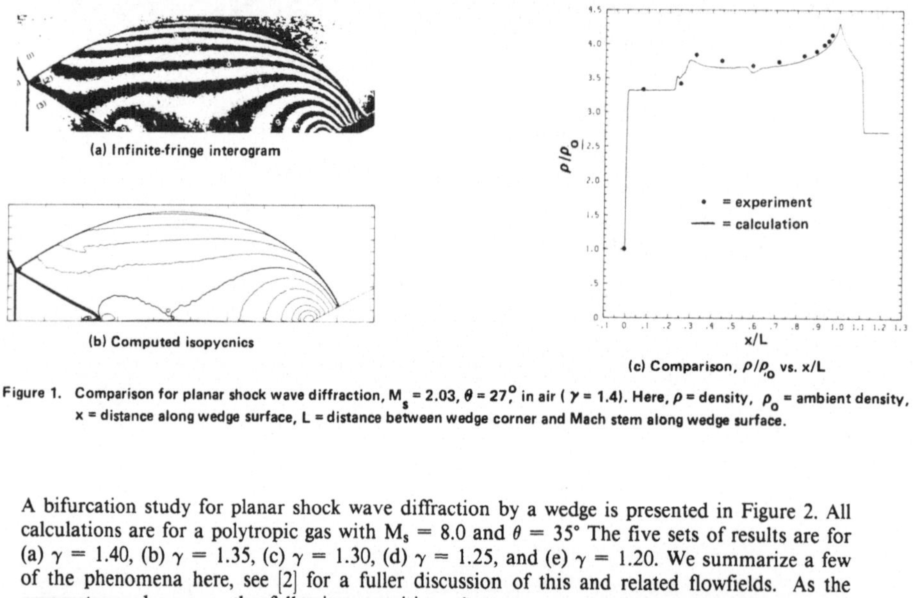

(a) Infinite-fringe interogram

(b) Computed isopycnics

(c) Comparison, ρ/ρ_o vs. x/L

Figure 1. Comparison for planar shock wave diffraction, $M_s = 2.03$, $\theta = 27°$, in air ($\gamma = 1.4$). Here, ρ = density, ρ_o = ambient density, x = distance along wedge surface, L = distance between wedge corner and Mach stem along wedge surface.

A bifurcation study for planar shock wave diffraction by a wedge is presented in Figure 2. All calculations are for a polytropic gas with $M_s = 8.0$ and $\theta = 35°$ The five sets of results are for (a) $\gamma = 1.40$, (b) $\gamma = 1.35$, (c) $\gamma = 1.30$, (d) $\gamma = 1.25$, and (e) $\gamma = 1.20$. We summarize a few of the phenomena here, see [2] for a fuller discussion of this and related flowfields. As the parameter γ decreases, the following transition phenomena occur in the double Mach stem flowfield: (1) the leading Mach stem "toes out" and then forms a new triple point ((c), T_1); for those values of γ, the flow is supersonic in a coordinate system moving with the self-similar solution, and the slip line from that triple point is swept up in the highly rotational flow brought about by the curvature of the shock below T_1, (2) the expansion and compression waves in the supersonic region between the main slip surface and the reflected shock ((d), R) become stronger, until the cross-flow becomes supersonic, causing the compression wave to form into a shock ((d), S_R), (3) the flow inside the jet along the wall becomes transonic, and a backwards facing shock forms at the wall ((e), S_w).

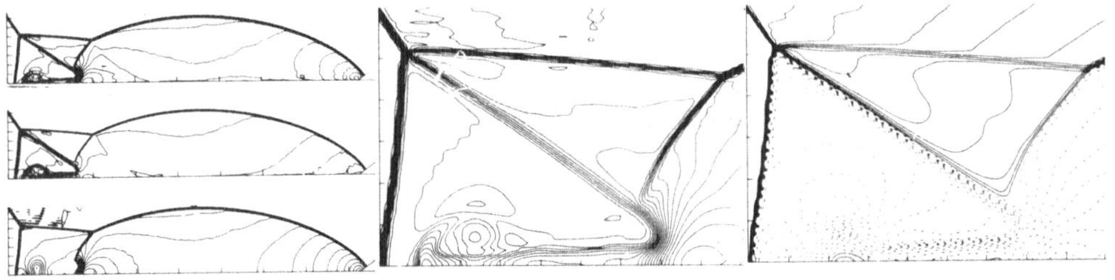

(a) $\gamma = 1.40$; (ρ, e, p) whole flowfield; $(\rho, \tilde{M})$ Mach stem region

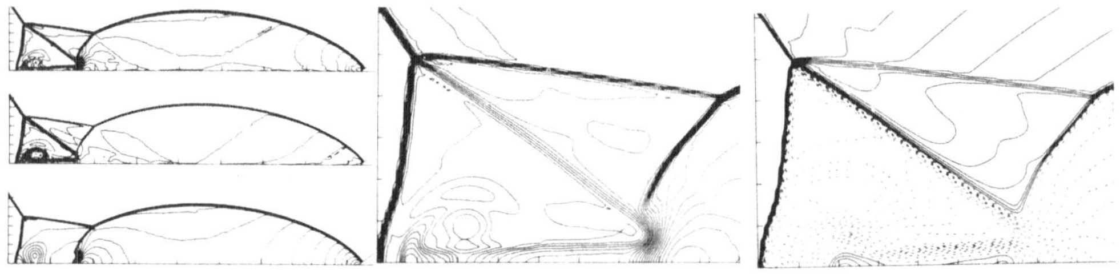

(b) $\gamma = 1.35$; (ρ, e, p) whole flowfield; $(\rho, \tilde{M})$ Mach stem region

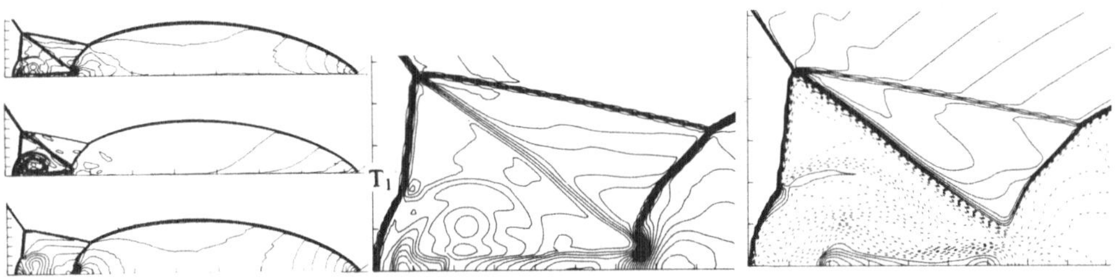

(c) $\gamma = 1.30$; (ρ, e, p) whole flowfield; $(\rho, \tilde{M})$ Mach stem region

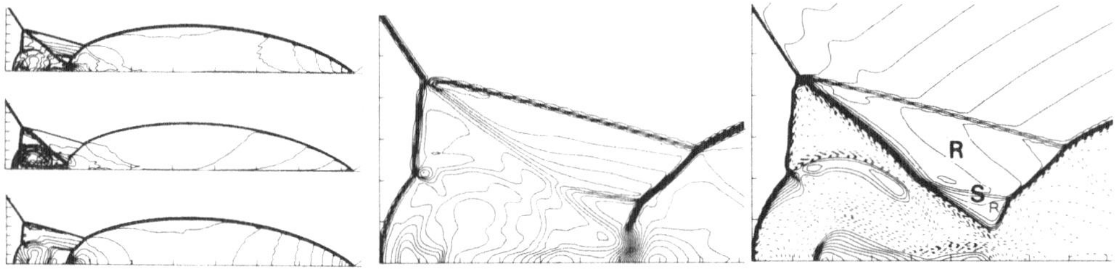

(d) $\gamma = 1.25$; (ρ, e, p) whole flowfield; $(\rho, \tilde{M})$ Mach stem region

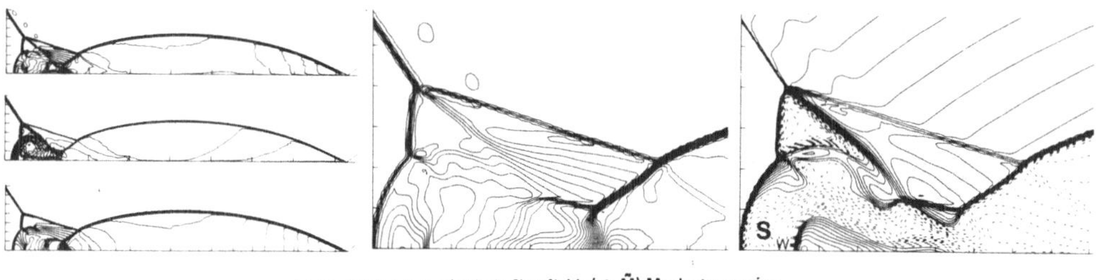

(e) $\gamma = 1.20$; (ρ, e, p) whole flowfield; $(\rho, \tilde{M})$ Mach stem region

Figure 2. Bifurcation study for planar shock wave diffraction, $M_s = 8.0$, $\theta = 35.^\circ$ Here, ρ = density, e = internal energy, p = pressure, $\tilde{M}$ = self-similar Mach number, where $\tilde{M}^2 = (\tilde{u}^2+\tilde{v}^2)/C^2$ and $\tilde{u} = u - \xi$, $\tilde{v} = v - \eta$, $(\xi, \eta) = (x/t, y/t)$ with the origin of coordinates at the wedge corner. In the plots of $\tilde{M}$, solid (dashed) lines correspond to $\tilde{M} > 1$ $(\tilde{M} < 1)$.

Although it would be possible to continue the series of calculations shown in figure 2 to lower values of γ, the problem is well-suited as an application of the mesh refinement algorithm proposed in [1] for use with the second-order Godunov schemes. Results for the case $M_s = 8.0$, $\theta = 35°$, $\gamma = 1.15$ are shown in Figure 3. For this case, the rarefaction fan emanating from the main contact surface impinges on the reflected shock, and creates a differential in the oblique jump conditions across this wave. As the figure indicates, this wave develops a new (but relatively weak) triple point; the overall configuration may be considered a "triple Mach stem".

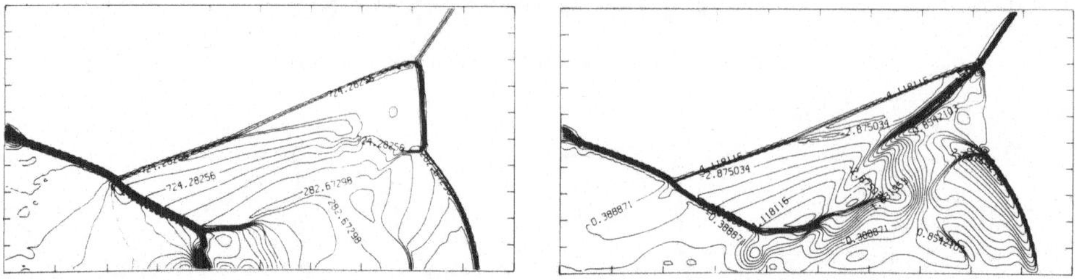

(a) Pressure   (b) Vertical velocity

Figure 3. Planar shock wave diffraction, $M_s = 8.0, \theta = 35^{o}, \gamma = 1.15$. The double Mach stem region is shown and it corresponds to the locally refined (fourfold relative to the coarse mesh) region; the grid shown is 208 x 104 points.

The flowfield resulting from the detonation of an 8 lb. sphere of chemical explosive (PBX-9404) 51.7 cm. above the reflecting surface is presented in Figure 4. Our calculation is hydrodynamic with the initial data provided by a similarity analysis of the detonation process (see [4]). The calculation is performed in cylindrical coordinates using a rectangular moving grid. the entire-grid is 100 cm. x 20 cm. and the fine grid is 8 cm. x 4 cm. The fine grid contains the point at which the incident shock wave intersects the surface. The coarse grid to fine grid ratio was 10 for this calculation. Transition from regular to double Mach reflection takes place in the calculation when the incident shock reaches about 57 cm. Since this is infinitesimal phenomenon, we expect a somewhat earlier transition for the grid strategies in [1],[2]. The similarities between the double Mach stem flowfield and the flowfields in Figure 2 may be noted.

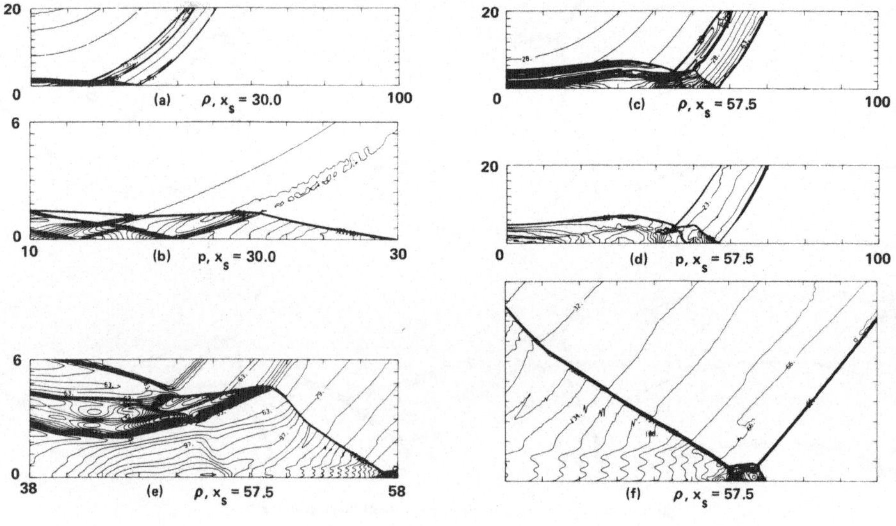

Figure 4. Continued on next page.

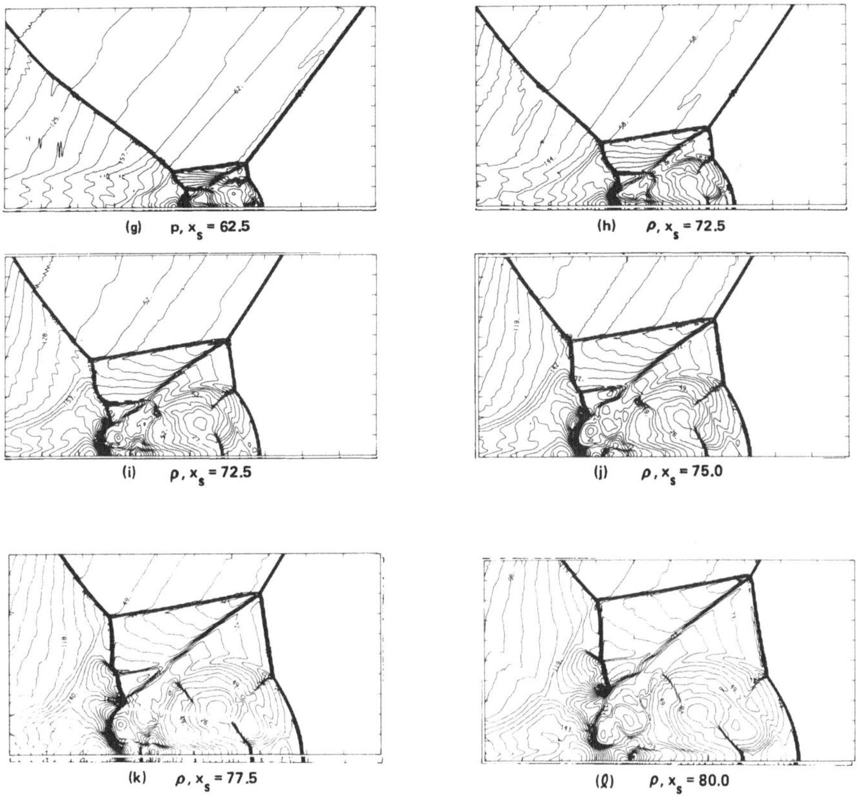

Figure 4. Blast wave flowfield. Plots (a), (c), (d) represent the entire grid; plots (b), (e) are 20 cm x 6 cm blowups; plots (f) - (ℓ) represent the entire 8 cm x 4 cm find grid. Here, ρ = density, p = pressure, x_s = location of intersection of incident shock and the surface. The grid contains 617 x 214 zones with 266 x 140 zones in the fine grid.

The interaction of a planar shock with a five zone thick layer of heated gas located five zones above a reflecting surface is shown in Figure 5. The gas is assumed polytropic with $\gamma = 1.4$. The solution is symmetric with respect to the centerline of the layer until waves from the lower boundary of the layer reach the surface. The interaction of the resulting reflected waves with the symmetric rollup leads to instabilities and the production of counterrotating vortices at later times.

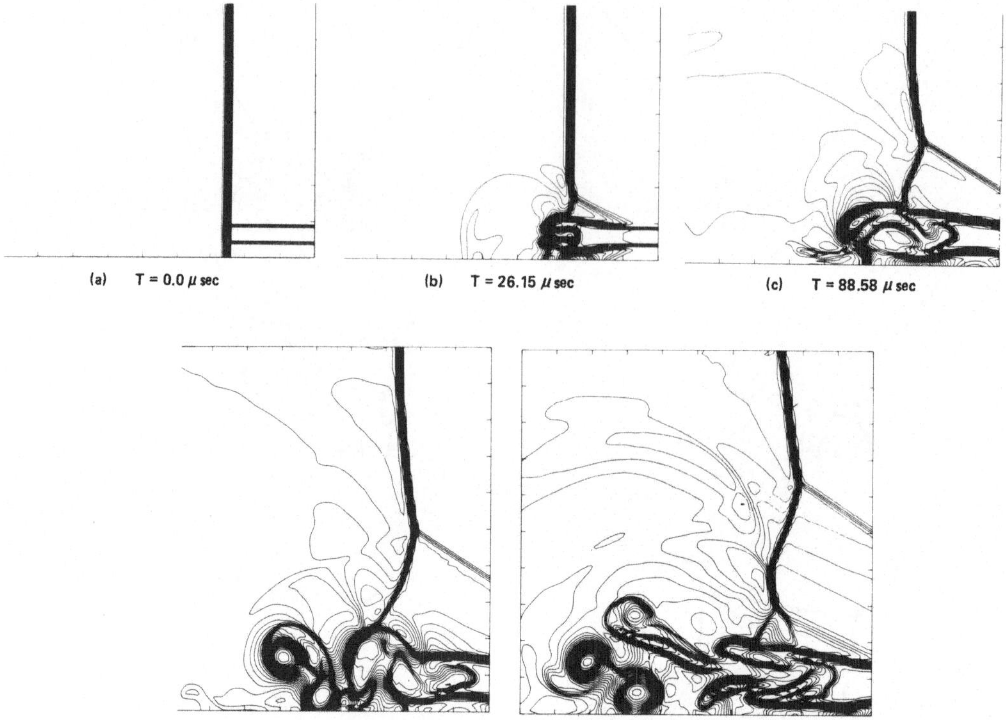

(a) T = 0.0 μsec (b) T = 26.15 μsec (c) T = 88.58 μsec

(d) T = 150.93 μsec (e) T = 217.86 μsec

Figure 5. Planar shock wave interaction with a layer of heated gas. The shock faces towards the right, T = 273°K in the ambient gas on the right above and below the heated layer, for which T = 2124°K. Pressure is atmospheric everywhere on the right and Δp = 30 bars across the shock ($P_L/P_R \sim 3$). $\Delta x = \Delta y \cong 0.05$ cm everywhere. Denisty contours are shown in the vicinity of the incident shock wave.

Acknowledgements. Discussions with M. Berger, I.I. Glass, and A. Kuhl are gratefully acknowledged. Portions of this work have been in collaboration with M. Berger, R. Deschambault, R. Ferguson, I.I. Glass, W. Glowacki, and A. Kuhl. The authors' work was supported by the U.S. Department of Energy at the Lawrence Berkeley Laboratory under contract DE−AC03−76SF00098, by the U.S. Defense Nuclear Agency under DNA task code Y99QAXSG, and by the Naval Surface Weapons Center Independent Research Fund.

References

1. Berger, M. and Colella, P. "Local Adaptive Mesh Refinement for Shock Hydrodynamics", forthcoming.
2. Berger, M., Colella, P., and Glaz, H.M., "Wave Bifurcations for Self-Similar Two-Dimensional Shocked Flows", forthcoming.
3. Colella, P. and Glaz, H.M., "Efficient Solution Algorithms for the Riemann Problem for Real Gases", J. Comput. Phys., in press.
4. Colella, P. , Ferguson, R., Glaz, H.M., and Kuhl, A., "Reflection of a Spherical, HE Driven Blast Wave from a Plane Surface", forthcoming.
5. Glaz, H.M., Colella, P., Glass, I.I., and Deschambault, R., "A Numerical Study of Oblique Shock Wave Reflections, with Experimental Comparisons", submitted for publication.

BOUNDARY LAYER MODELLING IN A NUMERICAL WEATHER PREDICTION MODEL

Jean Côté and Robert Benoit
Division de Recherche en Prévision Numérique, Dorval, P.Q., Canada

1. INTRODUCTION

The effectiveness of turbulence to transport momentum, heat and moisture makes it important to take into account the effect of the planetary boundary layer in a numerical weather prediction model. However there is a considerable discrepancy between the scales of interest in meteorology and those of atmospheric turbulence, and therefore a detailed description of the turbulence is out of the question. One therefore tries to model the effect of the turbulence on the large scale flow.

This paper is concerned with the introduction of a turbulence sub-model into the regional weather forecasting model soon to become operational at the Canadian Meteorological Center. Given that this model should run daily in quasi real-time the stability and the efficiency of the model will be highly emphasized.

To insure stability the forecast models usually run with time steps of about half an hour which renders the time truncation error much smaller than other errors. Consequently the time integration algorithms of the boundary layer model and the coupling with the host model were chosen so as not to introduce further constraints on the size of the time step.

2. THE MODEL

The large scale dynamical model is the finite-element regional model of Staniforth and Daley (1979). This model uses a polar stereographic grid in the horizontal and sigma, the ratio of pressure to surface pressure, as the vertical coordinate. The dependant variables are expanded in piece-wise linear finite-elements in each dimension. The differential operations and the products are performed using the Galerkin method.

The time integration uses the semi-implicit algorithm of Robert (1972). The stability constraint on the time step with this method is the CFL criterion due to the explicit treatment of the advection terms.

The turbulence sub-model is a 3D version of a finite-element model developed by Mailhot and Benoit (1982). The turbulence in this model is represented by the turbulent kinetic energy and the eddy mixing length which both satisfy 3D evolution equations. The various vertical transport or diffusion terms associated with the turbulence are then parametrized in terms of these quantities.

The model also assumes a surface layer between the ground and the first level (anemometer level). The state of the surface layer is diagnozed from the anemometer level fields (wind, temperature, moisture) and the surface fields (temperature, moisture, roughness). Continuity of the fluxes across the surface layer provides the set of boundary conditions needed to solve the diffusion equations. The integration of the evolution equations is performed using unconditionally stable methods and the time step is the same as that of the host model.

From the point of view of the large-scale host model, the boundary layer sub-model simply provides the vertical diffusion operators for the wind, the temperature and the moisture fields, along with appropriate boundary conditions. To integrate the resulting advection-diffusion equations the following split scheme is adopted:

(i) perform the usual semi-implicit step to obtain intermediate estimates of wind, temperature and moisture;
(ii) perform a fully implicit diffusion step to obtain the final values.

The interesting property of this algorithm is that the time step restriction remains the CFL criterion associated with the advection part of (i). The drawback is that the implicit (linear) diffusion step involved requires the solution of a not-so-sparse 3D set of linear equations. This is achieved using the generalised conjugate gradient method of Concus et al (1976). Since this method has a fast initial rate of convergence and the diffusion term is a small perturbation, it is found that 4 iterations are sufficient for a reasonably accurate solution.

3. RESULTS

The results are from an 18 hour forecast starting at 12 GMT on April 10 1979 with a time-step of 400 sec. The variable-resolution grid was 65 x 49 in the horizontal with a minimum spacing of 100 km centered over the American Midwest and 15 unequally spaced vertical levels, 10 of them below 3 km. This case was chosen because of the interesting meteorological evolution (a major tornado outbreak during the forecast period) and the availability of accurate data. All the results are for April 11 0 GMT.

Figs. 1 and 2 show a vertical cut of the potential temperature and specific humidity respectively, along a line extending approximately from El Paso to Nashville. Both exhibit a well defined mixed-layer. Note the maximum surface temperature and the minimum surface specific humidity occuring over the desert of New-Mexico. Fig. 3 shows the streamlines (dash) and the isolines of the y component (V) of the wind (contour interval, 10 knots) at 850 millibars. We see a strong jet bringing the moist air from the Gulf of Mexico over the Plains. This verifies well against observations as seen in Fig. 4, where we have plotted the predicted profiles of wind and temperature (bottom) and the observations (top) at Little Rock. Finally in Fig. 5 we show an empirical index of stability related to the occurrence of severe weather. Both its magnitude and structure are well correlated with the reported tornadoes and storms during the 24 h period starting at 12 GMT 10 April, as seen in Fig. 6 (from Hill et al.).

4. CONCLUSIONS

We have presented the coupling of a sophisticated boundary-layer model to a large-scale weather forecast model and verified the stability of the coupling strategy. The meteorological variables are realistically forecast in the planetary boundary-layer. Work is continuing towards operational implementation of the model for short-range weather forecasting.

REFERENCES

Concus, P., G.H. Golub and D.P. O'Leary, 1976: A generalized conjugate gradient method for the numerical solution of elliptic partial differential equations. Rep. No. STAN-CS-76-533, Computer Science Department, Stanford University.

Hill, K., G.S. Wilson and R.E. Turner, 1979: NASA's participation in the AVE-SESAME'79 Program, Bull. Am. Meteorol. Soc., 60, 1323-1329.

Mailhot J. and R. Benoit, 1982: A finite-element model of the atmospheric boundary layer suitable for use with numerical weather prediction models. J. Atmos. Sci., 39, 2249-2266.

Robert, A.J., J. Henderson and C. Turnbull, 1972: An implicit time integration scheme for baroclinic models of the atmosphere. Mon.Wea.Rev., 100, 329-335.

Staniforth, A.N. and R.W. Daley, 1979: A baroclinic finite-element model for regional forecasting with the primitive equations. Mon.Wea.Rev., 107, 107-121.

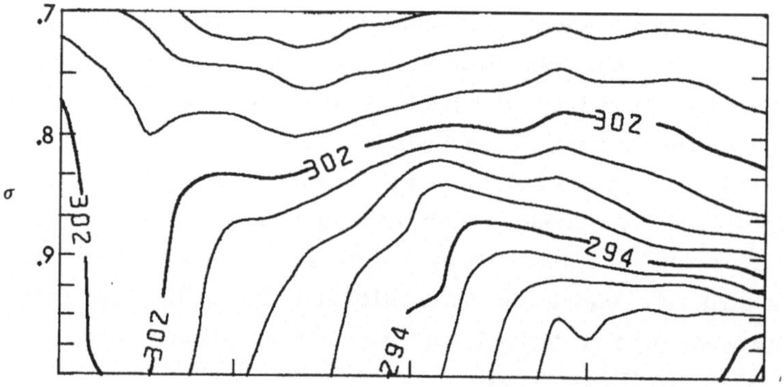

Fig. 1 Potential temperature (°K)

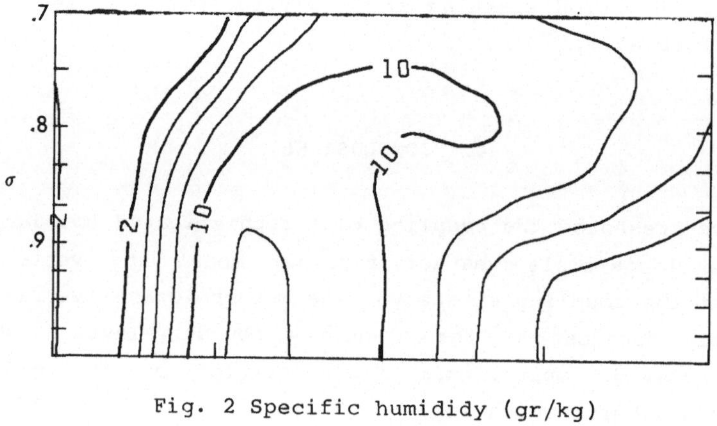

Fig. 2 Specific humididy (gr/kg)

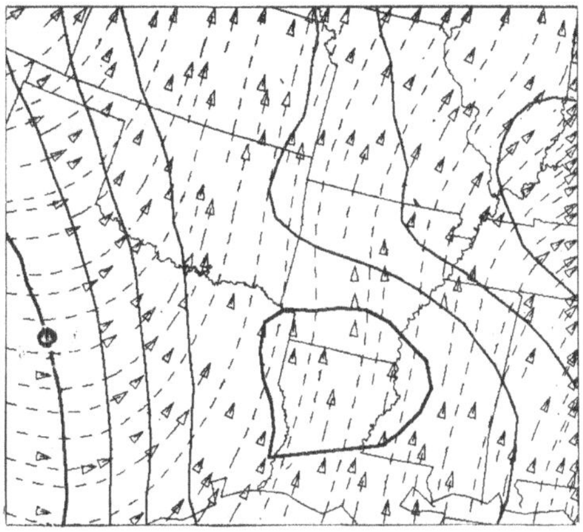

Fig. 3
Streamlines and V-isolines (knots)
at 850 millibars

Fig. 4
Wind (m/s) and potential temperature
(°C) at Little Rock

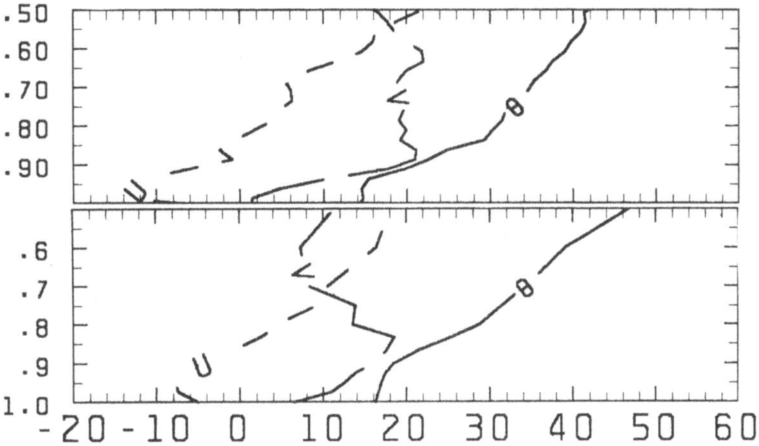

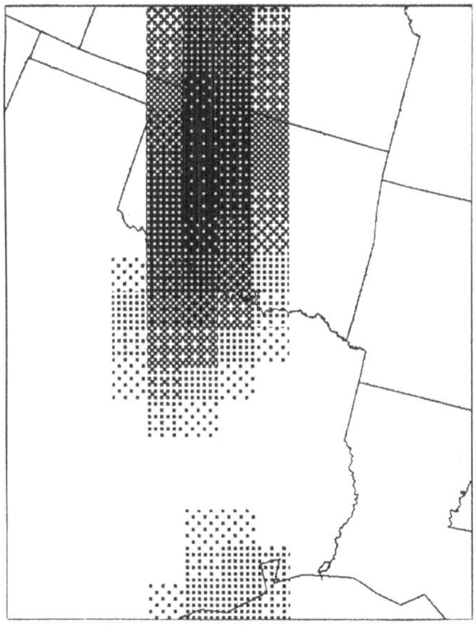

Fig. 5 Stability index

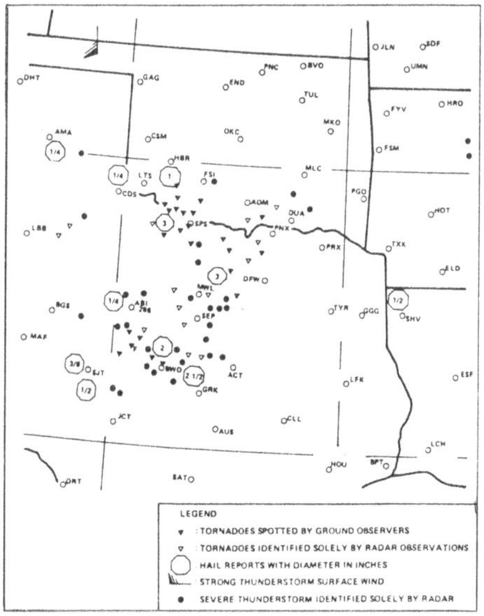

Fig. 6 Severe weather reports

MODELLING OF TWO-DIMENSIONAL BUBBLES IN VERTICAL TUBES

By B. Couët, G. S. Strumolo
Schlumberger-Doll Research, Old Quarry Road, Ridgefield, CT 06877
and A. E. Dukler
Department of Chemical Engineering,
University of Houston, Houston, TX 77004

INTRODUCTION

A problem related to gas-liquid slug flow in tubes arises in understanding the dynamics of the so-called Taylor bubbles preceding the slug. These bubbles occupy a significant cross-section of the tube and are often many tube-radii long. While for vertical flow the problem can be considered axisymmetric, we will consider a slightly simpler problem, namely, the flow past a two-dimensional bubble between two flat plates. As in the slug flow problem, the fluid runs around the outside of the bubble and remains as a layer down the surface of the tube falling freely under gravity. A characteristic quantity for this flow is the Froude number $F_r = U/\sqrt{(gb)}$, a dimensionless measure of the velocity U with which the gas bubble penetrates the liquid. g is the gravitational acceleration and b the two-dimensional tube's half width. The Froude number can be associated with the Reynolds number in the tube by the following relationship,

$$R_e \equiv \frac{2Ub}{\nu} = \left(\frac{2b^{3/2}g^{1/2}}{\nu}\right) F_r, \tag{1}$$

where ν is the kinematic viscosity of the liquid. For notation, we let x represent the vertical distance from the bubble nose, y the horizontal distance from the centerline, and the vector $\mathbf{r} = (x,y)$. We begin by computing, for each Froude number, the bubble shape needed to achieve a constant pressure along its surface in an inviscid, irrotational flow. A turbulent boundary-layer analysis is then applied in the region between the bubble and the tube. Results on the liquid film thickness at the intersection of the developing boundary layer and the bubble surface are presented.

POTENTIAL ANALYSIS

The irrotational flow field can be described by a stream function ψ written as a boundary integral of singularity distributions:

$$\psi(\mathbf{r}) = -\frac{1}{4\pi}\oint_C \gamma(\mathbf{r}') \ln(|\mathbf{r}-\mathbf{r}'|^2)\, ds' \, . \tag{2}$$

The boundary C of the fluid domain contained between the tube wall and the bubble surface is divided into N segments with γ constant on each segment and represented by N dipoles, each positioned at a given $\mathbf{r}$. The dipole strengths are obtained as the solution of the set of linear simultaneous equations:

$$\psi(\mathbf{r}_j) = -\frac{1}{4\pi}\sum_{i=1}^{N} \gamma(\mathbf{r}_i) \ln(\sigma + |\mathbf{r}_j-\mathbf{r}_i|^2)\, , \qquad j = 1,\ldots, N \, , \tag{3}$$

where σ is a small non-vanishing core radius for the dipoles and is included to handle the logarithmic singularity. It is assumed that upstream of the bubble the flow comes in uniformly. Similarly, for the outflow condition of the domain, the vertical (streamwise) component of the velocity is approximated to be a constant across the thin film. The stream function ψ is therefore prescribed as being a linear function of y for the inflow and outflow conditions while it is a constant along the tube ($y = \pm b$) and zero along the free surface of the bubble. An initial set of positions is picked for the dipoles describing the bubble and the rest of the boundary curve C is discretized using a well defined set of dipoles. Solving the linear system of equations (3) then provides their corresponding strengths which, in turn, give the velocity of the liquid along the surface of the bubble. Through Bernoulli's equation, a pressure is obtained which, if not zero, calls for an upgrade of the positions of the dipoles on the bubble. Iterating in this fashion finally provides the zero-pressure curve that describes the bubble.

It should be pointed out that the matrix associated with the linear system (3) is full. With about 200 dipoles to describe a bubble of a length equal up to 14 tube radii and another 300 dipoles to describe the rest of the boundary curve, $N = 500$. Since the system of equations (3) has to be solved for every non-linear iteration, it obviously turns out to be the costliest part of the computational effort. By observing that the matrix for the logarithmic coefficients of (3) is symmetric and that it does not change very much from a suitable initial guess for the bubble to the zero-pressure final curve, a preconditioned iterative technique to solve the system was used in lieu of the direct method (Cholesky factorization and back substitution). A conjugate gradient technique was the most convenient to implement. The preconditioning approach consists of premultiplying the system by an approximate inverse to the coefficient matrix so that the spectrum of the resulting matrix is closer to unity and thereby accelerate the convergence of the conjugate gradient technique. This inverse was determined and updated by using the direct method every tenth iteration. In that fashion, the computing time was cut by at least a factor of 4 from the direct method used alone.

The results of the computations appear in Figure 1 which displays some of the shapes satisfying the zero-pressure condition at the surface. Each shape is characterized by a particular Froude number. Garabedian (1957) also showed that the velocity of steady flow past a bubble in an infinitely long tube is not fixed by the tube diameter and the acceleration of gravity alone. Collins (1965) and Maneri (1970) conducted experiments on two-dimensional bubbles (between parallel plates of high aspect ratio so as to approximate a two-dimensional flow). They both observed that the bubbles have a circular shape at the nose. If we choose to satisfy this criterion in our potential flow simulations, a particular profile can now be selected from the family of curves. The computed result is that $F_r = 0.32$ and the ratio of the radius of curvature to the radius of the tube equals 0.64. This compared with 0.32 and 0.62 reported by Collins (1965) while Maneri (1970) measured values of 0.36 and 0.64 for parallel plates of the largest aspect ratio used. Figure 2 shows our calculated profile (solid line) compared with the profile obtained by digitization of Maneri's observation (dashed line). The agreement is very good indeed.

BOUNDARY-LAYER ANALYSIS

Although an inviscid approach is reasonably correct near the front of the bubble, it must break down further from the nose since it requires the surface to asymptote to the outer wall as $x \to \infty$ while observations seem to indicate an asymptotic, nonzero thickness well before then. This equilibrium thickness is conjectured to be the result of viscous forces acting in the thin liquid layer. To examine its effect, we will consider the development of a turbulent boundary layer between the bubble and the wall and its relation to a free falling film.

Taking a bubble rising with velocity U in a tube of radius b, we let the thickness of the liquid film at a distance x from the nose be $m(x)$ (see Figure 3). We will assume that the boundary layer begins at $x = 0$ and that its thickness is given by $\delta(x)$. Its subsequent development downstream is determined by the Von Karman momentum integral equation (Schlichting 1979),

$$\int_0^\delta \frac{d}{dx} \left(u(u_0 - u)\right) dy' + \frac{du_0}{dx} \int_0^\delta (u_0 - u)\, dy' = \frac{\tau_w}{\rho}. \tag{4}$$

Here u and u_0 are the velocities inside and outside the boundary layer, respectively, ρ is the liquid density, τ_w is the wall shear stress and $y' = b - y$. The velocity distribution inside the layer is given by the power law

$$\frac{u}{u_0} = \left(\frac{y'}{\delta}\right)^{\frac{1}{n}}, \tag{5}$$

where we allowed n to vary with the Reynolds number R_e as in Nikuradse's experiments, namely, $n \approx 3.51343 + 0.30254 \ln R_e$, for $R_e \leqslant 100{,}000$. Together with the transformations

$$X = x/b, \quad \Delta = \delta/m, \quad M = m/b, \tag{6}$$

and since $u \equiv u_0$ for $y' \geqslant \delta$, the momentum integral equation (4) then becomes a dimensionless nonlinear ODE:

$$\frac{d\Delta}{dX} = \left(\frac{\kappa \left(1 + \frac{n}{n+1} - \frac{2n}{n+2}\right)^{-1}}{M\Delta\left(\frac{R_e\,(n+1)\,\Delta}{2\,(n+1-\Delta)}\right)^{0.25}} + \frac{1}{M}\frac{dM}{dX} \right)(n+1-\Delta), \tag{7}$$

where κ is a function of n and the coefficient of the wall shear stress. Equation (7) is solved until $\Delta = 1$, or equivalently, until the boundary meets the potential bubble.

We postulate that beyond this intersection, the liquid film falls freely with uniform thickness. For turbulent flow in thin films, several relationships for film thicknesses can be found in the literature (Fulford 1964). One of the earliest showed that past a critical film Reynolds number,

$$\frac{d}{b} = \frac{1}{b}\left(\frac{3\,\nu^2 R_{eF}^2}{590\,g}\right)^{1/3} \cong 0.17196 F_r^{2/3}, \tag{8}$$

where d is the thickness of the falling film and $R_{eF} = d\,U_{LF}/\nu$ is the film Reynolds number with U_{LF} the liquid film velocity. If our postulate is correct, the film thickness at the intersection of the bubble with the developing boundary layer (δ/b) should equal that of a falling film (d/b).

RESULTS

Given any value of F_r, we calculate the potential bubble surface as described above. This then determines $m(x)$ and the boundary-layer equation can be integrated for a specific R_e. The solution provides the position where the boundary layer meets the bubble ($\Delta = 1$) along with the film thickness. Changing either R_e or F_r (another bubble shape) gives a different position for the matching. A set of computed film thickness values is thus obtained in the (R_e, F_r) plane as shown in Figure 4.

To test our postulate, we constructed analytic expressions for δ/b by performing a series of multiple linear regression analyses on the data display in Figure 4. The optimum two-variable model produced

$$\delta/b = 0.05209 + 0.17187\,F_r^{2/3} - 0.005453 \ln R_e, \tag{9}$$

which accounted for 99.7% of the variation. Note the agreement with the F_r dependence in (8). We then framed our results in a typical correlation plot for film thicknesses. Figure 5 is a plot of Nt against R_{eF}, where $Nt = d\,(g/\nu^2)^{1/3}$ is the dimensionless Nusselt film thickness parameter. Equation (8) indicates that a plot of Nt versus R_{eF}, on double-logarithmic coordinates should produce a straight line of slope 2/3 for turbulent film flow. The solid line in Figure 5 is derived from Equation (9) using (1). The dashed line comes from Dukler's theoretical investigation on free falling films (1960) and the data points are Reinius' experimental film thicknesses (1961) for the flow of water in a smooth channel. Our curve lies close to Dukler's theory in the low R_{eF} range, where both show excellent agreement with experiment, while (9) follows the experimental data more closely in the higher range.

In summary, using a highly accurate numerical technique for determining the potential bubble shape and a turbulent boundary layer analysis, it has been possible to describe the dynamics of a Taylor bubble from the nose to the free falling thin liquid film. This approach has succeeded in a domain where conventional numerics would most certainly fail.

REFERENCES

Collins, R. 1965 A simple model of the plane gas bubble in a finite liquid. *J. Fluid Mech.*, 22, 763.

Dukler, A. E. 1960 Fluid mechanics and heat transfer in vertical falling-film systems. *Chem. Engr. Progr. Symp. Ser.*, 30, 1.

Fulford, G. D. 1964 The flow of liquids in thin films, in *Advances in Chemical Engineering*, Vol 5 (eds Drew, T. B., Hoopes, J. W. & Vermeulen, T.). Academic Press.

Garabedian, P. R. 1957 On steady-state bubbles generated by Taylor instability. *Proc. Roy. Soc.* A, 241, 423.

Maneri, C. C. 1970 The motion of plane bubbles in inclined ducts. Ph. D. Thesis, Polytechnic Institute of Brooklyn.

Reinius, E. 1961 Steady uniform flow in open channels. *Trans. Roy. Inst. Technol. Stockholm*, 179, 1.

Schlichting, H. 1979 *Boundary Layer Theory*, McGraw-Hill.

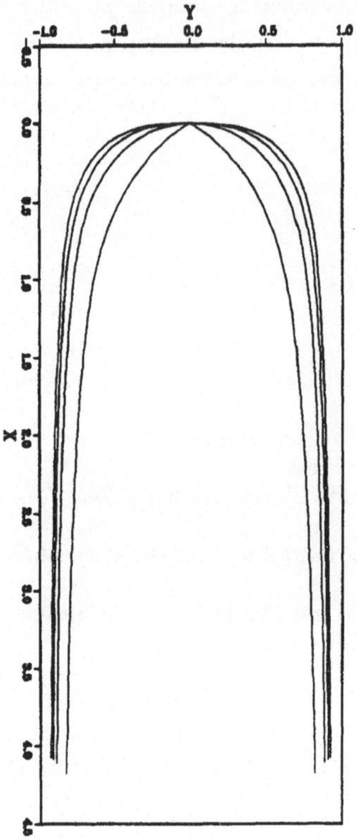

Figure 1.

Results of computations for zero-pressure condition at the surface

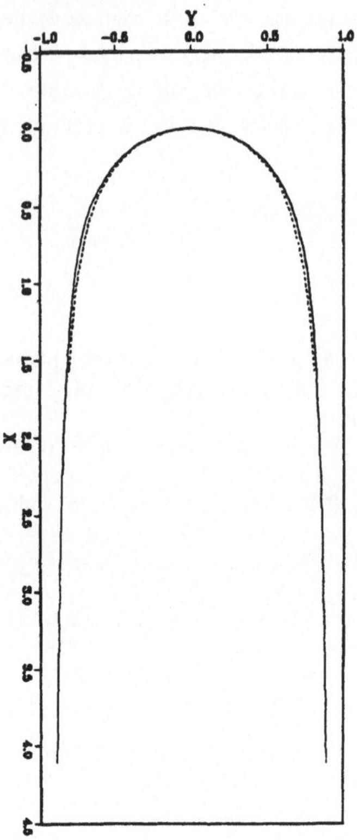

Figure 2.

Comparison of calculated and observed profiles

Figure 3.

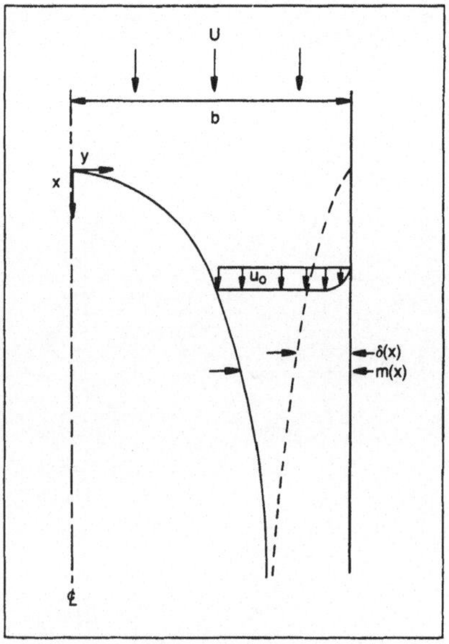

Thickness of the liquid film

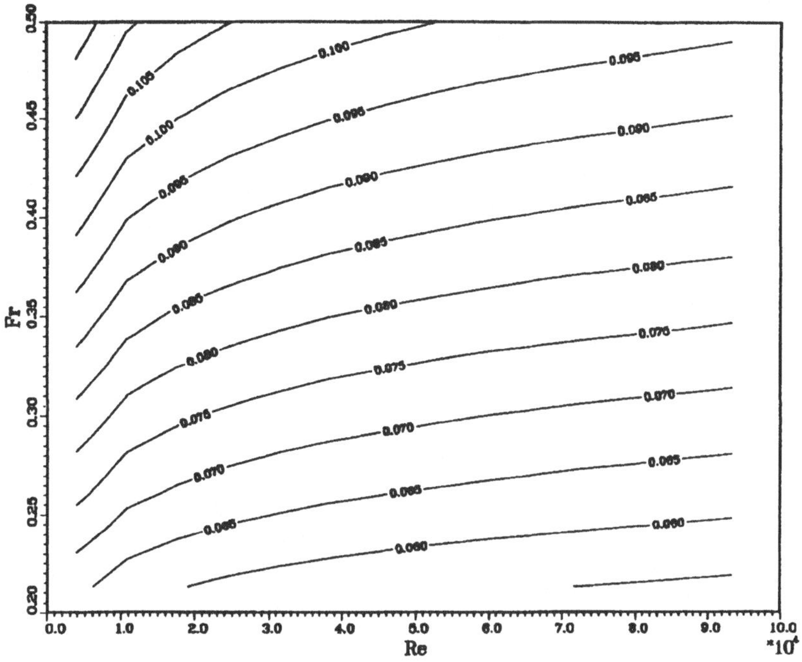

Figure 4.

Computed film thickness values

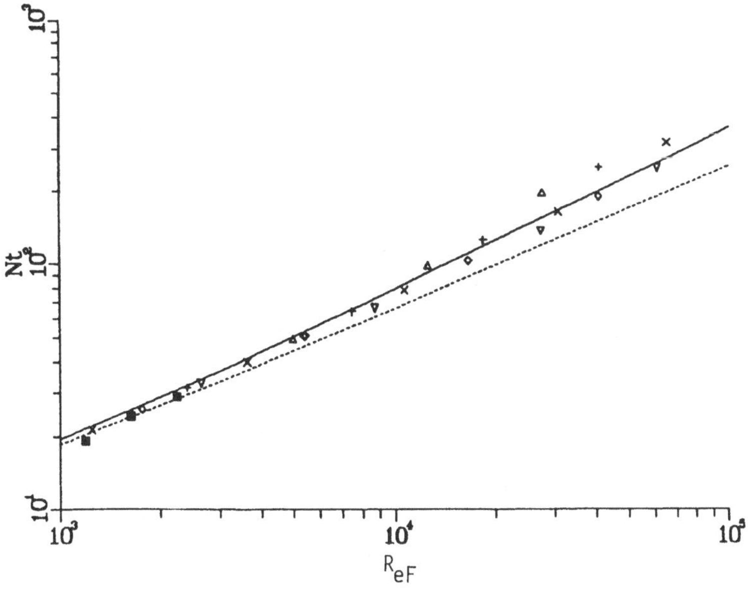

Figure 5.

Plot of Nt against R_{eF}

A TIME DEPENDENT FREE BOUNDARY GOVERNED BY THE NAVIER-STOKES EQUATIONS

C. Cuvelier
Delft University of Technology
Dep. of Mathematics
Delft, The Netherlands

In this paper we study a two dimensional linearized non-stationary capillary free boundary problem governed by the Navier-Stokes equations.

We consider a liquid in an open vessel V placed in the field of gravity g. V is given by $V = \{x = \{x_1,x_2\} \in \mathbb{R}^2 \mid x_2 > \phi_\Gamma(x_1)\ ,\ 0 < x_1 < L\}$ where ϕ_Γ is such that the container wall ∂V is smooth (see fig.) We assume that the fluid is isothermal, incompressible, newtonian with constant density ρ, viscosity μ and surface tension coefficient α.

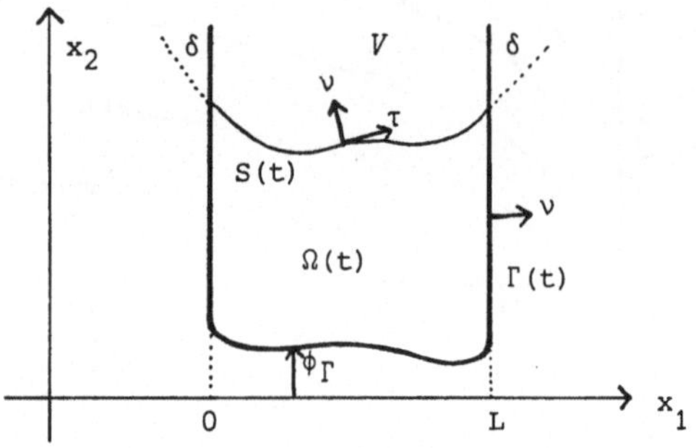

Under these assumptions the equations of non-stationary motion, which are the equations for conservation of momentum and mass, reduce to the Navier-Stokes equations for the velocity $u = \{u_1,u_2\}$ and the pressure p.

Let $\Omega(t)$ be the region occupied by the fluid at time t, then its boundary $\partial\Omega(t)$ can be subdivided into two parts: $\Gamma(t) = \partial V \cap \partial\Omega(t)$ = wet part of the container wall, and $S(t) = \partial\Omega(t) \setminus \Gamma(t)$ = free boundary.

To the Navier-Stokes equations we add the following boundary conditions: On $\Gamma(t)$ the no-slip (or free-slip) conditions. On S(t) the traction conditions of the continuity of the stresses:

$$\sigma_\nu = \frac{\alpha}{R_S} \qquad \sigma_\tau = 0$$

where σ_ν = normal stress, σ_τ = tangential stress. R_S denotes the radius of curvature of S(t) which is reckoned positive when the corresponding centre of curvature lies outside the liquid. The motion of the free boundary S(t) follows from the kinematic condition together with the conditions that, at the points of intersection of S(t) and $\Gamma(t)$, the angle between S(t) and $\Gamma(t)$ is equal to the given contact angle $\delta \in (0,\pi)$ (see fig.). The volume of the liquid contained in V is given. Finally we specify the initial conditions for the velocity and the initial position of the free boundary.

This problem will be studied within the context of small perturbations with respect to the static solution. We denote by $\Lambda = \Lambda(s,t)$ the deviation of the free boundary $S(t)$ from the static free boundary S_0 in the normal direction ν on S_0. By s we denote the length parameter on S_0. It can be proved that the linearized dimensionless form (L = length scale, $L^2\sqrt{\frac{\rho}{L\alpha}}$ = time scale, $\frac{\alpha}{L}$ = pressure scale) of the problem is:

(1)

$$\left|\begin{array}{ll}
\text{Find } u = u(x,t) = \text{velocity, } p = p(x,t) = \text{pressure, } \Lambda = \Lambda(s,t),\ x \in \Omega_0, & \\
s \in S_0,\ t \in [0,T) \text{ such that} & \\
\frac{\partial u}{\partial t} - Oh\ \Delta u + \nabla p = 0 & \\
 & \text{in } \Omega_0 \times (0,T) \\
\nabla . u = 0 & \\
u.\nu = 0 \qquad \sigma_\tau = 0 & \text{on } \Gamma_0 \times (0,T) \\
\sigma_\tau = 0 & \\
\sigma_\nu = \frac{\partial^2 \Lambda}{\partial s^2} + \left(\frac{1}{R^2_{S_0}} - Bo\ \frac{\partial x_2}{\partial \nu}\right) \Lambda & \text{on } S_0 \times (0,T) \\
\frac{\partial \Lambda}{\partial t} = u.\nu & \\
\frac{\partial \Lambda}{\partial s} = \frac{\Lambda \text{ cotan } \delta}{R_{S_0}} \text{ at } x_1 = 0 \qquad \frac{\partial \Lambda}{\partial s} = -\frac{\Lambda \text{ cotan } \delta}{R_{S_0}} \text{ at } x_1 = 1 & \\
u(x,0) = u_{init}(x) \qquad \Lambda(s,0) = \Lambda_{init}(s) &
\end{array}\right.$$

with Ω_0 the region occupied by the fluid in the static situation, $Oh = \mu\sqrt{\frac{1}{\rho L\alpha}}$ = Ohnesorge number, $Bo = \frac{g\rho L^2}{\alpha}$ = Bond number, R_{S_0} = radius of curvature of S_0.

Problem (1) admits the following weak formulation to which we have added an external force $f \in L_2(0,T;IL_2)$:

(2)

$$\left|\begin{array}{l}
\text{Find } u(t) \in IH^1 \text{ with } u_\nu = u.\nu = 0 \text{ on } \Gamma_0,\ u_\nu \in H^1(S_0),\ \text{div } u = 0 \text{ such that} \\
u(0) = u_{init} \text{ and} \\
\frac{d}{dt}(u(t),v)_{IL_2} + Oh\ a(u(t),v) + b\left(\int_0^t u_\nu(\xi)d\xi, u_\nu\right) = \\
\qquad = -\,b(\Lambda_{init}, v_\nu) + (f(t),v)_{IL_2} \quad \text{on } (0,T) \\
\text{for all } v \in IH^1 \text{ with } v_\nu = 0 \text{ on } \Gamma_0,\ v_\nu \in H^1(S_0),\ \text{div } v = 0.
\end{array}\right.$$

where $IL_2 = (L_2(\Omega_0))^2$, $IH^1 = (H^1(\Omega_0))^2$,

$$a(u,v) = \tfrac{1}{2} \sum_{i,j=1}^{2} \int_{\Omega_0} \left(\frac{\partial u_i}{\partial x_j} + \frac{\partial u_j}{\partial x_i}\right)\left(\frac{\partial v_i}{\partial x_j} + \frac{\partial v_j}{\partial x_i}\right) dx$$

$$b(\phi,\psi) = \int_{S_0} \left\{\frac{\partial \phi}{\partial s}\frac{\partial \psi}{\partial s} + \left(Bo\,\frac{\partial x_2}{\partial \nu} - \frac{1}{R^2_{S_0}}\right)\phi\psi\right\} ds + \frac{\phi \text{ cotan } \delta}{R_{S_0}}\,\psi\Big|_{x_1=0} + \frac{\phi \text{ cotan } \delta}{R_{S_0}}\,\psi\Big|_{x_1=1}$$

For $Bo > -Bo_{critical}$ and $Oh > 0$ existence and uniqueness of a solution $u \in L_2(0,T;IH^1)$ of problem (2) can be proved (see [1]). These results can be extended to the 3D case. Notice that the unknown position Λ of the free boundary and the pressure p have been eliminated.

* * *

Problem (2) will now be solved numerically using a finite element method in space and a finite difference method in time. One of the main difficulties is to handle the constraint div u = 0 in a correct way. We shall apply a finite element-penalty approach which consists of replacing a finite element discretization of the problem by a family of perturbation problems depending on a parameter ε. The solutions of these perturbation problems satisfy the incompressibility contraint in a global penalized way. This method has been proposed by several authors (see [1], [2] for references). The finite element we will use is the CROUZEIX-RAVIART extended quadratic element based on seven points per triangle. The incompressibility constraint is satisfied in the following global sense:

(3) $\int_K \text{div } u_h \, dx = 0$ (4) $\int_K (x_i - a_i^7)\text{div } u_h \, dx = 0 \quad i = 1,2$

for each triangle K, where $\{a_1^7, a_2^7\}$ denotes the coordinates of the seventh nodal point (barycentre) of K. The constraints (4) $i = 1,2$ are used to eliminate the velocity unknowns in the barycentre of each triangle. The resulting function space is called V_h. Constraint (3) will be satisfied approximately through the introduction of a penalty term with penalty parameter $\varepsilon \approx 10^{-6}$.
The numerical scheme defines the approximation u_h^n of the solution u of problem (2) at time $t = n\Delta t$, $\Delta t = \frac{T}{N}$. Let u_h^0 be the orthogonal projection of u_{init} onto V_h, then we define $u_h^{n+1} \in V_h$ by

$$(5)\left|\begin{array}{l} \frac{1}{\Delta t}(u_h^{n+1} - u_h^n, v_h)_{IL_2} + Oh\, a(u_h^{n+\theta}, v_h) + \Delta t\, b\left(\sum_{r=0}^{n-1} u_{h\nu}^{r+\theta} + \tfrac{1}{2} u_{h\nu}^{n+\theta}, v_{h\nu}\right) + \\ \quad + \frac{1}{\varepsilon}(\text{div}_h u_h^{n+\theta}, \text{div}_h v_h)_{L_2} = -\,b(\Lambda_{init}, v_\nu) + (f^{n+\theta}, v_h)_{IL_2}. \end{array}\right.$$

where generally $v_h^{n+\theta} = \theta v_h^{n+1}+(1-\theta)v_h^n$. The expression $div_h v_h$ in the penalty term denotes the projection of div v_h on the space of (not necessarily continuous) functions which are affine on each triangle. For $\frac{1}{2} \leq \theta \leq 1$ the scheme is unconditionally stable and convergent, while for $0 \leq \theta < \frac{1}{2}$ a condition $\Delta t < c\ \varepsilon\ h^2$, $h = diam(K)$, must be satisfied which makes the scheme useless in practice (see [1]).

* * *

When the Ohnesorge number is small (Oh=0.005), the results of numerical experiments can be compared with results obtained for perfect fluids. When $\Omega_0 = (0,1) \times (0,H)$ the linearized potential theory for perfect fluids proves the existence of vibration modes

(6) $\phi_k(x_1,t) = H + A \cos \tilde{\omega}_k t.\cos k\pi x_1 \qquad k = 1,2,\ldots$

(H = liquid height, A = amplitude) with frequency

(7) $\tilde{\omega}_k = (Bo\ k\ \pi + (k\pi)^3)\ \tanh(k\pi H)$

For our numerical experiments with $\delta = 90^o$, $H = 0.5$, $\theta = 1.0$, $Oh = 0.005$, $\Delta t = 0.005$, we chose

(8) $\begin{cases} \Lambda_{init} = -\ 0.1 \cos k\pi x_1 \qquad k = 1,2,3,4 \\ u_{init} = 0.0 \end{cases}$

In general the motion of the free surface showed an oscillatory damped behavior in time

(9) $\Lambda_h \sim e^{-\alpha_d t} \cos \omega t\ , \qquad \alpha_d$ = damping factor

$\sim real\ (e^{\lambda t})\ , \qquad \lambda = -\alpha_d + i\omega$

In table 1 we present some calculated values of ω for various situations as well as the (analytical) results from (7) for perfect fluids. Notice the excellent agreement of ω and $\tilde{\omega}$.

In table 2 some results for the 3D axi-symmetric case are shown: Ω_0 = cylinder, radius = 1.0, fluid height H = 1.0, Oh = 0.005, Δt = 0.005.

In table 3 we present some results for different values of Oh and Bo = 0.0. It can be seen that ω vanishes for large values of Oh. When Oh increases α_d increases as long as ω decreases and α_d decreases once ω is equal to zero. It can be proved that problem (2) has a countable set of eigenfrequencies λ which are all real, except possibly, for a finite number. When Oh is large enough all λ are real.

* * *

C. CUVELIER [1]: A capillary free boundary problem governed by the Navier-Stokes equations. Publication Marseille-Toulon 84/01. (To appear in Comp. Meth. Appl. Mech. Eng.).
C. CUVELIER [2]: On the solution of capillary free boundary problems governed by the Navier-Stokes equations. Report 82.09 Delft University.

* * *

k	Bo	ω	$\tilde{\omega}$
1	0.0	5.32	5.33
	5.0	6.52	6.55
	10^3	54.49	53.94
2	0.0	15.68	15.72
	5.0	16.63	16.68
	10^3	81.39	80.67
3	0.0	28.89	28.93
	5.0	29.57	29.73
	10^3	102.30	101.29
4	0.0	44.88	44.55
	5.0	45.65	45.25
	10^3	120.93	120.63

table 1.

k	Bo	ω	$\tilde{\omega}$
1	0.0	7.53	7.50
	10^3	63.00	62.36
2	0.0	18.64	18.58
	10^3	86.75	85.80
3	0.0	32.58	32.45
	10^3	108.87	105.96
4	0.0	50.79	48.63
	10^3	130.90	125.26

table 2

Oh	ω	α_d
0.005	5.32	0.87
0.01	5.24	1.04
0.05	4.72	2.38
0.1	3.2	3.98
0.5	0.0	0.71
1.0	0.0	0.35

table 3

* * *

A PERTURBATIVE LAMBDA FORMULATION

A. DADONE and M. NAPOLITANO
Istituto di Macchine, Università di Bari, via Re David 200, 70125 Bari, ITALY

ABSTRACT

The present paper provides a new perturbative lambda formulation for the numerical solution of compressible flows. The time-dependent Euler equations are recasted in terms of compatibility equations for perturbative bicharacteristic variables (which are the difference between the standard Riemann variables and those corresponding to an appropriate steady incompressible flow) and solved numerically by means of an ADI method. Results for subcritical and supercritical flows past a NACA 0012 airfoil are presented, which demonstrate the remarkable accuracy of the proposed methodology.

INTRODUCTION

This paper is concerned with the numerical simulation of compressible inviscid flows. It is the authors' belief that, when dealing with unsteady compressible flow phenomena, a very natural and convenient approach for solving the governing equations, numerically, is the so-called lambda formulation, recently developed by Moretti et al. /1, 2/: the time-dependent Euler equations are recasted in terms of compatibility equations for characteristic variables and discretized using upwind differences to account for the direction of wave propagation phenomena. As such, the lambda formulation combines the coding simplicity of finite difference methods with the intrinsic accuracy and physical soundness of the method of characteristics; and possesses a well documented /1-4/, although controversial, shock-capturing capability.

Since its first appearance as a working tool for solving compressible flows numerically /1/, the lambda methodology has undergone several improvements /3-7/. In particular, the authors have developed various implicit integration methods /3-5/ in order to overcome the CFL stability limitation of the original explicit lambda schemes /1, 2/. Also, when solving two-dimensional flows in a general orthogonal curvilinear coordinate system /4/, they have shown that a significant improvement in the accuracy of the solution is obtained by employing the incompressible potential flow net as the computational grid. However, when computing a compressible flow around an airfoil or inside a cascade by means of a rather coarse mesh, excessive spurious total pressure errors are usually obtained around the leading and trailing edges, where large gradients are present in the flow field. This paper addresses such a problem, specifically, by reformulating the governing lambda equations in terms of perturbative characteristic-type variables, which are the difference between the original variables and those corresponding to the mesh-generating incompressible steady flow. In this way, the geometry-induced gradients are mostly accounted for by the incompressible flow solution and the perturbative problem (i.e., the compressibility effects), which is smoother and better behaved than the full compressible one, can be solved very accurately even on a coarse mesh.

The proposed approach has been recently applied to solve subsonic and transonic internal flows in one and two dimensions; for the case of one-dimensional flows, in particular, transonic flows characterized by rather strong shocks have been computed very accurately and efficiently, using the shock-tracking procedure by Moretti /6/ and a fully implicit integration scheme /8/. Here, the great accuracy of the new methodology is demonstrated by computing subcritical and supercritical flows past an isolated airfoil.

GOVERNING EQUATIONS AND NUMERICAL TECHNIQUE

For the case of homentropic two-dimensional flows, the perturbative lambda-formulation equations are given in a general orthogonal curvilinear coordinate system /8/ as:

$$dC_t + dD_t + \frac{v_1 + a}{h_1}\frac{\partial\, dC}{\partial q_1} + \frac{v_2}{h_2}\frac{\partial\, dC}{\partial q_2} + \frac{v_1 - a}{h_1}\frac{\partial\, dD}{\partial q_1} + \frac{v_2}{h_2}\frac{\partial\, dD}{\partial q_2} = \frac{2\, dv_2}{h_1 h_2}\left(\frac{\partial h_2}{\partial q_1} dv_2 - \frac{\partial h_1}{\partial q_2} dv_1\right) -$$

$$\frac{2\,dv_1}{h_1}\left(\frac{\partial h_1}{\partial q_2}\frac{v_2'}{h_2}+\frac{\partial v_1'}{\partial q_1}\right)-\frac{2\,dv_2}{h_2}\left(\frac{h_2}{h_1}\frac{\partial v_2'}{\partial q_1}-\frac{\partial h_2}{\partial q_1}\frac{v_2'}{h_1}\right)-2\zeta\frac{\partial a'}{\partial q_1}\frac{da}{h_1} \tag{1}$$

$$dE_t+dF_t+\frac{v_1}{h_1}\frac{\partial\, dE}{\partial q_1}+\frac{v_2+a}{h_2}\frac{\partial\, dE}{\partial q_2}+\frac{v_1}{h_1}\frac{\partial\, dF}{\partial q_1}+\frac{v_2-a}{h_2}\frac{\partial\, dF}{\partial q_2}=\frac{2\,dv_1}{h_1 h_2}\left(\frac{\partial h_1}{\partial q_2}dv_1-\frac{\partial h_2}{\partial q_1}dv_2\right)-$$
$$\frac{2\,dv_2}{h_2}\left(\frac{\partial h_2}{\partial q_1}\frac{v_1'}{h_1}+\frac{\partial v_2'}{\partial q_2}\right)-\frac{2\,dv_1}{h_1}\left(\frac{h_1}{h_2}\frac{\partial v_1'}{\partial q_2}-\frac{\partial h_1}{\partial q_2}\frac{v_1'}{h_2}\right)-2\zeta\frac{\partial a'}{\partial q_2}\frac{da}{h_2} \tag{2}$$

$$(dC_t-dD_t+dE_t-dF_t)/2+\frac{v_1+a}{h_1}\frac{\partial\, dC}{\partial q_1}-\frac{v_1-a}{h_1}\frac{\partial\, dD}{\partial q_1}+\frac{v_2+a}{h_2}\frac{\partial\, dE}{\partial q_2}-\frac{v_2-a}{h_2}\frac{\partial\, dF}{\partial q_2}=$$
$$-\frac{2\,da}{h_1 h_2}\left(\frac{\partial h_2}{\partial q_1}dv_1+\frac{\partial h_1}{\partial q_2}dv_2\right)-\frac{2\,dv_1}{h_1}\left(\zeta\frac{\partial a'}{\partial q_1}+\frac{\partial h_2}{\partial q_1}\frac{a'}{h_2}\right)-\frac{2\,dv_2}{h_2}\left(\zeta\frac{\partial a'}{\partial q_2}+\frac{\partial h_1}{\partial q_2}\frac{a'}{h_1}\right)$$
$$-2\zeta\left(\frac{\partial a'}{\partial q_1}\frac{v_1'}{h_1}+\frac{\partial a'}{\partial q_2}\frac{v_2'}{h_2}\right) \tag{3}$$

$$dC-dD-dE+dF = 0 \tag{4}$$

In eqns(1-4), q_1, q_2, h_1 and h_2 are the two general orthogonal coordinates and the corresponding scale factors, v_1 and v_2 are the two components of the velocity vector $\underline{v}$, a is the speed of sound and the subscript t indicates partial derivative with respect to time; furthermore, C, D, E and F are the four bicharacteristic variables given as

$$C = v_1+\zeta a, \quad D = v_1-\zeta a, \quad E = v_2+\zeta a, \quad F = v_2-\zeta a \tag{5a, b, c, d}$$

with $\zeta = 2/(\gamma-1)$, γ being the specific heats ratio; finally, d in front of any variable (e.g., dC) indicates the corresponding perturbative variable, whereas primes (e.g., C') indicate the incompressible-flow variables, so that, for example, $dC = C - C'$. As far as the appropriate incompressible flow is concerned, its governing equations are given as:

$$\mathrm{div}\,\underline{v}' = \mathrm{curl}\,\underline{v}' = 0 \tag{6a, b}$$

$$v_1'^2+v_2'^2+\zeta a'^2 = \zeta a_o^2 \tag{7}$$

where a_o is the (known) stagnation speed of sound of the compressible flow under investigation. It is noteworthy that the perturbative lambda equations (1-4) preserve the basic structure of the standard (lambda) equations /4/; also, the coefficients of the bicharacteristic-type advection terms contain the full compressible flow variables v_1, v_2 and a, insofar as they provide the slopes of the physical bicharacteristic lines and, therefore, the direction of the upwind differences to be used for discretizing the associated spatial derivatives. Finally, in the right hand sides of eqns (1-3), the perturbative variables appear only as source terms, whereas spatial derivatives only act on the scale factors and the incompressible-flow variables; the latter ones are eliminated in favor of nonderivative terms by means of eqns (6) and of the spatial derivatives of eqn (7).

Equations (1-4) are discretized in time by means of a two-level Euler time stepping, using the incremental (delta) form /9/, with only the derivatives of the bicharacteristic variables in eqns (1-3) and all the terms in eqn (4) evaluated at the new time level. For example, eqns (1) and (4) become:

$$\frac{\Delta C}{\Delta t} + \frac{\Delta D}{\Delta t} + \frac{(v_1 + a)^n}{h_1}\frac{\partial \Delta C}{\partial q_1} + \frac{v_2^n}{h_2}\frac{\partial \Delta C}{\partial q_2} + \frac{(v_1 - a)^n}{h_1}\frac{\partial \Delta D}{\partial q_1} + \frac{v_2^n}{h_2}\frac{\partial \Delta D}{\partial q_2} = SSEQN(1)^n \qquad (8)$$

$$\Delta C - \Delta D - \Delta E + \Delta F = 0 \qquad (9)$$

where $SSEQN(1)^n$ is a shorthand notation for the steady state part of eqn (1) evaluated at the old time level t^n, $\Delta C = dC^{n+1} - dC^n$, etcetera. The resulting equations are then discretized in space using two-point first-order-accurate upwind differences for the incremental (Δ) derivatives and three-point second-order-accurate upwind differences in the right hand sides of the equations, so that the final steady state solution is second-order-accurate. After eliminating the ΔF unknowns at all of the gridpoints by means of eqn (9), a large 3×3 block-pentadiagonal linear system is obtained. Such a system is factorized by means of a two-sweep ADI method /4/, so that a 3×3 block-tridiagonal system needs to be solved along each row and column of the computational grid, to obtain the values of ΔC, ΔD and ΔE. All of the variables are then updated and the process is repeated until a satisfactory convergence is obtained.

As far as the boundary conditions are concerned, the computational grid is always chosen to coincide with the flow net of the incompressible potential flow, $\underline{v}'$. Therefore, at the body surface, which is aligned with the coordinate lines q_1, we have dv_2 equal to zero. The outer boundary is divided into an inlet and outlet boundaries, coinciding with incompressible-flow potential lines and into two far field boundaries (above and below the airfoil) coinciding with incompressible-flow streamlines. For the case of subsonic free-stream flows, of interest here, two boundary conditions need to be prescribed at the inlet boundary, namely the value of the total enthalpy (speed of sound) and the direction of the velocity vector $\underline{v}$, which is assumed to be parallel to $\underline{v}'$, so that $dv_2 = 0$. A single boundary condition is needed instead at the outlet boundary, where the static pressure (i.e., da) is assigned, and at the far field boundaries, where the impermeability condition $dv_2 = 0$ is enforced. At all boundary gridpoints, the appropriate boundary conditions (which are all "physical ones") are used to eliminate from the governing equations those derivatives whose upwind discretizations would involve gridpoints external to the flow domain /4/; for example, at the outlet boundary, the derivative of the variable D with respect to the coordinate q_2 is eliminated using the prescribed pressure condition.

The initial conditions for all of the calculations later presented in this paper is, of course, the incompressible-flow solution $\underline{v}'$, a', so that all of the perturbative variables are zero at the initial time level. The incompressible-flow solution, as well as the computational grid and its scale factors have been computed using the numerical integration of the Schwartz-Christoffel transformation due to Davis /10/.

RESULTS

The proposed formulation has been applied to compute subcritical as well as supercritical flows past a NACA 0012 airfoil. Symmetric flows at free stream Mach numbers M_o equal to 0.72 and 0.80 have been considered at first, using the rather coarse 27×11 nonuniform mesh depicted in figure 1. The present results are given in figure 2 as the distributions along the surface of the airfoil of both the Mach number and the pressure coefficient, for the $M_o = 0.72$ flow case, and of the pressure coefficient only, for the supercritical $M_o = 0.8$ flow case. The very accurate solutions due to Loch /11/ and Jameson /12/, also given as solid and broken lines, respectively, are seen to coincide with the present coarse-mesh solutions, for all practical purposes. It is noteworthy that the present total pressure errors never exceed 0.7%; furthermore, by considering that, for the supercritical flow case, Jameson /12/ solves the (full) Euler equations in conservation form on a very fine 128×32 mesh, the remarkable accuracy of the proposed approach, as well as its adequate "shock-capturing" capability are clearly demonstrated. Subcritical flows characterized by $M_o = 0.63$ and α (angle of attack) = 2° and $M_o = 0.5$ and $\alpha = 3°$ have then been considered, using a slightly finer $39 \times 14 \times 2$ mesh similar to the one of figure 1 (where only the upper half is given, due to the symmetry of the flow field). It is noteworthy that for such computations it has been necessary to extend the computational domain up to 3.5 chord-lengths above and below the airfoil in order to obtain results independent of the far-field boundary conditions. The pressure coefficient distributions on both sides of the airfoil, as computed by the present method, are given in figures 3 and 4 together with the very accurate solutions of Loch /11/ and Jameson /12/. For both flow cases, the agreement is excellent, demonstrating once more the remarkable accuracy of the proposed formulation.

As far as the efficiency of the calculations is concerned, a satisfactory convergence has always been obtained in less than 100 iterations (time steps), using a local CFL number of $10 \div 15$, for a total CPU time of about 5 and 18 minutes (on a HP 9000/9040A minicomputer) for the 27×11 and $39 \times 14 \times 2$ meshes, respectively.

CONCLUSIONS

A new perturbative lambda formulation has been provided, which uses an available incompressible flow solution and evaluates only the compressibility effects, so that extremely accurate results for the complete compressible flow problem are obtained even on a very coarse mesh. The validity of the proposed methodology has been verified by computing subcritical and supercritical flows past a NACA 0012 airfoil.

ACKNOWLEDGEMENTS

This research has been supported by M.P.I. and C.N.R.; the original idea of a perturbative lambda formulation was born during a very stimulating discussions among the authors and professor R. T. Davis, at the University of Cincinnati, in July 1982.

REFERENCES

1. Moretti, G., "The λ - scheme", Computers and Fluids, Vol. 7, 1979, pp. 191-205.
2. Zannetti, L. and Colasurdo, G., "Unsteady Compressible Flow: A Computational Method Consistent with the Physical Phenomena", AIAA J., Vol. 19, July 1981, pp. 851-856.
3. Dadone, A. and Napolitano, M., "An Implicit Lambda Scheme", AIAA J., Vol. 21, October 1983, pp. 1391-1399.
4. Dadone, A. and Napolitano, M., "Efficient Transonic Flow Solutions to the Euler Equations", AIAA Paper 83-0258, January 1983.
5. Napolitano, M. and Dadone, A., "Three-dimensional Implicit Lambda Methods", Fifth GAMM Conference on Numerical Methods in Fluid Mechanics, Rome, 5-7 October 1983.
6. Moretti, G. and Zannetti, L., "A New Improved Computational Technique for Two-dimensional Unsteady Compressible Flows", AIAA Paper 82-168.
7. Moretti, G., "Fast Euler solver for steady one dimensional flows", NASA-CR 3689, 1983.
8. Dadone, A. and Napolitano, M., "Accurate and Efficient Solutions of Compressible Internal Flows", AIAA Paper 84-1247, June 1984.
9. Beam, R. M. and Warming, R. F., "An Implicit Factored Scheme for the Compressible Navier-Stokes Equations", AIAA J., Vol. 16, April 1978, pp. 393-402.
10. Davis, R. T., "Notes on Numerical Methods for Coordinate Generation Based on a Mapping Technique", V.K.I., Lecture Series 1981-5, March 30 - April 3, 1981.
11. Lock, R. C., "Test Cases for Numerical Methods in Two-Dimensional Transonic Flows", AGARD-R575-70.
12. Jameson, A., "Solution of the Euler Equations for Two-Dimensional Transonic Flows by a Multigrid Method", Princeton University MAE Report n. 1613, June 1983.

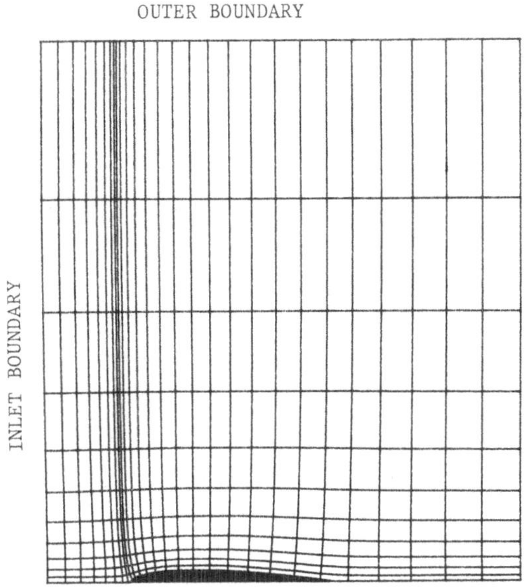

Figure 1. Symmetric flow computational mesh

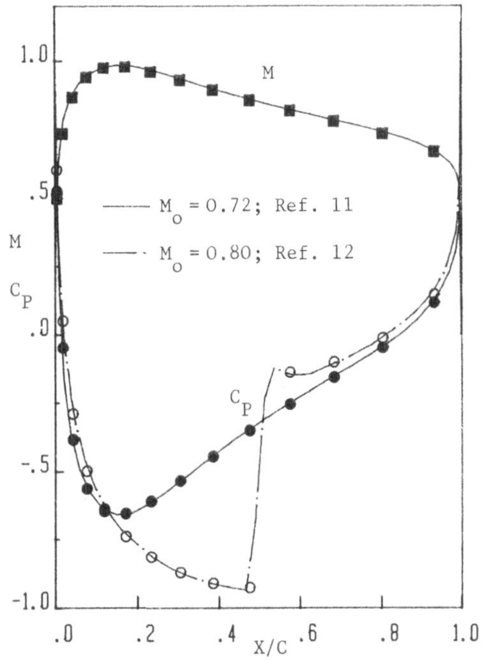

Figure 2. Symmetric flow results

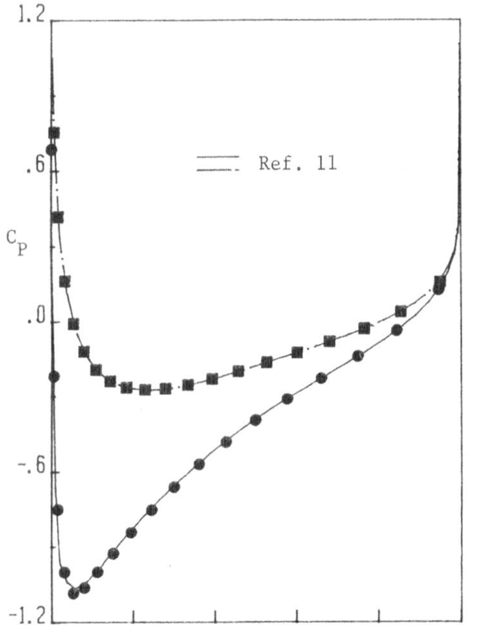

Figure 3. $M_o = 0.63$, $\alpha = 2°$ results

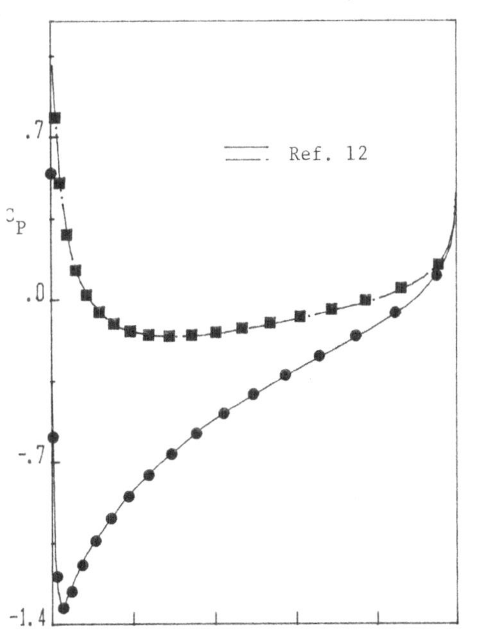

Figure 4. $M_o = 0.5$, $\alpha = 3°$ results

NUMERICAL MODELING OF VORTEX MERGING IN AXISYMMETRIC MIXING LAYERS

R. W. Davis and E. F. Moore
National Bureau of Standards
Gaithersburg, MD 20899, USA

I. Introduction

About a decade ago Brown and Roshko [1] observed that large vortical structures are the dominant features of mixing layers. Since then there has been much experimental and computational effort devoted toward improving the understanding of these coherent structures [2]. In particular, the experiments of Ho and Huang [3] have carefully elucidated the crucial role that subharmonic forcing plays in the vortex merging process inside a two-dimensional mixing layer. It is this merging process which is central to the downstream growth of the mixing layer. Thus, an understanding of it is essential if the flow is to be controlled.

It is the purpose of the present paper to present numerical solutions for spatially-developing axisymmetric mixing layers. The vortex merging inside these mixing layers is driven by small perturbations derived from linear inviscid stability theory. It is found that, as seen experimentally in the two-dimensional case [2, 3], the merging process is controlled by the frequency content of the forcing function. Thus it is possible to manipulate the downstream behavior of the mixing layer by altering the applied perturbation. Although the forced temporally-developing mixing layer with its simpler boundary conditions has been studied computationally [2, 4], this is not as desirable as studying the spatially-developing case which occurs in most physical situations.

II. Numerical Modeling

The axisymmetric mixing layer develops in the region between a jet with velocity U_j and a coflowing stream with velocity U_∞. Therefore, at the upstream boundary of the computational domain ($z = 0$), a velocity profile $U(r)$ characteristic of a coflowing jet is specified. This profile contains a shear layer centered at $r = R_j$ which develops into the mixing layer for $z > 0$. In an unforced physical experiment, random background noise will supply the perturbation necessary to trigger roll-up of this shear layer. The particular component of the background noise which is responsible for the initial vortex formation can be determined from linear inviscid stability theory as that frequency which exhibits the largest spatial growth rate [5]. Thus, for the computations, a perturbation of the form

$$\sum_n A_n(r) \exp \{i[R(\alpha_n)z - \beta_n t]\}$$

is applied to the vorticity over some region from $z = 0$ to $z = z_p \geq 0$. Here $A_n(r)$ are complex eigenfunctions, $R(\alpha_n)$ is the real part of a complex α_n, and β_n is real. When it is desired to match the situation in an unforced physical experiment, all parameters are chosen so as to minimize $I(\alpha_1)$ [3, 5]. In the case of a forced physical experiment, β_n for $n > 1$ must be subharmonics of β_1, i.e., β_1 is an integer multiple of β_2, another integer multiple of β_3, etc. [3]. Variations in the amplitude and domain of the applied perturbation have been found not to affect the resulting vortex dynamics in any fundamental way. For this study the maximum amplitude of the perturbation is about 0.01 U(r) and $z_p = 2R_j$. Reynolds number (Re) based on U_j and R_j has also been found not to affect the basic vortex dynamics although the vortices do smear out as Re decreases.

A finite difference method is employed in this study in order to solve the incompressible Navier-Stokes and continuity equations in primitive variables on a staggered mesh. This method utilizes quadratic upwind differencing for convection and an explicit Leith-type of temporal differencing [6]. This leads to effectively third-order accurate spatial differencing as $Re \to \infty$. The only relevant stability criterion is that the Courant number be less than one. This type of differencing scheme has been shown to perform well in computing vortex shedding from bluff bodies [6, 7]. At each time step a Poisson equation for pressure is solved by a direct method utilizing the FISHPAK package of FORTRAN subprograms for the solution of separable elliptic partial differential equations developed at NCAR [8].

The boundary conditions in the radial direction employed in this study are that radial derivatives are set to zero along $r = 0$ (axisymmetry), and a simple asymptotic analysis involving small perturbations about U_∞ is employed for large r. The free-stream velocity U_∞ is specified at the outflow boundary of the mesh by means of an infinite-to-finite mapping of the form $\zeta = K + K_1/z$, where K and K_1 are constants. This transform is employed for $z > 15\ R_j$.

The untransformed portion of the nonuniform mesh used in this study is shown in Fig. 1, where it can be seen that mesh points are concentrated in the region of the

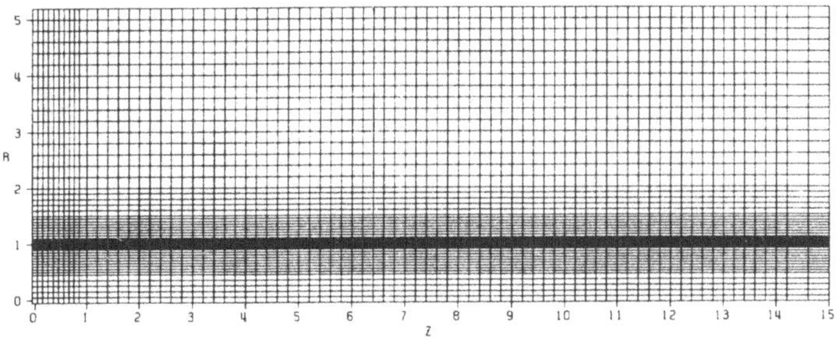

Fig. 1. The 79 x 52 nonuniform mesh.

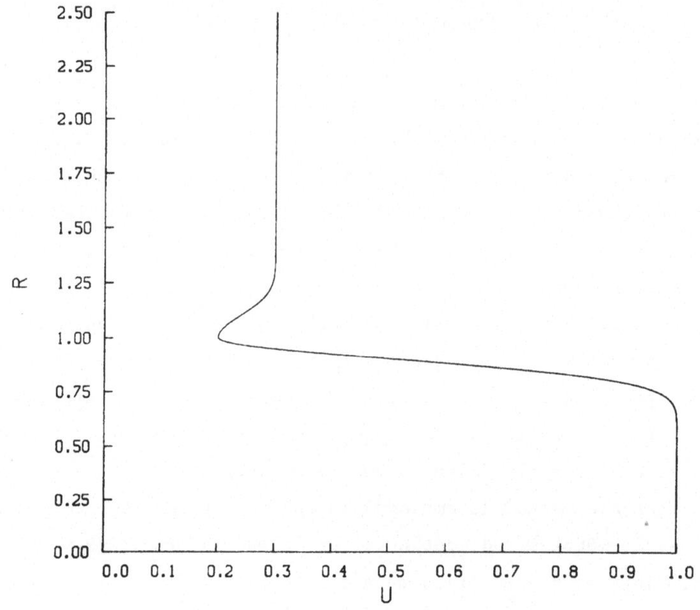

Fig. 2. Upstream velocity profile with $U_j/U_\infty = 3.33$.

initial shear layer near r = 1. The length scales in Fig. 1 are normalized with respect to R_j. All results in the next section are nondimensionalized with respect to R_j and U_j. Computation times on the NBS UNIVAC 1100/82 required to obtain a few unchanging cycles of vortex dynamics were typically about 3 hours.

III. Numerical Results

A plot of an upstream velocity profile with $U_j/U_\infty = 3.33$ is shown in Fig. 2. This profile, consisting of two Gaussians matched at r = 1, is typical of what is found downstream of a jet nozzle. A linear inviscid stability analysis reveals the most unstable frequency to be $\beta_1 = 3.48$. The results of perturbing this velocity profile with $\beta = \sum_n \beta_n = \beta_1$ (no subharmonics) are shown in Figs. 3 and 4 for Re = 1000. These two figures illuminate the same flowfield, the first by means of a streakline plot composed of passive marker particles and the second by means of isovorticity contours. What is seen here is the formation of vortices which shear as they move downstream but do not merge. Figures 5 and 6 show the effects of adding a subharmonic ($\beta = \beta_1 + \beta_1/2$) to the applied perturbation. A single vortex merging is now seen to occur. Adding a second subharmonic ($\beta = \beta_1 + \beta_1/2 + \beta_1/3$) results in two vortex mergings, as shown in Figs. 7 and 8. Thus, each subharmonic induces one merging, as seen experimentally in the two-dimensional mixing layer [3]. By applying only the fundamental and the second subharmonic ($\beta = \beta_1 + \beta_1/3$), three vortices merge into one (Figs. 9 and 10), a phenomenon also seen experimentally [3]. Changing the Reynolds number from 1000 to 333 results in the increased smearing of the vortices (with no change in the merging behavior) seen in Fig. 11 for $\beta = \beta_1 + \beta_1/2 + \beta_1/3$. For Re = 10^4 and this same β, the total streamwise energy content integrated across the jet as a function of axial distance downstream is shown in Fig. 12 for each of the three frequencies. The axial locations where each subharmonic saturates are, in fact, the same as the merging locations as seen from streakline and isovorticity contour plots. Also, the second subharmonic saturates at twice the distance from the upstream profile as the first. All this is, once again, in agreement with experimental results from the two-dimensional mixing layer [3].

IV. Conclusions

A computational model of the large-scale motions inside the forced axisymmetric mixing layer has been developed. The resulting vortex dynamics has been seen to be dependent on the subharmonic content of the forcing function in a manner analogous to that seen experimentally in the two-dimensional mixing layer. Thus, the behavior of the coherent structures inside the forced axisymmetric mixing layer can be predicted from knowledge of the forcing frequencies. Of course, since the modeling here does not account for either azimuthal instabilities or turbulence, these results may be applicable only in the near field, with the exact extent of this region dependent on the nature of the physical forcing.

Acknowledgment

This research was supported by the Air Force Office of Scientific Research.

References

1. Brown, G. L. and Roshko, A., J. Fluid Mech. 64, pp. 775-816 (1974).
2. Ho, C. M. and Huerre, P., Ann. Rev. Fluid Mech. 16, pp. 365-424 (1984).
3. Ho, C. M. and Huang, L. S., J. Fluid Mech. 119, pp. 443-473 (1982).
4. Corcos, G. M. and Sherman, F. S., J. Fluid Mech. 139, pp. 29-65 (1984).
5. Michalke, A. and Hermann, G., J. Fluid Mech. 114, pp. 343-359 (1982).
6. Davis, R. W. and Moore, E. F., J. Fluid Mech. 116, pp. 475-506 (1982).
7. Davis, R. W., Moore, E. F. and Purtell, L. P., Phys. Fluids 27, pp. 46-59 (1984).
8. Swarztrauber, P. and Sweet, R., Nat. Center Atmos. Res. Tech. Note IA-109 (1975).

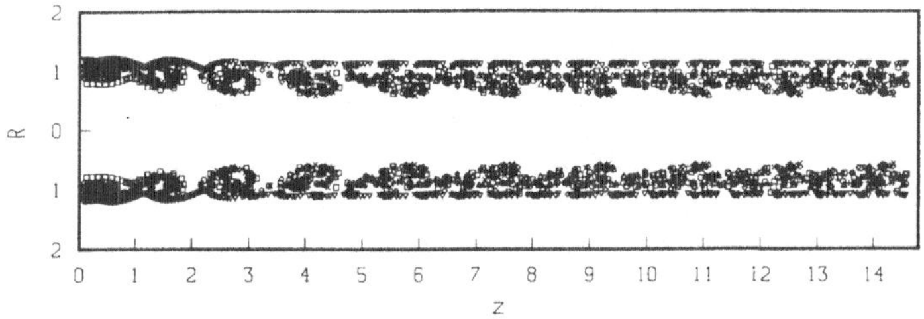

Fig. 3. Streakline plot: $\beta = \beta_1 = 3.48$.

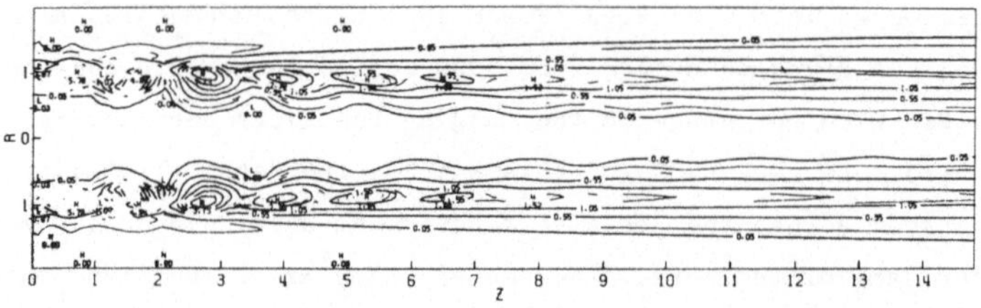

Fig. 4. Isovorticity contour plot: $\beta = \beta_1 = 3.48$.

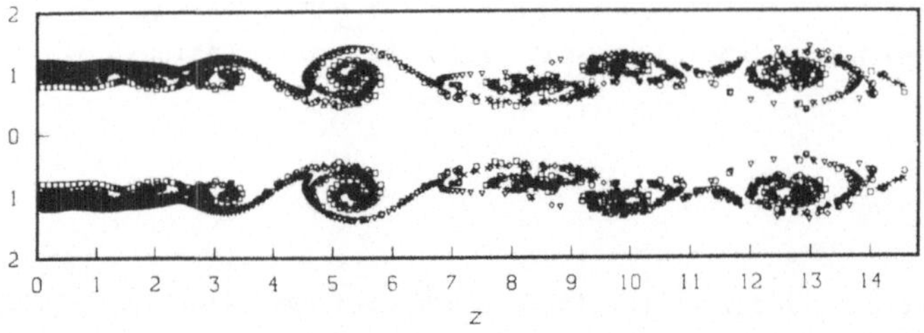

Fig. 5. Streakline plot: $\beta_1 = 3.48$, $\beta_2 = 1.74$.

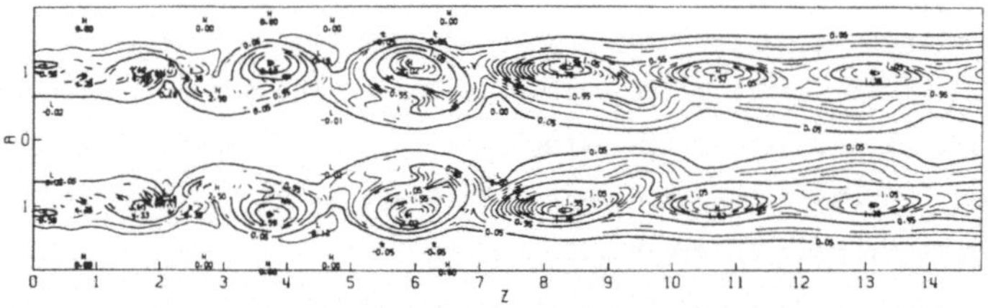

Fig. 6. Isovorticity contour plot: $\beta_1 = 3.48$, $\beta_2 = 1.74$.

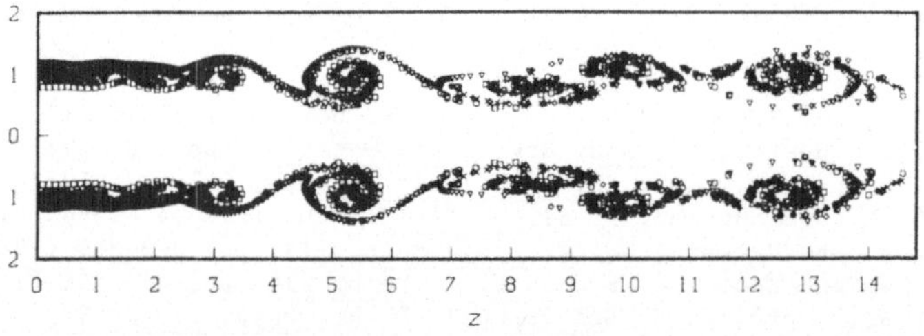

Fig. 7. Streakline plot: $\beta_1 = 3.48$, $\beta_2 = 1.74$, $\beta_3 = 1.16$.

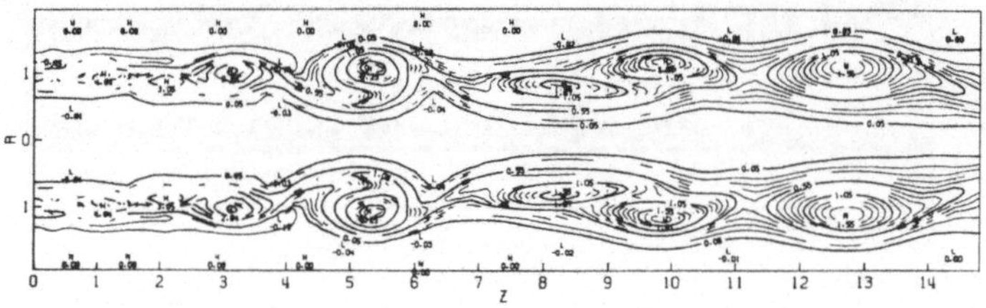

Fig. 8. Isovorticity contour plot: $\beta_1 = 3.48$, $\beta_2 = 1.74$, $\beta_3 = 1.16$.

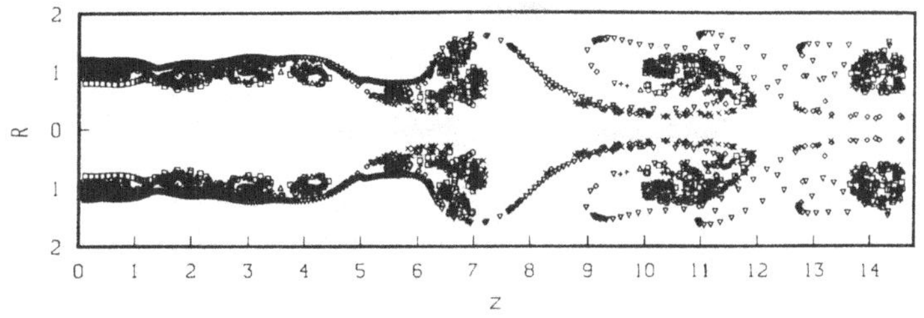

Fig. 9. Streakline plot: $\beta_1 = 3.48$, $\beta_2 = 1.16$.

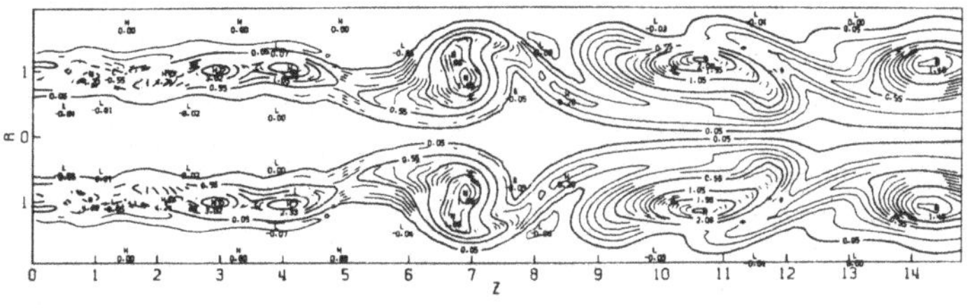

Fig. 10. Isovorticity contour plot: $\beta_1 = 3.48$, $\beta_2 = 1.16$.

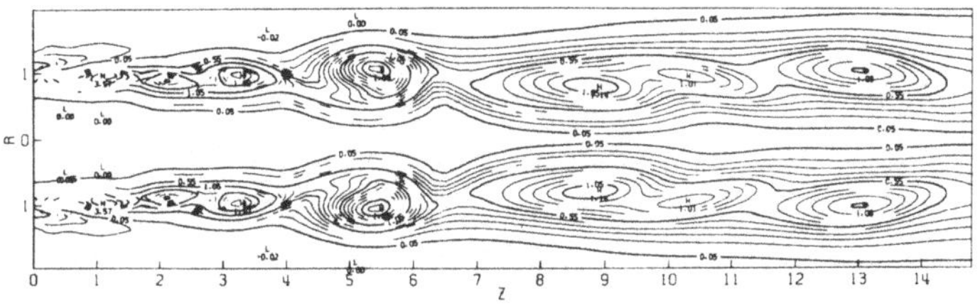

Fig. 11. Isovorticity contour plot for Re = 333: $\beta_1 = 3.48$, $\beta_2 = 1.74$, $\beta_3 = 1.16$.

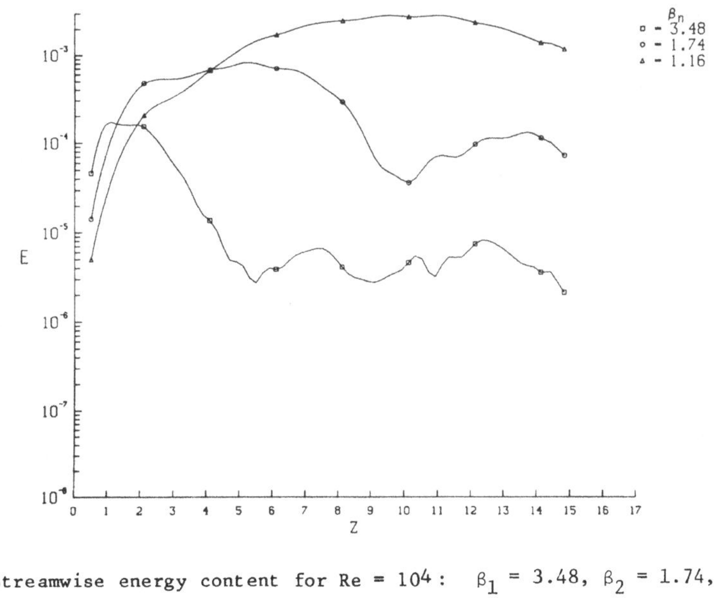

Fig. 12. Streamwise energy content for Re = 10^4: $\beta_1 = 3.48$, $\beta_2 = 1.74$, $\beta_3 = 1.16$.

A NEW MODIFIED SEMI-EXPLICIT DIFFERENCE SCHEME IN AERODYNAMICS

Fu De-Xun Ma Yan-Wen
BEIJING INSTITUTE OF AERODYNAMICS BEIJING CHINA

As it is known, explicit schemes are simple but time consuming. The implicit schemes can be carried out with larger time step but they are more complicated. In this paper we try to take advantage of superiorities of these two kinds of schemes and be free from their shortcoming. Two improvements on commonly used explicit schemes are made. One is scheme corrected with operator addition which may be used to solve both of steady state and time depend problems. The other is scheme corrected with operator amplification which is specially for solving steady state problems. The improved schemes are simple and easy to solve as explicit one and with large time increment like in the implicit schemes. The Navier-Stokes equations are approximated with this improved scheme to solve 2-D and 3-D shock wave boundary layer interaction problems. Numerical experiments show the corrected schemes are effective in improving the rate of convergence.

A. OPERATOR ADDITION

For simplicity consider the basic elements of the method applied to the following model equation with constant coefficients

$$\frac{\partial u}{\partial t} + c\frac{\partial u}{\partial x} = \gamma\frac{\partial^2 u}{\partial x^2} \qquad (1)$$

The difference scheme approximating the equation (1) can be written as

$$u_m^{n+1} = u_m^n + L(u_m^n) + L_{ad}(u_m^n, u_m^{n+1}) \qquad (2)$$

where L is a explicit operator and L_{ad} is a operator added correction. The following three conditions are required for the addition operator

a. The corrected scheme (2) has the same accuracy of approximation as without L_{ad};

b. the equation (2) is easy to solve;

c. stability condition can be improved.

For the equation (1) with c=0 the operator $L(u_m^n)$ and $L_{ad}(u_m^n, u_m^{n+1})$ can be taken as

$$L(u_m^n) = \gamma\frac{\delta_x^2 u_m^n}{\Delta x^2}\Delta t \qquad L_{ad}(u_m^n, u_m^{n+1}) = \frac{\beta_m^n}{2}(\delta_x^2 u_m^{n+1} - \delta_x^2 u_m^n) \qquad (3)$$

where $\delta_x^2 u_m^n = u_{m+1} - 2u_m + u_{m-1}$

The scheme (2) with (3) is stable for any $\Delta t/\Delta x$ if the parameter β is defined as

$$\beta_m^n = \max(\,0.0,\ \tfrac{1}{2}(\,2\gamma\Delta t/\Delta x^2 - 1\,)\,) \qquad (4)$$

The scheme (2) with (3) can be rewritten as

$$U_m^{n+1} = U_m^n + \frac{\beta_m^n}{2}\delta_x^2 U_m^{n+1} + \left(\gamma\frac{\Delta t}{\Delta x^2} - \frac{\beta_m^n}{2}\right)\delta_x^2 U_m^n \qquad (5)$$

In (5) the coefficient of $\delta_x^2 U_m^{n+1}$ is β instead of γ for commonly used six point implicit scheme. When (1) is a system of equations (c=0) the coeficient γ is a matrix. The corresponding parameter γ in (4) can be chosen, for some cases, as the maximum of eigenvalues of the matrix. In this case we have a tridiagonal matrix instead of block tridiagonal matrix for commonly used implicit schemes. For two step difference scheme the operator added correction can be introduced into each of predictor and corrector or into the corrector only. For 1-D hyperbolic system of equations the addition operator can be introduced into each of uncoupled equations in characteristic form. With special choice of parameter like β in (3) we can obtain much simpler system of difference equations approximating the original system of differential equations. For 2-D unsteady compressible form of Navier-Stokes equations $L(U_{i,j}^n)$ can be taken as explicit part of MacCormack scheme. In this case the addition operator in x direction can be given as

$$L_{ad}^{+}(\delta \overline{U_{i,j}^{n+1}}) = \frac{\Delta t}{\Delta x}\beta_1(\delta \overline{U_{i+1,j}^{n+1}} - \delta \overline{U_{i,j}^{n+1}})$$

for predictor and

$$L_{ad}^{-}(\delta U_{i,j}^{n+1}) = \frac{\Delta t}{\Delta x}\beta_1(-\delta U_{i,j}^{n+1} + \delta U_{i-1,j}^{n+1})$$

for corrector. The N-S equations can be numerically integrated in time by the following scheme corrected with simplified operator addition

$$P: \begin{cases} \Delta U_{i,j}^n = -\Delta t\left(\dfrac{\delta_x^+ F_{i,j}^n}{\Delta x} + \dfrac{\delta_y^+ G_{i,j}^n}{\Delta y}\right) \\ \left(1 - \dfrac{\Delta t}{\Delta x}\delta_x^+\beta_1\cdot\right)\left(1 - \dfrac{\Delta t}{\Delta y}\delta_y^+\beta_2\cdot\right)\delta \overline{U_{i,j}^{n+1}} = \Delta U_{i,j}^n \\ \overline{U_{i,j}^{n+1}} = U_{i,j}^n + \delta \overline{U_{i,j}^{n+1}} \end{cases} \qquad (6)$$

$$C: \begin{cases} \Delta \overline{U_{i,j}^{n+1}} = -\Delta t\left(\dfrac{\delta_x^- \overline{F_{i,j}^{n+1}}}{\Delta x} + \dfrac{\delta_y^- \overline{G_{i,j}^{n+1}}}{\Delta y}\right) \\ \left(1 + \dfrac{\Delta t}{\Delta x}\delta_x^-\beta_1\cdot\right)\left(1 + \dfrac{\Delta t}{\Delta y}\delta_y^-\beta_2\cdot\right)\delta U_{i,j}^{n+1} = \Delta \overline{U_{i,j}^{n+1}} \\ U_{i,j}^{n+1} = \frac{1}{2}\left(U_{i,j}^n + \overline{U_{i,j}^{n+1}} + \delta U_{i,j}^{n+1}\right) \end{cases} \qquad (7)$$

where $\delta_x^+ f_{i,j} = f_{i+1,j} - f_{i,j}$ $\qquad$ $\delta_x^- f_{i,j} = f_{i,j} - f_{i-1,j}$

$\delta_y^+ f_{i,j} = f_{i,j+1} - f_{i,j}$ $\qquad$ $\delta_y^- f_{i,j} = f_{i,j} - f_{i,j-1}$

The obtained system of difference equations is stable if

$$\beta_1 \geq \max_k\left(0.0,\ \tfrac{1}{2}\left(\lambda_k(A) + \frac{2\gamma}{\rho\Delta x}\right) - \tfrac{1}{2}\frac{\Delta x}{\Delta t}\right)$$
$$\beta_2 \geq \max_k\left(0.0,\ \tfrac{1}{2}\left(\lambda_k(B) + \frac{2\gamma}{\rho\Delta y}\right) - \tfrac{1}{2}\frac{\Delta y}{\Delta t}\right) \qquad (8)$$

where $\lambda_k(A)$ and $\lambda_k(B)$ are the eigenvalues of the Jacobian matrix A and B respectively. In the computation β_1 and β_2 are defined by

$$\beta_1 = \max\left(0.0, \tfrac{1}{2}(\lambda_a + \frac{2\gamma}{\rho\Delta x}) - \tfrac{1}{2}\frac{\Delta x}{\Delta t}\right) \qquad \lambda_a = |u| + c$$

$$\beta_2 = \max\left(0.0, \tfrac{1}{2}(\lambda_b + \frac{2\gamma}{\rho\Delta y}) - \tfrac{1}{2}\frac{\Delta y}{\Delta t}\right) \qquad \lambda_b = |v| + c \qquad (9)$$

The equation (6) and (7) are simple and easy to manipulate as the original explicit scheme because there is not any matrix operation. It is easier to solve than the implicit MacCormack scheme [1]. They are used to solve 2-D shock wave boundary layer interaction problem. The physical model is sketched in Fig.1. The inflow conditions are $M_\infty = 2$, $Re = 2.96\times10^5$. The total pressure increase is 1.4. Coordinate transformation is made in y direction in order to get fine solution in the viscous layer near the wall. 32x32 mesh points were used at first and then the mesh was rezoned to cover just the interaction region. The time step was successively reduced. The computer time required per step was less than that of implicit MacCormack scheme. The computed surface pressure and skin friction are given in the Fig.2 and 3. The comparison of results is also given there.

B. OPERATOR AMPLIFICATION

For steady state problems it is enough that the modified scheme has the same accuracy only for the steady state. In this case the supplementary operator may be simply taken as

$$L_{ad}(U_m^n, U_m^{n+1}) = -\beta\frac{\Delta t}{\Delta x}(U_m^{n+1} - U_m^n)$$

The modified scheme is obtained as

$$U_m^{n+1} = U_m^n + H\,L(U_m^n) \qquad (10)$$

where $H = 1/(1+\beta\frac{\Delta t}{\Delta x})$ can be considered as an amplification factor for $L(U_m^n)$. For L-W scheme approximating (1) with $\gamma=0$ the scheme (10) is stable for any $\Delta t/\Delta x$ if β is taken as $\beta = |c|$ because the Courant number $|c\frac{\Delta t}{\Delta x}/(1+|c|\frac{\Delta t}{\Delta x})|$ is always less than one. With large $\Delta t/\Delta x$ we have better uniform choice of Courant number for all mesh points where $c\neq0$. For system of equations the H can be defined as an specially chosen matrix. The equation (1) can be approximated by the following one step scheme with amplification factor [2,3]

$$U_m^{n+1} = U_m^n + H\frac{\Delta t}{\Delta x}\left(\gamma\frac{\delta_x^2 U_m^n}{\Delta x} - c\frac{\delta_x^+ U_m^n}{2} - c\frac{\delta_x^- U_m^{n+1}}{2}\right) \qquad (11)$$

$$H = 1/(1+\beta\frac{\Delta t}{\Delta x}) \qquad \beta = |c| + \frac{2\gamma}{\Delta x}$$

The simplified N-S equations are approximated by this scheme to solve 3-D shock wave boundary layer interaction problem. The flow field computed corresponds to flow of uniform supersonic stream at $M = 2.94$ over a plate with 10° half angle wedge standing vertically over the plate (Fig.4). Two cases with different Re based on the incoming boundary layer thickness have been computed. One with $Re_\delta = 687.5$, the other with $Re_\delta = 3000$. Coordinate transformation from (t, x, y, z) to (τ, ξ, η, ζ) is introduced to concentrate more grid points within the viscous layer next to the bottom wall y=0. The modified one step scheme (11) with following amplification factor

$$H = \left(1 + \lambda \frac{\Delta t}{\Delta x} \right)^{-1} \qquad \lambda = \lambda_1 + \lambda_2$$

$$\lambda_1 = |u| + k_1 \eta_y |v| + k_2 |w| + c\sqrt{1 + (k_1 \eta_y)^2 + k_2^2}$$

$$\lambda_2 = 2(1+(k_1 \eta_y)^2 + k_2^2)\frac{\gamma \mu}{Pr Re \Delta \xi}, \qquad k_1 = \frac{\Delta \xi}{\Delta \eta}, \qquad k_2 = \frac{\Delta \xi}{\Delta \zeta}$$

was used. The boundary conditions obtained from the Rankine-Hugoniot relations are given on the plane of symmetry z=0 instead of giving them on the wedge surface. 25x31x31 grid points were taken. The steady state solution is computed in about 300 time steps. Some of computed results are given in Fig. 5-7. Fig. 5 illustrates how the spanwise variation (along z) of pressure changes from the plate surface (j=1) to the inviscid field (j=30). When the computed results are projected in the plane normal to the inviscid shock and the flat plate y=0, we note that the flow is separated. V-U(n) plots at i=20 are illustrated in Fig.6 where U(n) is the projection of the velocity vector on the normal of the inviscid shock plane. In the Fig.7 U(s) is the projection of velocity vector in the shock plane in the direction parallel to the bottom wall. The profile U(s) as a function of y possesses break or departure form. This is because of spiral motion in the interaction region and getting thiner boundary layer thickness in the higher pressure region behind the shock.

References

1. MacCormack,R.W., AIAA paper 81-0110 , 1981
2. Ma Yanwen, Computational Mathematics, No. 2 , 1978 (in Chinese)
3. Ma Yanwen, Computational Mathematics, No. 1 , 1983 (in Chinese)

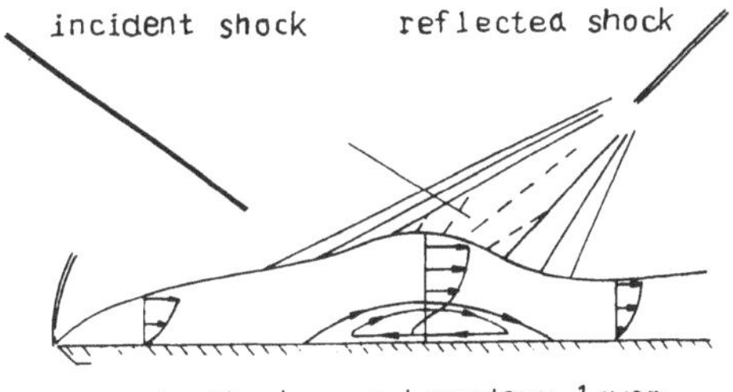

Fig.1 Shock wave boundary layer interaction

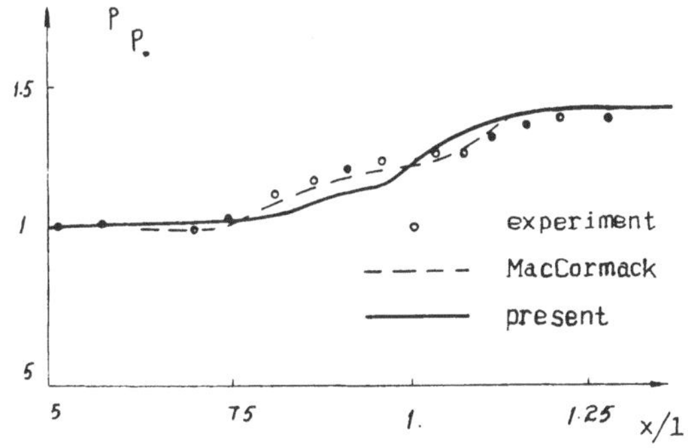

Fig.2 Surface pressure

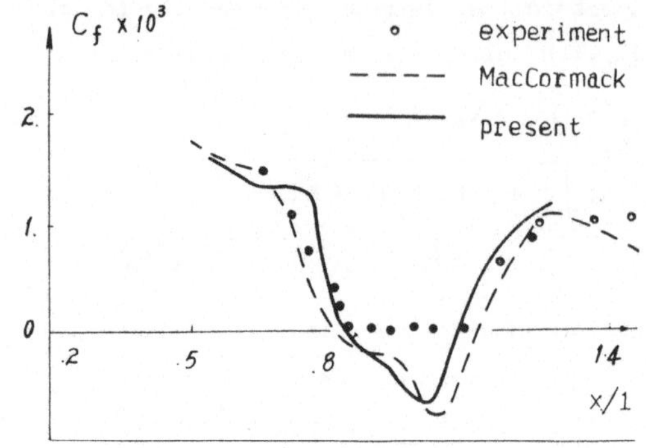

Fig.3 Skin friction

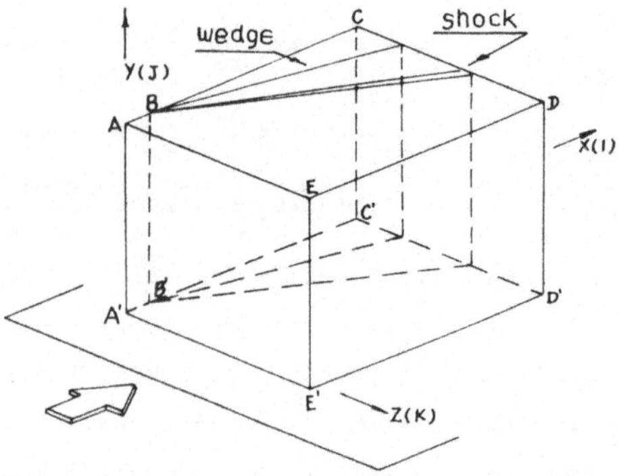

Fig.4 3-D computational field

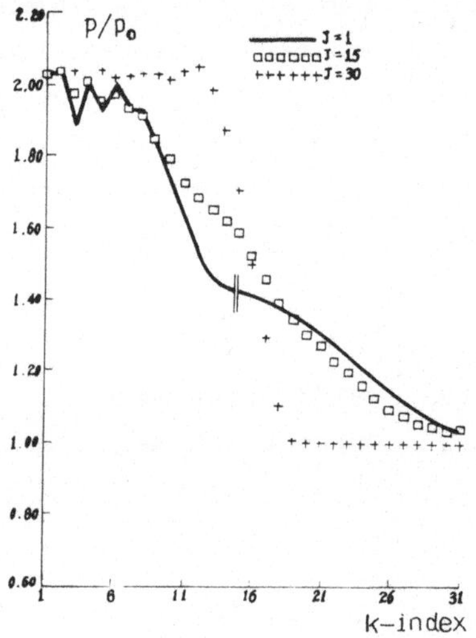

Fig.5 Pressure profile i=20

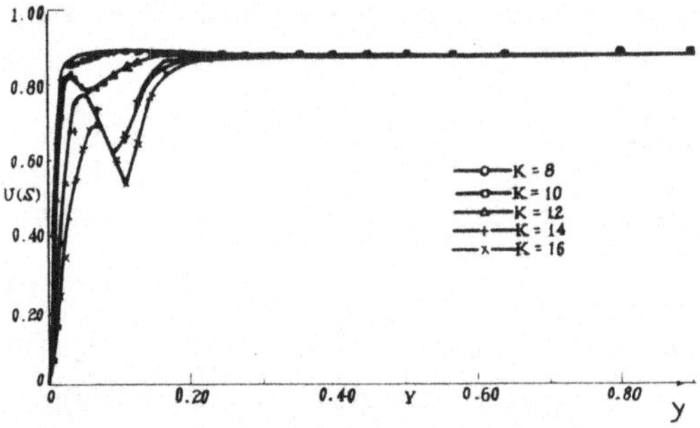

Fig.7a U_S profile at i=20

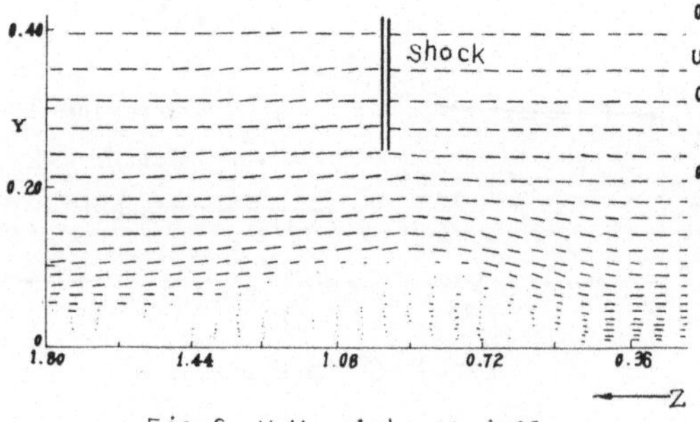

Fig.6 V-U_n plots at i=20

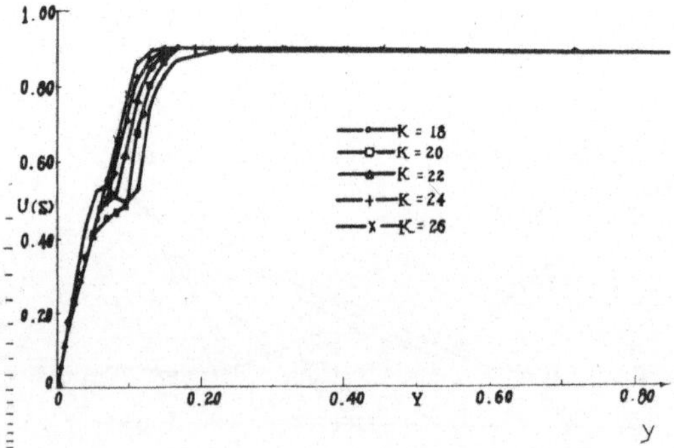

Fig. 7b U_s profile at i=20

THREE-DIMENSIONAL COMPUTATIONS OF NON-ISOTHERMAL WALL BOUNDED COMPLEX FLOWS

P. Dewagenaere - P. Esposito - F. Lana - P.L. Viollet
E.D.F. Laboratoire National d'Hydraulique
6, quai Watier - 78400 CHATOU FRANCE

Introduction

The practical computation of flows of engineering interest requires unsteady three-dimensional models able to take into account the turbulent properties of non isothermal flows as well as the complex shapes of the solid boundaries of industrial vessels. This paper describes an attempt to reach these requirements using a finite difference method, with examples of application.

Numerical methods

The algorithm is based upon the fractionary step technique, following Chorin [1], with primitive variables, in which the advective terms are treated in a separate step. A staggered cartesian or cylindrical grid is used to discretize the fields.

The three main steps of the algorithm are defined and solved as follows, inside a time step, the known fields f at time t^n being denoted as f^n.

a) advection :

$$\frac{\partial f}{\partial t} + \vec{C}^n \, \overrightarrow{\text{grad}} \, f = S \qquad (1)$$

where f is any scalar field to be transported (u_x, u_y, u_z, T, k, ε), and $\vec{C}$ is the convective field : velocity in the incompressible case, or momentum in variable density flows.
S is a source term including buoyancy effects, and production-dissipation terms for the case of the turbulent quantities.
This step is solved using a three-dimensional characteristic method whose Lagrangian nature allows to deal with high Reynolds number values. A third order interpolation is used, in order to reduce numerical diffusion [2].
The solution of this advection step leads to intermediate values of the unknown fields denoted as f_c.

b) diffusion :

$$\frac{\tilde{f} - fc}{\delta t} = \text{div}\ (K\ \vec{\text{grad}}\ f) \qquad (2)$$

where δ t is the time step, and K is an effective diffusivity, computed as the result of the turbulence model.

An implicit method is used for this elliptic operator.

c) continuity

$$\text{div}\ (\rho\ \vec{\text{grad}}\ P^{n+1}) = \frac{\rho}{\delta t}\ \text{div}\ (\rho\ \vec{\tilde{u}}) \qquad (3)$$

where $\vec{\tilde{u}}$ is the auxiliary velocity field resulting from the two previous steps.

This elliptic operator is solved with S.O.R. method, with Neuman boundary conditions for the pressure.

The final velocity field is then obtained through :

$$\frac{\vec{u}^{n+1} - \vec{\tilde{u}}}{\delta t} = -\frac{1}{\rho}\ \vec{\text{grad}}\ P \qquad (4)$$

Turbulence model and wall boundary conditions

The turbulence is modelled by means of the two equations k-ε model, with appropriate buoyancy induced terms [3].
The accurate description of the wall friction with practical mesh size makes it necessary to use the wall function techniques, which consists in boundary conditions for the shear stress rather than for the velocity, assuming a logarithmic velocity profile close to the wall.
The effective shape of the boundaries is discretized upon the finite difference grid, but general oblique boundaries are properly treated by means of the storage of the outer normal vector for each boundary cell.

Numerical performances

In order to deal with complex flows, a fine enough mesh must be used. For example the computation of figure 1 was made with 40 000 nodes (that is to say 280 000 unknowns). The computation of a 200 seconds transient in the plenum takes about 5 hours of CRAY 1 CPU time. Divided by the number of unknown, this CPU time turns out to be very close to the performances of many 2-D unsteady codes.

Application to the plena of pool-type fast breeder reactors

The liquid sodium flows, in the studied hot and cold plena, include steady recirculating situations as well as transient phenomena resulting from large temperature and flow rate variations (fast shut-down, power decay heat removal, ...). The large density effects, modelled using the Boussinesq approximation, may lead to stable thermal stratification and local damping of turbulence.
As an example, figure 1 shows the velocity field in the cold plenum of a fast reactor. From symetry considerations, 1/8 of the plenum is modelled. As it can be seen on the figure, the general shape of the

flow is strongly governed by large scale obstacles (pumps, heat exchangers, ...).
The main results of these studies are the temperature fields on the obstacles (structural mechanics point of view) and the caracteristics of the flows, for example near the pump intake of the cold plenum (hydrodynamics point of view). It can be added that, from a physical point of view, many kind of boundary conditions may be employed and the numerical model must be very adaptive. For instance, one of the last improvements of the model is the coupling with an external thermal field by means of temperature fluxes through the boundary.

Application to the mixing of a plasma jet in a pipe

The studied flow is the mixing of the subsonic high temperature (4000 °C) air jet emitted from an electric arc heater, in a cross-flowing air flow (20 °C to 1500 °C) in a pipe. The knowledge of the temperature field, in this cas, is very important to design the wall protection.
This kind of problem is characterized by small variations in the pressure, negligible gravity and radiation effects, but large variations of density.
These large density variations, lead to the use of mass-weighted (Favre) averaged values for the enthalpy, temperature, and velocity, defined as :

$$\hat{f} = \frac{\overline{\rho f}}{\rho} .$$

Preliminary two-dimensional computations [4], using mixed weighted averaging techniques in the modelling of density turbulent fluctuations, have shown that the difference between classical mean temperature and mass-weighted mean temperature is relatively small, especially close to the wall of the pipe.
In the algorithm, restricted here to steady-state situations, the convective field $\vec{C}$ in the first step is the momentum, and the variation of the mean density, for instance in the pressure equation, is taken into account.
The energy conservation equation is solved using the enthalpy ; the enthalpy-temperature law is non linear due to the chemical reactions in the plasma.

Figure 2 shows temperature and velocity fields computed in the vicinity of the discharge of a plasma jet in a pipe flow, with a 45° angle of incidence.

CONCLUSION

The presented results show that, through the use of an accurate treatment of the advective terms, a turbulence model allowing the turbulent length scales to vary with the flow, and an appropriate wall definition and treatment, it becomes possible to deal with complex three dimensional flows for applications to industrial equipments using finite difference methods. Among the remaining problems are the grid generation, which is at the present a quite time-consuming task, and the flow visualizations in complex three-dimensional geometries.

REFERENCES

[1] A.J. CHORIN "The numerical solution of the Navier-Stokes equations for an incompressible fluid". Bull. Amer. Math. Society 73, 1967.

[2] P. ESPOSITO "Résolution des équations de transport par la méthode des caractéristiques". Rapport EDF - HE 41/81.16.

[3] P.L. VIOLLET, J.P. BENQUE, J. GOUSSEBAILE "Two-dimensional numerical modelling of non-isothermal flows for unsteady thermal-hydraulic analysis". Nuclear Science and Eng. 84. 350-372. 1983.

[4] P.L. VIOLLET "Modélisation numérique du mélange d'un jet de plasma dans un écoulement froid en conduite". Rapport EDF - HE44/83.24.

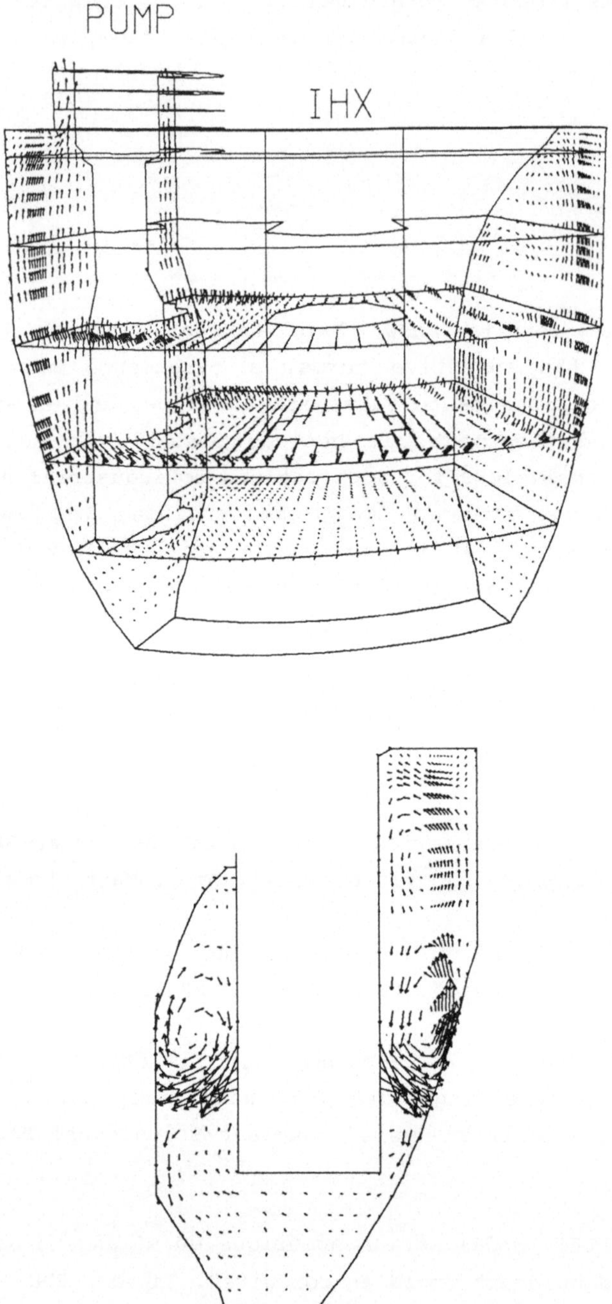

Figure 1
A GENERAL VIEW OF A VELOCITY FIELD IN THE COLD PRENUM OF A POOL TYPE FAST BREEDER REACTOR AND VERTICAL SECTION TROUGH THE HEAT EXCHANGER AXIS

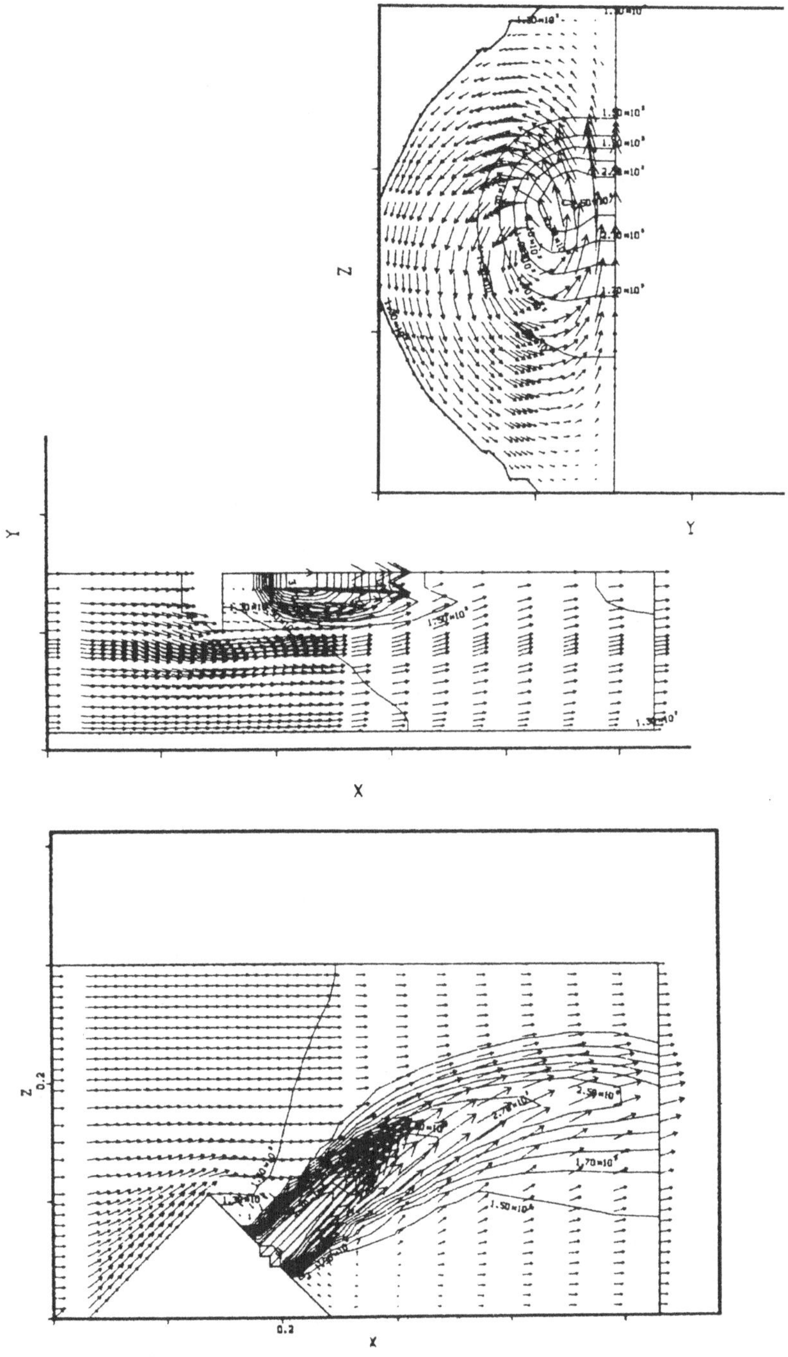

Figure 2

TEMPERATURE AND VELOCITY FIELDS COMPUTED IN THE VICINITY OF THE DISCHARGE OF A PLASMA JET IN A PIPE FLOW

A MULTIGRID TECHNIQUE FOR STEADY EULER EQUATIONS BASED ON FLUX-DIFFERENCE SPLITTING

DICK Erik

State University of Ghent

Department of machinery

Gent, Belgium

Recently, flux vector splitting techniques [1] and flux difference splitting techniques [2] were introduced to distinguish correctly between the influence of the forward- and backward moving waves in inviscid compressible flow, described by the Euler equations.

It was shown by Jespersen [3] that the flux vector splitting approach can be used succesfully to construct a relaxation operator usable in the multigrid technique. In this approach, the steady Euler equations are written as :

$$\frac{\partial F(Q)}{\partial x} + \frac{\partial G(Q)}{\partial y} = 0 \tag{1}$$

with $Q = \{\rho,\rho u,\rho v,\rho E\}^T$ (ρ is density, u and v are Cartesian velocity components, E is specific total energy).

Since the flux functions F and G are homogeneous of degree one in Q, one can write :

$$F(Q) = AQ \qquad\qquad G(Q) = BQ$$

A and B are diagonalizable with real eigenvalues.

With the left eigenvector matrix X : $X A X^{-1} = \Lambda = \Lambda^+ + \Lambda^-$ with $\Lambda = \text{diag}(\lambda_1,\ldots,\lambda_4)$

$$\Lambda^+ = \text{diag}(\lambda_1^+,\ldots,\lambda_4^+) \qquad\qquad \lambda_i^+ = \max(\lambda_i,0) \text{ (similar for B)}$$

(1) can be written as : $$\frac{\partial F^+}{\partial x} + \frac{\partial F^-}{\partial x} + \frac{\partial G^+}{\partial y} + \frac{\partial G^-}{\partial y} = 0 \tag{2}$$

with : $F^\pm = A^\pm\, Q$, $A^\pm = X^{-1}\, \Lambda^\pm\, X$ (similar for G).

In (2) backward differencing is used for the +terms while forward differencing is used for the -terms. Gauss-Seidel iteration is used on the resulting equations. This method works very well, but does not preserve the conservativity expressed by the primitive equations (1).

In this paper, it will be discussed how, on the basis of flux difference splitting, a similar algorithm can be constructed, which is fully conservative.

ONE-DIMENSIONAL EULER EQUATIONS

The discrete conservative equations for a one-dimensional channel can be written as :

Mass : $RUS(I+1) - RUS(I) = 0$ (3)

Momentum : $RUUS(I+1) - RUUS(I) + \frac{S(I+1)+S(I)}{2}\left(P(I+1)-P(I)\right) = 0$ (4)

Energy : $RHUS(I+1) - RHUS(I) = 0$ (5)

R = density, U = velocity, P = pressure, H = total enthalpy, S = section.

Using the splitting rule :

$$a(I+1)b(I+1) - a(I)b(I) = \frac{b(I+1)+b(I)}{2}\Big(a(I+1)-a(I)\Big) + \frac{a(I+1)+a(I)}{2}\Big(b(I+1)-b(I)\Big)$$

(3) can be written as :

$$\frac{S(I+1)+S(I)}{2}\left[\frac{R(I+1)+R(I)}{2}\Big(U(I+1)-U(I)\Big) + \frac{U(I+1)+U(I)}{2}\Big(R(I+1)-R(I)\Big)\right]$$

$$+ \frac{RU(I+1)+RU(I)}{2}\Big(S(I+1)-S(I)\Big) = 0 \qquad \text{or} \qquad \overline{U}\Delta R + \overline{R}\Delta U = -\frac{\overline{RU}}{\overline{S}}\Delta S \tag{6}$$

Similarly, (4) can be written as : $\overline{U}\Delta U + \frac{1}{\overline{\overline{R}}}\Delta P = 0$ with $\overline{\overline{R}} = \frac{\overline{RUS}}{\overline{U}\,\overline{S}}$ (7)

(5) can be written, through combination with (6) as : $\gamma\overline{P}\Delta U + \overline{U}\Delta P = -\frac{\gamma\overline{PU}}{\overline{S}}\Delta S$ (8)

The resulting system is :

$$\begin{pmatrix} \overline{U} & \overline{R} & 0 \\ 0 & \overline{U} & R^{-1} \\ 0 & \gamma\overline{P} & \overline{U} \end{pmatrix}\begin{pmatrix} \Delta R \\ \Delta U \\ \Delta P \end{pmatrix} = \begin{pmatrix} -(\overline{RU}/\overline{S})\Delta S \\ 0 \\ -(\gamma\overline{PU}/\overline{S})\Delta S \end{pmatrix} \qquad \text{or symbolically : } A\Delta q = b \tag{9}$$

The system (9) is exactly equivalent to the equations (3-5).

The matrix A has the eigenvalues : $\lambda_1 = \overline{U}$ $\lambda_{2,3} = \overline{U} \pm \overline{C}$ with $\overline{C} = \sqrt{\gamma\overline{P}/\overline{\overline{R}}}$

The matrix A can be splitted into a positive and a negative part (for subsonic flow) :

$$A^+ = X^{-1}\Lambda^+X \qquad A^- = X^{-1}\Lambda^-X \qquad \text{with} \quad \Lambda^+ = \begin{pmatrix} \overline{U} & & \\ & \overline{U}+\overline{C} & \\ & & 0 \end{pmatrix} \qquad \Lambda^- = \begin{pmatrix} 0 & & \\ & 0 & \\ & & \overline{U}-\overline{C} \end{pmatrix}$$

The system (9) can be splitted using the truth matrix approach of Lombard et al. [4]. Multiplicator matrices are defined by :

$$M_{A^+} = X^{-1}D^+X \qquad M_{A^-} = X^{-1}D^-X \qquad \text{with} \quad D^+ = \begin{pmatrix} 1 & & \\ & 1 & \\ & & 0 \end{pmatrix} \qquad D^- = \begin{pmatrix} 0 & & \\ & 0 & \\ & & 1 \end{pmatrix}$$

such that : $M_{A^+}.\,A = A^+$ $M_{A^-}.\,A = A^-$

The system $M_{A^+}.\,A\,\Delta q = M_{A^+}.\,b$ has two independent equations. (10)

The system $M_{A^-}.\,A\,\Delta q = M_{A^-}.\,b$ has one independent equation. (11)

The set of equations associated to node I can be composed by the two independent equations in (10) in the interval (I-1,I) and the one independent equation in (11) in the interval (I,I+1). In order to close the system, one boundary equation is to be prescribed at the outlet (pressure) and two boundary conditions at the inlet (stagnation properties). In practice, the surch for the independent equations in the A^+ and A^- systems is not necessary. An equivalent set of node equations can be found by assembling (i.e. adding equation per equation) the systems (10) in (I-1,I) and (11) in (I,I+1).

In order to make transonic calculations possible, the splitting of Λ can be done as :

$$\Lambda^+ = \begin{pmatrix} \overline{U} & & \\ & \overline{U}(1+\frac{1}{M}) & \\ & & \overline{U}(\frac{1}{M^*}-\frac{1}{M}) \end{pmatrix} \qquad \Lambda^- = \begin{pmatrix} 0 & & \\ & 0 & \\ & & \overline{U}(1-\frac{1}{M^*}) \end{pmatrix} \qquad \text{with} \quad \begin{matrix} M^* = M & \text{for } M < 1-\varepsilon \\ M^* = 1-\varepsilon & \text{for } M \geqq 1-\varepsilon \end{matrix}$$

The splitting for b has to be modified accordingly. The equations associated to node I are assembled in the same way as in the subsonic case. This leads to some inconsistency since for $M > 1-\varepsilon$, the A^+ equations in the interval (I-1,I) already form a set of independent equations. Yet they are assembled with A^- equations from the interval (I,I+1). This inconsistency is necessary to avoid the singularity of the node equations, which occurs by correct splitting of A in a sonic point (third eigenvalue is exactly zero). This inconsistency generates a conservation error at sonic points and at shocks.

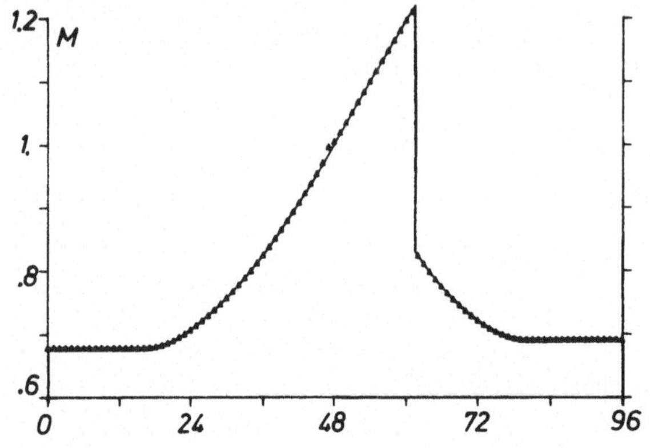

Fig. 1

Figure 1 shows the calculated result for a convergent divergent nozzle, discretized into 96 elements, with $\varepsilon = 0.007$. The solution was obtained by symmetric Gauss-Seidel relaxation (i.e. alternating the direction of sweep). Except for a small error at the sonic point, the solution cannot be distinguished from the exact solution. For small values of ε, the error in the sonic region becomes large due to the bad condition of the set of discrete equations for $M \cong 1$. For large ε the shock moves to a wrong position due to the conservation error at the shock.

TWO-DIMENSIONAL INCOMPRESSIBLE FLOW : CAUCHY-RIEMANN EQUATIONS

The foregoing flux difference splitting easily can be applied to Cauchy-Riemann equations :

$$\begin{pmatrix}1 & 0\\0 & -1\end{pmatrix}\frac{\partial}{\partial x}\begin{pmatrix}u\\v\end{pmatrix} + \begin{pmatrix}0 & 1\\1 & 0\end{pmatrix}\frac{\partial}{\partial y}\begin{pmatrix}u\\v\end{pmatrix} = 0 \qquad (12)$$

The matrices can be splitted into :

$$A^+ = \begin{pmatrix}1 & 0\\0 & 0\end{pmatrix} \qquad A^- = \begin{pmatrix}0 & 0\\0 & -1\end{pmatrix} \qquad B^+ = \begin{pmatrix}.5 & .5\\.5 & .5\end{pmatrix} \qquad B^- = \begin{pmatrix}-.5 & .5\\.5 & -.5\end{pmatrix}$$

The corresponding multiplicator matrices are :

$$M_{A^+} = \begin{pmatrix}1 & 0\\0 & 0\end{pmatrix} \qquad M_{A^-} = \begin{pmatrix}0 & 0\\0 & 1\end{pmatrix} \qquad M_{B^+} = \begin{pmatrix}.5 & .5\\.5 & .5\end{pmatrix} \qquad M_{B^-} = \begin{pmatrix}.5 & -.5\\-.5 & .5\end{pmatrix}$$

A complete conservative discretisation is obtained by a finite volume discretisation (on a square element) :

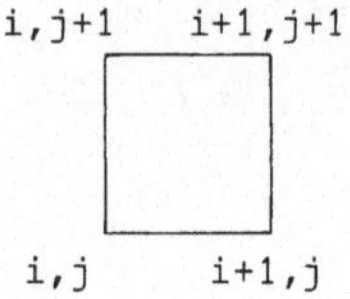

$$\begin{aligned} &u(i+1,j+1) - u(i,j+1) + u(i+1,j) - u(i,j) + v(i+1,j+1)\\ &\qquad + v(i,j+1) - v(i+1,j) - v(i,j) = 0\\ &u(i+1,j+1) + u(i,j+1) - u(i+1,j) - u(i,j) - v(i+1,j+1)\\ &\qquad + v(i,j+1) - v(i+1,j) + v(i,j) = 0 \end{aligned} \qquad (13)$$

The equations associated to a node are obtained by assembling 8 sets of equations.

NW		NE
	A^+B^-	A^-B^-
	A^+B^+	A^-B^+
SW		SE

From the south-west element an A^+ and a B^+ set are taken. These are obtained by multiplying (13) by M_{A^+} and by M_{B^+} respectively. The other elements give contributions as indicated in the sketch.

The resulting equations can be solved using coloured relaxation with 4 colours. Figure 2 shows the calculated lines of constant velocity for a square grid of 32 by 96 elements with boundary conditions : u = 1 at inlet, v = 0 at outlet, v = 0 at the upper boundary and v = αu at the lower boundary (α is positive in the first half of the boundary and negative in the second half).

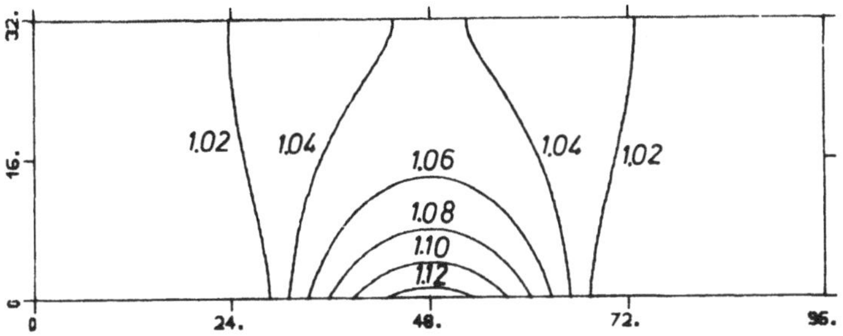

Fig. 2

The relaxation scheme possesses smoothing properties so that it can be used in the multigrid method. Figure 3 shows the convergence history for the single grid calculation with relaxation factor R = .45 (optimal acceleration) and the convergence history for the multigrid calculation using 4 grids : 32x96, 16x48, 8x24, 4x12. The optimal cycle is a sawtooth cycle with 4 relaxation steps on each grid before restriction to a coarser grid, 10 relaxation steps on the coarsest grid and direct prolongation from the coarsest to the finest grid. The cycle costs 6.71875 work units, i.e. : (4+1) x (1+1/4+1/16) + 10/64 (experimentally verified).

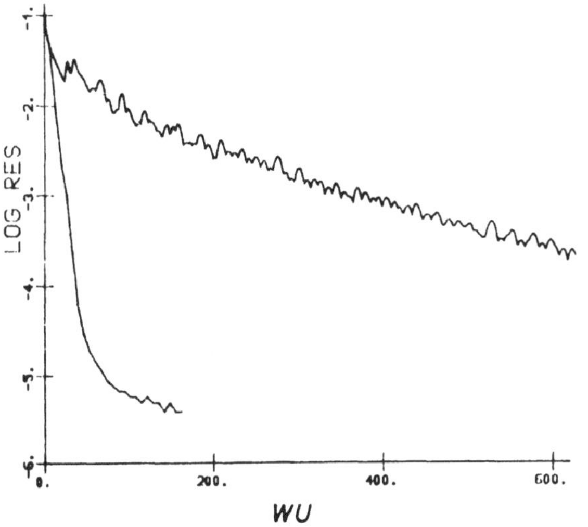

Fig. 3

TWO-DIMENSIONAL TRANSONIC FLOW : EULER EQUATIONS

The same procedure as described for Cauchy-Riemann equations can be applied to Euler equations if a suitable splitting method for the non-linear equations is used. The splitting method of Roe [2], based on the quadratic form of the equations with respect to $\sqrt{\rho}$, $\sqrt{\rho}$ u, $\sqrt{\rho}$ v and $\sqrt{\rho}$ H, can be applied to form a splitting of the finite volume equations, which, on a square grid, take the form :

$$A_1(q_{i+1,j}-q_{i,j}) + A_2(q_{i+1,j+1}-q_{i,j+1}) + B_1(q_{i,j+1}-q_{i,j}) + B_2(q_{i+1,j+1}-q_{i+1,j}) \doteq 0 \quad (14)$$

The splitting of the matrices has to be corrected for vanishing eigenvalues in a way similar to the one described in section 1. Of coarse, due to the slight differences between the matrices A_1 and A_2, respectively B_1 and B_2, in each finite volume, a correct simultaneous splitting of the system matrices in (14) is impossible. As a consequence, the ellipticity of the discrete equations is lost in sonic regions. This necessitates the use of artificial viscosity. Figure 4 shows a coarse grid result obtained on a square mesh, using first order extrapollations as numerical boundary conditions, together with v = 0 at the upper boundary, v = αu at the lower boundary, p equal to an imposed value at outlet and v = 0 at inlet with imposed stagnation conditions. For such a coarse grid, which gives an almost one-dimensional solution, artificial viscosity is not necessary.

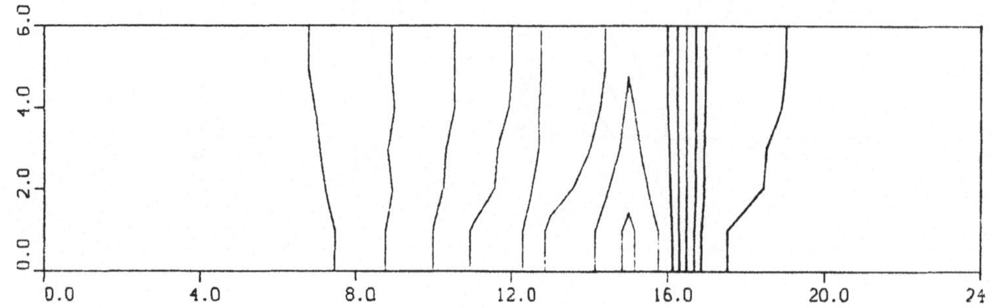

Fig. 4

On fine grids, usually a large amount of artificial viscosity is needed. This is completely unsatisfactory. Other drawbacks of the method are : a very complex splitting using square root of density, an assemblage of 8 sets of equations to form the node equations and additional complications in splitting for non-rectangular elements. Therefore, experiments are now done using the simple splitting, described for one-dimensional Euler equations, on rotated elements. This involves the assemblage of only 4 sets of equations to form the node equations, avoids the need of simultaneous splitting of matrices and, as a consequence, of artificial viscosity and can treat non-rectangular elements in a simple way.

ACKNOWLEDGEMENT

Part of this research was performed while the author was in residence at the NASA Ames Research Center, on a grant of the Belgian National Science Foundation (NFWO).

REFERENCES

1. Steger J.L. and Warming R.F. : "Flux vector splitting of the inviscid gasdynamic equations with application to finite-difference methods", J. Comp. Phys., Vol. 40, 1981, pp. 263-293.
2. Roe P.L. : "Approximate Riemann solvers, parameter vectors and difference schemes", J. Comp. Phys., Vol. 43, 1981, pp. 357-372.
3. Jespersen D.C. : "A multigrid method for the Euler equations", AIAA paper 83-0124, 1983.
4. Lombard C.K., Oliger J. and Yang J.Y. : "A natural conservative flux difference splitting for the hyperbolic systems of gasdynamics", AIAA paper 82-0976, 1982.

GENERATION OF FULLY ADAPTIVE AND/OR ORTHOGONAL GRIDS

H. A. Dwyer, Professor

O. O. Onyejekwe, Post Doctoral Researcher

Department of Mechanical Engineering
University of California
Davis, California 95616

Introduction

The need for the generation of adaptive grids for the numerical solution of the equations of fluid dynamics is very strong for those problems which contain regions of high gradient of an important dependent variable. Regions of high gradient are very common in fluid dynamics and are usually caused by the flow geometry and large values of the basic parameters which appear in the transport equations, such as Reynolds number. In general, the use of an adaptive grid which reflects the variation of a dependent variable in the partial differential equations improves both the accuracy and efficiency of a numerical method.[1] The present research paper is directed toward improving one of the possible major problems that can result from grid adaption, which is grid skewness or non-orthogonality. Grid systems which are highly skewed and non-orthogonal lead to many types of inaccuracies for both finite difference and finite volume applications. The most serious of these inaccuracies are usually large truncation errors and lack of convergence.

The problem of the formulation of adaptive and orthogonal grids is not a trivial one, and it can be shown that it is an overspecification of the mathematical problem to both adapt and orthogonalize a grid system simultaneously.[2] The approach that will be taken in the present two-dimensional research will be to form one family of grid lines based upon adaption of an important dependent variable and the other on the basis of orthogonality. Although this approach could be an attractive one for many problems, it sometimes cannot be utilized because of the need for adaption along both directions or because of very complex geometries. However, the basic method presented has many attractive features and can be useful in many problem areas in fluid dynamics.

Analysis

In the present paper the Navier-Stokes equations will be solved for the flow over a parachute-type body in axi-symmetric flow. The basic equations will be solved in streamfunction-vorticity variables (ψ, ω) and also in terms of generalized coordinates.[3,4] These equations in terms of the initial cylindrical coordinates (r,z) are

$$\frac{\partial \vec{Q}}{\partial t} + \frac{\partial \vec{E}}{\partial r} + \frac{\partial \vec{F}}{\partial z} = \frac{\partial \vec{R}}{\partial r} + \frac{\partial \vec{S}}{\partial z} + \vec{H} \tag{1}$$

where

$$\vec{Q} = \begin{pmatrix} 0 \\ \rho r \omega \end{pmatrix} \qquad \vec{E} = \begin{pmatrix} 0 \\ \rho r v \omega \end{pmatrix} \qquad \vec{F} = \begin{pmatrix} 0 \\ \rho r u \omega \end{pmatrix}$$

$$\vec{R} = \begin{pmatrix} \frac{1}{r}\frac{\partial \psi}{\partial r} \\ r\frac{\partial}{\partial r}[\mu\omega] \end{pmatrix} \qquad \vec{S} = \begin{pmatrix} \frac{1}{r}\frac{\partial \psi}{\partial z} \\ \frac{\partial}{\partial z}[\mu\omega] \end{pmatrix} \qquad \vec{H} = \begin{pmatrix} -\omega \\ -\frac{\mu\omega}{r} + \rho v \omega \end{pmatrix}$$

and where u and v are the velocity components in the z and r direction, respectively.

The adaptive method we will be using is based upon the use of an integral functional relationship which is applied along generalized arcs in space.[5] If f is the dependent variable used for adaption and s is the arclength location, then the adaptive relationship becomes

$$\xi(x,y,t) = \frac{\int_0^S \left(1 + b_1\left|\frac{\partial f}{\partial s}\right| + b_2\left|g\left(\frac{\partial^2 f}{\partial s^2}\right)\right|\right)ds}{\int_0^{S_{max}} \left(1 + b_1\left|\frac{\partial f}{\partial s}\right| + b_2\left|g\left(\frac{\partial^2 f}{\partial s^2}\right)\right|\right)ds} \tag{2}$$

where ξ is one of the generalized coordinates associated with the transformation

$$x, y, t \rightarrow \xi, \eta, t$$

S_{max} is maximum arclength distance, g is a function of the second derivative of f, and b_1 and b_2 are weighting functions which determine the relative importance of adaption criteria. The functions b_1 and b_2 have been discussed in Ref. [5] and are used to very accurately control the percentage change in a variable and also to limit the grid cell Reynolds number in regions of high gradient.

For an adaptive grid method it is still necessary to have a method to generate a starting grid and it is necessary to utilize a conventional grid generation technique.[4,6] In the present study the "hyperbolic" grid generation[6] method has been employed. This technique generates a body fitted orthogonal grid, which is generated as an initial value problem from the body surface. Shown in Fig. 1 are the results of the use of this method for an axisymmetric body (shown by white space) and for a flow from left to right. The grid has been expanded geometrically from the body surface and it is obvious that the orthogonality condition has been enforced.

The procedure which was followed after the generation of this initial grid was the following: (1) initial solution of the Navier-Stokes equations with an implicit windward difference method; (2) grid adaption based on velocity or vorticity; (3) grid orthogonalization with the use of a Green's function technique[7]; and (4) solution of the Navier-Stokes equations with an implicit central difference method for a final solution. Windward difference techniques were used initially to stabilize the convection terms with artificial viscosity and were not needed with the adaptive grid since the high gradient regions had been resolved. Some of the details of this procedure will be given in the results section of the paper.

Results

The initial grid and solution are shown in Figs. 1-3 for the parachute body and for a flow Reynolds number of 200. Because of the bluntness of the body, the flow separates at a very large angle with respect to the body (streamlines shown in Fig. 2) and the boundary layer vorticity is also ejected from the body surface, Fig. 3. Based upon this initial solution, adaptive grids are generated with the use of the integral equation presented previously. The grid lines leaving perpendicular to the body surface are retained and the grid points are moved along these lines. The results of this procedure are shown in Figs. 4 and 5 where the variable of adaption, f, has been the absolute value of vorticity and the absolute value of the total velocity. (Note: only first derivative adaption has been employed).

The grid generated based on vorticity, Fig. 4, accurately reflects the no-slip condition at the wall and the free shear layer which leaves the body. The other grid, Fig. 5, based on total velocity, sees the changes in both the inviscid flow as well as the viscous flow and is much more uniform. At the present time, the "best" variable for adaption is not known and the skill of the investigator must be employed to "pick" a final adaption variable. All variables have some strengths and weaknesses, but good grid adaption can dramatically improve the accuracy and efficiency of a numerical solution.

The final step in the sequence is to form an adaptive/orthogonal grid based on the use of a Green's function procedure which is applied from the body surface between successive lines around the body. The orthogonal trajectories between two lines are governed by the following relationship[7]

$$p(\vec{r}) = \frac{1}{\pi} \oint P \frac{\partial}{\partial n} \left\{ \ell n \frac{1}{\vec{r} - \vec{r}'} \right\} d\vec{r}' \tag{3}$$

The orthogonal trajectories can be defined from any arbitrary distribution of points and the adaptive lines are not changed in their location. Equal values of p on two adjacent lines are used with interpolation to generate the actual trajectories.

An example of this analysis is shown in Fig. 6 where the grid has been generated starting from Fig. 5. (Note: Fig. 4 could have been used equally as well). This new grid in Fig. 6 accurately follows the gradients in total velocity and the maximum percentage change in total velocity between node points has been limited to less than five percent. Therefore, we have a grid which will calculate solutions without artificial diffusion and which does not suffer from grid skewness problems. Almost all numerical methods will benefit from this grid and it will yield solutions which are both more accurate and efficient.

References

1. Dwyer, H. A., Smooke, M. D., and Kee, R. J., "Adaptive Gridding for Finite Difference Solutions to Heat and Mass Transfer Problems," Numerical Grid Generation, Ed. J. F. Thompson, North-Holland Pub., New York.

2. Brackbill, J. U. and Saltzman, "Adaptive Zoning for a Singular Problem in Two Dimensions," Los Alamos Scientific Labs, LA-UR-81-405, Los Alamos, NM, 1980.

3. Peyret, R. and Viviand, H., "Computation of Viscous Compressible Flows Based on the Navier-Stokes Equations," AGARD-AG-212, 1975.

4. Thompson, J. F., Thames, F. C., and Mostin, C. M., "Automatic Numerical Generation of Body Fitted Curvilinear Coordinate Systems for Fields Containing any Number of Arbitrary Two-Dimensional Bodies," Journal of Computational Physics, Vol. 15, July 1974, pp. 299-319.

5. Dwyer, H. A., "Grid Adaption for Problems with Separation, Cell Reynolds Number, Shock-Boundary Layer Interaction and Accuracy," to appear in AIAA Journal, 1984.

6. Stega, J., private communication.

7. Potter, D. E. and Tuttle, G. H., "The Construction of Discrete Orthogonal Coordinates," Journal of Computational Physics, Vol. 13, 1973, pp. 483-501.

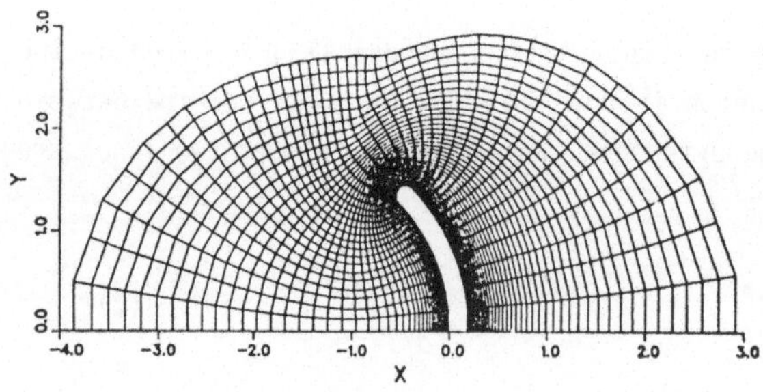

Figure 1. Grid Generated by the Hyperbolic Solver for the Parachute Body

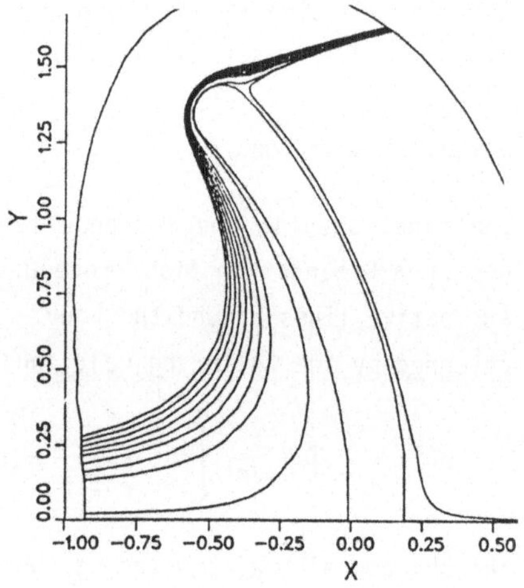

Figure 2. Streamlines Near the Body Surface, Re = 200

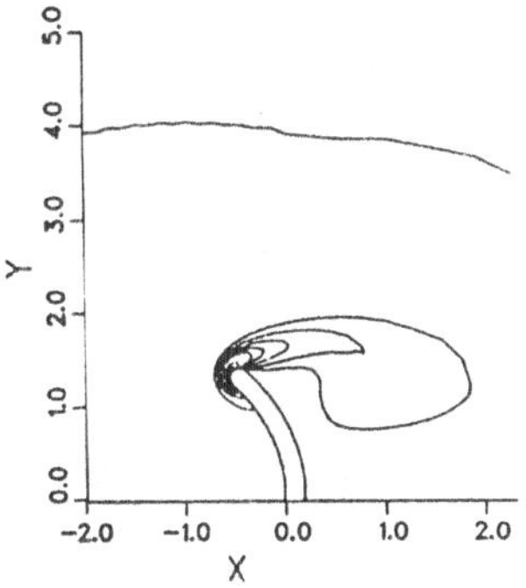

Figure 3. Vorticity Contours Around the Body

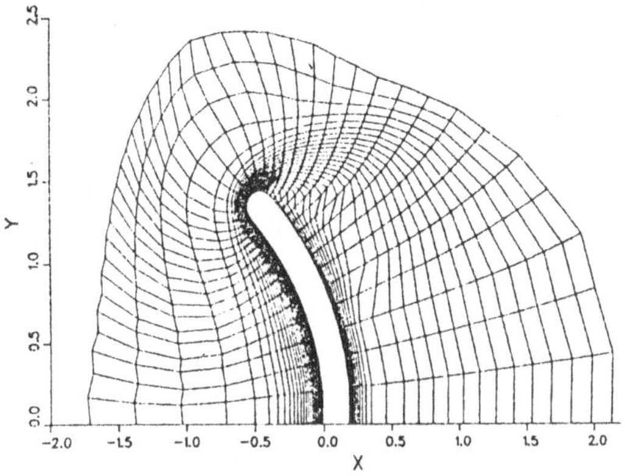

Figure 4. Adaptive Coordinates Based on Vorticity

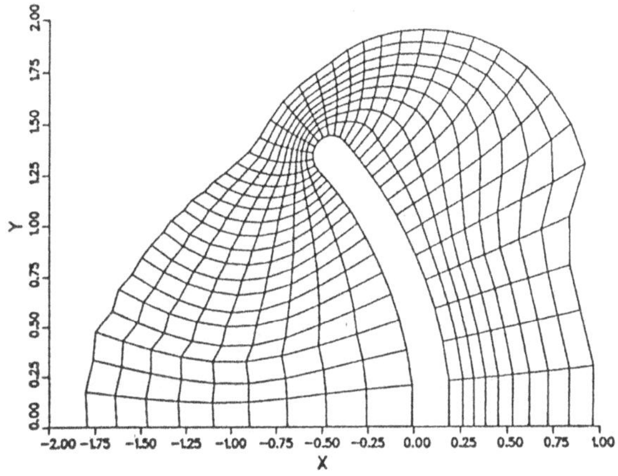

Figure 5. Adaptive Coordinates Based on Velocity

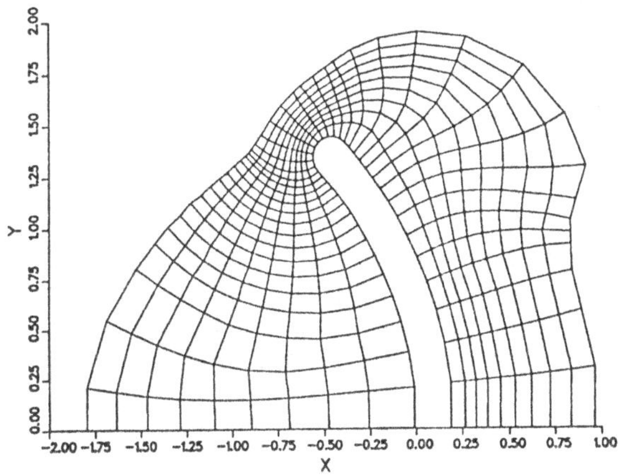

Figure 6. Adaptive/Orthogonal Coordinates Based on Velocity

COMPUTATION OF COMPRESSIBLE TWO-DIMENSIONAL TURBULENCE IN NON ROTATING AND ROTATING FLOWS

Marie FARGE
Laboratoire de Météorologie Dynamique
du C.N.R.S.
Ecole Normale Supérieure
24, rue Lhomond
75231 PARIS Cedex 5
Tel : 329.12.25 p. 3286

Introduction

The study of two-dimensional turbulence is essential for a better understanding of the dynamics of planetary flows. Its phenomenology is different from the three-dimensional case because, then, not only energy is conserved but also enstrophy (integral of the vorticity squared), which consequently may give rise to an inverse energy cascade. While the incompressible two-dimensional turbulence has already received a lot of attention, the compressible case has not yet been analyzed extensively.

1. *Equations*

We study the dynamics of compressible two-dimensional turbulence using Saint-Venant equations, i.e. Euler equations in the shallow water approximation.

Hypotheses :
1. incompressible barotropic fluid
   i.e. $\vec{\nabla}.\vec{V} = 0$ in three dimensions,
2. hydrostatic equilibrium
   i.e. $P = 4\rho(\vec{n}.\vec{g})$ and the horizontal components of the velocity field are independent of the vertical space variable,
3. plane periodic flow.

ρ	Density (ρ=1)
$\vec{g}$	Gravity
$\vec{n}$	Normal to the plane
$\vec{\nabla}_H$	Horizontal gradient
P	Pressure field
$\vec{V}$	Velocity field
f	Coriolis parameter

Equations (on the plane) :

$$P_t + \vec{\nabla}_H.(P\vec{V}) = 0$$

$$\vec{V}_t + (\vec{V}.\vec{\nabla})\vec{V} + \vec{\nabla}P + \vec{n} \times f\vec{V} = \vec{0}$$

In order to reduce the number of Fourier transforms needed for the spectral model we rewrite the equations introducing vorticity ($\zeta = \vec{\nabla}\times\vec{V}$) as a new variable :

$$\vec{V}_t + (\zeta + f)\,\vec{n} \times \vec{V} + \vec{\nabla}\left(P + \frac{\vec{V}.\vec{V}}{2}\right) = \vec{0}$$

2. *Invariants and spectra*

The problem has the following invariants :

Total mass $\quad M = \iint \frac{1}{-}\,P\,dxdy$

Total energy $\quad E = \iint \frac{1}{2}\,P(P + \vec{V}.\vec{V})\,dxdy$

Total potential enstrophy $\quad Z = \iint \frac{1}{2}\frac{(\zeta+f)^2}{P}\,dxdy$

Energy and potential enstrophy are no longer quadratic invariants, as for the incompressible case, which makes closure techniques (TFM, DIA, EDQNM) inadequate here. Therefore the only possible approach is direct numerical modelling.

In the compressible two-dimensional case the velocity field can be separated into a rotational part, related to the stream function ψ i.e. to incompressibility, and a divergent part, related to the velocity potential χ i.e. to compressibility :

$$\vec{V} = \vec{V}_{div} + \vec{V}_{rot} = \vec{n} \times \vec{\nabla}\psi + \vec{\nabla}\chi$$

Therefore it is interesting to study the following modal energy spectra (i.e. averaged energies in a given spectral band $k = |\vec{k}|$, $\alpha(k)$ being the number of modes in this band) :

Rotational energy $$E_{rot}(k) = \frac{1}{\alpha(k)} \sum_{k-\frac{1}{2}<|\vec{p}|<k+\frac{1}{2}} V_{rot}^{2}(\vec{p})$$

Divergent energy $$E_{div}(k) = \frac{1}{\alpha(k)} \sum_{k-\frac{1}{2}<|\vec{p}|<k+\frac{1}{2}} V_{div}^{2}(\vec{p})$$

Potential energy $$E_{pot}(k) = \frac{1}{\alpha(k)} \sum_{k-\frac{1}{2}<|\vec{p}|<k+\frac{1}{2}} P^{2}(\vec{p})$$

The eigenmodes of the Saint Venant equations, linearized around a resting state ($\vec{V} = \vec{0}$, $P = P_o$), can be classified into :

1. Rossby waves (stationary waves) corresponding to zero divergence,
2. Eastward and westward inertio-gravity waves (dispersive waves with phase speed $\omega = \pm \sqrt{f^2 + k^2 P_o}$) corresponding to zero potential vorticity ($\frac{\zeta+f}{P}$).

3. *Numerical model*

We have developed a pseudo-spectral model for which dissipation is represented by an interated Laplacian operator ($\nu\Delta^8$ i.e. in spectral νk^{16}), which dissipates enstrophy selectively near the smallest resolved scale. This technique has already been applied in most of the previous incompressible two-dimensional models [1][2]. The time integration is done with a 'leap-frog' scheme, resynchronized by an appropriate Laplacian operator in time when the odd and even tend to get desynchronized.

The model is run on a Cray 1-s using a 32x32 resolution with 1000 000 time steps (90 Kwords memory and 4 ms CPU per time step) for the study of statistical equilibria, and a 128x128 resolution with 100 000 time steps (400 Kwords memory and 66 ms CPU per time step) for decaying turbulence.

Our study is restricted to the case of subsonic flows, mainly large-scale geophysical flows, with periodic boundary conditions. We choose the parameters to be representative of some meteorological situation :

Periodic domain	6400 km
Coriolis parameter	$10^{-4}\ s^{-1}$
Mean atmospheric pressure	$P_o = 10^5 Pa$
Fluctuations in the pressure field	4% P_o
Speed of sound	$c = 300\ m\ s^{-1}$
Fluctuations in the velocity field	5% c

4. *Statistical equilibria*

The statistical equilibria of the inviscid truncated equations have been extensively studied for the incompressible case [3]. The compressible case has already been treated for initial conditions with no divergent energy : Sadourny [4], using a shallow water finite-difference model, has then found an enstrophy equipartition for rotational energy and an energy equipartition for divergent energy , while Errico [5], using a baroclinic multi-level spectral model, found that both rotational and divergent energies present energy equipartition. The results presented here are calculated in the case where, on the contrary, divergent

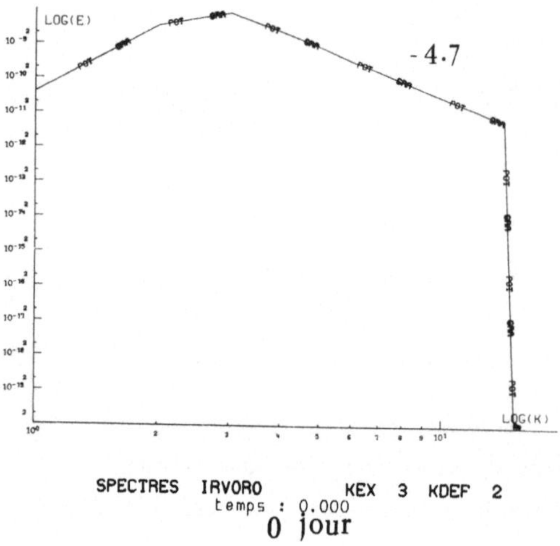

Figure 1

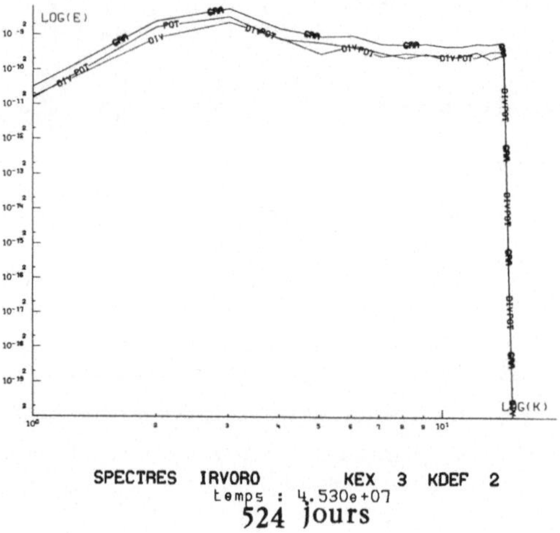

Figure 2

energy initially dominates rotational energy : we start from a random gaussian pressure field, whose potential energy is peaked at large scale (k_{ex} = 3), and with no initial velocity. In this case, flows without rotation (Fig. 1 and 2) or with rotation (Fig. 3 and 4) exhibit, after a very long time necessary for non linear interactions to develop (namely 1000 000 time steps), an equipartition of energy for both rotational and divergent energies. This seems to corroborate Errico's results, but it may well be that for stronger rotations we get an enstrophy equipartition of rotational energy instead.

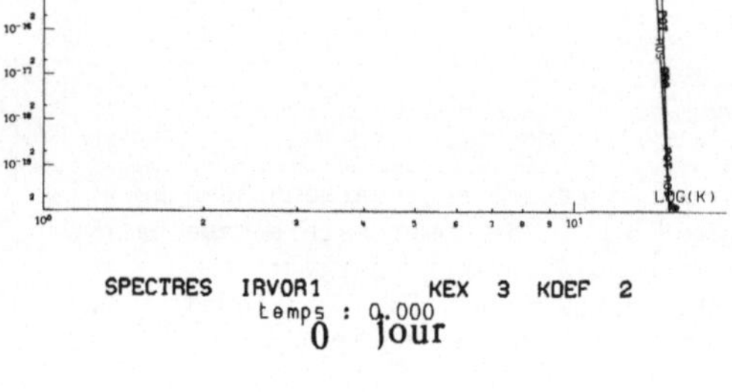

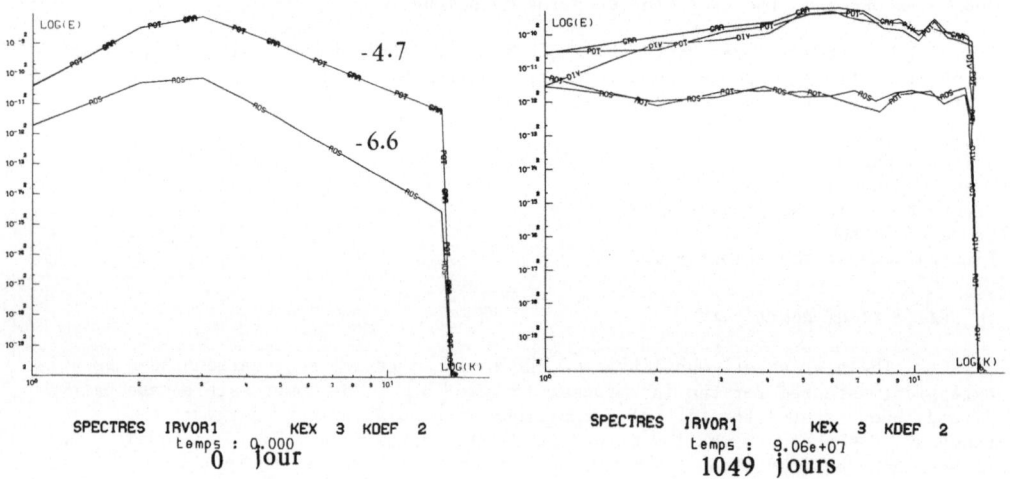

Figure 3

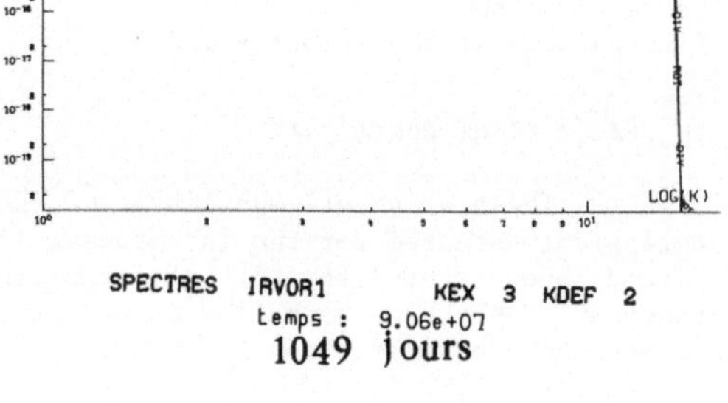

Figure 4

5. *Decaying turbulence*

We start from initial conditions whose energy is confined to large scales (k_{ex} = 3), with pressure and velocity fields balanced to insure the divergence and its tendency are initially zero. The radius of deformation, i.e. the scale above which rotation is felt by the flow, corresponds here to the wavenumber k_{def} = 2. Viscosity is calculated in such a way that enstrophy is significantly dissipated near the smallest scale k_{max} :

$$\nu = \frac{\left[\iint \zeta^2 dxdy\right]^{1/2}}{k_{max}^{16}}$$

After 100 000 time steps the spectra (Fig. 5 and 6) exhibit the following features :

1. the maximum of rotational energy is shifted from wavenumber 3 to 1, suggesting an inverse energy cascade,

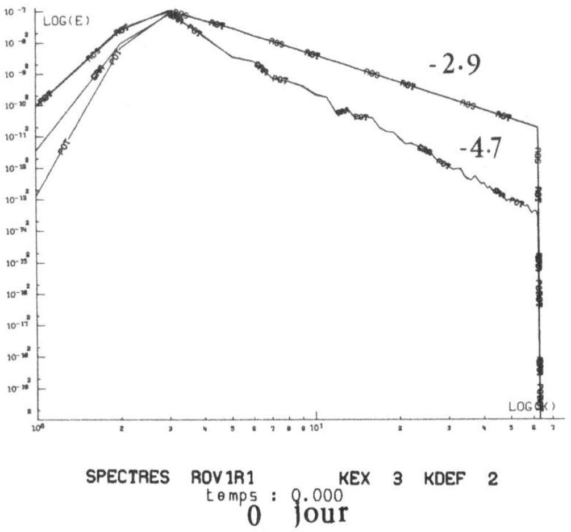

Figure 5

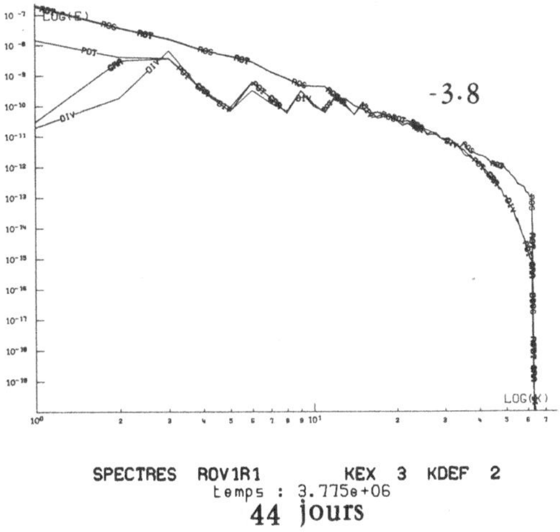

Figure 6

2. the modal spectra (i.e. energy per averaged mode) follow a $k^{-3.8}$ law, which corresponds to a $k^{-2.8}$ dependency for the unidimensional spectra (i.e. total energy per wave-number band) of rotational, divergent and potential energies. This seems similar to the incompressible case, where the Kraichnan phenomenology predicts a - 3 slope while different numerical studies [1][2] find steeper slopes, between - 3 and - 4.

3. the dissipation range of the spectrum is much broader for potential and divergent energies than for rotational energy. This implies that dissipation acts differently on divergent and rotational parts of the flow.

As spectra do not contain complete information on the structure of the flow, it is of interest to study the different fields in physical space. We present, as example, the evolution of the vorticity field : it is striking to see how, from a random gaussian distribution (Fig. 7), the field gets organized into coherent vortex-like structures (Fig. 8), which do not seem to be very much affected by dissipation and have therefore a very long life time. This behavior, already found by Basdevant,

Legras, Sadourny and Béland [1] and by Mc Williams [2], is due to the inverse energy cascade, characteristic of two-dimensional turbulence.

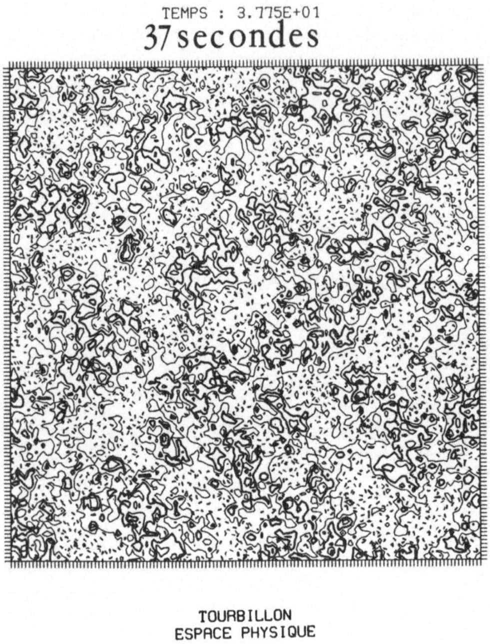

Figure 7

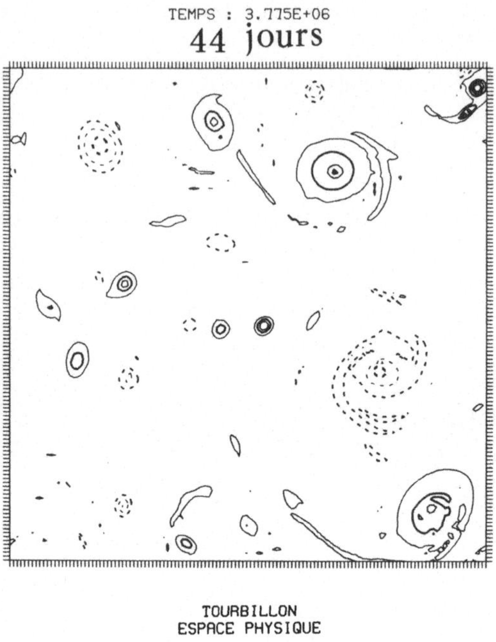

Figure 8

Conclusions

For the time being only preliminary experiments have been performed. It is therefore difficult to reach definite conclusions. We are now doing a systematic survey of different cases : inviscid statistical equilibria, decaying and forced turbulence, with different resolutions, parameters and initial conditions.

Acknowledgements

Calculations were performed on the Cray-1 of the Centre de Calcul Vectoriel pour la Recherche, Palaiseau.

References

[1] C. BASDEVANT, B. LEGRAS, R. SADOURNY and M. BELAND : A study of barotropic model flows : intermittency, waves and predictability. Journal of Atmospheric Sciences, Vol 38, n° 11, pp. 2305-2326 (1981)

[2] J. Mc WILLIAMS : The emergence of isolated, coherent vortices in turbulent flow. To be published in the Journal of Fluid Mechanics (1984)

[3] C. BASDEVANT and R. SADOURNY : Ergodic properties of inviscid truncated models of two-dimensional incompressible flows. Journal of Fluid Mechanics Vol. 69, part 4, pp. 673-688 (1975)

[4] R. SADOURNY : The dynamics of finite-difference models of the shallow-water equation. Journal of Atmospheric Sciences, Vol 32, n°4, pp. 680-689 (1975)

[5] R. M. ERRICO : The statistical equilibrium solution of a primitive-equation model. Tellus, Vol 36A, pp. 42-51 (1984)

COMPARISON OF THE FULL-POTENTIAL AND EULER FORMULATIONS FOR COMPUTING TRANSONIC AIRFOIL FLOWS

J. Flores, J. Barton, T. Holst, and T. Pulliam
NASA Ames Research Center
Moffett Field, CA 94035 USA

I. INTRODUCTION

Recently, much attention has been directed toward developing the Euler formulation for various applications in transonic aerodynamics. However, little effort has been made to compare the speed, accuracy, and robustness of these new Euler codes with the full-potential (FP) formulation. The purpose of this paper is to make such a quantitative comparison using a number of transonic airfoil cases.

The computed results are from four transonic airfoil computer codes: (1) TAIR [1,2]; (2) FLO36 [3]; (3) ARC2D [4,5]; and (4) FLO52R [6]. Codes (1) and (2) are FP codes, and codes (3) and (4) are Euler codes. The FP codes (TAIR and FLO36) use fully implicit iteration algorithms (AF2 and ADI, respectively); the convergence speed of FLO36 is further enhanced by a multigrid convergence acceleration process. The first Euler code (ARC2D) uses a fully implicit ADI iteration scheme; the second (FLO52R) uses an explicit Runge-Kutta time-stepping algorithm, which is enhanced by a multigrid convergence acceleration scheme.

The TAIR and ARC2D codes were each run using two types of grids. One grid was generated numerically, using an elliptic (Laplacian) solver [2], and the second was generated from an algebraic routine [7]. The FLO36 and FLO52R codes were run using an internally generated grid of the circle plane mapping variety.

The comments and conclusions reached in this study will be expressed generally, that is, in terms of FP versus Euler. The reader should bear in mind that these conclusions have been reached using the four specific codes mentioned above. We expect the results presented herein to be typical, but other codes that use different spatial or iteration algorithms may produce somewhat different results.

II. RESULTS

Figure 1 is a plot of lift coefficient versus the average mesh spacing on the airfoil. The airfoil is a NACA 0012, and the flow conditions are $M_\infty = 0.63$ and $\alpha = 2.0°$. As the grid was refined, the ratio of the number of grid points along the airfoil to the number of grid points away from the airfoil was held fixed. The outer boundary was placed at 12 chords from the airfoil. A study, which consisted of plotting the lift versus distance to the outer boundary, was conducted; it verified that this distance was sufficient to remove outer boundary effects. The TAIR and FLO36 codes produce lift asymptotes of 0.3326 and 0.3333, respectively, and the ARC2D and FLO52R codes produce asymptotic values of 0.3357 and 0.3342, respectively.

Theoretically, all results for the two formulations should reach the same asymptotic value for a subcritical case. The Lock solution (obtained through the Hodograph method and considered "exact") [8] yields 0.335 as the value of the lift. However, Lock extends the NACA 0012 airfoil to a sharp trailing edge at $x/c = 1.0089$, but does not normalize to unit length. In the present results, the NACA 0012 airfoil is both extended and renormalized to unit length. If the Lock result is renormalized, consistent with the present results, the lift coefficient would become 0.3321. This tends to suggest that the FP codes are in better agreement with the "exact" solution for this subcritical case.

Figure 2 is a plot of percent error in the lift coefficient versus CPU time in seconds on the Cray XMP computer for the conditions of Fig. 1 (NACA 0012, $M_\infty = 63$, $\alpha = 2.0°$). The timings from all codes are based on converging the lift to an accuracy of 10^{-4} (four decimal digits). The time-step and convergence acceleration parameters from all codes (in general) have been set at default values; that is, a minimal amount of "tuning" has been included. Thus, the convergence rates are not optimal, but are representative of the convergence rates that would be found in practical applications. Startup times, including initialization and grid generation, have been subtracted from each timing. The error is computed by first constructing the asymptotic values of the lift coefficient (as done in Fig. 1). Then the error is simply the absolute value of the difference between the asymptotic value and the value of the converged lift at a specific level of grid refinement. From Fig. 2 (also Fig. 1), it can be observed that the FP formulations are slightly more accurate than the Euler formulations, especially for the coarser grids. On the coarse grids, the Euler codes are more expensive than the FP codes by an average factor of about 17, based on CPU time. For the finer grids, this factor decreases to about 11.

Figure 3 displays a plot of lift coefficient versus the average mesh spacing for a transonic case with a moderate strength shock, NACA 0012, $M_\infty = 0.75$, and $\alpha = 1.0°$. No attempt was made to construct a lift error versus CPU time, as was done in Fig. 2, since, as can be seen in Fig. 3, some of the curves turn over on themselves, making the error measure potentially misleading. We point out here that the asymptotic characteristics of both the FP and Euler formulations are grid-dependent (also apparent in Fig. 1). The algebraic and Laplacian curves for both the FP and Euler formulations show different trends and levels of accuracy. The TAIR (algebraic) and TAIR (Laplacian) results approach their limits from different directions. The level of accuracy for the Euler results is typically less for the algebraic grids, whereas the reverse is true for the FP results. The FP results all approach the same asymptotic limit to within an error of about 1%. The Euler results also approach an asymptotic limit, but the error is significantly less. Another observation from Fig. 3 is that the level of accuracy owing to grid effects can be of the order of the differences in equation formulations (FP versus Euler) for these cases in which the FP is valid.

Utilizing the nonisentropic full-potential formulation [9] in TAIR yields the middle set of curves in Fig. 3. By adding entropy effects to FP formulation, the solutions were improved to within about 4% of the Euler formulation, which it is agreed is the more valid formulation for supercritical cases.

The CPU time at convergence versus the average surface mesh spacing is plotted in Fig. 4 for the conditions shown in Fig. 3 (NACA 0012, $M_\infty = 0.75$, $\alpha = 1°$). This yields a rough estimate of the cost of running each code for different grid sizes, without providing definitive information on the cost to obtain a desired level of accuracy. In general, the Euler codes are more expensive than the FP codes — by a factor of 10 based on CPU time and twice that based on operation count. An interesting observation is that both the ARC2D (Euler) and TAIR (FP) codes converge faster on the Laplacian grid than on the algebraic grid. In fact, the difference between TAIR (algebraic) and TAIR (Laplacian) convergence times is quite large (as much as a factor of 4). The cause for this behavior is not known for certain, but it may be that the stretching is too rapid in the algebraic grids. Because the FP formulation is based on a second-order PDE, it is more likely to be adversely affected by a grid that is nonsmooth or rapidly stretched.

Figure 5 illustrates the asymptotic lift behavior for a strong shock case (RAE 2822, $M_\infty = 0.75$, $\alpha = 3.0°$). Note that these conditions are considered to be beyond the valid range of the full-potential formulation, and only the TAIR and ARC2D codes were run for this case. The FLO codes were not used, a result of the difficulty of the case and the lack of user experience. It can be seen that the results for the TAIR code (algebraic and Laplacian grids) both reach the same asymptotic value of lift. The value obtained is about 1.69, which is grossly in error relative to the Euler results. Thus, the FP formulation is unacceptable for this calculation. The asymptotic values for the ARC2D code (algebraic and Laplacian grids) are in good agreement producing an asymptotic value of lift coefficient near 1.12. The effect of the FP entropy correction is seen to make a major difference in the FP solution, producing errors of a level comparable to those in the previously discussed case (NACA 0012, $M_\infty = 0.75$, $\alpha = 1°$). This improvement in lift is also reflected in a comparable improvement in the surface-pressure distribution, for the nonisentropic FP pressure distribution is in good agreement with the Euler pressure distribution.

Figure 6 presents a comparison of CPU time versus grid refinement for the RAE case. Again we note about an order of magnitude difference in CPU time for FP over Euler. For this case, which is admittedly difficult for isentropic FP, the convergence rates are strongly affected by the different grids. Again, the nonisentropic formulation helped improve the convergence speed of TAIR (Laplacian).

Figure 7 presents a plot of the convergence speed ratio (Euler to FP) versus the average surface mesh spacing for the NACA 0012, $M_\infty = 0.75$, $\alpha = 1°$ case. The convergence speed ratio is plotted based on two criteria: (1) CPU time, and (2) total operation count. Each data point plotted in Fig. 7 is obtained by means of a simple arithmetic average of the results for each formulation, three Euler and five FP (see Fig. 4). Although not monotonic, useful information can be obtained from these curves. The average convergence ratio based on total operations fluctuates from about 9 to 16, and based on CPU time the fluctuation is 4 to 8. The reason for the difference in average convergence speed ratio based on CPU time relative to total operation count is associated with vectorization efficiency. That is, the Euler codes are highly vectorized on the Cray XMP, but the FP codes are not. The Euler-to-FP speed ratio, based on CPU time, could be higher if the FP codes were more efficiently vectorized. However, the possible improvement in FP vectorization efficiency is difficult to estimate, since the AF2 algorithm in two dimensions cannot be vectorized as efficiently as the classical ADI-like implicit schemes or explicit methods. (Note that the AF2 algorithm in three dimensions does not have this disadvantage.)

In Fig. 8, an attempt is made to shed some light on an interesting controversy in which the Euler and FP formulations are involved: the proper level of solution convergence. Because of the differencing of the dependent variable ϕ to obtain the pressure distribution, truncation error is added to any FP solution. Since this error adds to the lack-of-convergence error (theoretically), the FP solution must be converged more tightly than the Euler solution for the same level of accuracy in the lift calculation. Figure 8 shows a plot of error in lift versus rms error in the dependent variable (E_{rms}), pressure for the Euler formulation, and ϕ for the FP formulation. The exact definitions for these two different types of error are displayed in Fig. 8. The two curves shown in Fig. 8 were produced from the NACA 0012, $M_\infty = 0.75$, $\alpha = 1°$ case. Initially, the test case was run until tight convergence was obtained. Then, the converged dependent variables and converged lift coefficient were saved and the case was rerun. The curves shown in Fig. 8 were obtained by plotting the lift error versus the rms dependent-variable error every 50 iterations. Convergence in this case for FP and Euler solutions were about 300 and 1600 iterations, respectively. This explains the difference in number of data points plotted for each code. For this case, the FP solution does need to be converged more tightly for the same error in lift. For a lift error of about 10^{-4}, the FP solution needs to be dropped about an order more in rms error.

III. CONCLUSIONS

A study involving four transonic airfoil computer codes, two FP and two Euler, has been performed. The major conclusions of the study are as follows: (1) the FP codes are faster than the Euler codes by about an order of magnitude based on CPU time on the Cray XMP; (2) the FP formulation loses accuracy as transonic flow develops, but entropy corrections yield FP solutions comparable to those of the Euler; (3) grid coarseness and type can be significant in affecting both accuracy and convergence characteristics; (4) the FP formulation must be more tightly converged than the Euler formulation for comparable levels of accuracy in the lift coefficient; and (5) in general, good accuracy for adequate meshes can be obtained with both formulations, irrespective of the solution method.

IV. REFERENCES

1. Holst, T. L., "A Fast, Conservative Algorithm for Solving the Transonic Full-Potential Equation," AIAA J., Vol. 18, No. 12, Dec. 1980, pp. 1431-1439.
2. Dougherty, F. C., Holst, T. L., Gundy, K. L., and Thomas, S. D., "TAIR – A Transonic Airfoil Analysis Computer Code," NASA TM-81296, May 1981.
3. Jameson, A., "Acceleration of Transonic Potential Flow Calculations on Arbitrary Meshes by the Multiple Grid Method," AIAA Paper 79-1458, July 1979.
4. Steger, J., "Implicit Finite-Difference Simulation of Flow about Arbitrary Two-Dimensional Geometries," J. Comp. Phys., Vol. 16, 1978, pp. 679-686.
5. Pulliam, T. H., "Implicit Finite-Difference Methods for the Euler Equations," Advances in Computational Transonics, Ed. W. Habashi, Pineridge Press Ltd., Swansea, U.K., 1983.

6. Jameson, A., Schmidt, W., and Turkel, R., "Numerical Solutions of the Euler Equations by Finite-Volume Methods Using Runge-Kutta Time-Stepping Schemes," AIAA Paper 81-1259, June 1981.
7. Pulliam, T. H., Jespersen, D. C., and Childs, R. E., "An Enhanced Version of an Implicit Code for the Euler Equations," AIAA Paper 83-0344, Jan. 1981.
8. Lock, R. C., "Test Cases for Numerical Methods in Two-Dimensional Transonic Flows," AGARD Report No. 575, 1970.
9. Hafez, M. and Lovell, D., "Entropy and Vorticity Corrections for Transonic Flows," AIAA Paper 83-1926, July 1983.

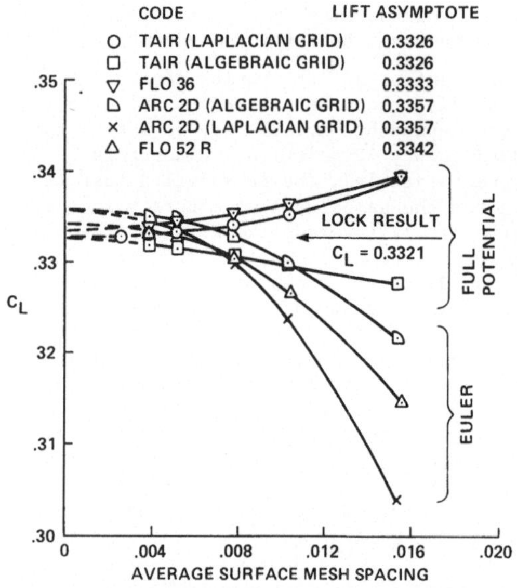

Fig. 1 Lift versus grid refinement: NACA 0012, $M_\infty = 0.63$, $\alpha = 2°$.

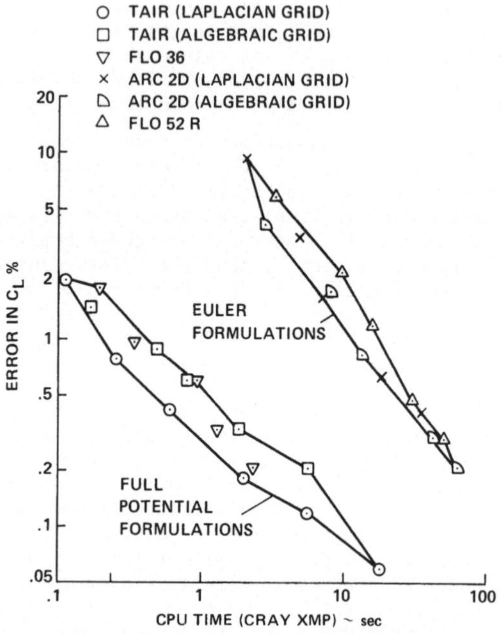

Fig. 2 Lift error versus CPU: NACA 0012, $M_\infty = 0.63$, $\alpha = 2°$.

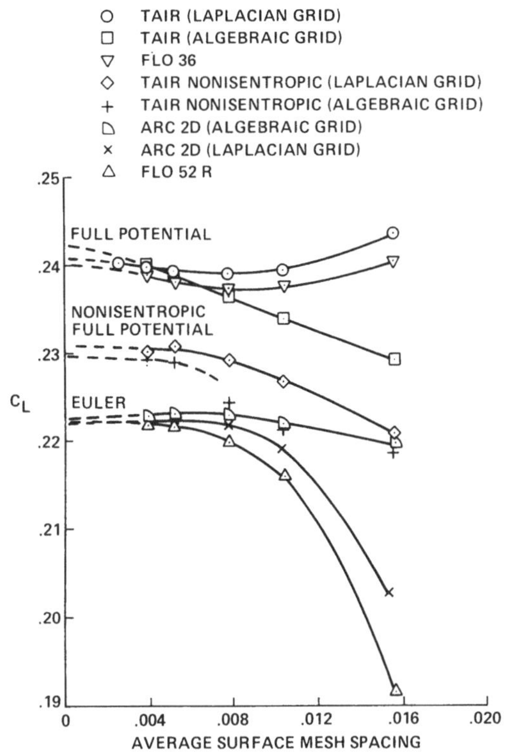

Fig. 3 Lift versus grid refinement: NACA 0012, $M_\infty = 0.75$, $\alpha = 1°$.

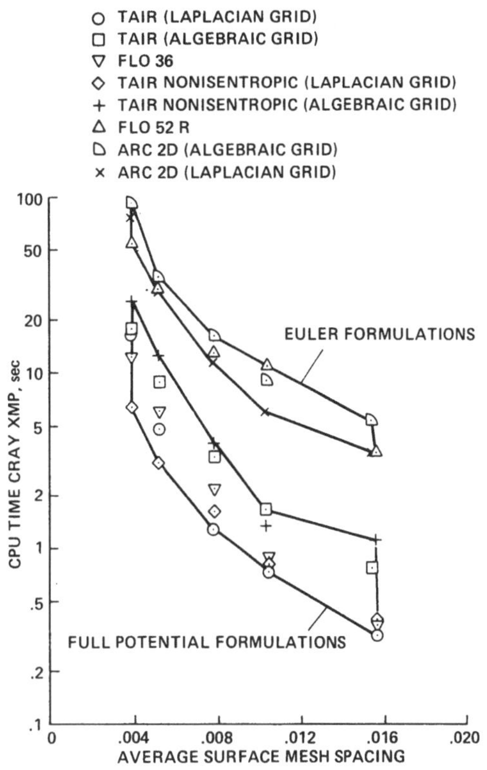

Fig. 4 CPU versus grid refinement: NACA 0012, $M_\infty = 0.75$, $\alpha = 1°$.

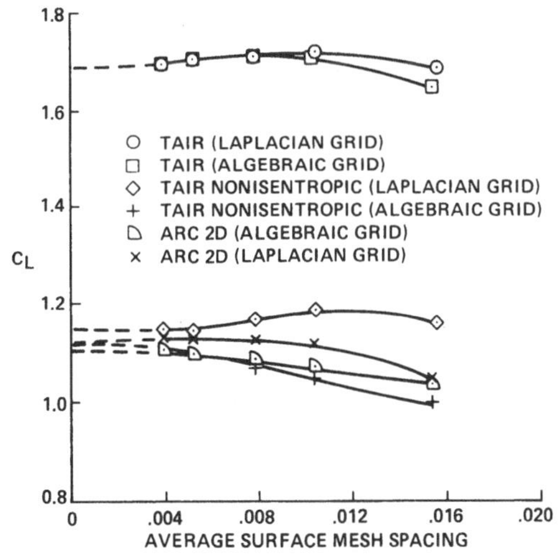

Fig. 5 Lift versus grid refinement: RAE 2822, $M_\infty = 0.75$, $\alpha = 3°$.

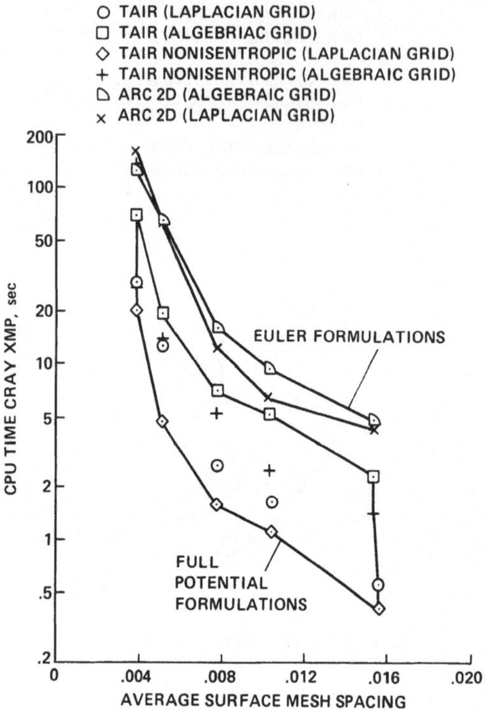

Fig. 6 CPU versus grid refinement:
RAE 2822, M_∞ = 0.75, α = 3°.

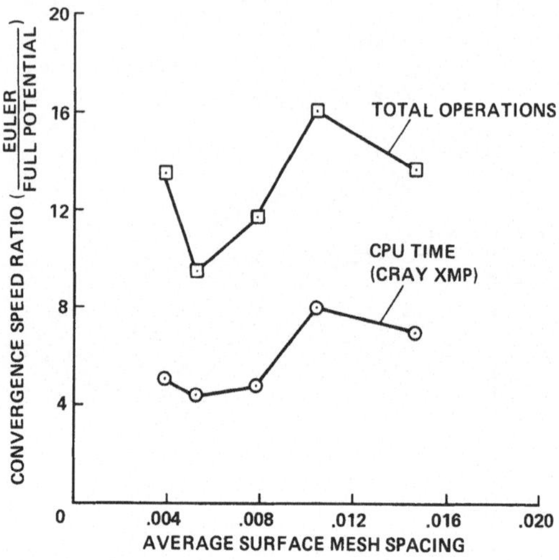

Fig. 7 Convergence speed versus grid
refinement: NACA 0012, M_∞ = 0.75,
α = 1°.

□ TAIR (ALGEBRAIC GRID)
   (FULL POTENTIAL)
△ ARC 2D (ALGEBRAIC GRID)
   (EULER)

$$E_{rms} = \left[\frac{\sum\limits_{ij} (r_{ij}^n - r_{ij}^{conv})^2}{\sum\limits_{ij} (r_{ij}^o - r_{ij}^{conv})^2} \right]^{1/2}$$

$$r = \begin{cases} P \text{ FOR EULER} \\ \phi \text{ FOR F. P.} \end{cases}$$

$$(C_L)_{ERR} = \frac{|C_L^{conv} - C_L^n|}{C_L^{conv}}$$

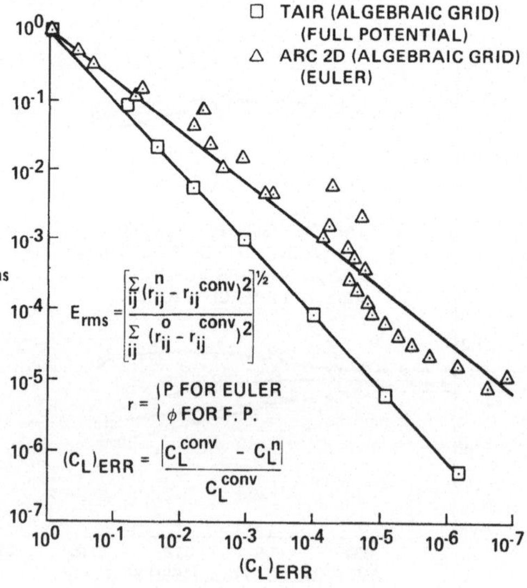

Fig. 8 Root-mean-square error versus lift error: NACA 0012,
M_∞ = 0.75, α = 1°.

NUMERICAL SIMULATIONS OF FUEL DROPLET FLOWS USING A LAGRANGIAN TRIANGULAR MESH

M.J. Fritts, D.E. Fyfe and E.S. Oran
Laboratory for Computational Physics
Naval Research Laboratory
Washington, D.C. 20375/USA

Particularly severe physical and mathematical approximations must be made to describe the detailed interactions between droplets and the external flow field in spray models (Williams, 1973; Faeth, 1983). For example, equivalent spheres approximate deformed droplets, the effects of droplet breakup are included by estimating breakup times and drop sizes after breakup, and wake effects and changes in the flow field due to droplet deformations are neglected. In general, the droplet concentration is dilute since little is known about droplet-droplet or droplet-wake interactions.

The need for these approximations arises directly from the difficulty in following several physically distinct regions as they distort and separate or merge through interactions with the external flow field. A Lagrangian technique is well suited to accurately modelling the transport of various fluid regions since it easily and naturally calculates the advection of material boundaries. However, unless the computational grid self-consistently adapts to the physical flows the accuracy of the numerical solution quickly deteriorates due to the flow-induced grid deformations.

The Lagrangian technique used in this study incorporates a fully conservative, finite difference method specifically designed for multiphase flow (Fritts and Boris, 1979). In this method a dynamically restructuring two-dimensional grid of triangles tessellates the interior of each fluid phase with the triangle sides aligned on material interfaces. Since vertex movement is Lagrangian, the interface sides accurately track the movement of the interface due to advection. A triangular grid can avoid the problems of mesh tangling through the Free Lagrange technique (Crowley, 1971) whereby individual mesh points are continually reconnected to account for the migration of fluid elements in the flow field. Since the number of grid lines meeting at a vertex is variable, the resolution can be altered where needed (e.g., enhancing resolution around a region of droplet distortion or lowering resolution in laminar flow regions) without affecting the resolution in other areas of the computation and without adding numerical diffusion. The technique allows the grid to evolve into multiply-connected regions, so that if conditions are appropriate, droplets can break up and shatter.

For incompressible and irrotational flow in two dimensions, the velocity field is specified by a velocity potential ϕ and stream function ψ which are specified through the conservation of vorticity and mass:

$$\nabla \cdot \bar{v} = \nabla \cdot (\nabla \times \psi) \equiv 0$$

and (1)

$$\nabla \times \bar{v} = \nabla \times \nabla\phi \equiv 0 \; .$$

Finite-difference operators can be defined for divergence, curl and gradient which have these continuum properties, but this requirement restricts the placement of variables. For example, for a scalar function ϕ specified at each vertex and assumed piecewise linear within each triangle, the vector gradient of ϕ (constant throughout the triangle j and discontinuous at the triangle sides) is given by

$$(\nabla\phi)_j = \phi_1 \frac{z\times(\bar{r}_3-\bar{r}_2)}{2A_j} + \phi_2 \frac{z\times(\bar{r}_1-\bar{r}_3)}{2A_j} + \phi_3 \frac{z\times(\bar{r}_2-\bar{r}_1)}{2A_j} \tag{2}$$

where z is the direction out of the page. With this placement of variables the dynamics of the flow, as well as the kinematics, behave properly. That is,

$$\nabla \cdot \bar{v} = \nabla \cdot \nabla\phi = \nabla^2\phi \tag{3}$$

is a general triangular grid Poisson equation which may be used to solve for the local pressure. In finite-difference form Eq.(3) becomes

$$A_c \langle \nabla\cdot\bar{v} \rangle_c = \sum_{i(c)} \left\{ \phi_i \frac{z\times(\bar{r}_c-\bar{r}_{i+1})}{2A_{i+1/2}} + \phi_{i+1} \frac{z\times(\bar{r}_i-\bar{r}_c)}{2A_{i+1/2}} + \phi_c \frac{z\times(\bar{r}_{i+1}-\bar{r}_i)}{2A_{i+1/2}} \right\} \times \frac{(\bar{r}_{i+1}-\bar{r}_i)}{2} \cdot z, \tag{4}$$

where A_c is the area of the vertex cell, defined as one third of the sum of the areas of all triangles including that vertex. The notation $\sum_{i(c)}$ is the sum over vertices i around a central vertex c. The quantity $A_{i+1/2}$ represents the area of the triangle having vertices (c,i,i+1).

The accuracy of the numerical algorithms is determined by both the local resolution and connectivity of the grid. For the approach used here, the local connectivity and resolution are both determined in part by the requirement that the matrix generated from the Poisson equation, Eq.(4), remains diagonally dominant. With this restriction, convergence of an iterative solver for Eq.(4) is assured.

Note that the coefficient of the ϕ_c term in Eq.(4),

$$a_c = -\sum_{i(c)} \frac{|\bar{r}_{i+1}-\bar{r}_i|^2}{4A_{i+1/2}}, \tag{5}$$

is always negative. The coefficient a_i of the ϕ_i term is

$$a_i = \frac{1}{2}(\cot\delta_{i+1/2} + \cot\delta_{i-1/2}), \tag{6}$$

where $\delta_{i+1/2}$ and $\delta_{i-1/2}$ are the angles in the (i+1/2)th and (i-1/2)th triangles opposite the line from c to i. Let the sum of $\delta_{i+1/2}$ and $\delta_{i-1/2}$ be θ. If θ is less than π radians for each i, the matrix is diagonally dominant. For any i, if θ is greater than π radians, then the line from c to i is reconnected to join (i+1) to (i-1). Since the sum of the angles in quadrilateral (c,i,i+1,i-1) is 2π radians, the angles opposite the new diagonal sum to less than π radians. Negative area triangles cannot form with an algorithm that requires that the sum of the opposing angles is greater than zero and less than π radians.

Since triangle sides aligned along interfaces cannot be reconnected, diagonal dominance cannot be preserved at interfaces in the same way. Instead, a vertex is added at the midpoint of the interface line. This scheme assures diagonal dominance while increasing the resolution in the neighborhood of the interface. Two additional restructuring procedures, vertex addition and deletion, are required to permit automatic grid restructuring and local alteration of grid resolution.

In a P-$\bar{v}$ formulation of the basic incompressible hydrodynamics equations the changes in vorticity are zero by construction since $\nabla\times\nabla P=0$. The new pressures are chosen to force the divergence of the velocity field to zero. The P-$\bar{v}$ algorithm specifies pressures, velocities and positions at full timesteps. A split-step algorithm is used to integrate the velocities forward half a time step, advance the grid a full time step, and then advance the velocities the remaining half time step.

Figure 1 illustrates a test of the grid restructuring algorithms in a calculation of the shattering of a droplet when the droplet and external fluid have a density ratio of 2:1 with no surface tension or viscosity present. The droplet is gridded into 28 triangular computational cells in a total system of 552 cells, as seen in the first frame. Boundary conditions are periodic at the sides and reflective at the top and bottom of the computational region. Early in the calculation a recirculation zone forms behind the droplet, compressing the droplet in the direction parallel to the flow. Flow within the droplet is initiated by this compression in a direction normal

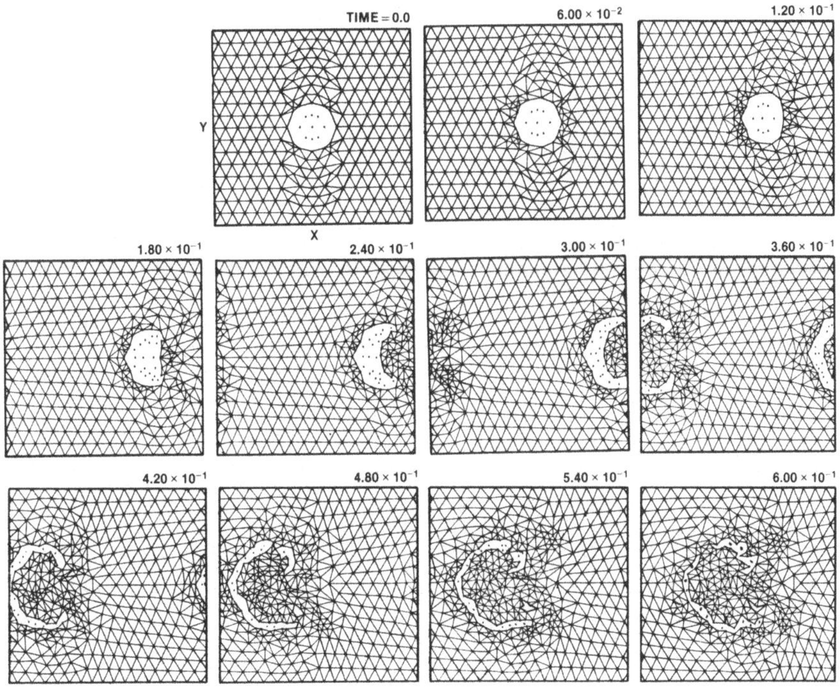

Figure 1.

Algorithms for grid restructuring

to the external flow. The bulges formed at the top and bottom of the distorted droplet are pulled around the recirculation zone by the shear flow which is a maximum at these points. Eventually the droplet is squeezed into a film coating the recirculation zone. The thinned film then shatters into several smaller pieces, first at the rear of the droplet and later in the more laminar flow toward the front of the droplet.

New algorithms for surface tension and viscosity have been added to the basic fluid dynamics model discussed above (Fritts, Fyfe and Oran, 1983). Surface tension is included as a jump in pressure across an interface by casting the surface tension forces in the form of a gradient of a potential. Since the pressure gradient forces are calculated in the same manner and on the same grid as those derived from the surface tension potential, exact balance can be achieved between the forces, and static pressure drops across the interface agree exactly with theory. Since the surface tension is normal to the interface and opposes the pressure drop, then the $\nabla P \times \nabla \rho$ terms which alter the vorticity are zero for the finite-difference algorithms.

The surface tension forces across an interface are

$$P_i - P_o = \sigma/R \tag{7}$$

where P_i is the pressure just inside the droplet at the interface, P_o is the pressure just outside the droplet at the interface, σ is the surface tension coefficient, and R is the radius of curvature of the cylindrical droplet, which is approximated by a parametric cubic spline interpolant to the interface vertices. These pressure jumps are included in the Poisson equation for the pressure. The average pressure, $(P_i + P_o)/2$, is computed at an interface vertex. From the average pressure and the pressure jump we can compute a pressure gradient centered on triangles, within and without the droplet, for inclusion in the momentum equation.

In the finite-difference formulation for viscosity, the coefficient of viscosity is centered on triangles and a vertex-centered velocity gradient is computed so that the divergence of the velocity gradient is also centered on triangles. This placement of variables puts the viscosity on the same footing as the density. Temporal changes in the triangle velocities are straightforward to compute, since now

$$\frac{d\bar{V}_t}{dt} \sim \nu_t \left(\nabla^2 \bar{V}\right)_t, \tag{8}$$

where the subscript "t" indicates that all quantities are triangle centered.

The surface tension algorithm was benchmarked by studying the oscillatory behavior of an n = 2 mormal mode. Comparisons with linear theory showed good agreement. All the difference between theory and the numerical result was consistent with second-order convergence to the theoretical frequency. The viscosity algorithm was tested by calculating the spreading of a viscous shear layer. The agreement with theory was excellent for both the width of the layer and the velocity profile despite significant distortion in the mesh induced by the flow field.

Figure 2 shows the pathlines of the internal and external flows of an oscillating and deforming kerosene droplet in an air jet. The jet is initially laminar (100m/s) about a 125 micron droplet. Surface tension and viscosity are included and the corresponding Reynolds number is about 1600. The first clear indication of the developing recirculation region behind the droplet is the pair of counter-rotating vortices seen in the fourth insert. By the last insert, another pair of vortices is forming near the droplet indicating that the original pair has been shed. Distortions in the face of the droplet are evident by at least the seventh frame, and are due to fluctuations in the external flow caused by the approaching wake of the preceding droplet.

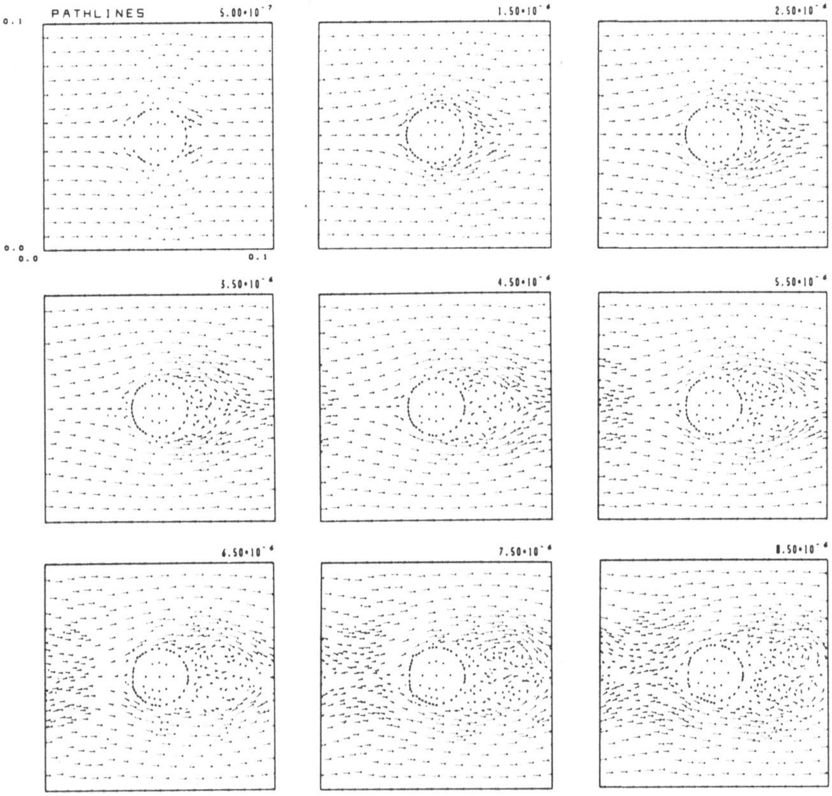

Figure 2

Pathlines of internal and external flows

Crowley, W. P., (1971) A Free-Lagrange Method for Numerically Simulating Hydrodynamic Flows in Two Dimensions, Proc. Second ICNMFD (Springer-Verlag, New York), p.37.

Fritts, M. J. and J. P. Boris, (1979) The Lagrangian Solution of Transient Problems in Hydrodynamics using a Triangular Grid, J. Comp Phys. 31, pp.173-215.

Fritts, M. J., D. E. Fyfe, and E. S. Oran, (1983) Numerical Simulations of Fuel Droplet Flows Using a Lagrangian Triangular Mesh, NASA CR-168263.

Faeth, G. M., (1983) Evaporation and Combustion of Sprays, Prog. Energy Comb. Sci 9, pp. 1-76.

Williams, A., (1973) Combustion of Droplets of Liquid Fuels, Comb. Flame 21, pp.1-31.

ON BOUNDARY CONDITIONS FOR INNER INCOMPRESSIBLE FLOWS

L. Fuchs
Department of Gasdynamics,
The Royal Institute of Technology,
100 44 Stockholm, SWEDEN

INTRODUCTION

We consider the flow of viscous incompressible fluids in confined two dimensional regions. The flow field is determined not only by the geometry and the properties of the fluid but largely by the conditions that are imposed at the inflow and the outflow boundaries. These conditions are usually not known explicitly. Often, one extends the physical domain so that analytically known free-stream conditions can be applied at 'infinity'. For numerical simulation the computational domain is redefined and in most cases in an arbitrary manner. The application of free-stream conditions, at finite distance, for channel and duct flows have been considered in [1] and [2]. It has been shown that by applying the free-stream velocity profile at some finite distance, errors appeared close to the outflow boundary. In several papers somewhat 'less restrictive' ([3], p.154) conditions were defined by assuming that the variations in the main flow directions are small. For both boundary conditions the computed flow approximates the physical one except in a thin region near the outflow boundary. To obtain uniform accuracy, parabolic boundary conditions have been developed [1,2]. These conditions assume that the flow has a main direction and that no separation occurs near the outflow boundary. Thunell [4] studied some of the effects of using free-stream velocity values (Dirichlet conditions) at the outflow boundary even when separation occurred. It was found that when such outflow boundary conditions were applied at a place where separation should take place, a distorted flow field, with most errors near the outflow boundary was obtained.

Here we investigate the effects of boundary errors on the solution. Such an investigation is important in those cases where the boundary velocity is determined experimentally. In such cases it is important to estimate how different boundary error components propagate into the flow field and to determine the required accuracy in measuring the boundary velocity. We also discuss the effects of boundary error when the continuity equation is replaced by the Poisson equation for the pressure. Our results show that it is important to have small amplitude low frequency Fourier-components in the boundary error, to ensure good global accuracy. This is the case even when the boundary is placed in a separated region.

GOVERNING EQUATIONS

We consider the flow of an incompressible viscous fluid in a 2-D region (Ω) bounded by rigid walls except at some segments (inflow and outflow) of the boundary. The dimensionless equations in cartesian coordinates are given by:

$$\nabla^2 u - p_x - Re(uu_x + vu_y) = 0 \qquad (1)$$

$$\nabla^2 v - p_y - Re(uv_x + vv_y) = 0 \qquad (2)$$

$$u_x + v_y = 0 \qquad (3)$$

The system (1) - (3) is elliptic of order 4 and requires 2 conditions on the boundary. Such conditions can be specified by either:

a. Dirichlet condition $\underline{q} = (u,v)$ on the whole boundary ($\partial\Omega$) provided that $\int_{\partial\Omega} \underline{q} \cdot d\underline{n} = 0$ where $\underline{n}$ is the unit vector normal to the boundary.

b. Neuman conditions (e.g. $u_x = g$; $v_y = f$).

c. Parabolic conditions [1,2].

In some numerical methods (such as the MAC-method [5])the continuity equation (3) is replaced by the Poisson equation for the pressure:

$$\nabla^2 p = Re\ J \qquad (4)$$

where $J = \text{div}(\underline{q} \cdot \text{grad}\underline{q})$.

The new system of equations (1), (2) and (4) is also elliptic of order 6 and therefore, for well posedness, 3 conditions must be specified on the boundaries. This implies that beside the conditions on the components of the velocity vector an additional condition, e.g. on the normal pressure derivative, should be specified. Usually, such a condition is not known explicitly and then the normal pressure derivative is computed from the momentum equations. By such a procedure errors in the velocity gradient near the boundary are spread out in the entire domain.

In the following we consider the effects of boundary errors for the flow in a rectangular 2-D geometry.

INFLOW AND OUTFLOW BOUNDARY ERRORS

As inflow (and outflow) boundary conditions we use the velocity profile which is obtained by assuming free-stream flow conditions.The parabolized equations are valid if no separation occurs and when the velocity gradients are small. The equation which describes the propagation of a boundary error, ε, is given by

$$Re_L \varepsilon_y = \varepsilon_{xx} \qquad (5)$$

where Re_L is the local Reynolds number and y is positive in the main flow direction. This equation is valid if the parabolizing approximation is not too bad. Each Fourier component of the error (ε^k) with a wave number k (in the x-direction) behaves as

$$\varepsilon^k = \exp\,[-k^2/Re_L + ikx] \qquad (6)$$

Thus, the slowest error component in the x-direction, would be convected the longest distance before it decays. The high frequency components decay at much shorter distances from the boundary.

The parabolization method, which guarantees the mass conservation, introduces mostly high frequency errors in the velocity components. For this reason the application of such boundary conditions would result in relatively small errors in the solution even for larger Reynolds numbers.

The momentum equation for an error ε near the outflow boundary (assuming almost parallel non-separated flow) is used to estimate the scales for error dissipation. If only the linearized equation is considered, it is found that the outflow boundary error can propagate upstream a distance proportional to 1/Re. When the non-linear error propagation is considered, the dissipation distance of the error is determined by rescaling the equation such that the convective term in the cross flow direction and the diffusion term on the upstream directions are of the same order. If this scaling is valid and by using the continuity equations one gets that

$$Re\varepsilon(\varepsilon_x + \varepsilon_y) = \varepsilon_{yy} \qquad (7)$$

The longest scale of viscous dissipation, for Re >> 1, is proportional then to $1/\sqrt{Re}$. It is clear that due to non-linearity of the system of equations, the outflow error region is not exactly proportional to the estimated scale. However, the qualitative error decay of exponential type could be confirmed by numerical experiments.

NUMERICAL RESULTS

Free stream inflow conditions were applied at a distance y = 5.0. A perturbation $\varepsilon_k = \sin(k\cdot\pi x)$ was added to the fully developed velocity profile. The propagation of the different error components (ε_k) was studied by computing the RMS of the difference between the perturbed and the unperturbed solutions at different y. The RMS-error was normalized by the mean velocity at the given cross-section. Figure 1.a shows the relative error propagation for different perturbation frequencies. Figures 1.b-1.e show the streamlines of the flow field with perturbed boundary conditions ($0\leqslant k\leqslant 3$, Re=150). The streamline pattern show that the perturbation in the boundary condition results in a separation buble which decreases in size as k increases. The amplitude of the error, at a given distance from the boundary depends on the wave-number of the error component and the Reynolds number. The field errors decrease as k increases (Fig. 1.a) and as Re decreases (Fig. 4).

As to the outflow section, the corresponding cases are displayed in Figures 2 and 5. Figures 3 show the case where free stream conditions are applied at distance where separation occurs (see Figure 1.b). The flow field (compared to Figures 2.b-2.e) is altered only near the outflow boundary. The errors decay exponentially.

From our numerical experiments we conclude that both the inflow and the outflow velocity profiles should be specified in such a way that the boundary error should contain small amplitude low frequency components. High frequency fluctuations are damped and do not effect the whole computational field. This fortunate situation enable the use of (not very accurate) measured boundary velocity profiles, or the use of some other proper approximation to it.

It is also noted that when the velocity field contains errors of the type shown here, there are large errors in the boundary pressure gradient. Such errors are spread into the flow field (even upstream) by the elliptic equation for the pressure (4). For this reason it is preferable to use the original system rather than the alternative system when boundary errors are unavoidable.

REFERENCES

1. L. Fuchs. Boundary Condition Effects on the Computation of Channel Flows. Proc. 2nd Asian Congress of Fluid Mech., 1983.
2. L. Fuchs, H.-S. Zhao. Solution of Three-Dimensional Viscous Incompressible Flows by a Multi-Grid Method. J. for Numerical Methods in Fluids, to appear, 1984.
3. P.J. Roache. Computational Fluid Mechanics. Hermosa Publishers, 1982.
4. T. Thunell. Numerical Simulation of Viscous Flows in a Separator. Report TRITA-GAD-6, 1984.
5. F.H. Harlow, J.E. Welch. Numerical Calculation of Time-Dependent Incompressible Flow. Phys. of Fluids, pp. 2182-2187, 1965.

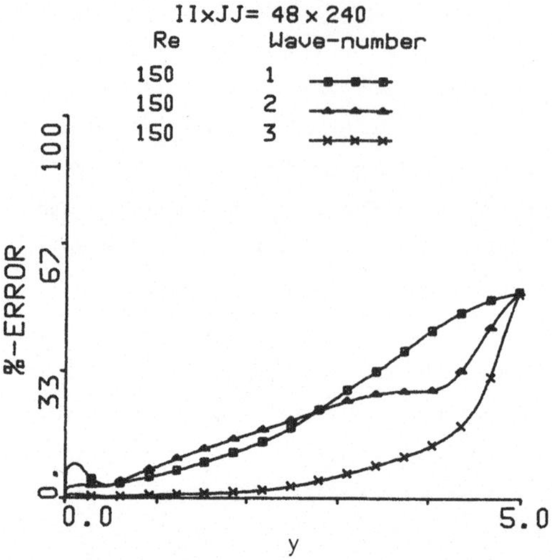

FIGURE 1.a: The mean error (%) vs. the distance from the inflow boundary (y=5.0) for different components of the boundary error.

FIGURE 1.b: The streamlines with unperturbed inflow B.C.

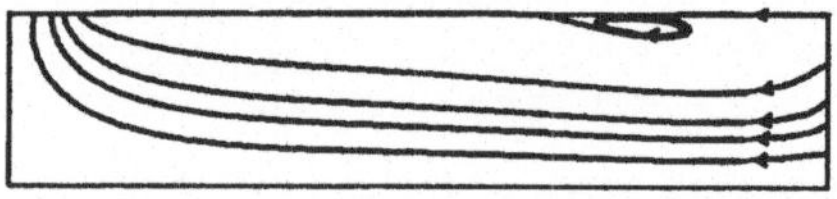

FIGURE 1.c: The streamlines with perturbed (k=1) inflow B.C.

FIGURE 1.d: The streamlines with perturbed (k=2) inflow B.C.

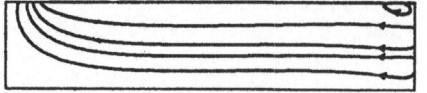

FIGURE 1.e: The streamlines with perturbed (k=3) inflow B.C.

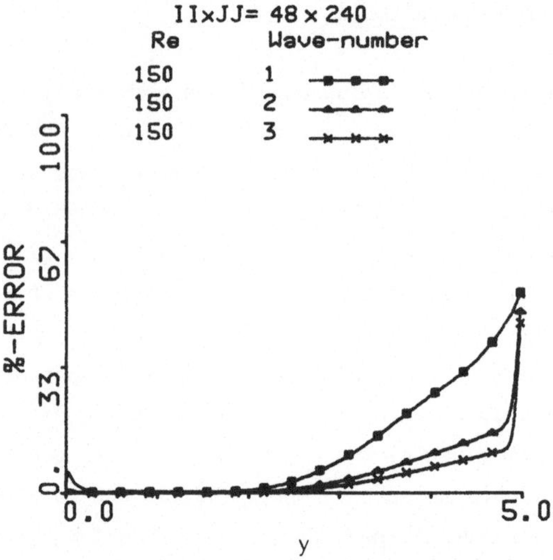

FIGURE 2.a: The mean error (%) vs. the distance from the outflow boundary (y=5.0) for different components of the boundary error.

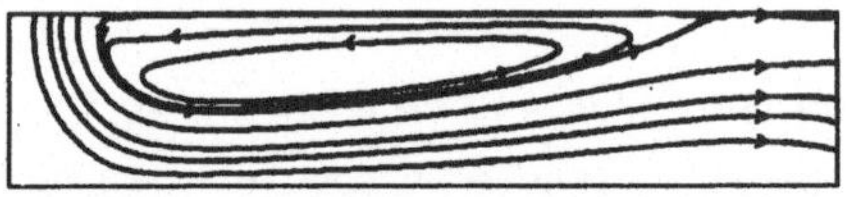

FIGURE 2.b: The streamlines with unperturbed outflow B.C.

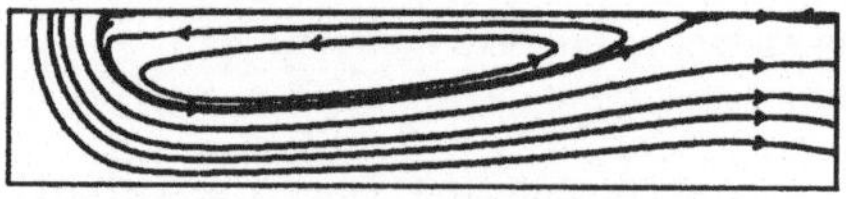

FIGURE 2.c: The streamlines with perturbed (k=1) outflow B.C.

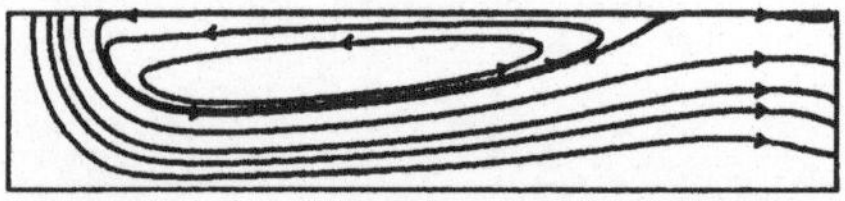

FIGURE 2.d: The streamlines with perturbed (k=2) outflow B.C.

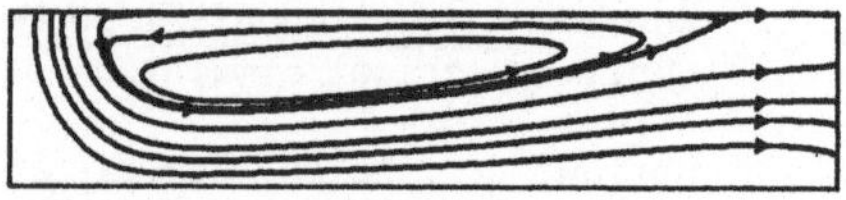

FIGURE 2.e: The streamlines with perturbed (k=3) outflow B.C.

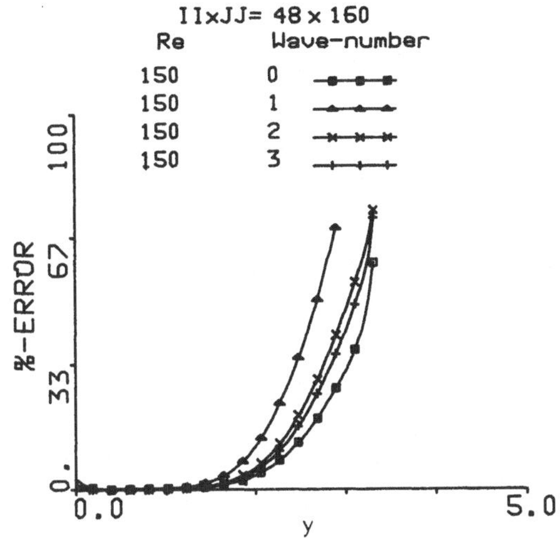

FIGURE 3.a: The mean error (%) vs. the distance from the outflow boundare (y=3.33) for different components of the boundary error.

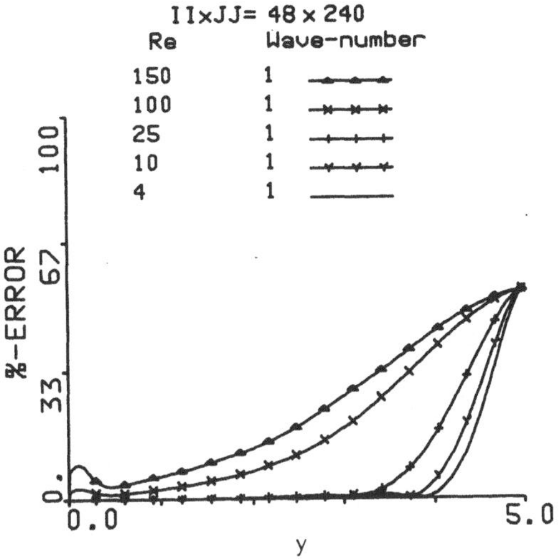

FIGURE 4: The mean error (%) vs. the distance from the inflow boundary (y=5.0) for different Reynolds numbers (k=1).

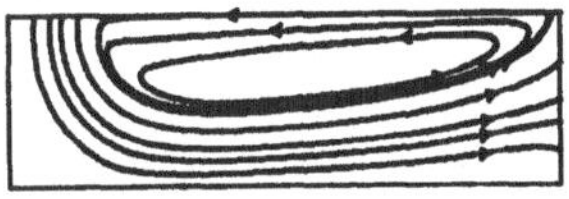

FIGURE 3.b: The streamlines with unperturbed outflow B.C.

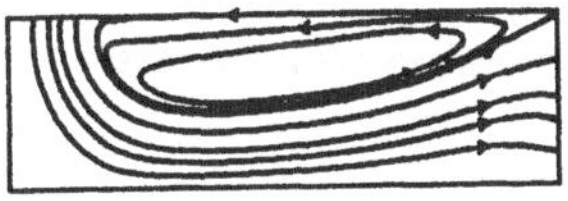

FIGURE 3.c: The streamlines with perturbed (k=1) outflow B.C.

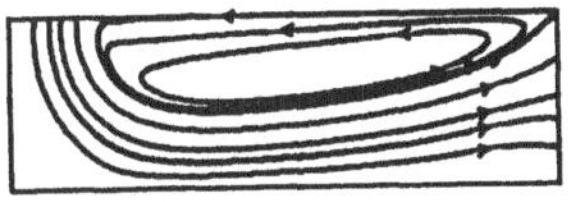

FIGURE 3.d: The streamlines with perturbed (k=2) outflow B.C.

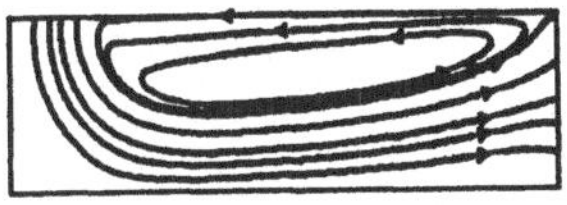

FIGURE 3.e: The streamlines with perturbed (k=3) outflow B.C.

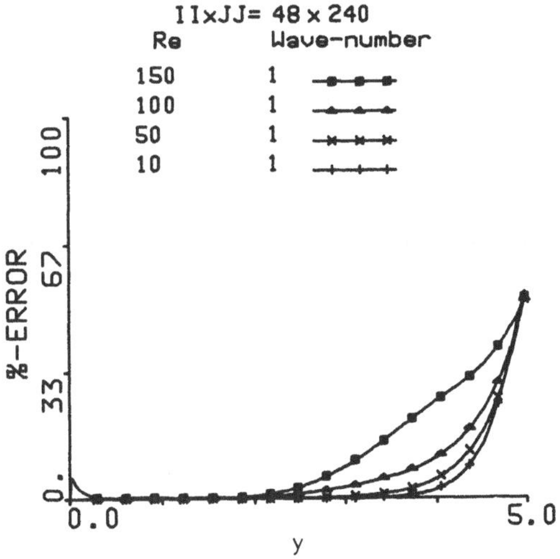

FIGURE 5: The mean error (%) vs. the distance from the outflow boundary (y=5.0) for different Reynolds numbers (k=1).

FAST THREE-DIMENSIONAL FLUX-CORRECTED TRANSPORT CODE FOR HIGHLY RESOLVED COMPRESSIBLE FLOW CALCULATIONS

David E. Fyfe, John H. Gardner, and Michael Picone
Laboratory for Computational Physics, Naval Research Laboratory
Washington, D.C. 20375
and
Mark A. Fry
Science Applications, Inc.
McLean, Virginia 22102

We describe FAST3D, a code which solves the three-dimensional hydrodynamic equations for the conservation of mass, momentum and energy of an ideal fluid in Cartesian geometry on a variably-spaced mesh using a new leapfrog FCT algorithm. The basic finite-difference technique is the three-dimensional leapfrog scheme with no time splitting. This scheme has very low dissipation, with a linear amplification matrix of unity on a uniform grid, but is subject to a weak (grid separation) instability and large dispersive errors at short wavelengths. These difficulties are overcome by using the Flux-Corrected Transport (FCT) technique, in which the final solution is made up of the weighted average of a high-order scheme (leap-frog) and a low-order scheme, in this case upwind differencing, designed to maintain positivity (monotonicity) where it is required physically. The FCT algorithm is implemented in the code by adding to the leapfrog convective terms a diffusive flux proportional to the absolute velocity, plus a small velocity-independent diffusive term. An antidiffusive step is then used in the algorithm in time-split fashion, removing most of the diffusion. The flux-correction part of the solution leaves just enough of the low-order scheme to guarantee monotonicity of the solutions.

The main region of the calculation employs equally spaced zones to maintain high accuracy. The grid may be stretched in regions of little activity or near the edges to reduce the influence of boundary approximations. Either reflecting or outflow boundaries may be applied in the transverse (y, z) directions, and a choice of reflecting, outflow, or periodic boundary conditions is available in the longitudinal direction. A variable time step is used based on the minimum CFL limit in each direction over the whole mesh. The momentum equation contains a gravity term and uses a real-gas equation of state routine to calculate pressure for strong shock calculations in air.

For some applications, the shocks and other unsteady phenomena being simulated are initially localized near one end of the system, conventionally taken as the origin of the x axis. At early times it is unnecessary to update variables far from this region, so the number of active zones in this direction is (NX)' < NXMAX. At late times (NX)' increases until the entire mesh is active.

Three-dimensional hydrodynamic calculations require large amounts of memory, typically more than is physically available on a given computer. This necessitates the use of auxiliary storage such as disk. The FAST3D algorithm uses only the neighboring values of the fluid variables to time-advance a given value. The data are stored on y-z planes, so that only three planes need be in memory at any given time during the computation (Fig. 1). Furthermore, by allowing two I/O buffers for transferring planes to and from the disk, computation can be overlapped with I/O to minimize overall computer time. This implementation of the algorithm requires two passes through the auxiliary disk file per time step: one for the leapfrog time advancement and one for the FCT correction. Other passes through the disk file which do not occur at every time step may be made as needed, for example, to produce diagnostics or restart/dump files.

The leapfrog FCT algorithm involves enough arithmetic to keep the transport part of the FAST3D code compute-bound. However, due to the explicit nature of the algorithm, the computations on the planes can be written to take advantage of vector hardware that exists on machines such as the Cray-1. When fully vectorized, FAST3D runs on the Cray-1 at about 20 μs/(zone-timestep).

After testing the numerical model with a variety of simple idealized calculations, we applied the code to the phenomenon of vorticity generation by a curved electric discharge. This problem is an important test of a mechanism proposed recently to explain the observed rapid cooling of discharge and laser channels in a gaseous medium and to predict the rate at which lightning produces nitrogen oxides (Picone et al., 1981). In air at STP a channel of radius $r_0 = 0.6$ cm is introduced, within which air at standard density is given a Bennett pressure profile with peak pressure equal to ~ 30 times the ambient value. The channel bends sinusoidally with wavelength $\ell_0 = 12$ cm and amplitude $a_0 = 0.5$ c.m

The x-axis is taken in the longitudinal direction, and the curved discharge axis is assumed to lie in the y = 0 plane. The boundary conditions are reflecting at x = 0, $x_{max} = \ell_0/2$, and y = 0, and transmissive on the other planes.

Two calculations were run with different resolutions. The first one used fine zones with $\Delta y = \Delta z = 1.2$ mm, 15 in the y direction and 29 in the z direction. Surrounding these was a layer of coarse zones, 18 at large y and 18 at each end of the z axis, stretched by 15 percent per zone. The total transverse extent was thus 12.2 cm × 24.4 cm. In the longitudinal direction there were 13 uniform zones with $\Delta x = 5$ mm, so the calculation involved NX × NY × NZ = 13 × 33 × 65 = 27,885 zones.

For the second calculation, the mesh was refined by approximately a factor of 2. Now the transverse plane comprised 29 × 57 zones with $\Delta y = \Delta z = 0.6$ mm in the fine-zoned region and a layer of 36 stretched zones around this, while there were 25 zones in the x direction with $\Delta x = 2.5$ mm. The mesh thus contained 209,625 zones.

The solution to this problem is important not only for research on electric discharges and lightning, but also for general studies of turbulence. As the channel expands, vorticity should be generated because $\nabla\rho \times \nabla p \neq 0$, where ρ is the density and p is the pressure (Picone and Boris, 1983). The evolution of the resulting vortices determines the rate of mixing, and therefore of cooling, of the hot channel. Although the channel only expands to a few centimeters in radius, it is necessary to keep the boundaries far away so that reflected waves do not contaminate the solution in the mixing region.

The calculation was carried out twice in order to use a technique similar to Richardson extrapolation to estimate the "exact" solution in the limit of vanishing zone size. Since the underlying leapfrog scheme in FAST3D is second-order accurate, we expect the errors to decrease by a factor of 4 in the second calculation. By plotting any calculated physical quantity against $(\Delta x)^2$, we should be able to draw a straight line whose intercept at $\Delta x = 0$ represents a better approximation to the correct answer than either of the individual calculations.

In addition, contour plots of vorticity, density and pressure show the evolution of the physical variables, and velocity vector plots display the flow field.

Regarding the theory of vorticity generation by energy deposition along a curved axis, this simulation has accomplished the following:

(1) The vortex strength predicted by the simulation and the general theory agree closely with each other and with experimental data; this verifies the mixing mechanism proposed above for the rapid cooling of such channels.

(2) The numerical simulation reveals a richness of vortex structure not readily predictable by an analytic approach. This structure

has led to a refinement of the theory, indicating the close relationship between points of inflection in the energy deposition and the positions of vortex centers. Figure 2 depicts the mass density in a transverse plane of the channel during expansion to pressure equilibrium. Notice that the density peaks are vertically displaced from y = 0 symmetry plane. In addition we have found, somewhat to our surprise, that the interaction of the pressure and density distributions in generating vorticity in the transverse planes is quite similar to the case of an uncurved discharge with an elliptical cross section. Figure 3 shows a velocity vector plot in a transverse plane after the channel has reached pressure equilibrium. A vortex pair is clearly visible near the center of the channel, and two oppositely directed vortices reside near the symmetry plane and the outer edges of the channel. These vortices remain in the same region, rapidly mixing cool ambient air with the hot channel gas. These numerical simulations have thus greatly enlarged our theoretical understanding of vorticity production by asymmetric energy deposition in a gaseous medium.

REFERENCES

1. Picone, J.M., Boris, J.P., Greig, J.R., Raleigh, M., and Fernsler, R.F., "Convective Cooling of Lightning Channels", Journal of the Atmospheric Sciences, 38 (9), 2056-62 (1981).

2. Picone, J.M., and Boris, J.P., "Vorticity Generation by Asymmetric Energy Deposition in a Gaseous Medium," Phys. Fluids 26 (2), 365 (1983).

ACKNOWLEDGEMENT

This work was supported by the Defense Nuclear Agency and DARPA.

THREE—DIMENSIONAL OUT-OF-CORE CALCULATIONS

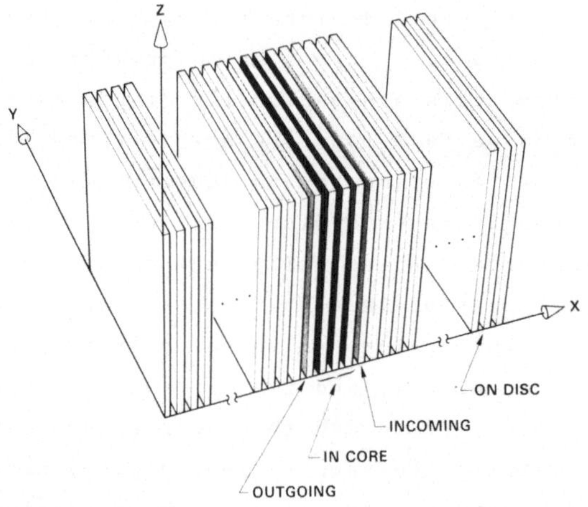

FIG. 1

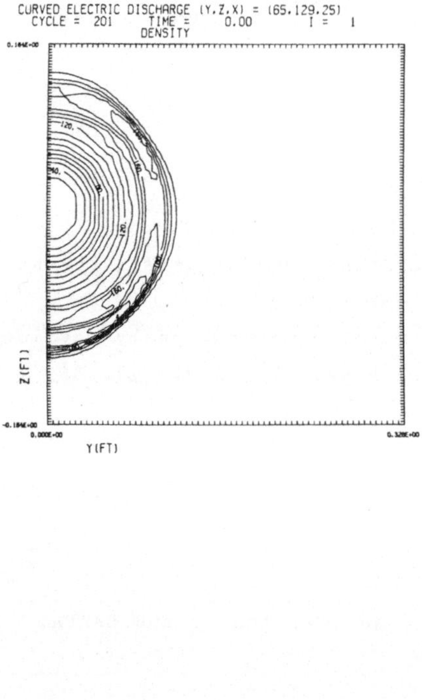

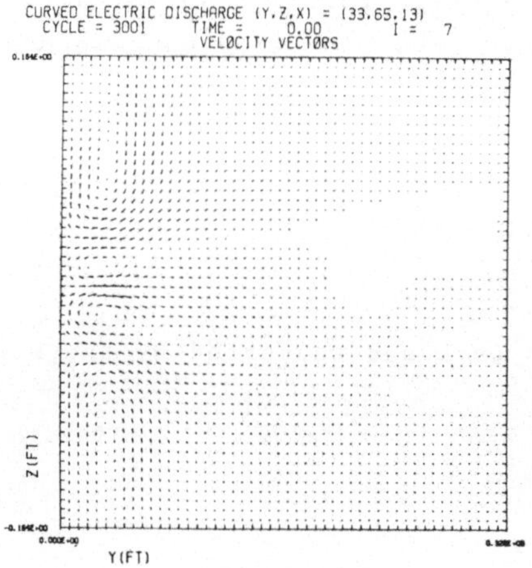

FIG. 3

A NUMERICAL STUDY OF THE TWO- AND THREE-DIMENSIONAL UNSTEADY NAVIER-STOKES EQUATIONS IN VELOCITY-VORTICITY VARIABLES USING COMPACT DIFFERENCE SCHEMES

T. B. Gatski
NASA Langley Research Center, Hampton, VA 23665
C. E. Grosch
Old Dominion University, Norfolk, VA 23508

Abstract

A compact finite-difference approximation to the unsteady Navier-Stokes equations in velocity-vorticity variables is used to numerically simulate a number of flows. These include two-dimensional laminar flow of a vortex evolving over a flat plate with an embedded cavity, the unsteady flow over an elliptic cylinder, and aspects of the transient dynamics of the flow over a rearward facing step. The methodology required to extend the two-dimensional formulation to three-dimensions is presented.

Introduction

The formulation of the Navier-Stokes equations in terms of velocity and vorticity is an alternate approach to the numerical solution of the Navier-Stokes equations. Previously, Dennis, Ingram, and Cook (1979) and Fasel (1980) have used this formulation as the basis of numerical calculations. Dennis et al treated the steady-state problem in three dimensions and Fasel treated the time-dependent problem in two dimensions. In both of these studies, Poisson equations for the velocity components were derived from the kinematic definitions of vorticity and used in the solution algorithm. In the numerical method developed by Gatski, Grosch, and Rose (1982) and used here, the kinematic definition of vorticity is used directly, along with the incompressibility condition of a divergence free-velocity field. These equations, coupled with the transport equation for the vorticity, form the basis of the algorithm.

Solution Method

The basic development and formalism for the two-dimensional solution method are described in Gatski, Grosch, and Rose (1982). It is desirable to extend this methodology to three-dimensional unsteady flows. As was the case in the two-dimensional problem, the continuity equation, kinematic definitions of vorticity and vorticity transport equations are used directly in the three-dimensional problem. The continuity equation and kinematic definitions of vorticity, constitute the velocity solver and are discretized using box-variables for the velocities. Such a formulation produces an over-determined system; however, Fix and Rose (1984) have shown that such a finite-difference approximation yields a least squares solution which is second order accurate. Before the vorticity transport equations can be brought into the solution sequence a modification to the form of the equations needs to be made. This is necessary because the three-dimensional vorticity transport equations contain a vortex stretching term which negates the direct use of the finite-difference basis set which was used in the two-dimensional case. Recall, however, that as the vorticity equations are solved over a time step Δt, the form of the equations allows for the introduction of an integrating factor (Rose, private communication) which transforms the vorticity transport equations into simple advection-diffusion equations. This system can then be solved for the three component vorticities in a completely analogous manner to the simple two-dimensional equation.

Computational Results

Consider first the evolution of a vortical structure over an embedded cavity. Such a flow is of interest since it serves as a qualitative model of a large scale vortical structure, omnipresent in wall bounded turbulent shear flows, evolving over an isolated surface roughness represented by an embedded cavity. In the present study, a Stuart vortex is introduced at the inflow boundary in a consistent mathematical manner. The vorticity contours are shown in Figure 1a. The figure shows the main vortical motion above the cavity as well as a remnant of an induced vortical region downstream of the main motion. Figure 1b shows an enlarged view of the motion in the cavity, as represented by the stream function contours, when the vortex is in the position shown in Figure 1a. The result shown in this figure indicates that the vortex in the boundary layer causes the cavity vortex to lift up. In Figure 1c is shown the pressure contours in the boundary layer. This figure shows the low and high pressure regions associated with this vortical flow.

A second example is the impulsive start of the flow over an ellipic cylinder. This is an example of the unsteady separation of an external flow. Some results are shown in Figure 2 at a time shortly after the beginning of separation. Here the Reynolds number, with the length scale based on the semi-major axis, is 100, the ratio of major to minor axis is 2, and the angle of attack is zero. The vorticity distribution, as shown in Figure 2a, is determined by a balance between diffusion and advection. Vorticity is produced at the boundary and is diffused away. Simultaneously it is advected towards the rear of the ellipse. In contrast, the stream function results given in Figure 2b indicate that the overall velocity field is only slightly different from that of the potential flow. A thin region of separation is also apparent in Figure 2, but at this early time a viscous wake has not yet formed.

Another example of practical relevance is the flow in a channel with a backward facing step. This is a simple prototype of a separating internal flow. Figures 3a and 3b show the steady-state distribution of the stream function and vorticity contours for this flow at Re = 300. The Reynolds number length scale is the height of the inflow channel and the velocity scale is the maximum speed at inflow. The geometry of Figure 3 has been considerably distorted in order to display the results clearly. As can be seen from the results, there is, at this Reynolds number, virtually no upstream effect of the step. The present result shows that the flow over the step generates a rather weak corner vortex, at this low Re. Due to the diffusive dominance in the cross-channel direction the effect of the step persists, in the lower half of the channel for very large distances downstream. In the upper half of the channel, there is a region of decelerating flow downstream of the step and over the recirculation zone. This deceleration, combined with the frictional drag of the upper wall gives a region of near separation on the upper wall. As is seen, this flow field is stable, but that does not prevent the formation of stable, that is decaying, shear waves. Any impulsive change in the flow will excite these transient waves. An example is shown in Figure 4. The flow field at Re = 300 was perturbed by decreasing the viscosity so that the Reynolds number was instantaneously changed from 300 to 500. Contours of the instantaneous values of stream function and vorticity, 20 time units after the viscosity was changed, are given in Figures 4a and 4b, respectively. Note that these waves are downstream of the separation zone behind the step; which is, in fact, where they were formed.

References

1. Dennis, S. C. R.; Ingham, D. B.; and Cook, R. N.: J. Comp. Phys., Vol. 33, (1979), pp. 325-339.
2. Fasel, H. F.: Lecture Notes in Mathematics, No. 771, Springer-Verlag, New York/Berlin, (1980), pp. 177-195.
3. Fix, G. J.; and Rose, M. E.: SIAM J. Numerical Analysis, (1984), to appear.
4. Gatski, T. B.; Grosch, C. E.; and Rose, M. E.: J. Comp. Phys., Vol. 48, No. 1, (1982), pp. 1-22.

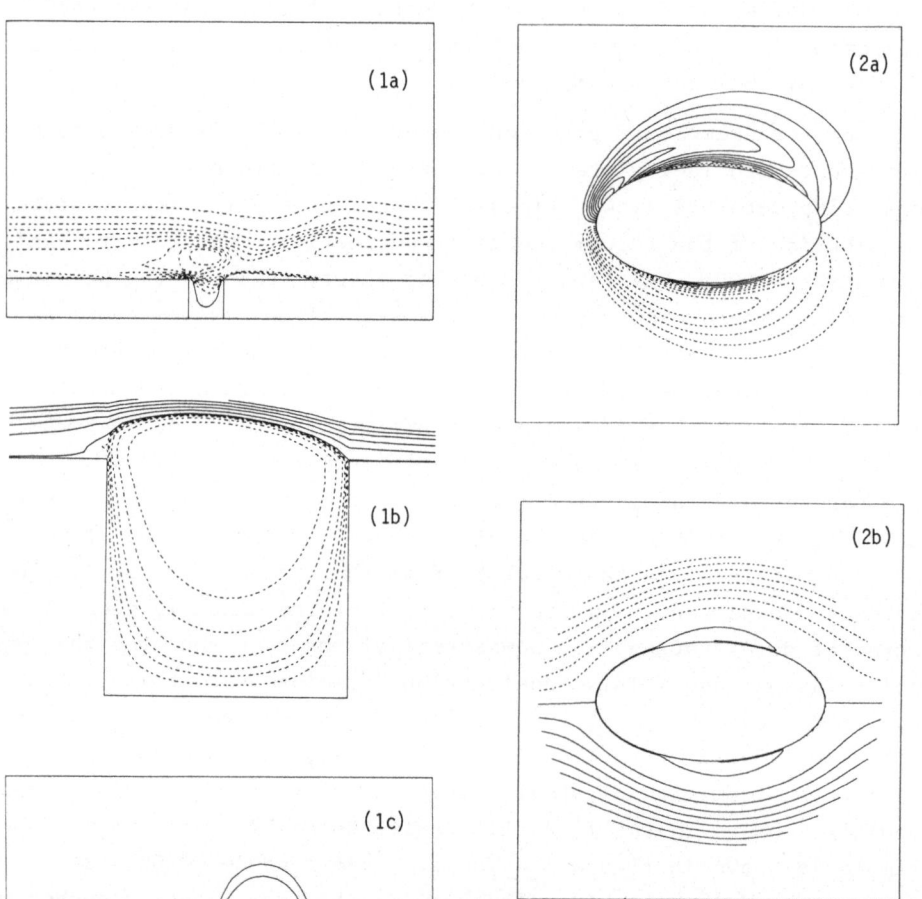

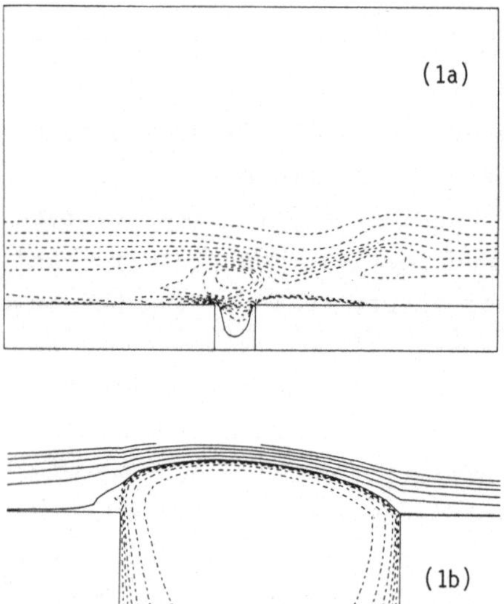

Figure 1: (1a) Vorticity contours in boundary layer (contour levels -1.8 to 0.0); (1b) Stream function contours in embedded cavity (contour levels -0.002 to 0.005); (1c) Pressure contours in boundary layer (contour levels -0.017 to 0.10).

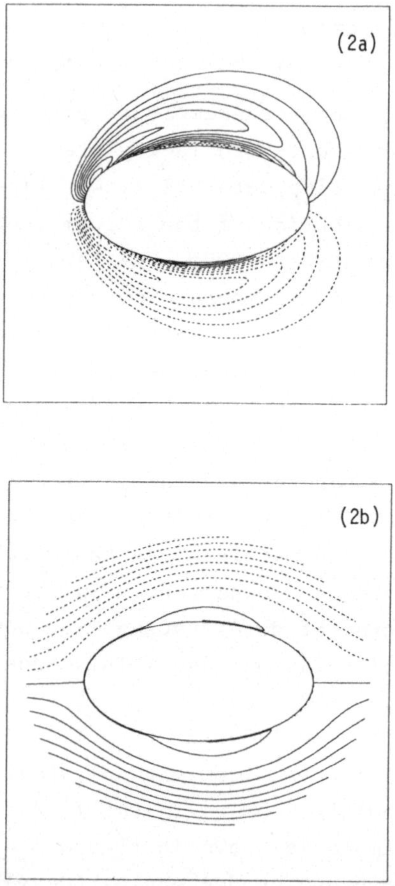

Figure 2: (2a) Vorticity contours in near field of elliptic cylinder (contour levels -6.40 to 6.40); (2b) Stream function contours in near field of elliptic cylinder (contour levels -0.56 to 0.56).

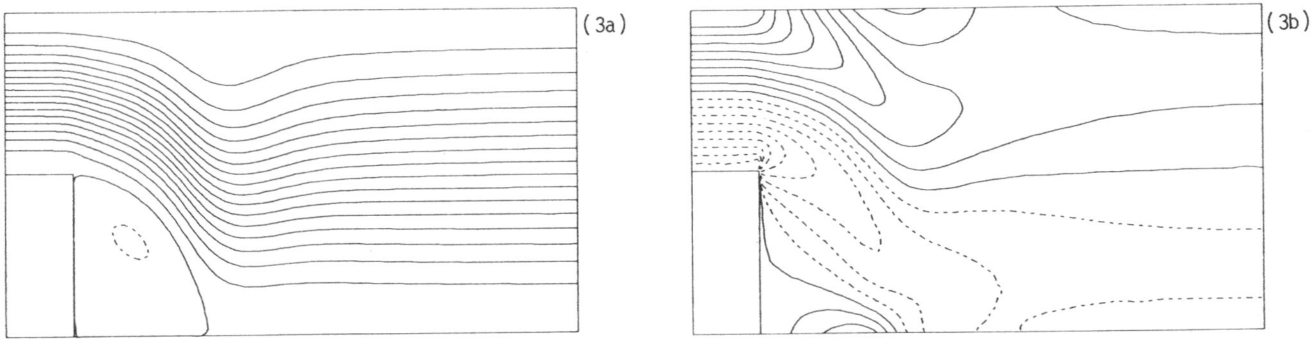

Figure 3: Channel flow with rearward facing step at Re = 300: (3a) Stream function contours (contour levels 0.0 to 2/3); (3b) Vorticity contours (contour levels -3.60 to 3.60).

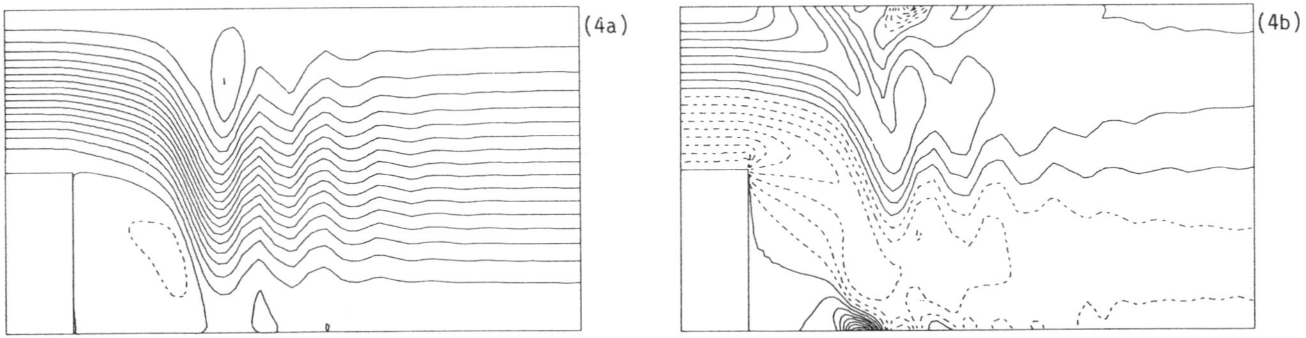

Figure 4: Channel flow with rearward facing step at Re = 500: (4a) Stream function contours (contour levels 0.0 to 0.75); (4b) Vorticity contours (contour levels -3.60 to 3.60).

IMPROVEMENTS IN THE ACCURACY AND STABILITY OF ALGORITHMS FOR THE SMALL-DISTURBANCE AND FULL-POTENTIAL EQUATIONS APPLIED TO TRANSONIC FLOWS

Peter M. Goorjian
NASA Ames Research Center
Moffett Field, CA 94035 USA

I. INTRODUCTION

This paper describes numerical techniques that improve the accuracy and stability of algorithms for the small-disturbance and full-potential equations used to calculate transonic flows. For the small-disturbance equation, the algorithm improvements are: 1) the use of monotone switches in the type-dependent finite-differencing, and 2) the use of stable and simple second-order-accurate spatial differencing. These improvements are for steady and unsteady transonic flows. For the steady, full-potential equation, the improvement is in the use of a monotone switch in the type-dependent finite-differencing of an approximate-factorization (AF2) algorithm. All these improvements can be implemented in present computer codes by making minor coding modifications.

II. SMALL-DISTURBANCE EQUATION IMPROVEMENTS

1. Monotone Implicit Algorithms

Most of the algorithms in transonic small-disturbance potential codes use the Murman-Cole [1] method of switching the differencing scheme for regions of supersonic and subsonic flow. A weakness in this method is that it allows stable solutions for flow fields containing entropy-violating expansion shocks in one-dimensional model cases, as shown in Ref. 2. In comparison, the implicit method presented here, which utilizes the monotone Godunov switch [3], does not allow such nonphysical solutions. In two-dimensional flows over airfoils [2], this weakness in the Murman-Cole switch allows numerical instabilities to develop in the calculations near the leading edges of the airfoils. In some cases of steady flows, converged solutions are obtained by the monotone approximate-factorization algorithm (MAF-G) (see Fig. 1 from Ref. 2), whereas calculations using the Murman-Cole switch (AF2) are unstable. In some cases of unsteady flows, the monotone method (LTRAN2-MG) allows the use of larger time-steps than the older method (LTRAN2), which uses the Murman-Cole switch. In Figs. 2(a) and 2(b) [2], a case is shown in which both methods use the larger time-step and the older method diverges (i.e., computer overflow); here the older method required a time-step 12 times smaller in order to agree with the monotone method. The jaggedness near the leading edge in the LTRAN2-MG results is due to the use of measured ordinates from an experimental model. This jaggedness provides a severe test case for the stability of any algorithm.

Two-Dimensional Steady Algorithm

Governing Equation

The MAF-G algorithm is for the low-frequency, unsteady, small-disturbance potential equation [2]

$$2kM_\infty^2\phi_{xt} = [(1 - M_\infty^2) - (\gamma + 1)M_\infty^m\phi_x]\phi_{xx} + \phi_{yy} \quad (1)$$

Although Eq. (1) is physically meaningful for unsteady flows, MAF-G is nonconservative in time for more efficient calculations of steady flows. The algorithm for unsteady flows, namely, LTRAN2-MG, is presented in detail in Ref. 2; it was used to generate the results shown in Fig. 2.

Algorithm

MAF-G is a modification of an approximate-factorization implicit algorithm, AF2, which uses the Murman-Cole switch. The MAF-G scheme uses the monotone switch of Godunov. It is given by the following two-step, finite-difference approximation to Eq. (1) at mesh point (i,j).

Step 1:

$$[\alpha - (A_{i,j}D_x)]f_{i,j} = [\alpha^2 \overleftarrow{\delta}_x + (A_{i,j}D_x)\delta_{yy}]\phi^n_{i,j} + \alpha(\omega - 1)R^n_{i,j} \tag{2}$$

Step 2:

$$(\alpha\overleftarrow{\delta}_x - \delta_{yy})\phi^{n+1}_{i,j} = f^n_{i,j} \tag{3}$$

where

$$A_{i,j}D_x = \tilde{G}_{i,j}\,\overrightarrow{\Delta}_x + \hat{G}_{i,j}\,\overleftarrow{\Delta}_x$$

$$\tilde{G}_{i,j} = (1 - \varepsilon_{i,j})[\tilde{A}_{i+(1/2),j} + \hat{A}_{i-(1/2),j}] + \tilde{A}_{i-(1/2),j}$$

$$\hat{G}_{i,j} = \varepsilon_{i-1,j}[\tilde{A}_{i-(1/2),j} + \hat{A}_{i-(3/2),j}] + \hat{A}_{i-(1/2),j}$$

$$\tilde{A}_{i-(1/2),j} = \frac{1}{2}C_1 + C_2\tilde{u}^n_{i-(1/2),j}\,, \qquad \hat{A}_{i-(1/2),j} = \frac{1}{2}C_1 + C_2\hat{u}^n_{i-(1/2),j}$$

$$\tilde{u}^n_{i-(1/2),j} = \bar{u} + [1 - \varepsilon_{i-(1/2),j}][u^n_{i-(1/2),j} - \bar{u}]$$

$$\hat{u}^n_{i-(1/2),j} = \bar{u} + \varepsilon_{i-(1/2),j}[u^n_{i-(1/2),j} - \bar{u}]$$

$$C_1 = 1 - M^2_\infty\,, \quad C_2 = -\frac{1}{2}(\gamma + 1)M^m_\infty\,, \quad \bar{u} = -C_1/2C_2$$

$$\varepsilon_{i+(1/2),j} = \begin{cases} 0 & \text{if} \quad u_{i+(1/2),j} \le \bar{u} \quad \text{(subsonic)} \\ 1 & \text{otherwise} \quad \text{(supersonic)} \end{cases}$$

$$\varepsilon_{i,j} = \begin{cases} 0 & \text{if} \quad u_{i+(1/2),j} + u_{i-(1/2),j} \le 2\bar{u} \text{ (possible upstream moving shock)} \\ 1 & \text{otherwise (possible downstream moving shock)} \end{cases}$$

$$u^n_{i-(1/2),j} = \overleftarrow{\delta}_x\phi^n_{i,j} = \frac{\phi^n_{i,j} - \phi^n_{i-1,j}}{x_i - x_{i-1}}\,, \qquad \alpha = 2kM^2_\infty/\Delta t$$

$$\overleftarrow{\Delta}_x f^n_{i,j} = \frac{f^n_{i,j} - f^n_{i-1,j}}{1/2(x_{i+1} - x_{i-1})}\,, \qquad \overrightarrow{\Delta}_x f^n_{i,j} = \frac{f^n_{i+1,j} - f^n_{i,j}}{1/2(x_{i+1} - x_{i-1})}$$

and where ω is a relaxation parameter, R^n is the residual given by

$$R^n_{i,j} = (A_{i,j}D_x\overleftarrow{\delta}_x + \delta_{yy})\phi^n_{i,j}$$

and $\bar{u}$ is the sonic value of ϕ_x. The Murman-Cole switch only uses the $\varepsilon_{i,j}$ switch [2]. The MAF-G scheme is identical to AF2 except in regions where the flow field changes type — near the sonic lines and shock waves.

2. Second-Order-Accurate Supersonic Spatial Differencing

Current methods for calculating transonic flows with the small-disturbance equation typically are only first-order accurate in the supersonic regions of the flow. However, calculations using the full-potential show significant improvements in accuracy when second-order methods are used. In this paper, a stable, simple algorithm [4] is described that is a second-order-accurate extension of the implicit monotone algorithm described above. For steady flow, Figs. 3(a)-3(c) show calculations of flow over a Korn airfoil, for coarse, medium, and fine grids. Note in Fig. 3(a) the improvement in resolution with the second-order method. Also, the convergence rates of the two methods are essentially the same. The improvement in the unsteady algorithm (LTRAN2-MG) and improvements in unsteady flow calculations are presented in Ref. 4.

Algorithm

The modifications to MAF-G to implement second-order accuracy are made by changing only the first step of MAF-G, given by Eq. (2). Let

$$A'_{i,j}D'_x = A_{i,j}D_x + (\Delta x)\overleftarrow{\delta}_x \hat{G}'_{i,j}\overleftarrow{\Delta}_x \quad (4)$$

where

$$\hat{G}'_{i,j} = \varepsilon_{i,j}\varepsilon_{i-(1/2),j}[\hat{A}_{i-(1/2),j} + \hat{A}_{i-(3/2),j}]$$

and

$$\alpha' = \alpha\,\Delta x\,\overleftarrow{\delta}_x\varepsilon_{i,j}\varepsilon_{i-(1/2),j}$$

Then the modified step 1 is given by the following:

Step 1':

$$[(\alpha + \alpha') - (A'_{i,j}D'_x)]f_{i,j} = [\alpha(\alpha + \alpha')\overleftarrow{\delta}_x + (-\alpha' + A'_{i,j}D'_x)\delta_{yy}]\phi^n_{i,j} + \alpha(\omega - 1)R'^n_{i,j}$$

Now the residual is given by

$$R'^n_{i,j} = (A'_{i,j}D'_x\overleftarrow{\delta}_x + \delta_{yy})\phi^n_{i,j}$$

III. FULL-POTENTIAL EQUATION IMPROVEMENT

Monotone Implicit Algorithm

Most of the algorithms in transonic, full-potential codes use type-dependent differencing that is a generalization of the Murman-Cole switch, such as the method of Jameson and the AF2 method of Holst and Ballhaus [5]. These methods suffer from a weakness similar to that of the Murman-Cole switch in dealing with nonphysical expan sions shocks. A monotone implicit method (AF2) has been developed [2] that eliminates this deficiency at sonic expansion points, as demonstrated by the calculations shown in Figs. 4(a) and 4(b) [2] for a two-dimensional case of flow over a modified, double-wedge profile. The full details of the MAG algorithm are given in Ref. 2. Here we describe the essence of the new switch. The AF2 method uses a nonmonotone switch, called upwind-density biasing, to stabilize the calculations in supersonic regions of the flow. That switch is implemented in the mass flux. An example, for $\rho\phi_x$, is

$$\bar{\rho}_{i+(1/2),j}\delta_x\phi^n_{i,j} = \{\rho_{i+(1/2),j} - \nu_{i,j}[\rho_{i+(1/2),j} - \rho_{i-(1/2),j}]\}\overrightarrow{\delta}_x\phi_{i,j} \quad (5)$$

MAF employs a monotone switch based on Godunov's ideas. An example, for $\rho\phi_x$, is

$$\bar{\rho}_{i+(1/2),j}\delta_x\phi^n_{i,j} = \{\rho_{i+(1/2),j} - \bar{\nu}_{i+(1/2),j}[\rho_{i+(1/2),j} - \rho^*]\}\overrightarrow{\delta}_x\phi_{i,j} + \{\hat{\nu}_{i-(1/2),j}[\rho_{i-(1/2),j} - \rho^*]\}\overrightarrow{\delta}_x\phi_{i,j} \quad (6)$$

The essence of the difference between MAF and AF2 is seen by comparing Eqs. (5) and (6). Whereas Eq. (5) uses the switch ν to smoothly shift the flow variables upwind as the flow becomes supersonic, Eq. (6) uses the switches $\tilde{\nu}$ and $\hat{\nu}$ to smoothly shift the flow variables to sonic values ρ^* and q^*, and to shift flow variables located upwind away from sonic values. Both methods use rotated differencing and treat shock waves identically [2], but the monotone method is more stable at sonic expansion points.

IV. REFERENCES

1. Murman, E. M. and Cole, J. D., "Calculations of Plane Steady Transonic Flow," AIAA Journal, Vol. 9, No. 2, 1971, pp. 114-121.

2. Goorjian, P. M., Meagher, M. E., and Van Buskirk, R., "Monotone Implicit Algorithms for the Small-Disturbance and Full-Potential Equations Applied to Transonic Flows," AIAA Paper 83-0371, Reno, Nev., 1983.

3. Godunov, S. K., "A Finite-Difference Method for Inviscid Transonic Flows with Embedded Shock Waves," Mat. Sb. 47, 1959, p. 271; also: Cornell Aeronautical Lab. (Calspan) Translation.

4. Goorjian, P. M., and Van Buskirk, R., "Second Order Accurate Supersonic Differencing for the Small-Disturbance Potential Equations Applied to Transonic Flows," AIAA Paper 84-0091, Reno, Nev., 1984.

5. Holst, T. L. and Ballhaus, W. F., "Fast, Conservative Scheme for the Full-Potential Equation Applied to Transonic Flows," AIAA Journal, Vol. 17, Feb. 1979, pp.

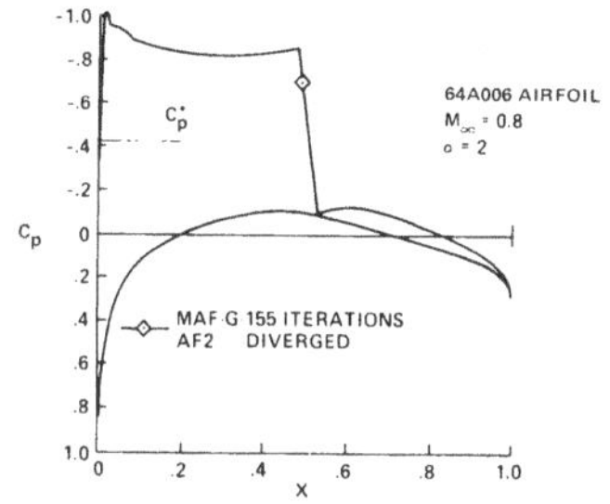

Fig. 1 Converged solution obtained from algorithm using the monotone Godunov switch MAF-G; algorithm using nonmonotones switch AF2 diverged.

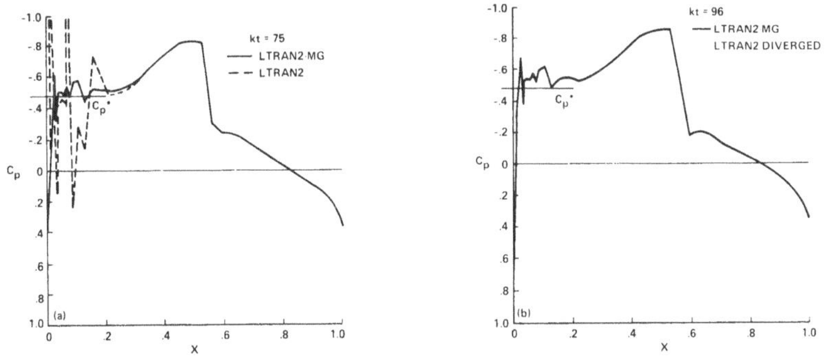

a) kt = 75°

b) kt = 96°

Fig. 2 Comparison of algorithms using the monotone Godunov switch LTRAN2-MG and nonmonotone switch LTRAN2. Plots of upper-surface pressure coefficients of a NACA 64A010 airfoil (experimental model ordinates) in pitching motion. Time given in degrees of motion kt and $M_\infty = 0.80$.

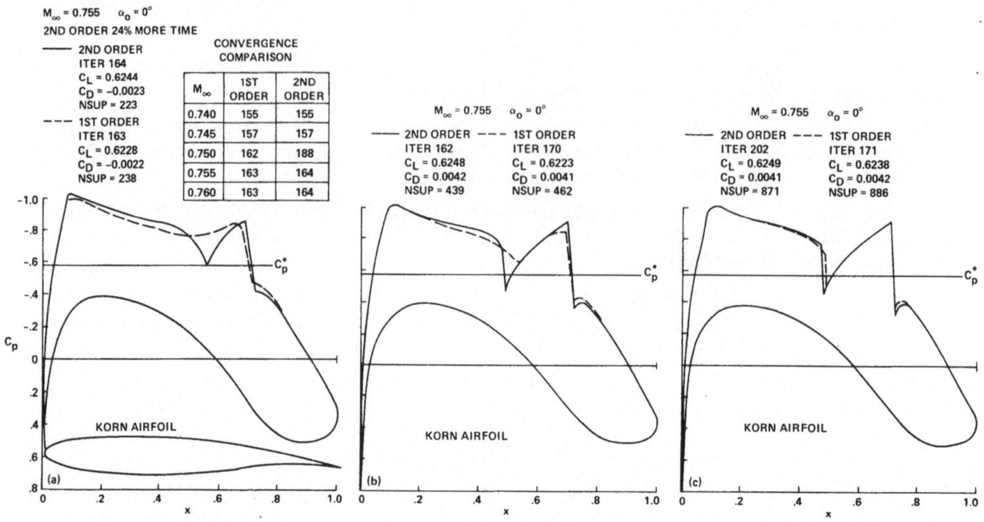

a) Coarse grid b) Median grid c) Fine grid

Fig. 3 Comparison of first- and second-order methods for steady flow over a Korn airfoil; pressure-coefficients plot: $M_\infty = 0.755$ and $\alpha_o = 0$.

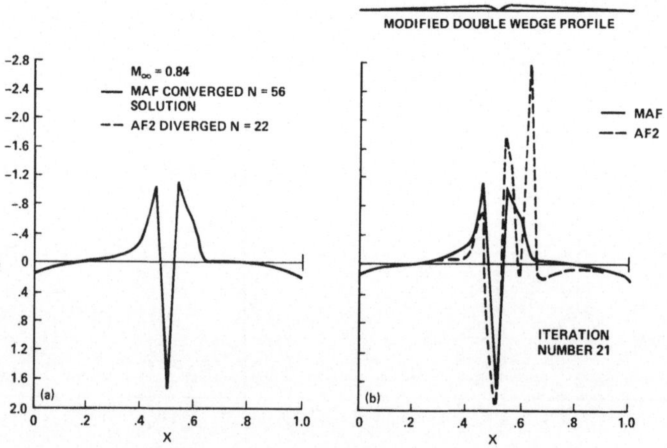

a) Converged solution using MAF, nonmonotone algorithm diverged. b) Iteration No. 21.

Fig. 4 Comparison of monotone MAF and nonmonotone AF2 algorithms for flow over a modified, double-wedge profile.

3D INDUSTRIAL FLOWS CALCULATIONS BY FINITE ELEMENT METHOD

GREGOIRE J.P.*, BENQUE J.P.**, LASBLEIZ P.* , GOUSSEBAILE J**

Electricité de France - Direction des Etudes et Recherches

* Mécanique et Modèles Numériques - Clamart (France)

* Laboratoire National d'Hydraulique - Chatou (France)

The Navier-Stokes equations solution, in case of 3D industrial uncompressible flows is characterized by a large number of unknows and complex geometries. For such problems, the non-linearity treatment inside the global velocity-pressure matrix, variable at each time step, is limited by the capabilities of the present computers. In order to overcome this difficulty, we separate, in the Navier-Stokes equations, the non-linear convection operator from the linear Stokes operator. The advantages of this approach are to replace non-linear matrix iterations by a characteristic curve algorithm and also to allow a velocity-pressure decoupling in the Stokes problem. The latter, in case of appropriate boundary conditions, can be split upon the three velocity components. In order to achieve this calculation, the direct Chorin method has been extended to the iterative Uzawa method, which improves the divergence constraint. Both numerical methods, previously tested, have been introduced in a 3D finite element code.

This paper presents on the one hand the algorithms used, on the other hand the results obtained by the code on a 3D industrial flow and a comparison between core requirements needed by several Stokes solvers.

I - NUMERICAL METHODS

The Navier-Stokes equations, treated by the splitting-up method, have the following form in the case of unsteady uncompressible flow :

Convection equation
$$\begin{cases} \dfrac{\partial \tilde{U}}{\partial t} + U^n . \Delta \tilde{U} = 0 \\ \tilde{U} = U^n \text{ on } \Gamma^- \end{cases}$$

Stokes problem
$$\begin{cases} \dfrac{U^{n+1} - \tilde{U}}{\Delta t} - \nu \Delta U^{n+1} + \dfrac{1}{\rho} \Delta p^{n+1} = F \\ \text{div } U^{n+1} = 0 \end{cases}$$

This scheme is of order one in time (and order two in space with quadratic elements). To obtain order two in time we are developping a new discretization of Navier-Stokes equations. This is done by using a weak formulation with test functions, variable in time and space. This new method, yet tested in 1D and 2D [2], [4], is now extended in 3D.

Convection equation

The solution $\tilde{U}$, at time t_{n+1} is given at each point M of the mesh by :

$$\tilde{U}(M) = U^n(P)$$

where $P = C(t_n)$ is the foot of the characteristic curve (C), passing through M and of equation :

$$\frac{d(C)}{dt} = - U^n$$

The calculation of each curve (C) is performed by Runge Kutta method.

Stokes problem

The finite element discretisation of the Stokes problem gives the matrix system :

$$\begin{cases} AU + B^t p = S & \left(S = F + \frac{1}{\Delta t} \tilde{U}\right) \\ BU = o & \end{cases}$$

a)Velocity-pressure decoupling.

This decoupling is done, by following Uzawa algorithm =

Let $p = p_o$

until $\|BU_i\| \leqslant \varepsilon$, repeat =

$$AU_i = S - B^t p_i$$

$$P_{i+1} = P_i - \rho BU_i \qquad (P_{i+1} = P_i - \rho E^{-1} BU_i \text{ if preconditionned}).$$

This algorithm is attractive only if we can find a good preconditionning matrix E. The natural way for calculating E is performed by discretization of the system :

$$\begin{cases} \Delta p = o \\ \frac{\partial p}{\partial n} = o \end{cases}$$

Then the preconditionned Uzawa algorithm (P.U.A.) can be considered as an iterative improvement of the Chorin method for the verification of the constraint div u = o.

b) Velocity components decoupling

In case of appropriate boundary conditions (imposed velocity or constraint) the A-matrix is formed by three identical diagonal blocks. This property is directly taken into account in the code for the solution of AU = S.

When velocity components are coupled, we have tested a new method which gives decoupled Dirichlet boundary conditions by solving an iterative boundary operator.

II - NUMERICAL IMPLEMENTATION

The previous algorithms have been implemented using the architecture of FIDAP code 6 . Then having several Stokes solvers, in the same code, we have set up comparisons between core and C.P.U. time requirements.

Implementing the convection solver

The determination of the characteristic curve (C) needs, at each intermediate point, the calculation of $U^n(M)$ given in the finite element method by :

$$U^n(M) = \sum_i U^n_i \varphi_i \, (\tau^{-1}(M))$$

where τ is the local mapping of the reference element into the real element containing M and φ_i the shape function in the reference element associated to the node i. The inversion of τ is very easy if τ is an affine mapping (case of plane faces tetrahedron or plane and parallel face hexahedron). In general, this operation must be done by a more expensive interative calculation. That is the reason why we use, at the moment, plane face tetrahedron meshes.

Implementing the preconditionned Uzawa algorithm

In order to illustrate our choice for a velocity-pressure decoupling Stokes solver, we give a comparison table in the case of the 3D problem presented in the results :

	Global Matrix	Matrix A	Matrix E	
Number of d.o.f.	12 318	4 128	702	
Number of words Sky-line mode	17 M	1,7 M	48 110	M = 1 million
Number of words Packing mode		200 000	8 000	
Time in seconds per STOKES P.	28*	70* for 20 iterations		* on CRAY-1

In the code, the linear solutions are computed by the incomplete Choleski conjugate gradient algorithm. When matrices are time dependent (turbulent flows) this solver will be very attractive because it avoids matrix factorisation (450 seconds for global matrix).

III - TESTS AND 3D RESULTS

The code have, yet, been tested on following problems : wall driven cavity [4] ,flow over a forward facing step [7], jet into a 3D-cavity.

In order to justify the interest of iterative Stokes solver, we are presently solving

the cooling tower out fall problem inside a crossing wind (474 hexahedrons to be subdivided into 2844 tedrahedrons ; 4653 nodes) :

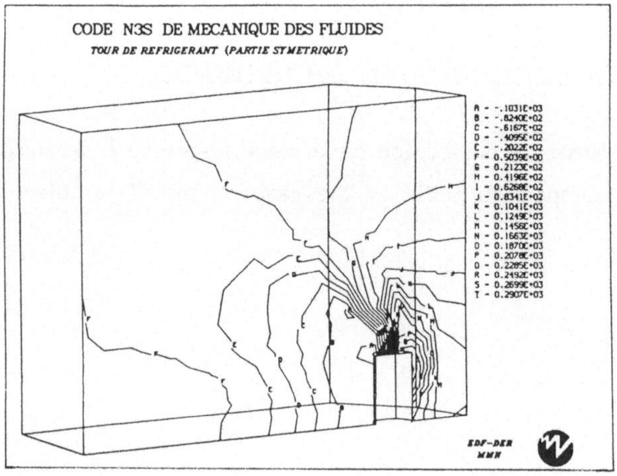

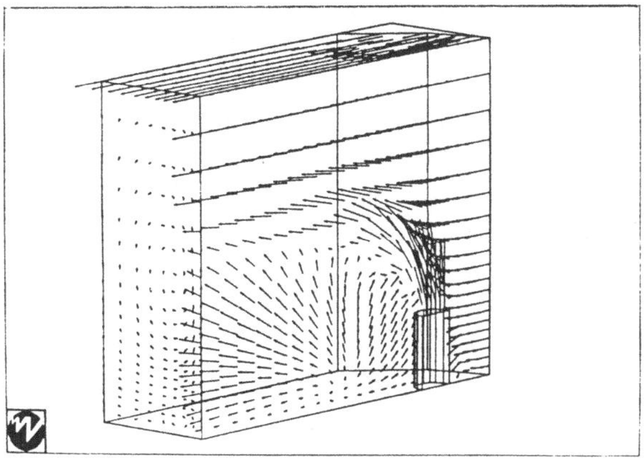

IV - CONCLUSION

The results presented here allow us to simulate more complex industrial flows. For this purpose we are now developping a more accurate modelization by taking into account temperature and turbulence phenomenae. In the same time we are improving code efficiency by a good vectorization and use of domain substructuring technics.

REFERENCES

[1] J.P. BENQUE, J. RONAT : "Quelques difficultés des modèles numériques en hydraulique". 5th International Symposium on Computing Methods in Applied Sciences and Engineering. Versailles (France) Décembre 1981.

[2] J.P. BENQUE, G. LABADIE, J. RONAT : "A new-finite element method for the Navier-Stokes equations coupled with a temperature equation". To be published in "International Journal of Numerical Methods in Fluids".

[3] J.P. BENQUE, P. ESPOSITO, G. LABADIE : "New decomposition finite element methods for the Stokes problem and Navier-Stokes equations" Numerical Methods in Laminar and Turbulent Flows, Seattle (USA) Août 1983.

[4] J.P. BENQUE, J.P. GREGOIRE, A. HAUGUEL, M. MAXANT : "Application des méthodes de décomposition aux calculs numériques en hydraulique industrielle". 6ème Colloque International sur les méthodes de calcul scientifique et technique, Versailles (France) Décembre 83.

[5] J. GOUSSEBAILE, J.P. GREGOIRE, A. HAUGUEL : "Iterative Stokes Solvers and Splitting Technics for Industrial Flows" - 5th International Conference on finite element methods in Flow Problems, AUSTIN (USA) January 84.

[6] M.S. ENGELMAN : "Fidap Theoretical Manual - Fidap Users Manual" Computer Applications International S.A. 1981.

[7] J. GOUSSEBAILE, H. HAUGUEL, P. HEMMERICH : "Numerical development of Navier-Stokes solvers for industrial flows" 6th IAHR working group on refined modelling of flows, Karlsruhe (R.F.A.) March 83.

TRANSONIC FLOWS THROUGH CASCADES

W. Haase
Dornier GmbH, Theoretical Aerodynamics
Postfach 1420, D-7990 Friedrichshafen 1, FRG

Introduction

In recent years, finite volume techniques have been developed in order to simulate complex flow phenomena about arbitrary geometries. Associated with Runge-Kutta time stepping schemes, efficient methods became available integrating the governing equations for inviscid, compressible, unsteady, rotational flows.

Based on the previously published Runge-Kutta stepping schemes and especially related to 4-stage schemes a new approach has been developed to solve the Navier-Stokes equations for two-dimensional, compressible, turbulent, time-dependent flow. This new method has been successfully applied to airfoil flow problems [1] and is now applied to transonic flows through cascades.

Method

Based on the physical laws for mass, momentum and energy conservation a finite volume method is used to integrate the Navier-Stokes equations as the governing equations for viscous, compressible and turbulent flows.

The corresponding set of ordinary differential equations is solved by a four-stage Runge-Kutta time-stepping scheme [2]. This scheme is of fourth order accuracy, it is stable up to CFL-numbers of about 2.8 and allows latitude in the introduction of dissipative (filter) terms. To prevent the appearance of oscillations it proves necessary - at least for high Reynolds number flows - to add artificial dissipation to the existing scheme; blended second and fourth order artificial dissipation terms are used in the present approach. The difference equations are solved by local time stepping based on the maximum allowable time step for each cell.

In order to simulate turbulent flows, an algebraic turbulence model is used to model the Reynolds stresses. The turbulent model [3] is a two layer model using the normal to the wall vorticity distribution to compute a scaling parameter for the other eddy viscosity instead of the boundary layer displacement thickness which is extremely difficult to obtain directly (and accurately) from Navier-Stokes results.

Geometry and Boundary Conditions

The flow through the DFVLR-cascade SKG 2.7 serves as a test case with transonic flow conditions. Navier-Stokes calculations are performed in a C-type mesh with 128 x 30 mesh points simply doubled in fig. 1 presenting better insight into the geometrical discretisation of the computational domain. The profile used for calculation has been closed for computational reasons resulting in a sharp trailing edge.
At the inflow boundary - the flow is coming from the left - all physical properties (pressure, density, velocity, energy) remain fixed during the calculation. At the outflow boundary only the pressure is prescribed and periodic conditions are used at the upper and lower boundary, respectively.

Results

Related to measurements, performed at the DFVLR [4] the initial calculation parameters are:

Mach number = 0.7737 , $p_{inflow}/p_{outflow}$ = 1.1608
Reynolds number = 0.769 million, T_{total} = 652 R

Based on these flow parameters the velocity field after 5000 iterations is given in fig. 2. As usual, the vector length corresponds to the magnitude of the velocity. The separated flow region on the upper surface, physically caused by a steep pressure gradient - no shock occurs in this supercritical flow case - can also be seen from the streamline pattern given in fig. 3. The thick solid lines represent a zero stream-function value, the interval is equal to 0.025.

Fig. 4 gives a contour plot of vorticity lines. The outermost lines represent a vorticity amount of less than 0.05 % of its maximum value. Therefore, these lines are assumed to define the boundary between viscous and inviscid flow domains. Furthermore, the increase in boundary layer thickness as well as the decrease of vorticity and skin friction regarding to the steep pressure gradient - no shock occurs in that supercritical flow case - can be easily recognized.

In fig. 5 contours of the pressure coefficient are plotted, showing some wiggles at the outer - periodic - boundaries due to a rather coarse mesh in this area. The thick solid lines represents the critical C_p-value of -0.514, the interval between the contour lines is equal to 0.1.

A better insight into the accuracy and efficiency of the present methods is gained from the following figures. Fig. 6 shows the surface pressure distribution compared with the measured data [4]. The pressure peak near the leading edge might be caused by geometrical problems. The agreement is encouraging but could be improved by means of a solution adaptive grid technique [5] which redistributes the surface oriented mesh based on the curvature of the surface pressure in order to get smaller step sizes in regions of large curvature.

To investigate the iterative behaviour of the numerical procedure, in fig. 7 the transient behaviour of the drag coefficient is given. One recognizes easily an oscillatory behaviour smeared out with respect to time (or iteration cycles). At 5000 iterations the calculation has been truncated with Δc_d and Δc_l less than 0.1 %; furthermore, no changes in the surface pressure distribution have been observed.

Conclusions

The present finite volume code for solving the Navier-Stokes equations for compressible, turbulent flows has been applied to cascade flow problems. The SKG 2.7 cascade serves as a test case with transonic - supercritical - flow conditions. The computed results seem to be in good agreement with the measurements.

Acknowledgement

This work was partially sponsored by the German Ministry of Defense.

References

[1] W. Haase, B. Wagner, A. Jameson:
Development of a Navier-Stokes Method Based on a Finite Volume Technique for the Unsteady Euler Equations.
Proceedings of the 5th GAMM-Conference on Numerical Methods in Fluid Mechanics, 1983, Notes on Numerical Fluid Dynamics, Vol. 7, M. Pandolfi/R. Piva (Ed.), Vieweg Verlag (1984)

[2] A. Jameson, W. Schmidt, E. Turkel:
Numerical Solutions of the Euler Equations by Finite Volume Methods Using Runge-Kutta Time-Stepping Schemes.
AIAA-81-1259 (1981)

[3] B. S. Baldwin, H. Lomax:
Thin Layer Approximation and Algebraic Model for Separated Flows.
AIAA-78-257 (1978)

[4] SKG-2.7-Measurements (Measurement-No. 147) provided by DFVLR, Institut für Antriebstechnik, Köln

[5] W. Haase, K. Misegades, M. Naar:
Adaptive Grids in Viscous Flow.
Third Int. Conference on Numerical Methods in Laminar and Turbulent Flow, University of Washington, Seattle (1983)

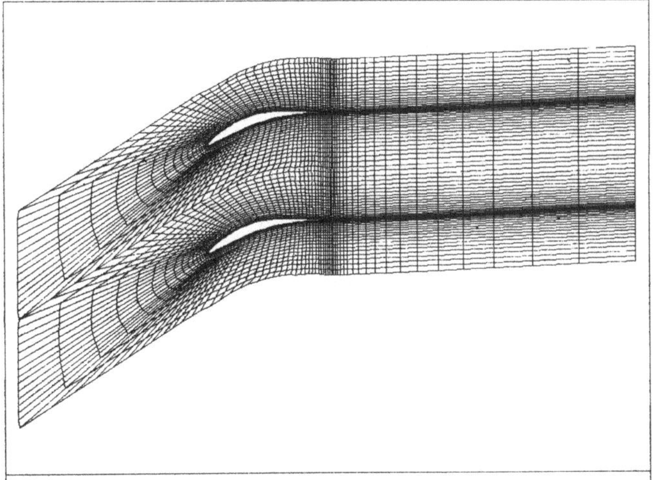

FIG. 1: SKG 2.7 CASCADE - MA=0.7737, RE=7.69E5
COMPUTATIONAL DOMAIN

I=2/126 J=2/30 XMIN=-1.491E+00 XMAX=3.338E+00 YMIN=-1.321E+00 YMAX=7.342E-01
ØFAK=4.500E-01 FLØNG=0.000E+00 FMAX= 0.000E+00 FMIN= 0.000E+00 CYCLES=0 20/06/84 13.09.24

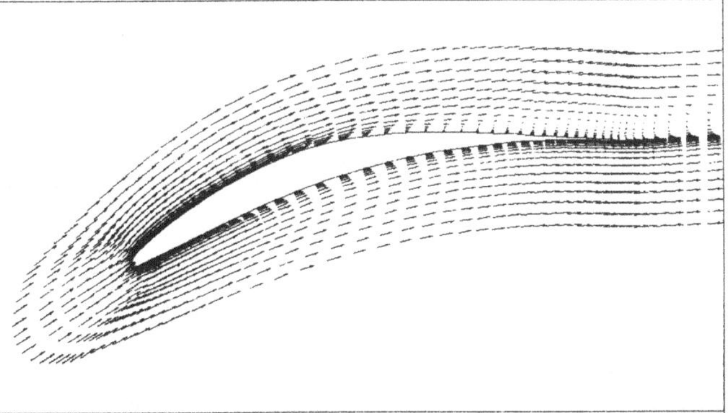

FIG. 2: SKG 2.7 CASCADE - MA=0.7737, RE=7.69E5
VELOCITY VECTORS

I=13/114 J=2/22 XMIN=-2.369E-01 XMAX=1.145E+00 YMIN=-1.929E-01 YMAX=4.088E-01
ØFAK=4.500E-01 FLØNG=8.000E-01 FMAX= 1.341E+03 CYCLES=0 20/06/84 13.05.49

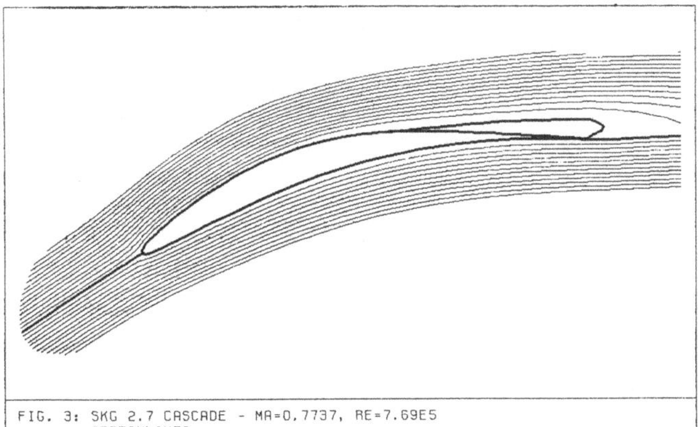

FIG. 3: SKG 2.7 CASCADE - MA=0,7737, RE=7.69E5
STREAMLINES

I=12/114 J=1/22 XMIN=-2.674E-01 XMAX=1.168E+00 YMIN=-2.190E-01 YMAX=4.207E-01
ØFAK=4.500E-01 FLØNG=0.000E+00 FMAX= 4.879E-01 FMIN= -5.097E-01 CYCLES=0 20/06/84 13.20.31

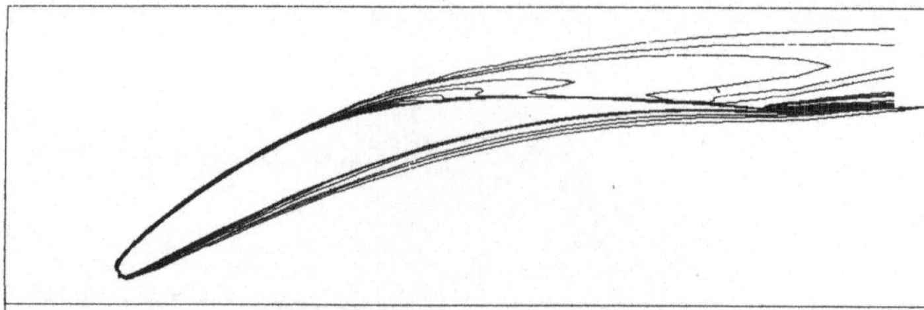

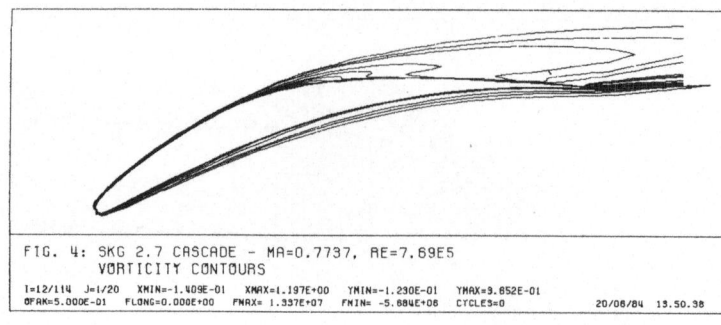

FIG. 4: SKG 2.7 CASCADE - MA=0.7737, RE=7.69E5
VORTICITY CONTOURS

I=12/114 J=1/20 XMIN=-1.409E-01 XMAX=1.197E+00 YMIN=-1.230E-01 YMAX=3.852E-01
OFAK=5.000E-01 FLONG=0.000E+00 FMAX= 1.337E+07 FMIN= -5.684E+06 CYCLES=0 20/06/84 13.50.38

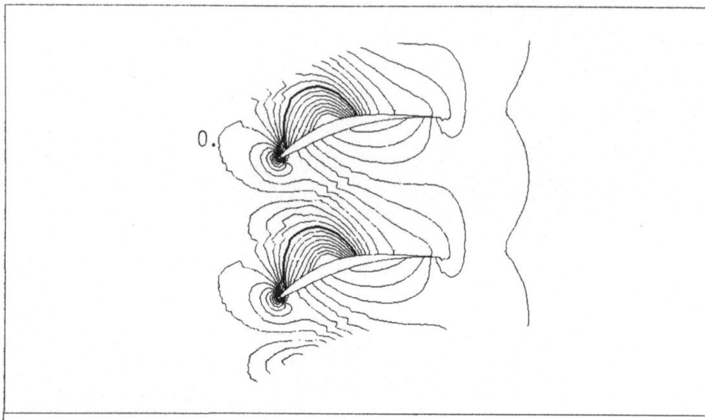

FIG. 5: SKG 2.7 CASCADE - MA=0.7737, RE=7.69E5
PRESSURE COEFFICIENT CONTOURS

I=1/126 J=2/90 XMIN=-1.000E+00 XMAX=2.000E+00 YMIN=-1.386E+00 YMAX=7.658E-01
OFAK=5.000E-01 SS FLONG=0.000E+00 FMAX= 1.817E+00 FMIN= -1.447E+00 CYCLES=0 20/06/84 13.51.33

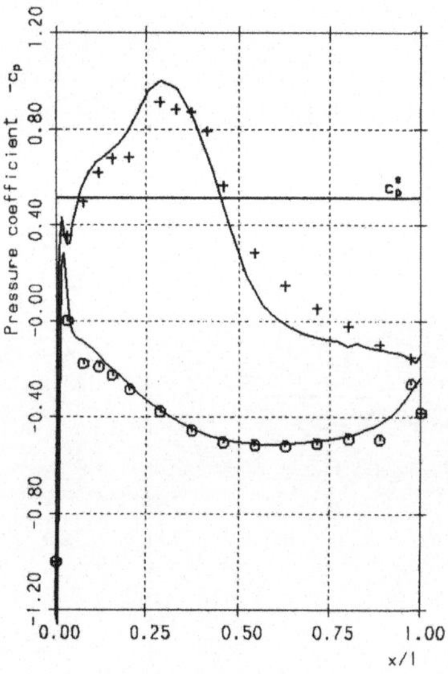

⊙ Lower surface: measurement
— present work
\+ Upper surface: measurement
— present work

Fig. 6: SKG 2.7 - Ma=0.7737, Re=7.69E5
Surface pressure distribution

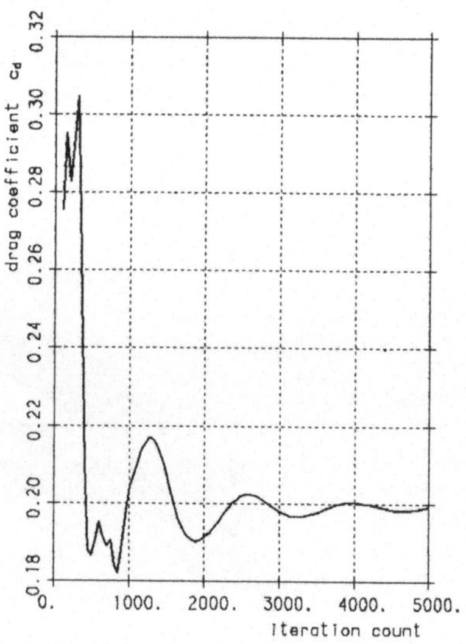

Fig. 7: SKG 2.7 - Ma=0.7737, Re=7.69E5
Transient behaviour of drag coefficient

TWO-DIMENSIONAL MODEL FOR THE TWO-PHASE FLOW SIMULATION IN A VIKING ROCKET ENGINE COMBUSTION CHAMBER

M. Habiballah and H. Monin
Office National d'Etudes et de Recherches Aérospatiales
BP 72, 92322 Châtillon Cédex, FRANCE

Introduction

High frequency instabilities in a first stage VIKING engine caused the loss of L02 Ariane launch vehicle. Both experimental and theoretical projects were then initiated to investigate the origin of such unexpected effects. SEP+ has overcome these difficulties for an engineering purpose. CNES++ had nevertheless charged ONERA with the development of a numerical model for two-phase flow applied to a VIKING rocket engine combustion chamber. Such a numerical simulation should explain and foresee these instabilities. At the same time, other studies are elaborated at ONERA, concerning the droplet size distribution in a liquid propellant engine and the droplet combustion modelling. These results will be implemented in the numerical model.

A two dimensional (x, y) model is now available for the two-phase flow simulation. This paper presents the main results of the computer program.

The gas-droplets model

The two-phase flow in the combustion chamber of a VIKING rocket engine is very complex. Two phases constitute it . The liquid one is composed of oxidizer (N_2O_4) and fuel (UDMH) droplets having different diameters, velocities and temperatures ; the gas one contains various chemical species. The two phases exchange mass, momentum and energy.

In the present model, the gas phase is supposed to be composed of species issued from droplets vaporization (UDMH and N_2O_4 gas) and burnt products (P). For the liquid phase then, the droplets are treated by groups : each group includes droplets of same chemical components, mean diameter, velocity and temperature. An eulerian description is used to solve the basic equations for each group.

+Société Européenne de Propulsion
++Centre National d'Etudes Spatiales

Governing equations

Gas phase.

$$\frac{\partial \rho}{\partial t}+\frac{\partial \rho u}{\partial x}+\frac{\partial \rho v}{\partial y}=\dot{w} \qquad (1)\ \text{continuity}$$

$$\frac{\partial \rho u}{\partial t}+\frac{\partial \rho u^2}{\partial x}+\frac{\partial \rho u v}{\partial y}=-\frac{\partial p}{\partial x}+\frac{\partial \sigma_{xx}}{\partial x}+\frac{\partial \sigma_{xy}}{\partial y}+f_x+S_{q_x} \qquad (2)\ \text{x-momentum}$$

$$\frac{\partial \rho v}{\partial t}+\frac{\partial \rho u v}{\partial x}+\frac{\partial \rho v^2}{\partial y}=-\frac{\partial p}{\partial y}+\frac{\partial \sigma_{xy}}{\partial x}+\frac{\partial \sigma_{yy}}{\partial y}+f_y+S_{q_y} \qquad (3)\ \text{y-momentum}$$

$$\frac{\partial \rho e}{\partial t}+\frac{\partial \rho e u}{\partial x}+\frac{\partial \rho e v}{\partial y}=-p\left(\frac{\partial u}{\partial x}+\frac{\partial v}{\partial y}\right)+\sigma_{xx}\frac{\partial u}{\partial x}+\sigma_{yy}\frac{\partial v}{\partial y}+\sigma_{xy}\left(\frac{\partial u}{\partial y}+\frac{\partial v}{\partial x}\right)+\frac{\partial q_x}{\partial x}+\frac{\partial q_y}{\partial y}+S_e \qquad (4)\ \text{energy}$$

$$\frac{\partial \rho_1}{\partial t}+\frac{\partial \rho_1 u}{\partial x}+\frac{\partial \rho_1 v}{\partial y}=\dot{w}_1+\frac{\partial}{\partial x}\left[\rho \mathcal{D}_1 \frac{\partial}{\partial x}(\rho_1/\rho)\right]+\frac{\partial}{\partial y}\left[\rho \mathcal{D}_1 \frac{\partial}{\partial y}(\rho_1/\rho)\right] \qquad (5)\ \text{UDMH gas transport}$$

$$\frac{\partial \rho_2}{\partial t}+\frac{\partial \rho_2 u}{\partial x}+\frac{\partial \rho_2 v}{\partial y}=\dot{w}_2+\frac{\partial}{\partial x}\left[\rho \mathcal{D}_2 \frac{\partial}{\partial x}(\rho_2/\rho)\right]+\frac{\partial}{\partial y}\left[\rho \mathcal{D}_2 \frac{\partial}{\partial y}(\rho_2/\rho)\right] \qquad (6)\ N_2O_4\ \text{gas transport}$$

$$\rho_3=\rho-(\rho_1+\rho_2)$$

The temperature is issued from the energy $T=(\rho e-\sum_{k=1}^{3}\rho_k e_{ok})/(\sum_{k=1}^{3}\rho_k C_{vk})$; e_{ok} is the internal energy of formation of specie k at $T = 0\,°K$, C_{vk} the specific heat of specie k at constant volume. The pressure results from the perfect gas equation of state $p=\sum_{k=1}^{3} r_k \rho_k T$, $r_k = C_{pk} - C_{vk}$; C_{pk} is the specific heat of specie k at constant pressure. $\vec{q}=\binom{q_x}{q_y}$ is the heat flux due both to conduction ($-k\,\overrightarrow{\text{grad}}\,T$) and enthalpy diffusion ($\sum_{k=1}^{3}\rho_k h_k \vec{V}_{dk}$) ; k is the thermal conductivity, h_k the enthalpy of specie k, $\vec{V}_{dk}$ the diffusion velocity of specie k given by Fick's law $\vec{V}_{dk}=-\mathcal{D}_k\,\overrightarrow{\text{grad}}\ln(\rho_k/\rho)$, $\mathcal{D}_k$ is the diffusion coefficient. σ_{xx}, σ_{yy}, σ_{xy} are the viscous stresses, a simple turbulence model yields the following expression for the effective viscosity $\mu_e=\mu+\mu_t$, $\mu_t = C_t\, l^2\left|\frac{\partial u}{\partial y}+\frac{\partial v}{\partial x}\right|$, l is the mixing length given by min $(\Delta x, \Delta y)$ and C_t a constant coefficient. The turbulent thermal conductivity and turbulent diffusion coefficient are then obtained from their laminar values : $\mathcal{D}_k=(\mathcal{D}_k)_{la}\times\frac{\mu_t}{\mu}$, $k=(k)_{la}\times\frac{\mu_t}{\mu}$.

Liquid phase : for each group m

$$\frac{\partial \rho_{\ell m}}{\partial t}+\frac{\partial \rho_{\ell m} u_{\ell m}}{\partial x}+\frac{\partial \rho_{\ell m} v_{\ell m}}{\partial y}=\dot{w}_m+\frac{\partial}{\partial x}\left(\mathcal{D}_{\ell m}\frac{\partial \rho_{\ell m}}{\partial x}\right)+\frac{\partial}{\partial y}\left(\mathcal{D}_{\ell m}\frac{\partial \rho_{\ell m}}{\partial y}\right) \qquad (7)\ \text{continuity}$$

$$\frac{\partial \rho_{\ell m} u_{\ell m}}{\partial t}+\frac{\partial \rho_{\ell m} u_{\ell m}^2}{\partial x}+\frac{\partial \rho_{\ell m} u_{\ell m} v_{\ell m}}{\partial y}=f_{mx}+S_{q_{mx}} \qquad (8)\ \text{x-momentum}$$

$$\frac{\partial \rho_{\ell m} v_{\ell m}}{\partial t}+\frac{\partial \rho_{\ell m} u_{\ell m} v_{\ell m}}{\partial x}+\frac{\partial \rho_{\ell m} v_{\ell m}^2}{\partial y}=f_{my}+S_{q_{my}} \qquad (9)\ \text{y-momentum}$$

$$\frac{\partial n_m}{\partial t}+\frac{\partial n_m u_{\ell m}}{\partial x}+\frac{\partial n_m v_{\ell m}}{\partial y}=S_{n_m}+\frac{\partial}{\partial x}\left(\mathcal{D}_{\ell m}\frac{\partial n_m}{\partial x}\right)+\frac{\partial}{\partial y}\left(\mathcal{D}_{\ell m}\frac{\partial n_m}{\partial y}\right) \qquad (10)\ \text{continuity for } n_m$$

We assume that any droplet of the same group has a constant temperature equal to the injection one Ti, except on its surface where it is equal to the vaporization temperature Tv. The energy transfers between the two phases are taken into account in the gas phase energy equation. The instantaneous diameter D_m is derived from $\rho_{1m} = \pi/6\ \rho_L\ n_m\ D_m^3$; n_m is the number of droplets per unit volume, ρ_L the density of the liquid composing the droplets. The second derivatives in eq (7) and (10) arise from the turbulent correlations modelling : $\overline{\rho'_{1m}\ u'_{1m}} = -\mathcal{D}_{1m}\ (\partial\rho_{1m}/\partial x)$, $\overline{\rho'_{1m}\ v'_{1m}} = -\mathcal{D}_{1m}\ (\partial\rho_{1m}/\partial y)$ (the same for n_m) ; $\mathcal{D}_{1m}$ is a turbulent diffusion coefficient. We can even now notice that the addition of these diffusion terms contributes to the stability of the numerical scheme. $Sn_m = \delta n_{m+1} - \delta n_m$, δn_m is the number of droplets leaving group m to group m - 1 per unit time.

Source terms : transfers between the two phases

- mass

The combustion model is issued from the quasi-steady analysis of a droplet combustion with a diffusion flame, without any forced convection [1]. This yields $\dot{w}_m^\circ = - n_m\ (2\pi k D_m/C_{pg})\ \ln\ (1 + B)$; B depends on the physical properties both of the surrounding gas and the liquid.

In the case of forced convection, a correction is proposed [2] : $\dot{w}_m = \dot{w}_m^\circ\ (1 + 0.276\ Re_m^{1/2}\ Pr^{1/3})$,

$$Re_m = \frac{\rho\,|\vec{V} - \vec{V}_{\ell m}|\,D_m}{\mu}\ ,\quad Pr = \frac{\mu C_{pg}}{k}\ ,\quad \dot{w} = -\sum_m \dot{w}_m$$

- momentum

The drag force is given by $\vec{f}_m = (\pi/8)\mu\, n_m D_m\ Cx_m\ Re_m\ (\vec{V} - \vec{V}_{1m})$, where the drag coefficient Cx_m is determined by the Wallis' analytical formula : $Cx_m = (24/Re_m)(1 + 0.15\ Re_m^{0.687})$ if $Re_m \leqslant 1000$, $Cx_m = 0.44$ if $Re_m > 1000$. $\vec{f} = -\sum_m \vec{f}_m$

The momentum loss due to droplets vaporization is $\vec{S}_{q_m} = \dot{w}_m\ \vec{V}_{1m}$; $\vec{S}_q = -\sum_m \vec{S}_{q_m}$

- energy

The analysis for the transfer terms between the two phases, based on the previous hypotheses yields the following expression $Se = -\sum_m \dot{w}_m\ (h_{mi} + 1/2\ |\vec{V} - \vec{V}_{1m}|^2)$, where h_{mi} is the specific enthalpy of droplets belonging to group m, at Ti.

- mass sources for gaseous species

$$\dot{w}_k = \dot{w}_k^{(v)} + \dot{w}_k^{(c)} \qquad k = 1,2$$

. vaporization : $\dot{w}_1^{(v)} = -\sum_m \dot{W}_m$ (UDMH droplets), $\dot{w}_2^{(v)} = -\sum_m \dot{W}_m$ (N_2O_4 droplets)

. combustion : the chemical reaction is assumed to be in the form

UDMH + 2 $N_2O_4 \longrightarrow$ P with a chemical velocity given by $\dot{\omega} = cc\left(\frac{\rho_1}{\mathcal{M}_1}\right)\left(\frac{\rho_2}{\mathcal{M}_2}\right)^2$

(cc is a constant). So $\dot{w}_1^{(c)} = -\mathcal{M}_1\dot{\omega}$, $\dot{w}_2^{(c)} = -2\mathcal{M}_2\dot{\omega}$, $\mathcal{M}_k$ is the molar mass of specie k.

Numerical method

The partial differential equations for the two-phase flow are solved by the method of finite differences, using an explicit predictor-corrector scheme [3]. The numerical integration uses a staggered grid : the momentum variables are defined on the mesh sides, and the other ones in the mesh center. Appendix A shows the discretization scheme of the equations for ρ and ρu. A similar treatment is applied to the other equations.

The well known explicit methods stability criterion-CFL criterion-valid in one-phase flow appears to be not sufficient in the case of multi-phase reacting flows where other characteristic time scales appear (e.g combustion time scale). The numerical tests we have performed showed that the scheme is stable when $\Delta t \leqslant (\Delta t)_{CFL}/5$.

Boundary conditions

A no-slip boundary condition is applied in the presence of a rigid wall for the two phases. At the exit area, a constant static pressure and zero gradients in the x-direction are imposed (Fig. 1).

The droplets are injected through fictitious cells located at the domain entrance. In those cells, the gas is supposed to be at rest (u = v = 0) and the injection parameters $|\vec{V}_{lm}(t)|$, $\theta_m(t)$, $\rho_{lm}(t)$, $D_m(t)$ are given for each group m of droplets.

Initial condition

At time t = 0, pressure, temperature, density are set uniform in the chamber. The calculation starts with the droplets injection.

Results

A simple calculation (one droplets size at the injectors) is presented for a better understanding of the results : it is performed using a 17 x 11 grid, in the case of two groups of UDMH droplets (m = 1,2) injected in the cell (1,5) and two groups of N_2O_4 droplets (m = 3,4) injected in the cell (1,7). The injection parameters imposed are : for all groups, $D_m = 100\,\mu m$, $Cv_m = (\rho_{lm}/\rho_L) = 0.035$, $u_{lm} = 40$ m/s ; $\theta_1 = -\theta_3 = 45°$, $\theta_2 = -\theta_4 = -30°$. The total injected liquid flow rate q_L is equal to 62.356 kg/s/m. At t = 0, the gas phase description is : $\rho_o = 5$ kg/m^3, $p_o = 40$ bar, $T_o = 2187.6°K$, $\rho_1 = 0.005\,\rho_o$, $\rho_2 = 0.1\,\rho_o$, $\rho_3 = 0.895\,\rho_o$. The integration parameters are $\Delta x = \Delta y = 0.01$ m and $\Delta t = 10^{-6}$ s.

Fig. 2 presents the temporal variation of the averaged pressure ($p_i(t) = \sum_{j=2}^{10} p_{ij}(t)/9$) at section 2. It's shown that $p_i(t)$ presents damped oscillations whose frequency f_L is equal to 1230 Hz and damping ratio equal to 11 %. These oscillations have to be compared, in the case of one-phase flow, with the first longitudinal mode of acoustic waves propagating in a cylinder closed at one side and with imposed pressure condition at the other one ; in such a case, the acoustic frequency ($f = a/4L$, a is the sound speed) should then be equal to 1500 Hz. In the case of a two-phase flow, the combustion effect is to excite the first longitudinal mode of the chamber that then damps down. The steady flow rate profiles of the two phases (x-direction) are shown in Fig. 3. The liquid flow rate is decreasing while the gas flow rate is increasing, so that the sum is nearly constant and equal to q_L (difference less than 5 %).

We can notice on the velocity gas field (Fig. 4) a gas acceleration from the injectors to the exit. We then observe that the flow is one dimensional in the second half of the chamber ; on the contrary, near the injectors, the transverse velocity is not negligible compared to the axial velocity. Fig. 5 presents the pressure contours at steady state. The pressure is almost constant in the y-direction, except near the injectors where the curves are distorted and show an over-pressure due to the great burnt gases production. Fig. 6 shows the mass fraction contours of the burnt products at steady state, ρ_3/ρ is maximum in the center region of the chamber where the production of (P) is high, it also corresponds to higher temperature. The volume fraction contours of the liquid phase are presented in Fig. 7. The volume fraction is maximum near the injectors and decreases downstream ; it becomes smaller than 10^{-3} in the second half of the chamber. Several calculations have been performed, the injection parameters being variable. Fig. 8 presents the influence of the injected droplets diameter on the damping ratio, while the injected flow rate is constant. The oscillations are more damped as the injected droplets diameter is large.

Conclusion

We have presented here a two-dimensional model for the two-phase flow simulation in a VIKING rocket engine combustion chamber. This model allows the transient flow description and shows the chamber unsteady response to combustion. Several calculations have been performed and the injection parameters influence on the flow pattern has been studied, as the concentration distribution of droplets, frequency and damping ratio of the oscillations. The numerical generation of undamped oscillations should enable us to a better understanding of this phenomenon. Therefore, numerical modellings of the main physical processes (e.g unsteady combustion) have to be improved. These problems are currently under development.

References

[1] Forman A. Williams : Combustion theory - ADISON WESLEY Publishing Company, inc.

[2] G.A. Agoston, H. Wise and W.A. Rosser : Sixth International Symposium on Combustion. Reinhold Publishing Corp., New York (1957), p. 708-717.

[3] Robert W. MacCormack : The effect of viscosity in hypervelocity impact cratering. AIAA Paper n° 69-354 (1969).

Appendix A.

Predictor (first step)

$$\tilde{\rho}_{i,j}^{n+1} = \rho_{i,j}^{n} + \Delta t \left[- \frac{(\rho u)_{i+1/2,j}^{n} - (\rho u)_{i-1/2,j}^{n}}{\Delta x} - \frac{(\rho v)_{i,j+\frac{1}{2}}^{n} - (\rho v)_{i,j-\frac{1}{2}}^{n}}{\Delta y} + \dot{w}_{i,j}^{n} \right]$$

$$(\tilde{\rho u})_{i+\frac{1}{2},j}^{n+1} = (\rho u)_{i+\frac{1}{2},j}^{n} + \Delta t \left\{ \frac{-\left[(\rho u^2)_{i+1,j}^{n} - (\rho u^2)_{i+\frac{1}{2},j}^{n}\right] - \left[P_{i+1,j}^{n} - P_{i+1/2,j}^{n}\right] + \left[(\sigma_{xx})_{i+1,j}^{n} - (\sigma_{xx})_{i+\frac{1}{2},j}^{n}\right]}{\Delta x/2} + \right.$$
$$\left. \frac{-\left[(\rho u v)_{i+\frac{1}{2},j+\frac{1}{2}}^{n} - (\rho u v)_{i+\frac{1}{2},j}^{n}\right] + \left[(\sigma_{xy})_{i+\frac{1}{2},j+\frac{1}{2}}^{n} - (\sigma_{xy})_{i+\frac{1}{2},j}^{n}\right]}{\Delta y/2} + (f_x)_{i+\frac{1}{2},j}^{n} + (S_{qx})_{i+\frac{1}{2},j}^{n} \right\}$$

Corrector (second step)

$$\rho_{i,j}^{n+1} = \frac{1}{2}\left(\rho_{i,j}^{n} + \tilde{\rho}_{i,j}^{n+1}\right) + \frac{\Delta t}{2}\left[- \frac{(\tilde{\rho u})_{i+1/2,j}^{n+1} - (\tilde{\rho u})_{i-\frac{1}{2},j}^{n+1}}{\Delta x} - \frac{(\tilde{\rho v})_{i,j+\frac{1}{2}}^{n+1} - (\tilde{\rho v})_{i,j-\frac{1}{2}}^{n+1}}{\Delta y} + \tilde{\dot{w}}_{i,j}^{n+1} \right]$$

$$(\rho u)_{i+\frac{1}{2},j}^{n+1} = \frac{1}{2}\left[(\rho u)_{i+\frac{1}{2},j}^{n} + (\tilde{\rho u})_{i+\frac{1}{2},j}^{n+1}\right] + \frac{\Delta t}{2} \left\{ \frac{-\left[(\tilde{\rho u^2})_{i+\frac{1}{2},j}^{n+1} - (\tilde{\rho u^2})_{i,j}^{n+1}\right] - \left[\tilde{P}_{i+\frac{1}{2},j}^{n+1} - \tilde{P}_{i,j}^{n+1}\right] + \left[(\tilde{\sigma}_{xx})_{i+\frac{1}{2},j}^{n+1} - (\tilde{\sigma}_{xx})_{i,j}^{n+1}\right]}{\Delta x/2} + \right.$$
$$\left. \frac{-\left[(\tilde{\rho u v})_{i+\frac{1}{2},j}^{n+1} - (\tilde{\rho u v})_{i+\frac{1}{2},j-\frac{1}{2}}^{n+1}\right] + \left[(\tilde{\sigma}_{xy})_{i+\frac{1}{2},j}^{n+1} - (\tilde{\sigma}_{xy})_{i+\frac{1}{2},j-\frac{1}{2}}^{n+1}\right]}{\Delta y/2} + (\tilde{f}_x)_{i+\frac{1}{2},j}^{n+1} + (\tilde{S}_{qx})_{i+\frac{1}{2},j}^{n+1} \right\}$$

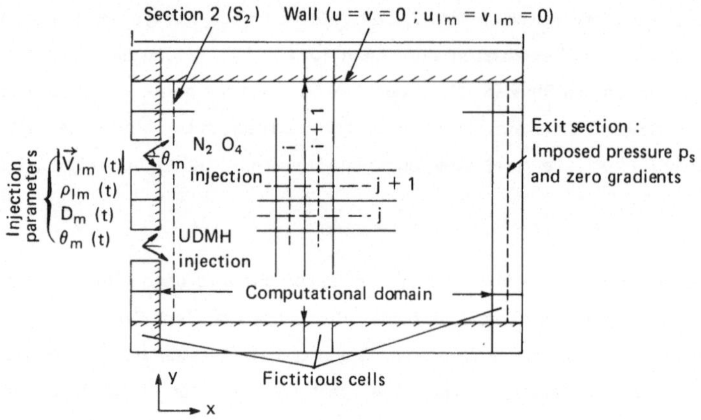

Fig. 1 — Boundary conditions.

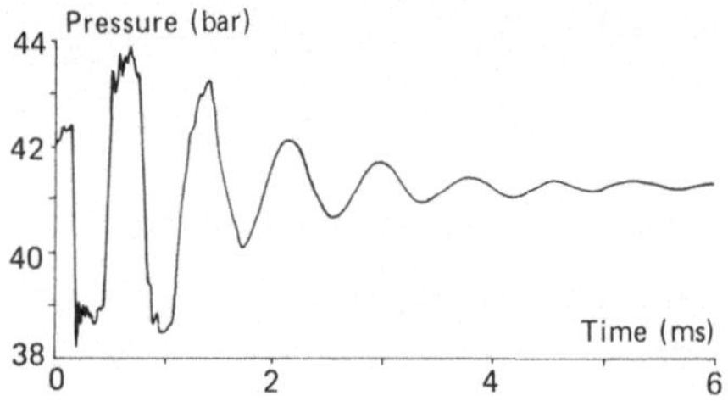

Fig. 2 — Temporal variation of the averaged pressure at S_2.

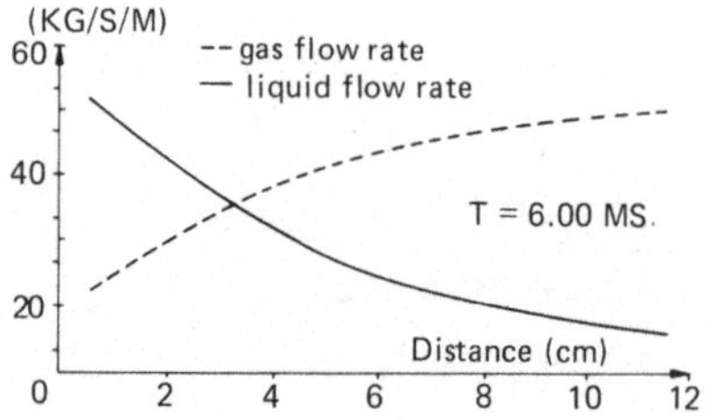

Fig. 3 — Liquid and gas flow rate profiles at steady state.

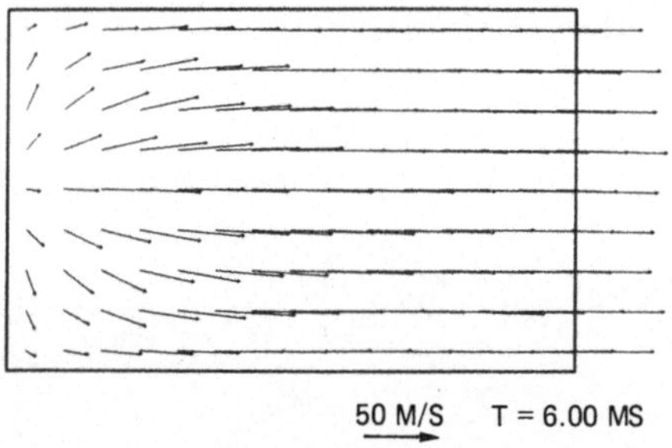

Fig. 4 — Velocity field of the gas at steady state.

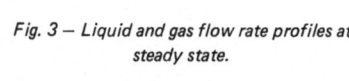

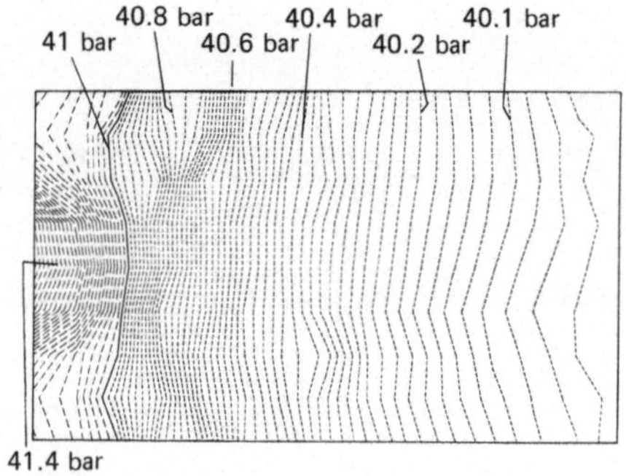

Fig. 5 — Pressure contours at steady state.

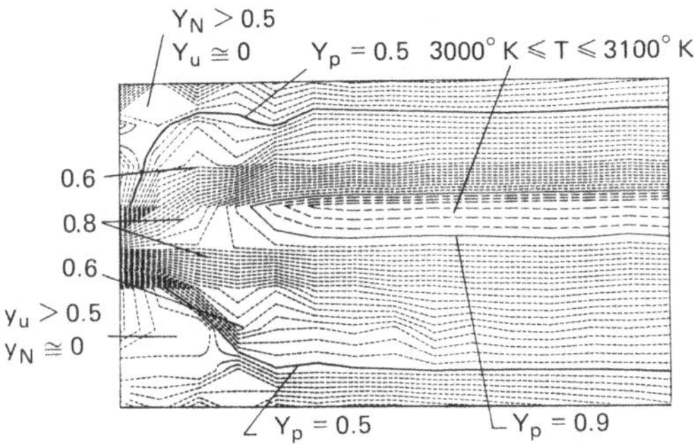

Fig. 6 – Mass fraction contours of the burnt products at steady state.

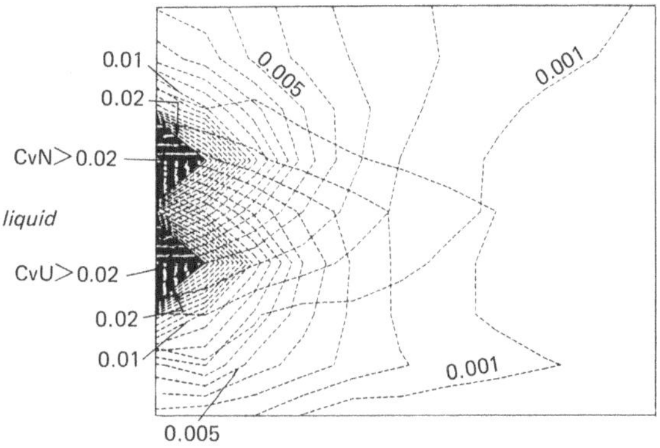

Fig. 7 – Volume fraction contours of the liquid phase at steady state.

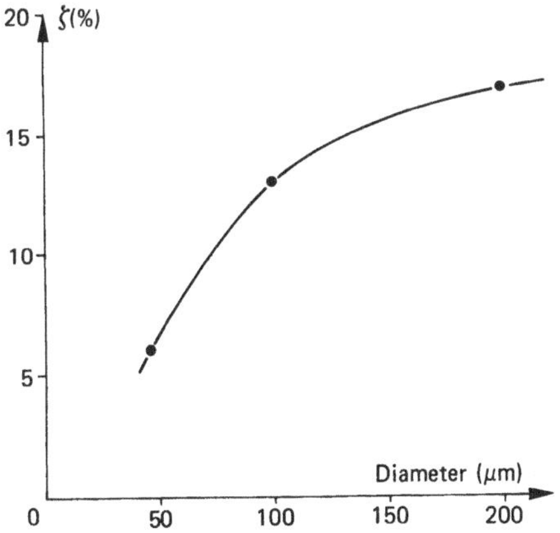

Fig. 8 – Damping ratio versus injected droplets diameter (q_L = 55.675 kg/s/m).

A NUMERICAL SOLUTION TO THE MOTION OF A LUBRICANT SQUEEZED BETWEEN TWO ROTATING COAXIAL DISKS

E.A. HAMZA
School of Mathematical Sciences
University of Khartoum
P.O. Box 321
Khartoum - Sudan

The behaviour of a thin film of liquid squeezed between two rotating parallel plane surfaces has received considerable attention due to its importance in the field of hydrodynamic lubrication. Here we are concerned with the numerical investigation of the impulsive motion of a fluid film squeezed between two rotating coaxial disks. The solution is obtained by employing a numerical technique due to Pearson [1] for the solution of the partial nonlinear differential equations. The objective of the study is to examine the way in which the velocity field, the normal force (load) which the fluid exerts on the upper disk and the torques exerted by the disks on the fluid change with time and to investigate the way in which they depend on the squeeze Reynolds number R_e^S and the rotation Reynolds number R_e^R.

Mathematical formulation

We consider the motion of a film of lubricant squeezed between two rotating coaxial disks each of which rotates in its own plane. Take cylindrical polar coordinates (r', θ', z') with corresponding velocity components (u', v', w') and suppose that the lower disk is at $z' = o$ and the upper disk is at $z' = h(t')$, where $h(o) = H$. Assume the lubricant is at rest for $t' < o$ and that at $t' = o$ the lower and upper disks are impulsively set in motion in such a way that they are respectively, rotating with constant angular velocities Ω_1 and Ω_2 and the upper disk is moving towards the lower one with constant velocity **V**. **(see fig,)**.

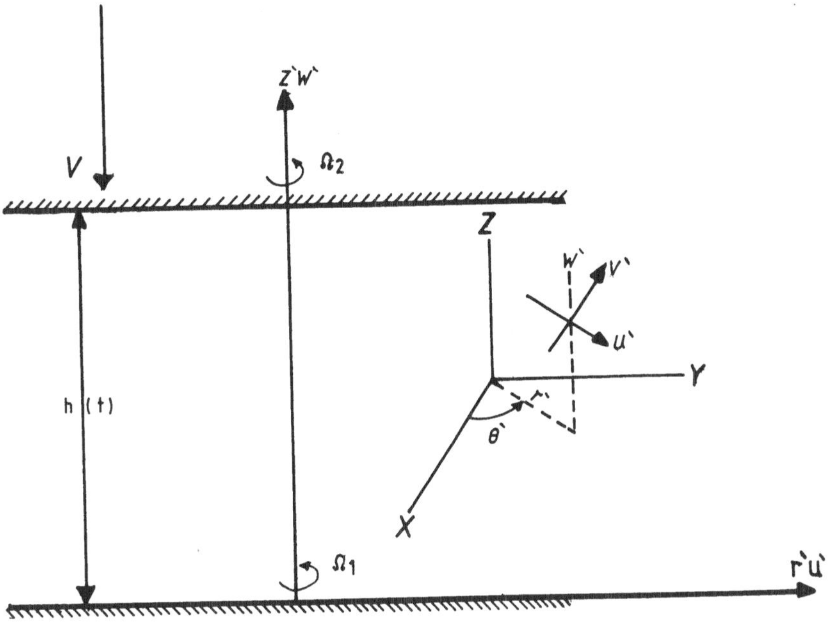

Fig. System configuration at time t', (co-ordinate scheme inset).

In the Navier-Stokes equations of motion, we non-dimensionalize all lengths with respect to H, u',w' with respect to V,v' with respect to HΩ, time t' with respect to H/V and pressure p' with respect to ρV^2, where Ω denotes a representative angular velocity and ρ denotes density. The corresponding dimensionless variables are those without the dashes.

We choose functions F(z,t) and G(z,t) such that

$$u = r\,F_z, \quad w = -\,2F\,, \quad v = rG$$

and the pressure p may be written in the form

$$p = \tfrac{1}{2}r^2\,p_1(t) + p_2(z,t),$$

Whence differentiation with respect to z and use of the transformation

$$y = z'/(1 - R^s e\tau)\,, \quad \tau = t/R_e^s$$

leads to the equations

$$T^2 M_\tau = M_{yy} - Re^S \left[yT M_y + 2M - 2TFM_y \right] + 2(Re^R)^2 (Re^S)^{-1} T^3 GG_y, \quad (1)$$

$$F_{yy} = M , \quad (2)$$

$$yRe^S TG_y + T^2 G_\tau + 2Re^S GF_y - 2Re^S T FG_y = G_{yy} \quad (3)$$

where $Re^S = \rho VH/\mu$, $Re^R = \rho\Omega H^2/\mu$ (μ denotes viscosity) and $T = 1 - Re^S\tau$.

The boundary conditions are

$$F = F'_y = 0, \; G = \alpha, y = 0, \; F = \tfrac{1}{2}, \; F_y = 0, \; G = \beta, y = 1, \quad (4)$$

where $\alpha = \Omega_1/\Omega$ and $\beta = \Omega_2/\Omega$

The conditions on τ which specify the impulsive start are

$$F = \tfrac{1}{2}y, \; G = 0, \; \tau = 0^+ , \; y \neq 0,1. \quad (5)$$

Numerical solution

Approximate analytic results can be obtained for the special cases where

(i) $Re^S \ll 1$ and (ii) $0 < Re^S\tau \ll 1$.

However, in general a numerical solution of the governing nonlinear equations is necessary. To do this we employ a highly efficient numerical technique., due to Pearson. By use of it we are able to present a much wider range of results than would otherwise be possible. Indeed, Pearson has claimed that the method will give accuracy to within a fraction of one percent with time steps greater than those permitted with conventional methods such as the Crank-Nicolson approach.

Since M is not specified at the boundaries y = 0,1, the finite difference formulation we adopt must hinge equations (1) and (2) together at the interior mesh points immediately adjacent to y = 0,1; this hinging procedure is iterative in nature.

We use an implicit finite-difference scheme in which the mesh spacing is h and the time step is k and we denote the values of G,M and F at the mesh point (y_i, τ_j) by G_i^j, M_i^j and F_i^j, where $y_i = ih$, $i = 0,1,\ldots,m+1$ $\left[(m\ 1)h=1\right]$ and $\tau_j = jk$, $j = 1,2,\ldots,(n-1)$, $nk = 1/\,R_e^S$. Equations (3), (1) and (2) are replaced by their corresponding finite-difference equations which constitute a system of 3m algebraic equations in (3m±2) unknowns. The system is closed by the addition of the two approximate equations

$$F_1^{(j\pm1)} = \tfrac{1}{2}h^2 \left[2M_1^{(j\pm1)} - M_2^{(j\pm1)}\right],$$

$$F_m^{(j\pm1)} = \tfrac{1}{2} + \tfrac{1}{2}h^2 \left[2M_m^{(j\pm1)} - M_{m-1}^{(j\pm1)}\right] \tag{6}$$

We select $10h = 1/2^{\ell}$, $\ell = 0,1,2,\ldots$ so that $3(10\times2^{\ell}-1)$ algebraic equations replace the original partial differential equations. The hinging equations (6) will be used t to couple the corresponding finite-difference approximations to equations (1) and (2). The coupling is iterative in nature and is very sensitive to the values of M at the boundaries. To avoid instabilities it is necessary during the iteration to use a smoothing formula. On the other hand the finite-difference equations (corresponding to equations (1), (2) and (3)) are in tri-diagonal forms and so may be solved by Gaussian elimination.

The computer program was so written that it could be used to give results for wide ranges of the paremeters R_e^S , R_e^R, α and β . However, for previty, we studied only the case where $\alpha = 1$ and $\beta = 0$. The calculations were performed on a CDC 7600 machine which retained 15 significant figures, for values of R_e^S and R_e^R in the ranges $0.1 < R_e^S < 10$ and $10 < R_e^R < 100$. The solution was started at $\tau = \tau_s$, where $0 < \tau_s \ll 1$, rather than at $\tau = 0$. The initial conditions which were employed in this case were derived from the approximate solutions obtained for the case where $0 < R_e^S\tau \ll 1$. For fixed R_e^S and R_e^R, accuracy was checked by comparing the results for two consecutive ℓ values.

Reference

(1) Pearson, C.E., "A computational method for viscous flow problems", Journal of Fluid Mechanics, Vol. 21, part 4, 1965, pp. 611.622.

NUMERICAL SIMULATION OF GAS MOTION IN PISTON ENGINES

H. Henke and D. Hänel
Aerodynamisches Institut, RWTH Aachen
Aachen, Germany

Introduction

The flow in piston engines is governed by complex physical processes, like turbulent transport or chemical reactions. The solution of this problem has been attempted by several investigators under various assumptions. A survey of the most important investigations is given in [1]. In the present paper the gas motion in the piston engine is simulated numerically by solving the time-dependent conservation equations for inviscid, compressible, non-reactive, plane or axisymmetric flow. The equations are formulated in a time-dependent, curvilinear coordinate system and solved with an implicit factorization method. The coupled system of difference-equations was transformed into an uncoupled diagonal form to reduce the computational time.

Governing Equations

Unsteady, inviscid, compressible, nonreacting flow in cylinders of piston engines can be described by the Euler equations. In time-dependent curvilinear coordinates $\xi = \xi(x,y,t)$, $\eta = \eta(x,y,t)$, $\tau = t$ the equations are written in the following compact form:

$$\bar{U}_\tau + \bar{F}_\xi + \bar{G}_\eta + q\bar{H} = 0 \tag{1}$$

where

$$\bar{U} = J^{-1}\begin{pmatrix} \rho \\ \rho u \\ \rho v \\ e \end{pmatrix}, \quad \bar{F} = J^{-1}\begin{pmatrix} \rho\tilde{U} \\ \rho u\tilde{U} + \xi_x p \\ \rho v\tilde{U} + \xi_y p \\ (e+p)\tilde{U} - \xi_t p \end{pmatrix}, \quad \bar{G} = J^{-1}\begin{pmatrix} \rho\tilde{V} \\ \rho u\tilde{V} + \eta_x p \\ \rho v\tilde{V} + \eta_y p \\ (e+p)\tilde{V} - \eta_t p \end{pmatrix}, \quad \bar{H} = \frac{J^{-1}}{y}\begin{pmatrix} \rho v \\ \rho u v \\ \rho v^2 \\ (e+p)v \end{pmatrix}$$

q is zero for plane and one for axisymmetric flows. In these equations $\tilde{U}$ and $\tilde{V}$ are the contravariant velocities and J is the Jacobian of the transformation.

Method of Solution

The numerical solution was first established by means of the factorization method of Beam and Warming [2]:

$$(I + \Delta\tau \frac{\partial}{\partial\xi}\bar{A}^n)(I + \Delta\tau\frac{\partial}{\partial\eta}\bar{B}^n)\Delta\bar{U}^n = -\Delta\tau\left(\frac{\partial\bar{F}}{\partial\xi} + \frac{\partial\bar{G}}{\partial\eta}\right)^n + O(\Delta\tau^2)$$
$$\Delta\bar{U}^n = \bar{U}^{n+1} - \bar{U}^n \tag{2}$$

The flux-vectors $\bar{F}$ and $\bar{G}$ were locally linearized in time and $\bar{A}$ and $\bar{B}$ are the corresponding Jacobian matrices.

Spatial derivatives of second order accuracy and an appropriate metric formulation preserve the conservative properties in the transformed domain of integration. Thereby a 4x4-block-tridiagonal system was obtained. The solution of this system required relatively long computational time.

A substantial reduction of computing time could be obtained by diagonalizing the matrices $\bar{A}$ and $\bar{B}$ with a similarity transformation of the form [3, 4]

$$\bar{A} = T_\xi \Lambda_\xi T_\xi^{-1} \quad , \quad \bar{B} = T_\eta \Lambda_\eta T_\eta^{-1} \tag{3}$$

yielding the system of 4 scalar equations

$$T_\xi^n \cdot (I + \Delta\tau \frac{\partial}{\partial \xi} \Lambda_\xi^n) N^n (I + \Delta\tau \frac{\partial}{\partial \eta} \Lambda_\eta^n)(T_\eta^{-1})^n \Delta\bar{U}^n = R.H.S.^n$$

with

$$N^n = (T_\xi^{-1})^n \cdot T_\eta^n \tag{4}$$

The matrices T_ξ^n, $T_\xi^{-1})^n$,... were taken outside of the brackets of the equations, so that the block matrices are diagonalized, whereby a decoupling is achieved.

In axisymmetric flow $\partial\bar{H}/\partial\bar{U}$ has eigenvalues which differ from those of $\bar{B}$. Therefore the source term was included only in the explicit part, so that the implicit part remained the same as for plane flows. Then the explicit part becomes:

$$R.H.S.^n = -\Delta\tau \left(\frac{\partial \bar{F}}{\partial \xi} + \frac{\partial \bar{G}}{\partial \eta} + q\bar{H} \right)^n - \Delta\tau \cdot q \cdot \Delta\bar{H}^{n-1}$$

with

$$\Delta\bar{H}^{n-1} = \bar{H}^n - \bar{H}^{n-1} \tag{5}$$

This formulation corresponds to a linear extrapolation in time, so that the overall accuracy is not impaired due to the source terms and the diagonalization for the plane flow can be retained.

To damp high frequency error components second order implicit and fourth order explicit dissipation terms were added to the equations (2) and (4).

The boundary conditions for the correction variable is assumed to be $\Delta\bar{U}^n = 0$ and the new values $\bar{U}^{n+1}$ on the boundary are found explicitly. This formulation being first order in time is easy to implement for all types of boundary conditions, as for example, fixed and moving walls or the flow through a valve. On rigid walls the tangential velocity component and the density are found by linear extrapolation and the pressure by integrating the normal momentum equation. For the flow through the valves during the intake stroke the mass flux, angle of injection and the density are prescribed. The pressure on the boundary is found by linear extrapolation from the interior.

For the exhaust stroke the mass flux is prescribed and the other variables are extrapolated from the interior.

The accuracy of the method of solution was tested by checking the overall conservation of mass and energy in closed cylinders.

Results

Numerical results for the flow in piston engines were obtained for several contoured pistons and different arrangements of valves.

In Figs. 1a, b vector plots of the velocity of plane flow in a cylinder with a moving step piston are shown. For this computation the gas is injected tangentially to the cylinder wall during the intake stroke. Two vortices are being formed as shown in Fig 1a. They are still present during the compression stroke (see Fig. 1b). If the step is removed only one vortex appears in the intake and compression stroke.

In other computations for plane flows the influence of the angle of which the gas is injected was investigated. For example, Fig. 2 shows lines of constant vorticity for an angle of injection of 60 degrees. If it is reduced to 30 degrees than two vortices occur.

Figs. 3a-d show a sequence of vector plots of the flow in an axisymmetric cylinder for the four strokes. During the intake stroke, Fig. 3a, two vortices are formed for an injection angle of 30 degrees. These vortices, which can clearly be seen at the end of the intake stroke, Fig. 3b, form a single vortex in the compression stroke. The flow pattern of the exhaust stroke is shown in Fig. 3d. In the Fig. 4 the lines of constant vorticity are plotted for a crank angle of 164 degrees. They clearly indicate position and extent of the double ring vortex developed during the intake stroke.

References

[1] Butler, T. D., Cloutman, L. D., Dukowicz, J. K., Ramshaw, J. D.: Multidimensional Numerical Simulation of Reactive Flow in Internal Combustion Engines. Prog. Energy Combust. Sci., Vol. 7, pp. 293-315, Pergamon Press Ltd., 1981.

[2] Beam, R., Warming, R. F.: An Implicit Finite-Difference Algorithm for Hyperbolic Systems in Conservation-Law-Form. Journal of Comp. Physics, Vol. 22, Sept. 1976, pp. 87-110.

[3] Warming, R. F., Beam, R., Hyett, B. J.: Diagonalization and Simultaneous Symmetrization of the Gas-Dynamic Matrices. Mathematics of Computation, Vol. 29, Oct. 1975, pp. 1037-1045.

[4] Pulliam, T. H., Chaussee, D. S.:A Diagonal Form of an Implicit Approximate-Factorization Algorithm. Journal of Comp. Physics, Vol. 39, 1981, pp. 347-363.

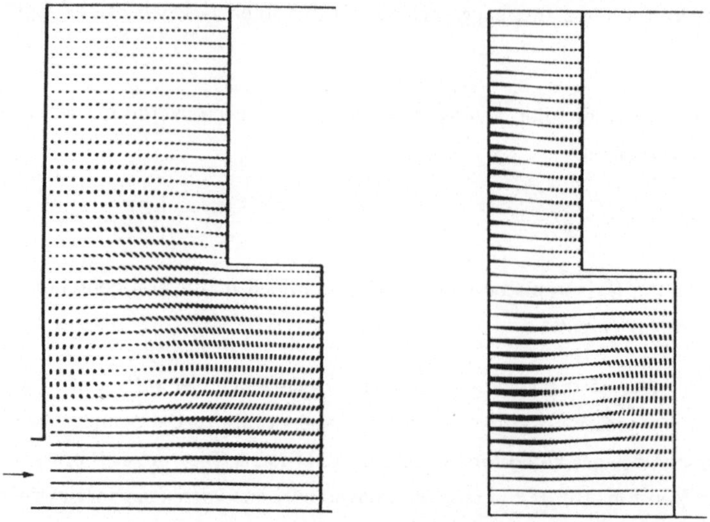

a) intake stroke, crank angle 52 degrees b) compression stroke, crank angle 360 degrees

Fig. 1: Velocity pattern for plane flow and a piston with a step

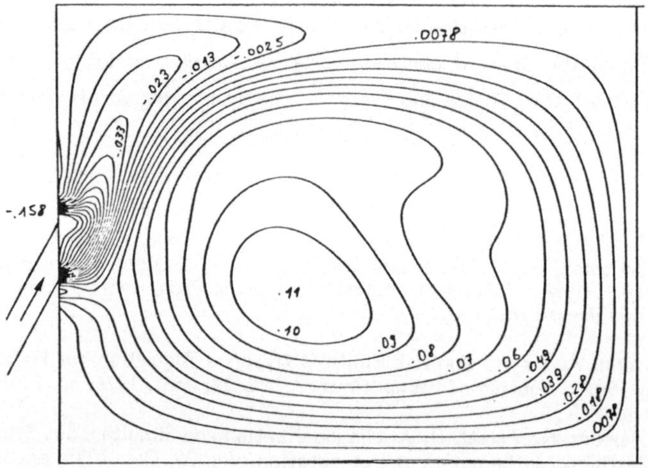

Fig. 2: Lines of constant vorticity for a plane piston, injection angle 60 degrees, crank angle 164 degrees

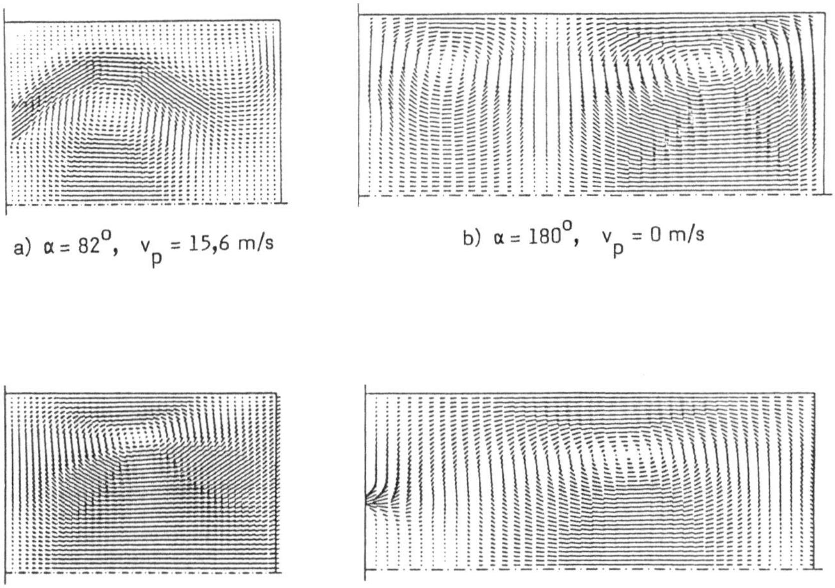

a) $\alpha = 82^o$, $v_p = 15{,}6$ m/s

b) $\alpha = 180^o$, $v_p = 0$ m/s

c) $\alpha = 440^o$, $v_p = 15{,}5$ m/s

d) $\alpha = 562^o$, $v_p = -6$ m/s

Fig. 3a-d: Velocity pattern for axisymmetric flow at different crank angles α, injection angle 30 degrees

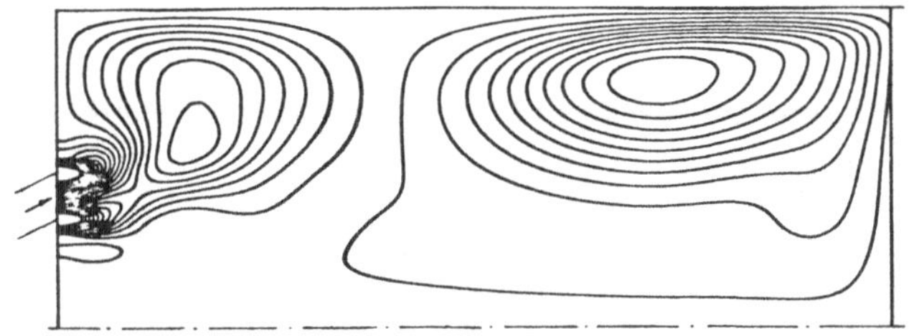

Fig. 4: Lines of constant vorticity for axisymmetric flow, injection angle 30 degrees, crank angle 164 degrees

MODELISATION NUMERIQUE DE LA SEPARATION CENTRIFUGE D'UN MELANGE

A. HOLCBLAT - P. BICHET - J. ALLAIS
FRAMATOME - Tour FIAT / CEDEX 16/92084 PARIS-LA-DEFENSE

Résumé

Le code SEPSOM étudie la séparation centrifuge d'un mélange liquide-vapeur dans un tuyau à l'aval d'une hélice. On emploie un modèle à 2 fluides. Les champs de pression, vitesses et concentration sont calculés pas à pas de l'amont vers l'aval. La méthode de résolution des équations est une adaptation de l'algorithme IPSA de D.B. Spalding. La divergence à fort couplage entre phases est évitée grâce à l'algorithme d'élimination partielle du même auteur.

1 - Introduction

Dans les générateurs de vapeur à recirculation des réacteurs à eau sous pression et la cuve des réacteurs à eau bouillante, on utilise des séparateurs centrifuges pour assécher la vapeur. Dans les modèles usuels, le mélange liquide-vapeur monte dans un tuyau contenant une hélice fixe. A sa traversée, il se met en rotation et la séparation se produit sous l'action des forces centrifuges. On a écrit le code SEPSOM pour prévoir la hauteur nécessaire à la séparation des deux phases.

2 - Modélisation

L'écoulement est permanent, adiabatique et sans changement de phase. Dans ce cas, un modèle à deux fluides ne contient que quatre équations:

- 2 équations de continuité (liquide : $l = 1$; vapeur : $l = 2$)

$$\operatorname{div} (\alpha_l \ \rho_l \ \vec{V}_l) = 0$$

- 2 équations phasiques de quantité de mouvement (liquide, vapeur)

$$\alpha_l \ \rho_l \ \frac{D\vec{V}_l}{Dt} = - \alpha_l \ \overrightarrow{\text{grad}}\, p \pm \rho_2 \ \frac{K}{2} \ \left\| \vec{V}_2 - \vec{V}_l \right\| \ (\vec{V}_2 - \vec{V}_l)$$

(le signe est + pour $l = 1$ et - pour $l = 2$)

où p est la pression et α_l et $\vec{V}_l$ sont respectivement la concentration et la vitesse de la phase l.

Comme on néglige l'effet de la gravité, il est assez naturel de supposer que les deux phases suivent la même surface de courant. La vitesse de chaque phase est la somme d'une vitesse radiale inconnue u_l et d'une composante dite principale W_l qu'on impose par une corrélation :

$$\vec{V}_l = W_l \ \vec{n} + u_l \ \vec{r}$$

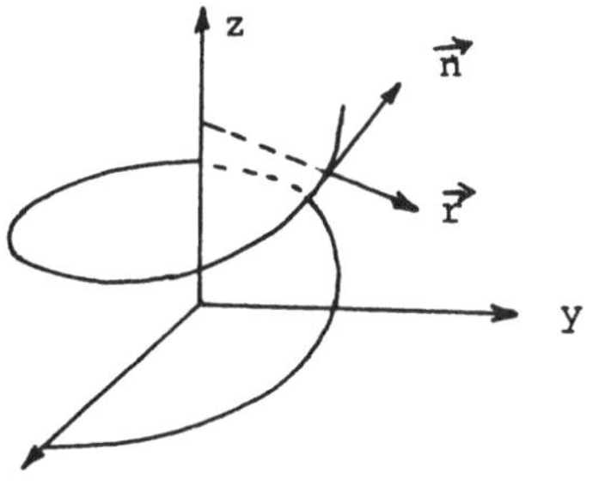

3 - Discrétisation

Le domaine de calcul est un tube de courant, mais on suppose que chaque grandeur est constante dans l'épaisseur du tube. On emploie deux maillages décalés dans le sens radial, l'un pour les équations de continuité, l'autre pour les équations de quantité de mouvement.

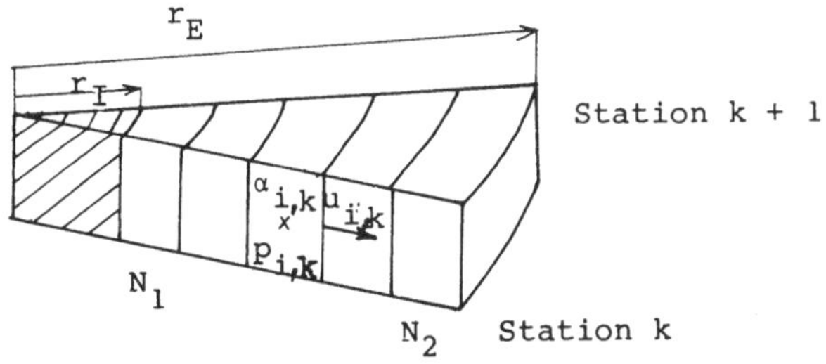

Pour réduire la diffusion numérique, on emploie en général un schéma de différence centrée pour écrire les flux aux interfaces entre cellules. On n'applique la méthode de la cellule donneuse qu'au liquide dans les équations de continuité (équation du liquide, équation du mélange), ce qui est d'ailleurs indispensable pour assurer la convergence.

4 - Algorithme

L'algorithme choisi est une adaptation pour un système de type parabolique de l'algorithme IPSA (Interphase Slip Algorithm) développé par D.B. Spalding (cf [1]).
On progresse d'une station à l'autre le long de la direction principale. Les valeurs des variables à la station k ne dépendant que de leurs valeurs à la station située à l'amont, le schéma est le suivant :

a. Attribuer une valeur à u_f et u_g à la station k

b. Résoudre l'équation de continuité du liquide au niveau k (inconnues : $\alpha_{i,k,f}$, $N_1 \leqslant i \leqslant N_2$)

c. Choisir une nouvelle distribution de pression au niveau k : $p^*_{i,k}$

d. A partir de $\alpha_{i,k,f}$ calculé à l'étape b et de la distribution de pression $p^*_{i,k}$, résoudre les équations de quantité de mouvement des cellules comprises entre les stations k - 1 et k, ce qui donne une solution provisoire.
Le système linéaire est résolu de façon itérative par la méthode de JACOBI par blocs, chaque bloc correspondant à une phase. Dans le cas de fort couplage (valeurs élevées du coefficient de frottement interfacial K), la convergence n'est assurée que si on emploie l'algorithme d'élimination partielle suggéré par le même auteur. Il consiste à combiner les 2 équations concernant le $i^{ème}$ élément de façon à éliminer du terme source la vitesse de l'autre phase dans le même élément.
Après résolution du système, on relaxe les vitesses.

e. A partir de $\alpha_{i,k,1}$ et $u^*_{i,k,1}$, dériver les équations de quantité de mouvement pour calculer les coefficients de correction de pression, qui sont les dérivées partielles des vitesses par rapport aux variables de pression adjacentes. On en déduit des variations de vitesses à la station k

$$\delta u_f = \frac{\partial u_f}{\partial p_i} \delta p_i + \frac{\partial u_f}{\partial p_{i+1}} \delta p_{i+1}$$

$$\delta u_g = \frac{\partial u_g}{\partial p_i} \delta p_i + \frac{\partial u_g}{\partial p_{i+1}} \delta p_{i+1}$$

f. L'équation de continuité globale est la somme des équations de continuité de chacune des phases. Dans chaque maille, elle s'écrit en fonction de $u^*_f + \delta u_f$ et $u^*_g + \delta u_g$

En les remplaçant par les expressions obtenues à l'étape e, on obtient un système linéaire en δp_i à la station k.

g. Résoudre le système en $\delta p_{i,k}$, $N_1 \leqslant i \leqslant N_2$

h. En déduire les corrections de vitesse $\delta u_{i,k,l}$
Relaxer les corrections de pression par le coefficient β

i. Corriger les vitesses et les pressions

$$u_{i,k,l} = u^*_{i,k,l} + \delta u_{i,k,l}$$

$$p_{i,k} = p^*_{i,k} + \beta . \delta p_{i,k}$$

j. Si les corrections δu_f et δu_g sont suffisamment petites, on arrête le calcul à la station k et on commence le calcul de la station k + 1. Ce calcul commence en éliminant éventuellement les mailles qui sont devenues monophasiques. (Dans ce cas N_1 augmente ou N_2 diminue).

Si le calcul à la station k n'a pas convergé, on repart à l'étape b

5 - Résultats

Les cas tests montrent que l'algorithme permet de poursuivre le calcul jusqu'à séparation totale dans une large gamme de variation des conditions d'entrée ainsi que du coefficient de frottement interfacial. Ce dernier dépend du régime d'écoulement. Si on emploie un modèle à gouttes dont le diamètre correspond à la valeur critique du nombre de Weber local, on obtient une hauteur de séparation totale physiquement réaliste.
La figure ci-dessous qui représente l'évolution du profil radial de concentration en liquide, permet de bien schématiser le processus de séparation : à l'entrée (k = 1), tout le rayon est diphasique; ensuite il se scinde en 3 zones distinctes, un film liquide sur la paroi du tube, un noyau de vapeur près de l'axe, entre les 2 précédents une zone diphasique de concentration à peu près uniforme, dont la largeur diminue peu à peu.

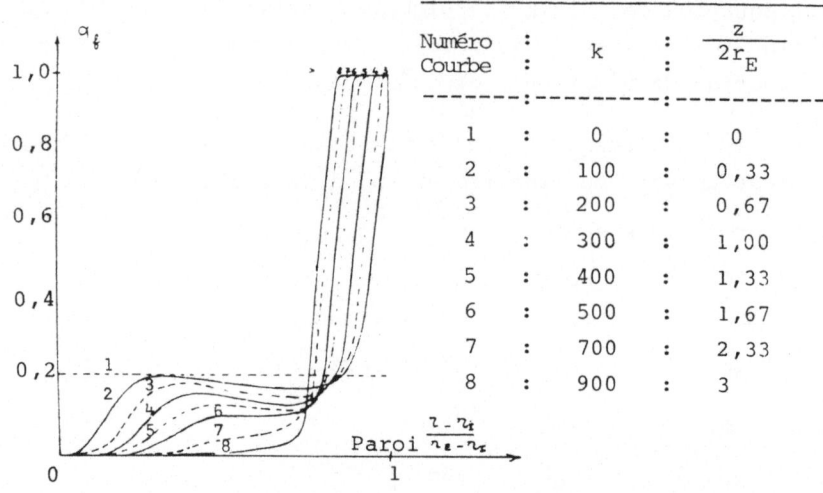

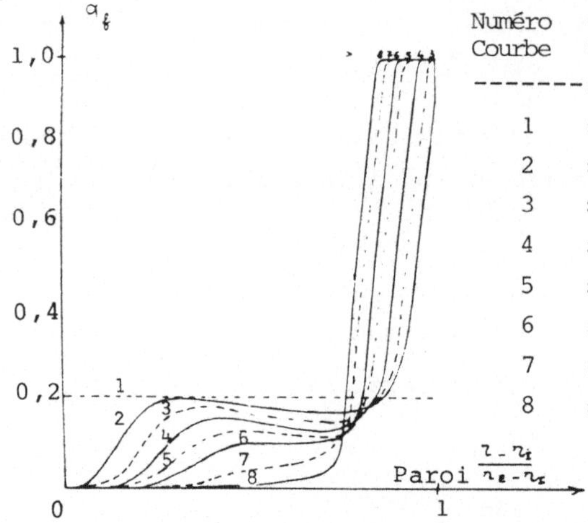

Numéro Courbe	k	$\frac{z}{2r_E}$
1	0	0
2	100	0,33
3	200	0,67
4	300	1,00
5	400	1,33
6	500	1,67
7	700	2,33
8	900	3

Profil radial de concentration en liquide
(Q_2 = 30 kg/s ; Q_1 = 96 kg/s ; P = 73 bar ; r_E = 0,266 m)

Un programme expérimental en cours permettra de qualifier les corrélations à introduire dans le code (coefficient de frottement interfacial, rapport de glissement et profil de vitesse dans le direction principale).

5 - Références

[1] D. BRIAN SPALDING - "Numerical computation of multi-phase flow and heat transfer" dans "Recent Advances in numerical mechanics" édité par C. TAYLOR, Université de Swansea.

[2] P.W. WYATT and al - Swirl Vane Separator Performance" ASME - IEEE Joint Power Generation Conference, St Louis, Missouri October 4 to 8, 1981.

Remerciements

Cette étude, dont les premiers résultats ont été présentés dans[2], a été menée dans le cadre de la coopération technique entre le Commissariat à l'Energie Atomique, Electricité de France, FRAMATOME et Westinghouse.

SUPERSONIC FLOW PAST CIRCULAR CONES AT HIGH ANGLES OF YAW, DOWNSTREAM OF SEPARATION

Maurice Holt and Mostafa Aghazadeh
Department of Mechanical Engineering
University of California, Berkeley

ABSTRACT

The calculation of viscous supersonic flow over circular cones at high angles of yaw was partially carried out by Fletcher and Holt (1976) and by Holt and Chan (1979). The flow field was calculated as the interaction between the outer inviscid flow and an inner conical boundary layer flow. The latter was treated by the orthonormal version of the Method of Integral Relations (M.I.R.) (Fletcher and Holt, 1975) and continued up to the cross flow separation line.

The present work deals with the boundary layer downstream of this separation line where the circumferential velocity component, w, is reversed. The orthonormal version of M.I.R. needs to be modified in this region to take account of a minimum point in w near the cone surface. In contrast to two-dimensional flow, this can be achieved by using polynomials to represent the normal gradient of w as a function of w, and square root factors, which seriously complicate the use of orthonormal M.I.R., are not needed. The extended calculation enables us to calculate the complete flow field over yawed supersonic cones, at different Mach numbers and yaw angles, including that in the far leeward region.

INTRODUCTION

The calculation of laminar boundary layer flow in regions of unfavorable pressure gradient was first considered in two dimensions, in the incompressible case, and was treated as a strict extension of the Prandtl boundary layer concept. For a given profile shape the inviscid pressure distribution was first calculated and then introduced into the boundary layer equations. This approach led to a singularity at the boundary layer separation point. The nature of the singularity was extensively investigated by a number of authors, cited in the review paper by Brown and Stewartson (1969).

An approach to the problem distinct from that originally made was proposed by Crocco and Lees (1952) for laminar boundary layers in supersonic flow. This takes account of the thickening of the boundary layer in retarded flow regions and requires the simultaneous calculation of pressure distribution at the edge of the boundary layer and boundary layer displacement thickness. Thus, in determining the inviscid pressure distribution, the height of the effective body is increased by the local displacement thickness. The Crocco-Lees approach was applied to typical supersonic boundary layer flows with separation by Lees and Reeves (1964) and Klineberg and Lees (1969). The Crocco-Lees concept was presented formally for low speed flow by Catherall and Mangler (1966). The latter case is more difficult to treat because of strong upstream influence near separation. In supersonic flow the displacement thickness is directly connected with pressure increase. In subsonic flow the pressure depends on displacement thickness distribution over a whole region.

The extension of this inviscid-viscous interaction approach to two-dimensional turbulent flow requires a very simple representation of the turbulent boundary layer behavior. This was first developed for the incompressible case in the Lag Entrainment method of Green (Green et al. 1977, East et al. 1977). Extension of this to the more difficult case of supercritical (transonic) airfoil flow was achieved by

Le Balleur (1978,1980) and applied by Wigton and Holt (1981).

The calculation of three-dimensional laminar boundary layer flow was investigated by K. C. Wang (1970-1974) using a finite difference approach. For conical flow Lin and Rubin (1973) developed a numerical method based on the parabolized Navier-Stokes equations.

THE PRESENT APPROACH

The methods previously referred to are mostly based on finite difference techniques. We now discuss the application of the Method of Integral Relations to boundary layer problems. A general description of this, both in the original Dorodnitsyn (1960) formulation and in the modified Fletcher-Holt orthonormal formulation (1975,1976) is given in Holt (1984). Both formulations need to be modified in retarded flow regions.

In the Dorodnitsyn approach a square root factor must be included in the representation of the transverse velocity gradient and this complicates the algebra required considerably. As a consequence, it has only been applied to separated flow problems in the lowest approximation, albeit with good results.

An orthornormal formulation for separated flows in two dimensions has been proposed but not yet applied. This is discussed in Holt (1983) and uses expansions in Chebycheff polynomials.

The Dorodnitsyn formulation was extended to three-dimensional laminar flow by Modarress and Holt (1977), and illustrated with a series of applications. It was applied to flow past ellipsoids by Modarress (1978).

CONICAL BOUNDARY LAYERS

The orthonormal version of M.I.R. was developed for laminar boundary layer flow over supersonic circular cones, at moderate and high angles of attack, by Fletcher and Holt (1976). It was successfully applied to calculate boundary layer behavior up to the cross flow separation line. The approach needs to be modified for calculations beyond this line and we conclude with an outline of the changes needed.

The Fletcher-Holt method is based on their orthonormal version of the Method of Integral Relations and requires the solution of the following equations

$$\frac{\partial}{\partial\zeta}\int_0^1 g_k \frac{w}{\tau}\,du = \sin\theta_b\Big[\ell_5\int_0^1 g_k\frac{w}{\tau}\,du - 1\cdot5\int_0^1 g_k\frac{u}{\tau}\,du - \ell_2\{g_k'\tau\}_{wall} - \ell_2\int_0^1 g_k''\tau\,du - \int_0^1 g_k'\frac{w}{\tau}\{u\frac{w_e}{u_e} - w\}du\Big] \quad , \qquad (1)$$

$$\frac{\partial}{\partial\zeta}\int_0^1 g_k\frac{w^2}{\tau}\,du = \sin\theta_b\;\ell_7\int_0^1 g_k\frac{w^2}{\tau}\,du - 2\cdot5\int_0^1 g_k\frac{uw}{\tau}\,du - \ell_2\int_0^1 g_k''\tau w\,du - 2\ell_2\int_0^1 g_k'\tau\frac{\partial w}{\partial u}\,du$$

$$- \int_0^1 g'_k \frac{w^2}{\tau} \{u \frac{w_e}{u_e} - w\} du + \ell_6 \frac{B}{u_e^2} \int_0^1 g_k \frac{(1-s)}{\tau} du$$

$$- \ell_2 \{g_k \frac{\partial w}{\partial u} \tau\}_{wall} + \ell_8 \int_0^1 g_k \frac{u^2}{\tau} du] \quad , \qquad (2)$$

$$\frac{\partial}{\partial \zeta} \int_0^1 g_k \frac{sw}{\tau} du = \sin \theta_b [\ell_5 \int_0^1 g_k \frac{sw}{\tau} du - 1{\cdot}5 \int_0^1 g_k \frac{su}{\tau} du$$

$$- \ell_2 \{g'_k s\tau\}_{wall} - \ell_2 \int_0^1 g''_k \tau s du - (\ell_2 + \ell_3) \int_0^1 g'_k \tau \frac{\partial s}{\partial u} du$$

$$- \int_0^1 g'_k \frac{sw}{\tau} \{u \frac{w_e}{u_e} - w\} du - \ell_3 \{g_k \frac{\partial s}{\partial u} \tau\}_{wall}$$

$$- \ell_4 \int_0^1 g'_k \tau \{u + \frac{\partial w}{\partial u} w\} du] \quad , \qquad (3)$$

where ζ is the coordinate along a cone generator, u is the radial velocity component, w is the circumferential velocity component, $\tau = \partial u/\partial \eta$ (η is the coordinate in the direction normal to the surface), s is a total enthalpy function and g_k are orthonormal weighting functions of u. The factors ℓ_2-ℓ_7 depend on the inviscid flow.

In the general Fletcher-Holt formulation $g_k(u)$ have the representations

$$g_k(u) = \sum_{k=1}^{j} b_{kj} (1-u)^k \quad , \qquad (4)$$

where b_{kj} are evaluated by the Gram-Schmidt orthonormalization process.

In the cross flow separation region, Eq. (4) has to be modified so that w/τ and sw/τ can be negative for small values of u and can have minimum values at a certain point $u = \beta$ near the cone surface. The simplest modification to w/τ meeting these requirements is

$$\frac{w}{\tau} = [a_o + (u-\beta)^2 \sum_{k=1}^{j} a_k (1-u)^{j-1}] \qquad (5)$$

with a similar change in sw/τ and w^2/τ.

We apply representation (5) in the third approximation and solve three ordinary differential equations for a_1, a_2 and β, the other seven coefficients having been eliminated with the use of seven algebraic relations derived from boundary conditions.

This formulation is currently being solved for the two sets of free stream conditions considered by Holt and Chan (1979): (i) Free stream Mach number $M_\infty = 7$, cone semi angle $\theta_b = 20°$, angle of incidence, $\alpha = 40°$; (ii) $M_\infty = 7$, $\theta_b = 10°$, $\alpha = 25°$. In both cases three values of Reynolds numbers are taken, $Re = 10^5$, 5.10^5, 10^6.

The calculations generate reverse cross flow velocity profiles for circumferential angle ϕ in the range $160° < \phi < 180°$ in case (i) and $155° < \phi < 180°$ in case (ii). The two lower angles correspond to minimum surface values in the respective cases. The pressure remains essentially constant at circumferential stations beyond these positions so that the reversed cross flow is not expected to reattach and in fact extends up to the farthest leeward generator. The separated boundary layer flow in the leeward region is not strongly dependent on the outer inviscid flow.

This work is supported by the Air Force Office of Scientific Research, Mechanics Branch, Dr. James D. Wilson, Program Manager.

REFERENCES

S. N. Brown and K. Stewartson, "Laminar separation," Annual Review of Fluid Mechanics, Vol. 1, Annual Reviews Inc., 1969.

D. Catherall and K. W. Mangler, Journal of Fluid Mechanics, 26, 163-182, 1966.

L. Crocco and L. Lees, Journal of Aeronautical Sciences, 19, 649-676, 1952.

A. A. Dorodnitsyn, Advances in Aeronautical Sciences, Vol. 3. New York, Pergamon Press, 1960.

L. F. East, P. D. Smith, P. J. Merryman, "Prediction of the development of separated turbulent boundary layers by the lag-entrainment method," British Royal Aircraft Establishment Technical Report 77046, 1977.

C. A. J. Fletcher and M. Holt, Journal of Fluid Mechanics, 74, 561-591, 1976.

J. E. Green, D. J. Weeks, J. W. F. Brooman, "Prediction of turbulent boundary layers and wakes in incompressible flow by a lag-entrainment method," British Aeronautical Research Council Reports and Memoranda No. 3791, 1977.

M. Holt, "The Changing Scene in Computational Fluid Dynamics," Computational Techniques and Applications Conference, Sydney, New South Wales, Aug. 1983.

M. Holt, Numerical Methods in Fluid Dynamics, Second revised edition, Springer-Series in Computational Physics, Springer-Verlag, 1984.

J. M. Klineberg and L. Lees, AIAA Journal, 7, 2211-2221, 1969.

J. C. Le Balleur, "Couplage Visqueux—Non Visqueux: Méthode Numérique et Applications aux Ecoulements Bidimensionnels Transsoniques et Supersoniques," La Recherche Aérospatiale, Mar.-Apr. 1978.

J. C. Le Balleur, "Calcul des Écoulements à Forte Interaction Visqueuse au Moyen de Méthodes de Couplage," AGARD Conference on Computation of Viscous-Inviscid Interactions, Preprint No. 291, Paper 1, Colorado Springs, 1980.

L. Lees and B. L. Reeves, AIAA Journal, 2, 1907-1920, 1964.

D. Modarress, Computer Methods in Applied Mechanics and Engineering, 14, 145-157, 1978.

D. Modarress and M. Holt, Proceedings of the Royal Society (A), 353, 319-347, 1977.

T. C. Lin and S. G. Rubin, Journal of Fluid Mechanics, 59, 593-620, 1973.

K. C. Wang, Journal of Fluid Mechanics, 43, 187-209, 1970.

K. C. Wang, AIAA Journal, 10, 1044-1050, 1972.

K. C. Wang, Proceedings of the Royal Society (A), 340, 1974.

L. B. Wigton and M. Holt, AIAA 5th Computational Fluid Dynamics Conference Proceedings, AIAA-81-1003-CP14, pp. 77-89, 1981.

A TWO-GRID METHOD FOR FLUID DYNAMIC PROBLEMS WITH DISPARATE TIME SCALES

M. Israeli and G. Enden
Computer Science Department
Technion-Israel Institute of Technology
Haifa, Israel

Many problems in fluid dynamics are driven by the interaction between convective and diffusive processes. The diffusive time scale T_D is L^2/ν (where L is a length scale, and ν is the kinematic viscosity and may represent any diffusive process), while the convective time scale T_C is L/U (where U is a characteristic velocity). For high Reynolds number flows, $T_D/T_C = LU/\nu = Re$ is large. On the other hand, the diffusive processes act quickly over the boundary layer scale $\delta = \sqrt{\nu L/U}$ as $\delta^2/\nu = T_C$.

We want to use this information in order to considerably speed up certain fluid dynamic computations without degrading the accuracy. In particular, we have in mind flows with recirculations, such as the driven cavity or the flow behind a step where the diffusion takes a long time to penetrate across the streamlines. Even more difficult and time consuming are cases where the diffusion is against the flow direction. We are interested both in the transient stages and in the approach to steady state. In many cases a quasisteady situation quickly develops over the convective time scale and is later modified slowly by the diffusion.

The spatial structure of high Reynolds number flows is also characterized by non-uniformity; there are large regions of slow variations and narrow regions where the variables undergo large changes. Mathematically, this behaviour is caused by the "singular perturbation" nature of the problem and can be used to generate approximate solutions by asymptotic techniques. Numerically, this behaviour requires very fine meshes for high Reynolds number flows and extremely small time steps for explicit marching in time dependent problems.

For certain problems we were able to incorporate asymptotic solutions in the numerical scheme to obtain considerable improvement in accuracy by using the "Booster Method" [1,2] or the "Asymptotic Finite Element Method" [3,4]. These schemes use the approximate solution to correct the truncation error in the basic numerical scheme and are therefore similar to Richardson correction schemes. The effect of the asymptotic corrections is to reduce the bounds for the error by a factor of ε^{n+1}, where ε is proportional to the relative thickness of the boundary layers and n is the order of the approximation. For complicated time dependent fluid dynamic problems it is difficult to obtain good approximate solutions. Here we propose to obtain an approximation by solving the problem on two grids, a fine grid and a coarse grid. The solution on the fine grid will enable us to correct the truncation error on the coarse grid, where large time steps can be used without the full penalty of reduced resolution.

The present method contains some of the elements of the multigrid "frozen τ" method [5] but the motivation and justification are different. The present method is simple to implement and follows the correct physical transient. It allows for arbitrary mesh ratios and does not require storage of all time levels [6].

Formulation

We assume that our problem is represented by a differential equation or system of equations of the form

$$\frac{\partial u}{\partial t} = Lu, \; u = u(x,t), \; x \quad R, \; t > 0 \; . \qquad (1)$$

R is a d-dimensional region, L is a differential operator, and u satisfies appropriate boundary conditions on the boundary of R and an initial condition at $t = 0$. (The following considerations apply equally well to approximations of the finite difference, finite element or spectral type, as long as the spatial operator is treated separately.)

Equation (1) is discretized on two grids, a fine grid denoted by the subscript f, and a coarse grid denoted by the subscript c. Thus, we have $\nabla_t U_f = L_f U_f$ for the numerical solution U_f on the fine grid, and respectively, $\nabla_t U_c = L_c U_c$. Here ∇_t approximates the time derivative and L_f and L_c approximate L on the fine and coarse grids respectively.

$$\nabla_t U = L_c U - (L_c U_f - L_f U_f). \qquad (2)$$

where U is expected to be more accurate than U_c. To estimate the accuracy of the new scheme we consider the truncation errors. The fine grid truncation error, $T_f(u)$ is defined by $T_f(u) = \nabla_t u - L_f u$. The coarse grid truncation error is, $T_c(u) = \nabla_t u - L_c u$, while for the new scheme we have

$$T(u) = \nabla_t(u-U_f) - L_c u + L_c U_f - (L_f U_f - \nabla_t U_f) \quad .$$

The last term is zero, therefore, we get in the linear case

$$T(u) = \nabla_t(u-U_f) - L_c(u-U_f) = T_c(u-U_f) \quad .$$

Since U_f is a fine grid approximation to u we expect that $|u-U_f| << |u|$ in some norm, and therefore, that $|T(u)|$ is much smaller than $|T_c(u)|$. In fact $T_c(u-U_f)$ can be much smaller than $T_f(u)$. Of course, we do not want to compute both U and U_f for all times. Therefore, we use the expected behaviour of u, namely that u (and therefore U_f) adjusts rapidly on a convective time scale and then changes slowly by diffusion. Excessive computations with small time steps can be avoided by using the following strategy. We compute U_f (for the fine grid) as long as it changes rapidly. When U_f settles down, we compute (on the course grid) the correction $L_c U_f - L_f U_f$, and substitute in the improved scheme (2), to compute U. We use the restriction of U_f on the coarse grid as initial condition for U. The scheme (2) will be accurate as long as $u-U_f$ is small, but since U_f is not updated, we monitor $|U-U_f|$ and switch back to the fine grid to recompute the correction when $|u-U_f|$ exceeds a predetermined magnitude. To restart the computation on the fine grid we use as initial condition the last available U_f updated by the interpolation of $U-U_f$.
We note that this is a non-iterative time dependent algorithm and cannot be expected to work efficiently unless the underlying assumptions hold during the computation.
To estimate the savings in computation we consider the simplest explicit scheme in d-dimensions, and a mesh ratio, M, between course and fine space increments. If the number of operations to integrate over a time interval, τ, using the fine grid is F, the work for the same τ using the coarse grid is F/M^{2+d}. The total work will be $\alpha F + (1-\alpha)F/M^{2+d}$, where $\alpha\tau$ is the sum of fine grid time increments. Since we expect to compute with the fine grid over convective time scales and with the new scheme over diffusive time scales, α should be small and therefore we should approach a situation of coarse grid work with fine grid accuracy.

Examples

To establish the feasibility of the method we first solved a linear 2-D problem of passive convection

$$\frac{\partial E}{\partial t} + \vec{V}\cdot\nabla E = \frac{1}{Re}\nabla^2 E, \quad E = E\,(x,y,t), \quad 0 < x,\ y < 1$$

$$\vec{V} = (u,v), \quad u = \frac{\partial\psi}{\partial y}, v = -\ \frac{\partial\psi}{\partial x}\ \psi = x(1-x)y(1-y),$$

$$E(x,y,0) = 0, \quad E(x,1,t) = 1$$

$$E(x,0,t) = E(0,y,t) = E(1,y,t) = 0 \quad .$$

We used central second order differences in space and simple Euler or improved Euler for the time integration. Results are shown in Figure 1.

The second example is that of a driven cavity in vorticity (Ω) stream function (ψ) formulation, i.e.:

$$\frac{\partial\Omega}{\partial t} + \vec{V}\cdot\nabla\Omega = \frac{1}{Re}\ \nabla^2\Omega\ , \quad 0 < x,y < 1$$

$$\vec{v}(u,v), \ u = \frac{\partial\psi}{\partial y}\ , \ v = -\ \frac{\partial\psi}{\partial x}\ , \quad \Omega = \nabla^2\psi \quad ,$$

with no slip boundary conditions on all boundaries. The velocity on the moving boundary is $u(x) = x(1-x)$. We used upstream derivatives for the convection terms and first order integration in time. The truncation error correction was also applied to the Poisson equation and the boundary conditions relating ψ and Ω. The vorticity time development is plotted in Figure 2 at two points on the boundary, for a fine solution, a coarse solution, and for the two grid method.

References

1. M. Israeli and M. Ungarish, Numerische Mathematik, 39, 309-324 (1982).
2. M. Israeli and M. Ungarish, IC7NMFD, Stanford, 1980. Lecture Notes in Physics 141, Springer-Verlag.
3. M. Israeli and P. Bar-Yoseph, Fifth GAMM Conference on Numerical Methods in Fluid Mechanics, Rome, October, 1983.
4. P. Bar-Yoseph and M. Israeli, "The Asymptotic Finite Element Method for Boundary Value Problems, Recent Advances in Numerical Methods in Fluids, Volume V (C. Taylor, ed.) Pentech Press 1984.
5. A. Brandt, in "Multigrid Methods" (V. Hackbush and U. Trottenberg, eds.) Springer-Verlag, 1981, 220-312.
6. A. Karlsson and L. Fuchs, in "Numerical Methods in Laminar and Turbulent Flow (C. Taylor, J. A. Johnson, W. R. Smith, eds.) Pineridge Press 1983.

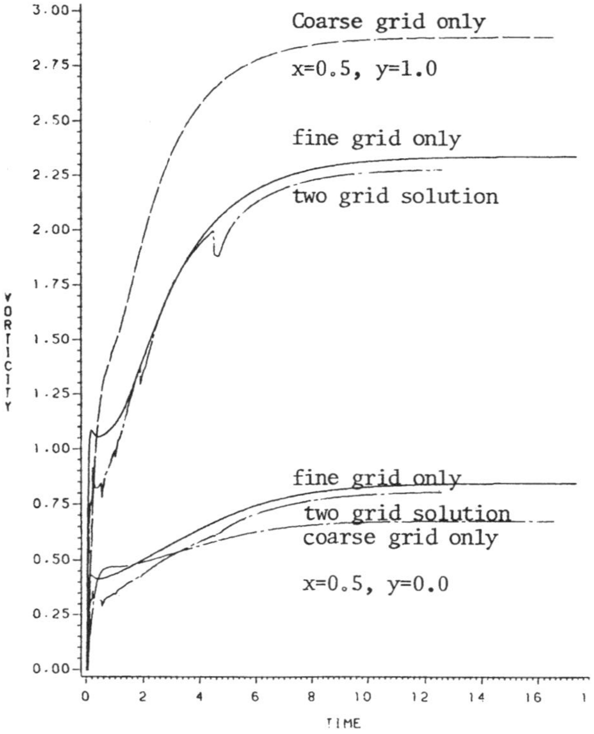

Figure 2
Re = 150
fine grid 20 x 20
coarse grid 10 x 10

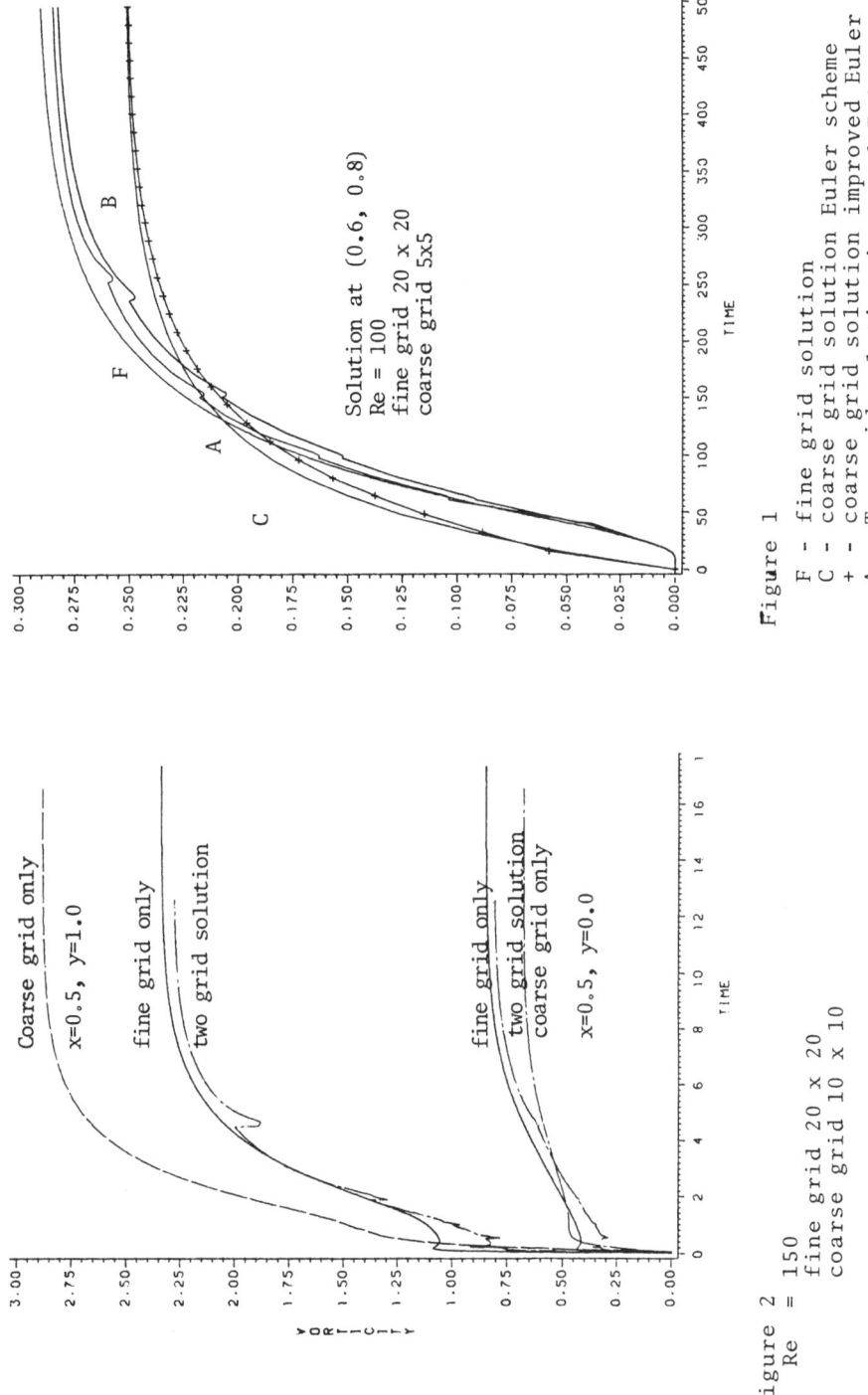

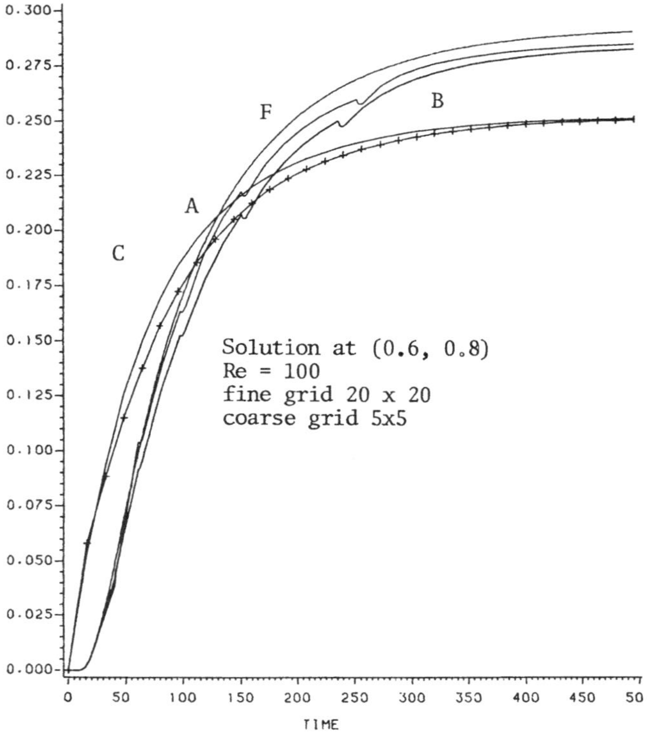

Figure 1

F - fine grid solution
C - coarse grid solution Euler scheme
+ - coarse grid solution improved Euler
A - Two grid solution improved Euler
B - Two grid solution simple Euler

MULTIPLE-GRID SOLUTION OF THE THREE-DIMENSIONAL EULER AND NAVIER-STOKES EQUATIONS

Gary M. Johnson and Julie M. Swisshelm
Institute for Computational Studies
Colorado State University
Fort Collins, Colorado, U.S.A.

SUMMARY

A procedure for the accelerated solution of the three-dimensional, compressible Navier-Stokes and Euler equations is described. The convergence of an explicit fine-grid scheme is enhanced through the use of a multiple-grid technique on a collection of coarser grids. The coarse-grid scheme is itself fully explicit and is independent of such details of the fine-grid problem as the formulation of viscous or damping terms and the specification of boundary conditions. Furthermore, this multiple-grid technique may be used, without modification, with a variety of fine-grid algorithms. Results are presented for the flow through a cascade of finite-span, swept blades. The Euler equations are solved for both subcritical and shocked, supercritical flows. The Navier-Stokes computations include laminar and turbulent, attached and separated flows. The procedure is vectorized for use on a CDC Cyber 205 computer and an algorithm version suitable for use on a multiple instruction - multiple data computer is mentioned.

INTRODUCTION

Motivated by the need to efficiently simulate complex, three-dimensional flow phenomena and encouraged by recent successes in two dimensions, attention has begun to focus on the development of algorithms for the rapid solution of the steady, three-dimensional Euler and Navier-Stokes equations. The method described here maintains the simplicity and low operations count of an explicit method while offering the accelerated convergence behavior usually associated with an implicit method. It uses a collection of successively coarser grids to accelerate the convergence of an explicit fine-grid solution procedure. In this general sense, it is a multiple-grid method. However, the details of its implementation are a good deal simpler than is the case with the conventional multigrid approach. The coarse-grid acceleration scheme presented here is quite modular. It may be used without modification in conjunction with a number of different fine-grid solution procedures and with any set of flow equations in the hierarchy ranging from the Euler equations to the full Navier-Stokes equations. It is also independent of the formulation of fine-grid viscous or damping terms and boundary conditions.

EQUATIONS OF MOTION

The nondimensional equations of motion may be written in conservation-law form as

$$q_t = -(F_x + G_y + H_z) \qquad (1)$$

where, for the full Navier-Stokes equations,

$$F = f - Re^{-1}p \qquad G = g - Re^{-1}r \qquad H = h - Re^{-1}s$$

while, for their thin-layer version,

$$F = f \qquad G = g \qquad H = h - Re^{-1}d$$

and, for the Euler equations,

$$F = f \qquad G = g \qquad H = h$$

Although, for simplicity, the equations of motion are presented here written in Cartesian coordinates, this does not imply any loss of generality. Note that the thin-layer approximation is implemented by using a body-fitted coordinate system and neglecting the viscous terms in the coordinate direction along the body. For Cartesian coordinates, with x and y representing the body-conforming coordinates, the thin-layer version of the Navier-Stokes equations is as given above. Turbulence is simulated by means of a two-layer algebraic eddy viscosity model due to Baldwin and Lomax [1].

SOLUTION PROCEDURE

Given a basic fine grid on which a numerical solution of Eqn.(1) is required and an explicit, conditionally stable solution procedure, the acceleration concept is to construct a collection of successively coarser grids by means of which the fine grid solution may be rapidly advanced, while respecting the stability limits on all grids. One cycle of the procedure consists of an application of a fine-grid integration step and one step of the coarse-grid scheme on each coarser grid. Grids may be updated either sequentially or simultaneously. The latter choice is particularly appropriate for computation on a parallel machine.

The fine-grid integration scheme used here is the forward predictor - backward corrector version of the two-step Lax-Wendroff method due to MacCormack [3]. First derivatives in the viscous terms are backward differenced in the predictor and forward differenced in the corrector. Any of its many variants could also be used, as could any other one- or two-step Lax-Wendroff scheme [3]. In fact the class of fine-grid schemes with which the coarse-grid scheme may be applied appears to be quite large, including schemes not of Lax-Wendroff type [4].

Given the fine-grid corrections, δq, we wish to use successively coarser grids to propagate this information throughout the computational domain, thus accelerating convergence to the steady state while maintaining the accuracy determined by the fine-grid discretization. A coarse-grid scheme may be derived from a one-step Lax-Wendroff method written in a coarse-grid setting. For example, if we choose a coarse-grid spacing which is double that of the fine grid we obtain:

$$\delta q_{i,j,k} = \frac{1}{8} \sum_{\substack{i\pm1 \\ j\pm1 \\ k\pm1}} \left[(I + \frac{\Delta t}{\Delta x} A + \frac{\Delta t}{\Delta y} B + \frac{\Delta t}{\Delta z} C) \Delta q \right] \tag{2}$$

where the indicated sum is taken over the eight diagonal neighbors of the central point and

$$A = \frac{\partial F}{\partial q} \qquad B = \frac{\partial G}{\partial q} \qquad C = \frac{\partial H}{\partial q}$$

Instead of discretizing the flux balance, Δq, on the coarse grid, we approximate it with a restriction of the fine-grid correction

$$\Delta q_{coarse} \simeq R(\delta q_{fine}) \ .$$

The effect of this restricted fine-grid correction is then distributed according to Eqn.(2) to obtain a coarse-grid correction. This is, in turn, prolonged to the fine grid to become the new fine-grid correction and update the fine-grid solution. Observe that in the basic integration scheme a correction at one grid point affects only its nearest neighbors, while in a k-level multiple-grid scheme the same correction affects all points up to 2^{k-1} mesh spacings distant. Since the coarse-grid scheme simply propagates the effects of fine-grid corrections, fine grid accuracy is maintained. Furthermore, a coarse-grid scheme based on the Euler equations may be employed, without modification, to accelerate the convergence of viscous flow computations based on the Navier-Stokes equations, the thin-layer equations, or any other viscous model equations which contain the full inviscid Euler equations.

An alternative coarse-grid scheme which does not require the use of flux vector Jacobian matrices has been developed:

$$\delta q_{coarse} = \Delta q - \frac{\Delta t}{2} \left[(F_x + G_y + H_z)^{n+1} - (F_x + G_y + H_z)^n \right] \tag{3}$$

In this flux-based coarse-grid scheme Δq is approximated by a restriction of the latest fine-grid value of δq and

$$F^n = F(q^n) \ , \qquad F^{n+1} = F(q^n + \Delta q) \ .$$

As both the fine-grid solution procedure and the coarse-grid shemes used here are explicit, the resultant multiple-grid algorithm may be readily vectorized. Such vectorization has been performed for computation on a CDC Cyber 205 machine.

RESULTS

Extensive preparatory testing of the multiple-grid procedure was carried out in two dimensions [5]. The speedups produced ranged from about 2 to 8 over a broad span of conditions, including subsonic and transonic inviscid flows and separated and attached, laminar and turbulent subsonic viscous flows with Reynolds numbers from 8.4×10^3 to 2.0×10^5. Three-dimensional computations have been carried out for the geometry illustrated in Fig. 1. It is a rectilinear cascade of finite-span, swept blades mounted between endwalls. The sweep angle ranges from 0 to 26 degrees. Results are available for both subcritical and supercritical inviscid flows and for laminar and turbulent, attached and separated, subsonic viscous flows. The speedups for these cases fall in the same range as obtained previously in two dimensions. Sample three-dimensional subcritical and supercritical inviscid results are shown in Fig. 2. Separated laminar and turbulent flow results using the thin-layer Navier-Stokes equations are shown in Fig. 3.

CONCLUSIONS

An explicit multiple-grid algorithm for the accelerated solution of the steady, three-dimensional Euler and Navier-Stokes equations has been presented. The coarse-grid scheme used to accelerate convergence is compatible with a variety of fine-grid algorithms and may be used with any set of flow equations in the hierarchy ranging from the Euler equations to the full Navier-Stokes equations. Furthermore, several variants of the coarse-grid scheme are possible. The multiple-grid procedure has been vectorized for use on a CDC Cyber 205 computer. Results have been obtained for subsonic and transonic inviscid flows and for separated and attached, laminar and turbulent subsonic viscous flows. Multiple-gridding yielded speedups comparable with those previously obtained in two dimensions.

REFERENCES

1. Baldwin, B.S. And Lomax, H.: AIAA Paper 78-257, 1978.
2. MacCormack, R.W.: AIAA Paper 69-354, 1969.
3. Johnson, G.M.: NASA TM-82843, 1982.
4. Stubbs, R.M.: AIAA Paper 83-1945, 1983.
5. Johnson, G.M.: NASA TM-83361, 1983.

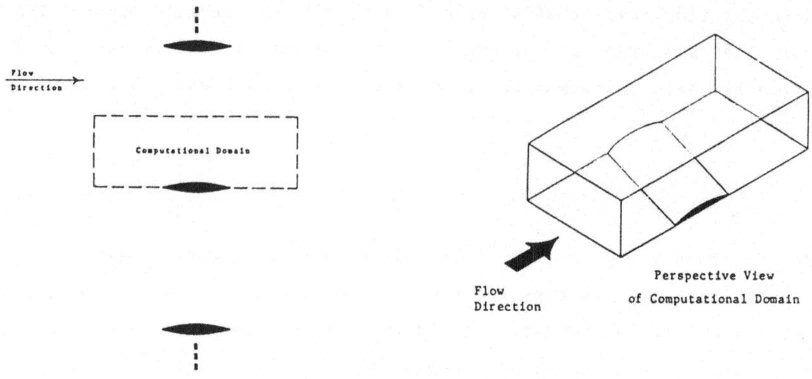

FIGURE 1. - Three-Dimensional Cascade

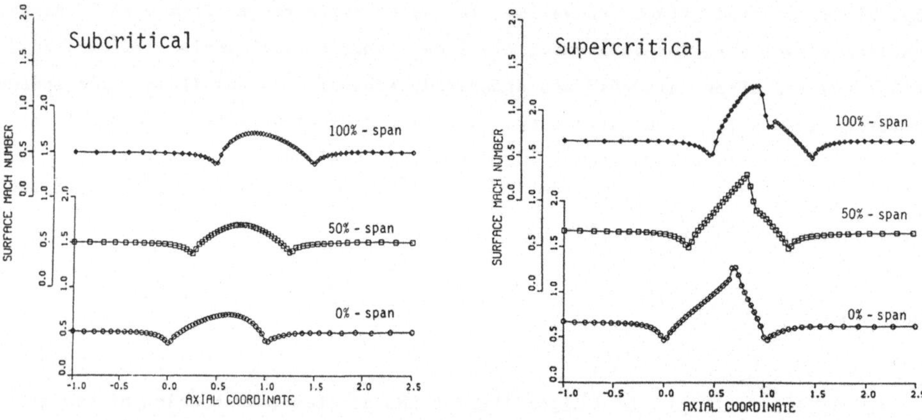

FIGURE 2. - Inviscid Flow, Sweep Angle = 26 Deg.

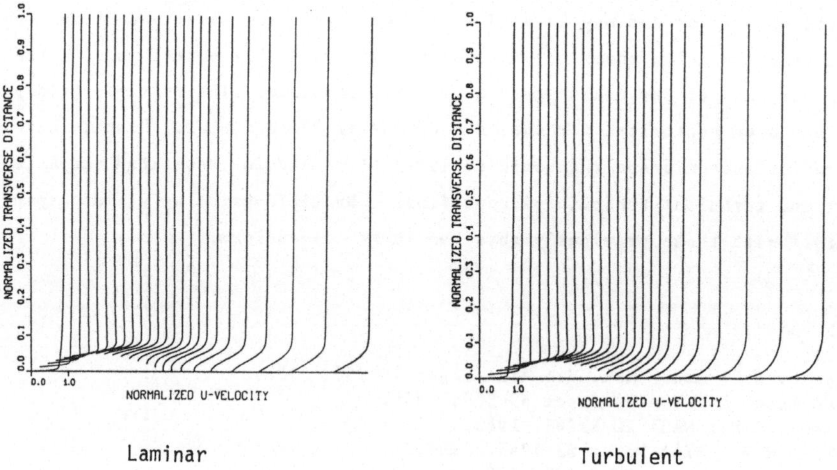

Velocity Profiles at 50% Span

FIGURE 3. - Viscous Flow, Re = 3.4×10^4, Sweep Angle = 26 Deg.

New Higher-Order Upwind Scheme for Incompressible Navier-Stokes Equations

Tetuya KAWAMURA, Hideo TAKAMI, *Kunio KUWAHARA
Department of Applied Physics, University of Tokyo, Hongo, Bunkyo-ku, Tokyo, Japan
*The Institute of Space and Astronautical Science, Komaba, Meguro-ku, Tokyo, Japan

Introduction

Computation of high-Reynolds number flow by finite-difference method has been severely hindered by numerical instability. The most widely used way to overcome this instability is to use some kind of upwind schemes. The stability of first-order upwind scheme is very good, but this has a strong diffusive effect similar to the molecular viscosity. The second-order upwind scheme is better in this sense but the stability is worse. It causes undesirable propagation of errors.

In this paper, a new upwind scheme is developed, which has third-order accuracy. The leading error term is a fourth-order derivative, this has a local diffusive effect, but the global effect is much smaller than the second-order derivative term.

Using this upwind scheme, three different flows are computed, i.e. a flow in a square cavity, a flow around a circular cylinder with surface roughness and a three-dimensional turbulent flow in a duct.

Computational scheme

Following MAC method,the basic equations of this scheme are the Navier-Stokes equations and Poisson equation for pressure and equation of continuity.

$$\frac{\partial \mathbf{v}}{\partial t} + (\mathbf{v}.\nabla)\mathbf{v} = - \text{grad } p + \frac{1}{Re}\Delta \mathbf{v} \ , \qquad (1)$$

$$\Delta p = - \text{div}(\mathbf{v}.\nabla)\mathbf{v} + R \qquad (2)$$

where

$$R = - \frac{\partial D}{\partial t} + \frac{1}{Re}\Delta D \ , \quad D = \text{div } \mathbf{v} \ . \qquad (3)$$

and Re is the Reynolds number and R is the corrective term in order to prevent the accumulation of numerical errors.

The Euler implicit scheme is used for the time integration. The nonlinear term is linearized for $\mathbf{v}^{n+1}$ as follows:

$$(\mathbf{v}^{n+1}.\nabla)\mathbf{v}^{n+1} = (\mathbf{v}^{n}.\nabla)\mathbf{v}^{n+1}, \qquad (4)$$

In the case of a turbulent duct flow, Euler explicit scheme is used.

All spatial derivatives except the nonlinear term are approximated by central difference. The nonlinear terms are approximated by a new third-order upwind scheme.

$$\left(f \frac{\partial u}{\partial x}\right)_{i,j} =$$

$$f_{i,j}(u_{i+2,j} - 2u_{i+1,j} + 9u_{i,j} - 10u_{i-1,j} + 2u_{i-2,j})/6\Delta x \quad \text{for } f_{i,j} \geq 0,$$

$$f_{i,j}(- 2u_{i+2,j} + 10u_{i+1,j} - 9u_{i,j} + 2u_{i-1,j} - u_{i-2,j})/6\Delta x \quad \text{for } f_{i,j} < 0. \qquad (5)$$

The finite-difference equations are obtained by discretizing Eqs.(1) - (2) using the relation (5). These equations are solved by point SOR method.

Flow in a Cavity

Computation of a flow in a square cavity is widely used to check numerical schemes, but even for this simple problem reliable computations at high Reynolds numbers are very scarce. One of the reliable computation is by Ghia et al.[1] at Reynolds number 5000. Using the above upwind scheme, the same flow was computed at Re=5000. The number of the mesh points is 40x40. The results is shown in Fig. 1(a). This results agree very well with that by Ghia et al. Their computation uses 257x257 mesh points. This means the present upwind scheme is very effective even if the number of the mesh points is very small.

Flow past a Circular Cylinder

The main purpose of this computation is to simulate the flow in the critical range and to investigate the mechanism of the sharp reduction of drag. It is well known from the experiments that the surface roughness around a circular cylinder makes the critical Reynolds number lower, that is, the boundary layer is thick enough for practical computations. Keeping this in mind, numerical calculation is carried out for the flow with surface roughness. Generalized coordinate system is used so that enough grid points are concentrated in the boundary layer and the wake. The main computations were performed over the Reynolds number range 10^3 - 10^5. Figures 2(a) and 2(b) show time developments of the computed flow patterns around a circular cylinder at Re=2000, and 40000, respectively. In these figures, only the stream lines which pass through the boundary layer are drawn. At Re=40000, small eddies separated from the boundary layer are clearly seen (see also Fig.2 (c));they make the width of the wake narrower. This indicates that the main reason for the drag reduction is the separation of small eddies. At Re=2000 on the other hand, the separation is very smooth in spite of the surface roughness.

Figure 3(a) shows the time histories of the drag and lift coefficients just after the shedding of Karman vortex begins at Re=2000 and Fig. 3(b) shows time histories from impulsive start at Re=40000.

Figure 4 shows the dependence of the mean drag coefficient on the Reynolds number, showing clearly the transition phenomena.

Turbulent Flow in a Duct

Three-dimensional turbulent flow in a duct is first studied numerically by Deardorff[2] and recently by Moin et al.[3] and Horiuti et al[4] by using large eddy simulation. Their results are very good and it has become widely believed that these turbulent flows are well simulated only by large eddy simulation.

However, from the above results, it has become clear that high-Reynolds-number flows are very well simulated by the present upwind scheme. This suggests that a turbulent flow in a duct also might be simulated by this scheme without turbulence model.

Keeping this in mind, a turbulent flow in a duct was simulated directly by integrating the Navier-Stokes equations without any turbulence model. The initial and boundary conditions are essentially the same as those by Deardorff, Moin et al. and Horiuti et al. The number of mesh points is the same as that used by Deardorff (24x14x20).

Figure 5 shows the contour lines of pressure and velocity components. These clearly indicate the large scale structures of the turbulent flow. The variation of the velocity components is shown in

Fig.6 which agrees well with experimental results and the computational results by Moin et al. Figure 7 shows the time lines at Re=25000. Burst is clealy seen in this figure.

In order to simulate the transition to turbulence of duct flow, above-mentioned flow is recomputed under the different initial conditions. The flow is assumed at rest initially and only pressure gradient is imposed. Figure 8 shows space averaged velocity profiles from impulsive start to turbulence. Up to non-dimensional time t=560, the flow is almost laminar and the velocity distribution becoming parabolic, but sometime around t=620 the asymmetry developes, which is the beginning of the transition. The transition to turbulence takes place during rather short time period as is shown in Fig.9. Figure 10 shows time averaged velocity profile after the transition. Reasonable agreement with Laufer's experiment[5)] is obtained.

These results indicate that turbulent flows can be directly simulated by the present scheme, and the number of the grid points need not be so large.

Present method has no adjustable constants and is much simpler and based on ordinary finite-difference method, and the boundary conditions are very easily imposed.

Conclusions

A new upwind scheme for computation of incompressible flow has been developed. It was found that this scheme works well at high Reynolds number even using limited number of mesh points.

By using this scheme, three different types of flows were computed. At first, a cavity flow was simulated at Re=5000, with 40x40 grid, and the results agree very well with the computation with 257x257 grid by Ghia et al.

Secondly, a flow past a circular cylinder was simulated. The sharp drag reduction in the critical Reynolds number range is clearly obtained.

Thirdly, a three-dimensional turbulent flow in a duct was simulated and large structures in the turbulent flow are captured by this scheme without any turbulence model. Agreement of statistical values with experiments and computations by Moin et al. are excellent. Moreover the transition to turbulence is well simulated.

From these results, we can conclude that this scheme is very widely applicable and robust and easy to program for any type of incompressible flow computations.

References

1) U. Ghia, K. N. Ghia and C. T. Shin; 1982 High-Re Solutions for Incompressible Flow Using the Navier-Stoke Equations and a Multigrid Method. J. comput. Phys. vol.48 pp.387-411

2) J. W. Deardorff; 1970 A numerical study of three-dimensional turbulent channel flow at large Reynolds numbers. J. Fluid Mech. vol.41, pp.453-480.

3) P. Moin and J. Kim; 1982 Numerical investigation of turbulent channel flow. J. Fluid Mech. vol.118, pp.341-377.

4) K. Horiuti and K. Kuwahara; 1982 Study of Incompressible Turbulent Channel Flow by Large Eddy Simulation. Proc. 8th ICNMFD (Springer-Verlag)

5) J. Laufer; 1950 Investigation of turbulent flow in a two-dimensional channel. NACA TN 2123.

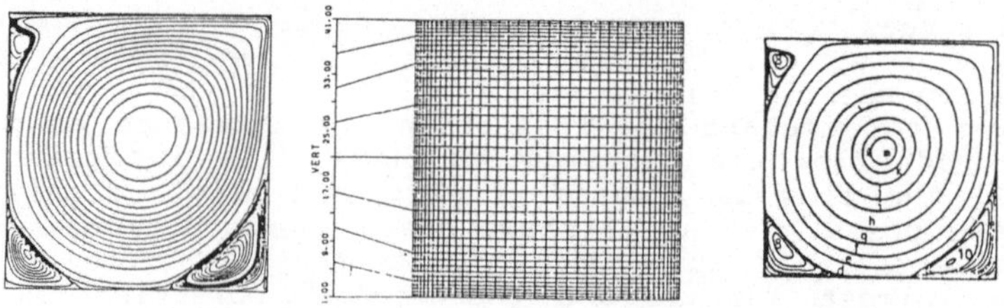

(a) (b) (c)

Fig.1 Flow in a cavity at Re=5000.
(a) 41x41 present method. (b) grid system. (c) 257x257 (Ghia et al.)

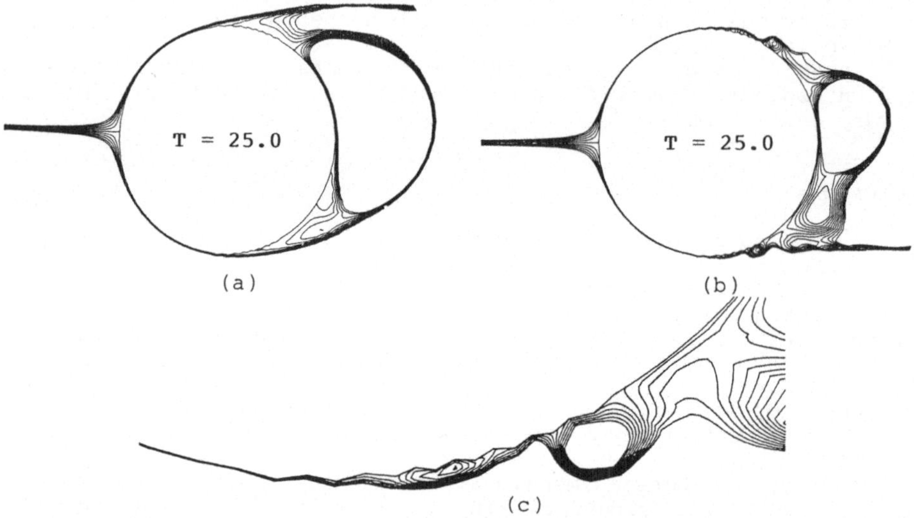

Fig.2 Flow past a circular cylinder with surface roughness.
(a) stream lines at Re=2000. (b) Re=40000.
(c) separation bubble near the separation point at Re=40000.

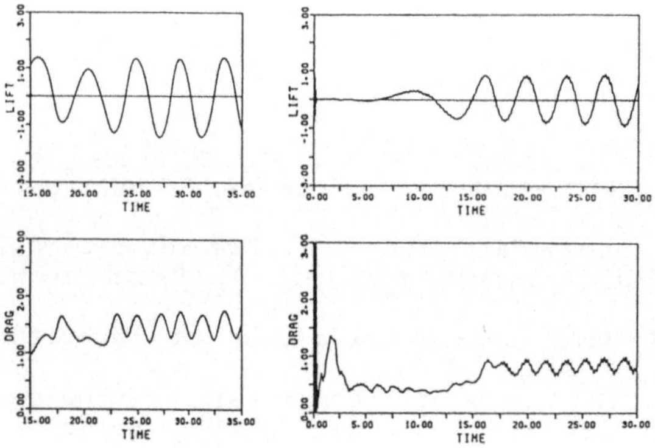

(a) (b)

Fig.3 Lift and drag.
(a) Re=2000. (b) Re=40000.

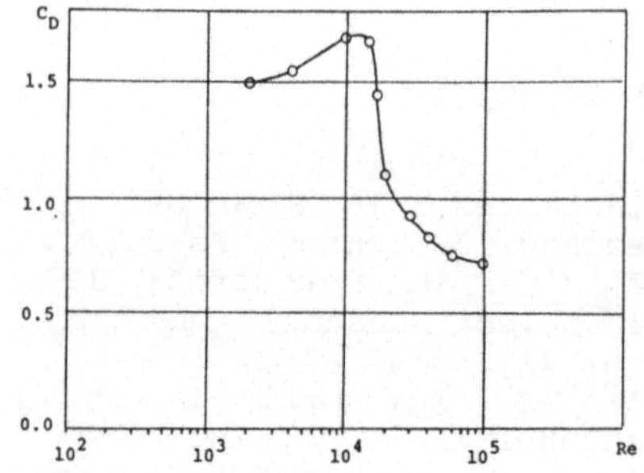

Fig.4 Dependence of the drag on Reynolds number.

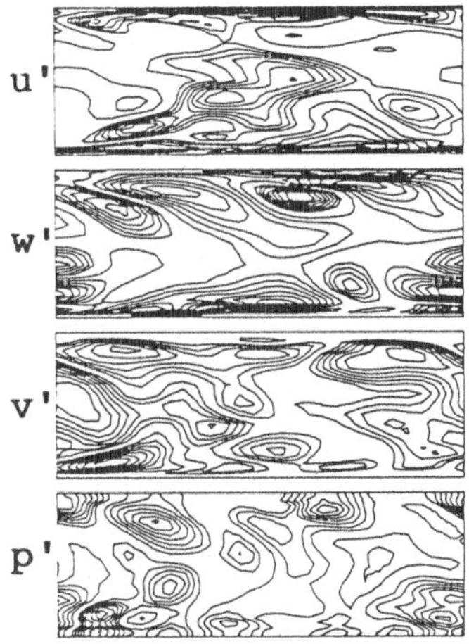

Fig.5 Contour lines for turbulent duct flow; the mesh is 32x20x40.

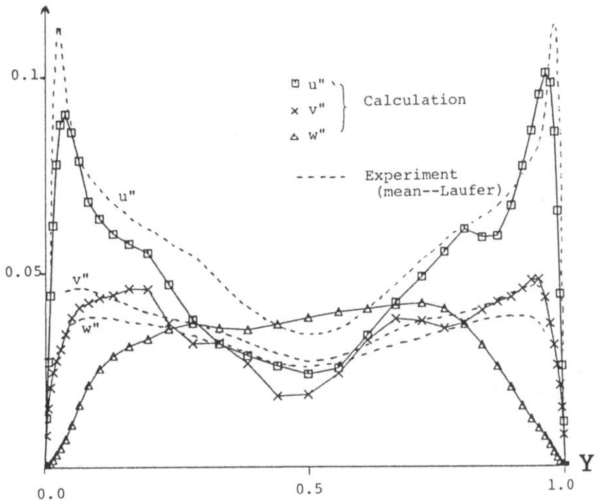

Fig.6 Turbulent intensities (space averaged, 30x20x40).

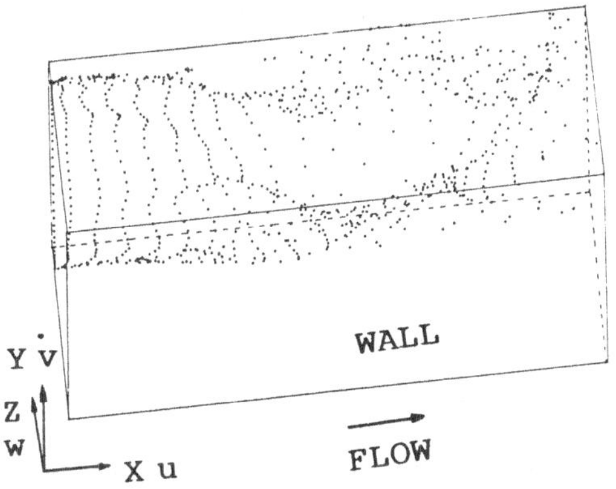

Fig.7 An example of the time lines at the time about the transition (25x15x20).

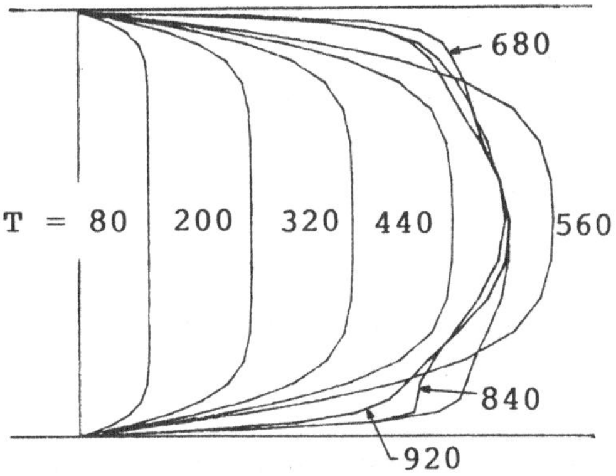

Fig.8 Space averaged velocity profiles from impulsive start to turbulence (25x15x20).

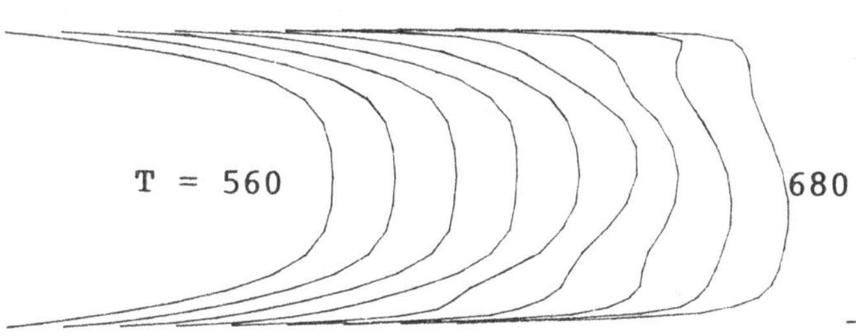

Fig.9 Velocity profiles showing the transition from laminar to turbulent flow (25x15x20).

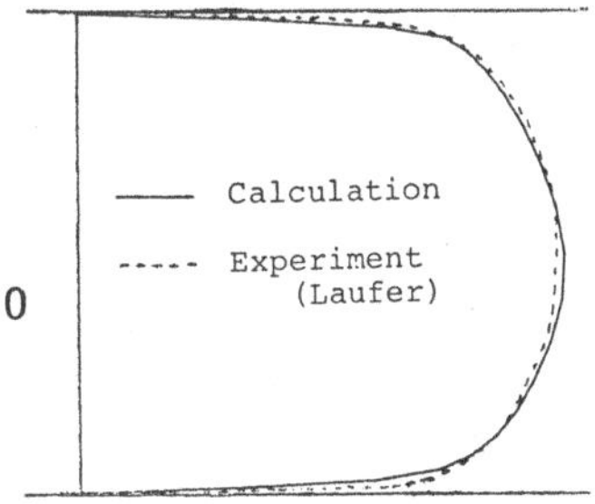

Fig.10 Time averaged velocity profile (25x15x20).

SOLUTION OF THE PARABOLIZED NAVIER-STOKES EQUATIONS FOR THREE-DIMENSIONAL INTERNAL FLOWS

P.K. Khosla and E.E. Bender
Dept. of Aerospace Engineering and Applied Mechanics
University of Cincinnati, Cincinnati, Ohio 45221

Introduction

The numerical solution of the 3-D Navier-Stokes equations for large Reynolds number flows requires prohibitively large computer resources and so are quite expensive to obtain. It has, therefore, been preferable to utilize some of the approximate formulations of the Navier-Stokes Equations. Classical and interacting boundary layer theory and the PNS approximation have found considerable acceptance for the computations of certain classes of viscous flows. Of these approximations, the PNS system, which is more appropriately termed the reduced Navier-Stokes (RNS) equations, contain more of the salient features of the Navier-Stokes equations for large Reynolds Number flows, yet allow for computational simplicity. The RNS equations contain an exact representation of the inviscid flow field as well as the viscous effects associated with the boundary-layer approximation and normal pressure gradients. For problems with strong viscous-inviscid interaction (e.g., separated flow), it has been shown that the RNS equations are capable of describing accurately this kind of flow. The earlier investigations tried to exploit the resemblance of these equations to a parabolic system. This then could be solved as an initial-value problem with efficient marching procedures accompanied by substantial reductions in computer time and storage [1-3]. It was soon realized that with complete pressure interaction, the marching technique exhibited exponentially growing behavior of the solution. This is known as the departure effect. These exponentially growing solutions are indicative of the presence of upstream influence, characteristic of elliptic systems. For supersonic flows, Lubard and Helliwell [1] presented a stability analysis to show that stable marching could be achieved with p_x backward differenced, but only when the marching step size was greater than a lower bound. Based on an eigenvalue analysis, Vigneron et al. [2] neglected that portion of p_x which is associated with upstream influence and eliminated the step size restriction. For incompressible or low subsonic flows, this amounts to neglecting the axial pressure gradient completely. Any attempt to incorporate p_x fully via iteration or a sublayer approximation in a single sweep procedure leads to the departure effect.

The phenomenon of pressure interaction and the associated upstream influence were first investigated by Lighthill [4]. Rubin [5] investigated the mathematical character of the RNS equations as related to the streamwise pressure gradient term. A significant ellipticity of the full Navier-Stokes equations is retained in the RNS equations through the pressure interaction. The departure solutions are an indicator of this ellipticity. Numerically, the RNS equations should be solved as a boundary-value problem taking due account of the downstream boundary condition. This consideration is necessary for flows that experience strong pressure interaction associated with the singular behavior; e.g., separated flows, or flows over geometries that have large axial variations. One step marching procedures that do not reflect the elliptic character of the RNS equations would yield inadequate solutions.

In order to incorporate the strong pressure interaction fully, Rubin and Lin [6] have investigated a relaxation method for the incompressible RNS equations. A stable marching procedure can be devised by using a simple "forward" or midpoint differencing of the axial pressure gradient term. Such treatment of p_x leads to a global computational procedure which includes a parabolic marching step imbedded in a global relaxation step for the pressure. Such a procedure has no limitation on step size [6,7]. This boundary layer relaxation procedure has been extended to subsonic and transonic flows by Khosla and Lai [8,9], Ramakrishnan and Rubin [10], and for two-dimensional internal flow by Rosenbaum [11].

The present investigation extends the procedure to 3-D internal flows with strong viscous-inviscid interaction. Incompressible flow through a converging-diverging channel of square cross-section will be investigated. The governing implicit system of difference equations are solved iteratively by an approximate explicit LU decomposition. The solution procedure allows for the coupling of all the flow variables in the cross-plane so that rapid convergence is achieved at each station. Furthermore, the coupling of the pressure allows for complete pressure interaction and eliminates any singularity at separation which might be encountered in the type of flows under investigation.

Governing Equations

For 3-D flows, the PNS equations can be written as:

$$u\frac{\partial u}{\partial x} + v\frac{\partial u}{\partial y} + w\frac{\partial u}{\partial z} = -\frac{dp_o}{dx} - \frac{\partial p'}{\partial x} + \frac{1}{Re}\left(\frac{\partial^2 u}{\partial y^2} + \frac{\partial^2 u}{\partial z^2}\right) \tag{1a}$$

$$\frac{\partial^2 v}{\partial y^2} + \frac{\partial^2 v}{\partial z^2} = \frac{\partial \Omega}{\partial z} - \frac{\partial^2 u}{\partial x \partial y} \tag{1b}$$

$$\frac{\partial^2 w}{\partial y^2} + \frac{\partial^2 w}{\partial z^2} = -\frac{\partial \Omega}{\partial y} - \frac{\partial^2 u}{\partial x \partial z} \tag{1c}$$

$$u\frac{\partial \Omega}{\partial x} + v\frac{\partial \Omega}{\partial y} + w\frac{\partial \Omega}{\partial z} + \frac{\partial u}{\partial z}\frac{\partial v}{\partial x} - \frac{\partial u}{\partial y}\frac{\partial w}{\partial x} + \Omega\left(\frac{\partial v}{\partial y} + \frac{\partial w}{\partial z}\right) = \frac{1}{Re}\left(\frac{\partial^2 \Omega}{\partial y^2} + \frac{\partial^2 \Omega}{\partial z^2}\right) \tag{1d}$$

$$-\frac{\partial^2 p'}{\partial y^2} - \frac{\partial^2 p'}{\partial z^2} = \frac{\partial u}{\partial y}\frac{\partial v}{\partial x} + \frac{\partial u}{\partial z}\frac{\partial w}{\partial x} + u\frac{\partial^2 v}{\partial x \partial y} + u\frac{\partial^2 w}{\partial x \partial z}$$

$$+ \left(\frac{\partial v}{\partial y}\right)^2 + 2\frac{\partial v}{\partial z}\frac{\partial w}{\partial y} + \left(\frac{\partial w}{\partial z}\right)^2 - v\frac{\partial^2 u}{\partial x \partial y} - w\frac{\partial^2 u}{\partial x \partial z} \tag{1e}$$

where $\Omega = \frac{\partial v}{\partial z} - \frac{\partial w}{\partial y}$

On the channel wall $u = v = w = 0$. The inflow is prescribed to be fully developed. The usual Neumann type boundary condition on p' is abandoned in favor of a Dirichlet boundary condition. The advantage of using a Dirichlet as compared to a Neumann boundary condition is that no compatibility condition is required in the former case. The details of this unique treatment of the pressure boundary condition are discussed in ref. [12].

At each axial location, $\frac{dp_o}{dx}$ is determined by satisfying global mass conservation. It should be noted from (1b,1c) that a pressure or velocity correction found in many other internal flow studies is not required to satisfy the continuity equation exactly. Since pressure is treated as unknown these equations include an internal mechanism for pressure interaction to circumvent singular behavior at separation.

Solution Algorithm

The governing equations are discretized using second order central differences. The x-derivatives are upwind differenced except that $\frac{\partial p'}{\partial x}$ is "forward" differenced. Quasi-linearization is used for the nonlinear terms. The algebraic system can be written in matrix form as $\mathbf{Au} = \mathbf{b}$ where $\mathbf{u} = [u,v,w,\Omega,p']^T$ is the solution vector. The coefficient matrix $\mathbf{A}$ is written as

$$\mathbf{A} = \mathbf{L} + \mathbf{D} + \mathbf{U}$$

where L is the lower triangular, U the upper triangular and D the diagonal components of $\mathbf{A}$. The iterative solution algorithm utilized is the coupled block SSOR

$$(L + D)D^{-1}(U + D)(u_{n+1} - u_n) = b - \mathbf{A}u_n$$

or

$$M(u_{n+1} - u_n) = r_n$$

This algorithm is a two step procedure and consequently requires the storage of an intermediate solution vector. Since each pass is equivalent to a Gauss-Seidel computation, the overall procedure is usually considered to be twice as fast as Gauss-Seidel. For a small number of grid points in the cross plane, say 11 x 11 with about 600 unknowns, this algorithm performs reasonably well. However, for finer meshes, the rate of convergence of SSOR slows down considerably. In a global computational mode, where each axial location has to be computed many times, such an algorithm can become prohibitively time consuming. In order to accelerate the convergence, a preconditioned conjugate residue (PCR) method is used. The procedure minimizes the Euclidean norm of the residual and requires that the symmetric part of the coefficient matrix be positive definite. The appropriate preconditioning is provided by the coupled block SSOR procedure. Such a solution procedure has been investigated by Axelsson [13], Eisenstat [14], and Petravic and Kuo-Petravic [15] for non-symmetric matrices. A number of variants of PCR exist in literature. The following version due to Eisenstat is utilized in the present investigation:

$$r_0 = b - \mathbf{A}u_0 \quad ; \quad r_0' = M^{-1}r_0 \quad ; \quad p_0 = r_0'$$

For k = 0 step 1 until convergence Do

$$\alpha_k = (r_k, r'_k)/(p_k, Ap_k) \quad ; \quad u_{k+1} = u_k + \alpha_k p_k$$

$$r_{k+1} = r_k - \alpha_k Ap_k \quad ; \quad r'_{k+1} = M^{-1} r_{k+1}$$

$$\beta_k = (r_{k+1}, r'_{k+1})/(r_k, r'_k) \quad ; \quad p_{k+1} = r'_{k+1} + \beta_k p_k$$

Preliminary results on the use of PCR are quite encouraging. Convergence was improved by a factor of three over the original SSOR algorithm for a 11 x 11 mesh. There should be a more significant increase for finer meshes.

Results

In order to test the algorithm, the flow in a curved duct of square cross-section has been investigated for a Dean number of 55. A uniform grid of 21 x 21 has been utilized in a single pass calculation. The fully developed solution is compared with the one obtained by Ghia et al [16], in Fig. 1. The agreement is very good.

Laminar flow in converging and converging-diverging ducts of square cross-section have been investigated to test the applicability of the global procedure. A simple shearing transformation is utilized to generate the grid. Typical results are shown in Fig. 2. The corresponding results for the diverging duct are shown in Fig. 3. The effect of the Dirichlet boundary conditions for p' at the wall is examined in Fig. 4. It can be seen that the Neumann boundary conditions usually applied for Poisson pressure solutions are in fact automatically satisfied.

The authors would like to thank Prof. S.G. Rubin for many helpful discussions. This research was supported by the office of Naval Research under contract No. N00014-79-C-849.

References

1. Lubard, S.C., and Helliwell, W.S. (1974), AIAA Journal, Vol. 12, No. 7, pp. 965-974.
2. Vigneron, Y.C., Rakich, J.V., and Tannehill, J.C. (1978), AIAA Paper 78-1137.
3. Davis, R.T., Rubin, S.G. (1980), Computers and Fluids, Vol. 8, pp. 101-132.
4. Lighthill, M.J., Proc. Roy. Sci. Lond., A, Vol. 217, pp.478-507, 1953.
5. Rubin, S.G. (1981), Symposium on Numerical and Physical Aspects of Aerodynamic Flows, Long Beach, California, Springer-Verlag Publ., Berlin, pp. 171-185.
6. Rubin S.G., and Lin, A. (1980), Israel Journal of Technology, Vol 18., pp 21-31.
7. Rubin S.G., and Reddy, D.R. (1983), Computers and Fluids, Vol. 11, pp. 281-306.
8. Khosla, P.K., and Lai, H.T. (1983), Computers and Fluids, Vol. 11, pp. 325-339.
9. Lai, H.T., and Khosla, P.K. (1984), AIAA paper 84-0458.
10. Ramakrishnan S.V., and Rubin S.G. (1984), Paper under preparation.
11. Rosenbaum, D., Private communication.

12. Bender, E.E., (1984), M.S. Thesis, Dept. of Aerospace Eng. and Appl. Mech., University of Cincinnati.
13. Axelsson, O. (1980), Linear Algebra Appl. Vol. 29, pp 1-6.
14. Eisenstat, Stanley C. (1981) SIAM J. Sci. State. Comp., Vol. 2 pp. 1-4.
15. Petravic M., and Kuo-Petravic, G. (1979), J. Comp. Phys., Vol. 32, pp 263-269.
16. Ghia, K.N., and Sokhey, J.S. (1976), Proc. of Numerical/Laboratory Computer Methods in Fluid Mechanics.

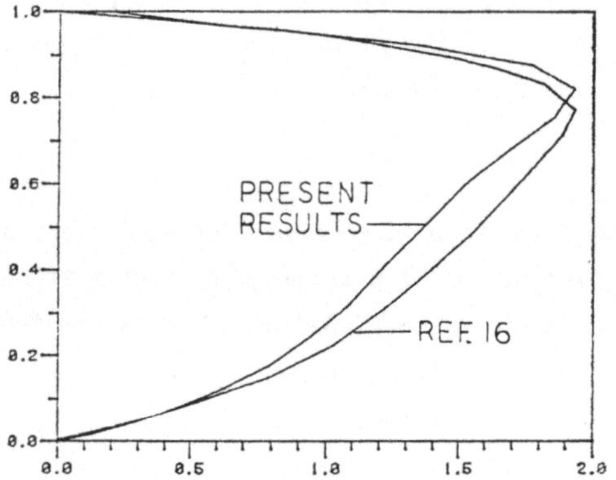

FIG. 1a FULLY DEVELOPED AXIAL VELOCITY PROFILE IN A CURVE DUCT, k = 55 (DEAN NO.).

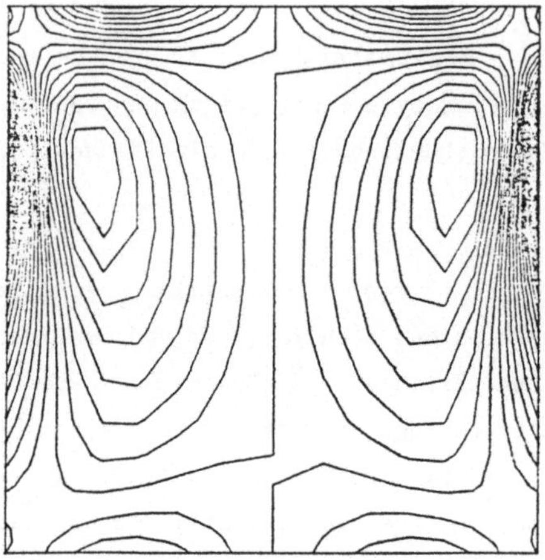

FIG. 1b AXIAL VORTICITY CONTOURS FOR FULLY DEVELOPED FLOW IN A CURVED DUCT, k = 55.

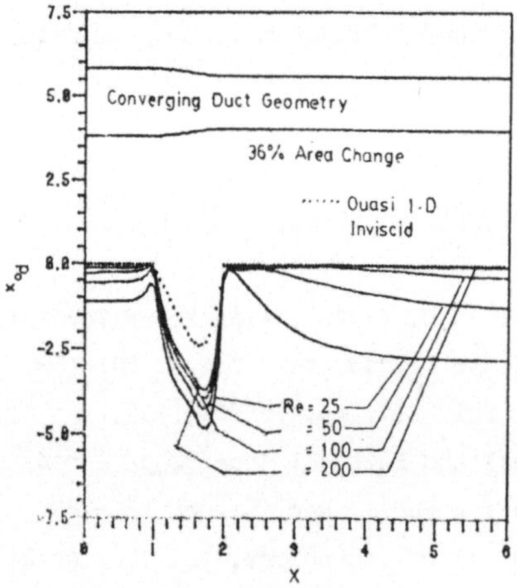

Fig. 2a - Average Pressure Gradient in a Converging Duct

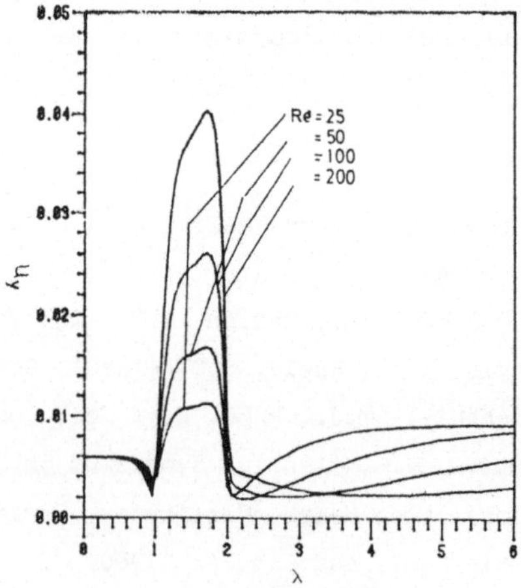

Fig. 2b - Skin Friction Parameter in the Corner of a Converging Duct

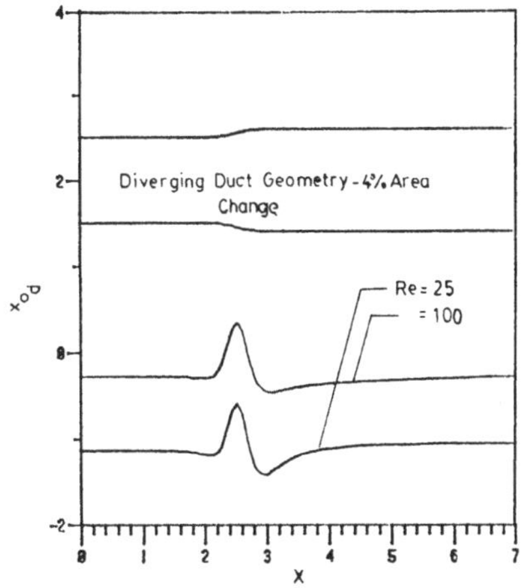

Fig. 3a - Average Pressure Gradient in a Diverging Duct

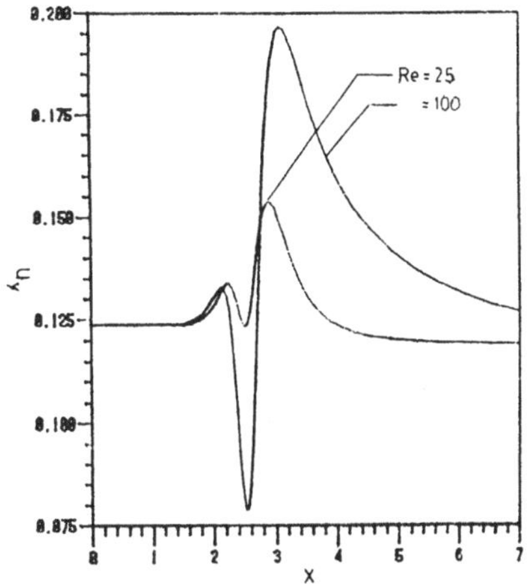

Fig. 3b - Skin Friction in the Corner of a Diverging Duct

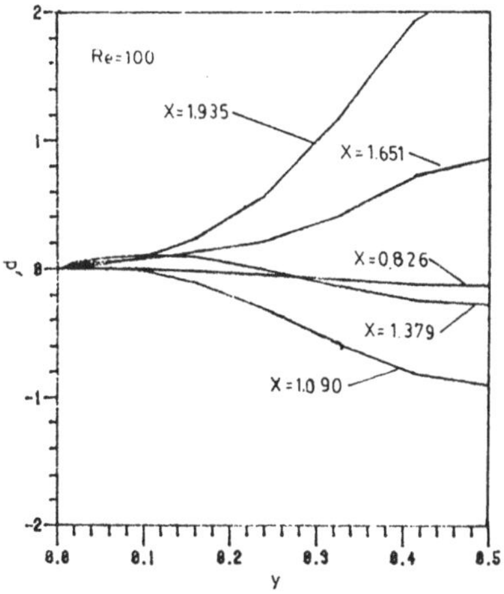

Fig. 4a - Centerline Pressure Profiles at Different Axial Locations (Converging Duct)

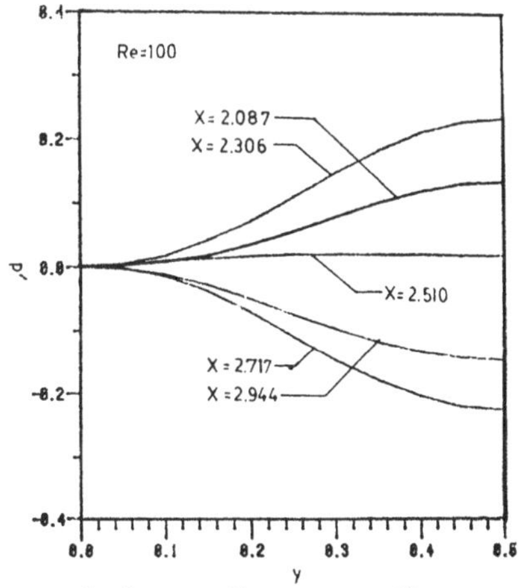

Fig. 4b - Centerline Pressure Profiles at Different Axial Locations (Diverging Duct)

IMPLICIT SOLUTION OF THE 3-D COMPRESSIBLE NAVIER-STOKES EQUATIONS FOR INTERNAL FLOWS*

K. R. Kneile
Sverdrup Technology, Inc./AEDC Group
Arnold AFS, TN 37389 USA

R. W. MacCormack
University of Washington
Seattle, WA 98195 USA

GOVERNING EQUATIONS

The governing field equations for compressible, viscous flow consist of the equations of conservation of mass, momentum, and energy (plus state equations) and are here collectively termed the Navier-Stokes equations. The equations in nondimensional conservation form can be written as

$$\frac{\partial U}{\partial t} + \nabla \cdot F = 0 \tag{1}$$

where the flux vector F is

$$F = \begin{bmatrix} \rho u \\ \rho uu + PI - \tau \\ (E + P)u - u \cdot \tau - \varkappa \nabla T \end{bmatrix} \tag{2}$$

and the dyadic τ represents the viscous portion of the stress tensor given by

$$\tau = Re^{-1}(2\mu \widehat{\nabla u} + \lambda \nabla \cdot uI) \tag{3}$$

The quantity U consists of the conservative variables, (ρ, ρu, E). The velocity vector u is normalized by the reference speed of sound a_r; pressure P and total energy per unit volume E are normalized by $\rho_r a_r^2$; density ρ and temperature T are normalized by their respective reference values. The Reynolds number Re is based on the reference sound speed and the viscous coefficients, μ and λ, are normalized by μ_r with λ set to $-2\mu/3$ using Stokes hypothesis. $\varkappa$ is a nondimensionalized thermal conductivity given by $\mu/RePr(\gamma-1)$. The dyadic $\widehat{\nabla u}$ is the symmetric part of ∇u. I is the unit dyadic.

EXPLICIT EULER CODE

The code used by the authors in this paper was obtained by modifying an existing code. This section contains a brief description of the finite volume explicit MacCormack Euler code (Ref. 1) which was converted to an implicit Navier-Stokes code. The primary motivation for starting with this code was its flexibility with respect to the geometries it can handle: external, internal, and combinations of the two.

The grid imposed on the computational flow region is topologically equivalent to a regular grid on a cube. The resulting volume elements are hexahedrons with quadrilateral faces. Degenerate (nonhexahedral) volumes may be used; e.g., two edges may coincide to form a distorted triangular prism typically found along the axis in axisymmetrical grids.

Flow variables are calculated at volume centers. Figure 1 shows a projection of a 2-D slice passing through volume centers defined by a value of k for the third subscript. The bold line represents the outer boundary of the flow region. The points outside the flow region are phantom points used for boundary conditions. For notational brevity, points like A and B will be described by i+,j,k and i,j-,k rather than the more conventional i+1/2,j,k and i,j-1/2,k.

The finite volume algorithm is obtained by integrating Eq. (1) over a grid volume and applying the divergence and mean value theorems

$$\frac{\partial U}{\partial t} = -\frac{1}{V} \oint n \cdot F dS \tag{4}$$

*The research reported herein was performed at the Arnold Engineering Development Center (AEDC), Air Force Systems Command. Work and analysis for this research were done by personnel of Sverdrup Technology, Inc., operating contractor for the AEDC propulsion test facilities. Further reproduction is authorized to satisfy needs of the U. S. Government.

Discretization yields

$$\Delta U = -\frac{\Delta t}{V} \sum An \cdot F \tag{5}$$

where the summation is over the six faces of the volume. The flow variables U and their corrections ΔU are evaluated at volume centers; the $An \cdot F$ terms are evaluated at face centers. V represents the cell volume and An represents outward pointing area vectors with n being a unit normal. The explicit MacCormack algorithm uses a two-step Runge-Kutta procedure where forward spatial differences are used for the first (predictor) step and backward differences for the second (corrector) step, or vice versa. For finite volume methods this is equivalent to evaluating the F for a given face using the values at either the forward or backward volume centers, e.g., $F_{i+,j,k} = F_{i,j,k}$ or $F_{i+1,j,k}$. For inviscid calculations F does not contain any derivatives and can be evaluated directly.

IMPLICIT NAVIER-STOKES MODIFICATIONS

This section describes the modifications for conversion to an implicit Navier-Stokes code. For viscous (Navier-Stokes) calculations F contains terms with first order derivatives. The terms are evaluated with central differences about face centers using a local coordinate transformation, e.g.,

$$\begin{aligned} \Delta_i T_{i+,j,k} &= T_{i+1,j,k} - T_{i,j,k} \\ \Delta_j T_{i+,j,k} &= T_{i+,j+1,k} - T_{i+,j-1,k} \\ \Delta_k T_{i+,j,k} &= T_{i+,j,k+1} - T_{i+,j,k-1} \end{aligned} \tag{6}$$

where $T_{i+,j+1,k} = \dfrac{T_{i,j+1,k} + T_{i+1,j+1,k}}{2}$ etc.

The Cartesian coordinate values are similarly differenced to obtain the Jacobian matrix of the transformation:

$$J = \begin{bmatrix} \Delta_i X_{i,j,k} & \Delta_j X_{i,j,k} & \Delta_k X_{i,j,k} \\ \Delta_i Y_{i,j,k} & \Delta_j Y_{i,j,k} & \Delta_k Y_{i,j,k} \\ \Delta_i Z_{i,j,k} & \Delta_j Z_{i,j,k} & \Delta_k Z_{i,j,k} \end{bmatrix} \tag{7}$$

The inverse of this matrix is then used to obtain viscous term derivatives with respect to the Cartesian coordinate system. Sutherlands law is used to obtain μ.

The implicit MacCormack algorithm (Ref. 2) can be represented as

$$C \, \delta U = \Delta U \tag{8}$$

where ΔU are the explicit corrections for a predictor or corrector step and δU are the implicit corrections for the corresponding step. As a result, the conversion from explicit to implicit can be easily accomplished by adding implicit routine(s) at the appropriate place in the explicit code. The implicit MacCormack algorithm uses an approximate factorization concept

$$C = C_K C_J C_I \tag{9}$$

where I, J, and K represent 1-D operators in the three grid oriented directions. Defining $(\Delta U)_I = \Delta U$, $(\Delta U)_J = (\delta U)_I$, $(\Delta U)_K = (\delta U)_J$, and $(\delta U)_K = \delta U$, Eq. (8) can be written as three 1-D equations to be solved successively:

$$C_I (\delta U)_I = (\Delta U)_I \tag{10a}$$

$$C_J (\delta U)_J = (\Delta U)_J \tag{10b}$$

$$C_K (\delta U)_K = (\Delta U)_K \tag{10c}$$

These 1-D implicit operators are forward or backward corresponding to the determination of the explicit ΔU. Further description of these equations will be limited to Eq. (10a), as the other two can be treated similarly. The I subscripts are dropped to simplify notation.

Define

$$B = n \cdot G \tag{11}$$

where $G = \partial F/\partial U$ is considered inviscidly (viscous terms set to zero). The eigenvalues Λ and eigenvectors S are

$$\Lambda = \begin{bmatrix} u_n - a & & \\ & u_n I & \\ & & u_n + a \end{bmatrix} \tag{12}$$

$$S = \begin{bmatrix} 0 & -\rho a n & 1 \\ n & I - nn & -n/a^2 \\ 0 & \rho a n & 1 \end{bmatrix} \begin{bmatrix} 1 & \emptyset & 0 \\ -u/\rho & I/\rho & \emptyset \\ \alpha\beta & -\beta u & \beta \end{bmatrix}$$

where $\alpha = u \cdot u/2$, $\beta = \gamma - 1$, $u_n = n \cdot u$, I is a 3x3 identity submatrix, and n is the unit area normal as a 3x1 submatrix. The 1-D implicit forward and backward operators can now be written as

$$\left(I + \frac{\Delta t A_{i-}}{V_i} \left|n_{i-} \cdot G_i\right|\right)\left(V_i \,\delta U_i\right) = \left(V_i \Delta U_i\right) + \frac{\Delta t A_{i+}}{V_{i+1}} \left|n_{i+} \cdot G_{i-1}\right| \left(V_{i+1} \,\delta U_{i+1}\right)$$

$$\left(I + \frac{\Delta t A_{i+}}{V_i} \left|n_{i+} \cdot G_i\right|\right)\left(V_i \,\delta U_i\right) = \left(V_i \Delta U_i\right) + \frac{\Delta t A_{i-}}{V_{i-1}} \left|n_{i-} \cdot G_{i+1}\right| \left(V_{i-1} \,\delta U_{i-1}\right) \tag{13}$$

where the j and k subscripts are dropped to simplify notation. The $|B| = |n \cdot G|$ terms are evaluated as

$$|B| = S^{-1} D^{+} S \tag{14}$$

where

$$D^{+} = (D + |D|)/2$$

$$D = |\Lambda| + \left(\frac{2\gamma\mu A}{Re\,Pr\,\rho V} - \frac{V}{2A\Delta t}\right) I$$

$|\Lambda|$ and $|D|$ are diagonal matrices of absolute values.

Computationally, these operators are block bi-diagonal requiring only a backward substitution algorithm. Furthermore, the transformation matrix S can be used to diagonalize the blocks to a scalar system (Ref. 2). The definition of D^{+} automatically reverts the algorithm to an explicit form when allowed by the local CFL criterion. The implicit calculations can be bypassed at these points resulting in further computational savings.

BOUNDARY CONDITIONS

In addition to the conventional boundary conditions, boundary conditions are also required for the implicit operator. The following is a description of the boundary conditions used for the nozzle flow test case (Fig. 2) discussed in the results section.

Inflow and outflow boundary conditions are imposed using a characteristic splitting method. ΔU is calculated for the interior cells along the boundary using the explicit MacCormack algorithm modified to use inward differences normal to the boundary. The transformation matrix S is used to obtain

$$\widehat{S}\,\widehat{\delta U} = \widehat{S}\,\delta U \tag{15}$$

where $\widehat{S}$ contains the rows of S for which the eigenvalues are nonnegative with respect to the outward direction defined by n. $\widehat{\delta U}$ represents the values after boundary conditions are imposed. Unless all eigenvalues are nonnegative, the system [Eq. (15)] is incomplete. Boundary conditions (as many as there are negative eigenvalues) must be supplied to complete the system.

For subsonic inflow only one eigenvalue, $u_n + a$, is nonnegative. Therefore four boundary conditions must be specified. Typically these are total temperature, total pressure, and two flow angles. For the test case, the incoming flow was assumed parallel ($v = w = o$) and the total conditions were held constant in time. Usually these conditions are applied at the first interior point by interpolating between the inflow plane and the second interior points. Because of the straight inflow section, these conditions were applied directly to the first interior points giving the simple expression

$$\Delta u = -\delta/(\rho u + \rho a) \tag{16}$$

where δ is the scalar value $\widehat{S\delta U}$. The remaining variables ρ, P, and E can be obtained from

$$\rho = \rho_T\left(1 - \frac{\gamma - 1}{2}\frac{u^2}{a_T^2}\right)^{\gamma/(\gamma-1)}$$

$$P = P_T\left(1 - \frac{\gamma - 1}{2}\frac{u^2}{a_T^2}\right)^{1/(\gamma-1)} \tag{17}$$

$$E = P/(\gamma - 1) + \rho u^2/2$$

For supersonic outflow all eigenvalues are positive and no boundary conditions need be imposed. $\widehat{S}$ is nonsingular and Eq. (15) becomes

$$\widehat{\delta U} = \delta U \tag{18}$$

Implicit inflow and outflow boundary conditions are not needed since the local explicit CFL criterion is satisfied in this direction and the implicit operator is not in effect.

At the wall, adiabatic no-slip conditions are specified. Phantom points are used as follows:

$$\begin{aligned} u_1 &= -u_2 \qquad & \rho_1 &= \rho_2 \\ v_1 &= -v_2 \qquad & E_1 &= E_2 \\ w_1 &= -w_2 \qquad & P_1 &= P_2 \end{aligned} \tag{19}$$

where the 1 subscript represents a phantom point and 2 represents its neighboring interior point. Because of the grid packing near the wall, the implicit operator is active at the wall in the normal direction. When the operator passes information away from the wall, incoming values of δU are set equal to zero. When the operator passes information toward the wall, the outgoing flux is mirrored about the wall plane and propagated back into the flow using the inward operator.

Reflections are used at centerlines. For example at the XZ plane,

$$\begin{aligned} u_1 &= u_2 \qquad & \rho_1 &= \rho_2 \\ v_1 &= -v_2 \qquad & E_1 &= E_2 \\ w_1 &= w_2 \qquad & P_1 &= P_2 \end{aligned} \tag{20}$$

Implicit boundary conditions are not required because of the local CFL criterion.

RESULTS

The code was verified using three check cases of Thomas (Refs. 3 and 4): boundary layer on a flat plate, 2-D transonic nozzle flow, and transonic nozzle flow for a rectangular converging-diverging nozzle. The wall boundary-layer calculations produced good agreement with the Blasius boundary-layer solution and with the calculations of Thomas. The 2-D nozzle calculations produced good agreement with measured wall pressures along the symmetry plane of a rectangular nozzle. The results were also in close agreement with the results of Thomas. Results are shown in this paper for the rectangular nozzle case. The Reynolds number based on stagnation conditions and throat half-height is 930,000. The flow is assumed laminar and the Prandtl number is 0.72. Figure 2 illustrates the nozzle computational grid. Two planes of symmetry (XY and XZ) were used. Figure 3 shows static pressure results along the nozzle wall at the vertical symmetry plane. Figure 4 shows static pressure along the nozzle sidewall at the horizontal symmetry plane. The open figures represent experimental data (Ref. 5), whereas the solid figures are the results of the method presented in this paper. The solid line shows the results of Thomas.

SUMMARY AND CONCLUSIONS

The implicit MacCormack algorithm is a simple and efficient method. Existing explicit codes can easily be modified to include the implicit operator. Its efficiency is due to: the use of a block bi-diagonal system instead of the more conventional tri-diagonal systems, the conversion to scalar systems using the transformation matrix, and the ability to bypass implicit operations where local conditions permit.

According to linearized 1-D theory the algorithm is unconditionally stable. In practice, however, an upper stability limit is prevalent. For the test case, a max CFL of 20 was used. Although this may seem small when compared with the CFL of 516 used by Thomas, continued running of the Thomas code at this CFL produced instabilities. Reduction of the CFL gave drastic changes in the velocity distribution, although the pressure distribution remained essentially the same. The CFL restrictions appear to be dominated by the approximate factorization. The instabilities occurred where the factorization has the largest error, in the corner where the grid is packed in two directions. Test cases with packed grids in only one direction ran with a significantly larger CFL. An advantage of the MacCormack implicit operator is that the approximate factorization can be eliminated without a significant increase in computer time. It is planned to modify the code to use an unfactored implicit operator.

REFERENCES

1. Jacocks, J. L. and Kneile, K. R., "Computation of Three-Dimensional Time-Dependent Flow Using the Euler Equations," AEDC-TR-80-49 (AD-A102463), July 1981.

2. MacCormack, R. W., "A Numerical Method for Solving the Equations of Compressible Viscous Flow," AIAA Paper 81-0110, January 1981.

3. Thomas, P. D., "Numerical Method for Predicting Flow Characteristics and Performance of Nonaxisymmetric Nozzles - Theory," NASA CR-3147, September 1979.

4. Thomas, P. D., "Numerical Method for Predicting Flow Characteristics and Performance of Nonaxisymmetric Nozzles. Part 2 - Applications," NASA CR-3264, October 1980.

5. Mason, Mary L., Putnam, Lawrence E., and Re, Richard J., "The Effect of Throat Contouring on Two-Dimensional Converging-Diverging Nozzles at Static Conditions," NASA TP-1704, 1980.

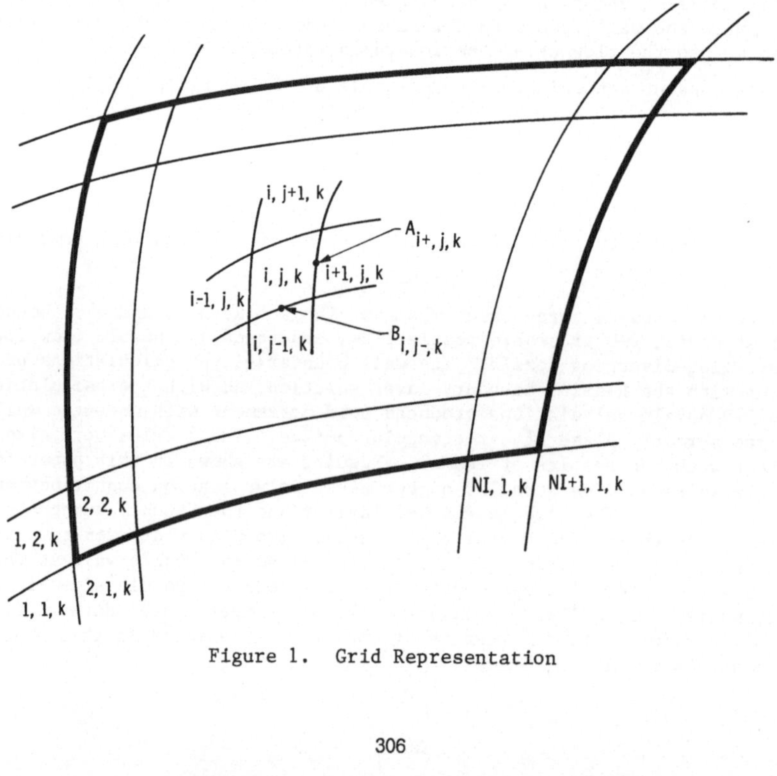

Figure 1. Grid Representation

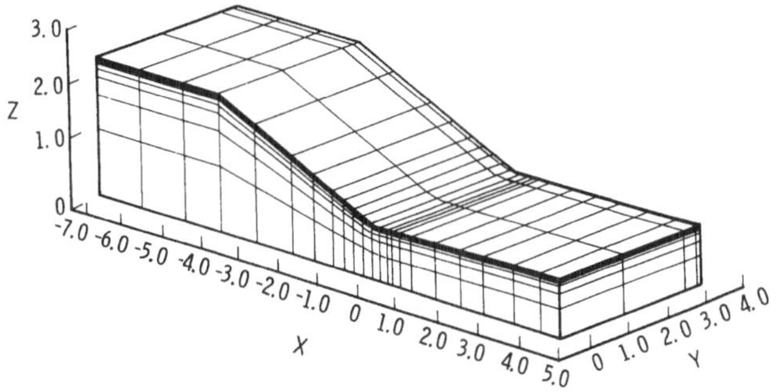

Figure 2. Grid for Rectangular Nozzle

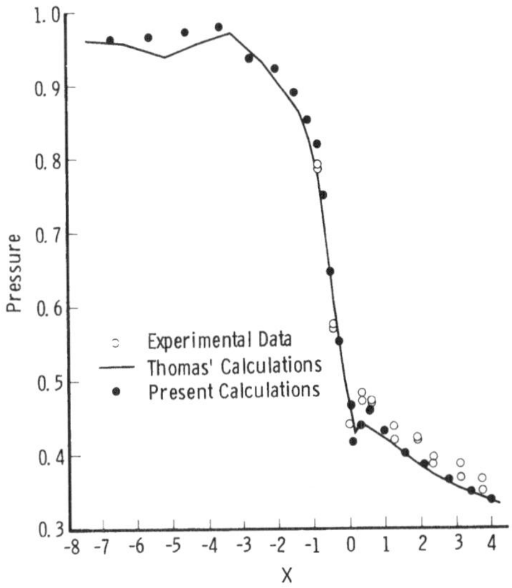

Figure 3. Upper Wall Static Pressure at Symmetry Plane

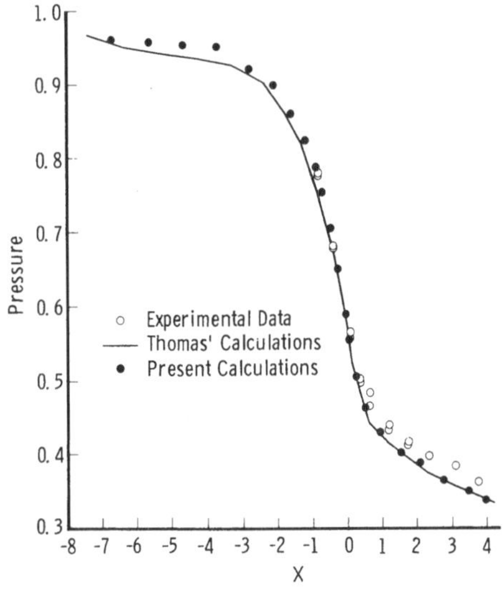

Figure 4. Side Wall Static Pressure at Symmetry Plane

COMPUTATION OF THREE-DIMENSIONAL VORTEX FLOWS PAST WINGS USING THE EULER EQUATIONS AND A MULTIPLE-GRID SCHEME

Ch. KOECK and J.J. CHATTOT
MATRA 78146 VELIZY-VILLACOUBLAY (FRANCE)

SUMMARY

A numerical method for solving the EULER Equations is presented, which is well suited to the computation of flows containing strong shock waves, and complex vortex structures. The EULER equations system is written in a pseudo-unsteady form, with constant total enthalpy ; it is integrated step by step in time with the explicit finite-volume scheme of Ron-Ho NI. Convergence speeding-up is achieved using NI's multiple-grid procedure. The farfield boundary conditions are treated with the compatibility relations technique. The computed examples presented concern the ONERA-M6 wing, and a sharp leading-edged delta wing, the DILLNER wing.

INTRODUCTION

Missile aerodynamic at high angles of attack is concerned mostly with non linear phenomena, such as shock waves and vortex sheets. Unlike methods based on a potential model, the EULER Equations method gives the exact jump relations at discontinuities ; furthermore, when the starting positions of the vortex sheets are known (trailing-edge, sharp leading-edge), the method yields solution with vortex rolling-up, emphasizing its ability to capture vortex sheets in the same sense as shock waves.

The EULER Equations are numerically solved using the scheme of Ron Ho NI[1]. It is an explicit one-step scheme, of the class of the LAX-WENDROFF schemes, with a finite-volume formulation (which is particulary convenient for 3D computations with mesh singularities). The scheme being an explicit one, it offers the main advantage of numerical simplicity, but, even when using the local time step technique, owing to the great number of iterations, it requires a large computational time to converge. The multiple-grid technique allows a noticeable reduction of the number of iterations, for a very small additional computational work per iteration (about 16 % in 3D with 3 grids). The multiple-grid procedure is based on a efficient damping of the unsteady waves (especially the low-frequency waves, whose convergence is very slow on the fine-grid) by cycling the calculation on several coaser and coarser grids, the accuracy of the solution depending only on the finest grid.

MATHEMATICAL MODEL

The EULER Equations are written as follows :

$$U_t + \operatorname{div} \overline{F}(U) = 0$$

with :

$$U = \begin{bmatrix} \rho \\ \rho u \\ \rho v \\ \rho w \end{bmatrix} \qquad \overline{F}(U) = \begin{bmatrix} \rho u & \rho v & \rho w \\ \rho u^2 + p & \rho v u & \rho w u \\ \rho u v & \rho v^2 + p & \rho w v \\ \rho u w & \rho v w & \rho w^2 + p \end{bmatrix}$$

the pressure p is a function of the density ρ and the velocity (u, v, w) :

$$p = \frac{\gamma - 1}{\gamma} \rho \left(H_i - (u^2 + v^2 + w^2)/2 \right)$$

Hi being the uniform total enthalpy.

The boundary conditions associated with the system are of three types ($\overline{N}$ is the outwards unit vector normal to the boundary) :

- Solid wall : $\overline{V} \cdot \overline{N} = 0$
- Inflow boundary : $\overline{V} \cdot \overline{N} < 0$
  if $\overline{V} \cdot \overline{N}$ is subsonic, the total pressure and the velocity direction are set to their free-stream values ; in the supersonic case, all the unknowns are set to their free-stream values.

- Outflow boundary : $\bar{V} . \bar{N} > 0$
  no condition is required for a supersonic flow ; otherwise, the following non-reflective condition is used: $\frac{\partial p}{\partial t} - \rho_o c_o \frac{\partial \bar{V}\cdot\bar{N}}{\partial t} = 0$
  where $c_o = \left[\left(\frac{\gamma-1}{2\gamma}\bar{V}\cdot\bar{N}\right)^2 + \frac{a^2}{\gamma}\right]^{1/2} - \frac{\gamma-1}{2\gamma}\bar{V}\cdot\bar{N}$

Outflow and inflow boundaries are treated with compatibility relation theory of ONERA[2]

FINE-GRID EULER SOLVER

The explicit finite-volume scheme of Ron - Ho NI is now presented. At a node ij, the correction $\delta U_{ij} \equiv U_{ij}^{n+1} - U_{ij}^{n}$ is obtained by adding contributions coming from the 4 cells having this node as a vertex (fig. 1). These contributions are given by the DISTRIBUTION FORMULAS (c is the cell $i+1/2\ ,\ j+1/2$):

$$(\delta U_{ij})_c = \frac{1}{4}\left(\Delta U_c - \frac{\Delta t}{v_c}\Delta F_{ic} - \frac{\Delta t}{v_c}\Delta F_{jc}\right)$$

$$(\delta U_{ij+1})_c = \frac{1}{4}\left(\Delta U_c - \frac{\Delta t}{v_c}\Delta F_{ic} + \frac{\Delta t}{v_c}\Delta F_{jc}\right)$$

$$(\delta U_{i+1\,j})_c = \frac{1}{4}\left(\Delta U_c + \frac{\Delta t}{v_c}\Delta F_{ic} - \frac{\Delta t}{v_c}\Delta F_{jc}\right)$$

$$(\delta U_{i+1\,j+1})_c = \frac{1}{4}\left(\Delta U_c + \frac{\Delta t}{v_c}\Delta F_{ic} + \frac{\Delta t}{v_c}\Delta F_{jc}\right)$$

The sum of contributions is written as :

$$\delta U_{ij} = (\delta U_{ij})_a + (\delta U_{ij})_b + (\delta U_{ij})_c + (\delta U_{ij})_d$$

ΔU_c is the first-order change occuring in the cell ; it assumes the following finite-volume expression :

$$\Delta U_c = (-\Delta t / v_c)\,(\delta_i . F_i + \delta_j . F_j)$$

The numerical fluxes Fi and Fj are given by (fig. 1) :

$$F_i = \bar{S}_i \cdot \mu_j . \bar{F}(U^n)$$
$$F_j = \bar{S}_j \cdot \mu_i . F(U^n)$$

The operators δ and μ stand for the difference and average operators ; $\bar{S}_i$ and $\bar{S}_j$ are the area vectors of the cell faces, v_c is the volume.
The fluxes ΔF_{ic} and ΔF_{jc} are defined by :

$$\Delta F_{ic} = (\mu_i . \bar{S}_i) . \Delta \bar{F}_c$$
$$\Delta F_{jc} = (\mu_j . \bar{S}_j) . \Delta \bar{F}_c$$

$\Delta \bar{F}_c = (\partial \bar{F} / \partial U)_c . \Delta U_c$; $(\partial \bar{F} / \partial U)$ is the jacobian of $\bar{F}$.
This way to calculate δU_{ij} is equivalent (provided the contributions are weighted by the volumes) to the following procedure :

$$\delta U_{ij} = \Delta t (U_t)_{ij} + \Delta t^2/2\, (U_{tt})_{ij} \equiv \Delta U_{ij} - \Delta t/2\, (\mathrm{div}\, \Delta \bar{F})_{ij}$$
$$\Delta U \equiv \Delta t\, U_t\ ,\ \Delta \bar{F} \equiv (\partial \bar{F} / \partial U) . \Delta U$$

and then compute $(\Delta U)_{ij}$ and $(\mathrm{div}\,\Delta \bar{F})_{ij}$ using the control-volumes described on fig. 1.
The artificial dissipative terms are of two types :

- Linear : $D_o(U) = Q_o \nabla^2 . U = 4\, Q_o (\mu_i \mu_j \mu_i \mu_j . U - U_{ij})$
- Nonlinear : $D_1(U) = D_i(U) + D_j(U)$

D_i is very similar to that of Jameson[3] and al : $D_i(U) = \delta_i . (\varepsilon_{i+1/2}\ \delta_i . U)$

with $\varepsilon_{i+1/2} = Q_1 \sup_{i,i+1} (|\delta_i^2 . \rho| / \rho)$

In our computations, we have used :

$$Q_o \leqslant 0.01$$
$$Q_1 \leqslant 0.1$$

COARSE-GRID SCHEME

A coarse-mesh is obtained from the previous one by deleting every other line. Each grid is denoted by it's typical meshsize : h, 2h, 4h.

The two-grids method reads as follows :

$U^* = U^n + \delta U^h$	Basic finite-volume scheme, dissipation and boundary conditions.
$\Delta U_c^{2h} = T_h^{2h} \cdot \delta U^h$	Transfer fine-to-coarse : Fine-grid residuals ⟶ coarse-grid changes.
$\delta U^{2h} = Q\left(\Delta U^{2h} - \beta \Delta t \, div \Delta \bar{F}_c^{2h}\right)$	Coarse-grid distribution formulas ⟶ coarse-grid corrections.
$U^{n+1} = U^* + I_{2h}^{h} \cdot \delta U^{2h}$	Interpolate and update the coarse-grid corrections ⟶ fine-grid solution at cycle n+1.

This procedure can be repeated on coarser grids.
We have introduced two scaling factors : Q (for the residuals) and β (for the time increments). These two factors have to be chosen in order to preserve the stability of the complete algorithm. The stability analysis is performed directly by computer, in the linear scalar case. For that, we compute several iterations of the scheme, with, as initial vector, the Fourier mode $V_L^0 = e^{i2\pi x/L}$ (L is the wave-length).

Due to the interpolation, two neighbor points do not have the same amplification factor, and the Fourier mode is not an eigenvector. A sufficient condition for stability is that :

$$\forall L \,,\; \forall n \,,\; \|V_L^n\| \leq \|V_L^0\| \qquad n = \text{number of iterations}$$

The result of this analysis is presented fig. 2. In all our computations, the transfert operator is the "full weighting" operator :

$$T_h^{2h} \cdot \delta U^h = \mu_i \mu_j \mu_i \mu_j \cdot \delta U^h$$

and residuals are interpolated using the bilinear interpolation operator, expressed in the computational plane. The CFL number of the fine-grid is taken equal to 0.9

COMPUTED EXAMPLES

ONERA M6 Wing, M_∞ = 0.84 , α = 3.06°

A C-O mesh system is used, with 53 x 49 x 17 points (fig. 3). Three grids are used in this calculation, h-2h-4h, and 700 cycles are performed (on the basic grid alone, more than 1500 it. are needed before reaching the same residuals, see fig. 4). The Cp distribution on the wing is presented on fig. 5 ; the present solution and a MacCormack's scheme solution[4] do not show notable differences.

DILLNER Wing, M_∞ = 0.7, α = 15°

For this Delta-wing, with a sharp leading-edge which is swept 70°, a C-O mesh composed of 45 x 57 x 17 nodes is used (fig. 6). The computed flow separates at the leading-edge, and rolls-up into a vortex (fig. 7 and 8). No condition is applied at the leading-edge. The Cp distribution on the wing (fig. 9) shows the suction peak due to the vortex ; we have found that the hight of this peak is very sensitive to the linear dissipation (we have taken Q_0 = 0.004).

CONCLUDING REMARKS

The EULER method enables the calculation of flow containing shock waves and vortex sheets, both of these discontinuities being captured by the scheme. However, the capture of a vortex sheet is more difficult than that of a shock; a bad use of the numerical dissipation may smear completely the vortical flow features.

REFERENCES

1 Ron-Ho NI "A multiple-grid scheme for solving the EULER Equations" AIAA Journal, november 1982.

2 H. VIVIAND "Pseudo-unsteady methods for transonic flow computation" Lecture note in physics, vol. 141 Springer Verlag, Berlin (1980).

3 A. JAMESON "Numerical solution of the EULER Equations for compressible inviscid fluids" 6 th ICCMAS Versailles, FRANCE december 1983.

4 C. KOECK and M. NERON "Computation of 3D inviscid flow on a wing by pseudo-unsteady resolution of the EULER Equations" 5 th GAMM conference ROMA October 1983 Vieweg Verlag.

5 M. BREDIF "A multigrid finite-element method for transonic potential flow" AIAA paper 83 - 0507.

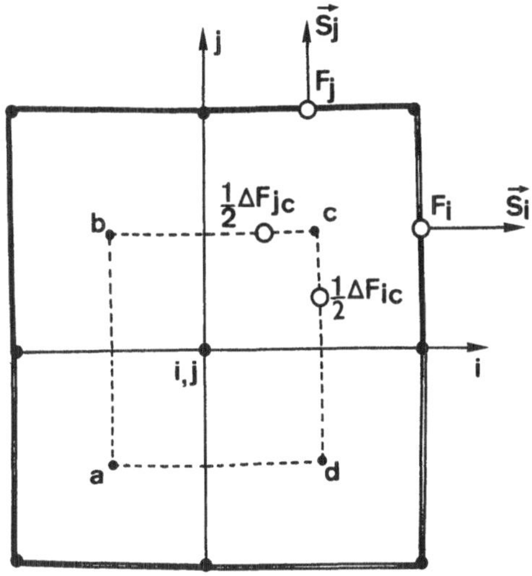

Fig. 1 Numerical integration-volumes : ═══ $(\Delta U)_{ij}$
- - - - - $(div\ \Delta\vec{F})_{ij}$

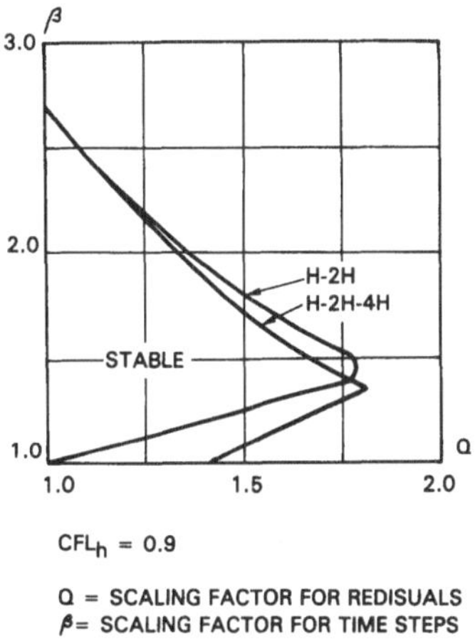

Fig. 2 Stability bound of the multiple-grid scheme

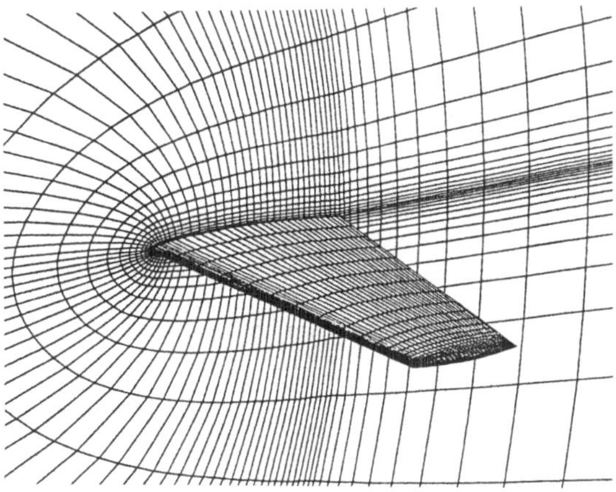

Fig. 3 ONERA M6 Wing

C-O mesh 53 X 49 X 17

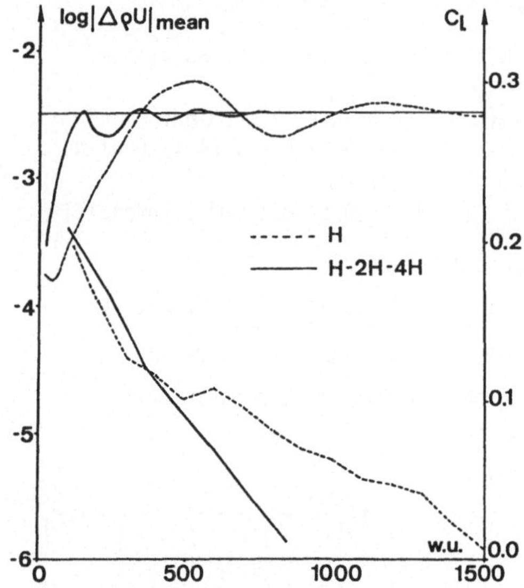

Fig. 4 M6 wing convergence history

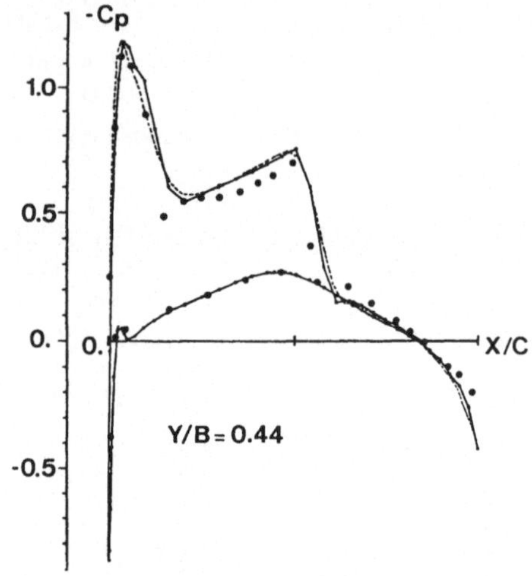

Fig. 5 M6 wing, M = 0.84, α = 3.06°

Cp distribution on the wing

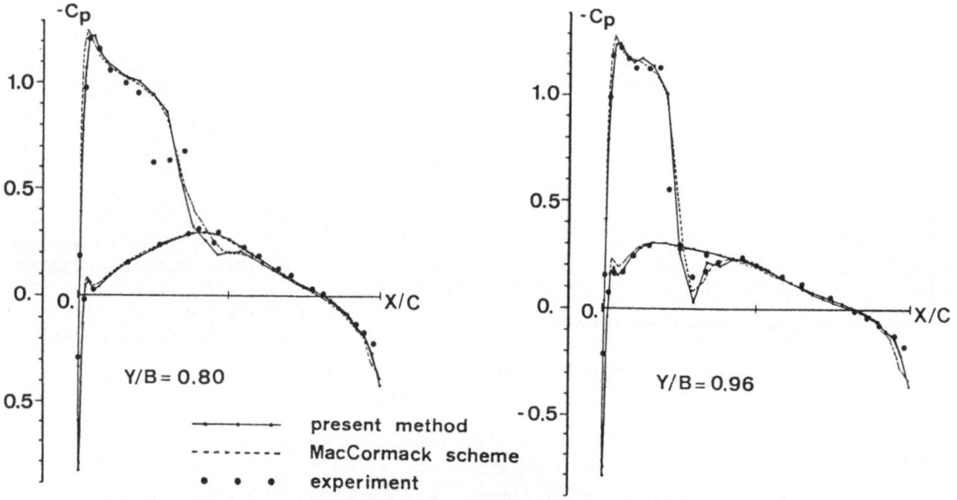

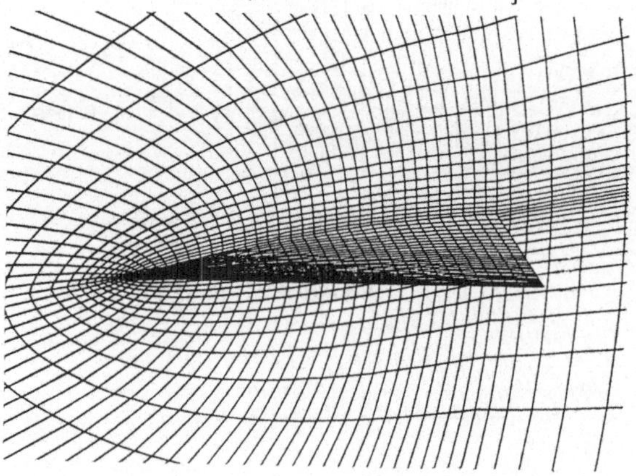

Fig. 6 DILLNER wing

C-0 mesh 45 X 57 X 17

DILLNER WING M = 0.7, α = 15°

a) X/C = 0.8

Wing coordinates

b) X/C = 1.15

free-stream coordinates

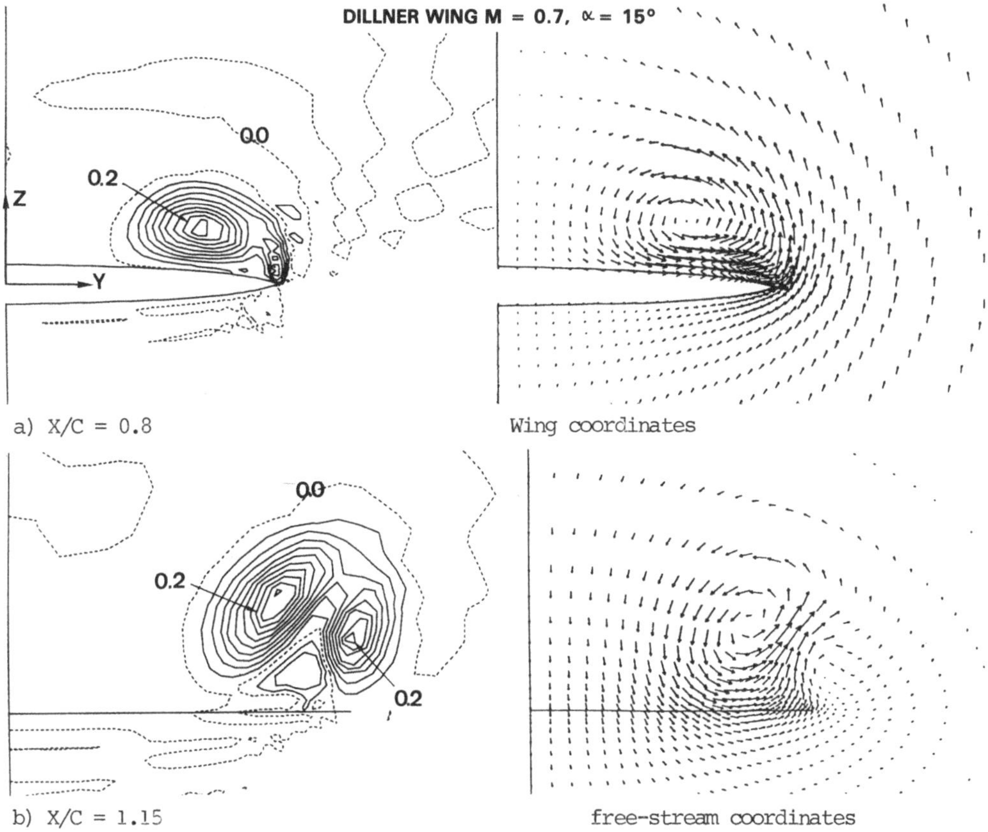

Fig. 7 Iso-total pressure loss and crossflow velocity field

DILLNER WING M = 0.7, α = 15°

-Cp

••• expérience
—— calcul

1.0

X/C = 0.8

Y/S

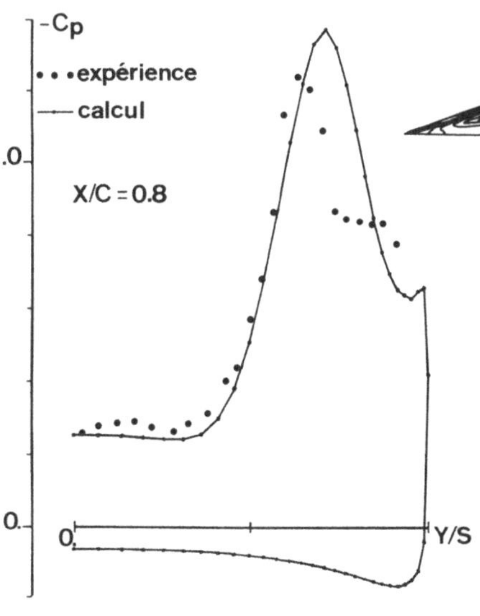

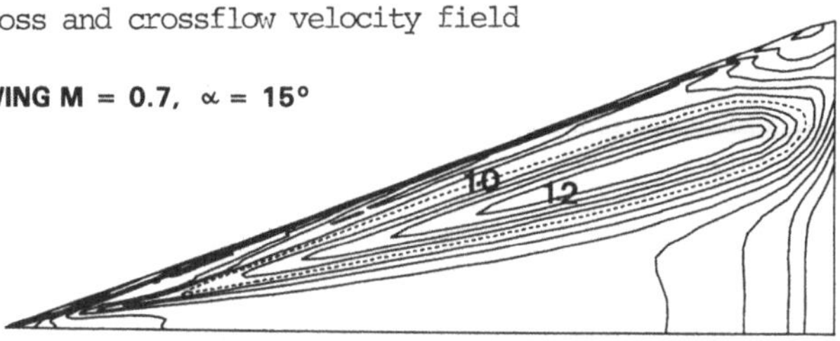

isomach on the upper-surface of the wing

x/c=0.8   x/c=1.15

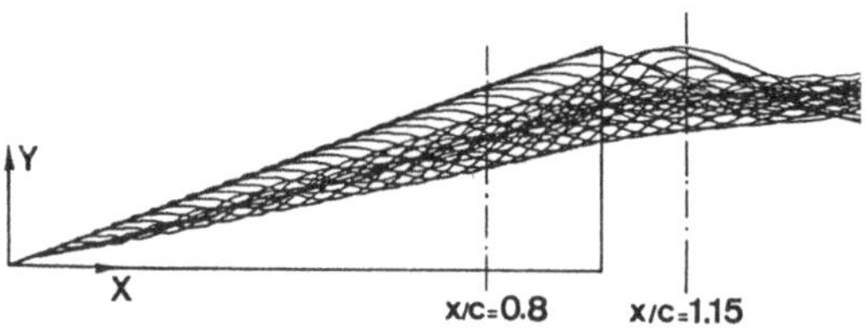

Fig. 8 Computed stream lines

Fig. 9 Cp distribution on the wing

A SPECTRAL ELEMENT METHOD APPLIED TO UNSTEADY FLOWS AT MODERATE REYNOLDS NUMBER

K. Z. Korczak and A. T. Patera
Department of Mechanical Engineering
Massachusetts Institute of Technology
Cambridge, MA 02139

ABSTRACT

The spectral element method is a high-order technique for solution of the incompressible Navier-Stokes equations which combines spectral expansions with finite element methodology to give high accuracy in general geometries. In the spectral element discretization, the computational domain is broken up into macro-elements, and the velocity and pressure in each element are represented as high-order Lagrangian interpolants. The nonlinear terms in the equation are then treated with explicit collocation, while the pressure and viscous contributions are handled implicitly with variational projection operators. Parallel static condensation applied to the implicit equations gives an operation count commensurate with that for low-order sub-structure techniques at the same resolution. A time-splitting technique is presented for solution of the Navier-Stokes equations, and results are given for linear and (three-dimensional) secondary spatial stability of plane Poiseuille flow, and for steady and unsteady separated channel flows at Reynolds numbers of several thousand.

INTRODUCTION

Spectral [1] and (low-order) finite element [2] methods represent the global and local limits, respectively, of weighted-residual techniques. Although high-order methods are clearly superior in sufficiently simple geometries, (e.g., a periodic channel), there has been no indication that they can be efficiently extended to handle the general geometries possible with low order techniques. In this paper, we present a general high-order technique for the Navier-Stokes equations, the spectral element method [3], [4], which does appear to be competitive with low-order methods if parallelism is effectively exploited.

The spectral element method is, in fact, nothing more than a high-order ("p-type") finite element technique, with projection operators and basis functions chosen so as to give an efficient and accurate solution. Inasmuch, the method relies heavily on existing weighted-residual (spectral and finite element) theory and practice. Similar techniques have been applied to classes of problems previously (e.g., the global element method for elliptic equations [5]) with apparently very good success.

In previous papers, the spectral element method was illustrated on one-dimensional wave, Poisson, and convection-diffusion equations [3], and applied to Navier-Stokes solutions of flow in a channel expansion [3] and oscillatory flow in a grooved channel [4]. We present here new results (at higher Reynolds numbers) for flows in regular and irregular (grooved) channels.

Section 1 Spatial Discretization - Passive Scalar

We consider here the equation for convection and diffusion of a (two-dimensional) passive scalar field $\theta(\vec{x},t)$, by an incompressible velocity field, $\vec{v}(\vec{x},t)$,

$$\frac{\partial\theta}{\partial t} + \nabla\cdot(\vec{v}\theta) = \alpha\nabla^2\theta . \tag{1}$$

The fact that the convective terms involve time-varying coefficients (and, furthermore, are nonlinear in the Navier-Stokes equations) suggests a semi-implicit treatment in time. In particular, we choose third-order Adams-Bashforth and Crank-Nicolson for the convective and diffusive pieces, respectively.

For the spatial discretization, the domain is broken up into elements, and the elements mapped into a square, $(x,y) \to (r,s)$, $-1 \leq r \leq 1$, $-1 \leq s \leq 1$. (Most work to date involves rectilinear elements, and correspondingly trivial mappings.) In a given element k, the field $\theta(r,s)$ is represented by a tensor-product basis,

$$\theta_N^k(r,s) = \sum_{i=0}^{N_r} \sum_{j=0}^{N_s} \theta_{ij}^k h_i(r) h_j(s) , \tag{2}$$

(N_r, $N_s \sim O(N)$), where the $h_i(z)$ are local Lagrangian interpolants through Chebyshev collocation points. For purposes of future discussion, we assume the domain is an M by M array of elements, each element containing (N+1) by (N+1) points.

To evaluate the convection terms we use standard collocation [1]. As in spectral methods, we choose collocation over Galerkin, as convolution sums are prohibitively expensive to evaluate (e.g., $O(N^6)$/element for Galerkin in nontrivial geometry vs. $O(N^3)$/element for collocation). Using collocation to evaluate the derivatives, the updated field will not be unique at points on element boundaries. The simplest way to proceed is to average the values from surrounding elements, which is, in fact, what the formal collocation operator gives. In Ref. [3] this is shown to result in exponential convergence and have satisfactory stability properties; however, the result is not necessarily conservative. A conservative spectral element scheme for solution of (conservative) hyperbolic systems can be constructed by averaging at element interfaces using the finite element distributed identity operator rather than the collocation procedure.

For solution of the (linear, time-independent coefficient) elliptic contributions of the form

$$\nabla^2\theta - \lambda^2\theta = f \tag{3}$$

which arise from the implicit step, collocation offers no advantages. A much better approach is to follow the standard finite element variational procedure for general Helmholtz equations. Recognizing that solution of (3) is equivalent to maximization of the functional

$$I = \int [-\frac{1}{2} \nabla\theta \cdot \nabla\theta - \frac{\lambda^2}{2}\theta^2 - \theta f] d\vec{x} \tag{4}$$

(with restriction to admissible variations), the elemental matrices are found by inserting (2) into the elemental version of (4) and requiring stationarity with respect to variations in the nodal values θ_{ij}^k. This process gives the elemental equations; the system equations are then obtained by direct stiffness (no patching is required).

Straight inversion of the system matrices is extremely inefficient due to the large bandwidth introduced by the high-order expansions. However, static condensation (also known as sub-structuring) [6] can be used to decouple the equations. In particular, consider the B-I (boundary-interior) node decomposition, where all points on element boundaries are in B, and all points in the interior of elements are in I. The static condensation algorithm exploits the fact that direct stiffness couple elements <u>only</u> through the equations for the $[{}^B\theta^k]$. Thus, once a "condensed" system is solved for the $[{}^B\theta^k]$, all the elemental calculations for the $[{}^I\theta^k]$ can be performed <u>in parallel</u>. The procedure is particularly effective when applied to high-order methods due to the larger number of internal degrees of freedom (i.e., I-points). In fact, parallel static condensation appears to make the spectral element method nearly as efficient as low-order finite element substructure solvers with similar resolution [6].

In Ref. [3], it is shown that for both hyperbolic and elliptic operators the spectral element method gives exponentially convergent solutions for sufficiently smooth problems. With respect to accuracy, we comment briefly on the choice of collocation points. The formulation (based on (4)) does not use a Chebyshev weighting, and in fact the choice of collocation points only enters into evaluation of the inhomogeneous terms, various polynomial representations being otherwise equal. Inasmuch, numerous point distributions are acceptable, and in fact Legendre might appear the best choice given the unity weighting implicit in (4). However, for truly general geometries using isoparametric elements, the point distribution is specified in (x,y), independent of the representation in (r,s), and therefore a strategy must be devised for correctly crowding the point distribution on complex curves in (x,y).

Section 2 Splitting Technique for the Navier-Stokes Equations

We consider here solution of the time-dependent incompressible Navier-Stokes equations,

$$\frac{\partial \vec{v}}{\partial t} = \vec{v} \times \vec{\omega} - \nabla\Pi + \frac{1}{R}\nabla^2\vec{v} \tag{5a}$$

$$\nabla\cdot\vec{v} = 0 \tag{5b}$$

in some domain D, with appropriate boundary conditions imposed on the boundary ∂D. Here $\vec{\omega}$ is the vorticity, Π is the dynamic pressure, and R is the Reynolds number, $R = \frac{UL}{\nu}$, where ν is the kinematic viscosity, and U and L are a characteristic velocity and length, respectively.

As for our model problem (1), we use a semi-implicit procedure, first updating the nonlinear terms explicitly using the third-order Adams-Bashforth scheme (we denote the result $\hat{\vec{v}}^{n+1}$). Note no boundary conditions are imposed at this point. We are then left with the Stokes problem, which is solved implicitly using a fractional-step method [7]. In particular, the problem is split into two steps, first a pressure step

$$\nabla^2\Pi = \nabla\cdot\left(\frac{\hat{\vec{v}}^{n+1}}{\Delta t}\right) \quad \text{in } D \qquad \frac{\partial\Pi}{\partial n} = 0 \text{ on } \partial D \tag{6a}$$

$$\frac{\hat{\hat{\vec{v}}}^{n+1} - \hat{\vec{v}}^{n+1}}{\Delta t} = -\nabla\Pi \tag{6b}$$

which imposes incompressibility, followed by a viscous step (Crank-Nicolson),

$$\left(\nabla^2 - \frac{2R}{\Delta t}\right)\vec{v}^{n+1/2} = \frac{-R}{\Delta t}\left(\hat{\hat{\vec{v}}}^{n+1} + \vec{v}^{n}\right) \quad \text{in } D \tag{7}$$

where $\vec{v}^{n+1/2} = \frac{1}{2}(\vec{v}^{n+1} + \vec{v}^{n})$. No-slip boundary conditions are imposed on the viscous step. The fully-discrete implementation follows directly from the model problems of the previous section. The non-linear terms are evaluated using collocation. The Poisson (6a) and Helmholtz (7) operators required in the implicit step are constructed as for (3), and solved using static condensation.

As always in incompressible problems, the pressure is known only to within an arbitrary constant. The corresponding solvability condition for the semi-discrete problem (6a) is simply

$$\int_D \nabla\cdot\hat{\vec{v}}^{n+1} d\vec{x} = \int_{\partial D} \hat{v}_n^{n+1} ds = 0 \quad \left(= \int_D \nabla^2\Pi d\vec{x} = \int_{\partial D} \frac{\partial\Pi}{\partial n} ds\right) \tag{8}$$

which is generally satisfied. It is simple to show that the fully discrete problem is also consistent if the conservative formulation indicated in the previous section is followed for evaluation of the divergence in (6a).

Section 3 Three-Dimensional Vibrating Ribbon Experiment

As a test problem for the spectral element method, we investigate spatial growth of infinitesimal disturbances in plane Poiseuille flow. In particular, we look at the stability of general (wavy) two-dimensional flow to (infinitesimal) three-dimensional disturbances,

$$\vec{v}(x,y,z,t) = (1 - y^2)\hat{x} + \vec{v}^{(2)}(x,y,t) + \varepsilon\vec{v}^{(3)}(x,y,z,t), \quad \varepsilon<<1 \qquad (9)$$

where the first term is the laminar parallel solution, $\vec{v}^{(2)}(x,y,t)$ is the specified two-dimensional non-parallel (wavy) flow, and $\vec{v}^{(3)}(x,y,z,t)$ is the three-dimensional disturbance,

$$\vec{v}^{(3)}(x,y,z,t) = \begin{Bmatrix} \tilde{u}(x,y,t)\cos\beta z \\ \tilde{v}(x,y,t)\cos\beta z \\ \tilde{w}(x,y,t)\sin\beta z \end{Bmatrix}. \qquad (10)$$

Here x(u), y(v), and z(w) are the streamwise, cross-stream, and spanwise directions (velocities) respectively, as shown in Fig. 1, where velocity is non-dimensionalized with respect to the laminar (parabolic) centerline velocity, U_c, and length scaled with the channel half-width, h. The resulting Reynolds number is $R = U_c h/\nu$. Note the form of the three-dimensional perturbation (10) follows from linearity ($\varepsilon<<1$) and separability and symmetry in z.

To obtain an equation for $\tilde{\vec{v}}(x,y,t)$, we insert (9), (10) into the three-dimensional Navier-Stokes equations, and neglect terms $O(\varepsilon^2)$ and higher. The boundary conditions on $\tilde{\vec{v}}(x,y,t)$ appropriate for study of spatial stability are

$$\tilde{\vec{v}}(x,y = \pm 1, t) = 0 \qquad (11a)$$

$$\tilde{\vec{v}}(x=0, y, t) = \mathcal{R}e\{\vec{v}_\ell^{(3)}(y)e^{-i\omega t}\} \qquad (11b)$$

$$\frac{\partial\tilde{\vec{v}}}{\partial x}(x=L, y, t) = 0 \qquad (11c)$$

where (11a) imposes no-slip on the channel walls, (11b) simulates the vibrating ribbon of experiment [8], and (11c) is the outflow condition imposed at the (computational) boundary $x = L$. We take the form of the inflow disturbance, $\vec{v}_\ell^{(3)}(y)$, from linear theory, $F(\vec{v}_\ell^{(3)}, \alpha^{(3)}; \omega, R, \beta) = 0$, where F represents the characteristic function of Orr-Sommerfeld equation. Here $\vec{v}_\ell^{(3)}(y)$ is the eigenfunction (appropriately symmetrized as in (10)), $\alpha^{(3)}$ is the complex spatial eigenvalue, ω the specified (real) frequency, and β the spanwise wavenumber.

The (linear) system for $\tilde{\vec{v}}(x,y,t)$ is very similar in form to the two-dimensional Navier-Stokes equations, and the numerical methods described in the previous two sections apply with little modification. The (x,y) computational domain is broken up into M elements, as shown in Fig. 2, the effective resolution in the x-direction being determined by M, the length of the computational domain, L, and the order of the elemental interpolant, N_x. (Note the technique is essentially spectral in y as implemented here; more recent tests break up the y-direction into multiple elements as well.) The time-stepping scheme used is an $O(\Delta t^2)$ Green's function boundary-divergence-free method [3].

The first case studied is parallel-flow linear theory (i.e., $\vec{v}^{(2)} \equiv 0$). Results at $R = 3000$, $\omega = .292$, $\beta = 1.0$ ($\alpha^{(3)} = .8258 + .03556i$) are shown in Fig. 2 for $\tilde{u}_m(x,t)$, where $\tilde{u}_m(x,t) = \tilde{u}(x,y_m^{(3)}, t)$, and $y_m^{(3)}$ is the location of the maximum of $|u_\ell^{(3)}|$. It it is seen that accuracy better than 1% is achieved with only six points/wavelength. The effect of the outflow boundary condition (11c) is virtually negligible; this is unfortunately not true if (11c) is applied to two-dimensional disturbances, as in this case the pressure effect upstream is much larger.

The second case studied is spatial secondary stability, corresponding to the controlled-transition (vibrating ribbon) experiments [8] performed in channels. Here we choose $\vec{v}^{(2)}$ to be non-zero. As to the exact form of $\vec{v}^{(2)}$, the best way to proceed is to solve for the finite-amplitude two-dimensional flow (in fact, equilibria exist for $R \gtrsim 2900$ [7]). However, the three-dimensional instability appears to be fairly insensitive to the exact form of the two-dimensional wave, and we therefore follow the simpler route of taking the form of $\vec{v}^{(2)}$ from linear theory

$$\vec{v}^{(2)} = A^{(2)} \Re e\{\vec{v}_\ell^{(2)}(y) e^{i(\alpha^{(2)}x - \omega t)}\} \qquad (12)$$

where $\vec{v}^{(2)}$, $\alpha^{(2)}$ are given by $F(\vec{v}_\ell^{(2)}, \alpha^{(2)}; \omega, R, \beta=0) = 0$. For our parameters ($R = 3000$, $\omega = .292$), $\alpha^{(2)} = 1.0018 + .02396i$. The amplitude, $A^{(2)}$, is chosen so that $\langle [u^{(2)}(x=0, y_m^{(2)}, t)]^2 \rangle^{1/2} = .07$, where $y_m^{(2)}$ is the location of the maximum of $|u_\ell^{(2)}(y)|$, and $\langle \cdot \rangle$ refers to time-averaging. In Fig. 3 we plot $\langle \tilde{u}_m^2 \rangle^{1/2}$ after the flow has reached an (approximately) asymptotic state. The three-dimensional disturbance (damped in Fig. 3) is now seen to grow exponentially on a convective time scale.

Section 4 Unsteady Grooved Channel Flow

In this section, we study two-dimensional flow in a periodically grooved channel; the geometry is shown in Fig. 4. The velocity is non-dimensionalized with respect to the mean velocity (in y and t), U, and lengths are scaled with the minimum channel height, H. The associated Reynolds number is $R = UH/\nu$. The results presented here are for the particular case of $h = 1/2$, $\ell = 3/4$, $L = 3/2$.

As regards the boundary conditions, the velocity $\vec{v}$ is taken to be zero on solid walls, and L-periodic in x. Consistent with these conditions, the pressure gradient, not the pressure, is taken to be L-periodic in x. To wit, we write $\Pi = \tilde{\Pi} - ax$, and require that $\tilde{\Pi}$ be L-periodic. The constant a indirectly determines the Reynolds number of the flow.

To solve the Navier-Stokes equations we use the time-splitting technique described in Section 2. The spectral element discretization is shown in Fig. 4; for the Reynolds numbers investigated here ($R \lesssim 1600$) converged results were obtained with $N \leqslant 6$ (i.e., a 7 by 7 grid in each element). For $R \lesssim 1125$ the solution is found to converge to a steady solution. The placement of the vortex center, concavity of the separating streamline, and thickness of the shear layer all agree with previous numerical simulations in similar geometries [9]. For $R \gtrsim 1125$, the steady solution undergoes a regular Hopf bifurcation to a steady-periodic flow. This behavior is demonstrated in Fig. 6 at $R = 1586$ by a sequence of plots indicating the streamline patterns during one full period. The frequency of the oscillations is found to be $\Omega = fH/U = .47$. Preliminary experimental work [10] verifies the bifurcation to oscillatory flow, and, in fact, indicates that transition to turbulence quickly ensues. Similar self-excited oscillations have been seen in numerical studies of flow in a channel expansions [11]. Currently, the analysis is being extended to include a (periodic) third direction (as in Section 3) to examine the stability of the steady and unsteady two-dimensional flow field to three-dimensional perturbations.

These results indicate the accurate solutions that can be obtained with moderate resolution for high Reynolds number flows in complex geometries. Together with the efficiency considerations given in Section 2, this suggests the possibility of high-order techniques proving useful not only in simple (model) geometries, but also in general configurations with complex resolution requirements.

ACKNOWLEDGEMENTS

This work was supported by a Rockwell International Assistant Professorship, and by the NSF under Grant MEA - 821249. The work was also supported in part by the ONR, contract N00014-81-C-0263 and the AFOSR, contract F49620-83-C-0064. A majority of computations were performed on a VAX 750 supplied in part by a DEC External Research Grant.

REFERENCES

1. D.O. Gottlieb and S.A. Orszag, "Numerical Analysis of Spectral Methods", SIAM, Philadelphia, 1977.

2. F. Thomasset, "Implementation of Finite Element Methods for Navier-Stokes Equations," Springer-Verlag, New York/Berlin, 1981.
3. A.T. Patera, J. Comput. Phys., 54, 1984.
4. N. Ghaddar, A.T. Patera, and B.B. Mikic, AIAA Paper No. 84-0495, 1984.
5. A. McKerrell, C. Phillips, and L.M. Delves, J. Comput. Phys., 40, 1981, 444.
6. L. Adams and R.G. Voigt, ICASE Nasa Contractor Report 172219, 1983.
7. S.A. Orszag and A.T. Patera, J. Fluid Mech., 128, 1983, 347.
8. M. Nishioka, S. Iida, and Y. Ichikawa, J. Fluid Mech., 72, 1975, 731.
9. U.B. Mehta and Z. Lavan, J. Appl. Mech., December, 1969.
10. M. Greiner, Ph.D. Thesis, Dept. of Mech. Engr., M.I.T., in progress.
11. G.A. Osswald, K.N. Ghia, and U. Ghia, in Proc. AIAA 6th Comp. Fluid Dyn. Conf., Danvers, 686 (1983).

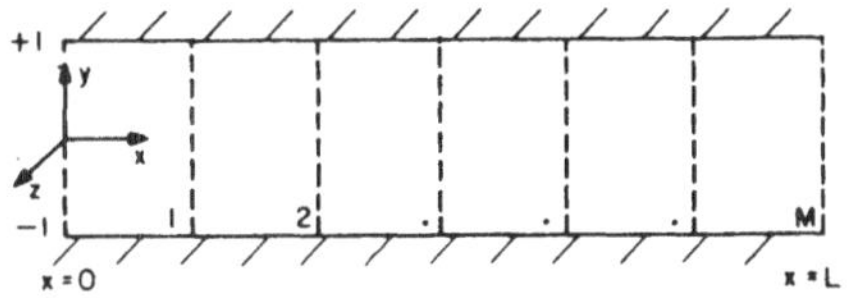

Fig. 1

Channel geometry for vibrating-ribbon experiments. Dashed lines indicate element boundaries.

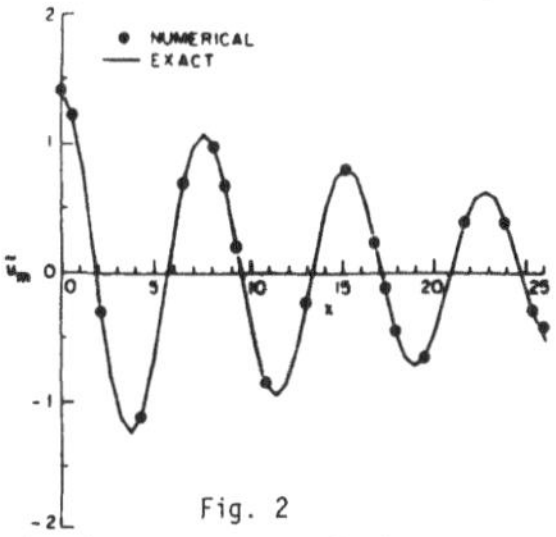

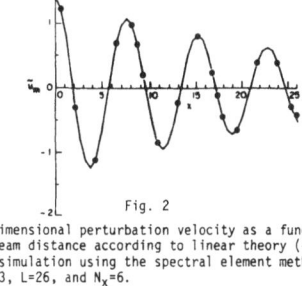

Fig. 2

Three-dimensional perturbation velocity as a function of downstream distance according to linear theory (—) and direct simulation using the spectral element method (•). Here M=3, L=26, and N_x=6.

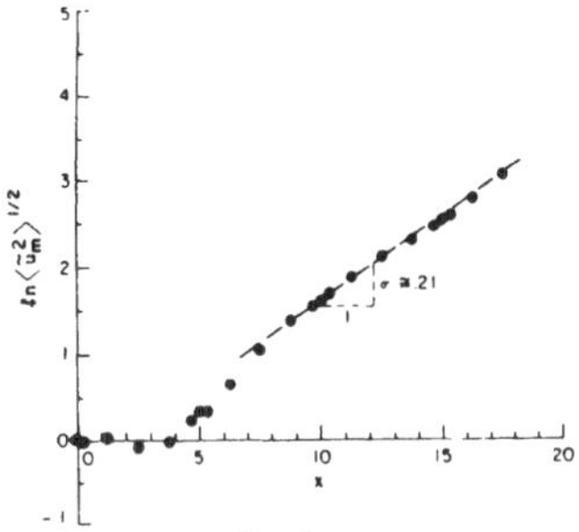

Fig. 3

Exponential growth of three-dimensional perturbation to a wavy two-dimensional flow. Growth rates compare well with previous temporal simulations. Here M=4, L=20, and N_x=6.

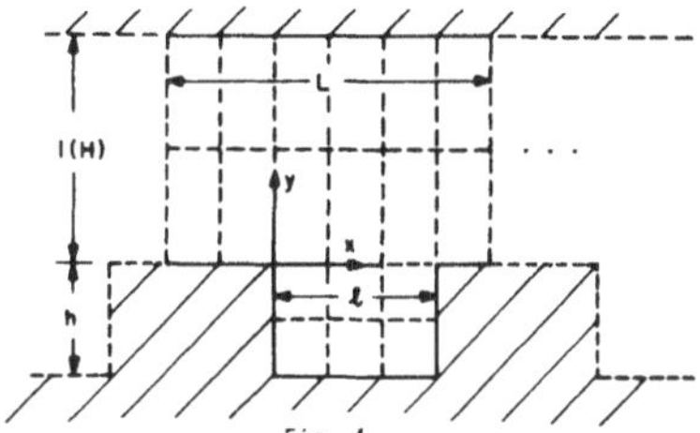

Fig. 4

Computational domain for grooved-channel simulations; dashed lines indicate element boundaries. Note the geometry is L-periodic in x.

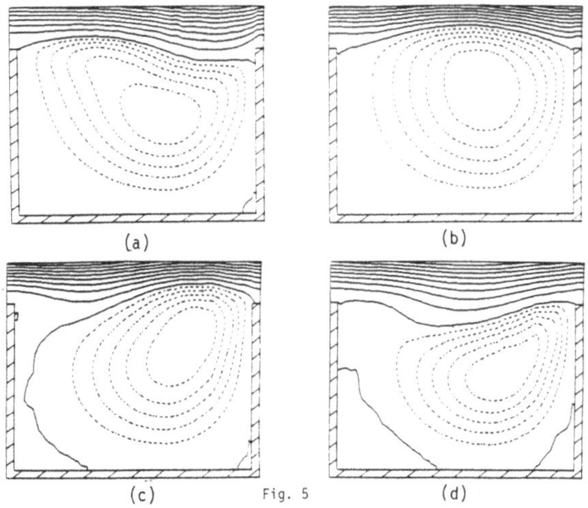

Fig. 5

Streamlines of the steady-periodic flow at R = 1586 at (a) t = 0, (b) t = T/4, (c) t = T/2, and (d) t = 3T/4, where T is the period of oscillation. The flow at t = T is virtually identical to that at t = 0. Only the flow in the vicinity of the groove is shown.

THE COMPUTATION OF THREE-DIMENSIONAL TRANSONIC VISCOUS FLOWS WITH SEPARATION

W. Kordulla
DFVLR-Institute for Theoretical Fluid Mechanics, Göttingen, FRG

INTRODUCTION

The investigation of three-dimensional flows with separation is rarely done on the basis of measurements of profiles of the governing flow quantities, and is often based on flow visualisation techniques (ref. 1). Moreover, surface flow visualisation techniques are preferred because of their relative simplicity. However, it is shown in (ref. 2) that there does not necessarily exist a unique relationship between the trace of the flow as observed on the surface of a body, and the structure of the flow field which encompasses that body. Therefore, in order to understand the structure of a flow and its impact on the aerodynamic behaviour of bodies, one must study the complete flow field. The topological structure of a flow with separation is characterized by singular points and the connections between them. Singular points are those locations within the flow field where the velocity is zero, and on surfaces where the shear stress vanishes, so that the local streamline direction is undetermined. Such locations are not easy to determine. The alternative to an experimental exploration of a flow field is to integrate numerically the governing equations of fluid flow, loosely called Navier-Stokes equations. Owing to the rapid development of both, numerical methods and supercomputers, in recent years, this alternative has become a reasonable means to simulate high Reynolds number viscous flows past simple three-dimensional configurations. In spite of the difficulties associated with modeling physics, in particular transition and turbulence, the success obtained in many cases is encouraging, see for example (ref. 3). However, in order to carry out meaningful investigations a large number of grid points is believed to be necessary, which cannot be handled in core of today's supercomputers. In (ref. 3), for example, 216000 points are used to simulate the flow past a body with jet using symmetry assumptions. Hence, either large in-core memories or efficient input/output devices are needed.

This paper reports on the progress in extending and applying the explicit-implicit MacCormack method (ref. 4) to three-dimensional flows on a CRAY/1S using large data bases. Due to the lack of space the scheme will only be sketched, see (refs.5,6,7). The method is used to simulate the laminar, transonic flow field past the geometrically simple hemisphere-cylinder configuration, investigated experimentally in (ref. 8) and numerically in e.g. (refs. 7,9,10) using coarse meshes. The ability to use fine meshes will allow to investigate the topology of such flows as a function of, e.g., angle of attack.

NUMERICAL METHOD

The present method is a finite-volume formulation of the explicit-implicit MacCormack scheme (ref. 4), see (refs. 5,6). The scheme is based on MacCormack's well-proven, second-order accurate, original explicit predictor corrector version of the Lax-Wendroff method. A bi-diagonal implicit procedure is incorporated into the predictor corrector sequences in order to get rid of the explicit stability condition on the time step, which is particularly restrictive in turbulent flows. While the explicit scheme approximates the integral form of the Navier-Stokes equations, the implicit scheme can best be viewed as an approximation to the differential equation, as obtained by differentiating the conservative partial differential form of the governing equations. However, the equations used are essentially based on the Euler terms , because the employed eigenvalues involve the absolute values of the Eulerian ones. Absolute values are taken because of stability reasons. The viscous terms are accounted for by adding the dominant corresponding eigenvalue. Note that MacCormack has shown for a scalar model equation that the implicit operation is a higher-order small perturbation of the explicit solution. The advantage of the bi-diagonal scheme versus fully implicit methods is the ability to skip the implicit step locally whenever the explicit stability conditions are fulfilled. To enhance stability a smooth blending of operations is performed at the boundary of purely explicit and purely implicit regions. Also, additional second-order numerical diffusion terms with variable coefficients are added in both, explicit and implicit, portions of the scheme. As in (ref. 5) the wall flux resulting from the implicit sweep towards the wall is reflected immediately, and not in the next step of the predictor corrector sequence. This procedure has not been presented previously:

$$(I+{}^3C_k^n)\delta U_k^+ = \Delta U_k + {}^3C_{k+1}^n \delta U_{k+1}^+, \quad k = k_{max}-1(-1)2,$$

$$(I+{}^3C_k^n)\delta U_k^{++} = {}^3C_{k-1}^n \delta U_{k-1}^{++}, \quad k = 2(1)k_{exp},$$

$$\delta U_k = \delta U_k^+ + \delta U_k^{++}, \quad k = 2(1)k_{exp},$$

$$\delta U_k = \delta U_k^+, \quad k = k_{exp.}+1(1)k_{max}-1.$$

Here, k is the mesh index for the wall (near-) normal direction, δU_k is the implicitly obtained change $\Delta U/\Delta t$, ΔU_k the explicit-implicit increment of the solution vector, and n indicates the known solution at the old time level. Hence the solution is being corrected for the effect of the wall flux. In order to be efficient it is necessary that the solution becomes explicit at some location k_{exp} not too far away from the wall. For a discussion of the implementation of the viscous terms the reader is referred to (refs. 6,11). As boundary conditions free stream conditions are assumed on the far field hull around the body, and extrapolation is used at the downstream exit plane. Numerical experiments are currently done based on quasi-one-dimensional characteristic variables.

RESULTS

The explicit-implicit scheme, which does not make use of the commonly employed thin-layer Navier-Stokes assumption, can be vectorised easily for use on a CRAY/1S computer (ref. 7). In (ref. 7) results were discussed for the hemisphere-cylinder configuration based on a coarse grid with 31×20×31 mesh cells. The corresponding computations could be performed in-core. In order to improve the resolution of the flow field a finer grid with 64×38×40 (i×j×k) cells, see Figures 1 and 2, is being used here. The nearly 85 thousand cells required to move data in and out of core, resulting in recoding the scheme in part. While the waiting time for input/output (I/O) is virtually zero for the in-core version, its ratio with respect to CPU-time ranges from 3 to 13, depending on the load of the computer. Note that the CRAY/1S (COS 1.12, CFT 1.11) used does not exhibit any dedicated I/O device such as a solid state device (SSD). The unfavourable ratio was obtained in spite of the following I/O strategy: use of random direct access files, unblocked buffered I/O and use of multiple parallel I/O streams. The typical CPU time per cell and time step is $8 \cdot 10^{-5}$ seconds which is comparable with the time for the in-core version, if one considers that the new version got rid of all subroutines which are not needed for the computation itself. The effect of the larger vector lengths seems to more than compensate the effect of the increase of the number of subroutine calls which is quite dramatic, because only four data planes are in-core at all times (this number being flexible). This plane concept, see (ref. 12), was preferred to the pencil concept (refs. 3,10) because the latter is well-suited for the thin-layer Navier-Stokes approximation but less efficient if the turbulent viscosity is needed for the solution sweeps in all three spatial directions. The pencil concept would then require at least one extra sweep through the data base per time step.

The flows chosen have the free stream conditions $M_\infty = 0.9$ and Re = 225000 with α ranging from 0° to 19°. Figure 2 shows a view of a typical computational surface j = constant (see Fig. 1) of the grid, which is twice as fine in the circumferential direction and employs twice as many cells in the axial direction as compared with the grid in (ref. 7). The first step size away from the wall is 0.00005 times the radius of the sphere. The downstream boundary is located at 10 radii, because a larger distance with a fine mesh would increase the computation time further. Laminar flow has been assumed for all cases, which is certainly valid on the nose portion on the body. Hence, due to the clustering of the grid cells near the wall and the fine resolution on the hemisphere good results can be expected there, as compared with experiment. Future investigations, however, should include some criterion for the location of transition, and use some turbulence modeling.

It is first noted that for $\alpha = 5°$ the results with the finer mesh confirm those obtained with the coarse grid in (ref. 7). The pressure distributions in windward, horizontal and leeward planes agree better with experimentally observed values, although the expansion continues too far downstream in the leeward plane, and the windward Cp distribution indicates a recompression near the hemisphere-cylinder junction. For $\alpha = 10°$ and 19° Figure 3 displays the predicted and measured Cp distributions in the leeward, horizontal and windward planes. Good agreement is achieved in spite of the laminar-flow assumption which may be responsable for the too large expansion region in the leeward plane. For both angles of attack Figure 4 compares computer plots with lines of constant Machnumber, with the sonic line denoted by crosses, and shadowgraph pictures (courtesy of T. Hsieh) of the flow in the symmetry plane. The shear layers coming off the nose in the leeward plane as well as the shock or no-shock situation in the windward plane are well simulated. It is interesting that the velocity profiles in these planes indicate that even for $\alpha = 19°$ the flow in the windward plane is still reversed with respect to the main flow direction within an extremely thin and very short layer. The complete evaluation of the 3-D flow field data, which is a non-trivial task, will allow to draw conclusions with respect to the topology of the above flows and its dependence on the angle of attack. The available information in form of lines of constant pressure, density and Machnumber in surfaces i, j and k = constant is not conclusive, but indicates interesting shapes of e.g. the sonic surface. Note that for $M_\infty = 1.2$ and $\alpha = 19°$ the numerical simulation yields excellent results concerning Cp distributions in windward, horizontal and leeward planes.

CONCLUDING REMARKS

The understanding of three-dimensional flow with separation can be enhanced by considering its topology, which requires a fine resolution of the flow. It has been indicated that the efficiency of the computation expressed in terms of computation time per cell and time step does not deteriorate, if the simulation of the flow cannot be done in core of a CRAY/1S. The lack of efficient input/output devices, however, results in unreasonably long turn-around times. The explicit-implicit MacCormack method has been shown to provide a useful numerical tool for simulating three-dimensional flows with separation, because the transonic flow past a hemisphere-cylinder at various angles of attack is well represented. It is assumed that the numerical quality (see jitters in Fig. 4) of the solution will improve when moving the downstream boundary conditions to 20 radii and refining the mesh in that and the radial direction. In order to reduce the turn-around time it will be attempted to improve the convergence and reduce the computation time by using characteristic-type boundary conditions and locally variable time steps. The investigation of the topological structure of flows requires more work with respect to the reduction of numerical data sets. More work is also needed with respect to criterions for transition and to turbulence modeling.

ACKNOWLEDGEMENTS

The author thanks T. Hsieh for the provision of his data in tabular form and for the shadowgraph pictures.

REFERENCES

1. Peake, D.J., and Tobak, M., 1980. Three-Dimensional Interactions and Vortical Flows with Emphasis on High Speeds. AGARDograph No. AG-252.

2. Dallmann, U., 1983. Topological Structures of Three-Dimensional Vortex Flow Separation. AIAA paper 83-1735.

3. Deiwert, G.S., 1983. Three-Dimensional Flow Over a Conical Afterbody Containing a Centered Propulsive Jet: A Numerical Simulation. AiAA paper 83-1709.

4. MacCormack, R.W., 1982. A Numerical Method for Solving the Equations of Compressible Viscous Flow. AIAA Journal, Vol. 20, pp 1275-1281.

5. Kordulla, W., and MacCormack, R.W., 1982. Transonic-Flow Computations Using an Explicit-Implicit Method. Proceedings, 8th Int. Conf. Numerical Methods in Fluid Dynamics. Lecture Notes in Physics, Vol. 170, Springer Verlag, pp 420-426.

6. Hung, C.M., and Kordulla, W., 1983. A Time-Split Finite-Volume Algorithm for Three-Dimensional Flow-Field Simulation. AIAA paper 83-1957.

7. Kordulla, W., 1983. The Computation of Three-Dimensional Transonic Flows With an Explicit-Implicit Method. Proceedings, 5th GAMM-Conf. Numerical Methods in Fluid Dynamics. Notes On Numerical Fluid Mechanics, Vol. 7, Vieweg Verlag, Wiesbaden, pp. 193-202.

8. Hsieh, T., 1976. An Investigation of Separated Flow About a Hemisphere-Cylinder at 0- to 90-Deg Incidence in the Mach Number Range From 0.6 to 1.5. AEDC-TR-76-112.

9. Pulliam, T.H., and Steger, J.L., 1978. On Implicit Finite- Difference Simulations of Three Dimensional Flow. AIAA paper 78-10.

10. Pulliam, T.H., and Lomax, H., 1979. Simulation of Three- Dimensional Compressible Viscous Flow on the Illiac IV Computer. AIAA paper 79-0206.

11. Kordulla, W., 1984. Three-Dimensional Viscous-Flow Simulations Based on Finite-Volume Formulations. DFVLR IB 221-84 A , Goettingen.

12. Shang, J.S., Buning, P.G., Hankey, W.L., and Wirth, M.C., 1979. The Performance of a Vectorized 3-D Navier-Stokes Code on the Cray-1 Computer. AIAA paper 79-1448.

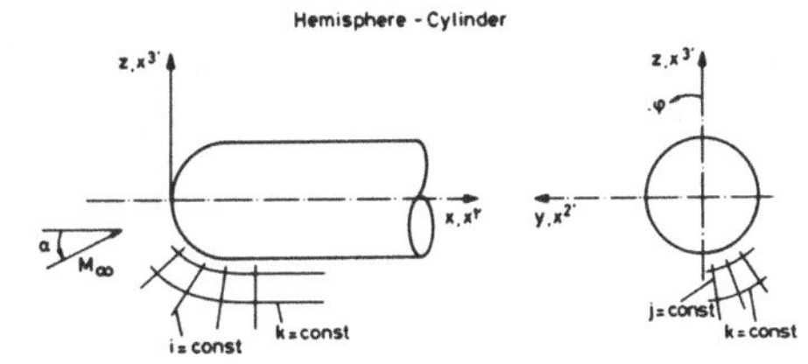

Fig. 1: Sketch of the hemisphere-cylinder configuration

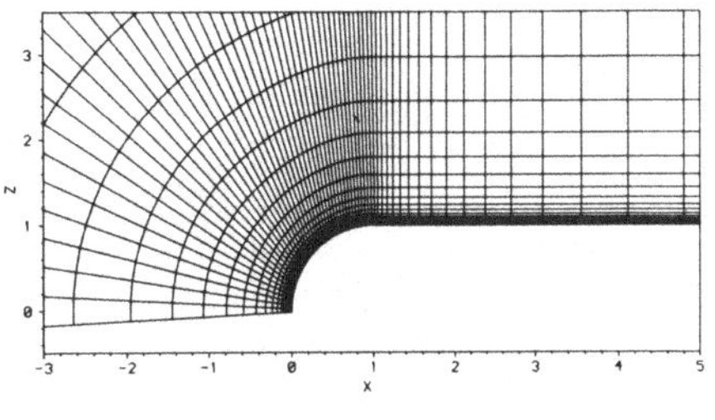

Fig. 2: Typical grid in surface j = constant (volume centers)

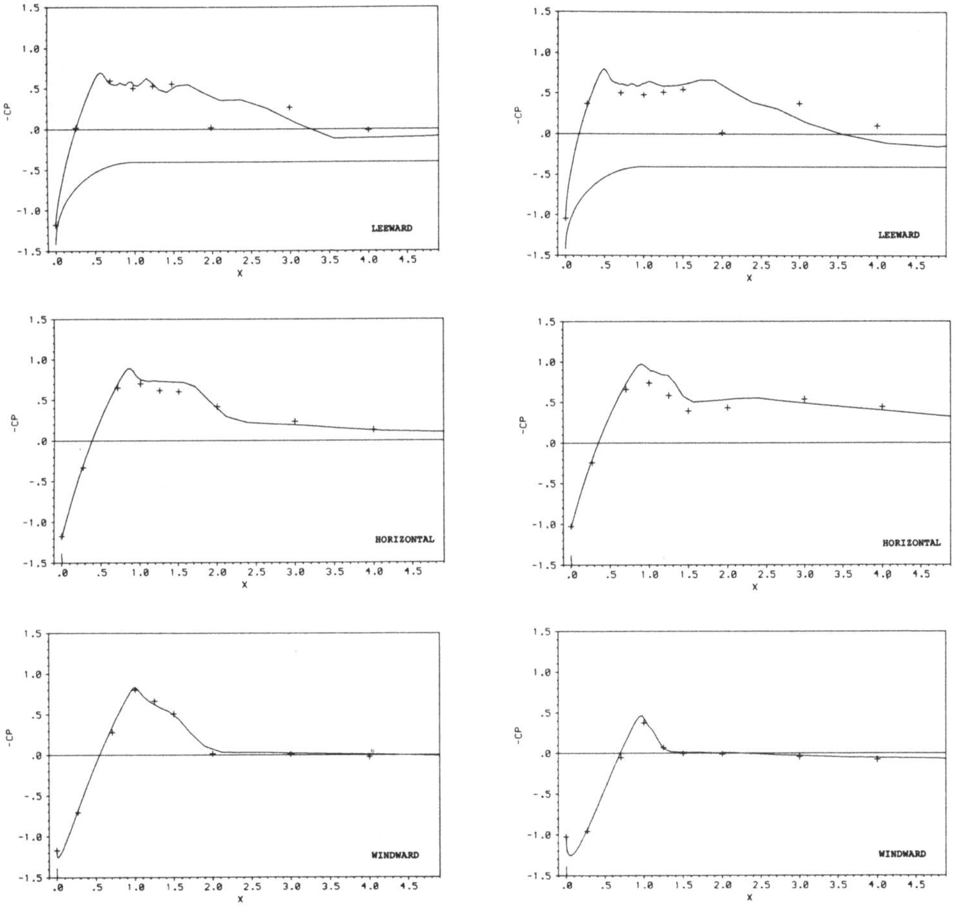

Fig. 3: Cp-distributions for $\alpha = 10°$ (left) and $\alpha = 19°$, $M_\infty = 0.9$, Re = 225 000 (—— Prediction, ++++ Measurement).

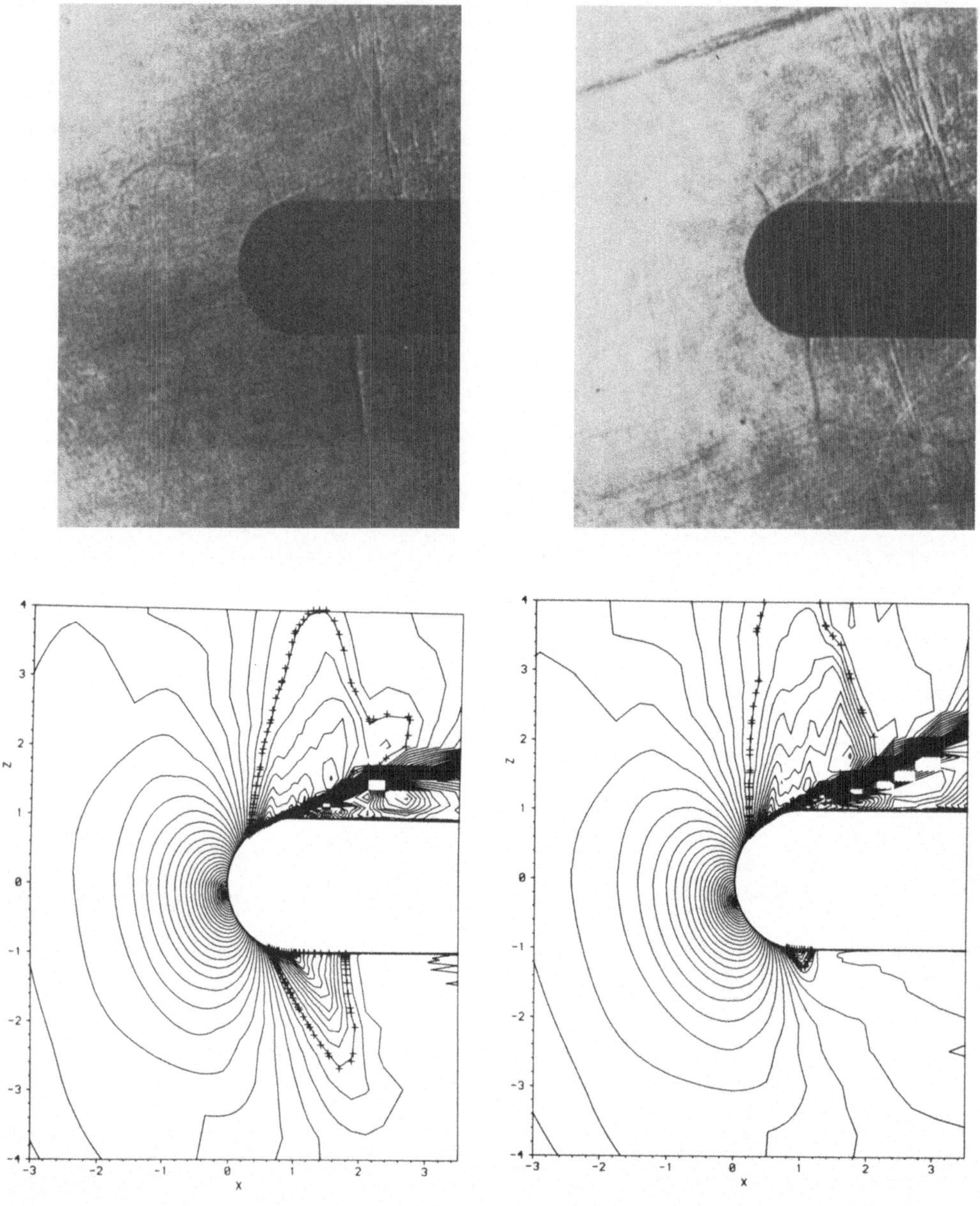

Fig. 4: Comparison of shadowgraph pictures and plots of lines of constant Machnumber for $\alpha = 10°$ (left) and $\alpha = 19°$, $M_\infty = 0.9$, Re = 225 000 (++++ sonic line).

A NUMERICAL METHOD OF SOLUTION FOR THE KELVIN-NEUMANN PROBLEM

C. Korving
Delft University of Technology
Department of Mathematics
The Netherlands

SUMMARY

A numerical procedure is described for the steady, inviscid flow of a moving disturbance at the free surface. The method is valid for application in the case of a moving pressure distribution at the free surface. The solution is obtained by collocation where Laplace's equation is transformed into a system of differential equations.
In the diagonal dominant structure one equation can be distinguished of the hyperbolic type and the other equations of the elliptic type.
The result involves a simplification for handling the radiation condition in a finite domain of calculation around the pressure distribution.
Computational results are presented, clearly demonstrating the Kelvin wave pattern downstream of the disturbance at the free surface.
An extension of the method to the case of a body at the free surface is currently being studied.

INTRODUCTION

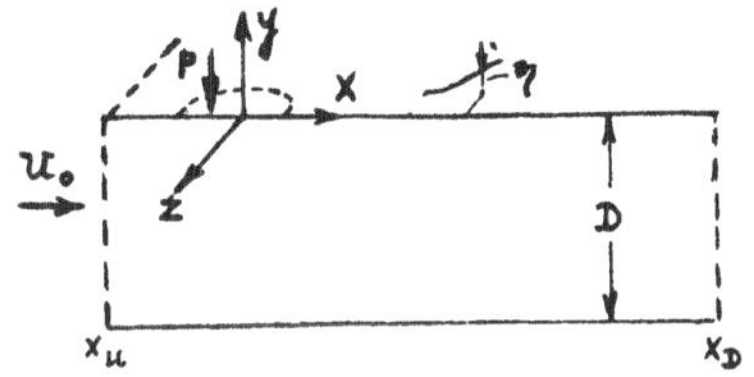

The problem of a moving pressure distribution at the free surface can be considered as steady with respect to a reference frame fixed to the pressure distribution.

Assuming frictionless, irrotational flow, the problem is governed by Laplace's equation for the velocity potential Φ. The dynamic and kinematic free surface conditions are respectively:

$$\eta + \frac{1}{2}F^2 \; \nabla\Phi \cdot \nabla\Phi = \frac{1}{2}F^2 - p^*$$

$$-\Phi_y + \Phi_x \eta_x + \Phi_z \eta_z = 0$$

The variables are written down in dimensionless form:

$x^* = \frac{x}{D}$, $p^* = \frac{p}{\rho g D}$, $\eta^* = \frac{\eta}{D}$ etc. Except for p^*, the star is omitted in the equations.

The Froude number is defined with respect to the depth D: $F = U_0/\sqrt{gD}$.

The addition of the conditions: undisturbed flow upstream, the radiation condition downstream and the bottom condition completes the formulation of the problem under consideration.

In Ref. [1] the method of solution is based on successive iteration of the computation of the free surface elevation on one hand, using the results and reversely obtained by the solution of the Neumann problem with fixed boundaries on the other hand. The iteration, however, has poor convergence, and even sometimes, the reason is not clear, there is no convergence at all. The method, presented in this paper, is based on iteration with respect to equations, following from collocation of Laplace's equation in combination with linearized free surface conditions.

The nonlinear problem is solved, if the formulation in curvilinear coordinates is used and Newton iteration is applied to the nonlinear system of equations.

The two-dimensional representation leads to a system of uncoupled equations. The solution is easily obtained, since the waves are found from one equation, involving a simplification in prescribing the boundary conditions.

The same holds for the three-dimensional case, where iteration is needed, because of the interaction of the equations in the cross-direction of flow.

3 METHOD OF SOLUTION

After the introduction of the perturbation potential $\phi = \Phi - x$ and the assumption of a small elevation of the free surface, the free surface conditions can be linearized, resulting into the condition:

$$\phi_y + F^2 \phi_{xx} = - p_x^*$$

In a subdomain j between the vertical coordinates $y = y_j$ and $y = y_{j+1}$ the potential function $\phi(x,y,z)$ is approximated by the expression:

$$\phi(x,y,z) = \sum_{i=0}^{1} \left[l_i(y)\, \phi^{j+i}(x,z) + m_i(y)\, \phi_y^{j+i}(x,z) \right]$$

This expression is valid for each subdomain j between the bottom $y = y_1$ and the undisturbed free surface $y = y_n$: $y_j \le y \le y_{j+1}$, $j = 1(1)\,n-1$. The polynomials l_i and m_i are determined by the condition that ϕ must be continuous up to and including the first derivative with respect to y (Hermitian approach).

A system of differential equations is obtained (with $\phi^j(x,z)$ as dependent variables), if Laplace's equation is satisfied at all collocation levels $y = y_j$. Adding the free surface condition and the bottom condition to this system, we obtain n equations in the n unknowns ϕ^l:

$$Z_{ml}\, \phi^l + I_{ml}\, \phi^l_{xx} + U_{ml}\, \phi^l_{zz} = Q_m \qquad m,l = 1(1)\,n \qquad (3.1)$$

Z_{ml} is a full matrix, I_{ml} is a unit matrix and U_{ml} is a diagonal matrix with one column of non-zero elements; Q_m contains p_x^*.
In order to get a diagonal dominant structure of the equations, the eigenvalues λ_l and eigenvectors E_k^l are determined from the matrix Z_{ml}.
Using the transformation $\psi^l = E_k^l \psi^k$ and multiplying the system of equations with the inverse matrix of eigenvectors, we arrive at a system of differential equations of the form:

$$\lambda_{(k)} \delta_{(k)l} \psi^l + \delta_{kl} \psi^l_{xx} + U^*_{kl} \psi^l_{zz} = Q_k^* \qquad l,k = 1(1)n \tag{3.2}$$

Calculations show a regular structure in the sign of the coefficients of the diagonal terms.
The system turns out to be of the following form:

$$\begin{aligned} \psi^1_{xx} + \alpha_1^2 \psi^1_{zz} &= Q_1^* + f_1 (\psi^2_{zz}, \psi^3_{zz}, \ldots .) \\ \beta_2^2 \psi^2 + \psi^2_{xx} - \alpha_2^2 \psi^2_{zz} &= Q_2^* + f_2 (\psi^1_{zz}, \psi^2_{zz}, \ldots .) \\ -\beta_j^2 \psi^j + \psi^j_{xx} + \alpha_j^2 \psi^j_{zz} &= Q_j^* + f_j (\psi^1_{zz}, \psi^2_{zz}, \ldots .) \qquad j = 3(1)n \end{aligned} \tag{3.3}$$

The first equation, belonging to the eigenvalue equal to zero, contains an arbitrary constant in the solution, which is in agreement with the original formulation of the problem. The second equation is of the hyperbolic type and the other equations are of the elliptic type.

The two-dimensional case leads to a system of uncoupled equations. So, the z-derivative terms don't occur in above system of equations. The frequency of the waves is known from the dispersion relation:

$$F^2 \nu = \tanh \nu$$

The frequency of the waves in above approximation follows from the second equation. The result is:

$$n = 2 \qquad \beta_2^2 = \frac{12(1 - F^2)}{4F^2 - 1} \qquad \frac{1}{4} < F^2 < 1$$

$$n = 3 \qquad \beta_2^2 = \frac{6(3\sqrt{P^2 + 16 F^4} - 5P)}{7 F^2 - 1}, \quad P = 4F^2 - 1, \quad \frac{1}{7} < F^2 < 1$$

The table below shows the differencies between the frequencies for several values of Froude.

	$F^2 = 1/4$	$F^2 = 1/2$	$F^2 = 3/4$
β_2 (n = 2)	—	2.45	1.22
β_2 (n = 3)	4.9	2.03	1.07
ν	4.0	2.00	1.05

The results show good agreement between β_2 and ν, if the number of collocation levels grows from n = 2 to n = 3. For small values of Froude, above example is a bad approximation, either with respect to the linearization of the free surface conditions (U_0 small), or because of the large depth being considered (Froude number is defined at the depth of the flow region). Is $F^2 \leq 1/4$, than it is impossible to calculate the waves for n = 2 and an unaccurate result is obtained for n = 3. A higher number of collocation levels is needed in this case.

The boundary conditions follow from the undisturbed flow upstream and the finiteness of the perturbation potential downstream of the pressure distribution.

The boundary conditions, expressed into the variables ψ^j, are at the upstream position:

$$\psi_x^j = 0 \quad j = 1(1)n; \quad \sum_{j=1}^{n} a_j \psi^j = 0 \quad \text{for equation 2 of (3.3)}$$

and at a downstream position:

$$\psi^j = \text{finite} \qquad j = 3(1)n.$$

It is remarked that the hyperbolic equation specifies a boundary condition containing the variables of the other equations. An analytical example shows that this condition must be replaced by the condition $\psi^2 = 0$. It involves small deviations of the free surface in the upstream region, nearly independent of the upstream position. This is not the case, when the formal condition is prescribed.

The example also shows that the downstream boundary condition may be replaced by $\psi^j = 0 \quad j = 3(1)n$. The further the downstream station is chosen away behind the pressure distribution, the smaller the error made in the free surface elevation.

In this formulation the solution of the linear two-dimensional problem is easily obtained.

The three-dimensional problem is more complex, because of the implicit structure of the equations (3.2) as a result of the coupling in the z-derivative terms. The matrix U^*_{kl} turns out to be only diagonal dominant, as far as the elliptic equations in the complete system (3.2) are concerned.

The hyperbolic equation can be made diagonal dominant with a suitable choice of the characteristic directions, determined by the coefficient α_2^2. The remaining part in relation to the z-derivative terms is expressed into the variables ϕ^j and the coefficient α_2 is calculated from the condition that the potential function of the free surface ϕ^n vanishes in the right-hand side of the hyperbolic equation. The variable ψ^2 mainly contains ϕ^n and in this way the diagonal dominant structure is achieved for the complete system (3.3).

The solution of (3.3) is obtained by a process of simultaneous iteration. The hyperbolic equation is solved by an implicit difference scheme and the elliptic equations with boundary conditions of the Dirichlet / Neumann type are solved by the Galerkin finite element method.

The boundary conditions of the two-dimensional case are supplied with the reflection condition at the boundaries parallel to the main stream direction.

RESULTS AND CONCLUSIONS

In Fig. 1 the results are presented of a two-dimensional case, clearly demonstrating the difference between the linear and nonlinear solution. The results are in agreement with the results presented in the References 1 and 2, but the solution is obtained by a much quicker and well founded method. Near the crest of the waves, the nonlinear solution turns out to be unstable, comparable to the real physical situation. This phenomenon does not affect the wave-resistance of the pressure distribution. It is a local effect and it is easily suppressed during the calculation.

A three-dimensional result on the basis of linearized free surface conditions is presented in Fig. 2, clearly showing the Kelvin wave pattern in downstream direction of the pressure distribution. The angle within the disturbances propagate, is sensitive with respect to the number of collocation levels, but the influence on the wave-resistance is small. If the number of collocation levels is increased, the coefficient of the diagonal dominant z-derivative term in the hyperbolic equation tends to a constant value. This value agrees with the angle of about 20°, within the Kelvin wave pattern is found for water of sufficient depth. The result of Fig. 2, carried out for $n = 4$, takes about 20 seconds on an Amdahl 470 V/7-B computer.

The method is currently extended to the case of a moving body at the free surface.

REFERENCES

1) Korving C., A Numerical Method for the Wave Resistance of a Moving Pressure Distribution on the Free Surface, Proceedings, Seventh I.C. on N.M. in F.D. 1980.

2) Kerczek, C. von and Salvesen, N., Nonlinear Free Surface Effects, The Dependence on Froude Number, Second I.C. on Num. Ship Hyd., Un. of Cal. Berkeley, 1977.

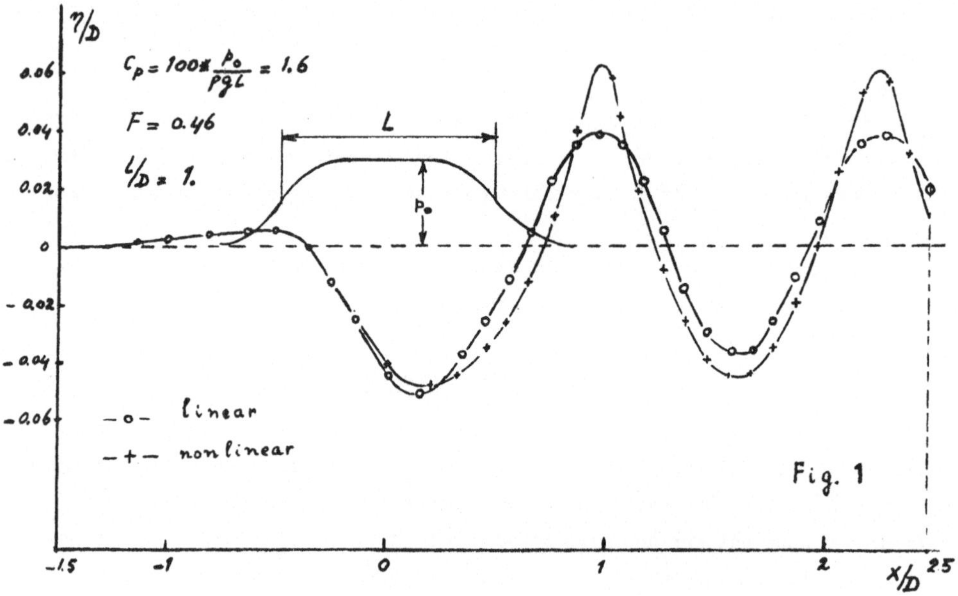

η/D
$C_p = 100 * \frac{p_0}{\rho g L} = 1.6$
F = 0.46
L/D = 1.
L
p_0
0.06
0.04
0.02
0
- 0.02
- 0.04
- 0.06
-o- linear
-+- nonlinear
-1.5
-1
0
1
2
2.5
x/D

Fig. 1

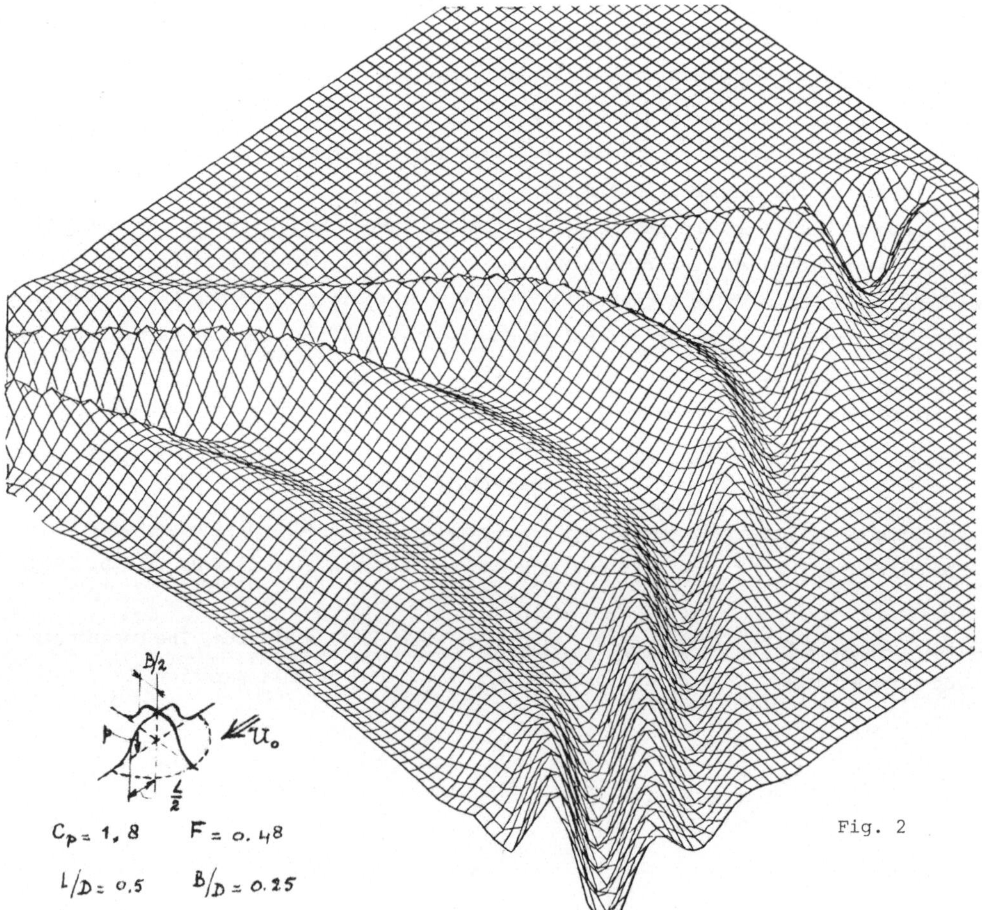

B/2
U_0
L/2
C_p = 1, 8
F = 0.48
L/D = 0.5
B/D = 0.25

Fig. 2

NUMERICAL SOLUTION OF UNSTEADY TRANSONIC FLOWS PAST THIN PROFILES

K.Kozel,M.Vavřincová

Dept.of Applied Mathem.

Faculty of Mech.Engineering

Technical University Prague

The work deals with a method of numerical solution of two-dimensional unsteady transonic flows past oscillating thin profile in a channel with $U_\infty=$ const. or $U_\infty = U_\infty(t)$.The problem is described by small disturbance equation and solved by finite difference method using a semiimplicit difference scheme.

Numerical results of transonic flows past thin non-symmetrical biconvex profile oscillating with a low frequency in a channel computed by presented method are compared with results computed by method of Ballhaus and Goorjian. Results of numerical simulation of flowfield properties for several values of reduced frequency $k \in (0,10)$ are presented also.

I. Formulation of the problem

The governing equation is small disturbance potential equation in the form (see [4])

$$A\varphi_{tt} + 2B\varphi_{xt} + C\varphi_x + D\varphi_t + E = V\varphi_{xx} + \varphi_{\tilde{y}\tilde{y}} , \qquad (1)$$

where

$A = k^2 M_\infty^2 \delta^{-2/3}$, $B = k M_\infty^2 \delta^{-2/3}$, $C = B[1 \;\; \frac{1}{2}(\varkappa-1)M_\infty^2][1+\frac{\varkappa-1}{2}M_\infty^2]^{-1} dM_\infty/dt$,

$D = -A\frac{\varkappa-1}{2}M_\infty^2[1+\frac{\varkappa-1}{2}M_\infty^2]^{-1} dM_\infty/dt$, $E = 2kM_\infty^2\delta^{-4/3}[1+\frac{\varkappa-1}{2}M_\infty^2]^{-1} dM_\infty/dt$,

$V = (1-M_\infty^2)\delta^{-2/3} - (\varkappa+1)M_\infty\varphi_x$, δ is thickness ratio, M_∞ is upstream Mach number, $k = \omega\bar{c}/U_\infty$ is reduced frequency, ω is frequency, $\bar{c}$ is lenth of chord, U_∞ is upstream velocity; $x = \xi/\bar{c}$, $y = \eta/\bar{c}$, $t = \tau/\omega^{-1} = \tau\omega$, where ξ , η , τ are space and time coordinates, $\tilde{y} = \delta^{1/3}y$ .Nonconservative form of equation (1) for M_∞=const. is

$$A\varphi_{tt} + 2B\varphi_{xt} = V\varphi_{xx} + \varphi_{\tilde{y}\tilde{y}} , \qquad (2)$$

conservative form

$$P_x + Q_{\tilde{y}} + R_t = 0 , \qquad (3)$$

$P = (1-M_\infty^2)\delta^{-2/3}\varphi_x - \frac{1}{2}(\varkappa+1)M_\infty\varphi_x^2$, $Q = \varphi_{\tilde{y}}$, $R = -(A\varphi_t + 2B\varphi_x)$ .A weak solution $\varphi(x,\tilde{y},t)$ of the problem is based on fulfilling the following integral relation

$$\oint\!\!\oint_{\partial D} \vec{F}\,d\vec{S} = 0 , \quad \vec{F} = (P,Q,R) , \qquad (4)$$

along the boundary ∂D of each closed domain $D \subset \Omega_T = \Omega \times \mathcal{T}$, $(x,\tilde{y}) \subset \Omega$, $\mathcal{T} \equiv \{t, t>0\}$ and next conditions:

a) $\varphi_x = 0$, $x \to \pm\infty$ ($x = -\bar{x}$, $x = 1+\bar{x}$, $\bar{x} > 0$) , $M_\infty < 1$ (5)

b) $\varphi_{\tilde{y}} = 0$, $\tilde{y} = \pm\tilde{L}$, $\tilde{L} = \delta^{1/3}L$, $L > 0$ (6)

c) $\varphi_{\tilde{y}}(x,0+,t) = M_\infty \delta^{-1}(\partial f_h/\partial x - \alpha(t))$, (7)

$\varphi_{\tilde{y}}(x,0-,t) = M_\infty \delta^{-1}(\partial f_d/\partial x - \alpha(t))$, $x \geq 1$, $t>0$, (8)

function $f_{h,d}(x,\tilde{y})$ describes upper(h) and lower(d)side of profile surface, $\alpha(t)$ is angle of attack

d) $\varphi_x(x,0+,t) = \varphi_x(x,0-,t)$, (9)

$\varphi_{\tilde{y}}(x,0+,t) = \varphi_{\tilde{y}}(x,0-,t)$, $x \geq 1$, $t>0$ (10)

e) $\varphi(x,\tilde{y},0)$ is solution of the steady problem. (11)

The Joukowski's condition(wake condition) is considered in (9),(10) and for a case $\gamma \neq 0$ this condition chooses a value of circulation fulfilling (9),(10); then

$$\gamma(t) = \varphi(x,0+,t) - \varphi(x,0-,t), \quad x \geq 1, \ t>0. \tag{12}$$

II. Difference problem

A. Low-frequency small disturbance equation

Consider unsteady transonic flow with $k \ll 1$, M_∞=const.Then $k^2 \ll k$, $A \ll B$, $C = D = E = 0$ and governing equation (1) has the form(see f.e. [1], [2])

$$2B\varphi_{xt} = V\varphi_{xx} + \varphi_{\tilde{y}\tilde{y}}. \tag{13}$$

The problem described by (13) was solved in [1] using semiimplicit difference scheme.The scheme used in [1] is conditionally stable with problems near the leading and trailing edges and for higher supersonic Mach numbers.The scheme published in [1] is possible to extend using (n+1,n,n-1)-time planes

$$2B\overleftarrow{\delta}_t\overleftarrow{\delta}_x\varphi_{ij}^{n+1} = (1-\mu_{ij})V_{ij}^n\delta_{xx}\left(\tfrac{3}{4}\varphi_{ij}^n + \tfrac{1}{4}\varphi_{ij}^{n-1}\right) + \mu_{i-1j}V_{i-1j}^n\overleftarrow{\delta}_{xx}\left(\tfrac{3}{4}\varphi_{ij}^n + \tfrac{1}{4}\varphi_{ij}^{n-1}\right) + \delta_{yy}\varphi_{ij}^{n+1}, \tag{14}$$

where $\overleftarrow{\delta}_t\varphi_{ij}^n = (\varphi_{ij}^n - \varphi_{ij}^{n-1})/\Delta t$, $\delta_{tt}\varphi_{ij}^n = (\varphi_{ij}^{n+1} - 2\varphi_{ij}^n + \varphi_{ij}^{n-1})/\Delta t^2$, $\overrightarrow{\delta}_x\varphi_{ij}^n = (\varphi_{i+1j}^n - \varphi_{ij}^n)/\Delta x$,

$\overleftarrow{\delta}_x\varphi_{ij}^n = (\varphi_{ij}^n - \varphi_{i-1j}^n)/\Delta x$, $\delta_{xx}\varphi_{ij}^n = \overrightarrow{\delta}_x\overleftarrow{\delta}_x\varphi_{ij}^n$, $\overleftarrow{\delta}_{xx}\varphi_{ij}^n = \overleftarrow{\delta}_x\overleftarrow{\delta}_x\varphi_{ij}^n$,

$\delta_{yy}\varphi_{ij}^n = (\varphi_{ij+1}^n - 2\varphi_{ij}^n + \varphi_{ij-1}^n)/\Delta\tilde{y}^2$, $V_{ij}^n = (1-M_\infty^2)\delta^{-2/3} - \frac{\varkappa+1}{2}M_\infty(\varphi_{i+1j}^n - \varphi_{i-1j}^n)/\Delta x$,

$\mu_{ij} = 0$ for $V_{ij}^n > 0$, $\mu_{ij} = 1$ for $V_{ij}^n < 0$.

Difference scheme (14) is also conditionally stable but with better condition for $\Delta t \leq F(\Delta x)$.The following difference scheme has better stability properties in comparison with (14) but lower order of approximation

$$2B\overleftarrow{\delta}_t\overleftarrow{\delta}_x\varphi_{ij}^{n+1} = (1-\mu_{ij})V_{ij}^n D_{xx}\varphi_{ij}^n + \mu_{i-1j}V_{i-1j}^n\overleftarrow{D}_{xx}\varphi_{ij}^n + \delta_{yy}\varphi_{ij}^{n+1} = N\varphi_{ij}^n, \tag{15}$$

$$D_{xx}\varphi_{ij}^n = \delta_{xx}\varphi_{ij}^n - \frac{1}{2}\frac{\Delta t^2}{\Delta x}\overleftarrow{\delta}_x\delta_{tt}\varphi_{ij}^n, \quad \overleftarrow{D}_{xx}\varphi_{ij}^n = \overleftarrow{\delta}_{xx}\varphi_{ij}^n + \frac{\Delta t^2}{\Delta x}\overleftarrow{\delta}_x\delta_{tt}\varphi_{ij}^n.$$

Difference scheme (15) is semiimplicit,conservative and unconditionally stable. Using combination of difference scheme (14),(15)in real computation we can find more accurate numerical results due to higher order of approximation of difference scheme (14).

The three-dimensional difference scheme for the problems with governing equation($k \ll 1$)

$$2B\varphi_{xt} = V\varphi_{xx} + \varphi_{\tilde{y}\tilde{y}} + \varphi_{\tilde{z}\tilde{z}} \tag{16}$$

is possible to use in the form

$$2B\overleftarrow{\delta}_t\overleftarrow{\delta}_x\varphi_{ijk}^{n+1} = N\varphi_{ijk}^{n} + D_{zz}\varphi_{ijk}^{n} = \mathcal{N}\varphi_{ijk}^{n} \tag{17}$$

where

$$D_{zz} = \delta_{zz}\varphi_{ijk} - \frac{1}{2}\frac{\Delta t^2}{\Delta z}\overleftarrow{\delta}_z\delta_{tt}\varphi_{ijk}^{n}$$

or in the form

$$2B\overleftarrow{\delta}_t\overleftarrow{\delta}_x\varphi_{ijk}^{n+1} = N\varphi_{ijk}^{n} + \delta_{zz}\varphi_{ijk}^{n+1} = \tilde{\mathcal{N}}\varphi_{ijk}^{n}. \tag{18}$$

Consider problem with $U_\infty = U_\infty(t)$ for $k \ll 1$. The governing equation is (1) with $A=D=0$. For numerical solution the following difference scheme

$$2B\overleftarrow{\delta}_t\overleftarrow{\delta}_x\varphi_{ij}^{n+1} + C\Delta_x\varphi_{ij}^{n} + E(t_n) = N\varphi_{ij}^{n} \tag{19}$$

has been used. Difference operator N is defined by (15) and $\Delta_x\varphi_{ij}^{n} = \overleftarrow{\delta}_x\varphi_{ij}^{n+1}$ for $C>0$, $\Delta_x\varphi_{ij}^{n} = \frac{1}{2}\overrightarrow{\delta}_x(\varphi_{ij}^{n}+\varphi_{ij}^{n-1})$ for $C<0$. Difference scheme (19) is unconditionally stable for $C>0$ and conditionally stable for $C<0$ ($\Delta t \le 2B\Delta x/|C|$).

B) Complete small disturbance equation

Consider a problem with M_∞=const. and complete two-dimensional small disturbance equation

$$A\varphi_{tt} + 2B\varphi_{xt} = V\varphi_{xx} + \varphi_{\bar{y}\bar{y}} \tag{20}$$

or three-dimensional small disturbance equation

$$A\varphi_{tt} + 2B\varphi_{xt} = V\varphi_{xx} + \varphi_{\bar{y}\bar{y}} + \varphi_{\bar{z}\bar{z}}. \tag{21}$$

Then the following difference approximation

$$A\delta_{tt}\varphi_{ij}^{n} + 2B\overleftarrow{\delta}_t\overleftarrow{\delta}_x\varphi_{ij}^{n+1} = N\varphi_{ij}^{n} \tag{22}$$

is possible to use for finite difference solution of the problems described by (20) and

$$A\delta_{tt}\varphi_{ijk}^{n} + 2B\overleftarrow{\delta}_t\overleftarrow{\delta}_x\varphi_{ijk}^{n+1} = \mathcal{N}\varphi_{ijk}^{n} \tag{23}$$

or

$$A\delta_{tt}\varphi_{ijk}^{n} + 2B\overleftarrow{\delta}_t\overleftarrow{\delta}_x\varphi_{ijk}^{n+1} = \tilde{\mathcal{N}}\varphi_{ijk}^{n} \tag{24}$$

for numerical solution of three-dimensional problems with governing equation (21). The system of difference equations for values φ_{ij}^{n+1}, φ_{ijk}^{n+1} is solved using a method similar to SLOR method for numerical solution of the steady problem.

III. Numerical results

A) Numerical results of transonic flow past oscillating 8% non-symmetrical biconvex profile ($f_h=-0,2(x^2-x)$, $f_d=0,12(x^2-x)$, $x \in \langle 0,1\rangle$) in the channel ($L=2$) with $M_\infty=0,77$ and $\alpha=0,5^\circ+1,6^\circ \sin t$ and $k=0,2$ using difference scheme (15) are presented in [3] and compared with numerical results of Ballhaus-Goorjian's method with good agreement.

B) Fig. 1 shows Mach number distribution along upper (full line) and lower profile surface of the numerical results of transonic flow past mentioned profile in the channel for the same $M_\infty=0,77$ and $\alpha=0,5^\circ+1,6^\circ.\sin t$ and different $k=0,1$ (fig. 1a), $k=0,2$ (fig. 1b) using difference scheme (22).

C)Numerical solution of transonic flow(using difference scheme (22)) past mentioned profile in the channel was used for simulation of flowfield properties for fixed M_∞=0,77 and α =0,5°+1,6°sin t and changing reduced frequency k(0,1;0,2;0,4;0,5;0,6;1;5;10).

Fig.2 shows numerical results of Mach number distribution along upper profile surface for t=1,99 and $k \in (0,1\rangle$.

Fig.3 shows"graph" $M_h = M_h(k), M_h = \max M(x,0+,t), x \in \langle 0,1\rangle$, $t \in \langle 0, 2\pi\rangle$.

Fig 4 shows "graph" $\Delta T = \Delta T(k)$, $\Delta T = T - \frac{\pi}{2}$. T is time when Mach number $M = M_h$ and $t = \pi/2$ is time when angle $\alpha(t)$ is maximal.

References

[1] Ballhaus,W.F.;Lomax,H.:The Numerical Simulation of Low-Frequency Unsteady Transonic Flow Fields,Lecture Notes in Physics,No 35,pp.57-62,Springer-Verlag,1975

[2] Ballhaus,W.F.;Goorjian,P.M.:Implicit Finite-Difference Computations of Unsteady Transonic Flows about Airfoils,AIAA Journal,Vol.15,No 12,1978

[3] Kozel,K.:Unconditionally Stable Difference Scheme for Calculation of Unsteady Potential Transonic Flows Past Thin Body,Report VZLÚ,1984

[4] Kozel,K.:Using Small Disturbance Theory for Description of Transonic Flows Past Thin Body,Strojnický časopis, V.34,1983 ,No 1-2(in Czech)

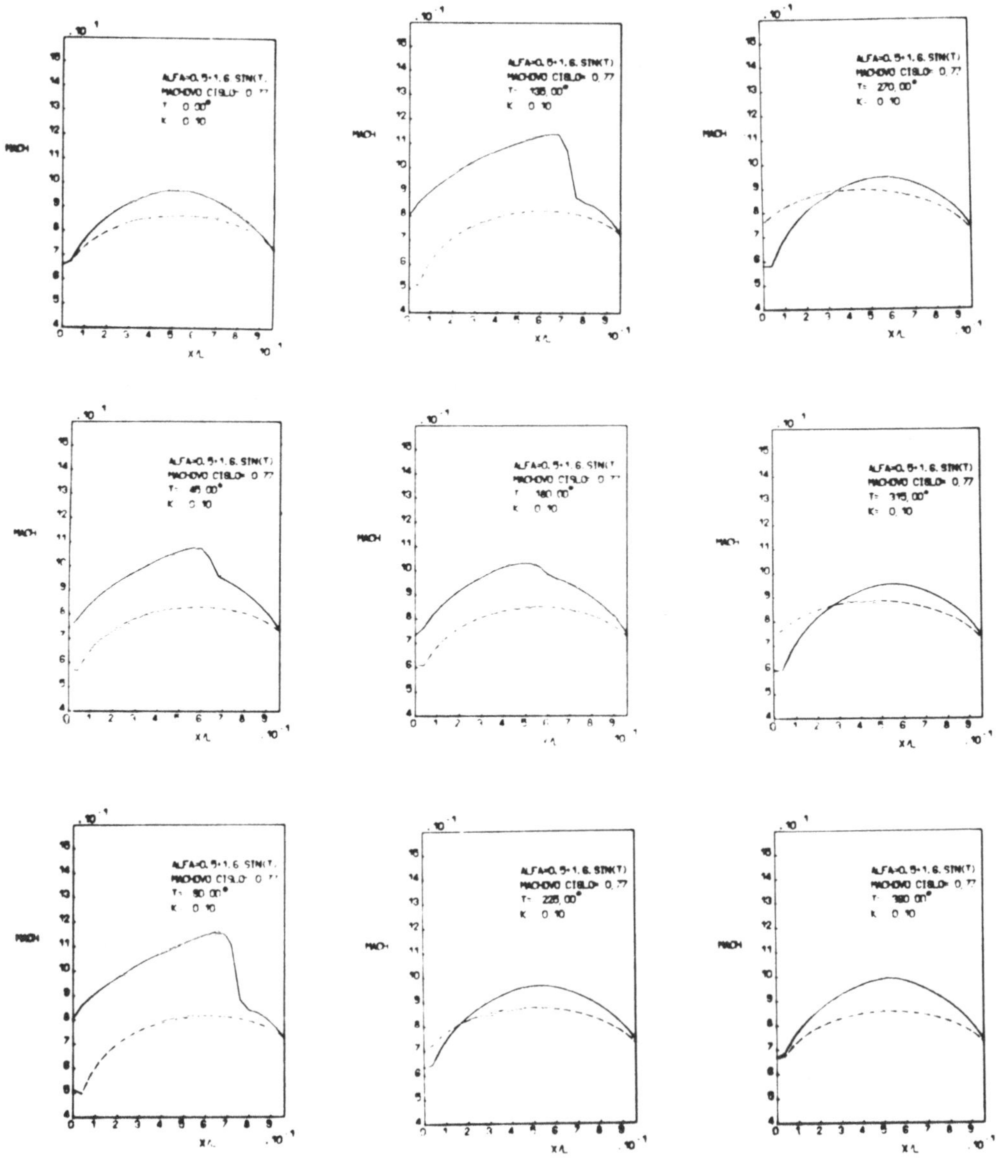

Fig.1a
Mach number distribution along upper(full line) and lower profile surface, $M_\infty = 0,77$; $\alpha(t) = 0,5^{\circ} + 1,6^{\circ} \sin t$; $k = 0,1$

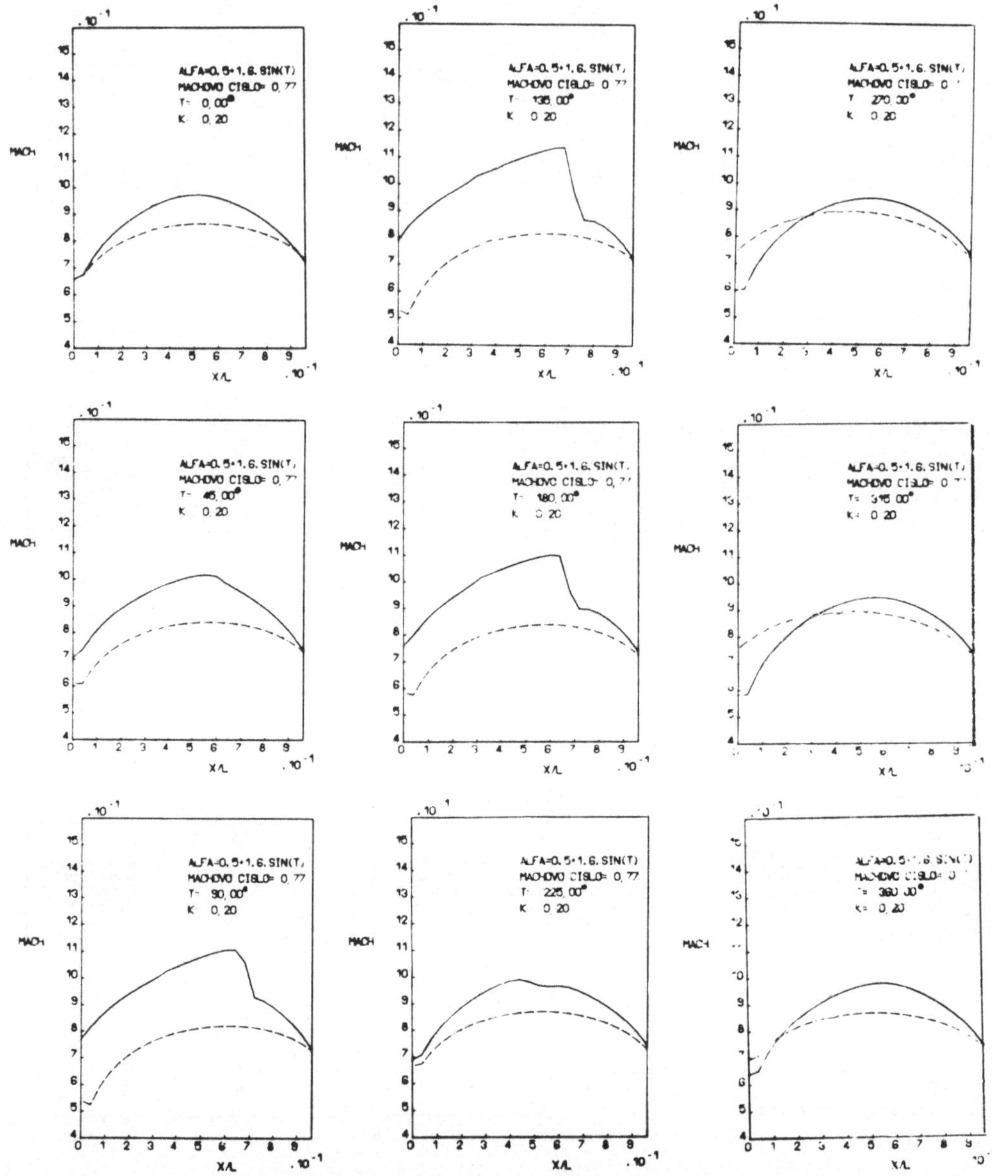

Fig.1b Mach number distribution along upper(full line) and lower profile surface, $M_\infty = 0{,}77$; $\alpha = 0{,}5^\circ + 1{,}6^\circ \sin t$; $k = 0{,}2$

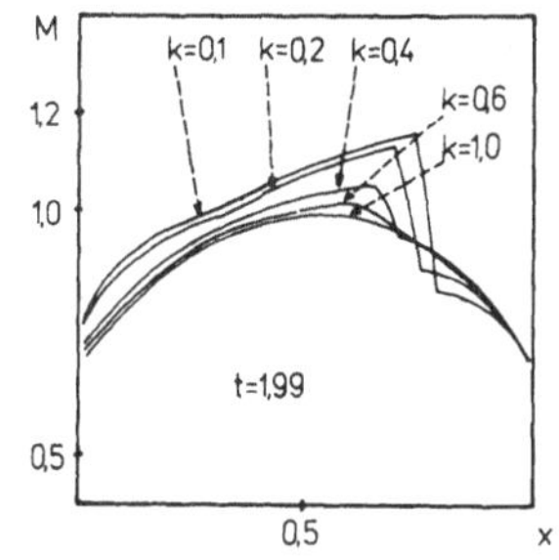

Fig.2 Mach number distribution along upper profile surface, $M_\infty = 0,77$; $\alpha = 0,5^o + 1,6^o \sin t$; $k \in (0,1\rangle$.

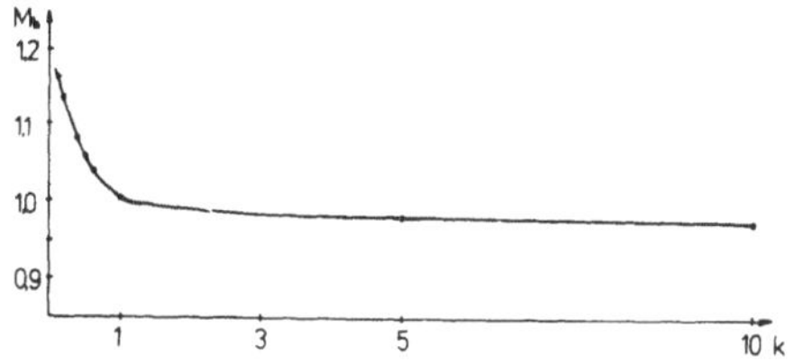

Fig.3 Graph $M_h = M_h(k)$, M_h is maximal Mach number along upper profile surface for $t \in \langle 0, 2\pi \rangle$ and fixed k.

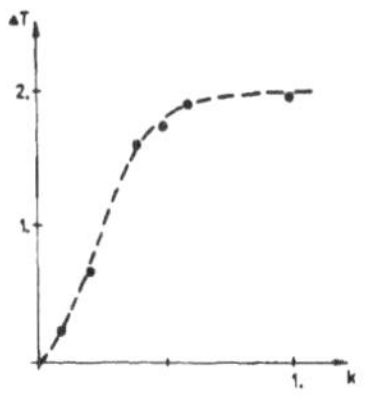

Fig.4 Graph $\Delta T = \Delta T(k)$, $\Delta T = T - \pi/2$, T is time when Mach number $M = M_h$ for fixed k.

POTENTIAL APPLICATION OF ARTIFICIAL INTELLIGENCE CONCEPTS TO NUMERICAL AERODYNAMIC SIMULATION

Paul Kutler, Unmeel B. Mehta, and Alison Andrews
NASA Ames Research Center
Moffett Field, CA 94035 USA

I. INTRODUCTION

The development of sophisticated computational fluid dynamic (CFD) tools for simulating the external flow field about complicated three-dimensional flight vehicles or internal flows within vehicle components requires vast expertise and enormous resources in terms of both human researchers and computer capacity (speed and storage). The creation of such simulation tools requires knowledge of the disciplines of numerical analysis, fluid dynamics, computer science, and aerodynamics, and the development of such tools takes an inordinate amount of time. Furthermore, the writing of software is becoming more expensive every year [1,2]. Shorter, less expensive development times resulting in more powerful, versatile, easy-to-use, and easy-to-interpret simulation tools are necessary if computational aerodynamics is to fulfill its potential in the vehicle design process. To this end, some of the concepts of artificial intelligence (AI) can be applied. It is the purpose of this paper, first, to briefly introduce these concepts and, second, to indicate how some of these concepts can be adapted to speed the numerical aerodynamic simulation process.

II. ARTIFICIAL INTELLIGENCE BACKGROUND

Artificial intelligence is a discipline of computer science concerned with the study of symbolic reasoning by a computer and symbolic representation of knowledge. The objective of applied AI is to design and construct computer programs that exhibit the characteristics normally associated with human intelligence (for example, performance, adaptability, and self-knowledge). The core elements of artificial intelligence are (1) heuristic search (rules of thumb to guide the search of the problem's solution space, as opposed to blind, exhaustive search, or an algorithmic solution procedure); (2) symbolic representation (representing knowledge by means of first-order predicate calculus or frames, for example); and (3) symbolic inference (methods of manipulating symbols to do reasoning). Research in these core areas is conducted through the study of such topics as natural language processing, formal theorem proving, concept learning, automatic programming, robot control, computer vision/perception, and problem solving and planning. Formal approaches to this research (using a formal, unambiguous language for representing facts and ideas, and a formal logic to reason about those ideas) have performed successfully on some rather simple problems. However, the elusiveness of high-level performance by formal methods on more difficult tasks led many AI researchers to an approach that emphasizes the importance of knowledge in expert problem solving [3]. That shift in approach has resulted in the emergence of expert systems technology.

Expert systems are knowledge-based AI programs which are capable of performing at the level of a human expert as a result of their emphasis on domain-specific knowledge and strategies. In addition to the characteristics of high-level performance and reliance on domain-dependent knowledge, expert systems are distinguished from other AI programs and computer programs in general by their ability to reason about their own processes of inference, and to furnish explanations regarding those processes [3]. These distinguishing characteristics are made possible by the underlying architecture common to most expert systems. There are two major components [4]: a knowledge base (domain-dependent facts, rules, heuristics) and a separate inference procedure. Knowledge acquisition and input/output components are usually included.

Expert systems are particularly well suited to two generic types of problems [4]. First, there are the problems in which pursuit of an exact or optimal solution would lead to a combinatorial explosion of computation; second, there are the problems that require interpretation of a large amount of data. In addition, the domains where application of expert systems technology is most appropriate are those fields in which "the difficult choices, the matters that set experts apart from beginners, are symbolic, inferential, and rooted in experiential knowledge" [4]. Expert systems have been constructed in such domains as medical diagnosis, chemistry, symbolic mathematics, geology, circuit design, structural engineering, and computer system configuration (Refs. 3-7 contain descriptions of these systems).

Expert system techniques are currently powerful enough to produce a few successful systems, such as MACSYMA, DENDRAL, and R1 [3]. But the state of the art still falls short of ideal intelligent behavior, or a mature technology. The domain of expertise must be very narrow, the problem representation languages and I/O languages are limited, there is little self-knowledge (which affects explanation capabilities and recognition of the system's own limitations), expertise is restricted to that from a single source, and much of the knowledge and problem-solving approach incorporated into an expert system is painstakingly hand-crafted, resulting in relatively long construction times [3]. Research continues to push the boundaries of capability of expert systems outward.

III. EXPERT SYSTEMS IN CFD

The design and application of computational aerodynamic simulation tools involves the synthesis of many facets (Fig. 1), each of which requires expertise and experience for its formulation, development, and use. It is conceivable that an expert system could be designed that would act as a flow-field synthesizer; that is, act on all of the facets depicted in Fig. 1 for a CFD computation. Present expert systems techniques could be used in at least five aspects of the CFD computation that would involve some of these facets (Fig. 2): three-dimensional grid generation (a pacing item in CFD [8]); flow problem definition and initialization; construction and analysis of numerical schemes; flow-solver selection and use; and data reduction, analysis, and display. Because grid generation has been identified as having the most promise, more detail is presented below.

One of the most important facets required to solve accurately a three-dimensional CFD problem using finite-difference procedures is the proper location of the nodal points in the flow region to be resolved. There are basically two decision stages and a feedback stage involved in the discretization process. The two decision stages involve (1) the grid topology and (2) the grid-generation scheme; the feedback stage involves an analysis and modification of the grid based on the geometric derivatives, the flow-solver algorithm employed, and the flow solution generated. Although grid generation is intrinsically complex, the elements of the decision stages are well understood by experts in the field and the feedback stage is currently receiving attention. Grid generation is, therefore, likely to offer the greatest potential for an early successful design of an expert system in computational fluid dynamics.

The schematic of an expert grid-generation system (EGGS) shown in Fig. 3 is based on some of the major components of an expert system; it depicts in detail the essential ingredients of some of those components. Input consists of three groups of information: (1) flow parameters such as the Mach number, angles of attack and yaw, and Reynolds number (these would determine, for example, whether planes of symmetry can be used, the position of the outer computational domain, and the nodal point clustering near surfaces); (2) geometric data (for external flows, the multiple, time-varying body coordinates at the inner boundary of the computational volume); and (3) qualitative program control information such as the level of accuracy required (e.g., calculations for understanding complicated fluid physics might require a fine grid, whereas those for performing preliminary engineering design might require a coarse grid), and the permissible level of expense to be incurred.

The knowledge base consists of facts (grid-generation schemes and grid-analysis theory) and heuristics (experience and good judgment regarding grid topology decisions, for example). Modern grid-generation philosophy concerned with three-dimensional discretizations dictates that some form of a zonal grid topology be employed. There currently exists no theory that can determine the zoning or grid patchwork for either two- or three-dimensional problems, so heuristics are used. It is hoped that theory can eventually replace many of these heuristics. Once the flow field has been zoned, each zone can then be discretized using the procedures denoted in Fig. 3; the procedures include either algebraic or differential approaches. With the flow region discretized, various levels of grid-analysis procedures can be used to judge the quality of the resulting grid. These vary in complexity from procedures that simply look at grid parameters, such as the transformation Jacobian, geometric derivatives, and ratio of the metrics, to procedures that combine these functions with the flow-solver algorithm and flow solution to yield an improved grid.

EGGS produces as output three pieces of information to be used by the flow solver: (1) the coordinate location of the nodal points, (2) the definition of each surface of the computational cube (e.g., plane of symmetry, body, shock wave), and (3) the zonal interface control parameters. The latter piece of information tells the flow solver which parts of the zonal grid boundaries are adjacent to each other. This is required by the boundary condition routines in the flow solver.

IV. RESOURCE REQUIREMENTS

Development of knowledge-based systems requires a significant investment of time and money, and requires a new kind of professional — the knowledge engineer. One time-estimate for building an expert system is anywhere from 7 months (for simple systems in a friendly environment with existing tools) to 15 yr (for complex systems in demanding environments where new tools must be researched and developed) [9]. Although the proposed expert grid-generation system would fall toward the simple end of the spectrum, a more comprehensive expert flow-simulation system will undoubtedly be more complex, and may require more powerful AI tools than are presently available. For a discussion of the issues involved in expert system development, see Refs. 3 and 9.

V. CONCLUDING REMARKS

The techniques of artificial intelligence, in particular those of expert systems, can be applied to most facets of the numerical aerodynamic simulation process. This paper describes some of the concepts underlying those techniques, and indicates the areas of aerodynamic simulation in which those techniques could play a significant role. A proposed expert grid-generation system is briefly described which, given flow parameters, configuration geometry, and simulation constraints, uses expert knowledge about the discretization process to determine grid-point coordinates, computational surface information, and zonal interface parameters. Additional details of this and other possible CFD expert systems can be found in Ref. 10. The potential payoff from the use of expert systems in the numerical aerodynamic simulation process is worthy of attention and warrants the allocation of resources as an investment in the future. Expert systems in CFD will promote the fusion, preservation, and distribution of aerodynamic knowledge, and will streamline research and design by managing the complexities of those processes. The users of these future systems will be freed from attending to the details of numerical simulation, and allowed to explore, innovate, and create at a higher level of abstraction.

VI. REFERENCES

1. Mueller, G. E., "The Future of Data Processing in Aerospace," Aeronaut. J., Apr. 1979, pp. 149-158.
2. Fleckenstein, W. O., "Challenges in Software Development," Computer, Mar. 1983, pp. 60-64.
3. Hayes-Roth, F., Waterman, D. A., and Lenat, D. B., Eds., Building Expert Systems, Teknowledge Series in Knowledge Engineering, Vol. 1, Addison-Wesley Publishing Co., Inc., 1983.
4. Feigenbaum, E. A. and McCorduck, P., The Fifth Generation, Addison-Wesley Publishing Co., Inc., 1983.
5. Gevarter, W. B., "Expert Systems — Limited but Powerful," IEEE Spectrum, Aug. 1983, pp. 39-45.

6. Gevarter, W. B., "An Overview of Artificial Intelligence and Robotics. Vol. I. Artificial Intelligence, Pt. A: The Core Ingredients," NASA TM-85836, June 1983.
7. Barr, A. and Feigenbaum, E. A., The Handbook of Artificial Intelligence, Vol. 2, William Kaufmann, Inc., 1982.
8. Kutler, P., "A Perspective of Theoretical and Applied Computational Fluid Dynamics," AIAA Paper 83-0037, Reno, Nev., 1983.
9. Hayes-Roth, F., "Codifying Human Knowledge for Machine Reading," IEEE Spectrum, Nov. 1983, pp. 79-81.
10. Kutler, P. and Mehta, U. B., "Computational Aerodynamics and Artificial Intelligence," AIAA Paper 84-1531, Snowmass, Colo., June 1984.

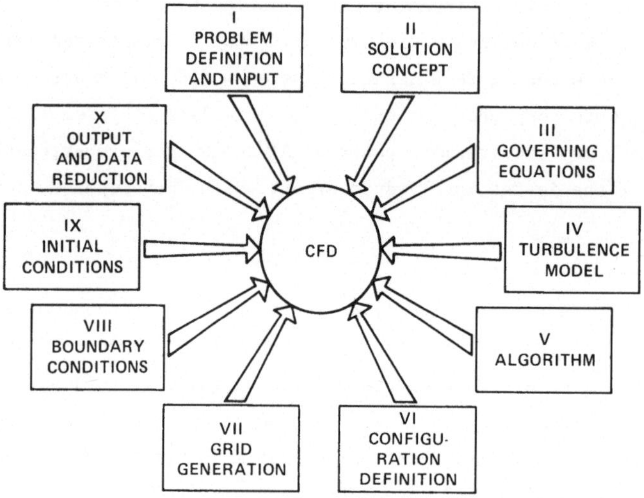

Fig. 1 Numerical aerodynamic simulation synthesizer (NASS).

$$\frac{\partial U}{\partial t} = \frac{\partial^2 U}{\partial x^2}$$

$$\frac{U_i^{n+1} - U_i^n}{\Delta t} = \frac{U_{i+1}^{n+1} - 2U_i^{n+1} + U_{i-1}^{n+1}}{\Delta x^2}$$

GRID GENERATION

- CONCEPTUAL OR TOPOLOGICAL DISCRETIZATION

- DEFINITION AND APPLICATION OF GRID-GENERATION PROCEDURE

- GRID-QUALITY ANALYSIS

- GRID-POINT LOCATION ADJUST-MENTS

FLOW-PROBLEM DEFINITION

- PHYSICAL ASPECTS OF FLOW
  SPEED REGIME
  VISCOUS OR INVISCID
  STEADY OR UNSTEADY
  THIN-, SLENDER- OR
  COMPLEX-SHEAR LAYERS
  SHOCK WAVES – FLOW
  DISCONTINUITIES

- SURFACE AND FIELD VARIABLES

- PHYSICAL BOUNDARY AND INITIAL CONDITIONS

- SUGGESTIONS AS TO SOLUTION METHODOLOGY

CONSTRUCTION AND ANALYSIS OF NUMERICAL METHODS

- DEVELOPMENT OF SCHEMES WITH SPECIFIC PROPERTIES

- STABILITY ANALYSIS

- ACCURACY ANALYSIS

FLOW SOLVERS

- NAVIER-STOKES SOLVER

- PNS SOLVER

- BOUNDARY LAYER SOLVER

- INVISCID SOLVER

DATA REDUCTION (ANALYSIS AND DISPLAY)

- ENGINEERING QUANTITIES

- FLOW VISUALIZATION

- ERROR ANALYSIS

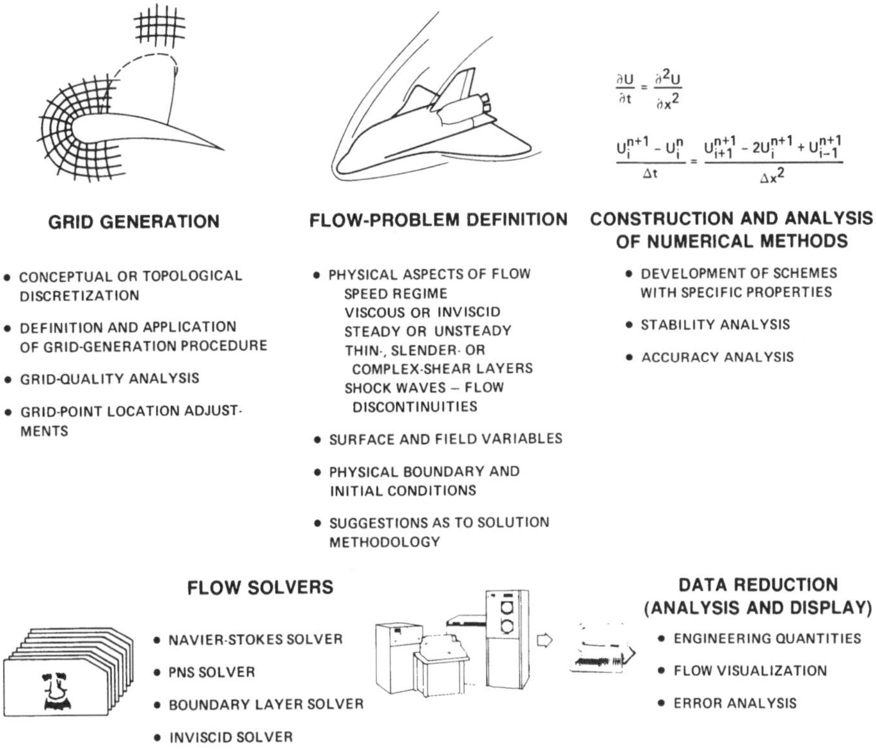

Fig. 2 Expert systems in numerical aerodynamic simulation.

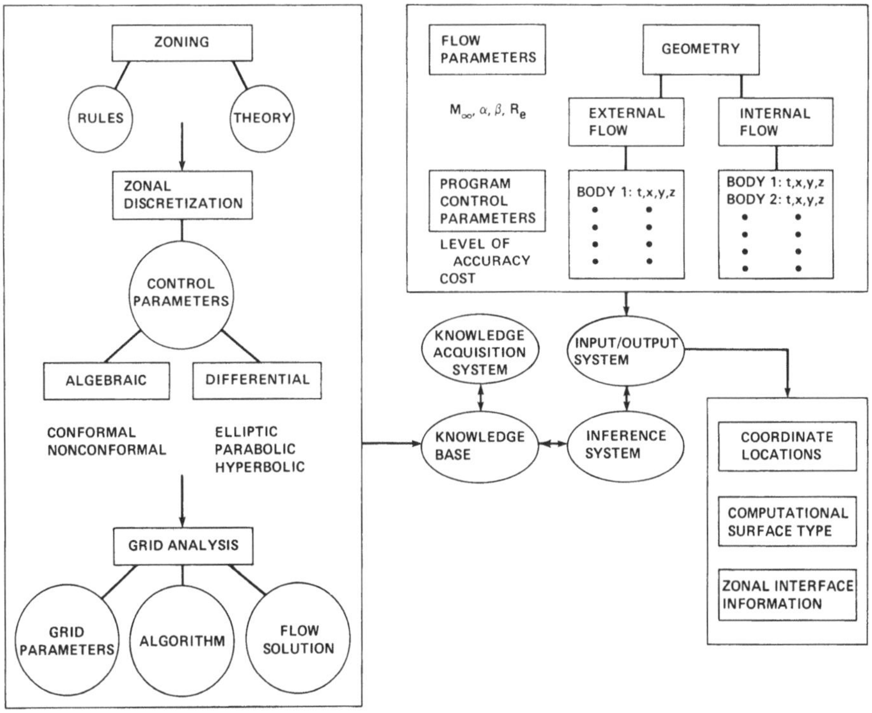

Fig. 3 Expert grid-generation system (EGGS).

A SOLUTION PROCEDURE FOR THREE-DIMENSIONAL INCOMPRESSIBLE NAVIER-STOKES EQUATION AND ITS APPLICATION

Dochan Kwak
NASA Ames Research Center, Moffett Field, California

James L.C. Chang and Samuel P. Shanks
Rocketdyne Division, Rockwell International, Canoga Park, California

I. Introduction

A major difficulty when solving the incompressible flow equations that use primitive variables is caused by the pressure term which is used as a mapping parameter to obtain a divergence-free velocity field. One commonly used approach is to solve the Poisson equation for pressure, which is derived from the momentum equations [1]. This approach can be very time consuming. To accelerate the pressure-field solution and alleviate the drawback associated with the Poisson equation approach, Chorin [2] proposed the use of artificial compressibility in solving the continuity equation. A similar method was adopted by Steger and Kutler [3] and Chakravarthy [4] using an implicit approximate-factorization scheme [5]. Based on this procedure, a pseudocompressible method has been developed for solving three-dimensional, viscous, incompressible flow problems cast in generalized curvilinear coordinates [6,7]. The purpose of the present paper is to show salient features of the pseudocompressible approach, which is primarily designed for obtaining steady-state solutions efficiently.

II. Description of the Method

In the present formulation, the three-dimensional, incompressible Navier-Stokes equations are modified to form the following set of governing equations written in dimensionless form :

$$\frac{1}{\beta}\frac{\partial p}{\partial t}+\frac{\partial u_i}{\partial x_i}=0 \tag{1a}$$

$$\frac{\partial u_i}{\partial t}+\frac{\partial u_i u_j}{\partial x_j}=-\frac{\partial p}{\partial x_i}+\frac{\partial \tau_{ij}}{\partial x_j} \tag{1b}$$

Here, t is time; x_i are the Cartesian coordinates; u_i are corresponding velocity components; p is the pressure; and τ_{ij} is the viscous stress tensor. The parameter $1/\beta$ is the pseudocompressibility. As the solution converges to a steady state, the pseudocompressibility effect approaches zero, yielding the incompressible form of the equations. In the present study, the approximate factorization scheme by Beam and Warming [5] is implemented to solve the finite-difference form of the governing equations written in general curvilinear coordinates (see ref. 6 for detail).

In the present formulation, waves of finite speed are introduced. And the system of modified equations given by equations (1a) and (1b) can be marched in time. The magnitude of the wave speed depends on β. To recover the incompressible phenomena, the physics requires that the pressure wave propagates much faster than the spreading of vorticity. From this, the following criterion for the lower bound on β is obtained [7]:

$$\beta > [1 + 4(x_{ref}/x_\delta)^2(x_L/x_{ref})/Re]^2 - 1 \tag{2}$$

where x_{ref} is the reference length, and x_δ and x_L are the characteristic lengths that the vorticity and the pressure waves have to propagate during a given timespan.

The upper bound on β depends upon the particular numerical algorithm chosen. In the present study, higher-order cross-differencing terms are added to obtain the approximately factored form of the governing equations. These added terms contaminate the momentum equations as well as the continuity equation, and therefore must be kept smaller than the original terms everywhere in the computational domain. This requirement leads to the following criterion for the upper bound of β:

$$\beta \Delta\tau < O(1) \tag{3}$$

where $\Delta\tau$ is the time-step used in the integration scheme.

III. Computed Results

Numerical experiments were performed to illustrate the present procedure. To represent an internal flow, the flow through a channel at Re=1,000 was chosen. The coordinate system and velocity vectors for a converged solution are shown in figures 1a and 1b. To change the ratio of the time scales required for the pressure waves and the vorticity to map the entire flow field, the channel length, L, is varied form 20 to 40. The recommended values of β for these cases using $\Delta\tau = 0.1$ are:

$$0.75 < \beta_{L=20} < 10, \quad 1.19 < \beta_{L=30} < 10, \quad 1.69 < \beta_{L=40} < 10$$

In table 1, the number of iterations for one roundtrip by the pressure wave (denoted by N_1) is tabulated for various values of β which include values outside the recommended range. In figure 2, root-mean-square (RMS) values of $(div\, u)$ are plotted to check the accuracy of the converged solutions. When the value of β is out of the range specified, the accuracy of the solution deteriorates.

To represent an external flow, the flow past a circular cylinder at a $Re = 40$ was chosen. To obtain the near-field solution only, the distance traveled by the waves and the spreading of the vorticity can be approximately the same in magnitude. In the present case, this leads to the range for β using $\Delta\tau = 0.1$ to be $0.1 < \beta < 10$. This indicates that the magnitude of β is less restrictive for external flows. In figures 3a and 3b, the stream-function contours and the pressure coefficient on the surface are shown for a steady-state solution. This solution agrees very well with that of Mehta who used a stream function and vorticity formulation in two dimensions (private communication, U. B. Mehta, 1983). In figure 4, in which the history of the pressure drag is shown for an impulsively started circular cylinder at $Re = 40$, four different values of β were compared with the time-accurate solution of Mehta. In all cases, the values of β are selected within the suggested range above, and the solutions converge rapidly.

To test internal flows further, an annular duct with a 180° bend is chosen. This configuration is similar to the turnaround duct of the hot-gas manifold in the Space Shuttle main engine (SSME). In figures 5a and 5b, the geometry and a laminar solution at Re=1,000 are shown, which reveals the formation of a large separated bubble after the 180° bend. For this geometry, the streamwise length normalized by the duct width is 20. The test problems presented here were treated using a 51 x 17 x 21 mesh for half-duct formulation and the computing time required was 1.1×10^{-4} sec per mesh point per time-step on the Cray X-MP computer at NASA Ames Research Center.

IV. Concluding Remarks

This paper presents salient features of the computational procedure developed for a three-dimensional, incompressible, Navier-Stokes code. This procedure has been applied to various geometrically complex flows, including a major application in analyzing the flow field in the SSME power head. The present algorithm has been shown to be very robust and accurate if the selection of β is made according to the guidelines presented here.

References

1. Harlow, F. H.; and Welch, J. E.: Numerical Calculation of Time-Dependent Viscous Incompressible Flow with Free Surface, Phys. of Fluids, vol. 8, no. 12, Dec. 1965, pp. 2182-2189.
2. Chorin, A. J.: A Numerical Method for Solving Incompressible Viscous Flow Problems, J. Comput. Phys., vol. 2, 1967, pp. 12-26.
3. Steger, J. L.; and Kutler, P.: Implicit Finite-Difference Procedures for the Computation of Vortex Wakes, AIAA J., vol. 15, no. 4, Apr. 1977, pp. 581-590.
4. Chakravarthy, S. R.: Numerical Simulation of Laminar Incompressible Flow within Liquid Filled Shells, Report ARBRL-CR-00491, U.S. Army Ballistics Research Laboratory, Aberdeen Proving Ground, Md., Nov. 1982.
5. Beam, R. M.; and Warming, R. F.: An Implicit Finite-Difference Algorithm for Hyperbolic Systems in Conservation-Law Form, J. Comput. Phys., vol. 22, Sept. 1976, pp. 87-110.
6. Kwak, D.; Chang, J. L. C.; Shanks, S. P.; and Chakravarthy, S.: An Incompressible Navier-Stokes Flow Solver in Three-Dimensional Curvilinear Coordinate Systems Using Primitive Variables, AIAA Paper 84-253, Reno, Nev., 1984.
7. Chang, J. L. C.; and Kwak, D.: On the Method of Pseudo Compressibility for Numerically Solving Incompressible Flows, AIAA Paper 84-252, Reno, Nev., 1984.

Table 1: Number of iterations required for one round-trip by pressure waves between in- and out-flow boundary of a channel: Re = 1000 and $\Delta\tau = 0.1$

	β	0.1	1	2	5	10	50
N_1	L = 20	4196	566	347	196	133	58
	L = 30	6293	849	520	294	199	86
	L = 40	8391	1132	693	392	266	115

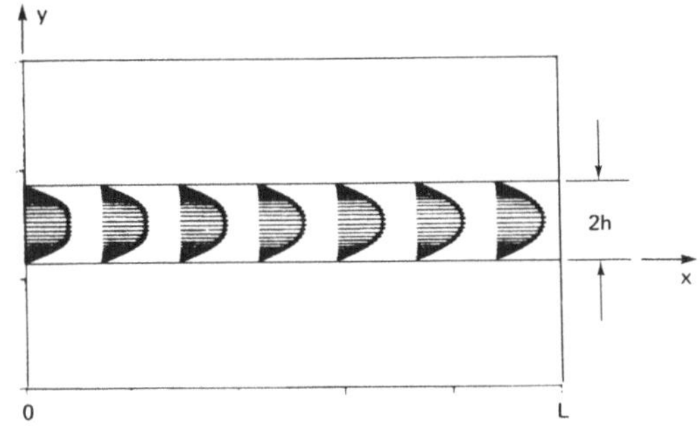

(a) Velocity vector

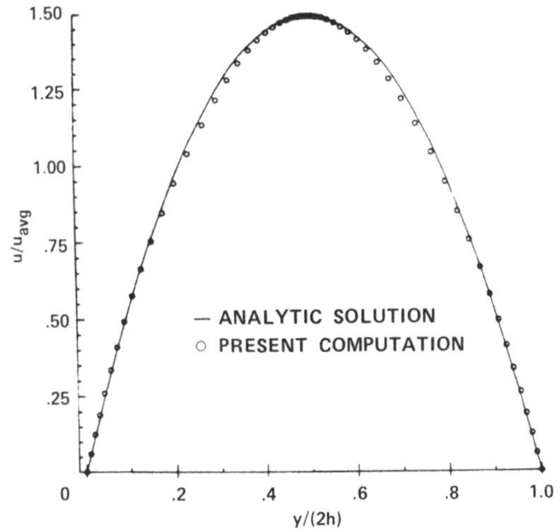

(b) Fully developed velocity profile

Figure 1.– Developing laminar channel flow at Re = 1,000 (Re based on channel width and average velocity).

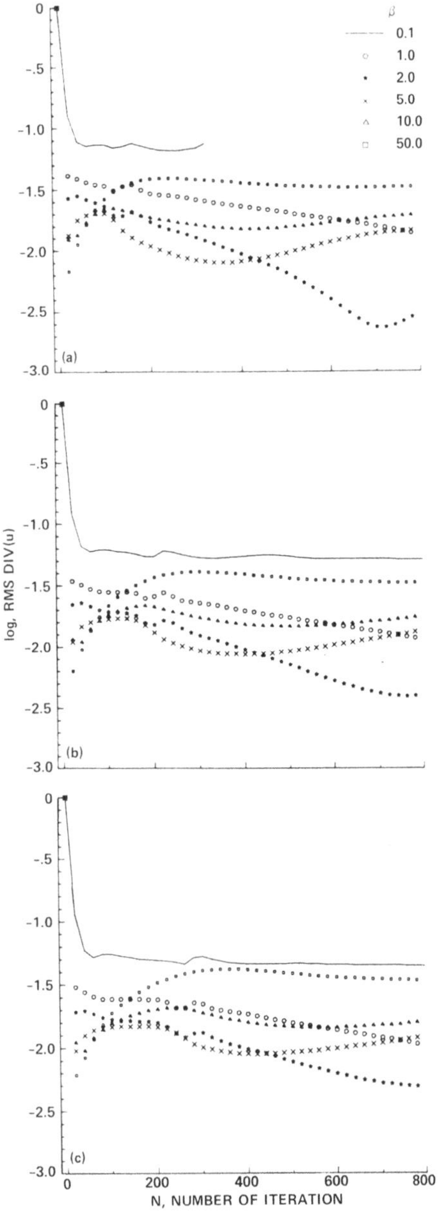

(a) L = 20
(b) L = 30
(c) L = 30

Figure 2.– RMS (divu) history of channel flow at Re = 1,000 and $\Delta\tau = 0.1$.

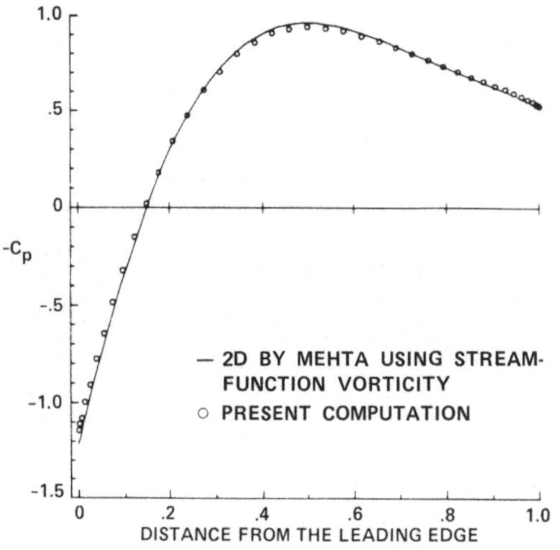

(a) Pressure coefficient on the surface

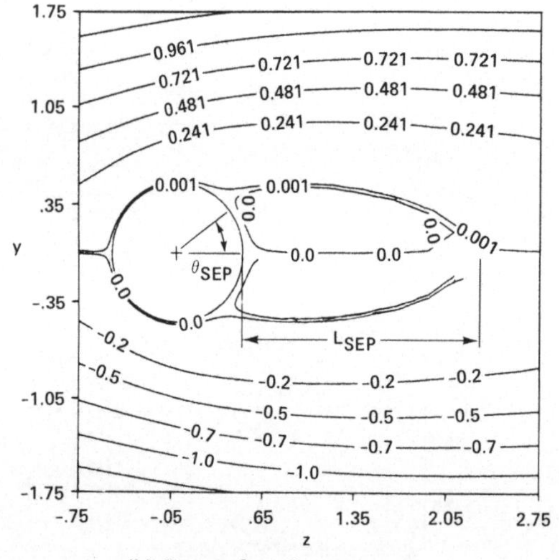

(b) Stream-function contours

Figure 3.– Steady-state solution for flow over a circular cylinder at Re = 40.

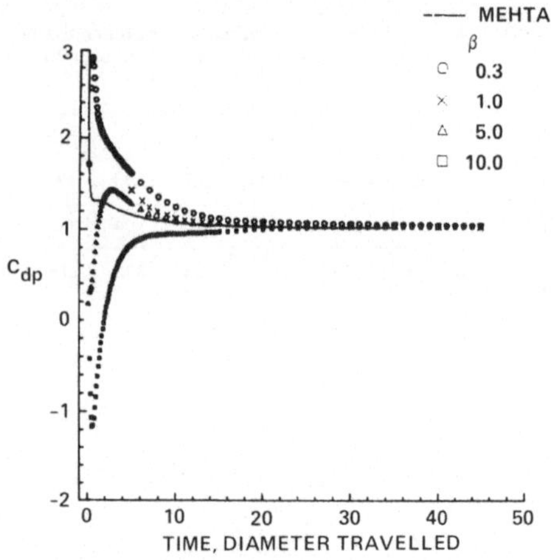

Figure 4.– Pressure drag history for flow over a circular cylinder at Re = 40.

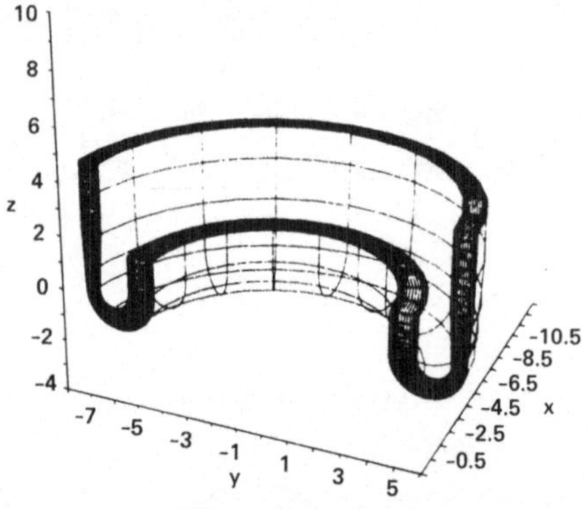

(a) Three-dimensional grid.

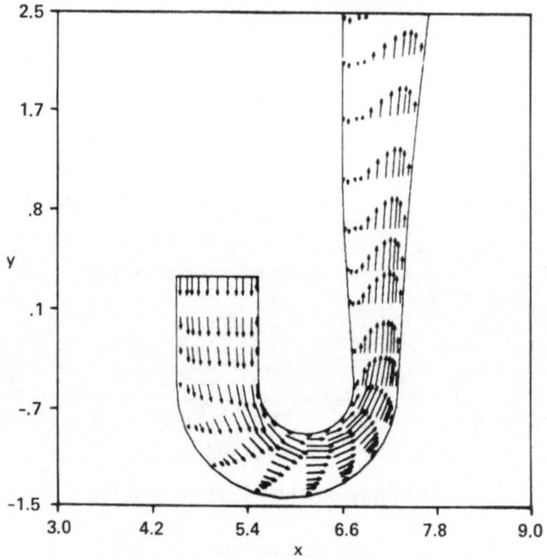

(b) Typical flow pattern with separation

Figure 5.– Flow through a turnaround duct.

A MULTI-ZONAL-MARCHING INTEGRAL METHOD FOR 3D-BOUNDARY LAYER WITH VISCOUS-INVISCID INTERACTION

J.C. LE BALLEUR and M. LAZAREFF

Office National d'Etudes et de Recherches Aérospatiales (ONERA)
92320 Châtillon (France)

INTRODUCTION

Much progress has been achieved, in two-dimensions, on "Viscous-Inviscid-Interaction" numerical methods, see [1 to 5] . The efficiency is based on iteratively coupling a viscous solver, which remains "fast" and well-conditioned at high Reynolds number, with a pseudo-inviscid field controlled through boundary conditions.

The "fast" viscous 2D-solvers (uncoupled) are provided by marching techniques and boundary-layer-like discretization schemes. Cost and grid requirements can be minimized successfully, in addition, for many strong-interaction flows, by solving "Defect" integral equations on the walls and wake-cuts [1 to 4] , marching in the free stream direction and using the wall-grid of the inviscid solver. Marching techniques can be maintained also in recirculating regions, by solving "inverse" integral methods, the viscous upstream influence being recovered from the strong coupling at any Mach number [1,4] .

The present method is a preliminary step to extend such viscous solvers in 3D-flows, with the following advances :

. A new 3D-integral numerical method of entrainment has been generated, using the 3D-extension of the turbulent modelling of Le Balleur [6] suggested for infinite swept wings [2-3] . The new 3D-numerical technique is based on a "Multi-Zonal-Marching" solution, and uses the curvilinear grid (i, j, k) of the interacting inviscid solver along the coordinate surface k = 1, which maps the body. This grid allows to control more easily the viscous-inviscid interaction, by computing the field of inviscid wall-transpiration velocities directly at the coupling nodes (i, j, 1).

. The turbulent modelling which closes the integral equations can describe the boundary layer 3D-velocity-profiles with reverse flow, but only the "Direct" method of solution, with the outer velocity prescribed from the inviscid solver, is considered. The calculation of "closed" 3D-separations is then only seen as a possible extension, accessible to an "Inverse" solution of the present integral method. The calculation of "open" 3D-separations, generating vortex sheets, would be tractable in principle, when obtaining convergence of the viscous-inviscid coupling and the vortex field.

. At the present time, solutions with viscous-inviscid interaction have been computed for transonic transport wings, where the vortex sheet is shed from the trailing-edge, and for prolate spheroids at incidence where the vortex sheets and the viscous recirculating zones are neglected.

MULTI-ZONAL-MARCHING INTEGRAL METHOD

Similar difficulties in 3D-flows are encountered in finite-differences or integral methods, to get first a satisfactory numerical technique for the "uncoupled" 3D-boundary layer, when using the steady equations and a prescribed grid. The main questions are the accessibility of each coupling node (i, j, 1) to a numerical integration based on a marching technique, and also the integration molecule on the (i, j) plane.

a) Viscous-Layer-Defect Equations : The present method solves the defect integral x- and y- momentum equations, the local momentum equation at the outer edge $z = \delta(x, y)$ of the layer (entrainment eq.), and the defect integral continuity equation, in non-orthogonal curvilinear coordinates. Knowing the contravariant inviscid velocity components at the wall $u_{i,j}$ and $v_{i,j}$, these four equations calculate the thickness $\delta_{i,j}$ of the layer, two free shape-parameters of the viscous velocity profiles, and also the coupling transpiration-velocity $w_{i,j}$. These equations are closed with a modelling of the 3D-turbulent velocity profiles, Fig. 2, suggested in previous publications [2 to 4,6] and which provides an equilibrium entrainment model.

The local characteristic cone of this hyperbolic set of integral equations is included inside the limiting outer- and wall-streamlines. This "Direct" velocity-prescribed solution may then be marched in principle along the inviscid streamlines coordinates, until an incipient reverse flow in streamwise direction or a wall-streamlines accumulation occurs.

b) Multi-Zonal-Marching integration : An equivalent integration capability is tentatively developed here with a "Multi-Zonal-Marching" new method (MZM) that uses the inviscid grid (i, j, 1), and maintains the low cost of a marching viscous solver.

The rectangular domain of integration in the computing plane (i, j) is dissociated into so many rectangular zones as necessary, possibly overlapping or degenerating into isolated lines. In each of them, the present MZM-solver may be numerically marched in the more appropriate grid-direction. Several crossed and overlapping zonal-sweeps may then be performed on a same rectangular zone of the computational plane. The new accessible nodes are updated during each zonal-sweep, possibly updating boundary conditions for the following zonal-sweeps. The MZM method of solution is used for example in the leading-edges regions, both for the ellipsoid and the transport wing, with the marching zonal-sweeps shown on Fig. 1.

c) Local integration scheme : In each marching zonal-sweep, the lateral step-size of the inviscid grid is used. Several steps are used in the (instantaneous) marching direction, between two inviscid stations, when it is locally needed for stability or accuracy of the integration, which is performed with an explicit scheme.

In order to select the scheme, a classical line-technique has been first experimented, based on first-order lateral differences schemes biased according to the local characteristic cone, and integrated with a fourth order Runge-Kutta technique in the marching direction.

However, the second-order explicit MacCormack's scheme has been selected, and found to be both less time consuming and more robust. The resulting improvement is shown for example on Fig. 3, with different views of the field of skin-friction directions for a laminar calculation on the ellipsoid at 10° of incidence, displaying an increased domain of accessibility. At the same incidence, the full laminar-turbulent multi-zonal calculation, with a prescribed transition line, Fig. 4, gives a very small inaccessible area.

VISCOUS-INVISCID INTERACTION

The computed wall-transpiration velocity $w_{i,j}$ at each node is prescribed to a coupled inviscid solver, using a panel method without vortex sheet for the ellipsoid, and the SLOR non-conservative method of Chattot [7] for the full potential equation, in the transonic supercritical-wing case.

The wall-source ratio $(w/q)_{i,j}^{n}$ is updated at each iteration n of the viscous calculation. The corresponding fixed point iteration for the vicous-inviscid coupling is here stabilized simply with an additional uniform-underrelaxation technique. The strong interaction is obtained only in the areas of regular attached boundary layer. The coupling relaxation n is embedded within the SLOR potential relaxation, with a coupling cycle each 20 (or 40) sweeps.

A converged wall-transpiration field (w/q) is seen on Fig. 5, in the case of the DVFLR-F4 wing at supercritical turbulent conditions, showing the rear-loading on the lower-side. The calculated skin-friction directions are also shown on Fig. 5 for the upper-side, lower-side, and leading-edge zone (computing plane projection). The viscous influence on the pressure field is seen on Fig. 6 at different spanwise sections, using a still rather coarse-grid calculation, and comparing with experimental data of Schmitt [8] .

REFERENCES

1 LE BALLEUR J.C. - Von Karman Institute Lecture Series 1982-04, Computational Fluid Dynamics (1982)

2 LE BALLEUR J.C. - Springer-Verlag, Proceed. Numerical and Physical Aspects of Aerodynamic Flows II, T. Cebeci ed., Chapter 13, p. 259-284, (1983-84)

3 LE BALLEUR J.C. - Pineridge Press, Recent Advances in Numerical Methods in Fluids, Vol 3 "Viscous flow Computational Methods", W.G. Habashi ed., (1984)

4 LE BALLEUR J.C. - AGARD-CP-351, Paper 1, (1983)

5 CARTER J.E., VATSA V.N. - Springer Verlag, Proceed. 8th ICNMFD, Lecture notes in Physics, (1982)

6 LE BALLEUR J.C. - La Recherche Aérospatiale n° 1981-3, English edition, p. 21-45, (1981)

7 CHATTOT J.J., COULOMBEIX C., TOME C. - La Recherche Aérospatiale n° 1978-4, p. 143-159 (1978)

8 SCHMITT V. - 8th Colloque Aero. Appl. AAAF, Poitiers (1981), ONERA TP-1981-122.

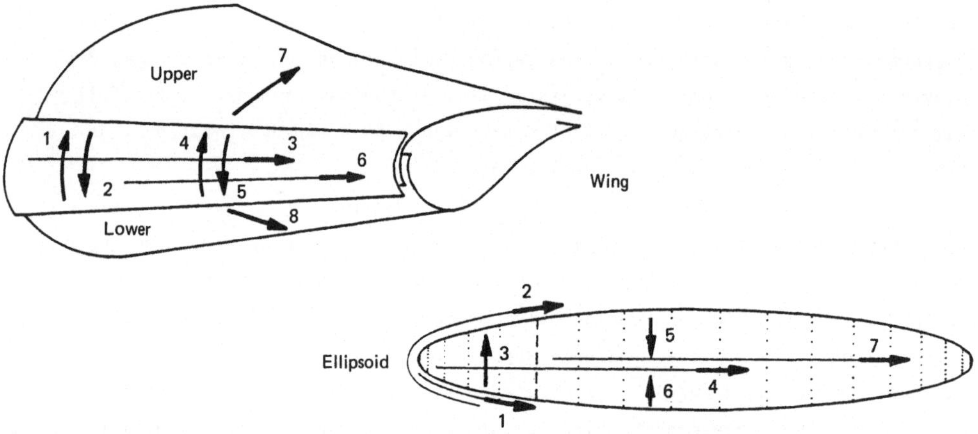

Fig. 1 — Multi-zonal-marching sweeps for 3D boundary layer.

Fig. 2 — 3D turbulent velocity profiles modelling.

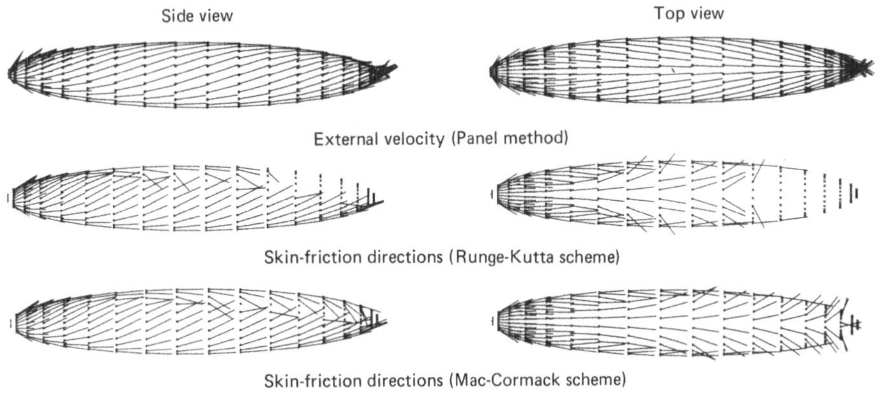

Fig. 3 – Explicit Mac-Cormack integration scheme (Laminar, $a/b = 6$, $U_0 = 10\,m/s$, $\alpha = 10°$, $R_a = 1.6 \times 10^6$).

Fig. 4 – Full multi-zonal laminar-turbulent calculation. Skin-friction directions ($a/b = 6$, prescribed transition line, $V_0 = 40\,m/s$, $\alpha = 10°$, $R_a = 6.4 \times 10^6$).

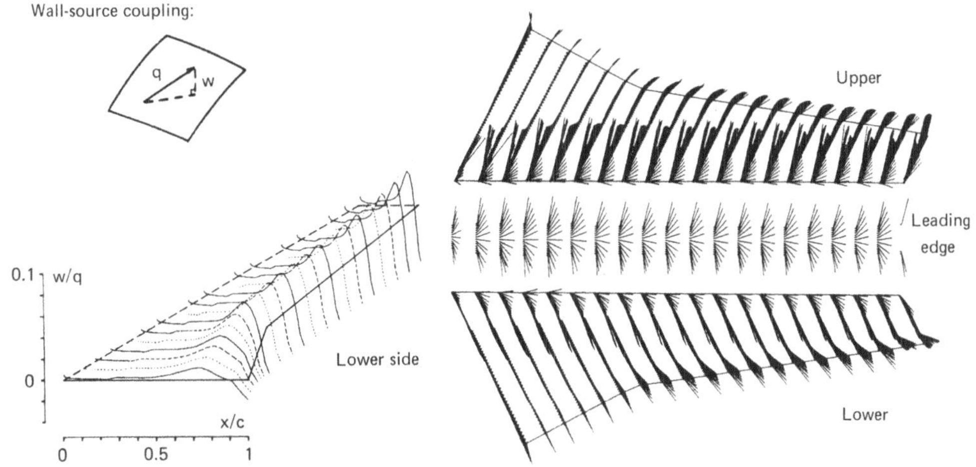

Fig. 5 – Viscous-inviscid interaction on transonic wings (DFVLR-F4 wing, $M = 0.75$, $\alpha = 0.10°$, $R = 2.6 \times 10^6$).

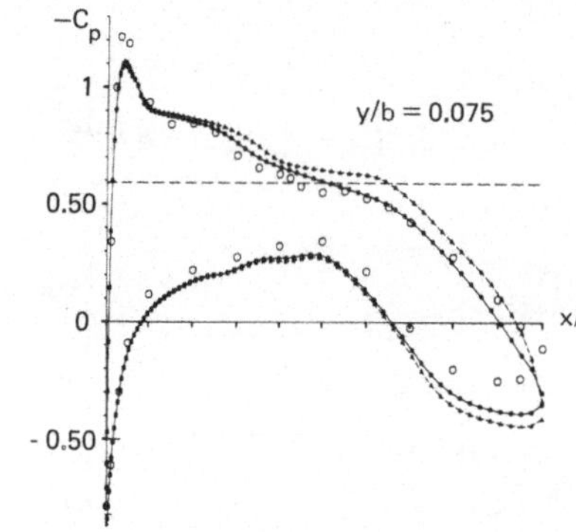

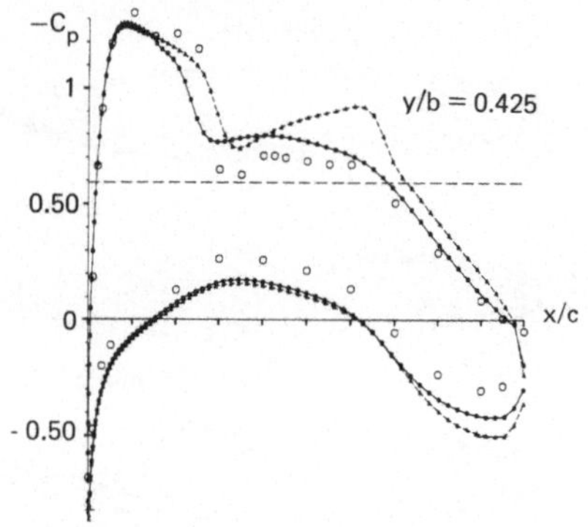

Fig. 6 – Pressure distributions along span (DFVLR-F4 wing, M = 0.75, $\alpha = 10°$, R = 2.6 x 10^6).

●–●–●–●–	Viscous calculation (wing)
–●–●–●–●–	Inviscid calculation (wing)
○ ○ ○	Experiment (wing + body)

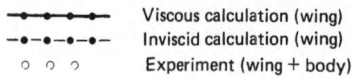

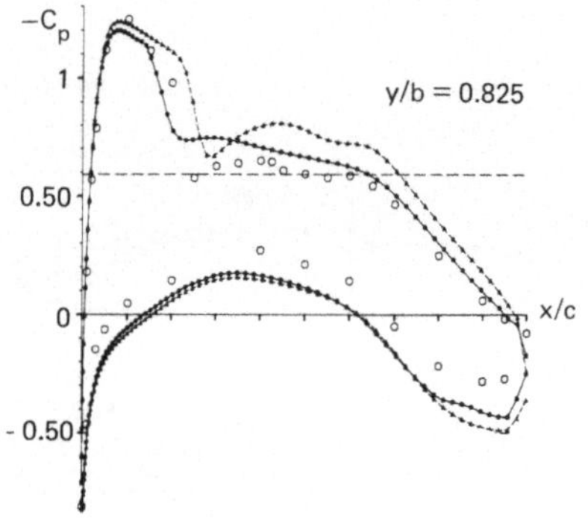

AN IMPLICIT METHOD FOR SOLVING FLUID DYNAMICS EQUATIONS

G. LE COQ EdF/DER Service IMA 1, Place Général de Gaulle
92141 CLAMART

P. RAYMOND CEA/IRDI DEMT-SERMA Bât.70 C.E.N. SACLAY
91191 GIF-sur-YVETTE CEDEX

R. ROY Collège St Jean Richelieu, Département de Mathématiques
C.P. 1018 St JEAN QUEBEC (Canada) J3B2A7

The Control Variable Method described in this paper is a fully implicit numerical method which solves conservation equations in fixed finite volumes for time dependant fluid flow problems. This method is a general technique allowing computations of compressible fluid flow with shock waves propagation, two phase flow transients, viscous and buoyancy driven flows.

The Control Variable Method

Consider a fixed finite volume V bounded by a surface S, conservations equations are expressed as :

$$\frac{\partial YV}{\partial t} = \oint_S \vec{y}.\vec{n}\, dS + \int_V F dV \qquad (1)$$

with

$$Y = \begin{bmatrix} \rho \\ \rho\,\vec{u} \\ \rho(e + \frac{u^2}{2}) \end{bmatrix} \qquad \vec{y} = \begin{bmatrix} \rho\,\vec{u} \\ \rho\,\vec{u}\otimes\vec{u} + P - \sigma' \\ \rho(e+\frac{u^2}{2})\;\vec{u} + P\,\vec{u}-\vec{q} \end{bmatrix}$$

F is a source terme
σ' is the viscous tensor
$\vec{q}$ is the heat flux

The solution to the problem (1) is obtained by Newton iteration method. At each step a flux prediction is performed and the conservation equations are strictly satisfied. A key point of the numerical method is to write the symetrie impulsion flux tensor under the following form : [1]

$$\bar{\bar{\Pi}} = \mathcal{P} \otimes \bar{\bar{I}} + \mathcal{N} \otimes \bar{\bar{I}} + \bar{\bar{\mathcal{M}}} \qquad (2)$$

where $\mathcal{P}$ is a scalar defined by (in cartesian coordinates)

$$\mathcal{P} = \frac{1}{n}\,(\text{trace}\ \bar{\bar{\Pi}}) = P + \frac{1}{n}\sum_{i=1}^{n} (\rho\, u_i\, u_i - (\lambda + 2\,\mu)\,\frac{\partial u_i}{\partial x_i}\,) \qquad (3)$$

$\vec{\mathcal{N}}$ is a vector such as $\sum_{i=1}^{n} \mathcal{N}_i = 0$

$$\mathcal{N}_i = \rho u_i u_i - 2\mu \frac{\partial u_i}{\partial x_i} - \frac{1}{n}\sum_{k=1}^{n} (\rho u_k u_k - 2\mu \frac{\partial u_k}{\partial x_k}) \quad (4)$$

$\bar{\bar{\mathcal{M}}}$ is a symetrie tensor such as

$$\mathcal{M}_{ij} = \{\rho u_i u_j - \mu (\frac{\partial u_i}{\partial x_j} + \frac{\partial u_j}{\partial x_i})\} \quad (1-\delta_{ij}) \quad (5)$$

It can be shown that this variables are solution of the following system.

$$\left[\begin{bmatrix} \frac{\partial^2}{\partial t^2} & & \\ & \frac{\partial}{\partial t} & \\ & & \frac{\partial}{\partial t} \end{bmatrix} + T - \begin{bmatrix} \alpha & & \\ & \mu & \\ & & \mu \end{bmatrix} - \begin{bmatrix} \Delta & \theta_1 & \theta_2 \\ \theta_1 & \Delta & 0 \\ \theta_2 & 0 & \Delta \end{bmatrix}\right] \begin{bmatrix} \mathcal{P} \\ \mathcal{N} \\ \mathcal{M} \end{bmatrix} = \begin{bmatrix} A_1 \\ A_2 \\ A_3 \end{bmatrix} \quad (5)$$

where :

$$\begin{bmatrix} \frac{1}{n}\frac{\partial}{\partial t}\,\vec{u}.\overrightarrow{grad} & \frac{\partial}{\partial t}(\frac{2}{n} u_i \frac{\partial}{\partial x_i} - \frac{1}{n}\sum_{j\neq i} u_j \frac{\partial}{\partial x_j}) & \frac{\partial}{\partial t}(u_i \frac{\partial}{\partial x_j} + u_j \frac{\partial}{\partial x_i}) \\ \frac{2}{n} u_i \frac{\partial}{\partial x_i} - \frac{1}{n}\sum_{i\neq i} u_j \frac{\partial}{\partial x_j} & \frac{2}{n}\vec{u}.\overrightarrow{grad} & \frac{2}{n} u_i \frac{\partial}{\partial x_i} - \frac{1}{n}\sum_{j\neq i} u_j \frac{\partial}{\partial x_j} \\ u_i \frac{\partial}{\partial x_j} + u_j \frac{\partial}{\partial x_i} & \frac{2}{n} u_i \frac{\partial}{\partial x_i} - \frac{1}{n}\sum_{j\neq i} u_j \frac{\partial}{\partial x_j} & \vec{u}.\overrightarrow{grad} \end{bmatrix}$$

$$\Delta = \sum_i \frac{\partial^2}{\partial x_i^2}$$

$$\theta_1 = \frac{2}{n}\frac{\partial^2}{\partial x_i^2} - \frac{1}{n}\sum_{j\neq i} \frac{\partial^2}{\partial x_j^2}$$

$$\theta_2 = \frac{2\,\partial^2}{\partial x_i\,\partial x_j}$$

A similar operator can be obtained for the total energy flux.

Numerical Results

1-D Results

This method has been used to compute solution of gas dynamics problems with discontinous initial conditions [2]. Figure 1 shows the propagation and the reflection of shock and expansion waves in a four meter closed ends pipe, the initial conditions are

$0 < x < 3$ m $\quad$ P = 12. MPa

$3 < x < 4$ m $\quad$ P = 0.1 MPa

A two phase flow instability calculation was performed. The aim of this test is to predict the transient resulting from heat addition to upward water flow in a vertical pipe with the homogeneous equilibrium model. The power (18 kW) applied a time t = 2 s produces a fast vaporization of the liquid, the pressure goes up in the pipe and a flow reversal occurs in the inlet section, then the flow reverses again and an oscillatory flow is established with a period of about two seconds (fig. 2) [3]. One can notice that for large courant number, oscillations are damped.

2-D Results

The first problem treated is the two dimensional flow induced by the in-plane motion of one wall for various Reynolds number, figures 3 and 4 shows the velocity and the pressure fields obtained for Re = 1 and $Re = 10^3$ [1].

For the second problem a viscous and heat-conducting fluid is considered, at initial time a heat flux is applied on the two vertical sides of the cavity and the fluid moves under effect of gravity (fig. 5, 6).

At last we present a 2-D calculation of the transient behaviour of the flow in the hot plenum of a fast breeder reactor due to a variation of the inlet température. Figure 7 shows the flow field at time t = 30 s and figure 8 gives the evolution of the outlet temperature.

Conclusion

In this paper we have shown the capability of the control variable method to compute various fluid problems. Until now 1-D or 2-D calculations was performed, but the method can easily generalized to 3-D flow problems. The implicitness of the method allows great time steps which is useful to obtain rapidly steady state flows. Further improvements have to be made in order to reduce numerical dissipation in order to obtain more precise solution.

References

[1] The Control Variable Method : A fully implicit Method for solving conservation equations for unsteady multidimensional Fluid Flow.
G. LE COQ, G. BOUDSOCQ, P. RAYMOND
CEA - CONF 6713

[2] Expansion and compression shock wave calculation in pipes with the C.V.M. numerical Method
P. RAYMOND, G. LE COQ, P. CAUMETTE, M. LIBMANN
CEA - CONF 6715

[3] The Triton Computer code, Finite Difference Methods for one dimensional single or two Fluid Flow Transient Computation.
G. LE COQ, M. LIBMANN, P. RAYMOND, Y. SOUCHET, J.P. SURSOCK
ANS-ENS Meeting MUNICH (1981)

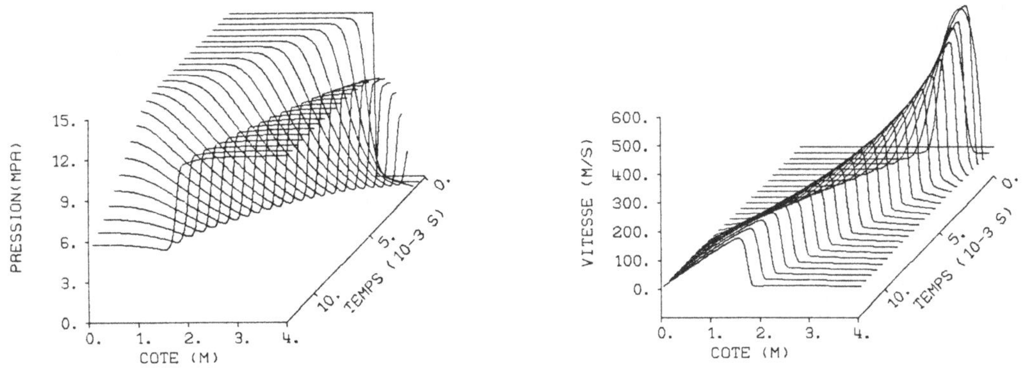

Figure 1 - Propagation and reflection of shock wave

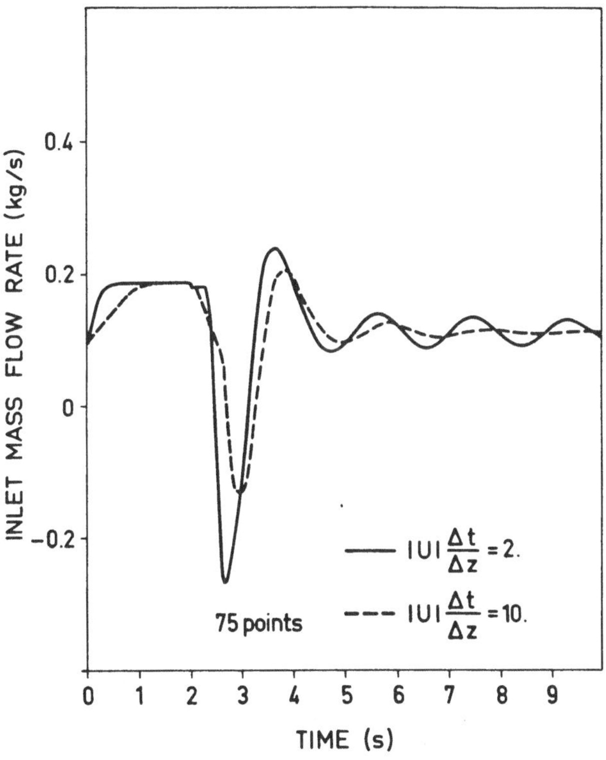

Figure 2 - Two phase flow instability calculation

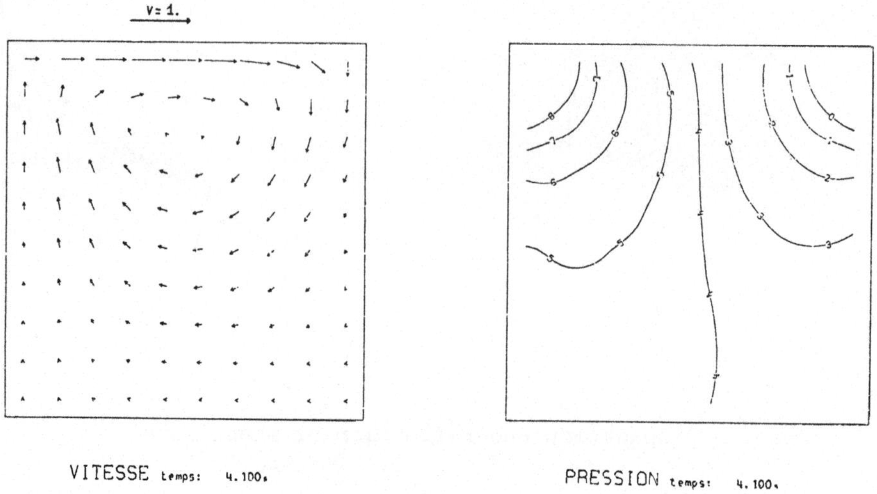

Figure 3 - Viscous Driven Flow calculation (μ = 1)

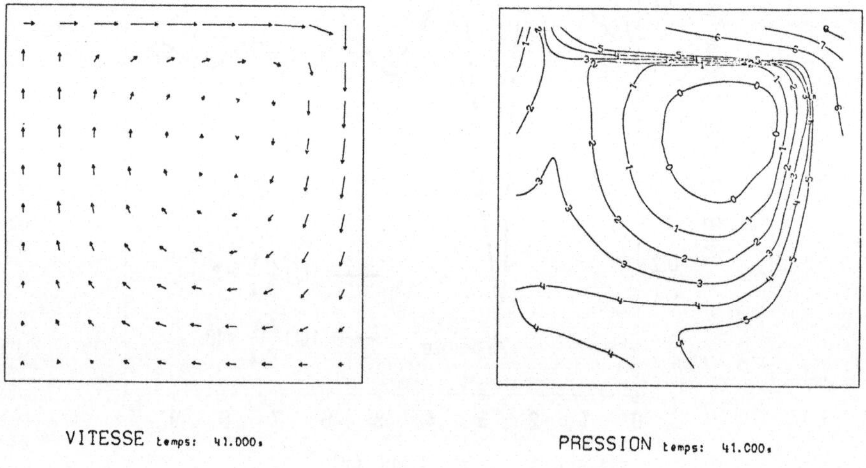

Figure 4 - Viscous Driven Flow calculation ($\mu = 10^{-3}$)

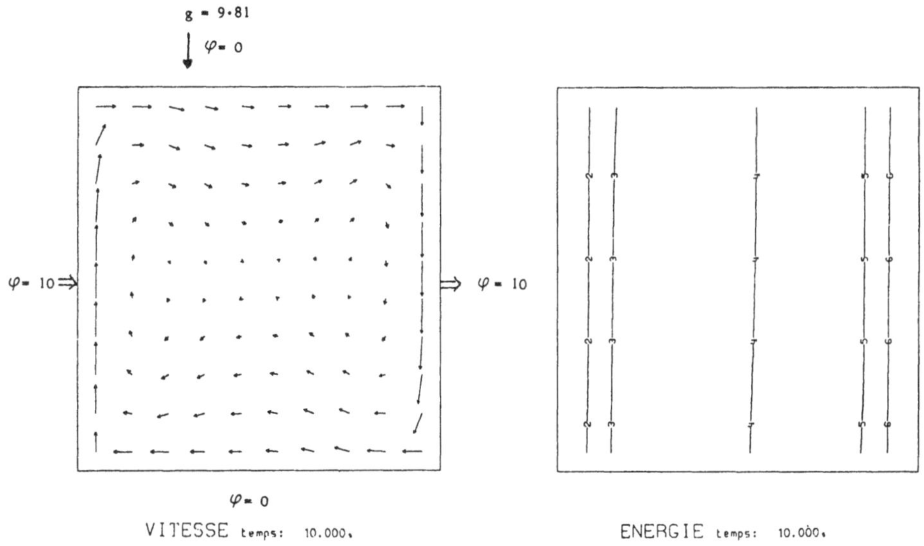

Figure 5 - Buoyancy Driven Flow, velocity-field and Iso-Energy curves at t = 10 s.

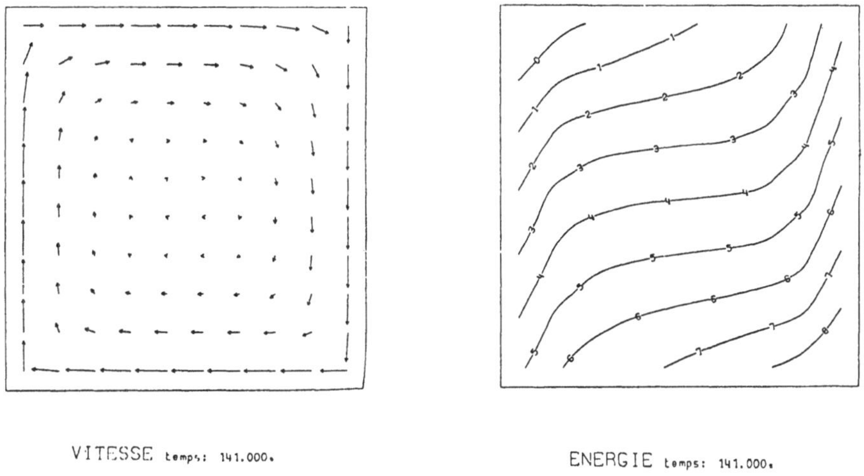

Figure 6 - Buoyancy Driven Flow : velocity Field and Iso-Energy curves for steady state.

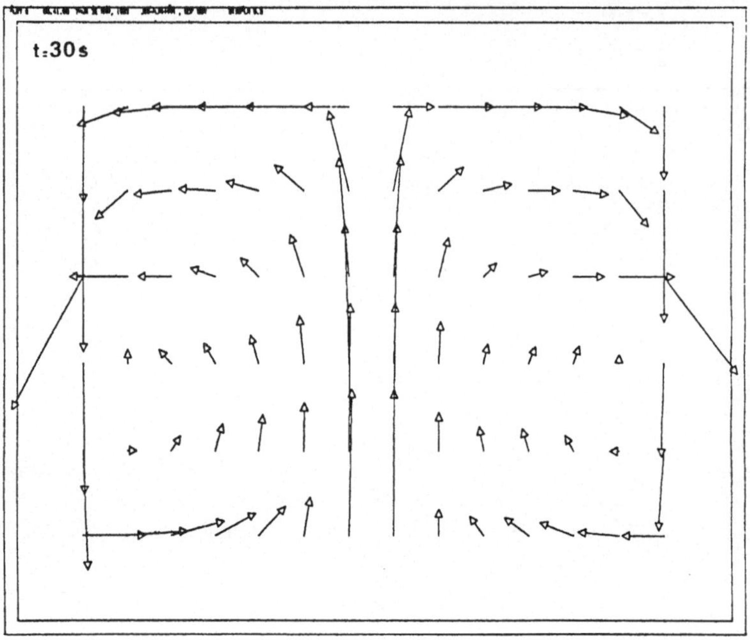

Figure 7 - L.M.F.B.R. Temperature Transient : hot Plenum velocity Field at t = 30 s.

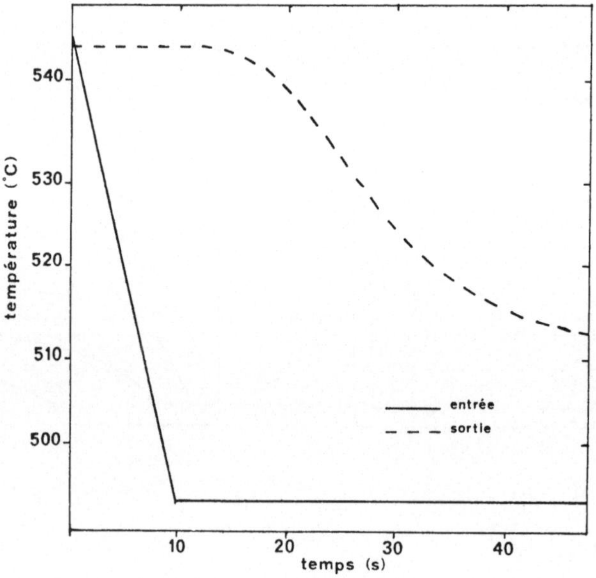

Figure 8 - L.M.F.B.R. Temperature Transient: Hot Plenum inlet and outlet Temperatures.

L.E.A.

UN CODE HYDRODYNAMIQUE MULTIFLUIDE BIDIMENSIONNEL

A.Y. LE ROUX (UNIVERSITE DE BORDEAUX I)

P. QUESSEVEUR (CENTRE D'ETUDES DE GRAMAT)

INTRODUCTION.

Nous proposons dans cette étude une méthode de résolution des problèmes d'hydrodynamique multifluide à deux dimensions d'espace en régime instationnaire. Les méthodes classiques pour résoudre ce type de problèmes peuvent être classées en deux catégories selon que l'on effectue les calculs dans un repère Lagrangien, ou dans un repère Eulérien. Dans la méthode proposée ici, la solution est calculée sur un maillage Lagrangien et projetée ensuite sur le maillage initial fixe, ceci à chaque itération. La phase Lagrangienne facilite le traitement des écoulements multifluides et permet d'obtenir une précision supérieure aux méthodes Eulériennes classiques. Les projections successives sur les différents maillages introduisent une diffusion numérique que l'on élimine par la suite selon une méthode de type flux corrigés, cf[1]. Dans une première approche on étudie la construction de ce schéma pour un problème monofluide.

I - PRESENTATION DU PROBLEME. Par la suite ρ représente la densité, u et v sont les composantes de la pression et e désigne l'énergie totale. L'énergie interne est notée par $\varepsilon = e - 1/2(u^2+v^2)$, et $p = f(\rho,\varepsilon)$ désigne la pression. Le système de l'hydrodynamique exprimé sous forme conservative s'écrit, avec $(x,y,t) \in \Omega \times]0,T[$, $T > 0$, et Ω un ouvert de $\mathbb{R}^2$:

$$(S1)\quad \begin{cases} \dfrac{\partial\rho}{\partial t} + \dfrac{\partial\rho u}{\partial x} + \dfrac{\partial\rho v}{\partial y} = 0 \\[2mm] \dfrac{\partial\rho u}{\partial t} + \dfrac{\partial\rho u^2}{\partial x} + \dfrac{\partial\rho uv}{\partial y} + \dfrac{\partial p}{\partial x} = 0 \\[2mm] \dfrac{\partial\rho v}{\partial t} + \dfrac{\partial\rho uv}{\partial x} + \dfrac{\partial\rho v^2}{\partial y} + \dfrac{\partial p}{\partial y} = 0 \\[2mm] \dfrac{\partial\rho e}{\partial t} + \dfrac{\partial\rho eu}{\partial x} + \dfrac{\partial\rho ev}{\partial y} + \dfrac{\partial pu}{\partial x} + \dfrac{\partial pu}{\partial y} = 0 \end{cases}$$

Par la méthode des directions alternées le système (S1) est décomposé en deux systèmes monodimensionnels. Ainsi le système à résoudre selon la direction OX s'écrit :

$$
(S2)\quad \left\{\begin{array}{l}
\dfrac{\partial\rho}{\partial t} + \dfrac{\partial\rho u}{\partial x} = 0 \\[2ex]
\dfrac{\partial\rho u}{\partial t} + \dfrac{\partial\rho u^2}{\partial x} + \dfrac{\partial p}{\partial x} = 0 \\[2ex]
\dfrac{\partial\rho v}{\partial t} + \dfrac{\partial\rho uv}{\partial x} = 0 \\[2ex]
\dfrac{\partial\rho e}{\partial t} + \dfrac{\partial\rho eu}{\partial x} + \dfrac{\partial pu}{\partial x} = 0
\end{array}\right.
$$

L'ouvert Ω est supposé de forme rectangulaire, et on introduit la discrétisation suivante :

$$x_0 < \ldots < x_i < x_{i+1} < \ldots < x_I$$

$$y_0 < \ldots < y_j < y_{j+1} < \ldots < y_J$$

avec $I \in \mathbb{N}$, $J \in \mathbb{N}$. Une itération comporte donc la résolution du système (S2) pendant un certain pas de temps Δt sur les bandes $[y_j, y_{j+1}]\times[x_0,x_I]$ pour tout $j \leqslant J-1$, suivie par la résolution du système analogue obtenue en direction y pendant le même pas de temps Δt. On expose à la suite la résolution du système (S2) pour indice j fixé.

II - CONSTRUCTION DE LA METHODE. On note $\Phi^n(.,t_{n-1})$, la solution à l'instant t_{n-1}, Φ valant respectivement ρ, u, v et e. Φ est à valeur dans l'espace V_h défini par :

$$V_h = \left\{\Phi \in L^\infty \ ;\ \Phi = \Phi_{i+1/2} \ \text{ sur }]x_i, x_{i+1}[\ ,\ \forall\, i \leqslant I-1\right\},$$

On introduit aussi les points du maillage intermédiaire $x_{i+1/2} = \dfrac{x_i + x_{i+1}}{2}$.

Dans [3] A.Y. Le Roux et E. Le Gruyer montrent qu'il est possible de décomposer un système analogue à (S2) en deux systèmes équivalents par une technique de splitting d'opérateurs. Ainsi on considère le système suivant qui traite uniquement les termes de convection :

$$
(S3)\quad \left\{\begin{array}{l}
\dfrac{\partial\rho}{\partial t} + \dfrac{\partial\rho u}{\partial x} = 0 \\[2ex]
\dfrac{\partial\rho u}{\partial t} + \dfrac{\partial\rho u^2}{\partial x} = 0 \\[2ex]
\dfrac{\partial\rho v}{\partial t} + \dfrac{\partial\rho uv}{\partial x} = 0 \\[2ex]
\dfrac{\partial\rho e}{\partial t} + \dfrac{\partial\rho eu}{\partial x} = 0
\end{array}\right.
$$

et le système (S4) qui traite la partie ondulatoire.

$$(S4)\begin{cases} \frac{\partial\rho}{\partial t} + \frac{\partial\rho u}{\partial x} = 0 \\ \frac{\partial\rho u}{\partial t} + \frac{\partial p}{\partial x} = 0 \\ \frac{\partial\rho e}{\partial t} + \frac{\partial pu}{\partial x} = 0 \end{cases}$$

Ces deux systèmes sont couplés par l'équation de conservation de la masse. Le premier système est traité par un schéma de type Lagrange-Euler, cf [4]. La solution est calculée sur le maillage Lagrangien déduit du maillage fixe par translation, puis projetée sur la grille initiale Eulérienne.

L'écriture de la phase Lagrangienne est particulièrement simple :

$$\overset{*}{\phi}{}_i^n = \delta_i^n \ \phi_i^n$$

où ϕ_i^n désigne la solution sur la maille $]x_{i-1/2}, x_{i+1/2}[$ à l'instant t_{n-1}, et δ_i^n la dilatation de cette maille pendant le temps Δt.

On note $\tilde{\phi}^{n+1}$ la solution obtenue après la projection Eulérienne, et on discrétise alors les équations de (S4) sur le maillage fixe.

$$\bar{\phi}_{i+1/2}^{n+1} = \tilde{\phi}_{i+1/2}^{n+1} - \frac{\Delta t}{(x_{i+1} - x_i)} \left[b_{i+1}^{n+1} - b_i^{n+1}\right],$$

avec $\phi = \rho u$, $b = p$ et $\phi = \rho e$, $b = pu$. La double projection introduit une diffusion numérique qu'il est possible d'éliminer à ce stade du calcul, ou bien après la projection Eulérienne. La solution des problèmes de Riemann étant constante le long des segments $x_i \times]t_{n-1}, t_n[$, les valeurs de la pression et de la vitesse au point x_i à l'instant t_n sont obtenues par une résolution approchée de ces problèmes.

L'extension à l'étude des problèmes multifluides et la complexité des lois de pression considérées nous ont amené à choisir un schéma prédicteur de type Lagrangien pour le calcul des valeurs u_i^{n+1}, p_i^{n+1}. Le pas de temps Δt est alors régi par une condition de Courant-Friedrichs Levy imposée par le schéma Lagrangien. Dans le cas de systèmes plus simples il est possible d'envisager des méthodes pour prédire la vitesse et la pression qui autoriseront des pas de temps importants.

Conclusion. Moyennant certains aménagements, la généralisation à l'étude des écoulements multifluides ne pose pas de problèmes majeurs, cf [2]. Les différents fluides sont alors représentés par leurs volumes partiels. De nombreux tests numériques ont été réalisés, [5], qui mettent en évidence les qualités de rapidité et de précision de la méthode L.E.A.

BIBLIOGRAPHIE

[1] J.P. BORIS, D.L. BOOK : Flux-Corrected Transport I : SHASTA, a fluid transport algorithm that works.

[2] C. COSTE, B. MELTZ, J. OVADIA : E.A.D. Un nouvel algorithme en hydrodynamique multifluide.

[3] E. LE GRUYER, A.Y. LE ROUX : A two dimensional Lagrange-Euler technique for gas dynamics.

[4] A.Y. LE ROUX, P. QUESSEVEUR : Convergence of an Antidiffusive Lagrange-Euler scheme for quasi linear equations.

[5] P. QUESSEVEUR : Analyse et génération d'un code hydrodynamique bidimensionnel. Thèse de 3ème cycle.

SPECTRAL SIMULATIONS OF 2D COMPRESSIBLE FLOWS

J. Léorat*, A. Pouquet**, J.P. Poyet*** and T. Passot*
*Univ. Paris VII, CNRS and Meudon Observatory, 92195 Meudon, France
**CNRS and Nice Observatory, 06003 Nice, France
*** CNRS and Toulouse Observatory, 31400 Toulouse, France

1) Introduction

Astrophysical flows, as found for example in the interstellar medium , are generally turbulent ; another consequence of the large scales which are available in such flows is the fact that compressibility effects may be important. This is particularly evident when dealing with the formation of stars, which is still an open problem, involving a lot of unanswered questions (concerning angular momentum, magnetic fields, molecular clouds, radiative heating and cooling etc...). Observations show that in the region where stars are forming, supersonic speeds are commonly encountered. To be able to predict for example mass spectrum of stars at the time of formation, it is first necessary to understand the basic properties of supersonic turbulence. Numerical simulations of this type of turbulence, which is very difficult to realize experimentally, represent a unique mean to study most astrophysical flows.

2) Equations

We choose to concentrate first on dynamical properties and follow the evolution of a single compressible fluid with a polytropic equation of state,

$$P \sim \rho^{\bar{\gamma}}$$

where ρ is the mass density, P the pressure and $\bar{\gamma}$ is a constant. The isothermal and adiabatic limits are thus easily recovered. In situations where the temperature field is important, it is necessary to take a thermodynamic state law and to add the second law of thermodynamics to the dynamical equations,

$$\partial\rho/\partial t + \nabla\rho\vec{u} = 0$$

$$\partial\rho\vec{u}/\partial t + \nabla(\rho\overline{\overline{uu}} + \overline{\overline{P}} + \overline{\overline{\sigma}}) = -\rho\nabla V$$

$$\dot{\vec{W}} = 4\pi G\rho$$

where $\vec{u}$ is the velocity, $\bar{\bar{\sigma}}$ the viscous stress tensor et V the gravitationnal potential.

Since in most astrophysical situations the influence of the boundaries of the flow may be neglected, we assume spatial homogeneity and take periodic boundary conditions : this assumption allows the use of a spectral method using Fourier space to deal with the spatial derivatives of the dynamical equations (Gottlieb and Orszag 1977). Time integration is performed in Fourier space using a second order Adams Bashforth scheme. With a Cray 1 Computer, the highest resolution which can be reached with 1M words in central memory is 256^2 grid points. Thus, 2-D flows with Reynolds numbers of the order of 600 and Mach numbers of the order of 3 may be simulated and each time step needs about 0.4s CPU. Taking the sound velocity (at the mean density) as unit, the characteristic dynamical time will be of the order of one (see the figures below for example) and will be reached in about 400s CPU.

3) Illustrations

The spectral code is particularly suited to study homogeneous turbulence. However, in presence of self gravitation, condensations ("clouds") may from and eventually are destroyed by the turbulence. Thus it is interesting to simulate basic events in compressible flows with self gravity : the gravitationnal collapse of a single cloud and the collision of two clouds. When the geometric scale of these events is small compared to the distance between them in the periodic flow which is computed, the simulation will give a satisfying approximation of an isolated event.

Figure 1 shows the result of a collision of two clouds with a non zero impact parameter. The initial relative velocity is of the order of the sound velocity and the density ratio between the cloud maximal density and the background density is 2.5. If absence of collision these two clouds would relax towards hydrostatic equilibrium with a smaller density ratio. The collision induces a gravitationnal collapse towards a single cloud : at final time (about half the box crossing time for a sound wave), the density ratio reaches a value greater than 100. . This simulation was performed with a small resolution (32^2 points only) and the CPU time needed is of the order of 5 seconds on a Cray 1, so that variations of the impact parameter are easy to explore. The zero impact case has been studied recently by Gilden (1984).

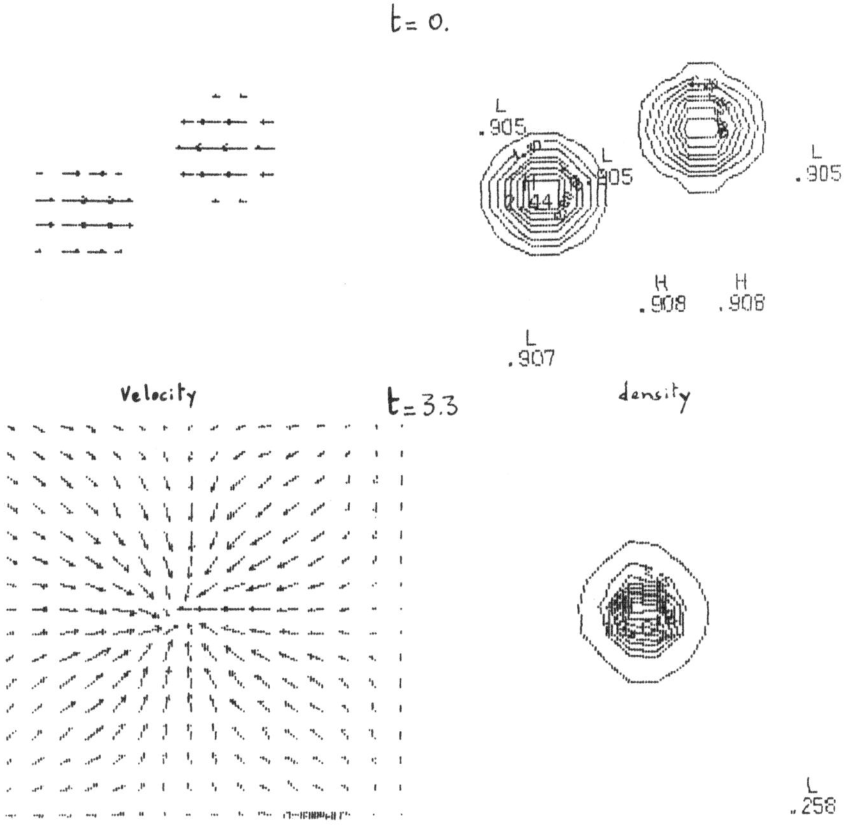

Fig 1 : Gravitationnal sticking of two clouds colliding with a relative velocity of the order of the velocity of sound. Velocity and density contours are shown. The final density contrast is above 100.

Another problem which can be studied in 2D is the gravitationnal collapse with angular momentum. The angular momentum of a collapsing cloud may be due for example to the initial angular momentum of two colliding clouds as seen above. Figures 2a and 2b compare the final state of collapse according to the value of the polytropic constant $\bar{\gamma}$. The initial state in both cases was a solid rotation velocity field with cylinder diameter nearby equal to the side of the box . The computation is performed using the highest resolution available (256^2 points). After the density reaches a maximum at the center, an annulus is formed, whose radius increases. The calculation breaks out when the density gradients become too large. Work is now in progress to control these gradients. The axisymetry in the final stage is striking ; fragmentation may eventually occur in the subsequent evolution of the annulus, since no axisymetry is introduced by the code (compare with Black and Bodenheimer 1976).

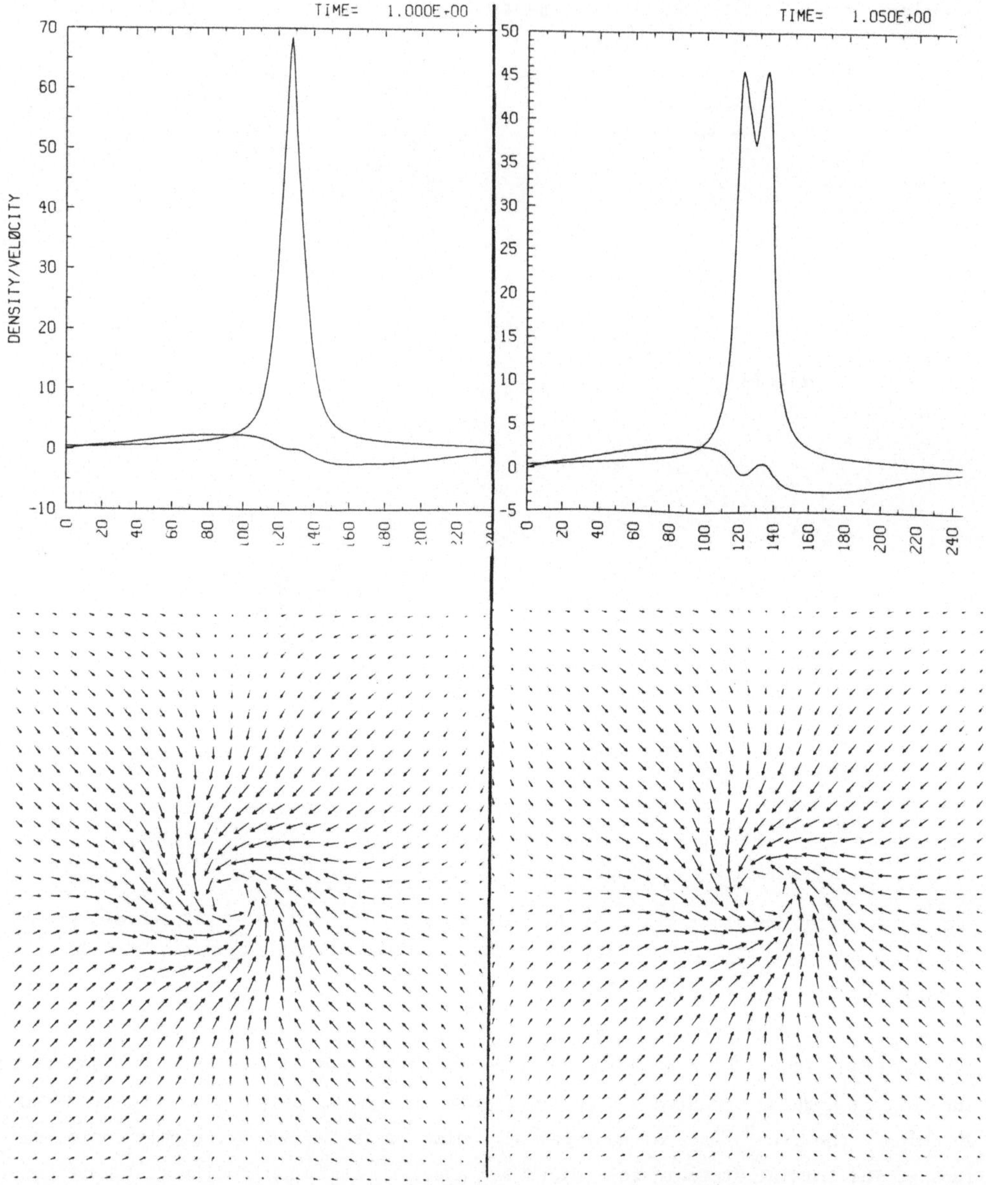

Fig 2a : formation of an annulus in the collapse of a cylinder with $\overline{\gamma} = 1$.
The velocity field and the density and radial velocity profiles are shown at $t = 1$ and 1.05.

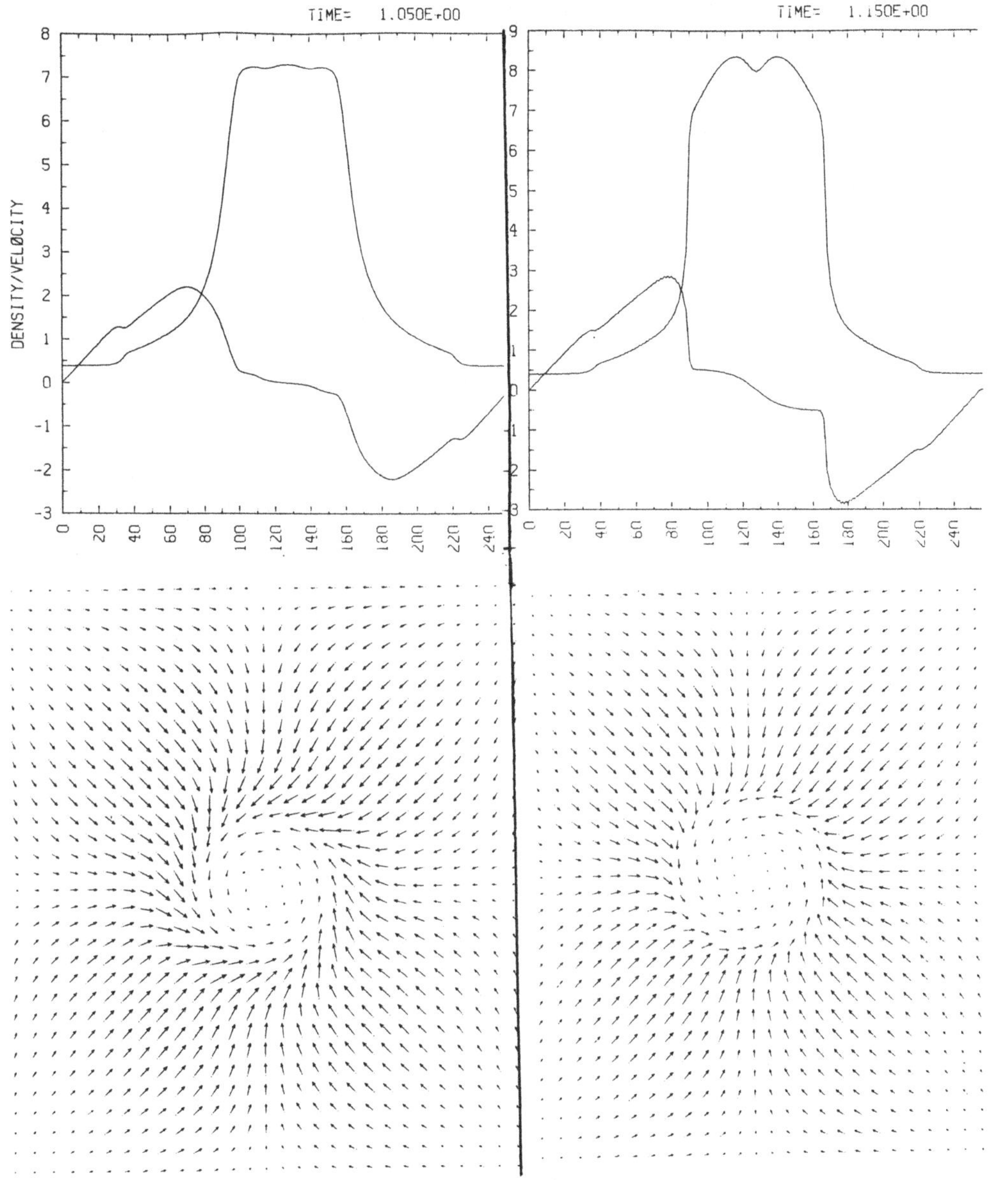

Fig 2b : formation of an annulus in the collapse of a cylinder with $\bar{\gamma} = 2$.
The velocity fields and the density and radial velocities are shown at $t = 1.05$ and $t = 1.15$.

References

Black and Bodenheimer, 1976, Ap.J., 206, 138.

Gilden, 1984, ApJ, 279, 335.

Gottlieb D.,Orszag S., 1977 Numerical Analysis of spectral methods, Siam Philadelphia.

Léorat J., Pouquet A., Poyet J.P., 1983, Numerical simulations of supersonic turbulent flows, to appear in "Problems of collapse and numerical relativity",Bancel and Signore Eds, Reidel

A MULTIGRID FACTORIZATION TECHNIQUE FOR THE FLUX-SPLIT EULER EQUATIONS

C. P. Li
NASA Johnson Space Center
Houston, Texas 77058

SUMMARY

The Euler equations formulated in characteristic components are solved by a time-like finite-difference method based on implicit multilevel grid sequencing. The conservative equations are made quasi-linear in metric coefficients in order to use upwind difference approximation of second order for the entire domain. Inside the computation region, the appropriate difference formula is automatically selected in accordance with the sign of the characteristics. When the flux components are originated outside the region, they are discarded and boundary conditions are imposed. Because the propagation path of signals is properly accounted for, higher accuracies of the solution and greater robustness of the numerical procedure are obtained. The implicit factorization procedure, which relies on the solution of four scalar matrices rather than of one block pentadiagonal matrix to save computation time, has removed severe restrictions on the time-step increments. Furthermore, the convergence rate is accelerated by a multigrid algorithm that switches the implicit procedure from fine to successively coarser grid levels. Newton's method is used to linearize the difference equations at the beginning of each step, then the correction vector is determined from the factorization technique applied to each grid level. Two-dimensional examples of a supersonic inlet flow and a transonic airfoil flow are considered in this study.

FORMULATION

The conservation-law form of the inviscid flow equations is given in generalized coordinates ξ and η,

$$U_t + \bar{F}_\xi + \bar{G}_\eta = 0 \tag{1}$$

Since the flux vectors of the Euler equation are the homogeneous function of the conserved variables U, they can be divided into two parts in accordance with the positive and negative eigenvalues of the Jacobians $\partial\bar{F}/\partial U$ and $\partial\bar{G}/\partial U$ as suggested by Steger and Warming,(1) or into three parts each associated with the individual eigenvalues as proposed by Reklis and Thomas.(2) Both versions can be approximated by one-sided differencing related to the sign of eigenvalues; however, the latter seems to have a slight advantage in treating the boundaries with a unified algorithm. The convective fluxes are split as follows.

$$\bar{F} = J\rho \sum_{\ell=-1}^{1} \gamma_\ell \lambda_{\xi\ell} \begin{bmatrix} 1 \\ u + \ell c\bar{\xi}_x \\ v + \ell c\bar{\xi}_y \\ q + \ell \bar{u}c/\bar{\xi} + |\ell|\gamma e \end{bmatrix}, \bar{G} = J\rho \sum_{\ell=-1}^{1} \gamma_\ell \lambda_{\eta\ell} \begin{bmatrix} 1 \\ u + \ell c\bar{\eta}_x \\ v + \ell c\bar{\eta}_y \\ q + \ell \bar{v}c/\bar{\eta} + |\ell|\gamma e \end{bmatrix} \tag{2}$$

where $\gamma_\ell = (1 - |\ell|)(\gamma - 1)/\gamma + \ell\gamma/2$, $\lambda_{\xi\ell} = \bar{u} + \ell c\bar{\xi}$, $\lambda_{\eta\ell} = \bar{v} + \ell c\bar{\eta}$, $\bar{u} = \xi_x u + \xi_y v$, $\bar{v} = \eta_x u + \eta_y v$, $\bar{\xi}_x = \xi_x/\bar{\xi}$, $\bar{\xi}_y = \xi_y/\bar{\xi}$, $\bar{\xi} = (\xi_x^2 + \xi_y^2)^{1/2}$, and $\bar{\eta}_x = \eta_x/\bar{\eta}$, $\bar{\eta}_y = \eta_y/\bar{\eta}$, $\bar{\eta} = (\eta_x^2 + \eta_y^2)^{1/2}$. Standard notation is used here for flow variables; viz: the density ρ, the pressure p, and the velocity components u and v in Cartesian coordinates; the total internal energy $\varepsilon = e + 0.5q$, and $q = u^2 + v^2$; and the internal energy $e = C_v T$, which relates to p and ρ by the equation of state. The sonic speed is $c = (\gamma p/\rho)^{1/2}$; γ is the ratio of specific heat. The integer ℓ is introduced to simplify the expression, in which each eigenvalue associates with $\ell = -1$, 0, or 1. The conventional matrix of eigenvalues consists of four components:

$$\Lambda_\xi = diag\ (\bar{u} - c\bar{\xi}, \bar{u}, \bar{u}, \bar{u} + c\bar{\xi}),\ \Lambda_\eta = diag\ (\bar{v} - c\bar{\eta}, \bar{v}, \bar{v}, \bar{v} + c\bar{\eta})$$

On the basis of the local eigenvalues, second-order one-sided differences are used upstream or downstream at each grid point for the three subflux vectors. The order of difference formula reduces to one for points adjacent to the wall. Furthermore, the subflux vectors are excluded from the calculation in case their corresponding characteristics originated externally. Appropriate boundary conditions are then assigned to the wall point. Details have been discussed by the author in Ref. 3.

AN UPWIND FACTORIZATION TECHNIQUE

The work done for flux-splitting and upwind schemes involves a larger amount of computation than for the usual flux vector and the central scheme. Hence, it is desirable to consider an implicit technique which will allow a greater time-step increment than required by the stability criterion. Some of the features of the implicit technique are highlighted here. Let Δv and Δu denote the unknown correction vectors abbreviated for $(\Delta\rho, \Delta u, \Delta v, \Delta e)^T$ and $(\Delta\rho, \Delta(\rho u), \Delta(\rho v), \Delta(\rho\varepsilon))^T$, respectively. The solution procedure for implicit calculation is implemented in four steps.

$$\begin{aligned}
&1.\ \Delta v_{i,j} = P^{-1}_{i,j}\, r_{i,j} \\
&2.\ (I + \Delta t \bar{\Lambda}_\xi \delta_\xi)\, \Delta w_{i,j} = T^{-1} \Delta v_{i,j} \\
&3.\ (I + \Delta t \bar{\Lambda}_\eta \delta_\eta)\, \Delta w_{i,j} = S^{-1} T\, \Delta w_{i,j} \\
&4.\ \Delta v^{k+1}_{i,j} = S\, \Delta w_{i,j}
\end{aligned} \tag{3}$$

where $r_{i,j}$ refers to a component in the matrix-vector product after replacing the derivatives of $\bar{F}$ and $\bar{G}$ by difference formulas. The subscripts i and j denote the spatial location of a grid network ranging from $i = 1$ to $Imax$ and $j = 2$ to $Jmax$. Since the difference approximations are type-dependent, overswitching from downwind to upwind schemes may occur when the magnitude of the characteristics is very small; viz $|\lambda_\ell| \ll c$. To alleviate the frequent change near the stagnation or sonic points and lower the error $\Sigma_{i,j}\, |r_{i,j}|^2$ or $max\ |r_{i,j}|$ efficiently, a stabilization mechanism is introduced such that the selection of a type-dependent scheme is controlled by both the orientation and the magnitude of the local characteristics. The notation P, T, S and the solution procedure for a pentadiagonal system of equations are given in Ref. 3. The step increment $\Delta t = CFL{*}min\ [1/|\bar{\Lambda}_\xi|_\ell,\ 1/|\bar{\Lambda}_\eta|_\ell]$, where CFL is the Courant number. For explicit second-order calculations, CFL must be equal to or less than 0.25.

AN IMPLICIT MULTIGRID ALGORITHM

Although the computation efficiency can be raised by a factor of 2 or more, depending on the nature of the problem and the distribution of grid points by solving Eq. (1) implicitly rather than explicitly, there is room for further improvement. One candidate method having potential to accelerate the convergence rate and consequently to reduce the number of iterative cycles is known as the multigrid technique advocated by Brandt.(4) Only recently has the application of multigrids found its way into the explicit solution of Euler's equations. For example, Ni(5) and Jameson(6) have successfully adopted it in their time-like iterative procedures, and Jespersen[7] has followed a strict relaxation concept. There have been questions on how to combine the multigrid and the implicit methods and on what the resulting advantages are. This is a separate subject and will be discussed in the balance of this paper.

The single-grid implicit procedure for solving Eq. (1) may be summarized as follows:

$$L\,\Delta v_{i,j}^{k+1} = r_{i,j}^{k} \tag{4}$$

$$v_{i,j}^{k+1} = v_{i,j}^{k} + \Delta v_{i,j}^{k+1} \tag{5}$$

where the implicit operator L represents the four steps in Eq. (3), and the updated vector $v_{i,j}^{k+1}$ is related to the correction vector by Eq. (5).

To intermix the multigrid algorithm of Brandt[4] with the factorization algorithm, one version of the coarse-grid correction scheme has been attempted. The calculation starts from the finest grid and determines $\Delta v_{i,j}$ from the explicit solution. Then, the implicit solution is sought for all levels until the coarsest grid level is reached. Thus, the solution accuracy is maintained at the finest grid level, but corrections are obtained at all allowable grid levels. Three steps constitute the basic algorithm. Let subscripts h and $2h$ denote the grid levels; then,

$$\begin{aligned} \Delta v_{2h} &= T_h^{2h}\,\Delta v_h \\ L^{2h}\,\Delta v_{2h} &= \Delta v_{2h} \\ v_h &= v_h + I_{2h}^{h}\,\Delta v_{2h} \end{aligned} \tag{6}$$

where the first equation performs a simple transfer function, i.e., $(T_h^{2h}\,r_h)_{i,j}$ = $1/4r_{2i-1,2j-2}$, for a grid system i = 1 to $Imax$ and j = 2 to $Jmax$, $Imax$ and $Jmax$ being, respectively, odd and even integers. Since metrics are calculated along with $r_{i,j}$, a factor 4 is introduced to compensate the greater spacing between grids h = 2** (level -1), with level = 1 denoting the finest grid. The operator L corresponds to the inverse of the distribution function, whereas the operator I performs bilinear interpolation between two levels. Eq. (6) is to be used as many times as the number of grid levels; the final solution designated at time step k + 1 is more accurate than and closer to the steady results than the single-level solution because error components in multiple wavelength can be eliminated. This is a time-like evolution process more like the multigrid scheme introduced by Ni[5] and Jameson[6] but different from the schemes used by Brandt[4] and Jespersen.[7]

DISCUSSION

The first test case was a supersonic flow of Mach 3 entering an inlet channel. An oblique shock and an expansion fan are generated on the lower wall and reflected from the upper wall. The computed pressure contours in Fig. 1 exhibit clearly the disappearance of the reflected shock due to its interaction with an expansion fan. Figure 1 also shows the pressure distributions on the plate and the midplane. The shock width is between 6 to 8 points and free of oscillations for a uniform grid 49 × 19. The solution accuracy is superior to that of a conventional Euler solver and comparable to other flux-split versions. A comparison of the rate of convergence given in Fig. 2 has shown that the maximum error ($\Delta\rho/\rho$) from multigrid calculations levels off quickly, whereas the error from the single-grid calculation decreases as iteration continues. After introducing the stabilization mechanism discussed earlier, a similar trend was achieved. The convergence histories of the maximum and minimum wall pressure in the field computed with various levels of grid are presented in Fig. 3. The multigrid sequences involving three grid levels are found to have the fastest rate of convergence. The work reduction factor is estimated to be

1.7. The Courant number was equal to 1; that is 2 times greater than the one required by the explicit MacCormack scheme. The explicit and implicit operators used approximately equal computation time during one iteration.

The second test case was a Mach 0.75 flow over the NACA-0012 airfoil at 2° angle of attack. The grid shown in Fig. 4 is a 65 × 22 O-grid generated by GRAPE.(8) The airfoil was taken to lie between 0 and 1 on the x-axis; computation domain was contained in a circle of radius equal to 6 and centered at $x = 0$. The common cut designated by $I = 1 = Imax$ connects the trailing edge and the downstream wake. The grid spacing along the airfoil was not found as influential to the accuracy as the spacing normal to the wall. Hence, $\Delta n = (x_\eta^2 + y_\eta^2)^{1/2}$ was selected to be 0.01 or less. Figure 5 shows the convergence for the maximum and minimum values of pressure obtained from both single-level and multilevel calculations. The multigrid solution has yielded higher accuracy at the leading edge and on the upper wall immediately upstream of the shock. The pressure coefficients in Figs. 6 and 7 have shown significant differences between the two methods, and both predict the shock location a bit downstream from the location computed by other finer grid Euler solutions.(6) However, the present results seem to agree well with the potential solution.(9) Some of the controversies may be caused by the relatively coarse grid spacing or by the geometry approximation of the trailing edge at $I = 1 = Imax$. The multigrid solution was obtained from levels 17 × 7, 33 × 12, and 65 × 22 using $CFL = 10$. The work reduction factor is about 2, after accounting for 30% increase of computation time due to additional implicit calculations. Finally, the convergence history of the multilevel calculations indicated that the maximum error stops decreasing as soon as the flow variables have converged. This difficulty remains to be resolved.

CONCLUSION

A new method for solving Euler's equations has been developed and tested on two problems of transonic and supersonic speeds. The coarse-grid solutions (65 × 22) are generally satisfactory and well-behaved near shock waves and around the leading edge. The corresponding conventional Euler's equation solved by central differencing often fails to converge despite the less costly computation effort. When the multigrid technique is incorporated with the upwind implicit scalar procedure, numerical stability, solution accuracy, and convergence rate are substantially improved. A work reduction factor as high as 3 is estimated for a fine-grid solution (161 × 32) of transonic airfoil flow. The present procedure should be useful in reducing the number of iterations for other implicit methods, and in solving the Navier-Stokes equations at high Reynolds number conditions.

REFERENCES

1. Steger, J. L. and Warming, R. F., *J. Comp. Phys.*, Vol. 40, 1981, pp. 263-293.
2. Reklis, R. P. and Thomas, P. D., *AIAA J.*, Vol. 20, Sept. 1982, pp. 1212-1218.
3. Li, C. P., Paper 83-0560, AIAA 21st Aerospace Sciences Meeting, 1983.
4. Brandt, A., *AIAA J.*, Vol. 18, No. 10, Oct. 1980, pp. 1165-1172.
5. Ni, R. H., *AIAA J.*, Vol. 20, No. 11, Nov. 1982, pp. 1565-1571.
6. Jameson, A. and Baker, T. J., Paper 84-0093, AIAA 22nd Aerospace Sciences Meeting, 1984.
7. Jespersen, D. C., Paper 83-0124, AIAA 21st Aerospace Sciences Meeting, 1983.
8. Sorenson, R. L., NASA TM-81198, 1980.
9. Holst, T. L., *Lecture Series 1983-04*, Von Karman Institute, Belgium, 1983.

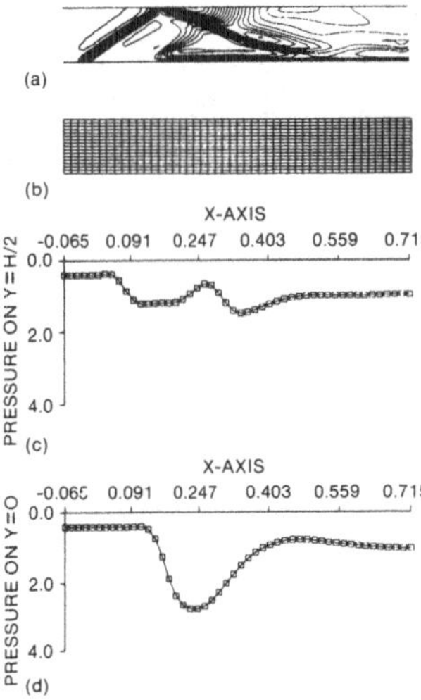

Fig. 1 Supersonic Mach 3 inlet flowfield results from implicit, multigrid calculations: (a) pressure contours; (b) 49 × 18 grid; (c) pressure distribution along the mid channel; (d) pressure distribution along the upper wall.

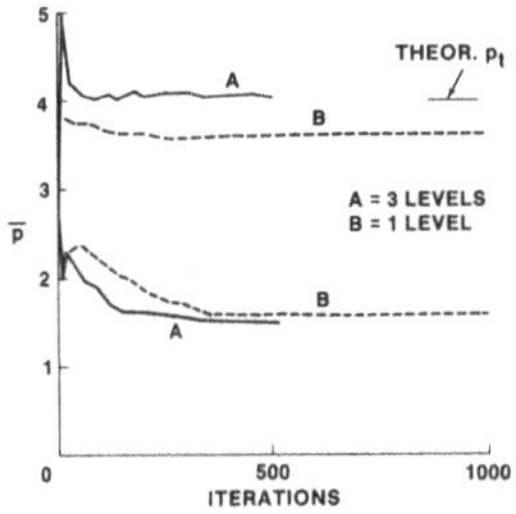

Fig. 5 History of convergence for the transonic $M = 0.75$, $\alpha = 2°$ airfoil problem: maximum and minimum local pressure.

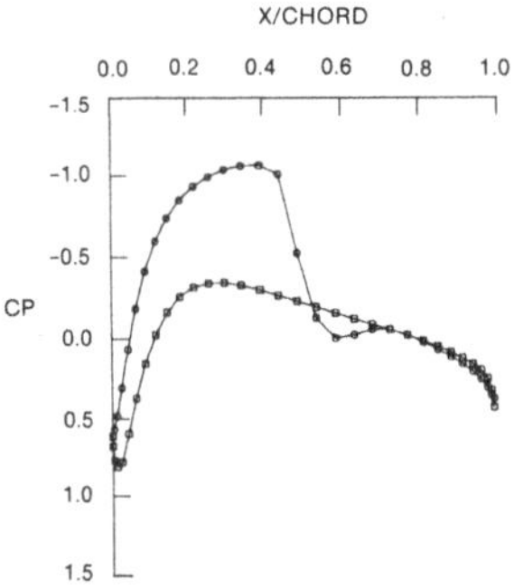

Fig. 6 Pressure coefficient distribution on the airfoil obtained from a single-level implicit calculation.

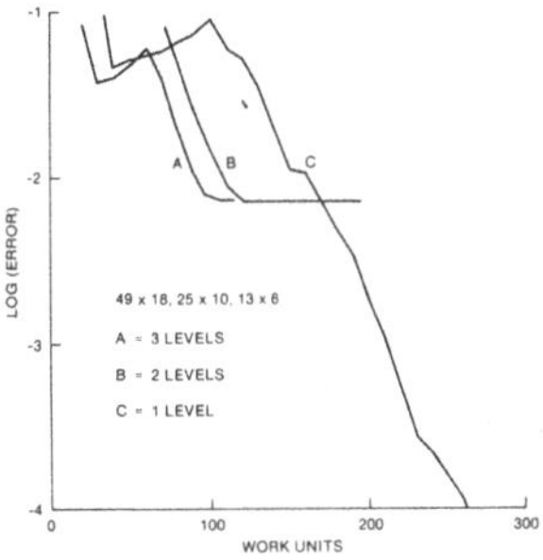

Fig. 2 History of convergence for the inlet problem: maximum local error.

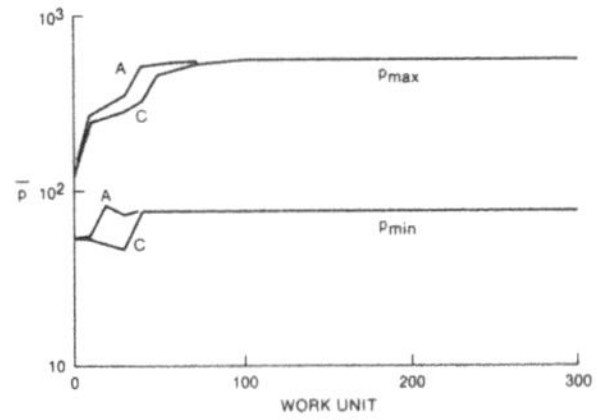

Fig. 3 History of convergence for the inlet problem: maximum and minimum local pressure.

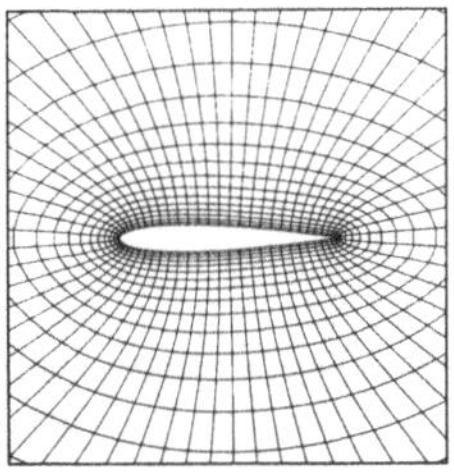

Fig. 4 An O-type 65 × 22 coarse grid for the NACA-0012 airfoil.

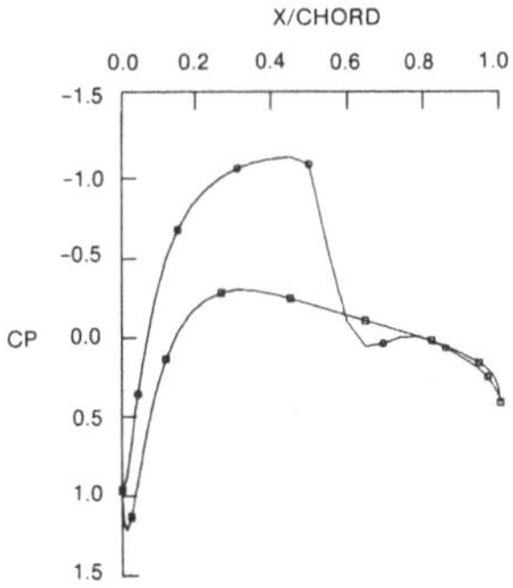

Fig. 7 Pressure coefficient distribution on the airfoil obtained from a multilevel implicit calculation.

Numerical Study of the Three-Dimensional Incompressible Flow Between Closed Rotating Cylinders .[1]

Avi Lin
Computer Science Department,
Technion - Israel Institute of Technology ,
Haifa , Israel , 32000 .
and
G. de Vahl Davis , E.Leonardi and J.A.Reizes
School of Mechanical and Industrial Engineering
The University of New South Walse, Kensington,
N.S.W. , Australia, 2033 .

Abstract

A new method for the solution of the vector potential - vorticity formulation of the equations of a fluid motion is presented in this paper. The fully coupled finite difference approximations to these equations have been solved using a general block tri - diagonal scheme. New boundary conditions for the vector potential are also presented. These conditions enables to satisfy exactly the conditions at the boundaries of the solution domain, like the mass flow through the boundaries. The method is found to converge more rapidly, and to be more accurate than previous solutions of the three dimensional vector potential - vorticity equations.

1. INTRODUCTION .

The three component velocity vector $\mathbf{U}$ and the scalar pressure P are the primitive variables describing three - dimensional incompressible flow at any point $\mathbf{r}$ in the three dimensional domain Ω with the boundary $\partial\Omega$. The flow field is governed by the following *continuity* and *momentum* equations

$$\nabla \cdot \mathbf{U} = 0 \tag{1}$$

$$(\mathbf{U} \cdot \nabla)\mathbf{U} = -\nabla p + \nu \nabla^2 \mathbf{U} \tag{2}$$

where ν is the flow kinematic viscosity coefficient. When solving these equations numerically, it is very difficult to satisfy the *continuity* equation iteratively especially where there is no explicit equation for the pressure. The three dimensional vector potential - vorticity method is one possible way to overcome these problems, since continuity is satisfied automatically and the pressure does not appear in the new formulation.

The variables of this method are the vector potential Ψ and the vorticity ω , defined by :

$$\mathbf{U} = \nabla \times \Psi \tag{3}$$

$$\omega = \nabla \times \mathbf{U} \tag{4}$$

---

[1] This research was partially supported by the Technion Research Foundation Grant No. 121-606 (1982) and by the National Energy Research, Development, and Demonstration Program of the Australian Commonwealth Department of Resources and Energy .

The governing equations therefore become:

$$\nabla^2 \Psi - \nabla(\nabla \cdot \Psi) + \omega = 0 \qquad (5)$$

$$\nabla \times (\omega \times \mathbf{U}) = \nu \nabla^2 \omega \qquad (6)$$

This six variables' (ψ , ω) scheme is in general much more complicated than eqs.(1,2). The main theoretical disadvantage is the correct definition of the boundary conditions for this system and the formulation of a stable scheme for solving these equations numerically, as will be discussed later.

Because of the definition of the vector potential the solution of eqs.(5) - (6) is not unique. Due to this flexibility, Ψ is usually chosen to be solinoidal [1,2] :

$$\nabla \cdot \Psi = 0 \qquad (7)$$

When imposing the last equation on the flow field may lead to wrong solutions of the flow. The present paper suggests a new approach for formulating the vector potential - vorticity boundary conditions and resents a numerical method which converges reasonably fast for the problems tested to date.

2. VECTOR POTENTIAL BOUNDARY CONDITIONS .

Although the governing equations seem to be quite simple and well defined, there has been considerable controversy about the precise boundary conditions which must be imposed on the vector potential [4] .

If the surface elements are piecewise smooth, they can be described locally by a mutually orthogonal right handed curvilinear coordinate system (x_1 , x_2 , x_n), where x_n denotes the outer normal and x_1 and x_2 the two tangential directions to $\partial\Omega$ and $\mathbf{e_i}$ is a unit vector along x_i , i = 1,2,n as is shown in figure (1a). Let us denote by s_i the three coordinates' scale factors. The original boundary conditions are defined by the *continuity* equation and the velocity vector given on $\partial\Omega$. Thus a two additional boundary conditions which do not contradict the other conditions can be freely chosen. In the present new approach $\mathbf{W}$ is a planar vector, which for the boundary normal to x_n is chosen as:

$$\mathbf{W} = w_1 \mathbf{e}_1 + w_2 \mathbf{e}_2 \qquad (8)$$

where :

$$w_1 = \alpha_n \frac{\partial}{\partial x_n}(s_1 \psi_1) \quad ; \quad w_2 = -\alpha_n \frac{\partial}{\partial x_n}(s_2 \psi_2) \qquad (9,10)$$

The $\mathbf{W}$ velocity is defined as " the induced velocity " due to the presence of a rigid boundary, and the parameter α_n is the " porosivity " or the " solidity " coefficient of this boundary. It turns out that α has to be defined as the ratio of the solid area of the boundary to the total boundary's area: $\alpha = 1$ is where the whole boundary is a solid wall, and is 0 for a free boundary, while for all other cases $0<\alpha<1$. In order to solve the system of equations uniquely, some relation between the $\mathbf{W}$'s components has to be assumed. In the present study the following relation has been used :

$$w_1 = w_2 = w \qquad (11)$$

With this assumption we get the following equations on the (x_1,x_2) boundary :

$$U_1 + w = \frac{1}{s_2 s_n} \frac{\partial}{\partial x_2}(s_n \psi_n) \qquad (12a)$$

$$U_2 - w = -\frac{1}{s_1 s_n}\frac{\partial}{\partial x_1}(s_n \psi_n) \qquad \textbf{(12b)}$$

$$\frac{\partial}{\partial x_n}(s_1 s_2 U_n) = -(\frac{\partial w}{\partial x_1} + \frac{\partial w}{\partial x_2}) \qquad (13)$$

and by combining eqs.(12a) and (12b) to the following Poisson equation for ψ_n is obtained:

$$\nabla_2^2(s_n \psi_n) = \frac{\partial}{\partial x_2}[s_2 s_n (U_1 + w)] - \frac{\partial}{\partial x_1}[s_1 s_n (U_2 - w)] \qquad (14)$$

where ∇^2_2 is the respective two dimensional Laplacian. Assuming for simplicity that this boundary is defined by four edges then a possible procedure for solving these boundary conditions' equations numerically is :

step [0] Assume values of w and ψ_n along one couple of neighboring edges.

step [1] Assume values $w(x_1,x_2), \psi_n(x_1,x_2)$ across the boundary domain.

step [2] Solve the inner field with the following boundary conditions for the vector-potential :

ψ_n - known, and

ψ_1, ψ_2 - have Newmann boundary conditions given by eq.(10).

step [3] Calculate the values of $\frac{\partial U_n}{\partial x_n}$ near the boundary.

step [4] Solve w using eq.(13).

step [5] Solve ψ_n using eq.(14).

step [6] If not converged go to step [2].

For the three dimensional field, it is necessary to specify in step [0] the values of w and ψ_n along three edges that are connected at one of the vertices of the domain's boundary. It turns out that there will be at least one edge which values will be calculated twice. Since

$$\int \mathbf{W} d\mathbf{s} = 0$$

along any close pass $\mathbf{s}$ around the boundaries, those values should be very close. It is not necessary to solve the inner field in step [2] till convergence is achieved with the given boundary conditions, since the relations between the boundary conditions, eqs.(12)-(13), are not necessarily fulfilled, as is discussed later in the paper.

3. FORMULATION OF THE TEST PROBLEM .

Usually, it is very comfortable to discuss the numerical procedure and the implementation of the boundary conditions in the context of the problem to be solved. In the present work a three dimensional flow field between two concentric cylinders has been considered. These finite length cylinders are closed at their both ends, while their axis of symmetry might have some inclination with respect to the direction of the gravity vector as is illustrated in figure 2. An isothermal fluid is assumed, where various parts of the solid boundaries are allowed to rotate, while the others are at rest.

The geometrical dimensions are also defined in this figure. R_o and R_i are the outer and the inner radii of the cylinders and L is their length. Let ω be the typical angular velocity, Δ be the reference length scale and V be the velocity used to undimensionalize the governing equations, where $V = \omega R_i$ and $\Delta \equiv R_o - R_i$. Thus the Reynolds number $R_e = \frac{V\Delta}{\nu}$, the aspect ratio of the configuration $l = \frac{L}{\Delta}$, and the radius ratio $\beta = \frac{R_o + R_i}{\Delta}$ are the three parameters defining the flow field configuration. The dimensionless co-ordinate system is r , φ , z and are also depicted on this figure where $\frac{\beta-1}{2} \le r \le \frac{\beta+1}{2}$, $0 \le \varphi \le 2\pi$ and $0 \le z \le l$.

It is convenient to present the variables of the vector potential - vorticity system by the following six functions' vector $\mathbf{q}$ defined as follows :

$$\mathbf{q} = (\psi_r , \psi_\varphi , \psi_z , \omega_r , \omega_\varphi , \omega_z) \tag{15}$$

Since the present numerical procedure is of the iterative type (artificial transient), the none steady version of the governing equations (3) - (4) can be expressed in terms of the vector variable $\mathbf{q}$ for the incompressible case (the normalized density is unity) as follows :

$$-\mathbf{T}\frac{\partial \mathbf{q}}{\partial t} + (\mathbf{A}_r + \mathbf{A}_\varphi + \mathbf{A}_z + \mathbf{K})\mathbf{q} + \mathbf{d} = 0 \tag{16}$$

where $\mathbf{A}_r$, $\mathbf{A}_\varphi$ and $\mathbf{A}_z$ are 6x6 differential operators acting in the r , φ and z directions respectively. $\mathbf{K}$ is a 6x6 algebraic operator. $\mathbf{T}$ is a 6x6 diagonal algebraic operator contains the artificial time derivative effects, where t is the time-like (iteration) co-ordinate. The detailed structure of these operators can be easily derived from the governing equations. The vector $\mathbf{d}$ depends only on the temperature gradients and it is assumed to be zero in the present study.

4. THE NUMERICAL PROCEDURE .

The numerical solution is obtained by spreading an equal spaced grid over the three-dimensional domain and approximating the partial differential operators at every inner grid point where standard second order central differencing mosels are assumed for all the derivatives. Hat $(\hat{\ })$ denotes the finite difference approximation of an operator, superscript denotes the iteration number and Δt denotes the artificial time step which really acts as an underrelaxation mechanism (that disappears as $\Delta t \to \infty$). Defining the finite difference operator $\hat{\mathbf{L}}$ as :

$$\hat{\mathbf{L}} = \hat{\mathbf{A}}_r + \hat{\mathbf{A}}_\varphi + \hat{\mathbf{A}}_z + \hat{\mathbf{K}} \tag{17}$$

the following iterative scheme is suggested:

$$\mathbf{T}\frac{\mathbf{q}^* - \mathbf{q}^{n-1}}{\Delta t} = \hat{\mathbf{L}}(\mathbf{q}^n) + \mathbf{d} \tag{18}$$

where $n-1 < * < n$. It is also assumed that

$$\mathbf{q}^n = \Sigma\mathbf{q}^* + (\mathbf{I}-\Sigma)\mathbf{q}^{n-1} \tag{19}$$

where $\mathbf{I}$ is the identity matrix and Σ is a 6x6 matrix which is usually diagonal. Denoting the correction in $\mathbf{q}$ by δ :

$$\delta = \mathbf{q}^n - \mathbf{q}^{n-1} \tag{20}$$

the following factorization is suggested :

$$(\mathbf{B}_1 - \mathbf{A}_r)(\mathbf{B}_2 - \mathbf{A}_\varphi)(\mathbf{B}_3 - \mathbf{A}_z)\ \delta = \hat{\mathbf{L}}(\mathbf{q}^{n-1}) + \mathbf{d} \tag{21}$$

so that

$$B_1 + B_2 + B_3 = \frac{1}{\Delta t}T\Sigma - K \qquad (22)$$

The rate of convergence of the solution depend very much on the choice of the **B**'s. The simple choice where all of the three **B**'s are equal to each another is not the best one, although a rapid rate of convergence is encountered for this case. Thus this choice was employed for most of the problems presented here, except for high Reynolds number fields where some priority was given to the r and z directions. For every time step (iteration) the solution for δ is obtained in three recursive steps defined by eq.(21).

Three major steps has been undertaken in order to increase the stability of the iterative technique as well as the rate of convergence:

(a) The matrix inversions of eq.(21) are done implicitly along the appropriate directions with the appropriate boundary conditions.

(b) The boundary conditions summarized in the next section are coupled implicitly into the inner field's finite difference system. coupled implicitly into the inner field's finite difference system.

(c) All the six algebraic equations at every grid point are solved simultaneously in a blocked coupled manner.

This numerical scheme is such that by an appropriate choice of the **B**'s and Σ one can reduce it to various known blocked ADI or Line Relaxation techniques.

5. NUMERICAL PRESENTATION OF THE BOUNDARY CONDITIONS.

Let us denote the four boundaries as follows : I - is the inner cylinder, II - is the outer cylinder, III - is the lower lid and IV - is the upper lid. The $\mathbf{W}$ variable has been defined initially along the three edges with the common vertice ($z=l, \varphi=0, r=R_o$)

Boundary I : Use here figure (1b). The normal component of the vorticity is always known and is given for the present boundary by:

$$\omega_r = \frac{1}{R_i}\frac{\partial U_z}{\partial \varphi} - \frac{\partial U_\varphi}{\partial z} \qquad (23)$$

which for most of the cases is zero. The other two components are obtained by using eqs.(3,4). By defining $\eta = h / R_i$, the following second order finite difference equation is obtained on the boundary:

$$\psi_{\varphi_2} - (1+\eta^2)(1-\eta)\psi_{\varphi_1} + \frac{h^2}{6}[\omega_{\varphi_2} + (2-\eta)\omega_{\varphi_1}] + S = 0 \qquad (24)$$

where

$$S = \frac{\eta}{2}(3\eta^2 - 3\eta + 2)\frac{\partial\psi_r}{\partial\varphi} + \frac{\eta}{2}(\eta+1)(\eta-2)rU_z \qquad (25)$$

$$-\frac{h}{3}\eta^2\frac{\partial^2}{\partial r\,\partial\varphi} + \frac{h^2}{2}[(1+\frac{\eta}{3})S_2 + \frac{h}{3}\frac{\partial S_2}{\partial r}]$$

Usually the cylinders do not move axially and $U_z = 0$. For the z vorticity component a similar equation can be obtained; but when using the condition given by eq.(11), the following relation can be generated:

$$\omega_\varphi + \omega_z = -\frac{C_i}{R_i^2} \qquad (26)$$

where C_i is a constant for the inner cylinder surface. The following relations for the vector potential, that can be obtained from eqs.(11), are used here:

$$\psi_\varphi + \psi_z = C_i \tag{27a}$$

$$w = \frac{\partial \psi_z}{\partial r} \tag{27b}$$

$$\frac{\partial w}{\partial z} + \frac{1}{R_i}\frac{\partial w}{\partial \varphi} = -\frac{\partial U_r}{\partial r} \tag{27c}$$

$$\frac{\partial^2 \psi_r}{\partial z^2} + \frac{1}{R_i^2}\frac{\partial^2 \psi_r}{\partial \varphi^2} = \frac{\partial}{\partial z}(U_\varphi + w) - \frac{1}{r}\frac{\partial}{\partial \varphi}(U_z - w) \tag{27d}$$

Boundary III : On the lower lid the normal vorticity component is known as before and the other vorticity components is given numerically by:

$$4\omega_{r_1} - \omega_{r_2} = \frac{6}{h^2}(\psi_{r_1} - \psi_{r_2}) + \frac{6}{h}S \tag{28}$$

where S can be obtained in a similar way to that given by eq.(25) [here $\eta \to \infty$]. ω_φ is governed by a similar equation to eq.(26).

The conditions for *Boundary II* and *Boundary IV* is similar to *I* and *III* respectively. Numerically the most important relations are those between the vorticity and the respective vector potential components, since by coupling them the scheme is stabilized and it converges very rapidly.

6. RESULTS AND SUMMARY.

Numerical results are reported here to serve as a simple illustration of the use of the present approach but no extensive investigation of the above properties and no detailed comparison with other methods attempted here.

The tested geometrical configuration is $l = 2$ and $\beta = 3$. Two test cases have been studied:

(a) only the outer cylinder rotates, and

(b) only one of the lids rotates.

where the Reynolds number range is $10 \le R_e \le 2000$ with $21 \times 21 \times 21$ equal space grid. The rate of convergence is shown in figure 3. The error E is defined as follows:

$$E = \underset{i,j,k}{MAX} \, ||\delta(i,j,k)||_2$$

where i,j,k define a grid point in the three dimensional domain. As it is expected, the rate of convergence becomes slower as the Reynolds number becomes larger, since the block diagonal dominance becomes weaker.However, in both runs the present method produces a much better rate of convergence compared to the iterative scheme used by RLD [3] for the present test cases. It is also shown that the case (b)' convergence is much more slower then that of case (a), while the superiority of the present method is more pronounced in case (b). Figure 4 presents the variation of the w value at the middle of the edge along the outer cylinder with the iterations. It turns out that w converges to is (near) final value faster than the other variabled in the field. One of the questions that has been raised before and the present paper does not suggest for it a definite answer, is when to stop the inner field iterations before updating again the source terms of the numerical boundary conditions formulation. Figure 5 present the rate of convergence for $R_e = 500$ when transferring back to the boundary conditions computer program modul after $k = 2,4,8$ and 16 inner field iterations. This question needs some more theoretical considerations. Figure 7 presents some results for $R_e = 1000$. Two basic solutions are acceptable for this high Reynolds number based on the initial guess for the vector potential field and the boundary conditions. One solution is a symmetric one and the other is not. Different solutions where obtained only for case (b), while for case (a) only one solution was always found.

In summary, the vector potential - vorticity formulation has been applied here successfully to a three dimensional flows in a rotating annulus. A new boundary condition formulation strategy has been presented together with a very powerful numerical technique to solve the governing and the boundary equations. Space has permitted the presentation of only simple basic results. The extension of this procedure is being finished in the present time.

REFERENCES .

[1] Hirasaki, G.J. and Hellums, J.D. , (1970) , "Boundary Conditions on the Vector and Scalar Potentials in Viscous Three-Dimensional Hydrodynamics" , *Quart.Appl.Math.* , XXVIII , 2 , pp. 293-296 .

[2] Richardson, S.M. and Cornish, A.R.H. , (1977) , "Solutions of Three-Dimensional Incompressible Flow Problems" , *J. Fluid Mech.* , *82* , 2 , pp.309-319 .

[3] Reizes, J.A. , Leonardi, E. and de Vahl Davis, G. , (1984) , "Problems with Derived Variable Methods for the Numerical Solution of Three-Dimensional Flows", *Techniques and Applications*, John Noye and Clive Fletcher eds., North-Holland Publ. Co. , pp.909-913.

[4] Pepper, D.W. and Copper, R.E. , (1979) , *ASME J. of Heat Transfer* , 101 , 4 , pp. 739-741.

[5] Lin, A. , (1984) , " A Primitive Variables Scheme For Solving the Navier Stokes Equations" , in preparation.

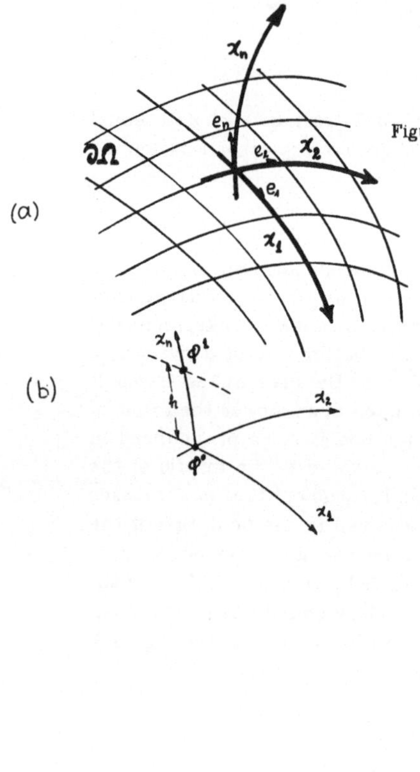

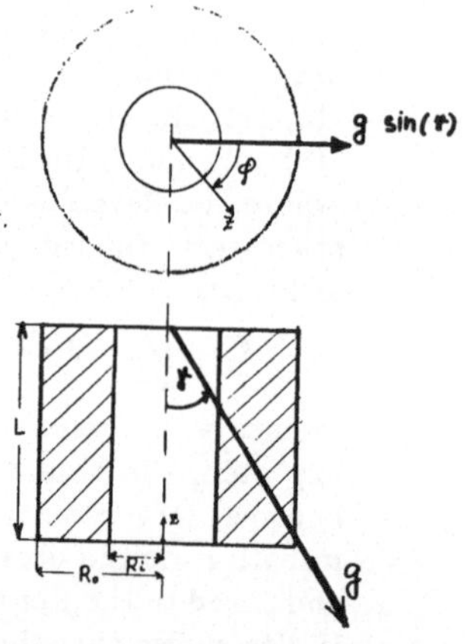

Figure 2 : Test problem geometry.

Figure 1 : (a) coordinate system; (b) finit difference grid on the boundary $\partial\Omega$.

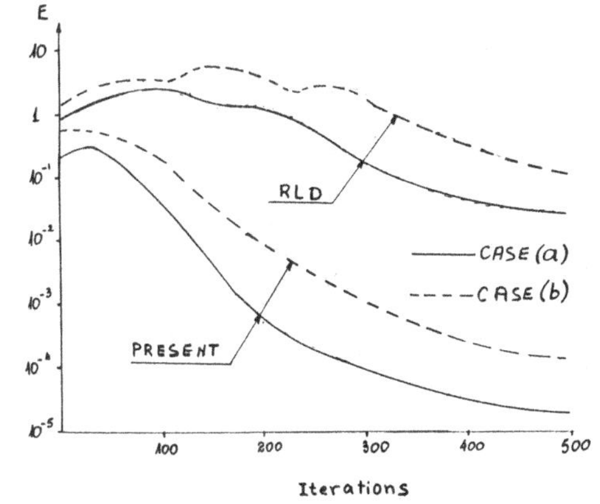

Figure 3 : Comparison of the maximum error .vs. ADI iterations

for R_e = 1000.

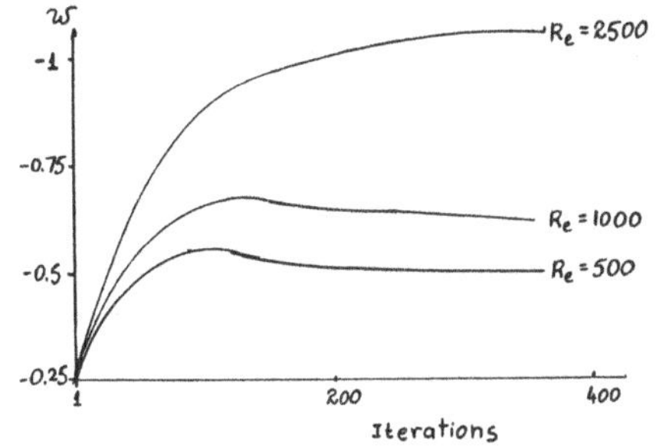

Figure 4 : A sample of the variation of w with iterations.

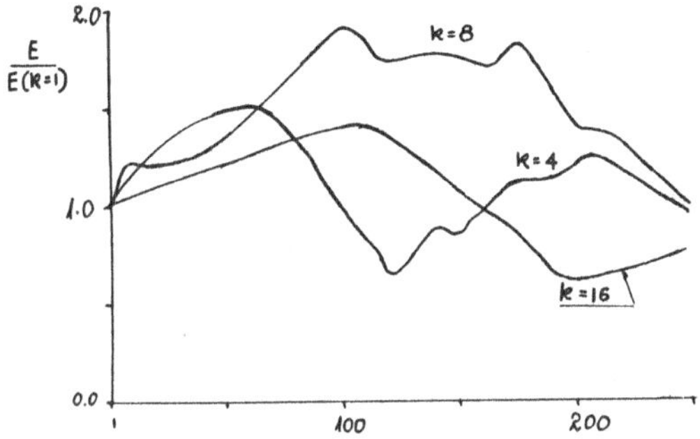

Figure 5 : Sensitivity of the rate of change of the explicit boundary conditions' contributions.

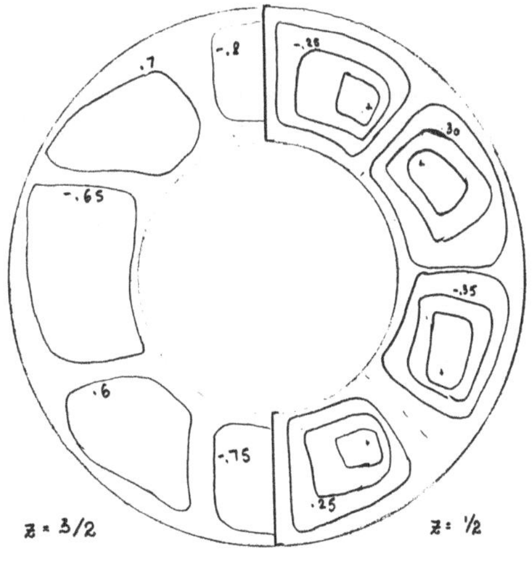

Figure 6 : Quasi Ψ_z lines for case (b) where the lower lid rotates with R_e = 1500.

AN ADAPTIVE FINITE ELEMENT METHOD FOR HIGH SPEED COMPRESSIBLE FLOW

R. Löhner, K. Morgan and O.C. Zienkiewicz

Institute for Numerical Methods in Engineering, University College of Swansea, Singleton Park, Swansea SA2 8PP, U.K.

Our aim is to solve large 2/3-D compressible fluid flow problems employing finite elements. Other numerical techniques, like finite difference or finite volume methods have reached a high degree of sophistication [1,2], but it is expected that the finite element method will make significant contributions due to its geometrical flexibility, a factor that is of importance for industrial applications.

The development has so far gone through the following stages:

(a) Definition of the basic algorithm: in order to be competitive, and at the same time to model correctly the physical properties of hyperbolic equations, it was decided to employ explicit time-marching schemes. As the straightforward Galerkin method is suboptimal [3], a modified or 'upwinded' Taylor Galerkin-type [4] procedure was implemented. The complete description may be found in [5], together with numerical examples. This one-step algorithm has now been superseded by an equivalent two-step method, the description of which may be found in [6]. This two-step scheme is 2-3 times faster than the one-step scheme on scalar machines (for two-dimensional problems) and has been vectorized in order to realise the full power of modern supercomputers.

(b) Domain splitting: it is well known that solutions of high speed compressible flow problems exhibit narrow regions of rapid change (e.g. shocks) which are embedded in larger regions where the solution is smooth. Accordingly, large variations in element size are expected in typical discretizations. However, the small elements might then require that a correspondingly small global timestep could be employed in larger elements. The remedy adopted here, and described in detail in [7], is to split the domain into regions in which different timestep-sizes can be used. The domain subdivision is performed completely automatically by the computer code at prescribed time intervals, and allows a time-accurate development of the unsteady solution.

(c) Adaptive mesh refinement: in general, an analyst will have no a priori knowledge of the location of those areas of the domain where more (i.e. smaller) elements should be employed. Therefore, usually, much more elements than necessary will be employed, leading to an inefficient overall procedure. An ideal computational algorithm would require the ability to refine the mesh where necessary as the solution proceeds. The geometric flexibility of the linear triangular element makes it ideally suited for refinement processes of this type. We adopted a posteriori methods [8], as they seem at present more economical, and for the same reason also did not implement hierarchical techniques [9], but the more classic enrichment of adding more elements. At present we are only considering steady state problems, the generalization to transient problems being an obvious extension. This means, that the timestepping scheme is utilized as a relaxation procedure. After a given number of timesteps the solution domain is analysed, and more elements are added where necessary. Generally speaking, there exist three possibilities for approximately determining the error

$$e = u - u^h \tag{1}$$

where u denotes the exact and u^h the discrete solution:

i) <u>Comparison with higher order schemes:</u> The significant derivatives of the partial differential equation (PDE) under consideration are evaluated twice, using in each case difference schemes of different order [10,11,12]. By determining the discrepancy of both approximations an estimation of the error can then be obtained. The problem with this kind of approach is that it fails near boundaries and at singularities or boundary layers (which are common in fluid dynamic problems). At the same time it is not extendable to FEMs, which operate on an element level.

ii) <u>Determination of the relative importance of further degrees of freedom:</u> Further degrees of freedom are introduced on an element by element basis, and the relative importance of these further degrees gives an error estimate [13,14,15]. The problem with this kind of approach is that it is relatively expensive in CPU-time requirements, so that for transient problems a considerable percentage of runtime will be spent on error estimation.

iii) <u>Use of error norms:</u> Here the classic theoretical error estimates are employed locally [16,17]. Thus, no further degrees of freedom are introduced and only first or second derivatives need to be evaluated. Our experience indicates that this type of error indicator works satisfactorily, and, as it is very economical, it is regarded as a good algorithm for transient problems as well. For elliptic problems the appropriate error norms appear naturally whereas for hyperbolic problems the theory is far from complete. Nevertheless one can assume

$$\|u - u^h\|_k < c\, h^{\ell-k} |u|_\ell, \tag{2}$$

where h is a representative element length. Using the L_2-norm (k=0) yields

$$\|u - u^h\|_0 < c\, h^{\ell} |u|_\ell. \tag{3}$$

The aim of any refinement is to obtain a reduction of errors according to some criterion, e.g. at a certain point, surface or evenly throughout the field. Particularly for hyperbolic problems the error at one point may influence the accuracy of the solution in the whole field (e.g. the root of an expansion fan), so that an even distribution of errors seems to be the only possible practical choice.

Therefore, at each refinement level all elements satisfying

$$h^{\ell}|u|_{\ell} > a \max_{e=1,nelem} h^{\ell} |u|_{\ell} \qquad (4)$$

are refined. Since the exact solution u is unknown, the practical requirement becomes

$$h^{\ell}|u^h|_{\ell} > a \max_{e=1,nelem} h^{\ell} |u^h|_{\ell}. \qquad (5)$$

Only the cases ℓ=1 or ℓ=2 appear to be of practical interest, and both have been studied (see examples). For the case ℓ=2 the first derivatives of u are evaluated inside the elements, and hereafter the nodal values for the second derivatives are recovered variationally as follows:

$$\int N^i N^j \, dV \, \hat{u}^j_{,xx} = - \int N^i M^k \, dV \, \bar{u}^k_{,x} \quad , \qquad (6)$$

where M^k is constant and $\bar{u}^k_{,x}$ is defined on an element basis. It has been found that a-values of the order

$$a = 0.6 - 0.9 \qquad (7)$$

yield the most effective refinement strategy. This is in contrast to [13], where the factor a = 0.1 was reported as optimal. A possible explanation for the discrepancy of these values may be found in the nature of the PDEs treated in both cases: whereas here the PDEs are hyperbolic - and this means th-t small disturbances propagate far into the field - , in [12] the effective solution of elliptic PDEs was pursued - and this means that small disturbances decay rapidly.

Results

(a) Supersonic flow past a wedge: the successive stages of the domain discretization as well as the solution obtained are shown in figure 1 . In this case the mesh was enriched according to equation (5) with ℓ=1 and a=0.6.

(b) Prandtl-Meyer expansion fan: the problem statement, as well as the successive stages of the domain discretization and the corresponding solutions are depicted in figure 2. The improvement in solution quality is readily seen. In this case the mesh was enriched according to equation (5) with ℓ=2 and a=0.8.

Acknowledgement

The authors would like to thank the Aerothermal Loads Branch of the NASA Langley Research Center for supporting this research under Grant Nr.NAGW-478, and especially A.R. Wieting and K.S. Bey for their continued interest and encouragement.

References

1. P. Woodward and P. Colella, J. Comp. Phys. 54, 115-173 (1984).
2. A. Jameson, J. Appl. Mech. 50, 1052-1070 (1983).
3. K.W. Morton and A.K. Parrott, J. Comp. Phys. 36, 249-270 (1980).
4. J. Donea, Int. J. Num. Meth. Eng. 20, 101-120 (1984).
5. R. Löhner, K. Morgan and O.C. Zienkiewicz, Int. J. Num. Meth. Fluids (1984) (to appear).
6. O.C. Zienkiewicz, R. Löhner and K. Morgan, Proc. MAFELAP-V Conf., May 1984 (to appear).
7. R. Löhner, K. Morgan and O.C. Zienkiewicz, Comp. Meth. Appl. Mech. Eng. 1984 (to appear).
8. J.T. Oden, TICOM Rep. 1983.
9. O.C. Zienkiewicz, J.P. de S.R. Gago and D.W. Kelly, J. Comp. Struct. 16, 53-65 (1983).
10. W. Schönauer, K. Raith and K. Glotz, Comp. Meth. Appl. Mech. Eng. 28, 327-359 (1981).
11. I. Babuska, personal communication.
12. A. Brandt, ICASE-Rep. 79-19 (1979).
13. A. Peano, M. Fanelli, R. Riccioni and L. Sardella, in Proc. Conf. Num. Meth. Fract. Mech., Swansea (1979).
14. O.C. Zienkiewicz and A.W. Craig, in Adaptive Comp. Methods for PDE's (I. Babuska et al. eds.), SIAM (1983).
15. A.W. Craig, J.Z. Zhu and O.C. Zienkiewicz, Proc. MAFELAP-V Conf., May 1984 (to appear).
16. A.R. Diaz, N. Kikuchi and J.E. Taylor, Comp. Meth. Appl. Mech. Eng. 1984 (to appear).
17. R. Löhner, Ph.D. Thesis, Univ. of Wales, 1984.

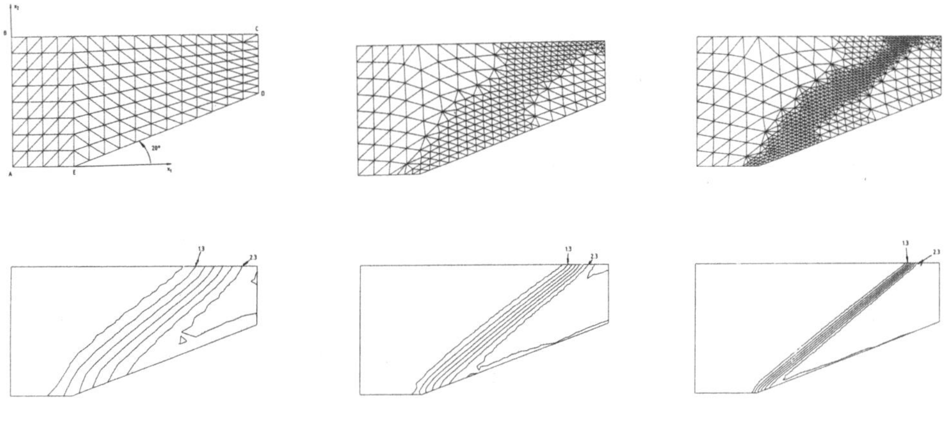

FIGURE 1

Discretization and solution for supersonic flow past a wedge

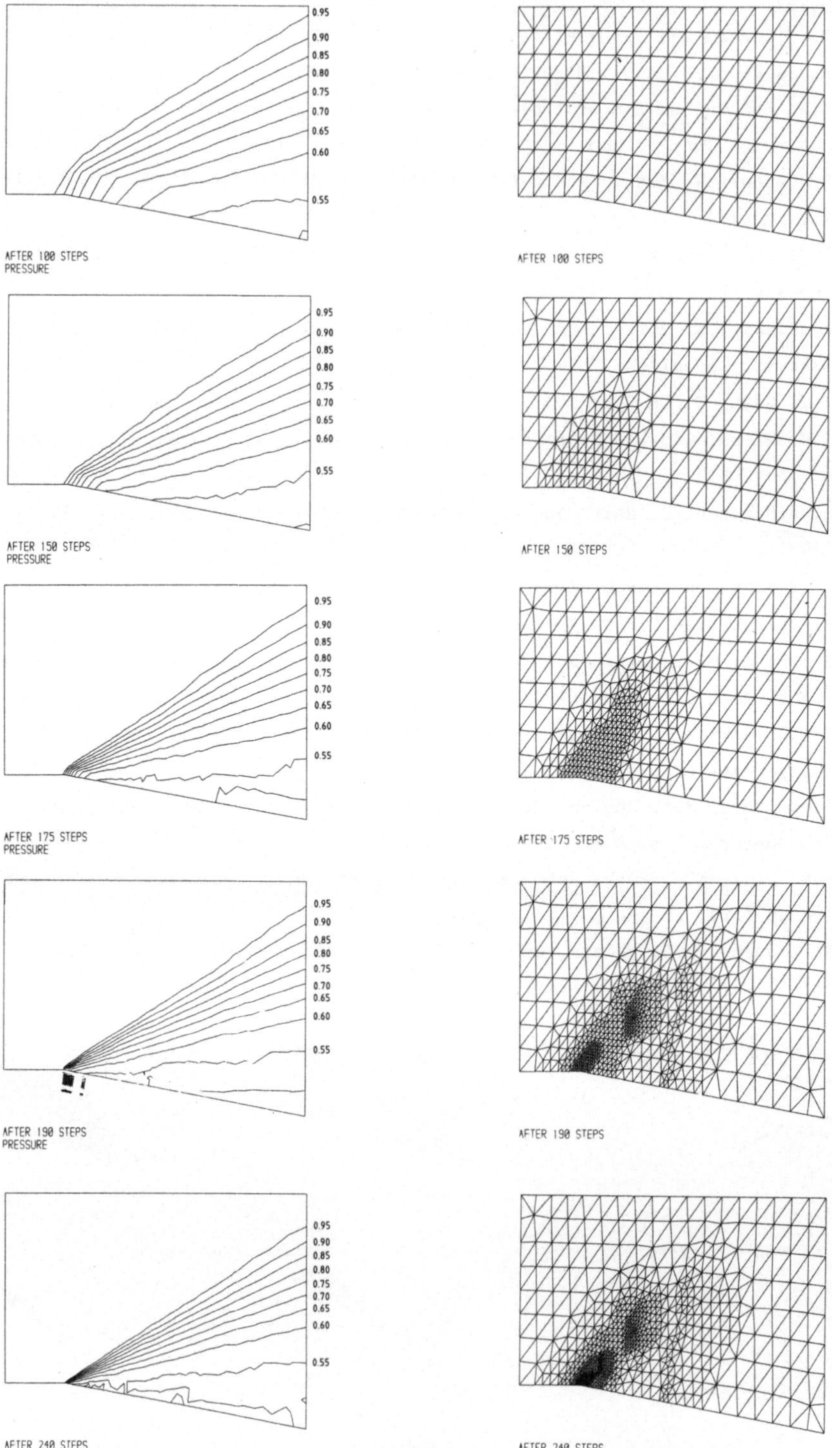

FIGURE 2

Discretization and solution for Prandtl-Meyer expansion fan

ANALYSIS OF SEPARATED FLOW IN A PIPE ORIFICE USING UNSTEADY NAVIER-STOKES EQUATIONS†

W.F. McGreehan, K.N. Ghia,
U. Ghia and G.A. Osswald
The General Electric Company
and University of Cincinnati
Cincinnati, Ohio, U.S.A.

Introduction

The loss characteristics for flow through complex geometrical restrictions in aircraft turbomachinery passages are often determined experimentally. A better understanding of the nature of profile losses can be obtained by accurately simulating the prevailing flow field. This poses a formidable task since many complex flow features are present in these configurations which consequently require solution of the unsteady Navier-Stokes equations. In the present study, emphasis is given to the analysis of separated flow with unsteadiness in a pipe-orifice using the solution of the Navier-Stokes equations in generalized coordinates. Flow through a pipe orifice, though functionally simple, is very rich in fluid mechanics phenomena. Of the published experimental results, those of Alvi et al. [1978] are the most comprehensive. These results were obtained from pressure measurements and are in the form of loss coefficients for various values of β (orifice to pipe diameter ratio) and Reynolds number Re_D based on the pipe diameter D, ranging from 10 to 10,000. Only recently, Mattingly and his associates [1983] have started extensively surveying orifice flow fields using laser-Doppler velocimetry and, like the present authors, they, too, have observed intrinsic unsteady phenomena in the orifice flow.

A number of numerical studies have been conducted by solving the steady and unsteady incompressible Navier-Stokes equations for axisymmetric flow through a pipe orifice. Mattingly and Davis [1977] employed a cylindrical grid which was restricted to square-edged orifices. Nigro et al. [1978] used a Schwarz-Christoffel transformation to generate a conformal grid which results in a more efficient grid distribution near the orifice plate where vorticity is generated. However, the inability to control grid-point location and distribution with this method does not permit the correct treatment of the vorticity singularities at the corners. Greenspan [1973] and Nigro et al. [1978], besides others, have attempted steady solutions for $Re_D > 100$ where the flow does not have truly steady behavior. The analysis of Mattingly and Davis [1977] and that of Coder and Buckley [1974] are the only analyses for unsteady flows but both of these studies were limited to $Re_D < 100$ where persistent unsteadiness does not appear.

† This research was supported, in part, by AFOSR Grant No. 80-0160

In light of this discussion, an unsteady flow analysis is developed in the present work to study separated flow in a pipe orifice using the method of Osswald and Ghia [1981]. This study is particularly important since it can simulate the persistently unsteady flow and analyze the nature of vortex shedding. Numerical results are obtained for Re_D ranging from 20 to 1000, with three different values of β, as shown in Table 1.

Governing Equations

The unsteady incompressible Navier-Stokes equations with cylindrical symmetry in the derived variables vorticity ω and stream function ψ are transformed from the physical plane (x,r) to the conformal plane (η^1,η^2) and finally to the computational plane (ξ^1,ξ^2). The equations used are in the conservative form and are as given by Osswald, K. Ghia and U. Ghia [1984]. The pipe radius a and the mean outflow velocity U_D are used in the nondimensionalization of the lengths and velocities. Thus, the Reynolds number Re is defined as $Re = \frac{U_D a}{\nu}$. For comparison of the results with the experimental data, Re_D based on the pipe diameter as well as Re_d based on the orifice diameter d are also used. Here, $Re_D = 2Re$ and $Re_d = U_d d/\nu$ where U_d is the mean velocity at the orifice.

The vorticity transport equation and the stream function equation, together with appropriate boundary and initial conditions, are solved numerically. The pressure field can be determined, thereafter, by integration of the ξ^1 momentum equation along the surface. It should be pointed out that, eventually, the pressure field will be obtained as the solution of a Neumann boundary-value problem.

Grid Generation For Orifice Configurations

Several factors need consideration in the generation of clustered surface-oriented grid distributions for the doubly infinite pipe-orifice configurations. The surface-oriented coordinates in the meridional (x,r) plane are obtained using a conformal transformation based on the technique developed by Davis [1981] and adapted to channels by Sridhar and Davis [1981]. This transforms the doubly infinite orifice-pipe configuration from the physical plane (x,r) to a doubly infinite pipe in the conformal plane (η^1,η^2). Grid clustering in the streamwise and normal directions is introduced to transform the doubly infinite pipe configuration in the (η^1,η^2) plane to a unit square in the computational plane (ξ^1,ξ^2); the details of many of these transformations are given by K. Ghia, Osswald and U. Ghia [1983]. The streamwise stretching is further improved to account for the large convective scales in the present problem by extending the work of U. Ghia et al. [1981]. Simultaneously, the vorticity singularities of the various corners are circumvented by mapping these corners to the grid mid-points. A typical grid distribution is illustrated in Fig. 1.

Some specific details are discussed here for the streamwise stretching. A unit distance along the streamwise computational coordinate ξ^1 is mapped to a doubly infinite length in an intermediate plane by a transformation of the form,

$$\xi^1 = \frac{1}{p}\left[\sum_{j=1}^{n} c_j \tan^{-1}\left(\frac{\eta^1 - b_j}{d_j}\right) + \frac{1}{\eta_\ell^1 - \eta_k^1}\int_{\alpha=\eta_k^1}^{\eta_\ell^1} C(\alpha)\tan^{-1}(\eta^1-\alpha)\,d\alpha\right] + \frac{1}{2} \tag{1}$$

where p is the summation of weighting coefficients c_j and of the integral of the clustering distribution coefficient $C(\alpha)$ from η_ℓ^1 to η_k^1. For a uniform distribution of grid points between η_ℓ^1 and η_k^1, $C(\alpha)$ will be constant, equal to, say, C, so that

$$p = \pi\{\sum_{j=1}^{n} c_j + C\}. \tag{2}$$

The transformation relation (1) then simplifies, by integration, to yield

$$\xi^1 = \frac{1}{p}\Bigg[\sum_{j=1}^{n} c_j \tan^{-1}\left(\frac{\eta^1-b_j}{d_j}\right) - \frac{C}{\eta_\ell^1-\eta_k^1}\{(\eta^1-\eta_\ell^1)\tan^{-1}(\eta^1-\eta_\ell^1)$$

$$- (\eta^1-\eta_k^1)\tan^{-1}(\eta^1-\eta_k^1) - \frac{1}{2}\ln(1+[\eta^1-\eta_\ell^1]^2) + \frac{1}{2}\ln(1+[\eta^1-\eta_k^1]^2)\}\Bigg] + \frac{1}{2} \tag{3}$$

Here, b_j denotes the location of clustering of strength d_j in the doubly infinite pipe plane (η^1,η^2). All the d_j's are solved for simultaneously to satisfy the grid-midpoint↔orifice-corner constraints which then permit determination of η^1 from the specified ξ^1 point locations using a Newton gradient method.

Comments on Numerical Method

A fully central differenced numerical method developed by Osswald and Ghia [1981] is used to solve the unsteady Navier-Stokes equations in generalized orthogonal coordinates for the axisymmetric flow through seven pipe-orifice configurations. A doubly infinite pipe is used, with inflow and outflow boundaries set truly at $\pm\infty$ in the physical plane. This is an important consideration because of the rather large convective length scales downstream of the orifice at higher Reynolds number; this was also observed by Greenspan [1973] and Alvi et al. [1978]. In the present study, the individual local scales are honored by the desired clustered grid distribution. The numerical method used solves the conservative form of the transport equation using the alternating-direction (ADI) method, whereas the stream function equation is solved by a direct block Gaussian elimination (BGE) method. This implicit ADI-BGE method offers an order-of-magnitude improvement in computational efficiency over existing iterative methods. The overall accuracy of the method is $O(\Delta t,(\Delta\xi^1)^2,(\Delta\xi^2)^2)$, and the computational effort parameter τ, which represents time required to advance the solution by one time step per spatial grid point, is 1.118×10^{-4} seconds for the AMDAHL V/7A computer.

Results and Discussion

Solutions have been obtained for the seven pipe-orifice configurations listed in Table 1. The configurations are so chosen as to provide transient results leading to steady state solutions as well as to persistently unsteady results. For the configurations with $\beta = 0.4$ and 0.6, Alvi et al. [1978] have provided experimental data which are mostly in terms of the loss coefficient and the discharge coefficient. Similar experimental data of Johanson and those of Tuve and Sprenkle for relatively lower range of Re_D is given in the work of Mattingly and Davis [1977]. The surface pressure obtained by integrating the ξ^1 momentum equation can give comparable values of the loss coefficient and discharge coefficient for $Re_D = 20$ and 100 as given by Alvi et al. [1978], but it was felt that comparison and improvement of these parameters will only be meaningful when the pressure field is computed rigorously from the solution of the Neumann boundary value problem; this will be reported in a forthcoming technical report by the authors. Typical flow results are now presented.

The transient flow development in terms of stream-function contours for $Re_D = 20$ and $\beta = 0.4$ is shown in Fig. 2. The computations were carried out using symmetry about the centerline and a (160,30) grid. The separation bubble aft of the orifice plate increases in size rapidly with characteristic time and reaches nearly a steady state by T = 1.0. No appreciable change takes place thereafter as the characteristic time progresses to T = 3.0. The reattachment length x/a = 2.7 is larger compared to that for the configurations with $\beta = 0.5$ and 0.6 at the same Re_D. Figure 3 shows the corresponding steady-state vorticity contours, with maximum vorticity being generated near the orifice plate and being convected downstream.

As listed in Table 1, for $Re_D = 100$, the reattachment length x/a = 13.2, represents a signficant increase as compared to that for $Re_D = 20$; also, it reduces to nearly half this value as β increases to 0.6. The (160,30) grid used earlier was found inadequate due to increase in the convective length scale in the streamwise direction for $Re_D = 100$ and, as such, a (220,24) grid was used for configurations II through VII. The transient stream-function contours are presented in Fig. 4. At T = 1.6, the separation bubble has grown considerably in length and shows the characteristics of an unsteady flow with two clockwise rotating eddies. By T = 4.0, the flow stabilizes to only a single separation bubble and reaches steady state at T = 7.0. The corresponding steady-state vorticity contours are presented in Fig. 5. There is considerable stretching and convection of the vorticity generated at the orifice plate.

The results of Alvi et al. [1978] show unsteady behavior in the flow for β = 0.6 at $Re_D = 270$. The results of the present analysis at $Re_D = 300$ for this geometry still show a steady-state solution. This observation parallels the findings of Ghia et al. [1983] who carried out comparisons of numerical and experimental results for the two-dimensional backstep geometry where three-dimensional effects were observed to initiate unsteadiness in the experiments at lower Re_D than would be predicted numerically from a two-dimensional analysis. Figure 6 depicts a persistently unsteady solution for the orifice with $\beta = 0.6$ at $Re_D = 1000$. As seen here, vortices originate from the free-shear layer just downstream of the orifice plate. As an individual vortex intensifies and grows, it quickly merges with its predecessor, eventually forming a street of toroidal vortices downstream of the orifice plate. At T = 35, five individual vortices have merged into one large train along the free shear layer. A large coherent vortex structure has just separated from the vortex train and is being convected downstream. At T = 40, seven individual vortices have merged together, with the last two vortices downstream being ready to separate and form discrete elements of the vortex street. Thus, the flow downstream of the orifice plate is highly unsteady as vortices are continuously convected downstream and eventually dissipated. Further analysis of these results will provide a better understanding of the nature of this unsteady phenomena and how it affects the primary function of the orifice, namely, flow measurement. Finally, Fig. 7 shows the vorticity contours corresponding to T = 40 and the origination and convection of vortices is very vividly seen here.

References

Alvi, S.H., Sridharon, K. and Lakshmana Rao, N.S., [1978], J. of Fluids Engr., Vol. 100, pp 299-306.

Coder, D.W. and Buckley, F.T., [1974], Comp. and Fluids, Vol. 2, pp 295-315.

Davis, R.T., [1981], VKI Lecture Notes, Springer-Verlag, New York, pp 1-44.

Ghia, K.N., Osswald, G.A. and Ghia, U., [1983], Second Symposium on Numerical and Physical Aspects of Aerodynamic Flows, Long Beach, CA.

Ghia, U., Ghia, K.N., Rubin, S.G. and Khosla, P.K. [1981], Comp. and Fluids, Vol. 9, pp 123-142.

Greenspan, D., [1983], Int'l J. for Num. Methods in Engr., Vol. 6, pp 489-496.

Mattingly, G.E., [1983], Private Communications.

Mattingly, G.E. and Davis R.W., [1977], ASME Paper No. 77-WA/FE-13.

Nigro, F.E.B., Strong, A.B. and Alpay, S.A. [1978], J. of Fluids Engr., Vol. 100, pp 467-472.

Osswald, G.A. and Ghia, K.N., [1981], Multigrid Methods, NASA CP-2202.

Osswald, G.A. Ghia, K.N. and Ghia, U., [1984], AIAA Paper No. 84-1854.

Sridhar, K.P. and Davis, R.T., [1981], Computers in Flow Predictions and Fluid Dynamics Measurements, Editor: K.N. Ghia et al., ASME Publication.

TABLE 1. VARIOUS PIPE-ORIFICE REYNOLDS NUMBERS AND REATTACHMENT LENGTHS

Configuration	β	Re_D $(\frac{U_D D}{\nu})$	Re $(\frac{U_D a}{\nu})$	Re_d $(\frac{U_d d}{\nu})$	x_ℓ/a	x_ℓ/D
I	0.4	20	10	50	2.7	1.4
II	0.4	100	50	250	13.2	7.6
III	0.4	1000	500	2500	*	*
IV	0.5	200	100	400	13.6	7.8
V	0.5	1000	500	2000	*	*
VI	0.6	200	100	333	6.2	3.1
VII	0.6	1000	500	1667	*	*

* For Re_D = 1000, the flow is unsteady and, hence, these reattachment lengths are not presented.

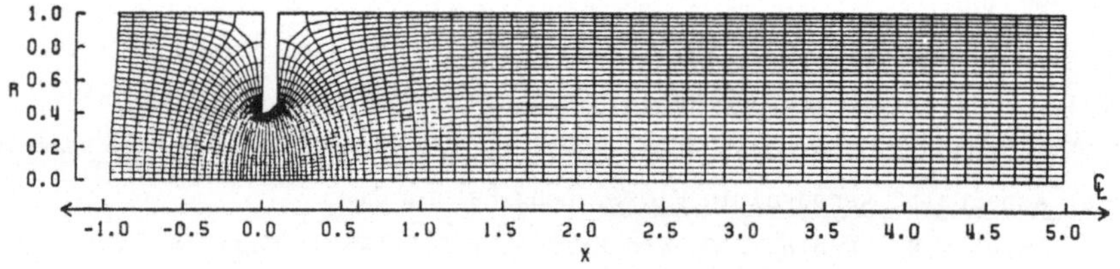

Fig. 1a Grid distribution for β = 0.4 with (160,30) mesh.

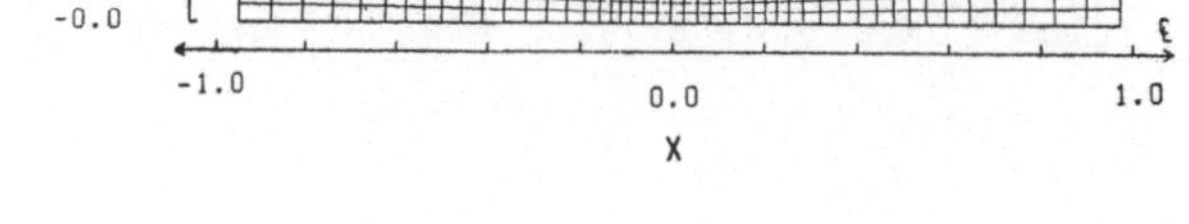

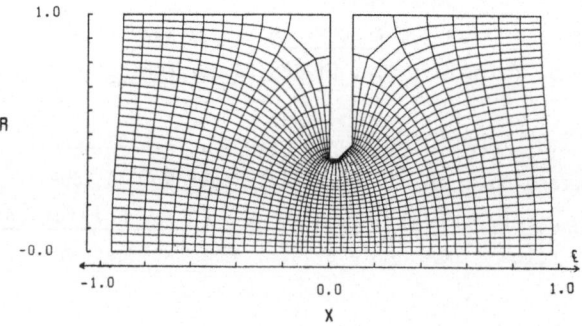

Fig. 1b Grid point distribution near the orifice plate with (160,30) mesh.

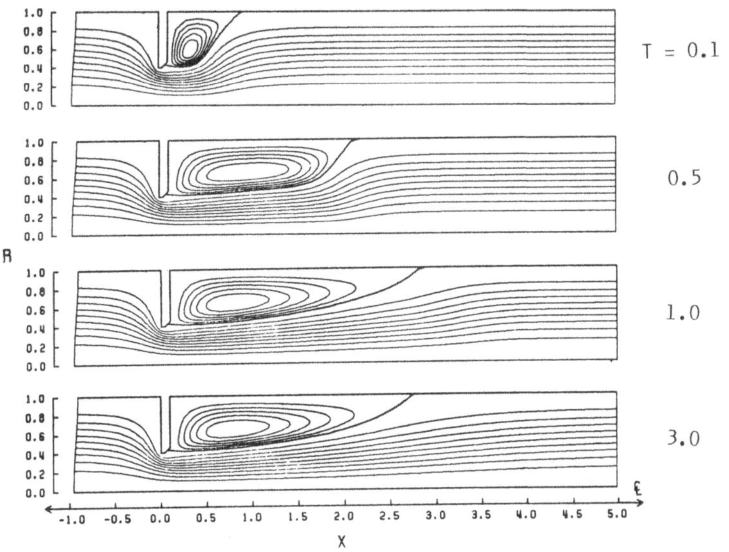

Fig. 2 Streamfunction contours for $\beta = 0.4$, $Re_D = 20$ ($\Delta\Psi = 0.05$).

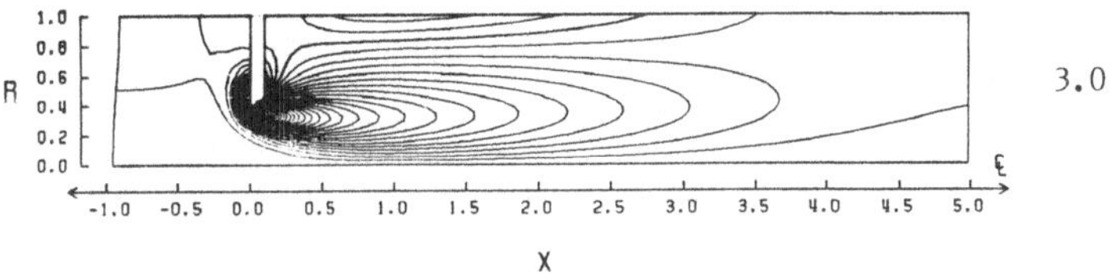

Fig. 3 Vorticity contours for $\beta = 0.4$, $Re_D = 20$ ($\Delta\omega = 2.0$)

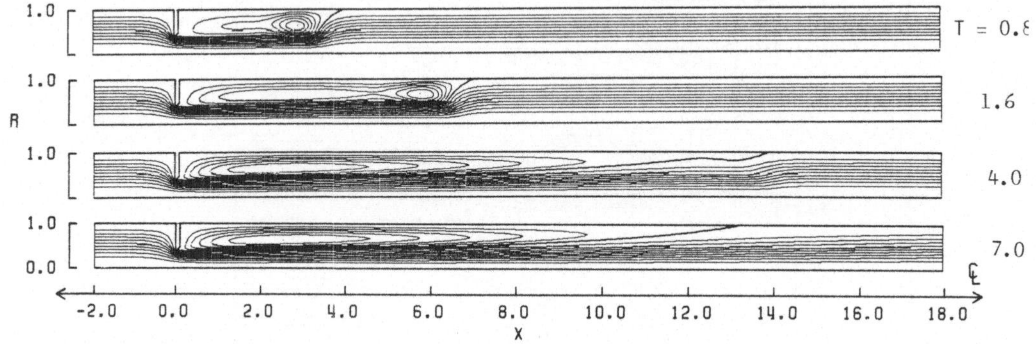

Fig. 4 Streamfunction contours for $\beta = 0.4$, $Re_D = 100$ ($\Delta\Psi = 0.05$).

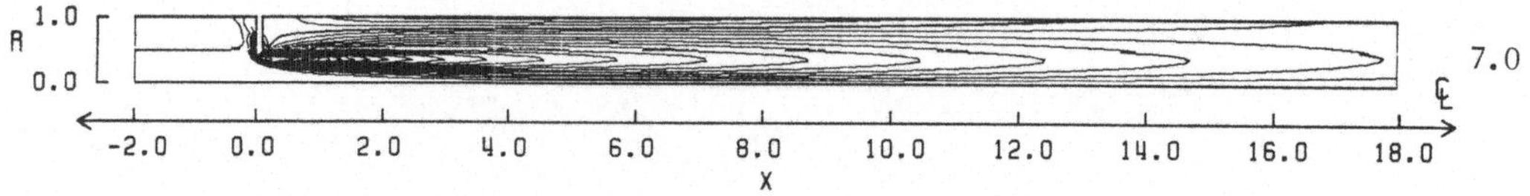

Fig. 5 Vorticity contours for $\beta = 0.4$, $Re_D = 100$ ($\Delta\omega = 2.0$).

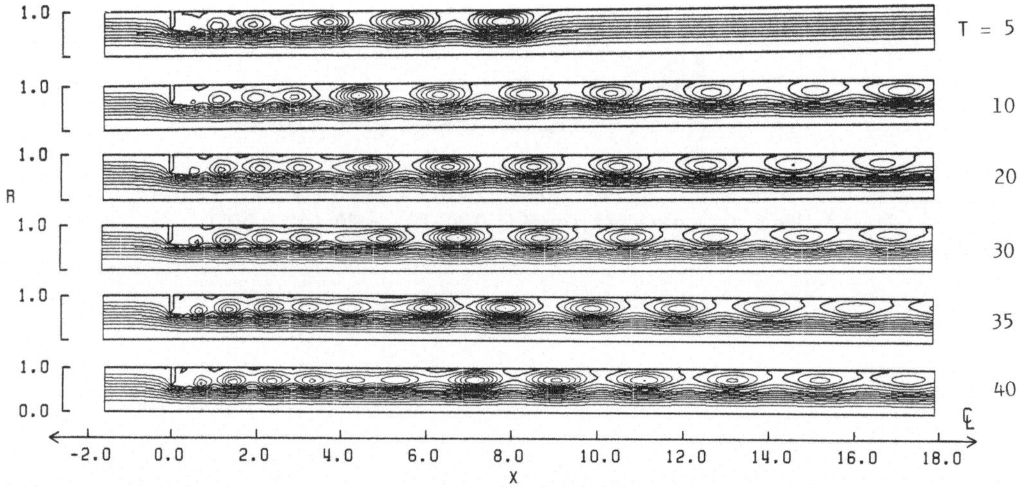

Fig. 6 Streamfunction contours for $\beta = 0.6$, $Re_D = 1000$ ($\Delta\Psi = 0.05$).

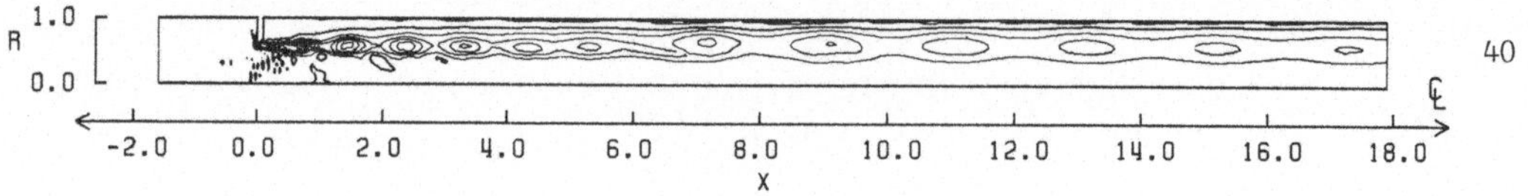

Fig. 7 Vorticity contours for $\beta = 0.6$, $Re_D = 1000$ ($\Delta\omega = 5.0$).

THE CONVECTIVE DYNAMO : A NUMERICAL EXPERIMENT

M.Meneguzzi[1] and A. Pouquet[2]

1 - Service d'Astrophysique, Centre d'Etudes Nucleaires de Saclay, 91191 Gif sur Yvette, and CNRS, France.

2 - CNRS, Observatoire de Nice, BP 139, 06007 Nice ,France

ABSTRACT

We report preliminary results obtained by high resolution numerical simulation of Boussinesq magnetoconvection in a flat infinite layer of fluid using a spectral method. The main result is that convection by itself, without rotation, can produce a dynamo effect for Rayleigh numbers about 100 times critical and magnetic Prandtl numbers larger than $\simeq 1$.

1 - THE DYNAMO PROBLEM

In a turbulent conducting fluid, an initially weak magnetic field can be considerably amplified. This requires that the stretching of the magnetic lines of force by the velocity gradients overcomes the Joule dissipation. In laboratory experiments on liquid metals, the magnetic Reynolds number R_m is in general too small to allow such a "dynamo effect" (R_m measures the ratio of the time scales of the afore mentioned physical processes). This is not the case for large scale industrial plants ($R_m \simeq 100$ for the liquid sodium cooling system of the SUPERPHENIX breeder reactor[1]) and in astrophisical turbulent flows ($R_m \simeq 10^6$ for the solar convective zone). Experiments to measure the ambient magnetic field will be set up in the summer of 1984 in the breeder reactors PHENIX and SUPERPHENIX. A similar experiment was performed in the USSR, and a magnetic field was observed[2] .

The turbulent dynamo effect can be studied in the "kinematical" framework[3] : the velocity field $\underline{v}$ is given, and one looks for the critical magnetic Reynolds number R_{mc} above which the magnetic field $\underline{B}$ grows indefinitely. On the other hand, one can study the full non linear problem,[4,5] where the reaction of the growing $\underline{B}$ on the $\underline{v}$ field through the Lorenz force is essential for the saturation of the magnetic field amplitude. Direct numerical simulation is presently the only way to deal with the non linear problem.

In a previous work[4] , we had used random forcing as a source of kinetic energy. With applications to geophysical and astrophysical flows in mind, we now solve the full 3D Boussinesq-MHD equations with the highest possible spatial resolution, for a flat infinite layer of fluid heated from below. The equations, in non dimensional form, are :

$$\frac{\partial \underline{v}}{\partial t} = \underline{v} \times \underline{\omega} + \underline{j} \times \underline{b} - \underline{\nabla} \Pi + R_a P \theta \underline{e}_3 + P \nabla^2 \underline{v} \quad (1)$$

$$\frac{\partial \underline{b}}{\partial t} = \underline{\nabla} \times (\underline{v} \times \underline{b}) + \frac{P}{P_m} \nabla^2 \underline{b} \quad (2)$$

$$\frac{\partial \theta}{\partial t} = - \underline{\nabla}.(\underline{v}\theta) + v_3 + \nabla^2\theta \quad (3)$$

$$\underline{\nabla}.\underline{v} = \underline{\nabla}.\underline{b} = 0 \quad (4)$$

We have taken as units of length, temperature and time respectively the heith h of the layer, the temperature difference between bottom and top $\Delta T = T_0 - T_1$ and the thermal diffusion time through the layer h^2/κ (κ is the themal diffusion coefficient). In these equations, $\underline{v}$ is the velocity field, $\underline{b}$ the Alfven speed (proportional to the magnetic induction $\underline{B}$), θ the temperature fluctuation respect to the conductive profile $T_0 - z\Delta T$, Π the total pressure $p + v^2/2$, R_a the Rayleigh number, P the Prandtl number and P_m the magnetic Prandtl number ν/η, where ν is the kinematic viscosity and η the magnetic diffusivity. $\underline{e}_3$ is the unit vector in the vertical direction z, $\underline{\omega} = \nabla \times \underline{v}$ and $\underline{j} = \nabla \times \underline{b}$

2 - NUMERICAL METHOD

The equations are integrated using a Fourier spectral method[6]. The computation of derivatives and time stepping are done in Fourier space. The non linear terms are computed in real space using Fast Fourier Transforms. The time stepping is done by leapfrog for the non linear terms and by an implicit scheme (Cranck-Nicolson) for the coupling terms between the equations of $\underline{v}$ and θ. For the dissipative terms, we use the following method[7] : Making the change of variables(in Fourier space)

$$\underline{v}' = \underline{v} \exp(Pk^2t) \quad ; \quad \underline{b}' = \underline{b} \exp(\frac{P}{P_m} k^2t) \quad ; \quad \theta' = \theta \exp(k^2t)$$

then discretizing the equations and returning to the original variables, one obtains the scheme

$$\frac{\underline{v}^{n+1} - \underline{v}^{n-1}\exp(-2Pk^2\delta t)}{2\ \delta t} = \underline{\underline{P}}(\underline{k}).(\underline{v}^n \times \underline{\omega}^n + \underline{j}^n \times \underline{b}^n - \underline{\nabla}\ \Pi^n)\exp(-Pk^2\delta t)$$

$$+ P\ R_a\ \underline{e}_3\ \frac{\theta^{n-1}\exp(-2Pk^2\delta t) + \theta^{n+1}}{2} \quad (5)$$

$$\frac{\underline{b}^{n+1} - \underline{b}^{n-1}\exp(-2Pk^2\delta t/P_m)}{2\ \delta t} = i\ \underline{k} \times (\underline{v}^n \times \underline{b}^n)\exp(-Pk^2\delta t/P_m) \quad (6)$$

$$\frac{\theta^{n+1} - \theta^{n-1}\exp(-2k^2\delta t)}{2\ \delta t} = i\underline{k}.(\underline{v}^n\theta^n)\exp(-k^2\delta t) + \frac{v_3^{n-1}\exp(-2k^2\delta t) + v_3^{n+1}}{2} \quad (7)$$

with $$P_{ij}(\underline{k}) = \frac{\delta_{ij} - k_i k_j}{k^2} .$$

Aliasing is removed by a method due to Patterson and Orszag[8] which requires no additional memory but doubles the integration time. Tests have shown that the scheme is strongly unstable at high Rayleigh numbers in the aliased version. The leapfrog is stabilized by mixing the last three time seps once in a while, according to

$$\theta^n = (\theta^{n-1} + 2\,\theta^n + \theta^{n+1}) / 4$$

We use periodic boundary conditions in the horizontal directions, and free-slip conditions in the vertical direction, i.e. we have at the top and bottom of the layer

$$v_3 = \partial_3 v_1 = \partial_3 v_2 = 0$$

and the same for the $\underline{b}$ field (perfect electrical conductors at the vertical boundaries) The spatial resolution is typically 48x48x24 on a 1 Megaword CRAY-1 computer, and 64x64x32 on a 2 Megaword CRAY-1. The correspondig time-steps take 5 and 12 seconds respectively (40% of which is time waiting for I/O).

Since we are looking for bifurcations from non magnetic to magnetic turbulent states, our method of investigation is as follows : For given Rayleigh and Prandtl numbers, we integrate the fluid equations ($\underline{b}$ = 0) in time until a statistical steady state is reached. A small magnetic seed (a few Fourier components with low amplitude) is then introduced, and the integration goes on until a new statistically steady-state is reached. If the magnetic Reynolds number R_m is grater than the critical value R_{mc} , an appreciable fraction of the energy of the new state will be in magnetic form.

3 - RESULTS

At moderate Rayleigh numbers (several times the critical Rayleigh number R_{ac} for the onset of convection), the flow preserves a cellular structure. As was previously seen in kinematical calculations, we observe the "flux expulsion" phenomenon[3] : the magnetic field is expulsed from the center of the cells and concentrated at their boundaries. As the Rayleigh number is increased to more than 100 times critical, the flow becomes more turbulent, and the above effect seems to disappear.

At $R_a = 120\, R_{ac} \simeq 10^5$, P=1 and a Rossby number $R_o = 0.5$ (defined as the ratio of the rotation to turnover times $v_0/\Omega h$, where v_0 is the rms velocity and Ω the rotation vector in the z direction), we obtain a dynamo effect for $P_m \geq 5$. Figure 1 shows the time evolution of the kinetic and magnetic energies. The ratio of magnetic to kinetic energy reaches 1% at $P_m = 10$. Figure 2 shows the corresponding results without rotation. One sees that there is very little difference between the two cases. This may come from the fact that ours is a solid body rotation around the vertical axis, while it is the combination of differential rotation and convection which is believed to generate helicity in the flow and therefore produce a more efficient dynamo[3] (The helicity of $\underline{v}$ is defined as $< \underline{v} \cdot \underline{\nabla} \times \underline{v} >$). Indeed, the helicity is low in both our runs with and without rotation (see figure 3). Note that a differential rotation would be inconsitent with our boundary conditions.

The same calculations have been repeated for P = 0.2, a value closer to the one of the liquid core of the Earth. The results are very similar to those of the P = 1 case, except that the ratio of magnetic to kinetic energy reaches a value of 10%, as shown in figure 4. The critical magnetic Reynolds number is found to be about 500. This is consistent to the value found in the random forcing case[4] as well as by closure calculations[1]. As also found in the random forcing case, the magnetic field exhibits a pronounced intermittency. This can be seen directly on plots of the fields, but also from the fact that the suprema of v and B are of the same order of magnitude while the kinetic and magnetic energies are quite different.

One could object that we have not integrated the equations long enough to ascertain that the magnetic field is not slowly decaying with a time scale of a few diffusion times. Indeed, we plan to continue at least one of our runs for two or three diffusion times. Note however that this has been done in the random forcing case[4], with a positive result. We therefore believe that these calculations strongly support the conclusion that turbulent non helical flows, and convection in particular, can produce a dynamo effect. A possible extension of the present work is the modification of the boundary conditions to allow for differential rotation in any direction, and also the inclusion of compressibility effects. The design of a code using the anelastic approximation is under way.

REFERENCES

1 - Léorat J., Pouquet A., Frisch U. (1981), J. Fluid Mech. **104**,419 .

2 - Kirko I.M., Kirko G.E., Telitchko M.T., Cheinkman A.G.(1982), Proceedings of the USSR Science Academy **266**,4,1384 .

3 - Moffatt H.K. (1978), Cambridge University Press : "Magnetic field generation in electrically conducting fluids".

4 - Meneguzzi M., Frisch U., Pouquet A.(1981), Phys. Rev. Letters **47**,1060 .

5 - Gilman P.A. (1983) Astroph. J. Supp. Ser. **53**, 243 .

6 - Gottlieb D., Orszag S.A. (1977), SIAM : "Numerical Analysis of Spectral Methods".

7 -Basdevant C., Legras B., Sadourny R., Beland B. (1981), J. Atmos. Sci. **38**, 2305.

8 - Patterson S., Orszag S. A. (1971) Phys. of Fluids **14**, 2538.

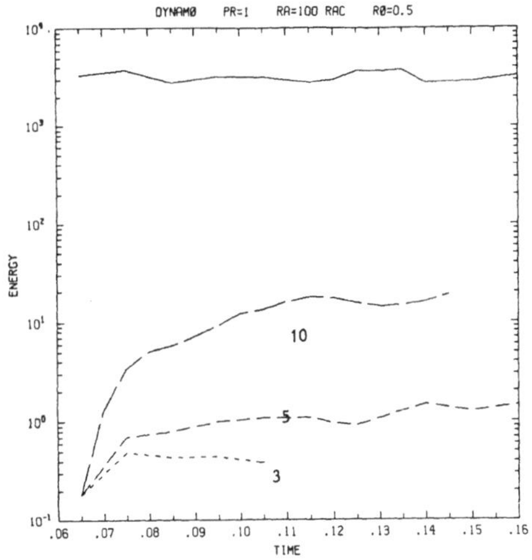

FIGURE 1

Time evolution of the kinetic (solid line) and magnetic (dotted lines) energies. The magnetic Prandtl number is indicated on the curves.

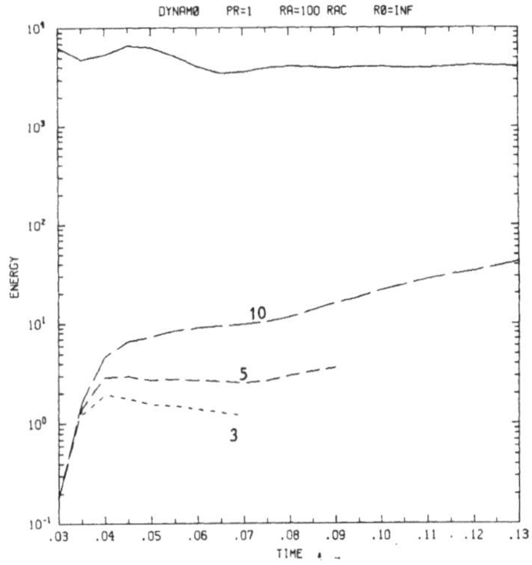

FIGURE 2

Same as figure 1 but without rotation.

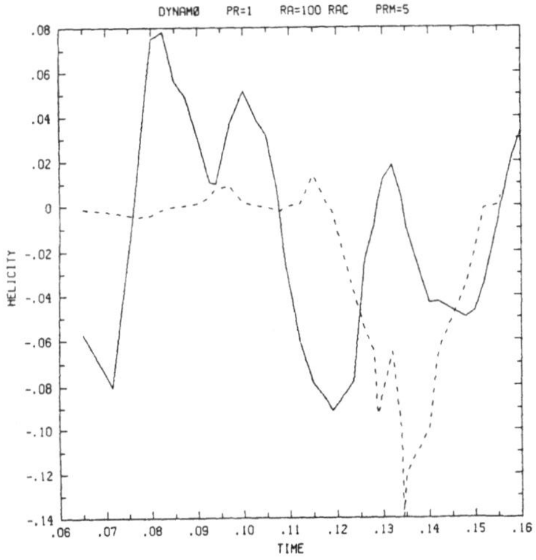

FIGURE 3

Temporal variation of the helicity of the flow for with rotation (solid line) and without (dotted line). The helicity is normalised so that its maximum possible value is 1.

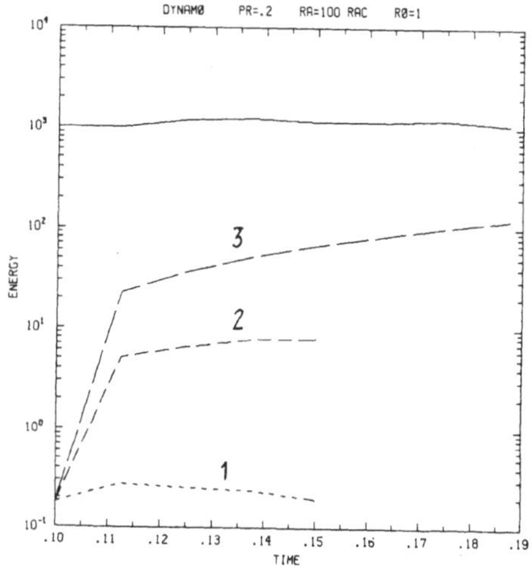

FIGURE 4

Same as figure 1 but for a Prandtl number of 0.2.

A SECOND-ORDER ACCURATE FLUX SPLITTING SCHEME IN TWO-DIMENSIONAL GASDYNAMICS

J.L. Montagné
Office National d'Etudes et de Recherches Aérospatiales
BP 72, 92322 Chatillon cedex, FRANCE

Recently, several authors have presented various improvements to the flux splitting or upwinding techniques for the numerical solution of hyperbolic equations. A systematic procedure to obtain the second order of accuracy, based on the definition of slopes in the one-dimensional spatial discretization has been introduced by Van Leer [4]. This approach is attractive because it appears as a straigthtforward modification of any first order accurate upwind scheme. Herein we propose to use this idea in two dimensions by introducing a spatial discretization with linear discontinuous elements in an explicit conservative Eulerian scheme.

I. Definition of the eulerian scheme

a) General description

We solve the system of Euler equations

$$\frac{\partial W}{\partial t} + \frac{\partial F(W)}{\partial x} + \frac{\partial G(W)}{\partial y} = 0 \qquad W = \begin{vmatrix} \rho \\ \rho u \\ \rho v \\ e \end{vmatrix} \qquad F(W) = \begin{vmatrix} \rho u \\ \rho u^2 + p \\ \rho u v \\ (e+p)u \end{vmatrix} \qquad G(W) = \begin{vmatrix} \rho v \\ \rho u v \\ \rho v^2 + p \\ (e+p)v \end{vmatrix}$$

$$p = (\gamma - 1)(e - \tfrac{1}{2}(u^2+v^2)\rho)$$

The scheme employs the following space discretization. The unknowns, which represents the components of W, are defined at the nodes i of a quadrilateral mesh. The elements are divided by their medians, and control volumes V_i are defined around each node i by joining together the quarters of elements surrounding the node. The discretized function W_h (x,y) at time $n\Delta t$ is discontinuous and piecewise linear on each control volume, with value w_i^n at the node i, and space derivatives p_i^n and q_i^n in x and y directions. The values W_i^n determine p_i^n and q_i^n through a regular operator and a monotonicity correction which are described below. The time integration comprises two stages :

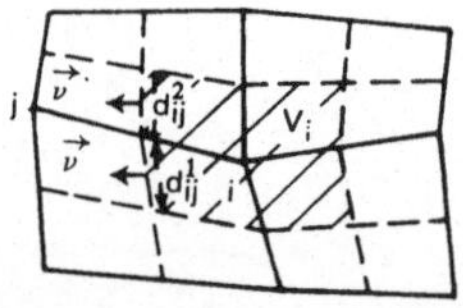

First stage : We compute an intermediate piecewise linear function W^* at the time $(n+½)\Delta t$ with the slopes p_i^n, q_i^n and new node values W_i^* given by :

$$W_i^* = W_i^n - \frac{\Delta t}{2} \cdot (A_i^n \cdot p_i^n + B_i^n \cdot q_i^n) \qquad A_i^n = \frac{dF}{dW}\bigg|_{W_i^n} \qquad B_i^n = \frac{dG}{dW}\bigg|_{W_i^n}$$

Second stage : We apply the conservative laws after the computation of fluxes at the boundaries of V_i from the intermediate function W ,

$$\frac{W_i^{n+1} - W_i^n}{\Delta t} \cdot \mathrm{Area}(V_i) + \sum_{j \in I} \left[H_{ij}^1(W^*) + H_{ij}^2(W^*) \right] = 0$$

$$H_{ij}^{k} = \int_{c_{ij}^{k}} \left[\nu_{ij_x}^{k} \cdot F(W^*) + \nu_{ij_y}^{k} \cdot G(W^*) \right] . ds = \int_{c_{ij}^{k}} h_{ij}^{k}(W_{(s)}^{*}) . ds = \left| c_{ij}^{k} \right| . h_{ij}^{k}(W_+^*, W_-^*)$$

I is the set of neighbours of the node i, H_{ij}^{k} is the integral of the flux on the boundaries d_{ij}^{k} of Vi. ν_{ij_x}, ν_{ij_y} are the components of the normal to the segment d_{ij}^{k}. The function $W^*(s)$ is double valued on d_{ij}^{k}. The integral is computed by a one point integration formula at the middle of each segment d_{ij}^{k}. We compute H_{ij}^{k} by using one of the upwind expressions $h_{ij}^{k}(W_+^*, W_-^*)$ defined for the first order schemes. These two stages describe the scheme for an inner node. At the boundaries of the domain, we have used the method of compatibility relations defined by Veuillot and Viviand I9I, which have given the best results in the case of slip conditions on a curved wall.

b) Accuracy and stability properties

We now give a summary of some results relating to the scheme without boundary conditions

Proposition a : a linear analysis shows that on a regular mesh, the two-dimensional scheme is second order accurate in space and time if the spatial approximation is of second order accuracy, and if the expression $h_{ij}^{k}(W^+, W^-)$ of the flux is consistent.

Proposition b : (Van Leer) I6I. In the one dimensional case, the scheme is linearly stable for L^∞ norm and total variation norm if we use an upwind expression of h (W^+, W^-), if the CFL condition (C0) : max $(|\partial f/\partial u| \cdot \Delta t/\Delta x)$, and a condition (C1) are fulfilled :

(C1) monotonicity condition : $0 \leqslant p_i . \Delta x/\Delta_+ u_i \leqslant 2$; $0 \leqslant p_i . \Delta x/\Delta_- u_i \leqslant 2$ $\quad \Delta_\pm u_i = \pm(u_{i\pm 1} - u_i)$

The non linear case needs a careful upwind computation of the fluxes. The basic first order accurate upstream differencing scheme is the Godunov scheme which has been proven to converge (see Leroux I6I). Using the proof given for the first order scheme I6I, we obtain.

Proposition C : For the one dimensional scalar equation, the second order scheme using the fluxes of Godunov scheme is stable for L^∞ norm and total variation norm under the GFL condition (C0), if $\partial f/\partial u$ is a Lipschitz function of u and if :

i) Condition (C1) is fulfilled away from sonic points

ii) at a sonic point, (C1) is replaced by (C2) : $0 \leqslant p_i . \Delta x/\Delta_+ u_i \leqslant 1$; $0 \leqslant p_i . \Delta x/\Delta_- u_i \leqslant 1$

Unfortunately, the author has not been able to verify the entropy condition without imposing too much restrictive conditions upon the slopes to preserve the second order. A complete proof of convergence is given by Osher I7I in the case of a convex flux.

In the case of two dimensions, we have only done a linear analysis on orthogonal meshes.

Proposition d : For the two-dimensional scalar linear equation, the scheme is stable in L norm if we use an upwind expression of the fluxes, if the CFL condition $\Delta t/\Delta x \, (|A| + |B|) < 1$ is fulfilled and if (C3) monotonicity condition : the approximation fulfils the condition (C1) in all the spatial directions, using a description of u made by a linear interpolation on the line joining the eight neighbours of each node i.

For a system, we define a global norm $|L^\infty|$ = sup (L^2 norm on R^m on $W = (w_i)$ i = i, m)

Proposition e : On an orthogonal mesh, the scheme with no intermediate time integration stage (first order accurate in time), applied to a linear system is stable for $|L^\infty|$ norm under the CFL condition ($\Delta t/\Delta x \cdot |A| < 1/2, \Delta t/\Delta y \cdot |B| < 1/2$), if the condition (C3) is fulfilled for each variable.

II. Numerical implementation

a) Computation of spatial derivatives

We use the previous results as indications to define a method for which : 1) The spatial approximation of each variable is done independently. 2) The linear monotonicity condition (C3) is applied. The leading idea of the method is to apply a second order accurate operator to interpolate a smoothed function $\tilde{W}(x)$ of the function $W(x)$ around each node :

We denote by $\mathcal{P}$ the operator which defines around each node i, a linear function $W_{\mathcal{P}}$ which has the value W_i at node i, and which minimizes the quadratic error $\sum_{j \in J} (W_j - W_{\mathcal{P}j})^2$ on the set J of the neighbours of i. The linear function is a second order approximation of W. Let us apply $\mathcal{P}$ to W. We compare on each node $j \in J$, the variation $a = W_j - W_i$ of W, and an estimated variation on the opposite direction : $b = W_i - W_j - 2(W_{\mathcal{P}j} - W_j)$. The value $\tilde{W}_j$ of the smoothed function W is computed in order to satisfy the condition (C2) in the direction of the node j : $\tilde{W}_j = \text{sign}(a).\max[\min(2|a|, 2.\text{sign}(a).b, (W_{\mathcal{P}j} - W_i).\text{sign}(a)), 0]$. We can also use the formula given by Van Leer and Van Albada [1] $\tilde{W}_j = (-(a^2 + \varepsilon)b + (b^2 + \varepsilon)a)/(a^2 + b^2 + 2\varepsilon)$, ($\varepsilon = O(\Delta x^3)$) which presents the advantage of a smooth dependency in terms of node values. The final approximation is given by applying the operator $\mathcal{P}$ to the values $\tilde{W}_j$. This spatial approximation does not satisfies exactly the condition (C2), but it does not smoothen too much a local extremum and it can be applied on a general finite element mesh. However, it can be greatly simplified on a structured mesh.

b) Upwind computation of the fluxes

Comparisons between several schemes, show that a linearized flux splitting seems to be sufficient for Eulerian gas dynamics, for which we propose a well adapted approximate Riemann solver. It can be described as follows : given states W_l, W_r on the left and right sides of a boundary between two control volumes, we compute an intermediate state $\tilde{W}$ and the flux $h(\tilde{W})$. The flux derivative matrix $H = dh/du$ can be diagonalized by $H = T \Lambda T^{-1}$, with $\Lambda = \{\lambda_k\}$ $k = i, 4$, the eigenvalues. We compute the matrices T^{-1}. The state $\tilde{W}$, ($\tilde{W}_k, k = 1,4$) is defined by the set of equations :

k^{th} component of $T^{-1}(\tilde{W} - W_l) = 0$ if $k > 0$

k^{th} component of $T^{-1}(\tilde{W} - W_r) = 0$ if $k < 0$

When expressed in terms of variables (ρ, u, v, p), the corresponding linear set of equations is quite easy to solve, and give good results for low transonic flows. Around a sonic expansion, the scheme is not isentropic. We apply an improvement proposed by Harten [3] which solves the Riemann problem with a linear interpolation between the left and right characteristic values. For a strong shock, the value $(\frac{\lambda^l + \lambda^r}{2})$ is no longer a good approximation of the shock velocity. One must use the Roe's method [8] to compute a matrix H such that $h(W^l) - h(W^r) = H.(W^l - W^r)$. The complete set of formula defining W is given below :

U_n = normal component of velocity, u_T = tangential component, c = speed of sound eigenvalues :

eigenvalues : $\lambda_0^* = u_n^*$, $\lambda_1^* = u_n^* + c^*$, $\lambda_2^* = u_n^* - c^*$

$\tilde{W} = (\tilde{\rho}, \tilde{\rho}\tilde{u}, \tilde{\rho}\tilde{v}, \tilde{e})$, $p = (\gamma-1)(\tilde{e} - \frac{1}{2}(u^2+v^2)\rho)$ verifies :

$$\tilde{\rho} - \frac{\tilde{p}}{c^{*2}} = \left(\frac{1+\mathrm{sign}(\lambda_0)}{2}\right).\left(\rho^l - \frac{p^l}{c^{*2}}\right) + \left(\frac{1-\mathrm{sign}(\lambda_0)}{2}\right).\left(\rho^r - \frac{p^r}{c^{*2}}\right)$$

$$\tilde{u}_T = \left(\frac{1+\mathrm{sign}(\lambda_0)}{2}\right).\, u_T^l + \left(\frac{1-\mathrm{sign}(\lambda_0)}{2}\right).\, u_T^r$$

$$\tilde{u}_n + \frac{\tilde{p}}{\rho^* c^*} = S_1.\left(u_n^l + \frac{p^l}{\rho^* c^*}\right) + (1-S_1).\left(u^r + \frac{p^r}{\rho^* c^*}\right)$$

$$-\tilde{u}_n + \frac{\tilde{p}}{\rho^* c^*} = S_2.\left(-u_n^l + \frac{p^l}{\rho^* c^*}\right) + (1-S_2).\left(-u_n^r + \frac{p^r}{\rho^* c^*}\right)$$

Roe's variables :

$$u^* = \frac{\sqrt{\rho^l}.u^l + \sqrt{\rho^r}.u^r}{\sqrt{\rho^l} + \sqrt{\rho^r}} \qquad \rho^* = \left(\frac{\sqrt{\rho^l} + \sqrt{\rho^r}}{2}\right)^2$$

$$h^* = \frac{1}{2}\left(\frac{p^l + e^l}{\sqrt{\rho^l}} + \frac{p^r + e^r}{\sqrt{\rho^r}}\right) \qquad c^* = \left(\frac{h^*}{\sqrt{\rho^*}} - \frac{1}{2}(u^{*2} + v^{*2})\right)$$

Harten's correction :

$$S_k = \begin{cases} \left(\dfrac{1+\mathrm{sign}(\lambda_k^*)}{2}\right) & \text{if } \lambda_k^l.\lambda_k^r > 0 \text{ or } \lambda_k^r < 0 \ \lambda_k^l > 0 \\ (\lambda_k^* - \lambda_k^l)/(\lambda_k^r - \lambda_k^l) & \text{if } \lambda_k^r > 0 ,\ \lambda_k^l < 0 \end{cases}$$

III. Numerical tests

We present three test problems (calculated with $\gamma = 1.4$). The first one, is the GAMM test case of a stationary flow at Mach 0.85 through a channel with a circular bump of 4,2 % (Fig. 1). The scheme has been used with a maximum CFL number of 0.8 and with a mesh of 72 x 21 grid points. It is a low transonic case, where no non-linear correction in the Riemann solver is needed. The use of compatibility relations on the walls was successful both in limiting the entropy production on the edges of the bump, and in damping an oscillation of entropy at the shock near the wall. The second case is the unsteady problem of a flow at Mach 1.6 past a wedge of 20 % slope in a channel. The computation is started under uniform flow conditions. We have done the calculation with two uniform meshes : (I) 51 x 151 (Fig. 2a-e), (II) : 26 x 76 (Fig. 2f), with a CFL number .8. The time scale is normalized with the sound velocity and the width of the channel at the inlet. If the entropic correction is not applied in the Riemann solver, some expansion shock can appear at the upper corner of the wedge. The third case is also an unsteady problem. This computation is likewise started under uniform flow conditions. The flow at Mach 3.0 in a channel hits ad step with height equal to 1/5 that of the channel. It is an example requiring Roe's method for the Riemann solver. Treating the upper corner of the step with variables on the nodes is somewhat difficult. The results at time 6.11 on a mesh 21 x 81 and a mesh 41 x 121 (Fig. 3), were obtained with a computation using a double valued state on the upper corner of the step. The results are not yet quite satisfactory, the eulerian scheme seems to yield a result slightly different from the one obtained by a lagrangian-eulerian computation (see Borrel and Morice [2]).

Conclusion

The domain of application of the proposed method is mainly unsteady flows with complex geometries. The scheme results from a combination of a spatial approximation, an upwind flux computation, and a time integration. We have proposed a Riemann solver for flux computation because it seems to be less dependent on the homogeneity property of the flux ; but many techniques are suitable [3, 5, 8, 10], and the flux splitting techniques proposed by Van Leer seem advantageous for their computational efficiency. The scheme could also be used on a triangular finite element grid and the time integration could be improved. A three-dimensional extension is under investigation.

References

[1] G.D. Van Albada, B. Van Leer, W.W. Roberts : "Comparative study of computational methods in cosmic gas dynamics". ICASE Reports N° 81-24, August 3, 1981.

[2] M. Borrel, Ph Morice : "A second order Lagrangian-eulerian method for computation of two-dimensional unsteady transonic flows". 5th GAMM Conf. Rome 1983.

[3] P. Harten, P. Lax, B. Van Leer : "On upstream differencing and Godunov type schemes for hyperbolic conservation laws". STAM Review, Vol. 25, N° 1 Jan. 83.

[4] B. Van Leer : "V. A second order sequel to Godunov method". JCP 23 276-299, 1977.

[5] B. Van Leer : "Flux vector splitting for the Euler equations". VKI 1983.

[6] A.Y. Leroux : "Approximation de quelques problèmes hyperboliques non linéaires". Thesis, Rennes University, April 1974.

[7] S. Osher : "Convergence of generalized MUSCL schemes". ICASE Report 172306, Feb. 1984.

[8] P.L. Roe : "Approximate Riemann solvers, parameter vectors and difference schemes". JCP 43 357.372 (1981).

[9] J.P. Veuillot, H. Viviand : "Méthodes pseudo-instationnaires pour le calcul d'écoulements transsoniques". ONERA publication N° 1978-4 (English translation, ESA-TT-561).

[10] G. Vijayasundaram : "Résolution numérique des équations d'Euler pour des écoulements transsoniques avec un schéma de Godunov en éléments finis". The., Paris VII, Oct. 1982.

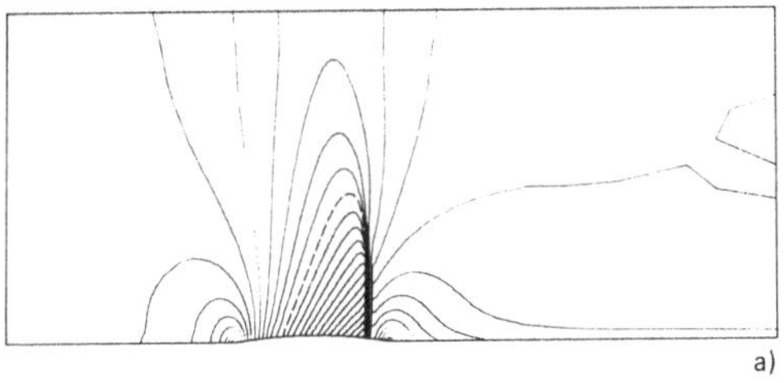

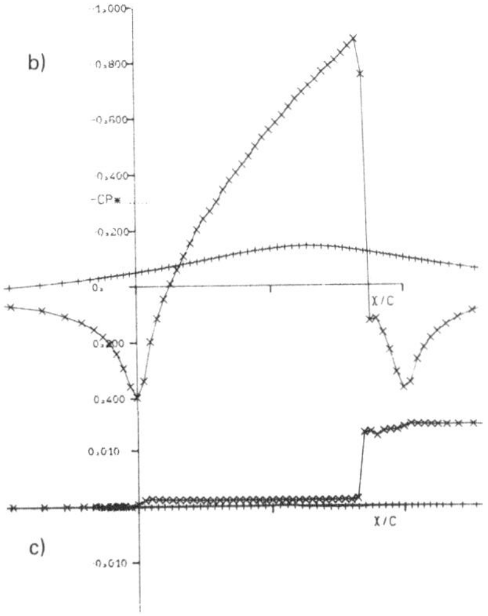

Fig. 1 – Flow past a circular bump in a channel at Mach 0.85 :
a) iso-mach lines
b) pressure curves on the wall
c) entropy curves on the walls.

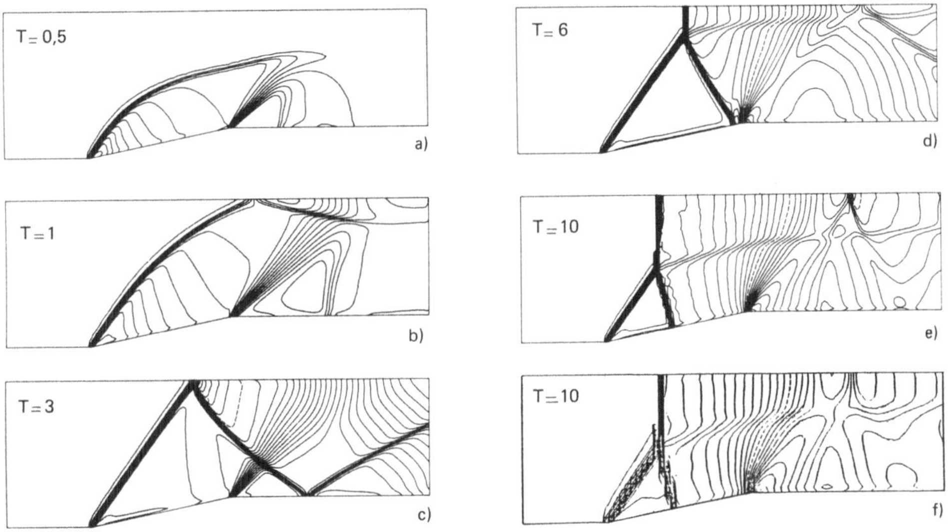

Fig. 2 – Iso-Mach lines at the times : 0.5, 1, 3, 6, 10, x t_{ref} ; for the wedge problem at Mach 1.6. t_{ref} x (width of the channel)/(speed of sound at Mach 1.6). (a)(b)(c)(d)(e) : mesh 51 x 151 (f) : Mesh 26 x 76

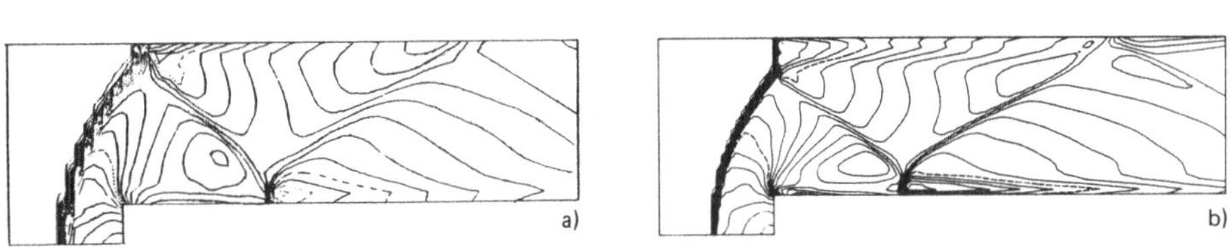

Fig. 3 – Iso-Mach lines at time T = 6,11 x t_{ref} for the flow over a step at Mach 3. (a) mesh 21 x 61 ; b) mesh 41 x 121.

A COMPARISON OF FINITE DIFFERENCE AND CHARACTERISTIC GALERKIN METHODS FOR SHOCK MODELLING

K.W. Morton and P.K. Sweby
Oxford University Computing Laboratory
8-11 Keble Road
Oxford

1. INTRODUCTION

In recent years many authors have devised high resolution, Total Variation Diminishing (TVD) finite difference schemes in order to obtain sharper profiles to represent discontinuities than is possible with first order schemes, whilst avoiding the spurious oscillations which plague the more classical second order schemes. One important class of techniques uses flux limiters [3,9,10,11,12,13] which, as with FCT (flux-corrected transport) methods [1], utilise a limited amount of anti-diffusive flux to add to a first order scheme.

More recently and in parallel with this work, finite element methods based on the Characteristic Galerkin formulation have started to be developed for shock modelling [6,7]. As shown in [6], piecewise constant elements lead to a first order scheme equivalent to the Engquist -Osher scheme [4] but data recovery techniques can be used to obtain higher accuracy.

In this paper we explore this idea using recovery techniques based on piecewise linear functions. We show that the recovery can be applied adaptively to ensure that monotonicity is preserved and has a similar role to that of flux limiters. Moreover, we show that the resulting update procedures have much in common with the difference schemes of Roe [9,10]. Comparisons on model problems will be made and distinguishing features of the methods discussed.

2. FLUX LIMITERS AND ADAPTIVE RECOVERY

Consider the scalar conservation law in 1D with convex flux f(u)

$$\begin{aligned} &\partial_t u + \partial_x f(u) = 0 \qquad t > 0,\ x \in \mathbb{R} \\ &u(x,0) = u^0(x), \ \text{given.} \end{aligned} \tag{2.1}$$

We suppose this is approximated by an essentially explicit scheme on a uniform $(\Delta x, \Delta t)$ mesh valid for CFL numbers less than unity. Then a convenient form in which to write the scheme in terms of nodal values U_k^n is

$$U_k^{n+1} = U_k^n - C_{k-\frac{1}{2}}\, \Delta_- U_k^n + D_{k+\frac{1}{2}}\, \Delta_+ U_k^n , \tag{2.2}$$

where $\Delta_- U_k := U_k - U_{k-1}$, $\Delta_+ U_k := U_{k+1} - U_k$ and the coefficients $C_{k-\frac{1}{2}}$, $D_{k+\frac{1}{2}}$ are functions of the set $\{U_k^n\}$: alternatively we may write this in terms of the flux differences $\Delta_+ f_k^n$, $\Delta_- f_k^n$ where $f_k^n := f(U_k^n)$, by introducing the CFL numbers $\nu_{k+\frac{1}{2}} := (\Delta t/\Delta x)\Delta_+ f_k^n/\Delta_+ U_k^n$. As shown by Sweby [11,12], the properties of many well-known difference schemes can be studied and compared by putting them in this basic form. Thus the TVD property, i.e. total variation diminishing or

$$\sum_k |\Delta_+ U_k^{n+1}| \leq \sum_k |\Delta_+ U_k^n| \quad , \tag{2.3}$$

follows from the inequalities

$$0 \leq C_{k+\frac{1}{2}}, \quad 0 \leq D_{k+\frac{1}{2}}, \quad 0 \leq C_{k+\frac{1}{2}} + D_{k+\frac{1}{2}} \leq 1. \tag{2.4}$$

This is a key property in eliminating spurious oscillations and in establishing convergence results [5].

To derive a scheme of the form (2.2), it is usual to start from a basic first order scheme and then to add limited contributions from the second order fluxes so as to maintain the TVD property. The framework used by Sweby [11] starts from the E-scheme fluxes of Osher [8], $h_{k+\frac{1}{2}}^n := h(U_k^n, U_{k+1}^n)$ satisfying

$$(U_{k+1}^n - U_k^n)\,[h_{k+\frac{1}{2}}^n - f(u)] \leq 0 \tag{2.5}$$

for all u between U_k^n and U_{k+1}^n: then positive and negative fluxes

$$(\Delta f_{k+\frac{1}{2}}^n)^+ := -[h_{k+\frac{1}{2}}^n - f_{k+1}^n] \;,\; (\Delta f_{k+\frac{1}{2}}^n)^- := [h_{k+\frac{1}{2}}^n - f_k^n] \tag{2.6}$$

are introduced for which the corresponding CFL numbers

$$\nu_{k+\frac{1}{2}}^{\pm} := (\Delta t/\Delta x)(\Delta f_{k+\frac{1}{2}}^n)^{\pm}/\Delta_+ U_k^n \tag{2.7}$$

are respectively positive and negative. The final flux-limited scheme has the form

$$U_k^{n+1} = U_k^n - (\Delta t/\Delta x)\Delta_-\{h_{k+\frac{1}{2}}^n + \tfrac{1}{2}\phi(r_k^+)(1 - |\nu_{k+\frac{1}{2}}^+|)(\Delta f_{k+\frac{1}{2}}^n)^+$$
$$+ \tfrac{1}{2}\phi(r_{k+1}^-)(1 - |\nu_{k+\frac{1}{2}}^-|)(\Delta f_{k+\frac{1}{2}}^n)^-\} \tag{2.8}$$

which can be written in the form (2.2) with

$$C_{k+\frac{1}{2}} = \nu_{k+\frac{1}{2}}^+ \{1 + \tfrac{1}{2}(1 - |\nu_{k+\frac{1}{2}}^+|)[\phi(r_{k+1}^+)/r_{k+1}^+ - \phi(r_k^+)]\} \tag{2.9}$$

and a similar expression for $D_{k+\frac{1}{2}}$. Here $\phi(\cdot)$ is the flux limiter: its arguments are the ratios

$$r_k^+ := d_{k-\frac{1}{2}}^+/d_{k+\frac{1}{2}}^+, \quad r_k^- := d_{k+\frac{1}{2}}^-/d_{k-\frac{1}{2}}^- \quad \text{where } d_{k+\frac{1}{2}}^{\pm} := \tfrac{1}{2}(1-|\nu_{k+\frac{1}{2}}^{\pm}|)(\Delta f_{k+\frac{1}{2}}^n)^{\pm}. \tag{2.10}$$

In [11] Sweby compared the effect of using different flux limiters.

For our present comparison we will concentrate on Roe's second order scheme [9,10]. In its original form it was based on Murman's upwind scheme in which the increments

$$g^n_{k+\frac{1}{2}} := -(\Delta t/\Delta x)\Delta_+ f^n_k = -\nu_{k+\frac{1}{2}}\Delta_+ U^n_k \qquad (2.11)$$

are added to either U^n_k or U^n_{k+1} according to whether the mean CFL number $\nu_{k+\frac{1}{2}}$ is negative or positive: this would admit entropy violating shocks across a sonic point but various devices have been used to split the flux difference then to satisfy the requirements of (2.5 - 2.7). The scheme is made second order by a further stage in which the increment

$$\tfrac{1}{2}(1 - |\nu_{k+\frac{1}{2}}|)g^n_{k+\frac{1}{2}} = -\tfrac{1}{2}\nu_{k+\frac{1}{2}}(1 - |\nu_{k+\frac{1}{2}}|)\Delta_+ U^n_k , \qquad (2.12)$$

modified by a flux-limiter, is transferred from one cell to the next against the direction of flow, i.e. from U^n_{k+1} to U^n_k if $\nu_{k+\frac{1}{2}} > 0$. This second stage operation corresponds to an anti-diffusive flux: and generally in these schemes the role of the flux-limiter is to ensure that, in pursuing a formally second order accurate scheme, one does not increase the total variation of the approximation $\{U^{n+1}_k\}$ as compared with that of $\{U^n_k\}$.

Now let us consider corresponding ECG (Euler Characteristic Galerkin) schemes for (2.1). Following Morton [6,7] we use only piecewise constant approximations for the basic scheme, relying on recovery procedures to obtain higher accuracy. In terms of the basis functions $\phi_k(x)$ [= 1 for $(k-\frac{1}{2})\Delta x < x < (k+\frac{1}{2})\Delta x$ and zero elsewhere], the basic ECG scheme is

$$\langle U^{n+1} - U^n, \phi_k\rangle + \Delta t \langle \partial_x f(U^n), \Phi^n_k\rangle = 0 \qquad (2.13a)$$

where

$$\Phi^n_k(x) := \frac{1}{a(U^n)\Delta t} \int_x^{x+a(U^n)\Delta t} \phi_k(z)dz \qquad (2.13b)$$

and $a(u) = \partial f/\partial u$ is the characteristic speed. The scheme in this form is unconditionally stable: but for CFL numbers less than unity it reduces to the Engquist-Osher first order scheme in which the flux difference $\Delta_+ f^n_k$ is broken up into increments

$$-(\Delta t/\Delta x)[f^n_{k+1} - f(\bar{u})] \quad \text{and} \quad -(\Delta t/\Delta x)[f(\bar{u}) - f^n_k] \qquad (2.14)$$

which are added to U^n_k or U^n_{k+1} according to whether $a(U^n_{k+1})$, and respectively $a(U^n_k)$, are negative or positive. Here $\bar{u}$ corresponds to the sonic point.

The nodal parameters $\{U_k^n\}$ are regarded as representing as closely as possible the projection of the true solution onto the space of piecewise constant functions. Then (2.12) is the result of tracing the evolution of this projection through one time step, by following the characteristics and using overturned manifolds, before projecting again at the new time level. To obtain greater accuracy we can (i) carry out the evolution more accurately and/or (ii) deduce more information regarding the true solution at time level n by combining neighbouring values U_k^n and strengthening the hypotheses on its properties. The recovery procedure in (ii) involves finding a function $\tilde{u}^n(x)$ whose projection is $U^n(x)$, and which is therefore given by

$$\langle U^n - \tilde{u}^n, \phi_k \rangle = 0 \qquad \forall k, \tag{2.15}$$

and then using $f(\tilde{u}^n)$ and $a(\tilde{u}^n)$ in (2.13). For example, in [7] it was seen that recovery with quadratic splines in the case of linear advection reproduced the highly accurate (3rd order) scheme obtained from the basic scheme with continuous piecewise linear approximations. At the other extreme of smoothness, in [6] a shock recovery algorithm was given in which a shock in cell k, involving a jump from U_{k-1}^n to U_{k+1}^n, is moved with the shock speed $\Delta_0 f_k^n / \Delta_0 U_k^n$ where $\Delta_0 := \frac{1}{2}(\Delta_+ + \Delta_-)$.

Here we pursue an alternative of recovering with continuous piecewise linears which spread the discontinuity at $(k+\frac{1}{2})\Delta x$ over $\frac{1}{2}\theta_{k+\frac{1}{2}}\Delta x$ either side. The intermediate recovered levels $\tilde{u}_k^n$ satisfy

$$\theta_{k-\frac{1}{2}}\tilde{u}_{k-1}^n + (8 - \theta_{k-\frac{1}{2}} - \theta_{k+\frac{1}{2}})\tilde{u}_k^n + \theta_{k+\frac{1}{2}}\tilde{u}_{k+1}^n = 8U_k^n . \tag{2.16}$$

For linear advection and smooth data, choosing $\theta=1$ gives a 2nd order scheme reducing to

$$U_k^{n+1} = U_k^n - \nu\Delta_0\tilde{u}_k^n + \tfrac{1}{2}\nu^2\delta^2\tilde{u}_k^n \tag{2.17}$$

for $|\nu| \leq \frac{1}{2}$, but stable for $\nu^2 \leq \frac{1}{2}$: here $\delta^2 := \Delta_+\Delta_-$ and the scheme is shifted for other ν. However (2.16) plus (2.17) is not TVD: so how best to choose θ ? We can rely on evolution plus projection to maintain key properties from $\tilde{u}^n$ to U^{n+1}, so we adopt as our main criterion

$$\Delta_+U_k^n = 0 \Rightarrow \Delta_+\tilde{u}_k^n = 0 \quad \text{and} \quad (\Delta_+\tilde{u}_k^n)/(\Delta_+U_k^n) \geq 0 \tag{2.18}$$

to be applied at the recovery stage. This can be iterated but a normally reliable choice is

$$\theta_{k+\frac{1}{2}} = \min(1, 3|r_k|, 3/|r_{k+1}|), \quad r_k := (\Delta_-U_k^n)/(\Delta_+U_k^n). \tag{2.19}$$

The evolution algorithm (2.16) with (2.13) can be given explicitly for $\theta_{k+\frac{1}{2}} > 0$ in two steps: first find the ordered set $\{x_{k+\frac{1}{2}}^{(i)}, i=1,2,..m\}$ satisfying

$$x_{k+\frac{1}{2}} = x^{(i)}_{k+\frac{1}{2}} + a(\tilde{u}^n(x^{(i)}_{k+\frac{1}{2}}))\Delta t, \quad |x^{(i)}_{k+\frac{1}{2}} - x_{k+\frac{1}{2}}| \leq \tfrac{1}{2}\theta_{k+\frac{1}{2}}\Delta x \qquad (2.20)$$

plus the end points; then the increment

$$-(\Delta t/\Delta x)[f(\tilde{u}^n(x^{(i+1)}_{k+\frac{1}{2}})) - f(\tilde{u}^n(x^{(i)}_{k+\frac{1}{2}}))], \quad i = 0,1,\ldots,m \qquad (2.21)$$

is added to U^n_k or U^n_{k+1} according to the sign of $a(\tilde{u}^n)$, while a transfer of $\frac{1}{2}[\tilde{v}^{(i)}]^2 \Delta_+\tilde{u}^n_k/\theta_{k+\frac{1}{2}}$ is made from U^n_{k+1} to U^n_k, or back, according to the sign of $1 + da(\tilde{u}^n(x))/dx$.

In comparative tests, linear advection of $\sin^2 x$ with (2.19) gives results very similar to standard flux-limiters, including the flattened top: use of (2.18) gives rather better results. For inviscid Burger's equation, these recovery procedures remove the dog-leg at a rarefaction sonic point typical of the first order Engquist-Osher scheme while shock recovery gives just a single intermediate point. Work is proceeding to carry these studies over to the Euler equations in 1D and a scalar law in 2D. The interplay of the ideas from the two viewpoints is already proving fruitful as progress is made towards shock modelling on an irregular mesh in 2D.

[1] BORIS, J.P. & BOOK, D.L., JCP 11, pp. 38-69, 1973.

[2] BRENIER, Y., INRIA Rapports de Recherche No. 53, 1981.

[3] CHAKRAVARTHY, S. & OSHER, S., AIAA 6th Comp. Fluid Dynamics Conf. 1983.

[4] ENGQUIST, B. & OSHER, S., Math. Comp. Vol. 34, No. 149, pp.45-75, 1980.

[5] HARTEN, A., JCP 49, pp. 357-393, 1983.

[6] MORTON, K.W., Proc. IC8NMFD, Lect. Notes in Physics 170, pp. 77-93, 1982.

[7] MORTON, K.W., Proc. 5th GAMM Conf., pp. 243-250, 1983.

[8] OSHER, S., SIAM Journal on Num. Analysis 21 No. 2, 1984.

[9] ROE, P.L., Numerical Methods for Fluid Dynamics, Academic Press, 1982.

[10] ROE, P.L. & BAINES, M.J., Proc. 4th GAMM conf., pp. 281-290, 1981.

[11] SWEBY, P.K., SIAM Journal on Num. Analysis 21, No. 5, 1984.

[12] SWEBY, P.K., Proc. AMS-SIAM Summer Seminar, La Jolla, California, 1983.

[13] VAN LEER, B., JCP 14, pp. 361-370, 1974.

[14] VAN LEER, B., JCP 32, pp. 101-136, 1979.

MULTIGRID RELAXATION FOR THE EULER EQUATIONS

W.A. Mulder
University Observatory, P.O. Box 9513
2300 RA Leiden, The Netherlands

1. Introduction

Implicit time-discretization, combined with upwind space-differencing, yields a fast and robust method for finding stationary solutions of the Euler equations. Particularly successful is the switched evolution/relaxation (SER) scheme, which provides a smooth switching between explicit time-integration and Newton's method for finding zero values of a given function. For one-dimensional problems quadratic convergence can be obtained, as shown in an earlier paper [1]. In two dimensions the exact inversion of the linear system arising in the implicit formulation is too costly. Various approximate solvers are described in [2].

In this paper an efficient approximate solver based on the multigrid method for the solution of large linear systems is described. An outline of the basic multigrid concepts can be found in [3]. The method is applied to compute the two-dimensional transonic flow through a channel with a circular bump at one wall. Numerical results are presented for single-grid and multigrid, with first- and second-order spatial accuracy.

2. Method

Let the system of hyperbolic equations in two dimensions be given by:

$$\frac{\partial \mathbf{w}}{\partial t} = -\frac{\partial \mathbf{f}}{\partial x} - \frac{\partial \mathbf{g}}{\partial y} \equiv \mathbf{r}(\mathbf{w}) \tag{1}$$

Here $\mathbf{f}(\mathbf{w})$ and $\mathbf{g}(\mathbf{w})$ are the fluxes, $\mathbf{w}$ is the vector of conserved state quantities and $\mathbf{r}(\mathbf{w})$ is the residual, the function that must be made to vanish. The implicit scheme of our choice is the linearized "backward Euler" scheme:

$$\mathbf{L}^n \, \Delta_t \mathbf{w}^n = \left[\frac{\Delta x \Delta y}{\Delta t^n} - \Delta x \Delta y \left(\frac{d\mathbf{r}}{d\mathbf{w}}\right)^n\right] \Delta_t \mathbf{w}^n = \Delta x \Delta y \; \mathbf{r}^n(\mathbf{w}^n) \tag{2}$$

The superscript n denotes values at a time t^n, while $\Delta t^n = t^{n+1} - t^n$ and $\Delta_t \mathbf{w}^n = \mathbf{w}^{n+1} - \mathbf{w}^n$. The discrete values $\mathbf{w}_{ij}$ of the state quantity are obtained by volume-averaging. The local residual $\mathbf{r}_{ij}(\mathbf{w})$ is computed, for the present purpose, by a first-order upwind-difference scheme on a 5-point stencil: $\mathbf{r}_{ij}(\mathbf{w}) = \mathbf{r}(\mathbf{w}_{i-1,j}, \mathbf{w}_{i,j-1}, \mathbf{w}_{ij}, \mathbf{w}_{i+1,j}, \mathbf{w}_{i,j+1})$. The timestep Δt^n is determined by:

$$\Delta t^n = \varepsilon / RES^n ,$$
$$RES^n = \max_{ijk}(|r^n_{ijk}| / |w^n_{ijk}| + h^n_{ijk}) . \qquad (3)$$

Here h is a bias to prevent division by zero; in case of the isenthalpic Euler equations used here, $h_{ij1}=0$ and $h_{ij2}=h_{ij3}=\rho c$. If Δt^n is small, the implicit scheme behaves very much like an explicit time-accurate scheme. Once the solution is getting closer to the steady state, Δt^n becomes larger and the scheme automatically switches to Newton's method. The constant ε controls the relative variation of w and is usually taken to be 1.

The inversion of the linear system (2) can be carried out efficiently by a multigrid scheme. Its basic ingredients are: (i) relaxation, (ii) restriction and (iii) prolongation. Symmetric Gauss-Seidel relaxation is used here for its excellent short-wave damping. Restriction is carried out by adding the values of the neighbouring zones and placing them on the coarser grid. The matrix L is restricted by addition of the corresponding blocks. For non-uniform grids the multiplication by the local cell-volume $\Delta_i x \; \Delta_j y$ ensures the proper weighting. Since restriction only involves additions, its cost is but a small fraction of that of a relaxation sweep. Finally, prolongation is carried out by distributing the coarse grid solution uniformly over the fine grid.

The multigrid strategy used in this paper is a simple V-cycle. Before every restriction and after every prolongation symmetric Gauss-Seidel relaxation (consisting of 2 sweeps) is carried out. On the coarsest grid an exact inversion is applied.

The usual quantity "work" is computed here by adding the number of relaxation sweeps, weighting each grid-total with respect to the finest grid. Thus, the amount of work for a single-grid iteration is 2, and for a V-cycle about $5\frac{1}{3}$.

3. Results for a test problem

The method is tested on the two-dimensional problem of transonic flow through a straight channel. The flow runs along the x-direction and is obstructed by a circular arc on the lower wall. The channel has an x-coordinate running from -1.5 to 2.5 and a y-coordinate running from 0.0 to 2.0. The circular arc between x = -0.5 and 0.5 has a maximum thickness equal to 4.2% of the chord. Thin-airfoil theory is used to transfer the boundary conditions at the arc to the flow. For simplicity a uniform square grid is adopted. In this setting the isenthalpic Euler equations in conservation form are solved for an ideal gas with $\gamma = 1.4$. The free-stream values are chosen to be: $\rho_\infty = 1$, $u_\infty = 0.85$, $v_\infty = 0$, $c_\infty = 1$. For the unchoked case two boundary conditions at the inlet and one at the outlet should be specified. At the inlet the direction of the flow and the total pressure are given, at the outlet the static pressure is specified; these parameters are computed from the free-stream values. The fluxes on the boundaries are computed by using differences of characteristic variables, with the appropriate upwind-switching to determine between extrapolation or direct computation. Boundaries at the lower and upper wall are simulated by an extra zone with reflected state quantities.

For the upwind differencing of the internal flow the split fluxes as proposed in [4] are used, as they can be easily linearized. for second-order accuracy an incomplete linearization is adopted to give $L^n(\mathbf{w})$ the same structure as for the first-order scheme (see [1]). This will obviously lead to some loss in convergence speed, but greatly simplifies the computation and inversion of $L^n(\mathbf{w})$.

The linearization L^n is frozen now and then to save cpu-time, just as in [2]. Furthermore, the inverses of the main-diagonal blocks and the restricted blocks are stored in memory, so that a multigrid cycle can be carried out must faster during freezing.

Convergence histories for a 16x8, 32x16 and 64x32 grid are shown in Fig. 1, both for the first-order and second-order accurate solutions. In all cases the multigrid scheme is faster. The second-order runs are somewhat slower than the first-order ones, due to the incomplete linearization. Fig. 2 shows the pressure coefficient on bottom and top wall as computed from the first-order and second-order accurate solution, respectively.

4. Concluding Remarks

It has been demonstrated that the multigrid technique can be successfully applied to compute a stationary transonic solution of the Euler equations. For the two-dimensional test problem the gain in efficiency with respect to a single-grid scheme, both in terms of work and cpu-time, is of the order $N^{0.4}$, where N is the total number of zones. Consequently, the number of iterations required to obtain a converged solution, whether first- or second-order accurate, increases only slowly with N. This result certainly justifies the additional effort of coding the multigrid scheme.

Finally, the power of the method presented here is demonstrated by Fig. 3, showing the stationary solution and convergence history for the flow in a rotating galaxy.

Acknowledgement. Due to a sudden illness of the author, the oral presentation of this paper was prepared by Bram van Leer and carried out by Piet Wesseling.

References

1. Mulder, W.A., Van Leer, B.: "Implicit Upwind Methods for the Euler Equations", AIAA paper 83-1930, Danvers, Massachusetts, July 1983.
2. Van Leer, B., Mulder, W.A.: "Relaxation Methods for Hyperbolic Equations", in Proceedings of the INRIA workshop on Numerical Methods for the Euler Equations for Compressible Fluids, Le Chesnay, France, 7-9 Dec. 1983, to be published by SIAM.
3. Brandt, A.: "Guide to Multigrid Development", in Multigrid Methods, Proceedings of the conference held at Köln-Porz, Nov. 1981 (Lecture Notes in Mathematics 960, Springer-Verlag).
4. Van Leer, B.: "Flux-Vector Splitting for the Euler Equations", in Proceedings of the 8th International Conference on Numerical Methods in Fluid Dynamics, Aachen, June 1982 (Lecture Notes in Physics 170, Springer-Verlag).

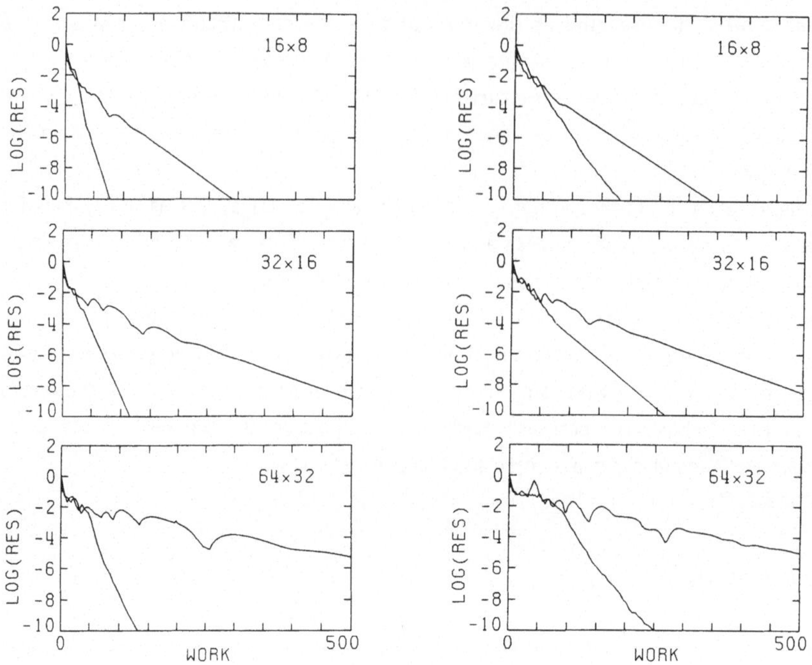

Fig. 1. Convergence histories for the first-order (left) and second-order (right) accurate solutions on 3 different grids. The residual RES is normalized by the initial value at t=0. In all cases the multigrid scheme is faster than the single-grid scheme. For the multigrid scheme, the total amount of work increases only slowly with the number of points.

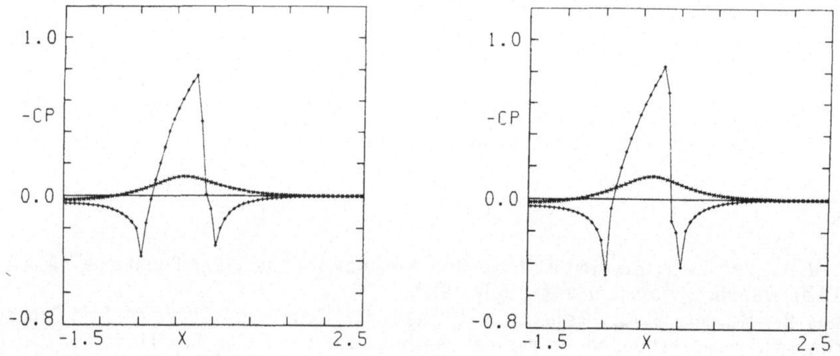

Fig. 2. Pressure coefficient on bottom (+) and top wall (x) for a 4.2% circular arc and a free-stream Mach number 0.85, as computed from the first-order (left) and second-order (right) solution on a 64x32 grid. Thin-airfoil theory is applied to transfer the boundary conditions to the flow.

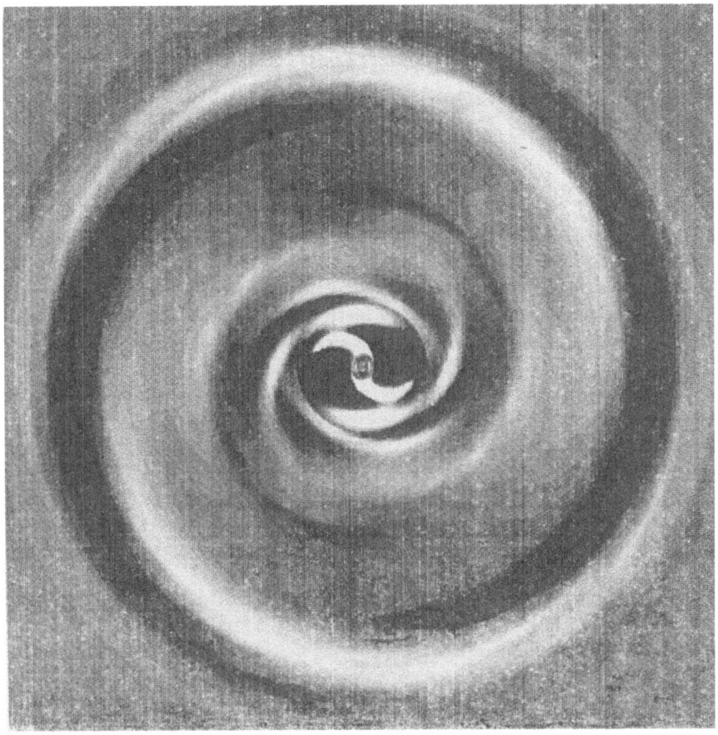

Fig. 3. Stationary spiral pattern in the co-rotating frame of a weakly barred galaxy. The gravitational potential consists of an axisymmetric part and a rotating $\cos(2\phi)$ perturbation, which ends at a co-rotation radius 8.36 . The polar grid (64x64 zones) covers one half-plane, with a radius R running from 0.3 to 30 and ΔR varying from 0.03 to 1.7. The computation is carried in single precision. Shown is the density divided by the average density per ring; the size of the figure is 40x40. Convergence histories for two different grids are shown below. The peaks in the initial phase are due to successive grid-refinement starting from a 4x4 grid.

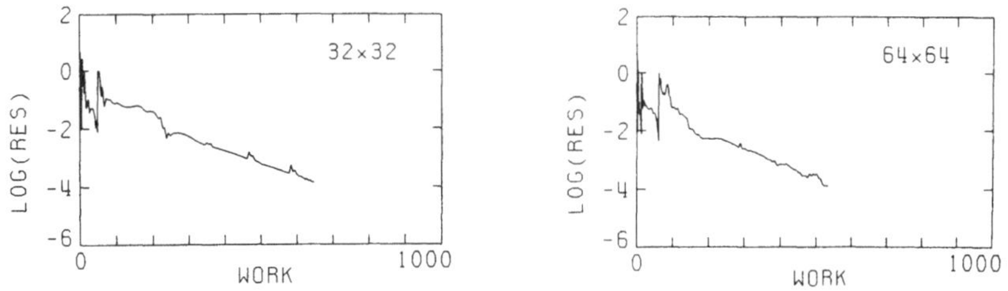

A Practical Adaptive-Grid Method for Complex Fluid-Flow Problems

KAZUHIRO NAKAHASHI AND GEORGE S. DEIWERT

NASA Ames Research Center
Moffett Field, California 94035, U.S.A.

INTRODUCTION

Adaptive-grid methods are an important subject of study in computational fluid dynamics because of their potential for improving the efficiency and accuracy of numerical methods. Various papers on grid adaptation have been presented in recent years, but few methods have been applied in practical applications.

In this paper, a practical solution-adaptive-grid method utilizing a tension and torsion spring analogy is proposed for multidimensional fluid-flow problems. The tension spring, which connects adjacent grid points to each other, controls grid spacings so that clustering is realized in regions containing shock waves and shear layers. The torsion spring, which is attached to each grid node, controls inclinations of coordinate lines and prevents excessive grid skewness. A marching procedure is used that results in a simple tridiagonal system of equations at each coordinate line to determine grid-point distribution. Multidirectional grid adaptation to the flow solution is achieved by successive applications of one-directional adaptation. Examples of applications for axisymmetric afterbody flow fields and two-dimensional transonic airfoil flow fields are shown.

GRID ADAPTATION IN ONE COORDINATE DIRECTION

For simplicity of illustration, consider grid adaptation to a flow field in which grid points are free to move along each η-coordinate line whose configuration is fixed. Let a grid point A in figure 1 be connected to its adjacent points, B and C, by tension springs whose spring constants are $K_{i,j-1}$ and $K_{i,j}$. To distribute grid points along the η_i-coordinate line in proportion to the gradient of selected flow properties, the relationship between the spring constant K and the gradient of the dependent variable f

$$K_{i,j} = 1 + C_1|f_{i,j+1} - f_{i,j}|/(s_{i,j+1} - s_{i,j}) \tag{1}$$

is used, where C_1 is a constant and $s_{i,j}$ is the arc length calculated from point (i,1) along the η_i-coordinate.

Using equation (1), the distribution of grid points along the η_i-coordinate line, namely, new values of $s_{i,j}$, is determined by

$$K_{i,j}(s_{i,j+1} - s_{i,j}) - K_{i,j-1}(s_{i,j} - s_{i,j-1}) = 0 \tag{2}$$

The idea of a spring analogy represented in equation (1) was introduced by Gnoffo(ref. 1). In his model, the distribution of points along the η_i-coordinate is determined only by the gradients of flow properties along that coordinate line and is not affected by the distribution on adjacent η-coordinates, η_{i-1} and η_{i+1}. This can lead to excessive skewness of grid lines, especially when applied to complex flow fields, and this lack of control of grid inclination makes it difficult to extend the scheme to more than one family of coordinates.

A force to control inclinations of ξ-coordinates in addition to that of grid spacings on η-coordinates will correct this deficiency. This control force can be given by considering torsion springs attached to nodes along the η_{i-1} line. The torsion spring enforces the inclination of line $\overline{DA}$ to that of a reference line. If the spring constant of the torsion spring is denoted by H, a mathematical statement of the force is

$$F_{torsion} = -H_{i-1,j}(\theta_{DA} - \phi) \tag{3}$$

where θ_{DA} is a inclination of line $\overline{DA}$ and ϕ the inclination of the reference line. The reference line can be chosen as an extension of $\overline{FD}$ to avoid kinks in the ξ-line at point D, as a line normal to the η_i-coordinate to make the grid quasi-orthogonal, or as a streamline, and so forth. In practical calculations, a combination of these reference lines is used. The torsion spring constant H can be prescribed for each coordinate line.

A balance equation for the complete spring system is

$$K_{i,j}(s_{i,j+1} - s_{i,j}) - K_{i,j-1}(s_{i,j} - s_{i,j-1}) - H_{i-1,j}(\theta_{i-1,j} - \phi_{i-1,j}) = 0 \tag{4}$$

To facilitate solutions to equation (4), the third term is rewritten

$$H_{i-1,j}(\theta_{i-1,j} - \phi_{i-1,j}) \Rightarrow \overline{H}_{i-1,j}(s_{i,j} - \overline{s}_{i,j}) \qquad (5)$$

where $\overline{s}_{i,j}$ is arc length to the intersection of reference line $\overline{DA'}$ with the η_i-coordinate as depicted in figure 2. The $\overline{H}_{i-1,j}$ term is set equal to $H_{i-1,j}$ divided by length of $\overline{DA'}$. Finally, equation (4) reduces to the following equation:

$$K_{i,j-1}s_{i,j-1} - (K_{i,j} + K_{i,j-1} + \overline{H}_{i-1,j})s_{i,j} + K_{i,j}s_{i,j+1} = -\overline{H}_{i-1,j}\overline{s}_{i,j} \qquad (6)$$

This is a tridiagonal system of equations for $s_{i,j}$ and can be readily solved.

In this analysis only the torsion force on the upstream side (η_{i-1}) influences the distribution at η_i. This permits simple marching schemes to be used, without any loss of generality, and contributes to the simplicity and robustness of the method. Conversely, if the influence from both sides (η_{i-1} and η_{i+1}) is considered simultaneously, the computational effort is increased considerably without any additional benefit. Note, too, that the downstream influence, (η_{i+1}), could be used instead of the upstream, without any additional complexity or loss of generality.

GRID ADAPTATION IN MULTICOORDINATE DIRECTION

A model using tension and torsion springs for a two-directional adaptation can be depicted as in figure 3(a). Since each grid point is connected to its four adjacent points, the procedure for the grid movements of this model is more complicated and requires more computational effort. To minimize this complexity, a split model, which is a combination of one-directional adaptation, is used(fig. 3(b)). Grid movement is achieved by successive applications of the one-directional adaptation method. This is analogous to ADI schemes for partial differential equations. It is not necessary to achieve convergence in this adaptation procedure and, in fact, one iteration is sufficient in practical applications of the scheme.

The solution field is interpolated onto the newly adapted grid, using second-order, one-dimensional Lagrange interpolation after each one-directional adaptation. Before this interpolation, it is possible to add or delete grid points at the users discretion, thus enhancing the method without any loss of accuracy or increased complexity. The successive application of one-directional adaptation also enhances the applicability of the method, in that unidirectional adaptation can be used in regions where gradients are large in only one direction. Extension of this scheme to three-dimensions is straightforward.

RESULTS

Shown in figures 4-8 are examples of applications of the method to axisymmetric, plume flow fields. These complex flow fields exhibit oblique shocks, barrel shocks, and slip surfaces, where grid points should be clustered for adequate resolution. The locations of these discontinuities are not known apriori, and a solution-adaptive grid is particularly useful in realizing an efficient and accurate simulation.

The initial grid(fig. 4) was generated by an algebraic method and the flow field determined using a code(ref. 2) for the thin-layer Navier-Stokes equations. The free-stream Mach number is 2.01, the jet-exit Mach number is 2.5, and the static pressure ratio of the exhaust jet to the free-stream is 1. The density gradient was chosen as a reference variable for the grid clustering, and figure 5 shows a solution-adapted grid which clearly has clustered points to the oblique shock, the slip surface, and the barrel shock. Shown in figure 6 are computed density contours using the grid shown in figure 5.

Shown in figure 7 is an adapted grid for a high-jet-pressure case (pressure ratio of 6), and density contours are shown in figure 8. The same initial grid(fig. 4) used in the previous case was used here.

The grids were adapted to the flow-field solutions periodically (typically two or three times) during the course of reaching a steady-state solution. In this way, the time required for adaptation is a negligible fraction of the total time required for flow-field solution.

Examples of the application to transonic flow fields past a NACA0012 airfoil are shown in figures 9-12. The free-stream Mach number is 0.8 and the angle of attack is 1.25°. Two-dimensional Euler equations were solved, using the code ARC2D(refs. 3,4). Figure 9 shows an initial O-grid generated by an algebraic method. Grid-point distributions along ξ-coordinate lines, which are parallel to the airfoil surface, were adapted to the density gradients in the flow-field solution. Shown in figure 10 is an adapted grid showing appropriate clustering of grid points to the shocks. Computed Mach contours obtained using the adapted grid are shown in figure 11. Figure 12 shows a comparison of pressure coefficients obtained using nonadapted and adapted grids. This figure shows that shock waves are crisply resolved using the adapted grid.

An example for supersonic flow is shown in figures 13-15. The free-stream Mach number is 1.2 and the angle of attack is 1.25°. In this case, both ξ and η-coordinate lines were adapted to the density gradients.

SUMMARY

The principal features of the proposed adaptive-grid method are as follows:

1. It is a simple concept consisting of a tension and torsion spring analogy. The combination of these springs produces a suitable adaptive grid without excessive grid skewness.
2. A marching-type of calculation procedure minimizes the computation time.
3. The split-solution procedure for multidirectional adaptation is simple and practical.
4. The method can be applied independently to selected parts of the entire grid.
5. The controllability of grid inclinations with the torsion spring makes it possible to generate a nearly orthogonal adaptive grid. Also, the inclinations of grid lines near boundary can be specified arbitrarily.

These features make the proposed adaptive-grid method practical and robust, and enhance its applicability. The method can be applied to two- and three-dimensional fluid-flow problems without any difficulty and requires little computational time and effort.

REFERENCES

1. Gnoffo, P. A.: A Finite-Volume, Adaptive Grid Algorithm Applied to Planetary Entry Flowfields, AIAA J., vol.21, no.9, 1983, pp. 1249-1254.
2. Deiwert, G. S.; Andrews, A. E.; and Nakahashi, K.: Theoretical Analysis of Aircraft Afterbody Flow, AIAA Paper 84-1524, 1984.
3. Pulliam, T. H.; Jespersen, D. C.; and Childs, R. E.: An Enhanced Version of an Implicit Code for the Euler Equations, AIAA Paper 83-0344, 1983.
4. Pulliam, T. H.: Euler and Thin Layer Navier-Stokes Codes : ARC2D, ARC3D, Notes for Computational Fluid Dynamics User's Workshop, The University of Tennessee Space Institute, Tullahoma, Tenn., Mar. 1984.

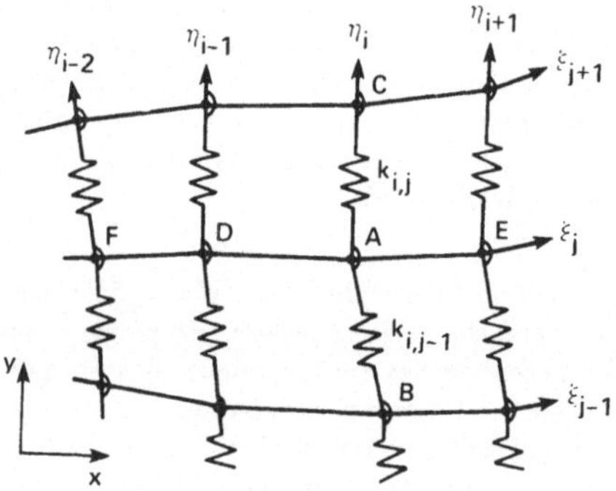

Fig.1 Schematic of adaptive-grid algorithm with tension and torsion spring analogy.

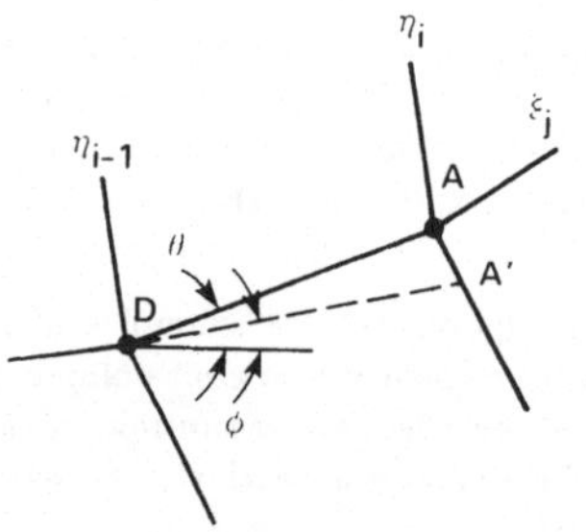

Fig.2 Notations for torsion spring analogy.

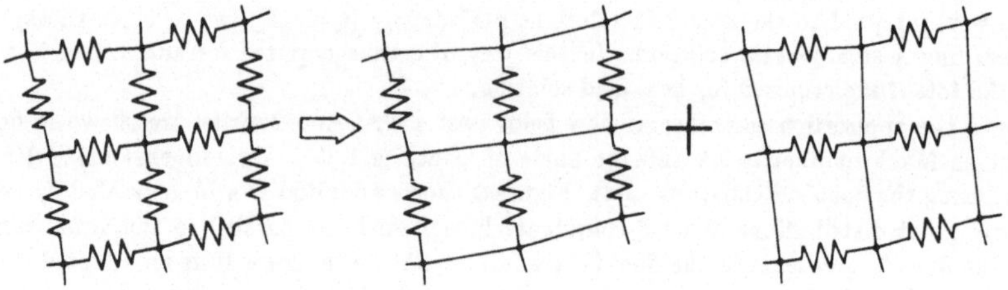

(a) Two-dimensional model.

(b) Split model.

Fig.3 Multidirectional adaptation.

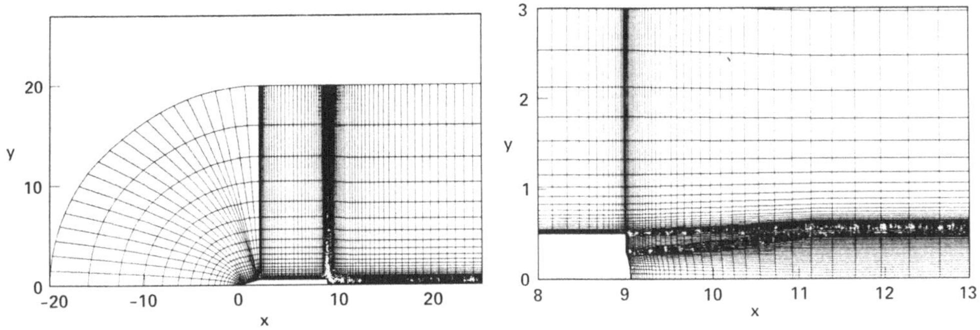

(a) Complete configuration.

(b) Base region detail.

Fig.4 Initial grid for afterbody flow field.

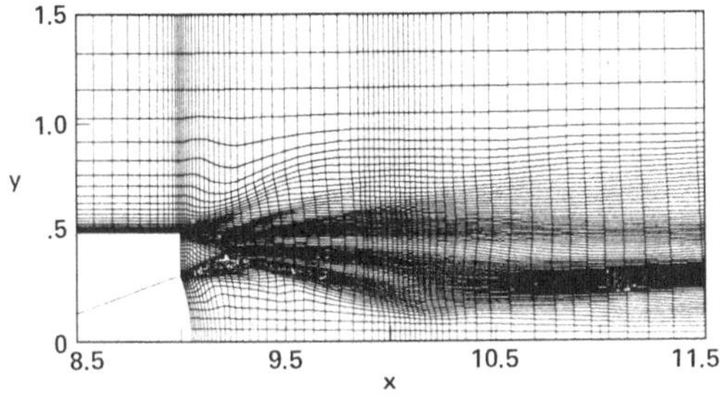

Fig.5 Adapted grid for afterbody flow field: $M_\infty = 2.01, M_J = 2.5 : P_J/P_\infty = 1.0$.

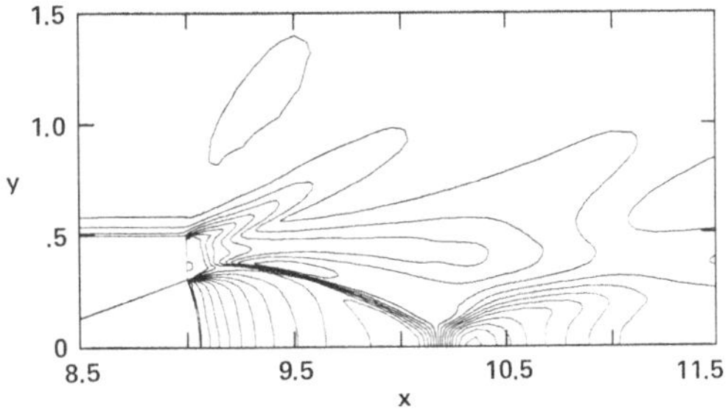

Fig.6 Computed density contours with adapted grid: $M_\infty = 2.01, M_J = 2.5 : P_J/P_\infty = 1.0$.

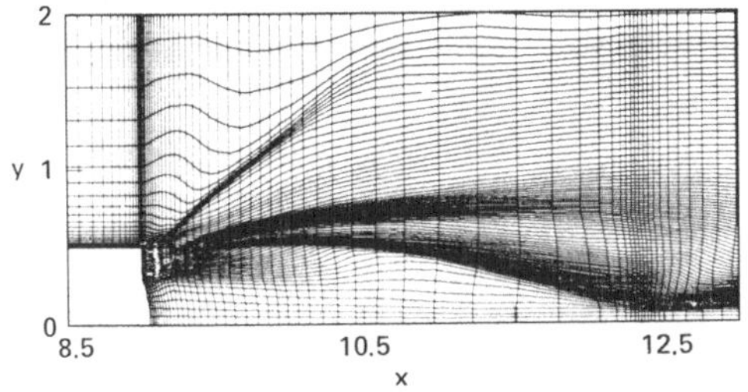

Fig.7 Adapted grid for afterbody flow field: $M_\infty = 2.01, M_J = 2.5 : P_J/P_\infty = 6.0$.

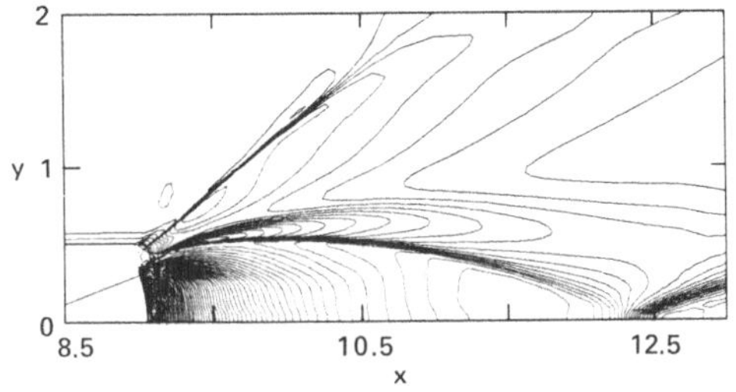

Fig.8 Computed density contours with adapted grid: $M_\infty = 2.01, M_J = 2.5 : P_J/P_\infty = 6.0$.

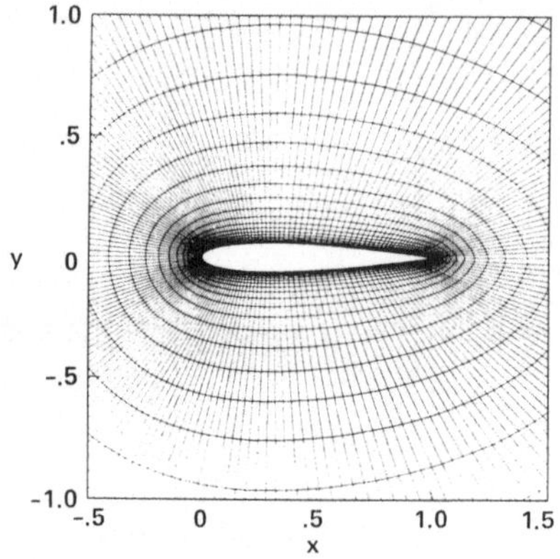

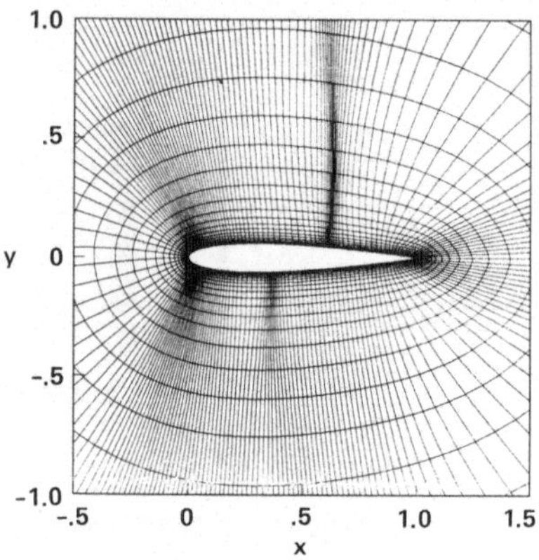

Fig.9 Initial O-grid for NACA0012 airfoil.

Fig.10 Adapted grid: $M_\infty = 0.8, \alpha = 1.25°$.

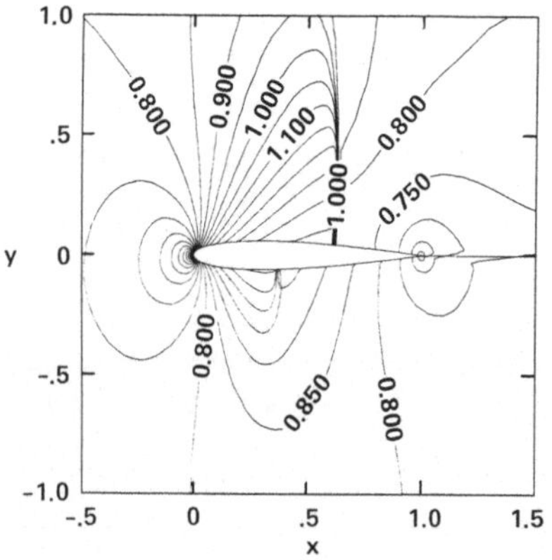

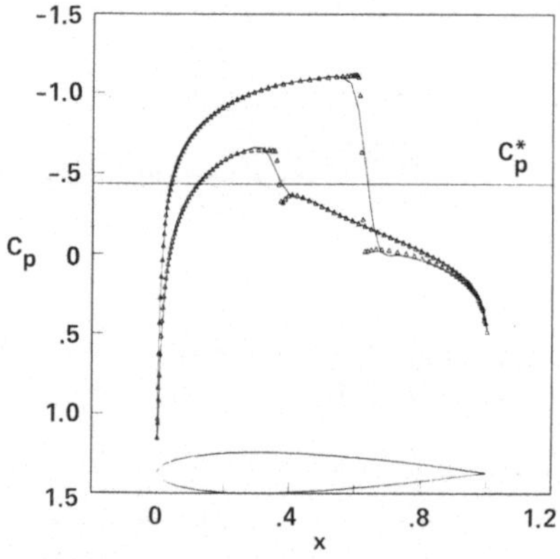

Fig.11 Computed Mach contours with adapted grid: $M_\infty = 0.8, \alpha = 1.25°$.

Fig.12 Cp for solutions with(dot) and without(line) adapted grid.

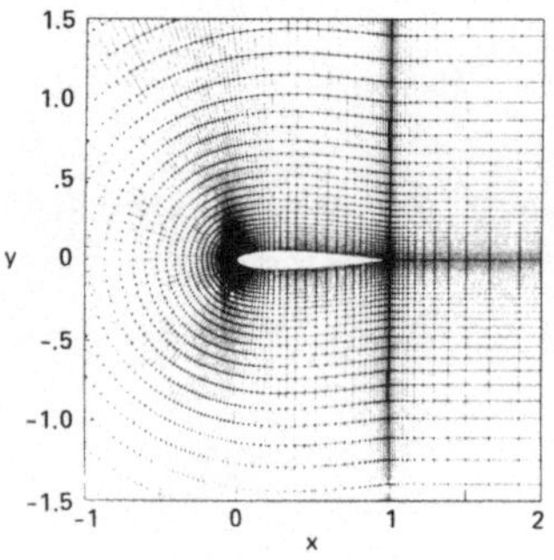

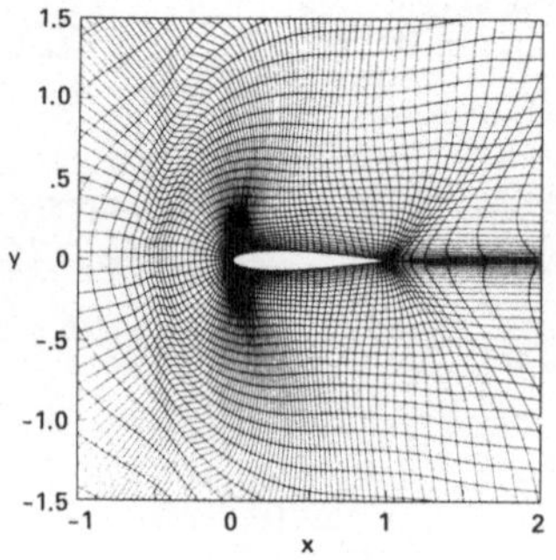

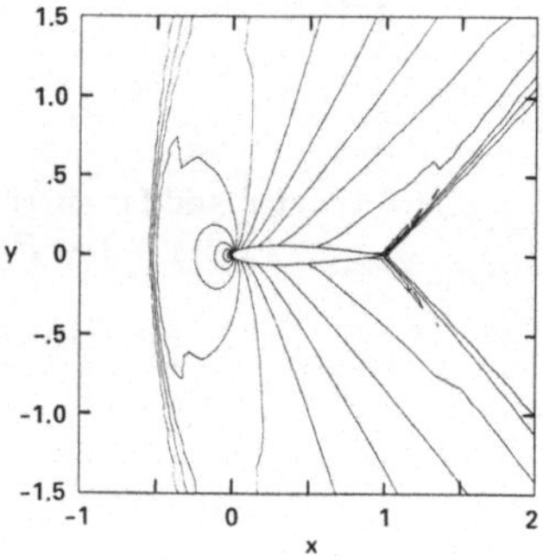

Fig.13 Initial C-grid.

Fig.14 Adapted grid: $M_\infty = 1.2, \alpha = 1.25°$.

Fig.15 Computed Mach contours: $M_\infty = 1.2, \alpha = 1.25°$.

ORTHOGONAL GRID GENERATION BY BOUNDARY GRID RELAXATION ALGORITHMS+

S. Nakamura
The Ohio State University
Mechanical Engineering Department
Columbus, Ohio 43210, USA

Two elliptic grid generation schemes, UGBR and FBGR, that generate orthogonal grids are presented. Each scheme is different from the original elliptic grid generation scheme developed by Thompson, et al[1], in the treatment of boundary points. With the UBGR scheme, all the grid points on the flow boundaries are automatically determined by the algorithm, while with the FBGR scheme at most one half of the boundary grid points are prespecified and the remainder of boundary grid points are determined automatically. Numerical examples show their capability of easy stretching, clustering and shock fitting while maintaining orthogonality of the grid. The present method can be implemented in existing elliptic grid generation programs with relatively minor modifications. The present method is more versatile than the conformal grid generation [2] because clustering and stretching of grids are easy, adapting to the flow solution is possible, orthogonal grids may be generated even when one half of the boundary grid points are fixed, and the algorithm may be extended to the control of orthogonality of grids in three-dimensional elliptic grid generation.

GRID GENERATION EQUATIONS

Denoting x-y and ξ-η as physical and computational coordinates respectively, let us consider two families of curves, $\xi(x,y)=c$ and $\eta(x,y)=d$, on the physical domain, where c and d are parameters. The orthogonality between $\xi=c$ and $\eta=d$ requires

$$\eta_x/\xi_y = -\eta_y/\xi_x = a\,k(x,y) \qquad (1)$$

where a is constant, and k(x,y) is a continuous function which may be arbitrarily specified. The transformation between x-y and ξ-η is conformal if ak=1. Otherwise the transformation is orthogonal but not conformal.

Since Eq.(1) cannot be directly used to generate grids, a set of elliptic partial differential equations and boundary conditions that are equivalent to Eq.(1) is derived. By eliminating ξ or η from Eq.(1) one at a time, we obtain the elliptic PDEs as follows

$$(k\xi_x)_x + (k\xi_y)_y = 0, \qquad (\tfrac{1}{k}\eta_x)_x + (\tfrac{1}{k}\eta_y)_y = 0 \qquad (2)$$

+*Financial support for this work was provided by NASA Ames research Center. For details of the work, see reference 3.*

Suppose that the whole boundary of the x-y domain is divided into four parts, G_g, g=1 to 4, each of which is transformed respectively to the left, top, right and bottom sides of the rectangular computational domain, $0<\xi<\xi_{max}$ and $0<\eta<\eta_{max}$. Thus, the boundary values of ξ and η are specified as $\xi=0$ along G_1, $\eta=\eta_{max}$ along G_2, $\xi=\xi_{max}$ along G_3, and $\eta=0$ along G_4. The remainder of boundary conditions are derived from Eq.(1) as

$$\begin{aligned} &\partial\eta(x,y)/\partial n = 0 \quad \text{along } G_1 \text{ and } G_3 \\ &\partial\xi(x,y)/\partial n = 0 \quad \text{along } G_2 \text{ and } G_4 \end{aligned} \tag{3}$$

Equation (2) is transformed onto the computational domain as

$$\begin{aligned} Lx &= [\alpha\gamma + 2x_\xi x_\eta \beta\ ] \\ Ly &= [\beta\gamma + 2y_\xi y_\eta \alpha\ ] \end{aligned} \tag{4}$$

where

$$L = A\partial^2/\partial\xi^2 - 2B\partial^2/\partial\xi\partial\eta + C\partial^2/\partial\eta^2$$

$$\alpha = (y_\eta k_\xi - y_\xi k_\eta), \quad \beta = (x_\eta k_\xi - x_\xi k_\eta), \quad \gamma = (x_\xi y_\eta + x_\eta y_\xi)$$

The boundary condition, Eq.(3), is transformed to

$$\begin{aligned} -y_\xi \ell_x + x_\xi \ell_y &= 0 \quad \text{along } G_1(\xi=0) \text{ and } G_3(\xi=\xi_{max}) \\ y_\eta \ell_x - x_\eta \ell_y &= 0 \quad \text{along } G_2(\eta=\eta_{max}) \text{ and } G_4(\eta=0) \end{aligned} \tag{5}$$

where ℓ_x and ℓ_y are directional cosines of the outward normal on the boundary. Equation (5) is combined with the finite difference form of Eq.(4) as written in more detail in the next section. The set of difference equations is solved by a relaxation scheme.

BOUNDARY GRID RELAXATION SCHEMES

Two different schemes of implementing boundary conditions are proposed: UBGR and FBGR. In the UBGR scheme, all the grid points on the boundary are undetermined, while up to one half of the boundary grid points may be prespecified in the FBGR scheme and the remaining boundary grid points are undetermined. The elliptic equations for x and y may be written as

$$As_{\xi\xi} - 2Bs_{\xi\eta} + Cs_{\eta\eta} = R(s), \quad At_{\xi\xi} - 2Bt_{\xi\eta} + Ct_{\eta\eta} = R(t) \tag{6}$$

where variables s and t are local Cartesian coordinates as shown in Fig. 1. Since the value of t is constrained by the shape of the boundary, we use only the first equation of Eq.(6).

The difference form of the third term of Eq.(6) may be written as

$$s_{\eta\eta} = [s_\eta]_{i,j} - 2(s_{i,j} - s_{i,j-1}) \tag{7}$$

The orthogonality boundary condition requires that the first term on the right side of Eq.(7) be zero, $[s_\eta]_{i,j} = 0$. Thus, the difference approximation for the first equation in Eq.(7) becomes

$$A(s_{i-1,j} - 2s_{i,j} + s_{i+1,j}) + 2C(-s_{i,j} + s_{i,j+1}) = R(s)_{i,j} \tag{8}$$

where B in Eq.(6) is set to zero because of orthogonality at the boundary. Then $t_{i,j}$ is calculated by introducing the value of $s_{i,j}$ into the equation that defines the boundary, $f(s,t) = 0$. Control of grid spacing for clustering and stretching is achieved by the following two algorithms: (1) variable grid spacing on the intermediate computational grid (once a desired grid is generated, it is used as if the grid spacing on the computational domain is uniform), and (2) space dependent coefficient, $k(x,y)$.

The FGBR scheme is based on the following concept: (a) one end point of a grid line is fixed, (b) the other end point of the grid line is left to be variable, and (c) grid spacings adjacent to the corresponding grid line on the computational domain are variable. This scheme can be implemented by modifying the UBGR scheme only in the treatment of the fixed boundary grid points. Suppose that the grid points with black circles in Fig. 2 are obtained after an iteration step with a given set of grid spacings

$$\delta\xi_i = \xi_i - \xi_{i-1}, \quad \delta\xi_{i+1} = \xi_{i+1} - \xi_i. \tag{9}$$

First, the hypothetical coodinates to satisfy Eq.(8) are calculated (open circle marked as B') as if the grid point (i,j) on the boundary is an undetermined point. Second, the following two chord length ratios are calculated:

$$r' = AB'/(AB' + B'C), \quad r = AB/(AB + BC)$$

where AB, BC, AB' and B'C are chord lengths. Third, the computational grid spacings $\delta\xi_i$ and $\delta\xi_{i+1}$ are replaced by

$$\delta\xi_i' = ar^2 + br, \qquad \delta\xi_{i+1}' = (\delta\xi_i + \delta\xi_{i+1}) - \delta\xi_i \tag{10}$$

where

$$a = \delta\xi_i/r' - \delta\xi_{i+1}/(r'-1), \quad b = (\delta\xi_i + \delta\xi_{i+1}) - a$$

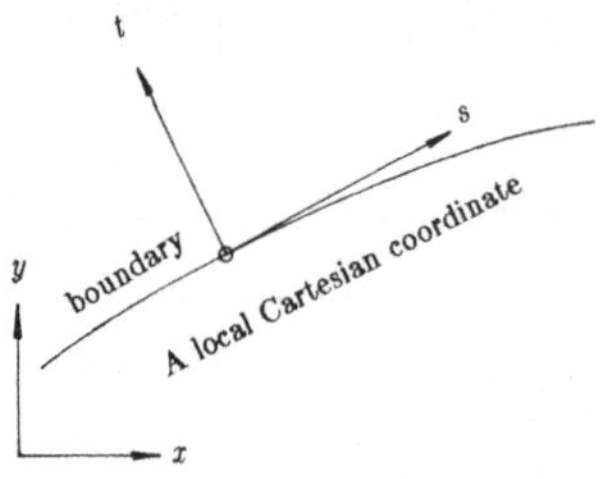

Figure 1

Local Cartesian coordinates

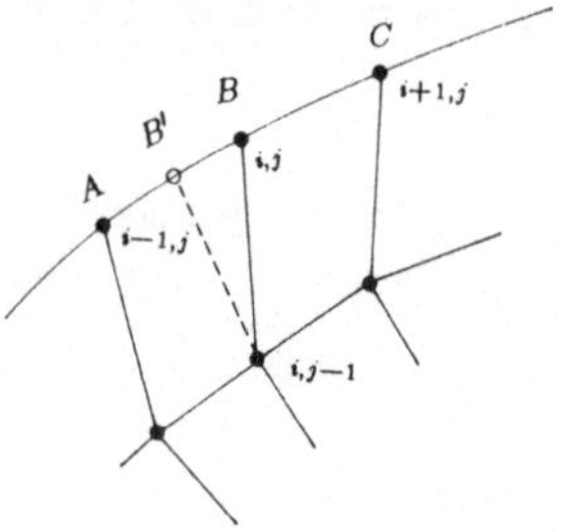

Figure 2

Grid points after iteration

NUMERICAL ILLUSTRATIONS

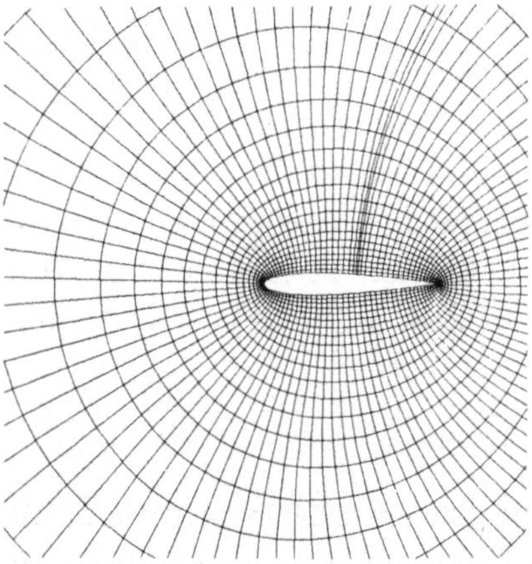

Figure 3

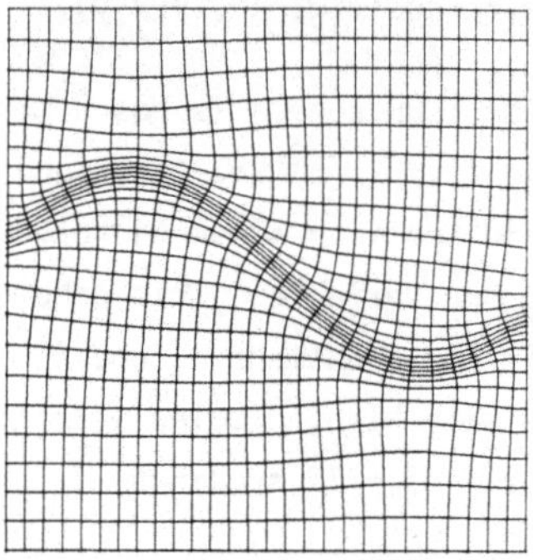

Figure 4

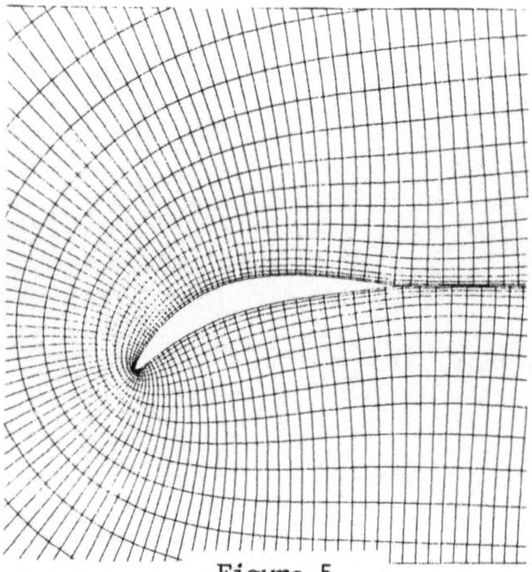

Figure 5

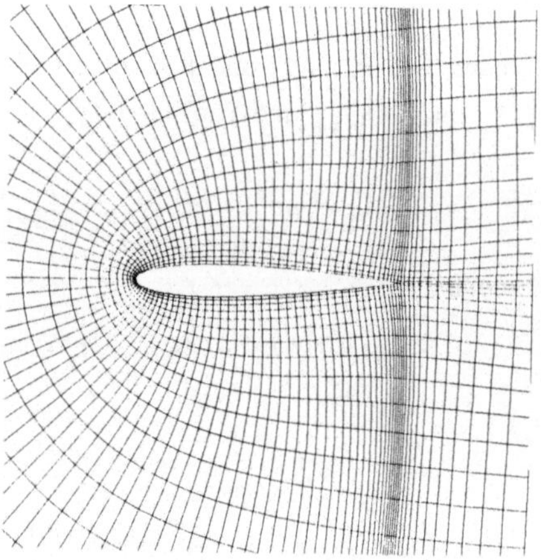

Figure 6

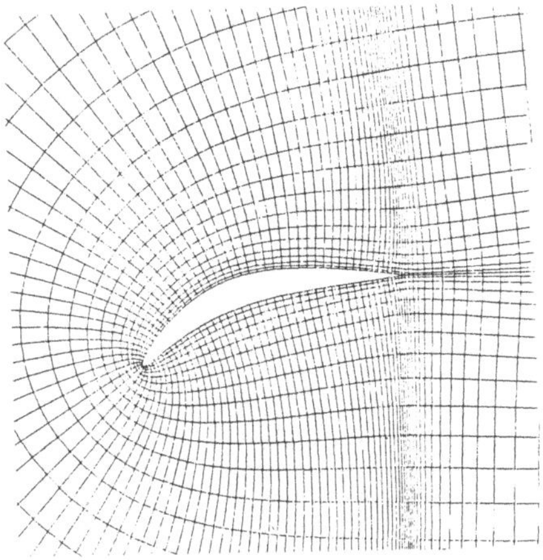

Figure 7

Figure 3 shows the grid generated for a NACA-0012 airfoil by the UBGR scheme with a clustering effect created by a variable grid spacing on the intermediate computational domain.

Figure 4 shows the grid points clustered around a prescribed curve to illustrate the capability of solution adapting. This is done by setting $k(x,y)$ in Eq.(1) as $k(x,y)=c_1+c_2/[1+h^2(x,y)]$ where c_1 and c_2 are positive constants and $h(x,y)=0$ is the curve (for example, a shock profile) to which grid points are to be clustered. The function $h(x,y)$ used in the present example is $h(x,y) = \sin(x)-y$.

Figure 5 shows the C-type grid generated by the UBGR scheme for a cambered airfoil. The grid points on the Kutta cut do not match because the points above and below were determined independently.

Figure 6 shows the C-type grid generated by the FBGR for a NACA- 0012 airfoil. The grid points on the airfoil surface and the Kutta cut are fixed, while the grid points on the outer boundaries are all determined by the UBGR scheme.

Figure 7 is for the same configuration as Figure 6 except the airfoil is cambered.

References

[1] Thompson, J. F., Thames, F. C. and Mastin, C. W., "Automatic Numerical Generation of Body-Fitted Curvilinear Coordinate System for Field Containing Any Number of Arbitrary Two-Dimensional Bodies," J. Comp. Physc., Vol. 15 (1974)

[2] Ives, D. C., "Conformal Grid Generation," Numerical Grid Generation, North-Holland (1982)

[3] Nakamura, S., "Generation of Orthogonal Grids by Boundary Grid Relaxation," NASA-CR-166523 (1983)

A New LU Factored Method
for the Compressible Navier-Stokes Equations

Shigeru Obayashi*, Kunio Kuwahara** and Yoshimasa Yoshizawa
Institute of Engineering Mechanics, Tsukuba University,
Sakura-mura, Ibaraki, Japan,
**The Institute of Space and Astronautical Science,
Meguro-ku, Tokyo, Japan.

§1. Introduction

The development of efficient methods for the compressible Navier-Stokes equations is desired. Numerical simulations by the popular Beam-Warming-Steger method/1,2/ is bounded both by the CPU time and by the temporary storage to invert block-tridiagonal matrices. Various factorization of the implicit procedure can be applied to improve convergence rates for steady-state problems.

In this paper, a new LU factored method is developed by splitting the implicit procedure into inversions of the lower and upper bidiagonal matrices. This approximate LU factorization is derived from the concepts of the flux vector splitting technique/3/ and the implicit MacCormack method/4/. The resulting method reduces the CPU time and the temporary storage, and retains the reliability such that a steady-state solution is independent of time increments.

§2. Algorithm Development

The new method is considered in Cartesian coordinates/5/ for brevity. The two-dimensional compressible Navier-Stokes equations are written in the conservation-law form,

$$U_t+F_x+G_y=Re^{-1}(R_x+S_y), \tag{1}$$

The Beam-Warming-Steger method/1,2/ applied to Eqs.(1) results in the following approximate factorization,

$$\begin{gathered}(I+hD_i(A+P))(I+hD_j(B+Q))\Delta U_{ij}{}^n=-hLr_{ij}{}^n,\\ A=(\frac{\partial F}{\partial U})_{ij}{}^n,\ B=(\frac{\partial G}{\partial U})_{ij}{}^n,\ Lr_{ij}{}^n=D_i(F_{ij}{}^n-R_{ij}{}^n/Re)+D_j(G_{ij}{}^n-S_{ij}{}^n/Re),\end{gathered} \tag{2}$$

where $U_{ij}{}^n=U(i\Delta x,\ j\Delta y,\ n\Delta t)$, D_i and D_j are the central finite-difference operator for i and j, respectively, and $h=\Delta t/\Delta x=\Delta t/\Delta y$.

---

*Present address: Dept.Aeronautics, Univ.Tokyo, Bunkyo-ku Tokyo Japan.

The viscous terms R and S are linearized by P and Q, respectively/2/. This algorithm requires inversions of the block-tridiagonal matrices.

The Jacobian matrices A and B are diagonalized/6/ as,

$$E_A=\mathbf{diag}(u,u,u+c,u-c) \text{ and } E_B=\mathbf{diag}(v,v,v+c,v-c), \tag{3}$$

where $A=XE_AX^{-1}$, $B=YE_BY^{-1}$, respectively, and X and Y are the eigenvector matrices. The diagonal matrix E_A can be split along the sign of each eigenvalue as $E_A=E_A^{+}+E_A^{-}$. The operators in Eqs.(2) can be replaced by using the flux vector splitting/3/, as $D_iA=D_{i-}XE_A^{+}X^{-1}+D_{i+}XE_A^{-}X^{-1}$. The operators D_{i+} and D_{i-} denote the forward and backward one-sided differences, respectively. The LU factored form can be obtained as,

$$I+hD_i(A+P)=(I+hD_{i-}(XE_A^{+}X^{-1}+\hat{P}))(I+hD_{i+}(XE_A^{-}X^{-1}-\hat{P})), \tag{4}$$

if D_iP can be rewritten as $D_{i-}\hat{P}-D_{i+}\hat{P}$. The eigenvalues of the block matrix $\hat{P}$ are related to the stability for the discretized viscous terms. The parameter k can be chosen so as to maintain the stability of the viscous terms and added as a weight of upwind differences/3,4/:

$$\hat{E}_A^{\pm} = E_A^{\pm} + k\ \mathbf{sign}(E_A^{\pm}),\ k=\frac{\nu}{Re\rho\Delta x},\ \nu=\mathbf{max}(2\mu,\lambda+2\mu,\frac{\gamma\mu}{Pr}), \tag{5}$$

where ν is identically set to 2μ if $\lambda=0$, $\gamma=1.4$ and $Pr=.7$. Finally, the LU factored scheme is described as,

$$(I+hD_{i-}\hat{A}^{+})(I-hD_{i+}\hat{A}^{-})(I+hD_{j-}\hat{B}^{+})(I-hD_{j+}\hat{B}^{-})\Delta U_{ij}^{n}=-hLr_{ij}^{n}, \tag{6}$$

$$\hat{A}^{\pm} = X|E_A^{\pm} + kI^{\pm}|X^{-1},\ \hat{B}^{\pm} = Y|E_B^{\pm} + kI^{\pm}|Y^{-1},\ k=\frac{2\mu}{Re\rho\Delta x}=\frac{2\mu}{Re\rho\Delta y}.$$

The usual fourth-order dissipation/1,2/ is added to the right-hand-side of Eqs.(6). The implicit smoothing factor is added to the prameter k in Eqs.(5). The resulting scheme is unconditionally stable for linearized analysis. It is efficient, and needs less temporary storage because no inversion of block-tridiagonal matrix is required.

§3. Results

A. Test Problem in Cartesian Coordinates

The interaction problem of shock wave with laminar boundary layer for $Re=.296\times10^{6}$, $M=2$ and a shock angle $\theta=32.6$ degrees/1,4,5/ was solved as shown in Figs.1 and 2. Molecular viscosity was calculated by Sutherland's formula. The mesh at first contained 32x32 points, and successively increased to 256x96 points.

The present algorithm is suitable for a vector machine. The vectorized ratio on Hitachi S810 is .98. The ratio of CPU times of the vectorized code to the original one is about one tenth.

B. Grid Generation

Nearly orthogonal mesh system is iteratively generated by algebraic technique. Subject to given boundary points, the inner points are initially determined by linear interporation. Then the mesh is iteratively obtained by interporating a grid point with a location along a normal vector of a lower grid line. The lack of smoothness overcomes by averaging grid points. The mesh is concentrated to the boundary layer by employing a double-exponential function so as to control the concentration both in the strength and in its increment. The typical mesh system for 81x40 points around an NACA0012 airfoil is shown in Fig.3 where the CPU time was about .2 seconds on Fujitsu M380.

C. Test Problems in the Generalized Coordinates

The transonic flows about an NACA0012 airfoil were solved. For invisid case, the mesh contains 321x40 points. The numerical results at 500 steps for M=.8 are shown in Fig.4. The CPU time on Fujitsu M380 is about $1.1x10^{-4}$ seconds per step per grid point. The L_2 residual reaches order of 10^{-8} with the maximum CFL number of 60. The oscillation before the shock wave in the coefficient of pressure is due to the explicit central difference. For viscous case, the mesh contains 161x40 points. The numerical results for the laminar flow, $Re=10^4$ and M=.2, are shown in Fig.5. The CPU time is $1.4x10^{-4}$ in which the implicit procedure is comparable to the explicit one. The results for the fully turbulent flow, $Re=10^6$, M=.75 and an angle of attack $\alpha=2$ degrees, are obtained at 2500 steps in Fig.6 by using the Baldwin-Lomax model. The CPU time is $1.6x10^{-4}$.

The present method is twice as fast as the Steger code and requires less computer storage. Furthermore, it is suitable for a vector machine and thus enables to precisely simulate flow fields with a large amount of grid points.

References

1) R. M. Beam and R. F. Warming, AIAA J. **16** (1978), 393.
2) J. L. Steger, AIAA J. **16** (1978), 679.
3) J. L. Steger and R. F. Warming, J. Comput. Phys. **40** (1981), 263.
4) R. W. MacCormack, AIAA J. **20** (1982), 1275.
5) S. Obayashi and K. Kuwahara, AIAA paper 84-1670
6) R. F. Warming, R. M. Beam and B. J. Hyett, Math. Comput. **29** (1975), 1037.

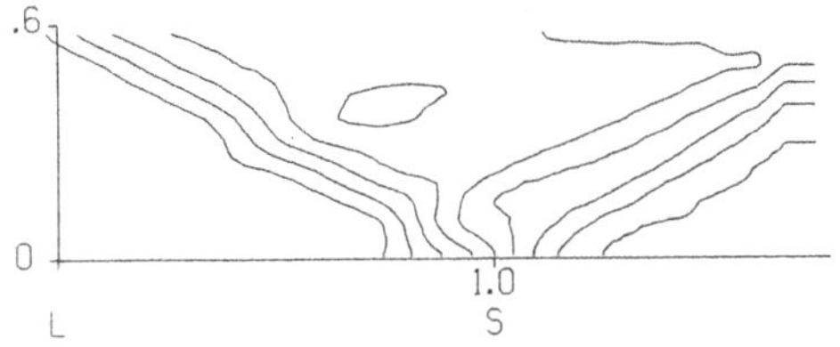

a) 32x32 points

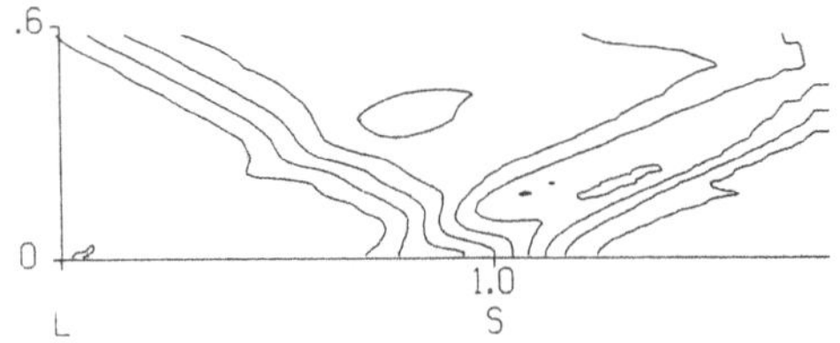

b) 64x32 points

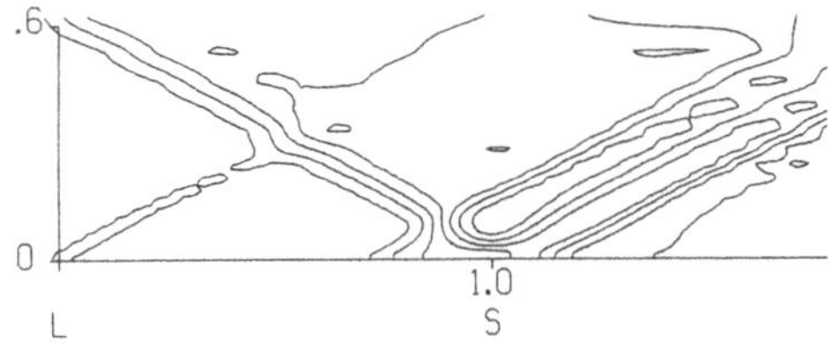

c) 128x64 points

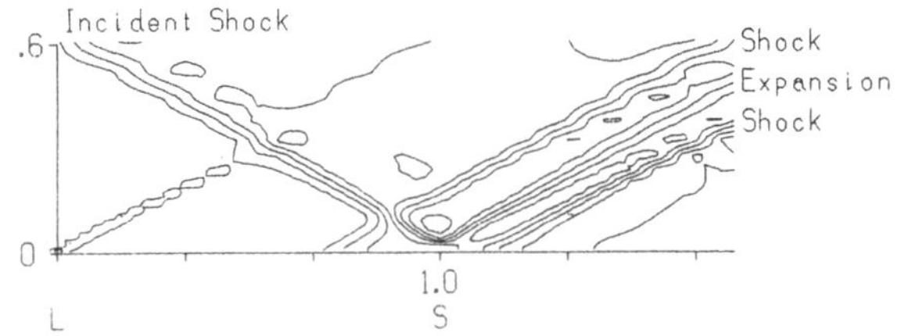

d) 256x96 points.

Fig.1 Numerical pressure contours. Shock angle $\theta=32.6$, Mach number $M=2.0$, Reynolds' number $Re=.296 \times 10^6$. L; the leading edge, S; the incident point of shock wave.

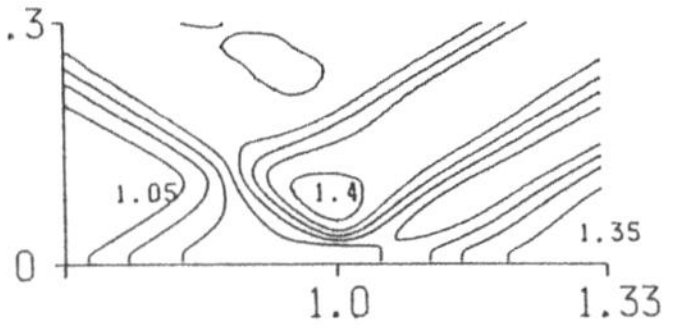

a) pressure contours

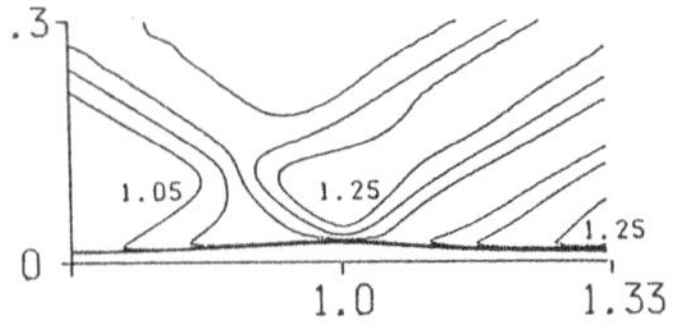

b) density contours

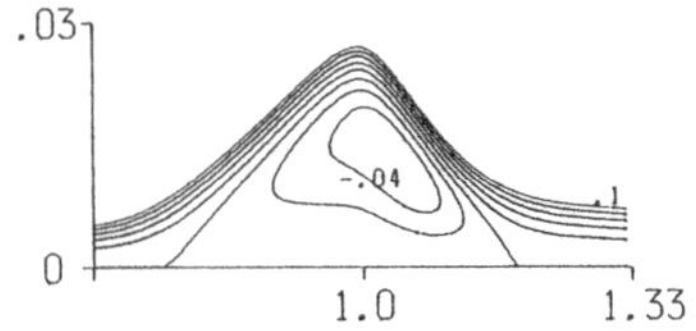

c) stream function contours

Fig.2 Numerical contour maps near the incident point of shock wave for 256x96 points.

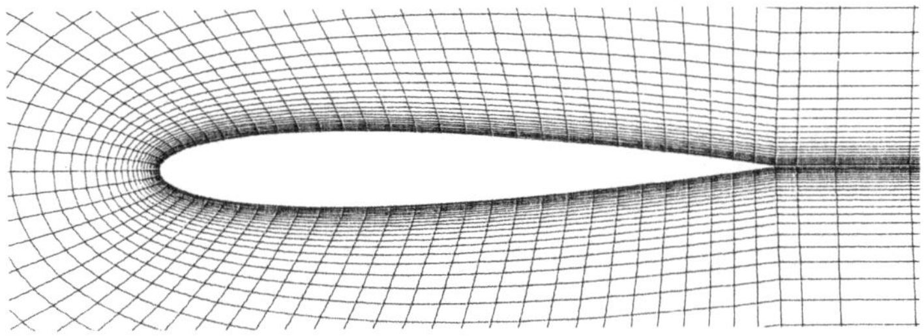

Fig.3 Near field view of grid for 81x40 points.

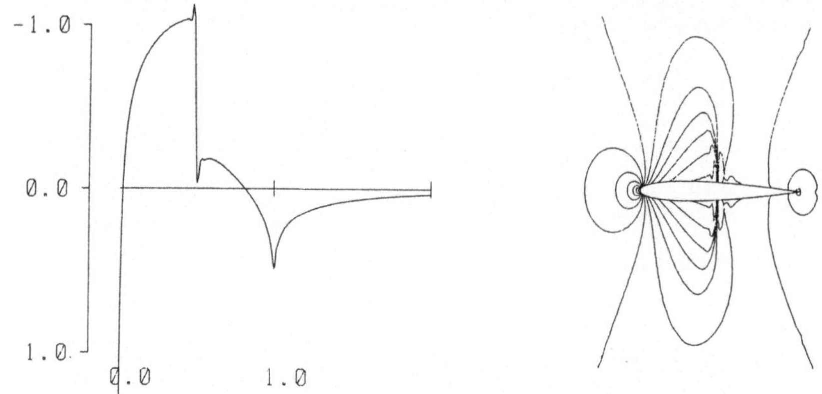

a) coefficient of pressure b) density contour map

Fig.4 Numerical results of inviscid flow for M=.8 and α=0.

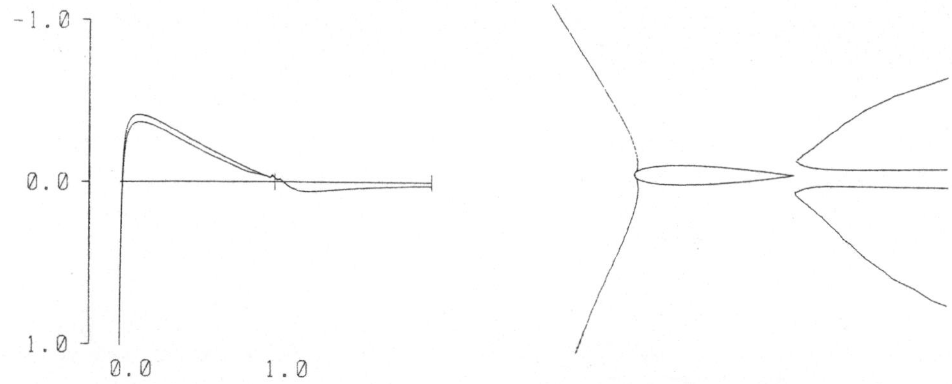

a) coefficient of pressure b) density contour map

Fig.5 Numerical results of laminar flow for Re=10^4, M=.2 and α=0.

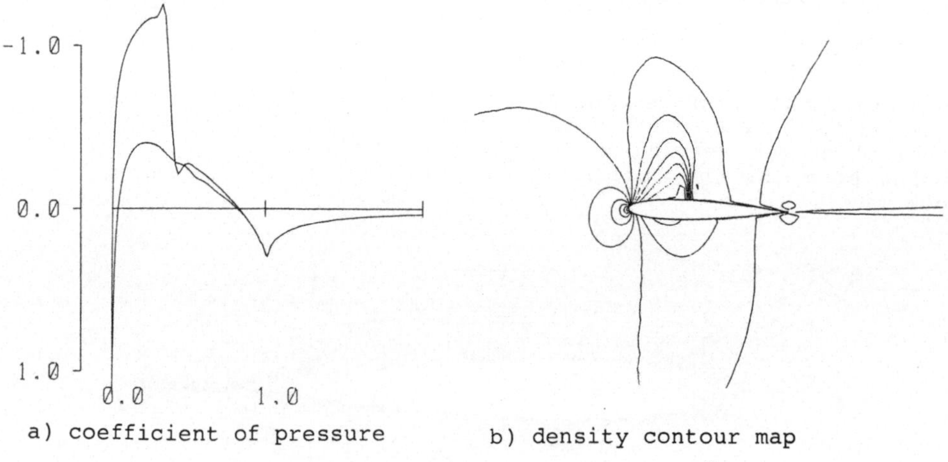

a) coefficient of pressure b) density contour map

Fig.6 Numerical results of turbulent flow for Re=10^6, M=.75 and α=2.

TIME-DEPENDENT NON-UNIFORM GRIDS FOR PARABOLIC EQUATIONS

P. Orlandi

Dipartimento di Meccanica e Aeronautica
Università degli Studi di Roma "La Sapienza"

ABSTRACT

A method based on a time dependent coordinate transformation has been applied to solve Burgers' equation. The numerical solution has been compared to the analytical solution of Lighthill [1]. The numerical results agree completely with the analytical ones. They have been obtained with very large time steps because a fast implicit non-iterative algorithm has been employed for the time discretization. The number of grid points in the space discretization is very low because the coordinate transformation gives a steady shock in the "new" coordinate.

INTRODUCTION

Often parabolic partial differential equations describe physical phenomena involving large gradients of the unknown quantities which move in time and space. The desirable numerical simulation of this behavior is that which employs the smallest possible number of grid points. In order to test a numerical model it is worthwhile to apply it to a one-dimensional equation, provided with analytical solutions. The Burgers' equation investigated by Lighthill [1] is one of the best examples in literature. Moreover, numerical solutions obtained using a predictor corrector scheme for the time discretization and both the finite difference scheme [2] and the cubic spline technique [3] for the space discretization are available. In both papers [2,3] the same non-uniform grid spacing was employed together with a mesh rezoning once every few time steps. Both these methods required an excessive number of mesh points, and very small time steps. Further, a considerable part of computing time was spent in the mesh rezoning.

The numerical method presented here was previously used to solve laminar and turbulent unsteady boundary layers [4]. By employing the following procedures our method overcomes the disadvantages of the others. A non-iterative implicit scheme is used in the time direction, based on the original idea of Briley and McDonald [5], and developed by Beam and Warming [6] for compressible Navier-Stokes equations. The implicit scheme has been modified to take into account moving grids. Thus the calculation no longer requires a mesh rezoning. The "real" coordinate is related to a "new" coordinate through a quantity similar to a boundary layer thickness. In the "new" variable representation the "shock" location is steady. Such an idea was suggested by Chong [2], who however was not able to find a transformation which kept the "shock" steady. Further, the coordinate transformation should provide an unequally spaced distribution of grid points.

NUMERICAL SCHEME

Burgers' equation is

$$\frac{\partial U}{\partial t} + U\frac{\partial U}{\partial x} = \frac{\delta}{2}\frac{\partial^2 U}{\partial x^2} \tag{1}$$

Lighthill [1] obtained the analytical solution of Eq.(1), it is

$$U(x,t) = \frac{x/t}{1 + (t/t_0)^{\frac{1}{2}} \exp(x^2/2\delta t)} \quad ; \quad t_0 = \exp\left(\frac{1}{4\delta}\right) \tag{2}$$

Since the solution is antisymmetric in x, previous authors [2,3] solved Eq.(1) only for $x > 0$, with one of the boundary conditions as $U(0,t) = 0$. In the present paper Eq.(1) has been solved in the entire field, because it was evident from a previous paper [7] that if the numerical scheme was inadequate or the mesh size too large the solution lost the antisymmetric property.

The use of a uniform grid makes the numerical solution of Eq.(1) at very low δ very onerous, mainly because the velocity profiles take a very sharp gradient in a region of thickness $(\delta t)^{\frac{1}{2}}$. Supposing this region is described only by a few grid points, the entire field requires a very large number of grid points N_T, making the solution impossible at very low values of δ. Moreover the number of mesh points must increase in time because the "shocks" move in opposite directions. To reduce substantially the number of mesh points a time dependent coordinate transformation can be used. This transformation, to be useful, must satisfy the following requirements:

A) The location of the "shock" in the "new" variable representation must be stationary.

B) Iterative procedures or mesh rezoning must be avoided.

The following coordinate transformation fulfills the above requirements

$$x = x_G(\eta)\, L(t) \tag{3}$$

where the function $x_G(\eta)$ of the "new" coordinate η allows for a large number of mesh points in regions where high gradients are located. The function $L(t)$ allows the grid distribution to follow the moving shock. If $x_G(1) = 1$, $L(t)$ is the value of x where the boundary conditions must be imposed. The boundary conditions necessary to obtain the analytical solution given by Eq.(2) are $U = 0$ for $x = \pm\infty$. Contrary to this in our calculation we assumed the almost equivalent boundary conditions

$$x = \pm L(t) \qquad \frac{\partial U}{\partial x} = 0 \tag{4}$$

The function $L(t)$ can be assigned by analytical expressions if the "shock" moves according to a law which is known. This does not occur in the general case. Therefore $L(t)$ is a function of the solution itself. If the function $L(t)$ is evaluated by the solution for the same time step, then an iterative procedure must be employed. To avoid the iterative scheme, $L(t)$ can be calculated by the solution at the previous time step. If $S(t)$ is the value of x where $U(x,t)$ reaches a very small value ε, e.g. $\varepsilon = 10^{-4}$, $L(t)$ can be expressed by

$$L(t) = S(t - \Delta t)C_E \tag{5}$$

The value of $S(t - \Delta t)$ is obtained by a linear interpolation of the values of $U(x,t)$ at two grid points where $U(x_{I-1},t) < \varepsilon < U(x_I,t)$. C_E is a constant which must be greater than 1 both because $L(t)$ has been related to the velocity at the previous time step and because the boundary conditions given by Eq.(4) have been assumed. As will be shown later, the variations of the constant C_E are reflected on the number of grid points in the region of large velocity spatial gradients.

Introducing the trasnformation of Eq.(3) in Burgers' equation this becomes

$$\frac{\partial U}{\partial t} + \frac{1}{L(t)}\frac{d\eta}{dx_G}\left[U - x_G\frac{dL(t)}{dt}\right]\frac{\partial U}{\partial \eta} = \frac{\delta}{2L^2(t)}\frac{\partial}{\partial \eta}\left(\frac{\partial U}{\partial \eta}\cdot\frac{d\eta}{dx_G}\right)\frac{d\eta}{dx_G} \tag{6}$$

The strongly non-uniform grid distribution, necessary to describe numerically the large velocity gradients, can be assigned by a combination of hyperbolic tangents in order to have low values of the coordinate derivatives in the shock region. Thus the truncation errors of the second derivative discretization are reduced.

To further reduce the computing time a very fast method must be employed to solve the non-linear Eq.(6) by the smallest possible number of time steps. The non-linear Eq.(1) can be discretized in the time marching direction by the Padé formulation as explained in previous papers [4,5,6,7]. The discretization of the spatial derivatives by second order accurate schemes brings to a system of algebraic equations of tridiagonal form.

RESULTS

The numerical method described above has been tested taking into consideration some of the cases analyzed in Ref. [2] and [3]. These authors emphasized the comparison between the exact Reynolds number and the numerical one

$$R = \frac{1}{\delta}\int_0^\infty U\,dx = \lg\left[1 + \left(\frac{t_0}{t}\right)^{\frac{1}{2}}\right] \tag{7}$$

In the present paper we have preferred to plot the velocity error distribution rather than the Reynolds number. The former show that errors can be antisymmetrically distributed around the shock, in which case they compensate each other and a very satisfactory Reynolds number prediction is obtained. For the exact and the numerical solution to agree it is important to have the greatest possible number of grid points around the shock region. This can be represented very well by analyzing the velocity profiles in the "new" coordinate η rather than in the physical coordinate x.

Fig. 1 shows, at $\delta = 10^{-2}$, the behavior of the initial velocity profile ($t = 1$) when the constant C_E is varied. The value of $C_E = 1.25$ describes the shock with a larger number of grid points, thus yielding a better error distribution. Fig. 2 shows the error distribution versus the coordinate x. The errors obtained with $C_E = 1.25$ are one order of magnitude lower than the errors obtained with $C_E = 1.50$. The Reynolds number error $E_R = (R_{EX} - R_N)/R_{EX}$ for the three cases at $t = 7$ are respectively $(ER)_{C_E=1.25} = 1.65 \cdot 10^{-3}$, $(ER)_{C_E=1.33} = 1.16 \cdot 10^{-3}$ and $(ER)_{C_E=1.5} = 5.58 \cdot 10^{-3}$. It follows that the best Reynolds number prediction is obtained by a coordinate transformation which does not give the best velocity prediction. Fig. 3 shows the velocity profiles versus the "new" coordinate η at different times. It can be argued that the best predictions obtained with this limited number of grid points ($N_T = 81$ for $-1 \le x \le +1$) is due to the fact that the coordinate transformation of Eq.(3) permits a steady shock in the η plane, located around $\eta = 0.7$. Not much worse a solution was obtained with $N_T = 41$ grid points, while with $N_T = 31$ grid points the solution does not converge. Also worth analyzing are the errors introduced by the time discretization. This has been done by using the largest time stem ($\Delta t = 0.1$) allowable by one of the three schemes ($\theta = 0.5$, $\xi = 0.$). Fig. 4 shows that for $\delta = 10^{-2}$ the second order accurate schemes give errors which do not differ too much between them, while the first order scheme gives unreliable predictions. The large errors due to the Euler implicit scheme depend mostly on the fact that the numerical solution predicts a growth of $L(t)$ in time faster than expected. This is analogous to having a C_E increasing in time, which reduced the number of grid points in the shock region.

At $\delta = 10^{-3}$ a sharper shock arises and a large number of grid points are needed to describe it. At $\delta = 10^{-3}$ previous authors [2,3] used about 400 grid points to find the solution in half field ($0 < x$). The coordinate transformation used here yields the solution with $N_T = 81$ grid points, in the whole field ($-1 \le x \le +1$), but the constants in equations (5) and in the $x_G(\eta)$ must be accurately tuned. Doubling the number of grid points a wide variation of the same constants was possible. At this δ a smaller change of C_E than for $\delta = 10^{-2}$ gave rise to large variations on the velocity behavior in the "new" variable representation. Also at this δ the same considerations made above about the error distribution hold true. Fig. 5 shows that the coordinate transformation giving the largest number of grid points in the shock region introduces the lowest errors. The other trasnformations show positive and negative error distributions around the shock. The errors in the Reynolds number calculations related to the error distributions of fig. 5 vary between $2 \cdot 10^{-4}$ and $8 \cdot 10^{-4}$. The best prediction is obtained by assuming $C_E = 1.475$. The time evolution of the velocity profile for $C_E = 1.475$ shows the shock in the η coordinate field is steady and centered around $\eta = 0.7$. Owing to the larger number of grid points and the enhanced coordinate stretching required, the time step Δt should be smaller than the Δt used in the case of $\delta = 10^{-2}$.

CONCLUSIONS

The method presented here solves non-linear partial parabolic equations with large unsteady gradients by a very limited number of grid points and time steps. Previously, this method was used to solve efficiently laminar and turbulent unsteady boundary layers [4]. The boundary layer case is not a very critical one because the large gradients are located in a well defined position and the moving edge is located where low velocity gradients occur. In this paper a more critical case has been investigated to show the method has a much larger range of applicability. It requires a much simpler programming work than the tedious and time consuming mesh rezoning. A further advantage lies in the fact that it can be easily applied to multidimensional problems.

ACKNOWLEDGEMENTS

This work was supported by the "Consiglio Nazionale delle Ricerche". The author is grateful to Mr. Giuseppe De Caro for the editing of this paper.

REFERENCES

1. M.J. Lighthill, "Surveys in Mechanics" (G.K. Batchelor and R.M. Davis, Eds.), p. 250, Academic Press, New Yrok & London, 1956.
2. T.H. Chong, SIAM J. Numer. Anal., 15 (1978), 835.
3. B.L. Lohar and P.C. Jain, J. of Comp. Physics, 39 (1981), 433.
4. P. Orlandi and J.H. Ferziger, AIAA J., 19 (1981), 1408.
5. W.R. Briley and H. McDonald, Lecture Notes in Physics, vol. 35, 105-110, Springer Verlag, New York, 1975.
6. R.M. Beam and R.F. Warming, AIAA J., 16 (1978), 393.
7. P. Orlandi, "Direct Simulation of Burgulence", Proceedings of CTAC-83, J. Noye and C. Fletcher, North-Holland Publ. Company, p. 641, 1983.

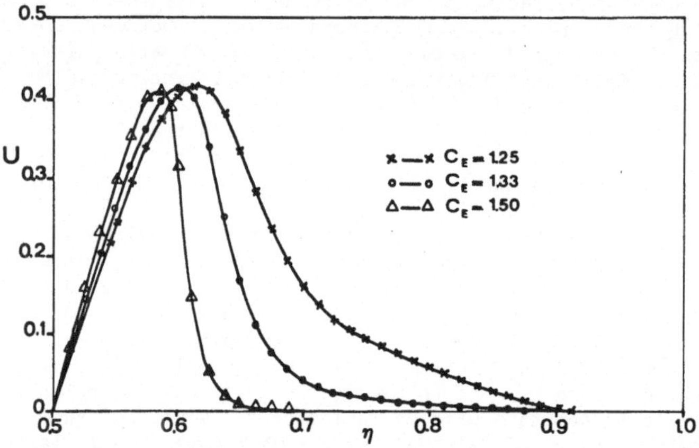

Fig. 1 - Velocity distribution versus the "new" variable η at $t = 1$ for $\delta = 10^{-2}$.

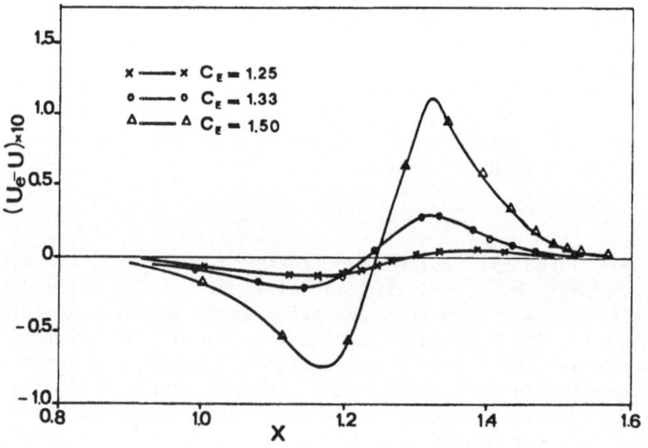

Fig. 2 - Error distribution versus the real coordinate x at $t = 7$ for $\delta = 10^{-2}$, $\Delta t = 10/500$.

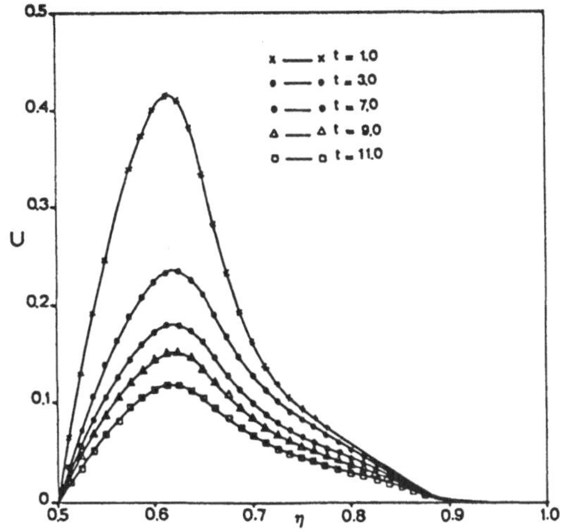

Fig. 3 - Time history of velocity profile versus the "new" variable η for $\delta = 10^{-2}$ and $C_E = 1.25$, $\Delta t = 10/500$.

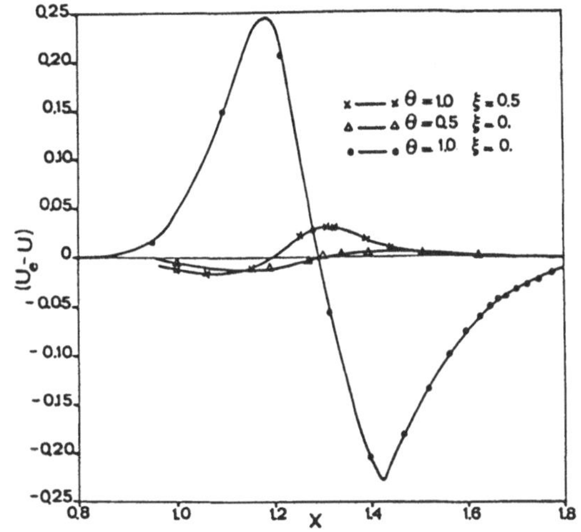

Fig. 4 - Errors distribution versus the real coordinate x at t = 7 for $\delta = 10^{-2}$, $\Delta t = 10/100$.

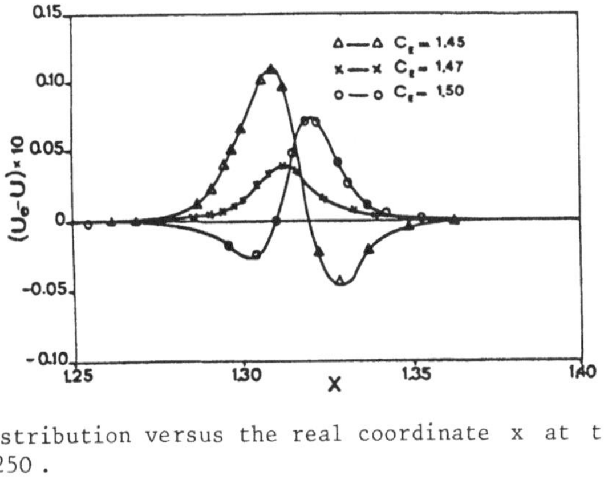

Fig. 5 - Errors distribution versus the real coordinate x at t = 7 for $\delta = 10^{-3}$, $\Delta t = 10/1250$.

NUMERICAL SIMULATION OF DYNAMICS OF AN AUTOROTATATING AIRFOIL

K.Oshima*, Y.Oshima** and N.Izutsu*

* The Institute of Space and Astronautical Science
Komaba, Meguro-ku, Tokyo 153, Japan
** Dept. Physics, Ochanomizu University
Ohtsuka, Bunkyo-ku, Tokyo 112, Japan

I. Introduction

The autorotation phenomenon has been extensively studied by many researchers, such as E.H.Smith(1), H.J.Lugt and S.Ohring (2), J.D.Iversen (3) and H.J.Lugt (4). An extensive review of autoratation was given by H.J.Lugt (5). In a previous study (6), flow field around a two-dimensional, elliptic cylinder rotating around its center axis fixed perpendicularly to the uniform flow has been studied, and it was concluded that the autorotation occurs when the aerodynamic moment of rotation averaged over one cycle of rotation is equal to the mechanical friction torque. The autorotating frequency thus predicted coincided with the experimentally observed value.

In this study, the aerodynamic forces calculated at each time instant by discrete vortex method are directly fed into the equation of motion of the airfoil. Integrating this equation timewise, the motion of the airfoil is traced.

II. Numerical Procedure

Discrete vortex method was applied for the flow around a rotating elliptic airfoil, in which the vortex blob model including the viscous dissipation effect was applied. The ellips in the physical plane is mapped onto a unit circle in the complex potential plane. At each time instant two vortex blobs are created at the both edges of the ellips. Intensity of the newly generated vortices at each time step is determined by applying the non-slip condition over a certain portion of the ellips surface, containing each edge point. These points are determined such manner that the potential difference over these two points given by the potential flow solution of a rotating airfoil in the still air is to be the same as the moving surface velocity difference of the corresponding points. The position of these points depends on the ellips shape but not on the rotating angular velocity. It is noted, that the very thin airfoil has the degenerated points to the each edge, that is, it reduces to the Kutta condition.

The aerodynamic forces thus obtained is directly fed into the equation of motion of the airfoil, and it is integrated timewise starting at the zero angle of attack position. In the case with very large value of the moment of inertia of the airfoil, the rotating angular velocity of the airfoil tends to be constant, which is the condition of forced rotation treated in the previous study (6).

III. Results and discussions

Experimental study has been carried out in a water tunnel and in a wind tunnel, for the conditions of the free rotation and the forced rotation with constant angular velocity externally driven by a motor. The flow pattern around the model was visualized as seen in Fig.1, which is in good agreement with the numerically obtained flow pattern shown in Fig.2, for the corresponding conditions. These cases have a constant rotation speed of S=0.14, where S is the reduced rotating period (the chord length/(the uniform flow velocity * the rotating period)) or is in forced rotation with a nondimenionalized value of the moment of inertia I* (the moment of inertia of the airfoil / the mass of the fluid of the same volume) of 0.5.

The variations of the angular velocity observed experimentally and numerically are presented in Figs.3 and 4, respectively. Fig.5 gives the relation of the autorotating frequency and the moment of inertia. With large value of it, the rotation tends to be constant. and as decreasing it, it ceases to autorotate suddenly. This also agrees with the experimentally observed value.

The wake field was simulated over rather far field behind the airfoil, and the vorticity flux which passed through a perpendicular plane at a wake position was calculated, and is shown in Fig.6. The Fourier analysis of them gives the wake Strouhal number St, which are plotted in Fig.7 against the forced rotation frequency. This shows that the merging and splitting of the vortices created by the rotation of the airfoil eventually results in the constant Strouhal number flow. This value is about 0.18, which is close to the Strouhal number of a fixed cylinder.

References

1. E.H.Smith; J.Fluid Mech. vol 50 p513, 1977
2. H.J.Lugt & S.Ohring; J.Fluid Mech. vol 79 p127, 1977
3. J.D.Iversen; J.Fluid Mech. vol 92 p327, 1979
4. H.J.Lugt; J.Fluid Mech. vol 99 p817, 1980
5. H.J.Lugt; Ann.Rev.Fluid Mech. vol 15 p123, 1983
6. Y.Oshima & K.Oshima; AIAA Paper 83-0130, 1983

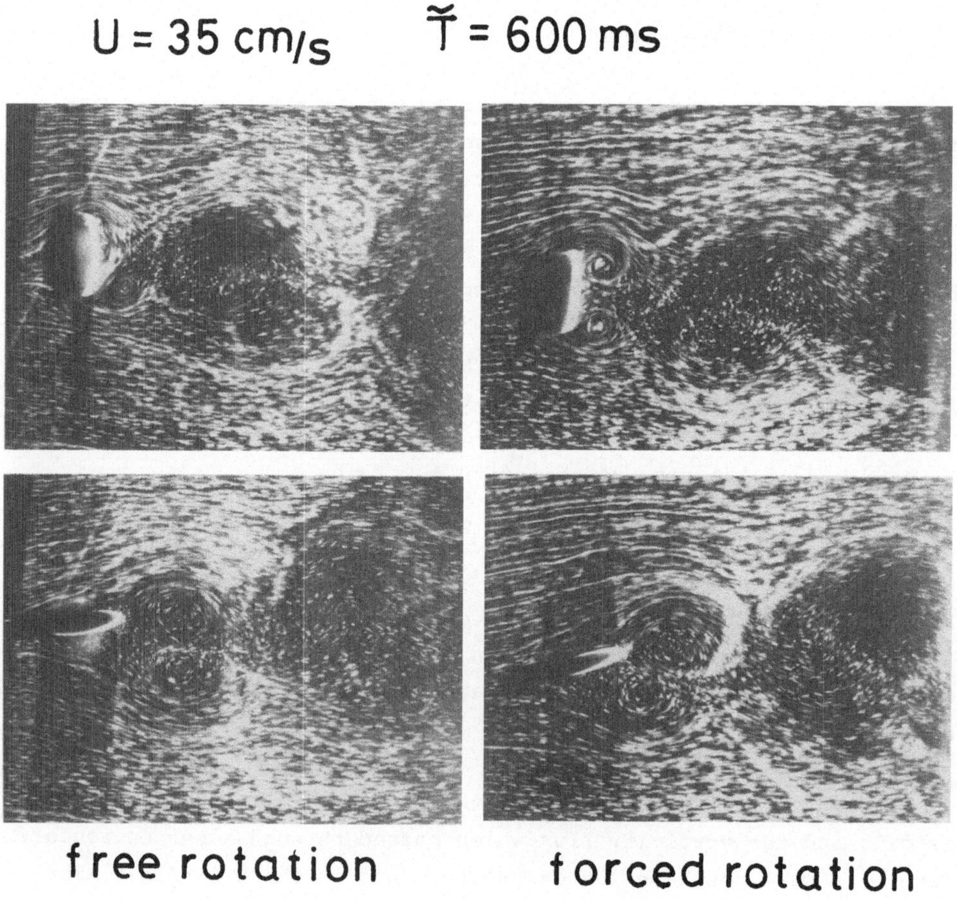

Fig.1 Visualized flow pattern

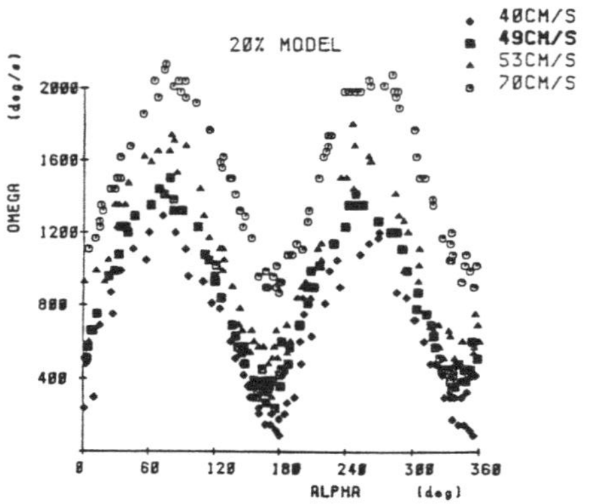

Fig.3 Variation of the angular velocity (Experimental)

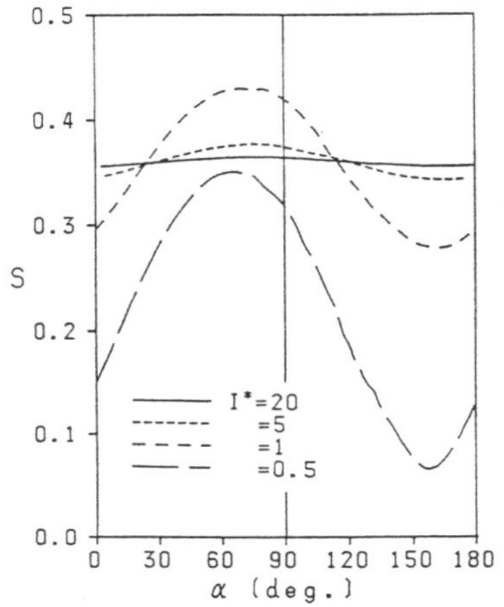

Fig.4 Variation of the angular velocity (numerical)

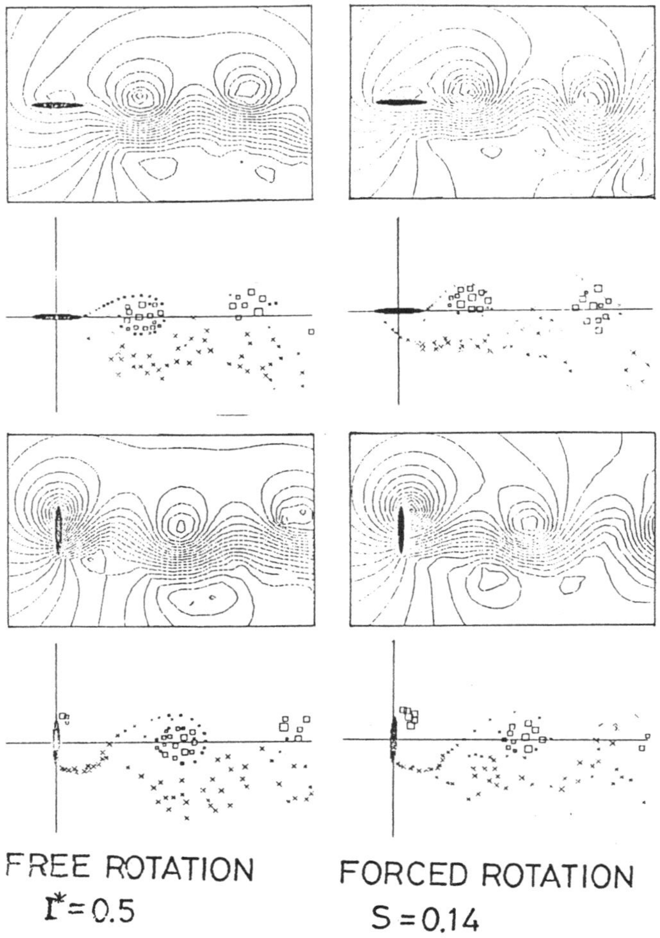

Fig.2 Numerically simulated flow pattern

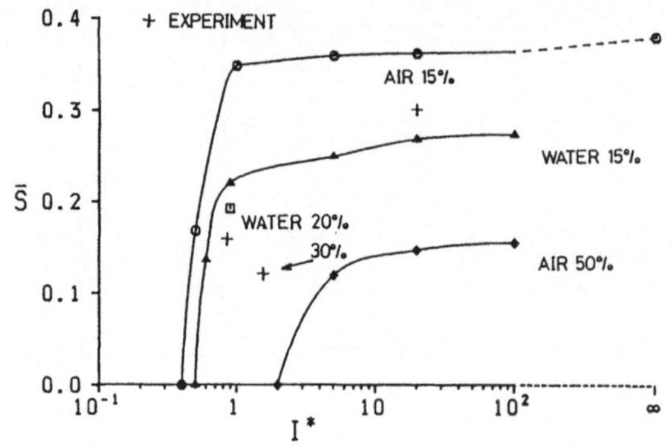

Fig.5 Autorotation frequency against the moment of inertia

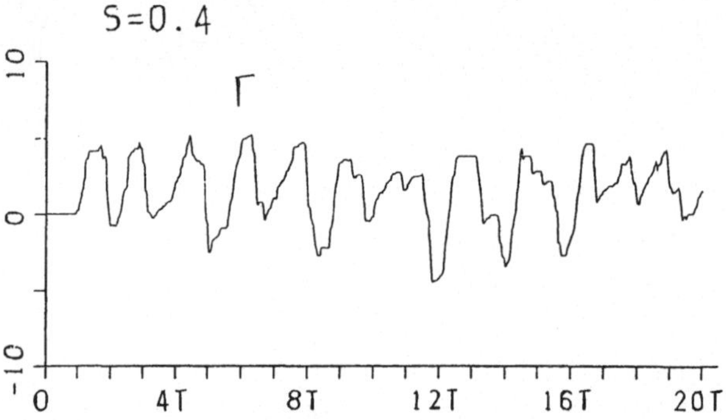

Fig.6 Vorticity flux passing trough a fixed plane

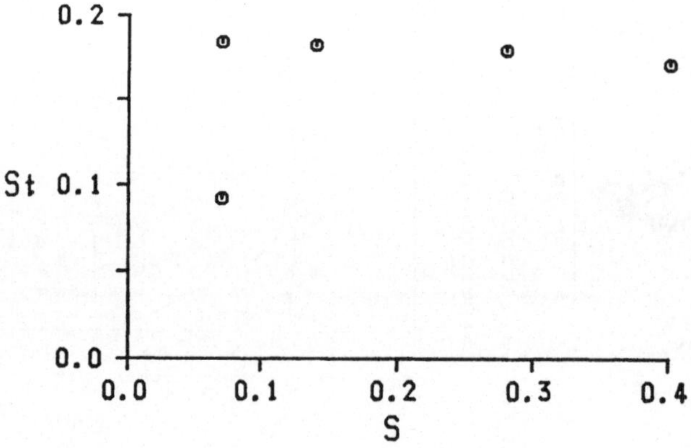

Fig.7 Wake Strouhal number

TRANSIENT MULTIPLE WAVE NUMBER CONVECTIVE INSTABILITY IN A 2-DIMENSIONAL ENCLOSED ROTATING FLUID

Charles Quon
Bedford Institute of Oceanography
Dartmouth, N.S. B2Y 4A2 Canada

Statement of the problem and method of solution

The classical Benard instability problem has traditionally been investigated in a thin layer of fluid in either a stationary or a rotating frame of reference (Chandrasekhar, 1961). When the layer is thin, the dominant unstable mode has vertical wave number one. The aspect ratio of the unstable cells is always 0(1). In this paper, we shall numerically study the transient, finite amplitude instability problem first studied by Daniels and Stewartson (1977, 1978). These authors had derived the critical Rayleigh number for multiple vertical wave number instability of a fluid in an enclosure of aspect ratio unity. They also did some weakly nonlinear analysis at supercritical Rayleigh number. The finite amplitude transient states are beyond the reach of classical analysis.

The essence of the problem is as follows. Consider a cavity as illustrated in Fig. 1, which can be considered as the cross section of an annulus with large inner radius, or of an infinite channel of fluid. The top and side boundaries are insulating, while the bottom boundary is assumed to be perfectly conducting and is maintained at some temperature as function of x, the horizontal coordinate. The whole system rotates about a vertical axis. The objective is to describe the evolution in time of the fluid motion and temperature distribution for an imposed rotation rate and a given temperature on the bottom boundary.

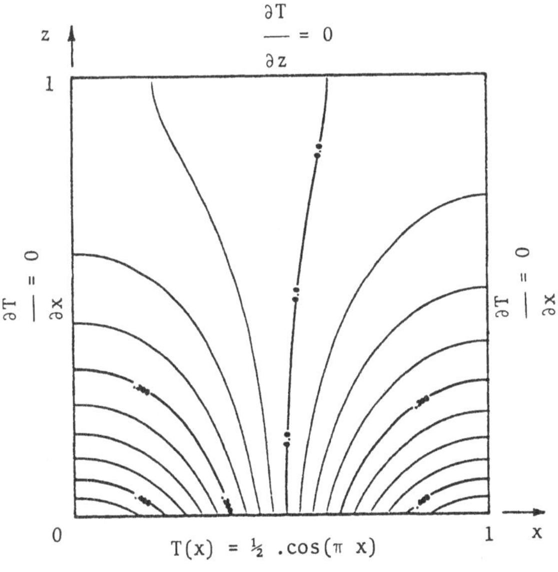

Figure 1. Fluid cavity with Cartesian coordinate. Contours of temperature distributive for almost purely conduction state. Contour interval = 0.05.

We shall impose a Cartesian coordinate $\underline{X} = (x, y, z)$ as shown in Fig. 1, and designate a velocity vector $\underline{V} = (u, v, w)$. After non-dimensionalization with characteristic velocity $U = 2\Omega L$, length L, and temperature ΔT, where Ω is the rotation rate, L the width and height of the square cross section, and ΔT the maximum imposed temperature, the 2-dimensional governing equations with Boussinesq approximation are:

$$\frac{\partial u}{\partial t} + u\frac{\partial u}{\partial x} + w\frac{\partial u}{\partial z} - v = -\frac{\partial p}{\partial x} + E\nabla^2 u \qquad (1)$$

$$\frac{\partial v}{\partial t} + u\frac{\partial v}{\partial x} + w\frac{\partial v}{\partial z} + u = E\nabla^2 v \qquad (2)$$

$$\frac{\partial w}{\partial t} + u\frac{\partial w}{\partial x} + w\frac{\partial w}{\partial z} = -\frac{\partial p}{\partial z} + \beta T + E\nabla^2 w \qquad (3)$$

$$\frac{\partial T}{\partial t} + u\frac{\partial T}{\partial x} + w\frac{\partial T}{\partial z} = \frac{E}{\sigma}\nabla^2 T \qquad (4)$$

$$\text{div}\ \underline{v} = 0 \qquad (5)$$

with the following boundary conditions:

$\underline{V} = 0$ on all boundaries
$\partial T/\partial n = 0$ on all boundaries except the bottom (6a, b, c)
$T = T(x) = \frac{1}{2}.\cos(\pi x)$ on the bottom

where $\nabla^2 = \partial^2/\partial^2 x + \partial^2/\partial^2 z$, and $\partial/\partial n$ is the normal gradient on the boundaries. The dimensionless parameters are:

$E = \nu/(2\Omega L^2)$	Ekman number
$\beta = \alpha g \Delta T/(4\Omega^2 L)$	External thermal Rossby number
$\sigma = \nu/\kappa$	Prandtl number

where ν and κ are respectively the kinematic viscosity and thermal diffusivity, α is the coefficient of volumetric expansion, g the earth's gravity.

Equations (1) to (6) have been solved by finite difference method on a transformed coordinate which has more grid points in the boundary regions than in the interior of the cavity. A detailed exposition of the finite difference procedure is given in Quon (1976). The transform functions used for this problem are symmetric about $z = \frac{1}{2}$, and asymmetric in x (Quon, 1984). Similar numerical algorithms have been used to solve successfully a number of convection problems with or without rotation (Quon, 1976, 1977, 1980, 1981, 1983).

The basic physics of the problem

At low values of β, the leading order of temperature is essentially distributed by conduction as shown in Fig. 1, where the isotherms are almost symmetric about the line at $x = \frac{1}{2}$. We note that the hot (left) side is thermally unstable because it is top heavy. When β increases, instability similar to Benard instability will set in. The steady states, if they exist, depend on the values of the non-dimensional parameters: E, β, and σ. Specifically, we shall discuss the transient states of a computation for $E = 1.5\times10^{-4}$, $\sigma = 10$, and $\beta = 1$.

The case $\beta \to 0$ can be interpreted either as $\Delta T \to 0$, i.e. the imposed temperature differential very small, or as $\Omega \to \infty$, i.e. the rotation rate very large. In either case, the convective motion does not affect the basic conduction state of the temperature, say, $T_o(x, z)$, as shown in Fig. 1. Suppose we increase β. Gentle flows will be created by $T_o(x, z)$ which in turn will be modified by the flows. If we denote the perturbation in velocity and temperature by $\underline{U} = (\Psi_z, v, -\Psi_x)$, and $T(x, z)$, then the linearized equations for T, v and Ψ are:

$$\frac{\partial v}{\partial z} = \beta \frac{\partial T}{\partial x} - E\nabla^4 \Psi \tag{7}$$

$$\frac{\partial \Psi}{\partial z} = E\nabla^2 v \tag{8}$$

$$\sigma \left(\frac{\partial \Psi \partial T_0}{\partial z \partial x} - \frac{\partial \Psi \partial T_0}{\partial x \partial z} \right) = E\nabla^2 T \tag{9}$$

The side boundary layer equation becomes

$$\Psi_{6\zeta} + \lambda T_{oz} \Psi_{\zeta\zeta} + \Psi_{zz} = 0 \tag{10}$$

where $\zeta = E^{-1/3} x$, $E^{1/3}$ being the boundary layer length scale, and $\lambda = \sigma\beta E^{-2/3}$.

If $\beta = 0$, as for a homogeneous rotating fluid, or if T_{oz} is zero, i.e. if the isotherms are nearly vertical on the side walls, such as in the classical rotating annulus problem where the side wall temperatures are kept constant, (10) represents the $E^{1/3}$ Stewartson layer. (Note that we cannot consider the limit $\sigma = 0$ because then we cannot linearize the momentum equations.) If T_{oz} is positive, the solution for (10) is still the Stewartson layer which decays rapidly with distance from the side wall. Physically, a positive T_{oz} means a gravitationally stable thermal stratification which damps vertical motion. The crux of the problem lies in the fact that in the warm side of the fluid cavity, T_{oz} is negative, and hence thermally unstable. We have then a potentially very interesting situation (Daniels and Stewartson, 1977, 1978). In order to understand the effect of the middle term in Eqn. (10), we shall define an internal Rayleigh number with a length scale $\ell = E^{1/3}$ and an internal temperature gradient $|\partial T_0/\partial z|$ as:

$$Ra(i) = Ra.|\partial T_0/\partial z|\ell^4 = \sigma\beta E^{-2}.|\partial T_0/\partial z|.E^{4/3} = \lambda.|\partial T_0/\partial z|$$

where $Ra = \alpha g \Delta T L^4/(H\kappa\nu) = \sigma\beta E^{-2}$ is the external Rayleigh number based on $\Delta T/H$, and L^4.

Hence λT_{oz} in (10) is the internal Rayleigh number Ra(i). Daniels and Stewartson found the critical Rayleigh number for vertical wave number n as:

$$Ra_c(i) = 8.6956\ (2n)^{4/3}, \qquad n = 1, 2, 3, \ldots$$

for equation (10) to hold. If $Ra(i) < Ra_c(i)$, the solution decays into the interior. However, if $Ra(i) \geqslant Ra_c(i)$, then the solution becomes oscillatory for large ζ, and consequently the boundary layer approach is no longer valid. Other approach must be taken.

The transient states

Due to page limitation, we can only explore the transient states to show that (a) unstable modes indeed contain multiple vertical wave numbers, and (b) quasi-steady states after the transient instability are attainable. This may also indicate that the instability is an exchange instability, and not a catastrophic one. Fig. 2 shows contour maps of the isotherms, stream lines, and isotachs of v at four different instances at 200 time step intervals (1 rotational period = 100π = 314.16 time steps). A typical early state of development is depicted at $t = 200\Delta t$ in Fig. 2 (2/3 of a rotation after start). The temperature shows a conduction profile, and the flow fields consist of a buoyant Ekman layer near the bottom boundary where $\partial T/\partial x$ is non-zero. Ekman suction entrains and detrains fluid from the interior in and out of the bottom boundary layer. The non-buoyant Ekman layer near the top is very weak at this instance. It is in fact invisible in the top frame of Fig. 2c. A very careful scrutiny

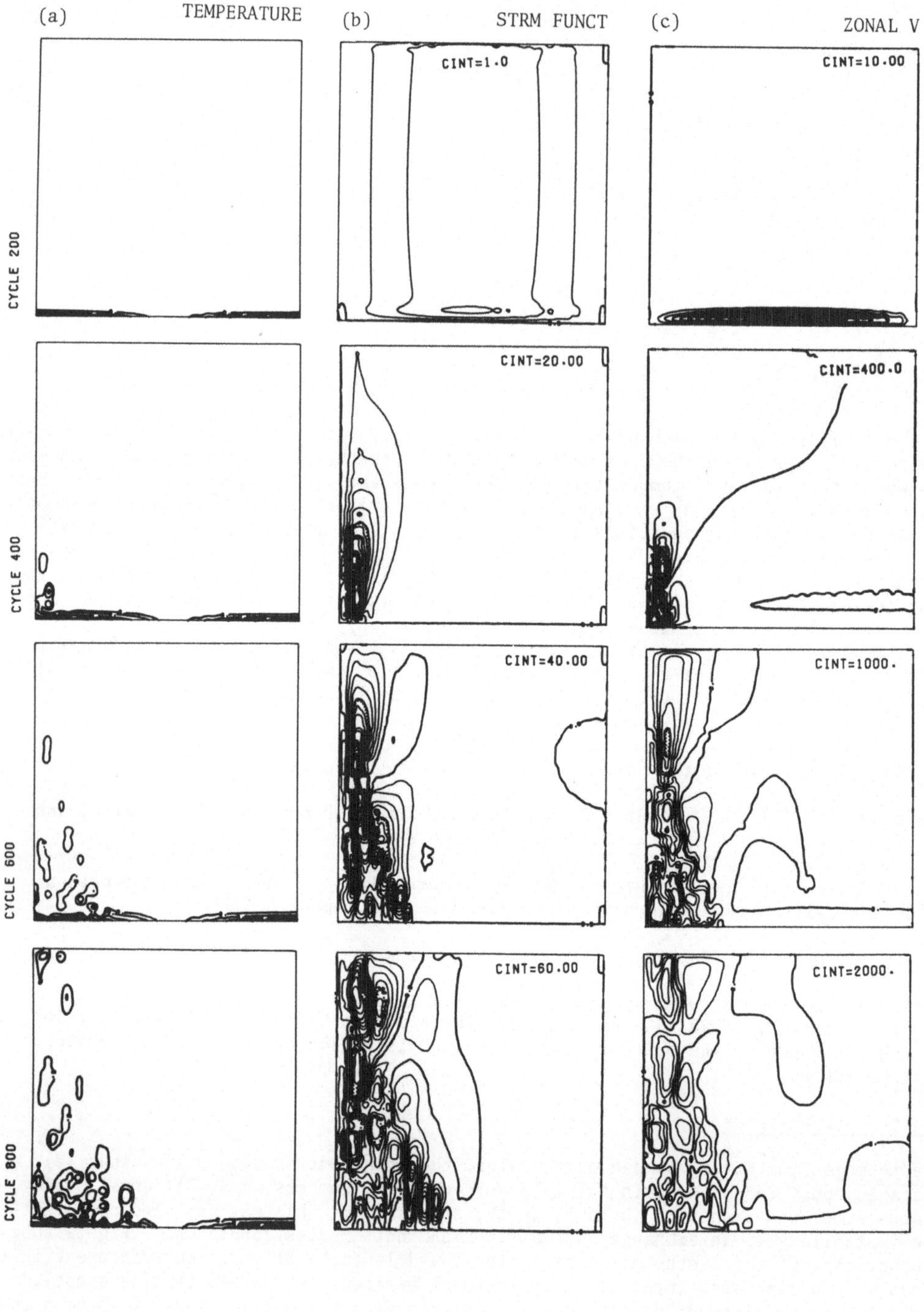

Figure 2. Time evolution of temperature, stream function x 10^5 and along channel velocity x 10^5. Cint = contour intervals. Cint = 0.1 for temperature. The last 2 frames of temperature have additional isotherms at 0.03 and 0.05.

of plots of temperature and velocity as functions of time at various space points shows that the most probable time for the onset of linear instability is slightly before 200Δt, and that for finite amplitude instability is around 250Δt. By t = 400Δt as shown in the second row in Fig. 2, all fields have already been severely affected.

The instability clearly starts from the lower left corner where the local Rayleigh number is the largest. Although the stream function looks like a single plume at 400Δt, there are a number of cells at its base near the bottom boundary. Multiple vertical wave numbers are not yet evident. At t = 600Δt as shown in the contour plots in the third row, instability has been spreading horizontally towards the center along the base while the thermal boundary layer thickens with time. Thus the local Rayleigh numbers reach the critical $Ra_c(i)$ successively at increasing values of x along the base. At t = 600Δt, wave numbers 3 and 4 are much in evidence while the thermal layer has been badly eroded already, although a portion at the extreme right of the unstable layer keeps its conduction profile intact. However, by t = 800Δt, cells have covered the whole unstable layer. Note that vertical penetration and the number of vertical cells decrease with increasing x from the side wall. At the peak of convective activity as inferred from the energy plots (not shown here), half the cavity is filled with cells which are not necessarily well aligned with one another, and their dominant vertical wave number is 4. The largest number of horizontal cells over the unstable half of the base varies from 8 to 10. The stable thermal layer maintains its conduction profile at all times. It is, however, suppressed by Ekman suction and cannot penetrate very far into the interior. A quasi-steady state is finally realized when conduction is exactly balanced by advection due to almost uniform downwelling resulting from Ekman suction as shown in Fig. 3. All the vigorous cells shown earlier have disappeared, and there is a distinct vertical boundary layer on the left which spreads out gradually with height from the base. The upward mass transport in the entire cavity is confined to this narrow jet while the return downward motion covers the rest of the interior. Ekman layers are maintained near the top and bottom boundaries. We cannot explore the dynamics here in detail for lack of space. We should, however, point out that for this highly nonlinear case study, initial instability seems to have taken place extremely early, within one rotation from t = 0. Although the inhibitive effect of rotation must be relevant to the onset of instability, the Daniels and Stewartson mechanism, i.e. instability of the side wall layer, cannot be important until later, when the multiple vertical cells begin to appear. The numerical results give the length scale at the onset of linear instability to be approximately 0.023 (cf.$E^{1/3}$ = 0.053, and $E^{1/2}$ = 0.012), the Rayleigh number is about 2,700 which is close to an estimate given by Howard (1964) for the onset of instability for high Rayleigh number convection in a deep fluid without rotation. A more thorough investigation will be reported elsewhere (Quon, 1984).

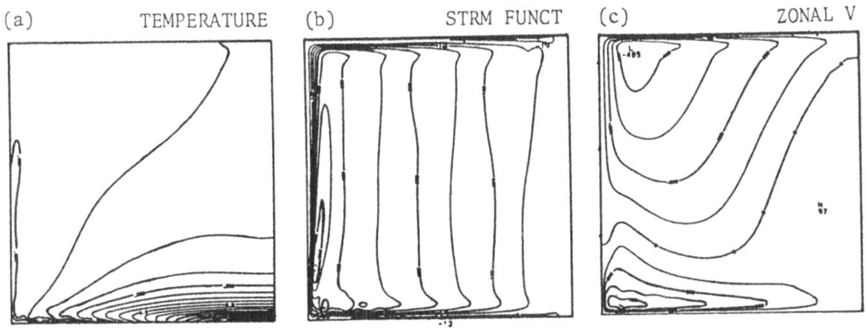

Figure 3. Contours of temperature, stream function and along the channel velocity at quasi-steady state after 30 rotations.

References

Chandrasekhar, S., 1961, Hydrodynamics and Hydromagnetics Instability, Oxford Press, pp 654.

Daniels, P.G., and K. Stewartson, 1977; On the spatial oscillations of a horizontally heated rotating fluid. Proc. Camb. Phil. Soc., 81, pp 325-349.

Daniels, P.G., and K. Stewartson, 1978; On the spatial oscillations of a horizontally heated rotating fluid II. Q.J. Mech. Apl. Math., XXXI, pp 113-135.

Howard, L.N., 1964; Convection at high Rayleigh number, in Applied Mechanics, proceedings of the 11th international congress of applied mechanics, Munich, Ed. H. Gortler, Springer-Verlag, Berlin, Heilelberg, New York, 1966.

Quon, C., 1976; A mixed spectral and finite difference model to study baroclinic annulus waves. J. Comp. Phys. 20, pp 442-479.

Quon, C., 1977; Axisymmetric states of an internally heated rotating annulus. Tellus, 29, pp 83-96.

Quon, C., 1980; Quasi-steady symmetric regimes of a rotating annulus differentially heated on the horizontal boundaries. J. Atmos. Sci., 37, pp 2407-2423.

Quon, C., 1981; In search of symmetric baroclinic instability in an enclosed rotating fluid. J. Geophy. Astrophy. Fluid Dyn., 17, pp 171-197.

Quon, C., 1983; Effects of grid distribution on the computation of high Rayleigh number convection in a differentially heated cavity, in Numerical Properties and Methodologies in Heat Transfer. Ed. T.M. Shih, Hemisphere Publishing Corp., New York, and Spring-Verlag, Berlin, pp 554.

Quon, C., 1984; Non-linear response of a rotating fluid to differential heating (in preparation).

NUMERICAL COMPUTATION OF 3-D FIRE-INDUCED FLOWS AND SMOKE COAGULATION

R. G. Rehm and H. R. Baum
National Bureau of Standards
Gaithersburg, Maryland 20899

Introduction and Mathematical Model

A major challenge in computational fluid dynamics is to describe a flow field over a wide range of dynamically active length and time scales within available computer resources. When additional physical phenomena, such as combustion, smoke dynamics and radiation, are important, as is the case in fires, the description becomes even more difficult.

In the previous conference the authors proposed a convective model which eliminates all small scale phenomena and concentrates on predicting the large scale buoyant convection without empirical parameters (such as occur in turbulence models). The numerical scheme, computational results and comparison of the large-scale features with experiments were presented for a thermally expandable fluid in the two-dimensional case. In the present paper, the numerical scheme is generalized to three-dimensional, time-dependent flows and specialized to a Boussinesq fluid. It is also shown how the small-scale phenomenon of smoke coagulation can be imbedded in the large-scale flow field. We suggest that this simple model, coupling a small-scale phenomenon to large-scale convective features, can be regarded as a prototype for more general couplings which occur during combustion, and other physical phenomena, including turbulence.

The equations used for the results presented in this paper are the inviscid, Boussinesq equations driven by a prescribed volumetric heat source Q(x,t). These equations can be derived from the acoustically filtered equations, determined in (1) and for which two-dimensional results were presented in (2) by introducing a velocity scale U and a density perturbation parameter ζ based upon the maximum total rate of heat release Q_0:

$$U = (Q_0/P_0H^2)^{1/3}\,(g\,H)^{1/3} \quad , \quad \zeta = (Q_0/P_0H^2)^{2/3}\,(g\,H)^{1/3} \qquad [1]$$

Here P_0 is the ambient pressure, H is the height of the enclosure and g is the acceleration of gravity. When the dimensionless parameter ζ is considered small, the acoustically filtered equations become the Boussinesq equations.

$$\nabla \cdot \vec{w}\,(\vec{r},\tau) = 0 \qquad [2]$$

$$\frac{\partial \vec{w}}{\partial \tau} + \nabla\,(1/2\;w^2) - \vec{w} \times \vec{\Omega} + \nabla p^*\,(\vec{r},\tau) + \rho^* \vec{k} = 0 \qquad [3]$$

$$\frac{\partial \rho^*}{\partial \tau}(\vec{r},\tau) + \vec{w}\cdot\nabla\rho^* = -\frac{\gamma-1}{\gamma} f(\tau)\,(H/\ell)^3\, Q^*\,(\vec{r}H/\ell) \qquad [4]$$

$$\vec{\Omega}\,(\vec{r},\tau) = \nabla \times \vec{w} \qquad [5]$$

The quanity $\vec{k}$ in Eq. [3] is a unit vector in the vertical direction.

In these equations, lengths have been made dimensionless using H, time using U and H and pressure deviation from ambient using the ambient density ρ_0 and U. The dimensionless density deviation ρ^* and the dimensionless heat source Q^* are related to the density ρ and the heat source Q through $\rho = \rho_0\,[1 + \zeta\rho^*]$, $Q = (Q_0/\ell^3)\, f(\tau)\, Q^*$, where Q^* is normalized so that $(H/\ell)^3 \int Q^*\, d^3r = 1$ and $f(\tau)$ determines the time history of heat release and ℓ its spatial extent. These equations contain only dimensionless parameters characterizing the location, shape, and time history of the source. For a closed room, the only necessary boundary condition is $w\cdot n = 0$. Initially, the fluid is assumed to be at rest at uniform ambient conditions P_0, ρ_0, T_0.

The major simplification obtained by using the Boussinesq equations comes from the calculation of the pressure perturbation p^*. Taking the divergence of Eq. [3] and using Eq. [2], p^* satisfies the following equation:

$$\nabla^2 p^*\,(\vec{r},\tau) = \nabla\cdot\{\vec{w} \times \vec{\Omega} - \nabla\,(1/2\; w^2) - \vec{k}\rho^*\} \qquad [6]$$

At solid boundaries, from Eq. [3] and the condition of no normal flow:

$$\frac{\partial p^*}{\partial n} + \vec{k}\cdot\vec{n}\,\rho^* = 0 \qquad [7]$$

The relative absence of parameters in both equations and boundary conditions is a direct consequence of the assumption that ζ is small. For a height H of three meters and a one hundred kilowatt fire; $\zeta \cong .08$. Thus, away from the combustion zone, solutions to the Boussinesq equations should be valid.

The acoustically filtered equations have been numerically integrated by the methods described in (2) to obtain fully three-dimensional, time-dependent results. However, the filtered equations and the numerical methods have been specialized to the Boussinesq equations in the present study to conserve computer resources while allowing adequate resolution.

Numerical Methods

The procedures used to numerically integrate Eqs. [2]-[5] are, with the exception of the pressure computation, almost identical with those described for the two-dimensional case in the previous ICNMFD, Baum and Rehm (2). Further details of the two-dimensional methods used in that work may be found in Baum et.al. (3) and Lewis and Rehm (4); extension to three dimensions is straightforward. Since w and ρ^* can be updated in time using the explicit discretized forms of Eqs. [3] and [4], at any given time the right hand side of the pressure equation, [6] is known. Thus, at each discrete time interval, the problem for p^* is the solution of a Poisson equation in a rectangular geometry with Neumann boundary conditions. The solution is obtained using discrete Fourier transforms on the horizontal coordinates, and solving the remaining independent tridiagonal linear equations for the Fourier coefficients. Finally, the discretized pressure is determined from the inverse Fourier transform. The procedure used to solve Eqs. [6] and [7] was developed by Wilhelmson and Erickson (5). This implementation was designed and programmed by Dr. Roland Sweet of NBS.

The computer program implementing the finite difference solutions was tested in a variety of ways to ensure that the computed flows are reasonable approximations to solutions of Eqs. [2]-[5]; see (6), (7), (8) and (2). While the results presented below have not been compared directly with experiments, we are confident that they are approximate solutions to the Boussinesq equations.

Particle Tracking and Computed Results

The hydrodynamic model has been chosen to capture large-scale features of the buoyant flow, whereas smoke dynamics are described on a small-scale using a frame of reference moving with the macroscopic fluid flow. A "particle tracking" procedure is used to obtain the desired Lagrangian information. Marker particles are introduced at each time step in the vicinity of the heat source. The initial particle locations are random, but consistent with the form chosen for the heat source, to provide an initial randomness apparent in real fires.

The coordinates of each particle is found by integrating the ordinary differential equations for particle trajectories. The velocity at the particle location is found by linear interpolation in the hydrodynamic grid. A second order Runge-Kutta method is used to integrate the differential equations. The particle injection frequency is adjusted to maintain the desired particle flux. The particle tracking technique can be used directly to visualize the large-scale features of the flow, in addition to providing a convenient reference frame for small-scale analyses. This is particularly true for three dimensional, time-dependent flows.

Calculations will be reported elsewhere for different enclosure configurations (11). The only results shown here are for a cube with the heat source located at the center of the floor. The time history of the source is $f(\tau) = \tanh(\tau)$, and the source has a Gaussian distribution in the horizontal coordinates and decays exponentially in the vertical.

Figure 1 shows a sequence of photographs of a dynamic graphics device capable of representing phenomena in three space dimensions. Figure 1a shows the plume at about the time it reaches the top of the enclosure. The necking in at the base of the plume due to a strong inflow at the enclosure bottom is evident. The plume has spread across much of the ceiling in Figure 1b. A pulsing motion has begun near the ceiling creating an apparent "ring" of particles moving out from the plume impingement region. The pulsations move rapidly down the plume as the hot gas moves downward along the enclosure boundaries in Figure 1c. Approximately twelve thousand particles were used to generate these displays. The device can show successive time frames rapidly enough for the motion to appear continuous.

Aerosol Coagulation

The aerosol particles generated in fires fall generally in a size range between .01 and 10 micrometers diameter. Such particles are too large to diffuse and too small to settle out over the time needed to fill most enclosures. Hence, if each marker particle is taken to represent a specified mass of smoke aerosol particles, then the solution to the particle trajectory equations also solves the particles mass conservation equation.

It is also possible to consider the size distribution within each "blob" of smoke. Let f(v,t) dv be the number of particles in the volume size range dv about volume v in such a blob. Note that t is the physical time. Then, if size changes occur under the action of sticking collisions between particles brought into close proximity by Brownian motion within the blob, the f(v,t) evolves according to the Smoluchowski equation

$$\frac{\partial f}{\partial t} = \Gamma \{ \int_0^v f(v') f(v - v') dv' - 2f(v) \int_0^\infty f(v') dv'\}. \qquad [8]$$

Here Γ is the coagulation frequency, taken to be constant. This equation has been solved analytically by Mulholland and Baum (9), and applied to the prediction of smoke detector performance in Baum et.al. (10). A derivation of this result by Laplace transform methods is given by Baum and Rehm (11).

The blob containing the smoke is introduced into the fluid at time t_0 with a truncated Junge distribution. The observational evidence for this choice of initial distribution is discussed in Mulholland and Baum (9). The physical significance of the analytical solution for aerosol physics is discussed in Mulholland and Baum (9). Two points are worth noting. First, the calculated size distributions do reproduce measured coagulation effects reasonably well. Figure 2, taken from Baum et.al. (10) shows a comparison between the analytical solution and data taken from a wood crib fire by Helsper et.al. (12). Second, the combination of particle tracking with this analytical result represents information depending on five independent variables. Any attempt to solve Eq. [8] directly on an Eulerian grid would be hopeless at this time.

It is possible to make some assessment of the feasibility of large eddy simulations in fire research. There are two main issues: the computer resources needed to calculate the large-scale processes, and the research needed to model the small-scale processes in an appropriate manner. Only the first of these is considered here. It is concluded that simulation of the hydrodynamics of fires in small to medium size enclosures of simple shape (i.e., enclosures of less than 120 M^3 volume composed of unions of rectangles with rectangular openings) is probably feasible now. With the advent of more powerful computers, this estimate will change rapidly to encompass both larger and more complex geometries.

References

1. Rehm, R. G. and Baum, H. R., NBS J. Res. 83, p. 297 (1978).

2. Baum, H. R., and Rehm, R. G., 8th International Conference on Numerical Methods in Fluid Dynamics, Springer-Verlag.

3. Baum, H. R., Rehm, R. G., Barnett, P. D. and Corley, D. M., SIAM J. Sci. Stat. Comput. 4. p. 117 (1983).

4. Lewis, J. and Rehm, R. G., NBS J. Res. 85, p. 367 (1980).

5. Wilhelmson, R. B. and Ericksen, J. R., J. Comp. Phys. 25, p. 319 (1977).

6. Baum, H. R. and Rehm, R. G., SIAM J. Sci. Stat. Comput., in press.

7. Rehm, R. G., Baum, H. R., Barnett, P. D. and Corley, D. M., submitted for publication.

8. Baum, H. R., Rehm, R. G. and Mulholland, G. W., 19th International Symposium on Combustion, The Combustion Institute, Pittsburgh, PA, p. 921 (1982).

9. Mulholland, G. W. and Baum, H. R., Phys. Rev. Letters 45, p. 761 (1980).

10. Baum, H. R., Rehm, R. G., and Mulholland, G. W., Luck, H. (Ed.), 8th International Symposium on Problems of Automatic Detection, University Duisberg, Duisberg, Germany, p. 259 (1982).

11. Baum, H. R. and Rehm, R. G., to appear in Combustion Science and Technology.

12. Helsper, C., Fissan, H. J., Muggli, J. and Scheidweiler, A., J. Aerosol Sci. 11, p. 439 (1980).

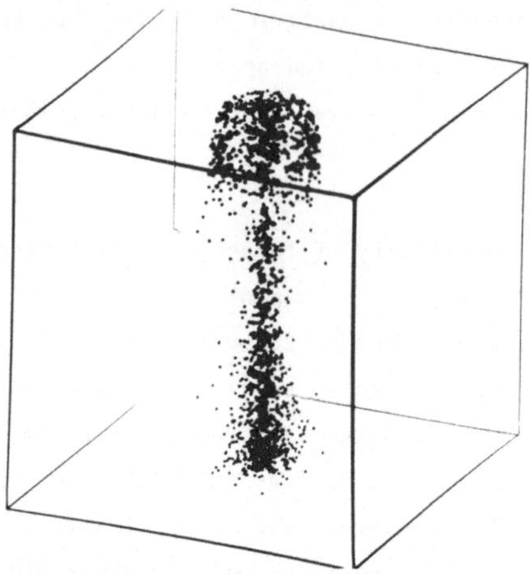

Fig. 1a Dynamic graphic display of particles in a plume, near time of arrival of the plume at the top of a cubic enclosure.

Fig. 1b The plume has spread across much of the ceiling.

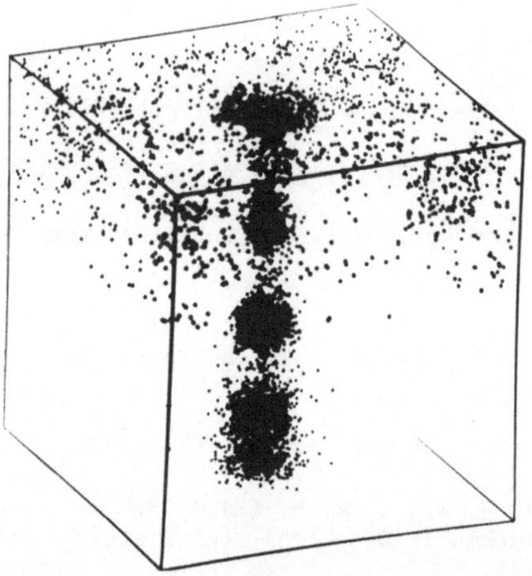

Fig. 1c Pulsations traverse the length of the plume and the enclosure begins to fill from the top with particles.

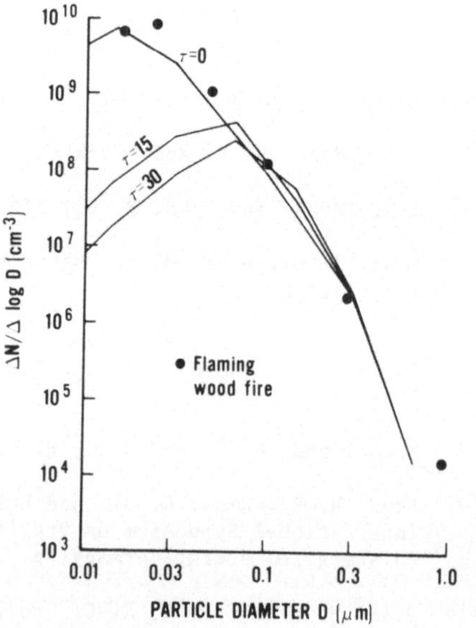

Fig. 2 Calculated smoke aerosol size distributions showing the evolution of the distribution under the influence of coagulation (from Baum et.al. (10).

CYBER 205 DENSE-MESH SOLUTIONS TO THE EULER EQUATIONS FOR FLOWS AROUND THE M6 AND DILLNER WINGS

Arthur Rizzi
FFA The Aeronautical Research Institute of Sweden, S-161 11 BROMMA, Sweden

INTRODUCTION

Methods to solve numerically the Euler equations now abound, all with different techniques and different implementations. But in trying to evaluate them all, it is difficult to judge the merits of one approach over another. Aware of the need for guaging the methods most commonly used today to solve the Euler equations, AGARD set up Working Group 07 to organize a competition open to all to obtain the best possible solutions to a number of representative 2D and 3D test cases[1]. The goal was to produce a body of definitive solutions against which other methods could be measured.

The accuracy of a computed solutions is related directly to the number of grid points that discretize the problem. And comparing a given solution with another one obtained with the same numerical method but on a denser mesh is one of the most certain ways to judge how near the given solution is to the ultimate accuracy of a converged sequence of solutions using successively refined meshes. In evaluating the contributed solutions the Working Group was able to use this measure for the two-dimensional results but not the three-dimensional ones because the computer resources (primarily memory) of even supercomputers (1 M or 2 M words of memory) do not allow refinement beyond the standard mesh. Recently, however, Control Data Corporation has built a CYBER 205 with 16 M 64-bit words of memory and very generously gave me time on the machine to compute some of the M6 wing and Dillner wing test cases using meshes that contain double and even triple the number of cells in each of the three coordinate directions of the standard meshes that I used in my solutions for the Working Group. Effective use of this machine requires vector coding of the computational algorithm so that it processes data in vector lengths as large as possible but less than 2^{16}-1. My previous computer code for the standard mesh cases is vectorized over the entire 3D data structure obtaining vectors whose lengths are the total number of grid points in the mesh[2]. The programming, however, had to be modified for the large-memory machine. It now vectorizes the operations over a 3D subset (whose size is free to choose on input) of the total data structure, and repeats these operations in a DO LOOP for each successive subset until the entire data set is processed--a programming technique commonly termed "strip mining". In this way vector lenths of between 40,000 to 60,000 are obtained easily for any given dense-mesh dimension. The program executes with 32-bit precision at an average rate of about 125 M flops sustained over the entire computation.

Three dense-mesh solutions are presented here. The first was carried out for the M6 wing at conditions M_∞=0.84 and α=3.06 deg. and shows only small improvements in accuracy compared to my earlier standard-mesh solution. But the other two for the Dillner wing cases reveal much richer detail in the vortex-shock-wave structure of these flowfields than was seen before. These computations confirm what, based on wind tunnel measurements, aerodynamicists recently have claimed must exist in real flowfields[3,4].

ONERA M6 WING ; M_∞=0.84 , α=3.06 deg.

With double the number of cells in all the directions of my standard mesh[5], I had 192 cells around the chord, 40 on the span, and 40 outward from the wing to the farfield. The variation in computed lift and drag with density given in Table 1 indicates the very small overall change achieved with the last mesh refinement.

The solution is presented in Figs. 1-7 which show that the suction peaks are nearer to the measured values, the pressure on the lower surface is in better agreement with the experiment, the shock waves are sharper, and the losses in total pressure are more localized to the shocks. The isobars on the upper run together at the trailing corner of the tip exactly as in the standard-mesh solution, ratifying our earlier belief in the reality of this strking feature[5].

Table 1. Variation in lift and drag with mesh density M6 wing. $M_\infty=0.84$ $\alpha=3.06$ deg.

MESH	C_L	C_D
coarse 49×11×11	0.266	0.0167
standard 97×21×21	0.285	0.0113
dense 193×41×41	0.283	0.0111

70 DEG. SWEPT DILLNER DELTA WING

The two solutions obtained with a much denser grid for subsonic and supersonic flows around the Dillner wing reveal flowfields of much more complex expansion and compression phenomena than the standard-mesh solutions do. In both cases weak and local shock waves interact with the vortex over the wing. Tables 2 and 3 show the sequence of lift and drag coefficients for the three levels of grid density.

Table 2. Variation in lift and drag with mesh density. Dillner wing. $M_\infty=0.7$ $\alpha=15$ deg.

MESH	C_L	C_D
coarse 33×11×15	0.563	0.147
standard 65×21×29	0.639	0.161
dense 161×49×81	0.681	0.171

Table 3. Variation in lift and drag with mesh density. Dillner wing. $M_\infty=1.5$ $\alpha=15$ deg.

MESH	C_L	C_D
coarse 33×11×15	0.482	0.135
standard 65×21×29	0.525	0.146
dense 193×57×97	0.532	0.147

Subsonic Case, $M_\infty=0.7$ $\alpha=15$ deg.

The solution computed for this case on a mesh of 160 cells around the semi-span, 80 on the chord, and 48 from the wing outward to the farfield is presented in a variety of views in Figs. 8 to 13. The most striking feature--not seen in the standard-mesh solution but clearly visible here in the isobars, Mach number, and total pressure contours--is the small shock wave situated between the vortex and the upper surface of the wing just outboard of the vortex core. From observations of wind tunnel experiments Wendt[3] recently has suggested the existence of just such a shock wave. In overall comparison with my previous solution, the suction peak under the vortex is markedly stronger and produces the shock wave, the smearing of vorticity across the vortex sheet at the leading edge is significantly reduced, and the total pressure losses are confined to a tighter region centered on the core of the vortex. In the wake the interaction of the leading and trailing edge vortices also is represented better.

Supersonic Case, M_∞=1.5 α=15 deg.

A similar set of Figs. 14 to 19 display the solution computed for this case on a mesh of 192 cells around the semi-span, 96 on the chord, and 56 from the wing outward to the farfield. The variation in pressure throughout this supersonic flowfield is not very great, but the distortion in the isobars and the coalescing of the Mach contours indicates the presence of a small shock wave intersecting the coiling vortex sheet just inboard of and above the core of the vortex. This feature of the flowfield is consistent with the experimental findings of Miller and Wood (see Fig. 12 of Ref. 4). Furthermore we see in the Mach number and total pressure contours at least two distinct shock waves between the suction peak and leading edge, forming a complex system of shocks and expansion waves being reflected from the apex to the trailing edge. Such phenomena should be expected because the flow, which is supersonic, has to turn abruptly where the vortex sheet leaves the leading edge. And accompanying these phenomena are heavy losses in total pressure.

CONCLUSIONS

The dense-mesh solutions presented here have shown that the flowfield around a trapezoidal wing like the M6 can be represented with reasonable accuracy on a standard-size mesh of say 50,000 grid points. But if the wing is one of low aspect ratio, like the Dillner delta wing, a much denser grid is required to capture the rich structure of rotating flow interacting with shock waves.

Acknowledgement

Many thanks to the Control Data Corporation for its interest in and support of this work, especially to Chuck Purcell, who invited me to Minneapolis, and Mike Hodous, who put in many hours helping me to produce these solutions.

REFERENCES

1 Fluid Dynamics Panel Working Group 07.: Test Cases for Steady Inviscid Transonic and Supersonic Flows, AGARD Pub., in preparation, 1984.

2 Rizzi, A.: Vector Coding the Finite-Volume Procedure for the CYBER 205, in Lecture Series Notes 1983-04, von Karman Institute, Brussels, 1983.

3 Vorropoulos, G. and Wendt, J.F.: Laser Velocimetry Study of Compressibility Effects on the Flow Field of a Delta Wing, AGARD-CP-342, 1983.

4 Miller, D.S. and Wood, R.M.: An Investigation of Wing Leading-Edge Vortices at Supersonic Speeds, AIAA Paper No. 83-1816, 1983.

5 Rizzi, A.W. and Eriksson, L.-E.: Computation of Flow Around Wings Based on the Euler Equations, Journal Fluid Mechanics, in press.

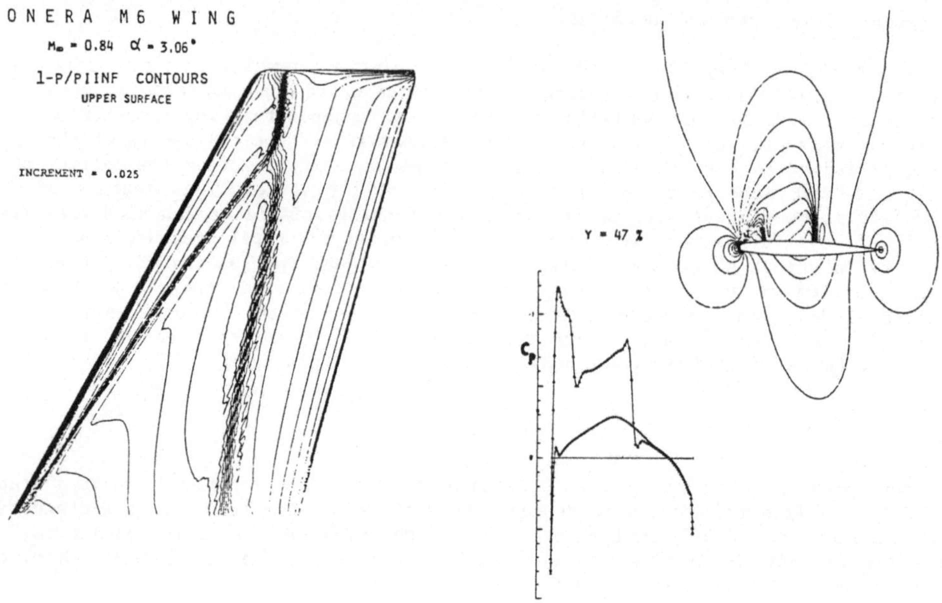

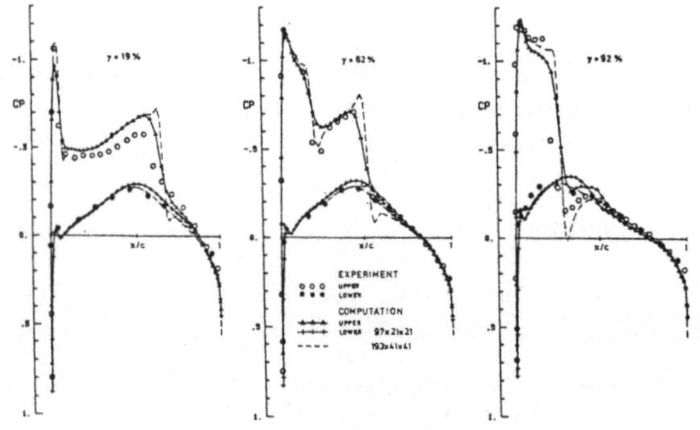

Fig 1 Pressure and total pressure fields computed on 193x41x41 mesh compared with solution from 97x21x21 mesh. M6 wing.

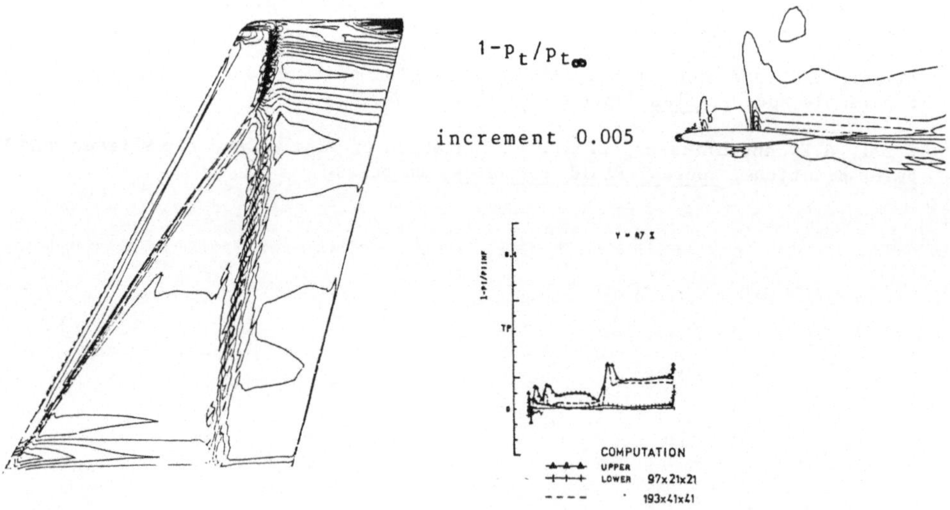

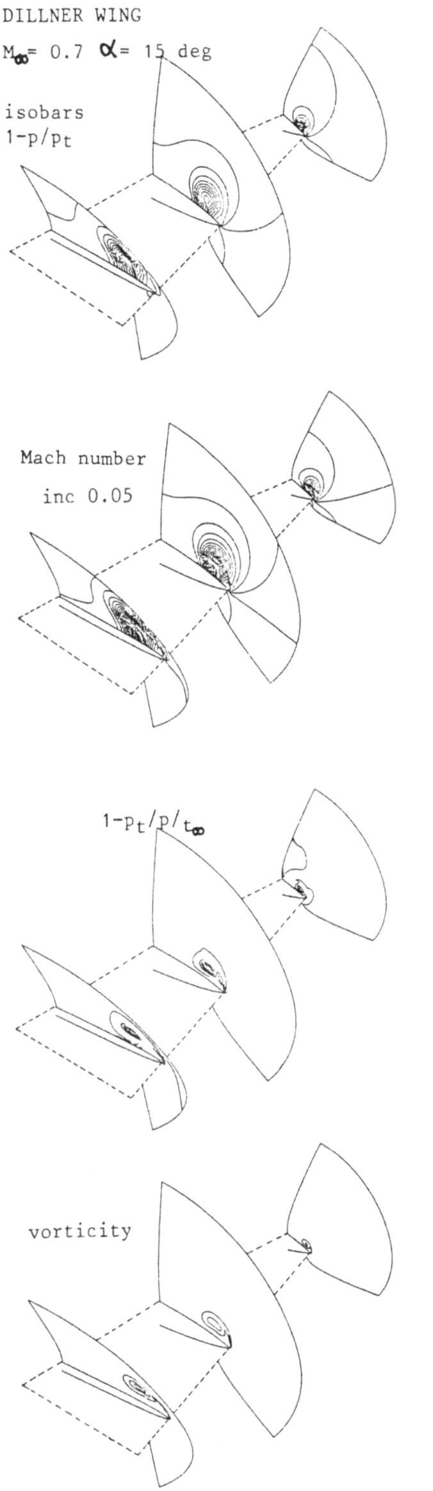

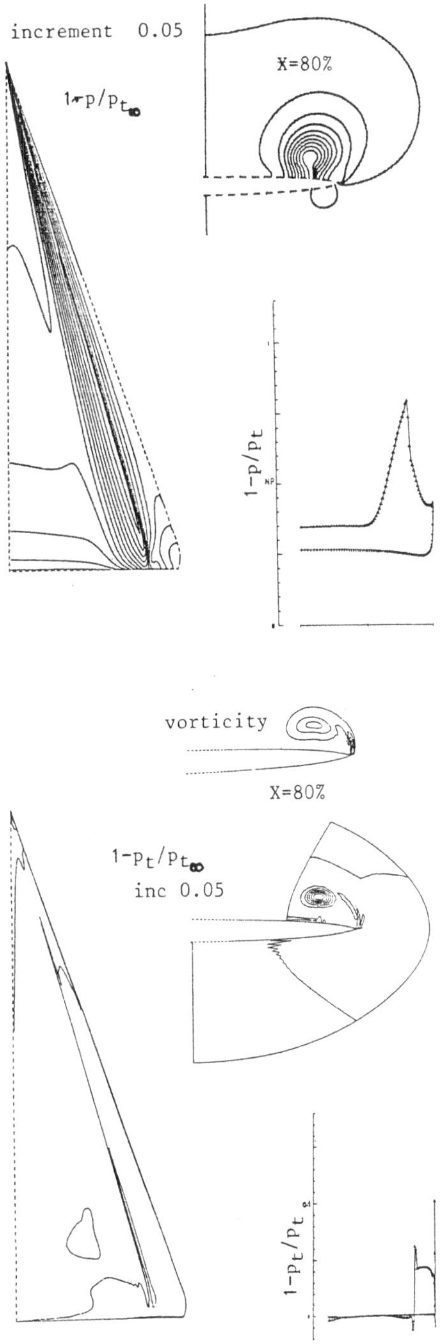

Fig 2 Delta wing flowfield computed on an 0-0 mesh of 161x49x81 points. Dillner wing M_∞=0.7 α=15 deg.

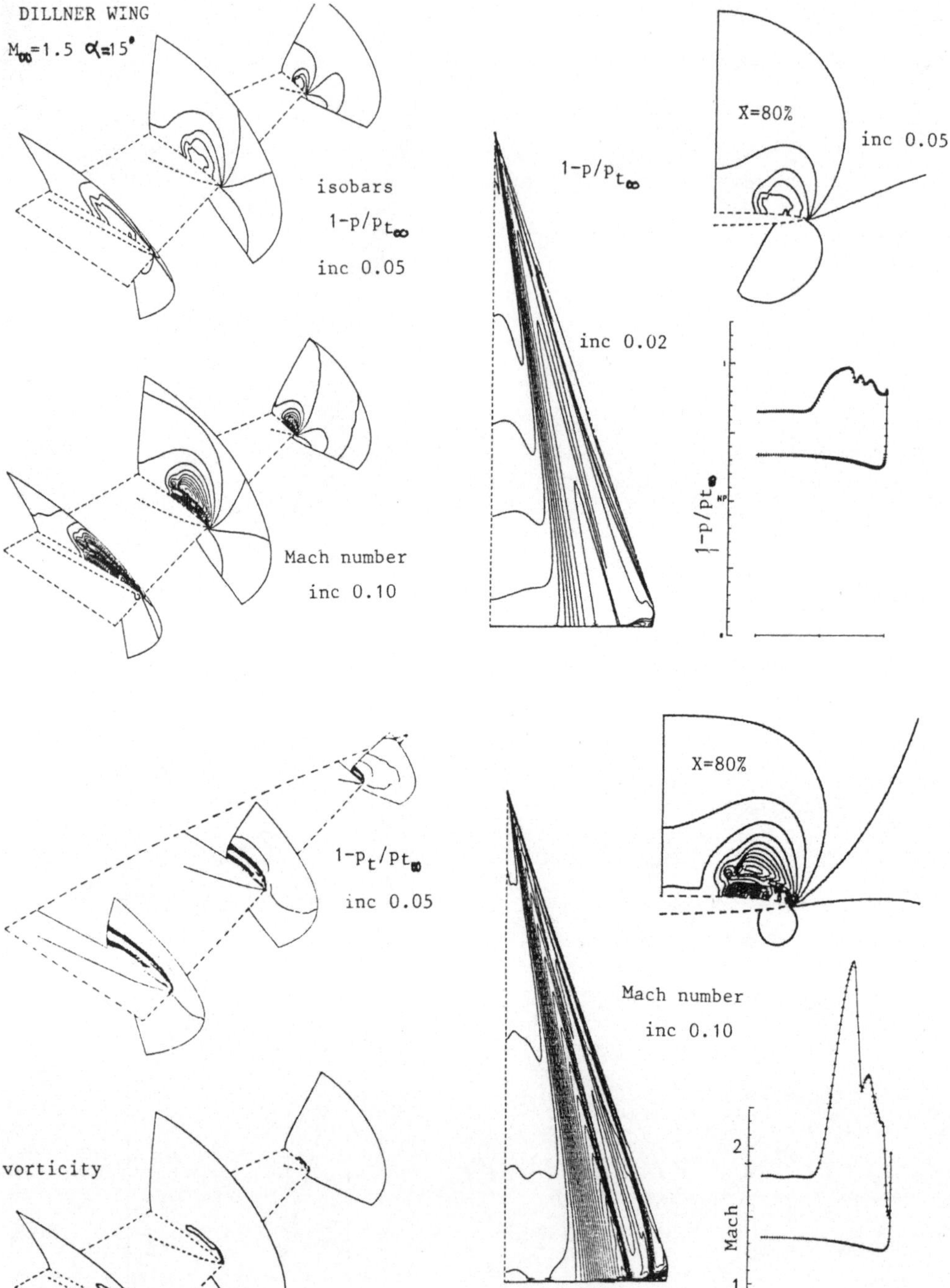

Fig 3 Delta wing flowfield computed on an O-O mesh of 193x57x97 points. Dillner wing M_∞=1.5 α=15 deg.

NONCONFORMING 3D ANALOGUES OF CONFORMING TRIANGULAR FINITE ELEMENT METHODS IN VISCOUS FLOW

V. Ruas
Pontifícia Universidade Católica
Departamento de Informática
22453 Rio de Janeiro/BRAZIL

INTRODUCTION AND NOTATION

This work deals with the numerical solution of three-dimensional Stokes' equations by mixed finite element methods related to non standard degrees of freedom. The elements, which are of the velocity-pressure type, can be viewed as 3D analogues with a nonconforming velocity, of classical piecewise (conforming) [quadratic velocity]×[constant pressure]and [(quadratic+bubble function)velocity]×[linear pressure] triangular elements, as the same kind of polynomials are used and the same error estimates are obtained.

We refer to [3] and [2]respectively for a detailed description of these 2D elements. Let us here just briefly recall their structure.

Given a polygon Ω of $\mathbb{R}^2$ partitioned into a set $\mathcal{T}_h$ of triangles, in the usual way of the finite element method, each velocity component and the pressure are approximated in finite dimensional spaces V_h and Q_h, respectively defined as follows:

1st element: $V_h=V_h^o=\{v/v_{/K}\in P_2 \quad \forall K\in \mathcal{T}_h, \quad v \text{ continuous over } \Omega\}$

$Q_h=Q_h^o=\{q/q_{/K}\in P_o \quad \forall K\in \mathcal{T}_h\}$

2nd element: $V_h=V_h^1=\{v/v_{/K}\in P_2\oplus\{\psi\} \quad \forall K\in \mathcal{T}_h, \quad v \text{ continuous over } \Omega\}$

$Q_h=Q_h^1 = \{q/ \quad q/_K\in P_1 \quad \forall K\in \mathcal{T}_h\}$

where P_k is the space of polynomials in two (or more) variables of degree less than or equal to k, and ψ is the cubic function that vanishes on the three edges of K, called the bubble-function [2].

The degrees of freedom associated with V_h^i and Q_h^i, i=0,1 at element level are functional values at the nodes indicated in Figure 1.

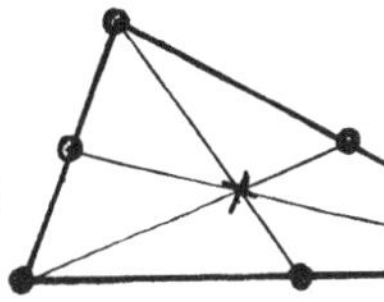

1st element $V_h^o\times Q_h^o$

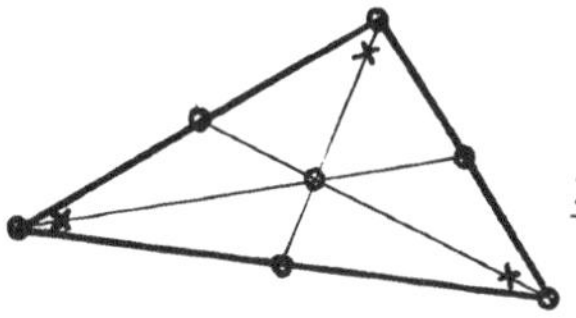

2nd element $V_h^1\times Q_h^1$

o Velocity node (functional value) × Pressure node (func. value)

Figure 1: 2D elements

In the following we denote by $H^k(\Omega)$, $k\in\mathbb{N}$, the Sobolev space of functions defined over a bounded domain Ω of $\mathbb{R}^n$, $n=2$ or 3, whose partial derivatives up to the k-th order are square integrable over Ω. Any function which is square integrable over Ω is said to belong to $L^2(\Omega)$. If $\vec{f}$ and $\vec{g}$ are two vector fields of $[L^2(\Omega)]^m$ we denote their inner product by $(\vec{f},\vec{g})_\Omega = \int_\Omega \vec{f}\circ\vec{g}\,dx$, where $\vec{f}\circ\vec{g} = \sum_{i=1}^{m} f_i g_i$, with corresponding norm $\|\vec{f}\| = \sqrt{(\vec{f},\vec{f})_\Omega}$

A function of $H^1(\Omega)$ that vanishes almost everywhere on the boundary of Ω is said to belong to $H^1_o(\Omega)$, and a function f of $L^2(\Omega)$ such that $\int_\Omega f dx=0$ is said to belong to $L^2_o(\Omega)$.

THE PROBLEM TO SOLVE

Consider the stationary Stokes problem defined in a flow region Ω of $\mathbb{R}^3$ with boundary Γ assumed to be polyhedric, written in variational form:

$$(\mathcal{P})\begin{cases} \text{Find } \vec{u}\in[H^1_o(\Omega)]^3 \text{ and } p\in L^2_o(\Omega) \quad \text{such that} \\ (\text{grad } \vec{u},\text{grad } \vec{v})_\Omega - (p,\text{div } \vec{v})_\Omega = (\vec{f},\vec{v})_\Omega \qquad \forall\vec{v}\in[H^1_o(\Omega)]^3 \\ (q,\text{div } \vec{u})_\Omega = 0 \qquad \forall q\in L^2_o(\Omega). \end{cases}$$

We are given a regular family of partitions $\{\mathcal{T}_h\}_h$ of Ω into tetrahedrons (see e.g. [1]). For each $\mathcal{T}_h$, h represents the maximum edge length of all tetrahedrons of $\mathcal{T}_h$.

As usual, we define the finite element analogue of $(\mathcal{P})$ by considering only test velocities v and pressures q that belong to finite dimensional spaces $[V_h]^3$ and Q_h, approximating $[H^1_o(\Omega)]^3$ and $L^2_o(\Omega)$, respectively, associated with $\mathcal{T}_h$. Then instead of $(\mathcal{P})$ we solve.

$$(\mathcal{P})\begin{cases} \text{Find } \vec{u}_h \in[V_h]^3 \text{ and } q_h\in Q_h \quad \text{such that} \\ \sum_{K\in\mathcal{T}_h} [(\text{grad } \vec{u}_h, \text{grad } \vec{v})_K - (p,\text{div } \vec{v})_K] = (\vec{f},\vec{v})_\Omega \qquad \forall\vec{v}\in[V_h]^3 \\ \sum_{K\in\mathcal{T}_h} (q,\text{div } \vec{u}_h)_K = 0 \qquad \forall q\in Q_h. \end{cases}$$

THE 3D ELEMENTS AND ERROR ESTIMATES

Refer to Figure 2 and let P_i, $i=1,2,3,4$ be the vertices of a tetrahedron $K\in\mathcal{T}_h$, and P_{ijk} be the centroid of the face whose vertices are P_i, P_j and P_k. Let also ℓ_{ij} be the edge of K whose ends are P_i and P_j, and P_{ij} be its midpoint.

We consider the following two choices V^i_h and Q^i_h of V_h and Q_h, $i=0,1$, analogous to the two-dimensional spaces:

The space V_h^o is the set of functions whose restriction to each $K \in \mathcal{T}_h$ is a quadratic function w associated with the set $\sum$ of ten degrees of freedom listed below:

- Functional value $w(P_{ijk})$, $1 \leq i<j<k \leq 4$
- The linear combination $[w]_{ij}$ of $w(P_{ij})$ and the mean value of w over ℓ_{ij}, $1 \leq i<j \leq 4$ given by:

$$[w]_{ij} = \frac{9}{5} \int_{\ell_{ij}} w \, ds/\text{length}(\ell_{ij}) - \frac{4}{5} w(P_{ij}) \quad (1)$$

For any function of V_h^o, each one of these degrees of freedom is assumed to coincide at interelement boundaries, and to vanish if it is attached to a node or an edge lying on Γ.

One can verify that this set of degrees of freedom is P_2-unisolvent by exhibiting the ten associated basis functions. We refer to [4] for further details.

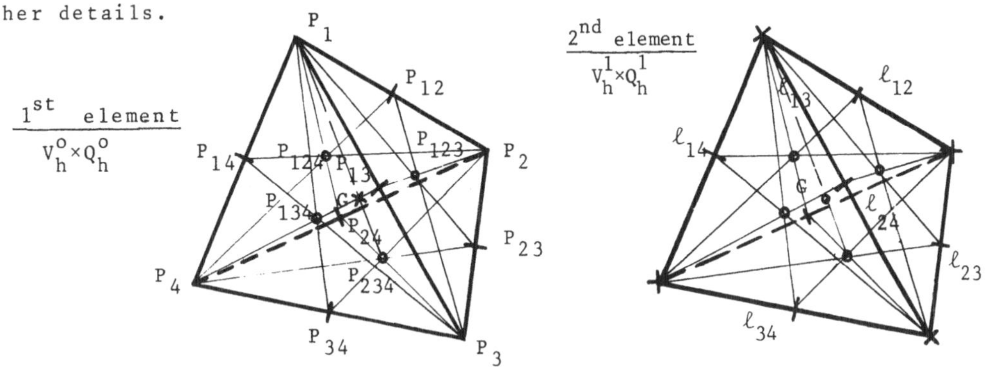

Velocity d. of f.: —+— linear combination (1); ∘ functional value

Pressure node: × functional value

Figure 2: The 3D elements

V_h^1 in turn is the space of functions whose restriction v to any $K \in \mathcal{T}_h$ belongs to $P_2 \oplus \{\psi\}$, i.e., is a function of the form $v = w + \alpha\psi$, where $w \in P_2$, α is a real coefficient and ψ is the bubble-function of K, namely, the quartic function that vanishes on the four faces of K.

The set of degrees of freedom of V_h^1 restricted to each K is the same set $\sum$ above, plus the functional value at the centroid G of K. These degrees of freedom are also required to coincide at interelement boundaries and to vanish on Γ.

The corresponding basis functions at element level are given in [4].

Likewise the 2D case Q_h^o and Q_h^1 are the spaces of functions whose restriction to each $K \in \mathcal{T}_h$ belong to P_o and P_1, respectively.

Now, as for error estimates, let $\vec{u}_h^i$ and p_h^i be the approximations $\vec{u}_h$ and p_h of $\vec{u}$ and p, when one sets in $(\mathcal{P}_h)$ $V_h = V_h^i$ and $Q_h = Q_h^i$, i=0 or 1.

We can prove [5] the following result ,

$$\| \vec{grad}(\vec{u}-\vec{u}_h^i) \| + \| p-p_h^i \| \leq C_i h^{i+1} \qquad (2)$$

assuming that $\vec{u}$ and p belong to the Sobolev spaces $[H^{i+2}(\Omega)]^3$ and $H^{i+1}(\Omega)$ respectively, C_i being constants depending on $\vec{u}$ and p but not on h. Note that (2) also applies to the corresponding 2D elements, under the same regularity assumptions.

The key to the proof is the fact that, although a function of V_h^i is necessarily continuous only at the centroids of the common faces of two neighboring tetrahedrons, a continuity property in a certain weak sense is satisfied at such interfaces, according to the following:

Lemma: Let K'and K" be two elements of $\mathcal{T}_h$ such that $K' \cap K'' = F$, F being a triangular face. For every $v \in V_h^i$, v'and v" denote the restriction of v to K' and K" respectively. Then we have:

$$\int_F v' p ds = \int_F v'' p ds \qquad \forall p \in P_1 .$$

The proof of this lemma, together with detailed proofs of error estimate (2) can be found in [5].

FINAL REMARK

The properties of the two elements described in this work confirm that, in the framework of viscous flow problems, the right generalization of conforming finite elements are nonconforming for the velocity. We had already pointed out this fact in [6], where another nonconforming 3D analogue of the $P_2 \times P_o$ element with an incomplete eight node quadratic velocity had been introduced.

REFERENCES

[1] CIARLET, P.G., The finite element method for elliptic problems, North Holland, 1978.

[2] CROUZEIX, M. &RAVIART, P.A., Conforming and nonconforming finite element methods for solving stationary Stokes' equations(I) , RAIRO R-3, pp. 33-76 (1973).

[3] FORTIN, M., Calcul numérique des écoulements des fluides de Bingham et des fluides newtoniens incompressibles par la méthode des éléments finis, thèse de Doctorat d'Etat, Université Paris VI, 1972.

[4] RUAS, V., A complete three-dimensional version of the quadratic velocity-constant pressure finite element method for fluid flow problems, Revista Brasileira de Computação, 3 nº 2,(1983/84).

[5] —— , Finite element solution of 3D incompressible flow problems using nonstandard degreesof freedom, to appear.

[6] —— , Méthodes d'éléments finis en élasticité incompressible non-linéaire et diverses contributions à l'approximation des problèmes aux limites, Thèse de Doctorat d'Etat, Université Paris VI, 1982.

ON THE NON-UNIQUENESS OF THE SOLUTION OF THE PROBLEM ON FLOW FIELD ABOUT A CONE AT INCIDENCE

V.V. Rusanov, V.A. Karlin
Keldysh Institute of Applied Mathematics
USSR Academy of Sciences
Moscow 125047, USSR

The problem of an uniform steady supersonic gas flow with an attached bow shock about a circular cone at incidence is of considerable interest in gas dynamics. At the zero angle of attack the problem is known to have two solutions differing primarily by the shock half-angle (Busemann, 1929). The shock is called strong in the case of the larger half-angle and weak in the case of smaller one.

From the continuity considerations the cone flow problem at non-zero angle of attack may also be expected to have two solutions. However only one solution has been found so far. This solution corresponds to the flow parameters observed in experiment and as the angle of attack is tending to zero, converges to the weak shock solution. The reason why the second solution cannot be calculated with available stabilization methods seems to be in the solution instability with respect to the corresponding nonstationary **models** (Rusanov and Sharakshanae, 1980).

In this paper the second solution to the problem on a flow about a cone at incidence is solved by the stabilization method with fictitious unknowns. The method allows one to determine stationary solutions of non-steady problems irrespective of the solution stability in a nonstationary model.

The essence of the stabilization method with fictitious unknowns may be outlined as follows.

If the stationary problem in question is described by, in general, a nonlinear operator equation

$$A(u) = 0 \tag{1}$$

in the abstract Hilbert space H, then under some restrictions impozed on operator A we may use as its solution an asymptotic of the solution component $U(t)$ as $t \to \infty$ in the following problem:

$$\frac{dU}{dt} + \kappa U - (A'_U)^* Q = \kappa u^{(0)}, \qquad \frac{dQ}{dt} + A(U) = 0, \qquad t \geq 0; \tag{2}$$

$$U(0) = u^{(0)}, \qquad Q(0) = 0.$$

Here $(A'_v)^*$ is the operator ajoint to the operator A'_V , the latter being the result of the linearization of the operator A when $u = U(t)$, $K>0$ is a real number, and $u^{(0)}$ is a good enough approximation to the sought solution u_0 of equation (1). The conditions sufficient for convergence of $U(t)$ to u_0 are established by Rusanov and Karlin (1983).

The idea of using fictitious unknowns in a stabilization method was earlier discussed by Essers (1977) and Wirz (1977).

In the formulation of the stationary problem on the supersonic flow about a cone at incidence used here the bow shock was treated as discontinuity surface where the Rankine-Hugoniot conditions were set. The condition of the normal velocity component vanishing was imposed on the cone surface.

If the conical shock-wave is attached to the solid cone at their common vertex, it is known that all parameters of the flow field will be constant along an arbitrary ray for which the origin is the cone vertex. Hence, all parameters of the flow field depend only on the angles of latitude θ and longitude φ in the spherical coordinate system in which the origin is the apex of the cone and $\theta=0$ coincides with the cone axis. It is expedient to transform the spherical coordinates into the coordinates (η , φ), where $\eta=(Ctg\theta-F(\varphi))/(Ctg\chi-F(\varphi))$, χ is the cone half-angle, $F(\varphi)=Ctg\sigma(\varphi)$, and $\sigma(\varphi)$ is the shock-wave half-angle.

To simplify the computational algorithm the system of differential equations was completed with an obvious identity $\partial F/\partial\eta=0$ and $F(\eta,\varphi)$ is interpreted as an unknown function. As a result, the problem in question can be written in a symbolic form:

$$\frac{\partial u}{\partial \eta}+B\left(u,\frac{\partial F}{\partial \eta},\eta,\varphi\right)\frac{\partial u}{\partial \varphi}+f\left(u,\frac{\partial F}{\partial \varphi},\eta,\varphi\right)=0, \quad (\eta,\varphi)\in[0,1]\times[0,\pi];$$

$$\tilde{u}=\psi(F,\partial F/\partial\varphi,\varphi) \quad \text{at } \eta=0, \quad v_\theta=0 \quad \text{at } \eta=1, \tag{3}$$

$$\partial u_i/\partial\varphi=v_\varphi=0 \quad \text{at } \varphi=0 \text{ and } \varphi=\pi \text{ for } i\neq 2,$$

where $u=(p, v_\varphi, v_\theta, v_r, F)^T=(\tilde{u}^T, F)^T=(u_1, u_2, u_3, u_4, u_5)^T$, p the pressure, v_r, v_θ, v_φ the velocity components. The density is excluded using the Bernoulli's integral.

The nonstationary problem constructed according to (2) for the stabilization method with fictitious unknowns is a nonlinear one for a set of partial differential equations with three independent variables and both initial and boundary conditions:

$$\frac{\partial}{\partial t}\begin{pmatrix}U\\Q\end{pmatrix}+\begin{pmatrix}0&E\\E&0\end{pmatrix}\frac{\partial}{\partial \eta}\begin{pmatrix}U\\Q\end{pmatrix}+\begin{pmatrix}0&E\\B&0\end{pmatrix}\frac{\partial}{\partial \varphi}\begin{pmatrix}U\\B^TQ\end{pmatrix}+\begin{pmatrix}g\\f\end{pmatrix}=\begin{pmatrix}0\\0\end{pmatrix},\ t\geq 0,(\eta,\varphi)\in[0,1]\times[0,\pi];$$

$$\tilde{U}=\psi,\quad \Psi^T\tilde{Q}+Q_5=\partial(\Phi^T\tilde{Q})/\partial\varphi \qquad \text{at } \eta=0,$$

(4) $$U_3=Q_1=Q_2=Q_4=Q_5=0 \qquad \text{at } \eta=1,$$

$$\partial U_i/\partial\varphi=\partial Q_i/\partial\varphi=U_2=Q_2=0 \text{ at } \varphi=0 \text{ and } \varphi=\pi \text{ for } i\neq 2,$$

$$U(0,\eta,\varphi)=u^{(0)}(\eta,\varphi),\quad Q(0,\eta,\varphi)=0,\quad (\eta,\varphi)\in[0,1]\times[0,\pi],$$

where $Q=(\tilde{Q}^{\mathrm{T}},Q_5)^{\mathrm{T}}=(Q_1,Q_2,Q_3,Q_4,Q_5)^{\mathrm{T}}$, $E=\mathrm{diag}(1,1,1,1,1)$,

$$g=\kappa(U-u^{(0)})-[\partial(f+Bu_\varphi)/\partial u]^TQ+\partial\{[(\partial B/\partial u_\varphi)u_\varphi+\partial f/\partial u_\varphi]^TQ\}/\partial\varphi,$$

$$\Psi=\partial\psi/\partial F,\quad \Phi=\partial\psi/\partial F_\varphi,\quad u_\varphi=\partial u/\partial\varphi,\quad F_\varphi=\partial F/\partial\varphi.$$

In order to solve this problem an implicit approximate factorization difference scheme was employed:

(5) $$(\mathcal{E}I+\tau\mathcal{B}_{n,\ell}^{(m)})(\mathcal{E}T+\tau\mathcal{A}R)\frac{\mathcal{W}_{n,\ell}^{(m+1)}-\mathcal{W}_{n,\ell}^{(m)}}{\tau}+(\mathcal{A}R+\mathcal{B}_{n,\ell}^{(m)}T)\mathcal{W}_{n,\ell}^{(m)}+\mathcal{F}_{n,\ell}^{(m)}=0.$$

Here I is the identity operator in the space of gridfunctions $\mathcal{W}_{n,\ell}^{(m)}=\mathcal{W}(m\tau,nh_\eta,\ell h_\varphi)=((U^{\mathrm{T}})_{n,\ell}^{(m)},(Q^{\mathrm{T}})_{n,\ell}^{(m)})^{\mathrm{T}}$, $\mathcal{F}_{n,\ell}^{(m)}=((g^{\mathrm{T}})_{n,\ell}^{(m)},(f^{\mathrm{T}})_{n,\ell}^{(m)})^{\mathrm{T}}$;

$$\mathcal{A}=\begin{pmatrix}0&E\\E&0\end{pmatrix},\quad \mathcal{B}_{n,\ell}^{(m)}=\begin{pmatrix}0&Y(B^T)_{n,\ell}^{(m)}\\B_{n,\ell}^{(m)}Y&0\end{pmatrix},$$

$T=(P_\eta+I)/2$, $R=(P_\eta-I)/h_\eta$, $Y=(P_\varphi-P_\varphi^{-1})/2h_\varphi$, $\mathcal{E}=\mathrm{diag}(E,E)$, and P_η, P_φ are the transition operators, i.e., $P_\eta\mathcal{W}_{n,\ell}^{(m)}=\mathcal{W}_{n+1,\ell}^{(m)}$, $P_\varphi\mathcal{W}_{n,\ell}^{(m)}=\mathcal{W}_{n,\ell+1}^{(m)}$.

At the boundaries $\varphi=0$ and $\varphi=\pi$ difference equations (5) are used and the mirror symmetry of the problem (4) is taken into account. The initial conditions and boundary conditions at $\eta=1$ are set exactly. The boundary conditions at $\eta=0$ (on the shock wave) are approximated with the second order of accuracy.

Difference scheme (5) is stable under the weak restrictions on the step τ and has the second order of accuracy for steady-state solutions.

In this paper in addition to the well known solution with the weak shock a numerical solution converging to the one with the strong shock as the angle of attack tends to zero was found. It was possible due to the specific properties of the employed method. The solutions were obtained on the 21x21 grid. The stabilization was considered to be achieved when the relative increments of the unknowns become less than $\tau\times10^{-6}$.

Figure 1 shows the intersections of the shocks with the plane P which is normal to the cone's symmetry axis. The incidence angles $\alpha = 0°$ and $\alpha = 4°$, the freestream Mach number $M_\infty = 2$, the cone half-angle $\chi = 40°$, the ratio of the specific heats $\gamma = 1,4$. The upwind side of the cone corresponds to the lower part of the figure. The figure shows that the solutions of different types are converging when the angle of attack is being increased. The speed of stabilization was found to slow down when $\alpha \to 4°$. These results put together seem to justify the suggestion that the problem in question has no solution with attached bow shock for $\alpha > \alpha_0 \approx 4°$.

Figures 2 and 3 show streamline's projections of both types solutions on the plane P .

REFERENCES

Busemann A. ZAMM, 1929, b. 9, h. 6, s. 496-498.
Rusanov V.V., Sharakshanae A.A. Comp. and Fl., 1980, v. 8, p. 243-250.
Rusanov V.V., Karlin V.A. Doklady AN SSSR, 1983, v. 268, n. 5, p. 1058-1062. (Translated in: Soviet Math. Dokl., v. 27, n. 1).
Essers J.A. Proc. Second GAMM - Conference on Numerical Methods in Fluid Mech., Köln, 1977, p. 20-27.
Wirz H.J. Ibid., p. 238-245.

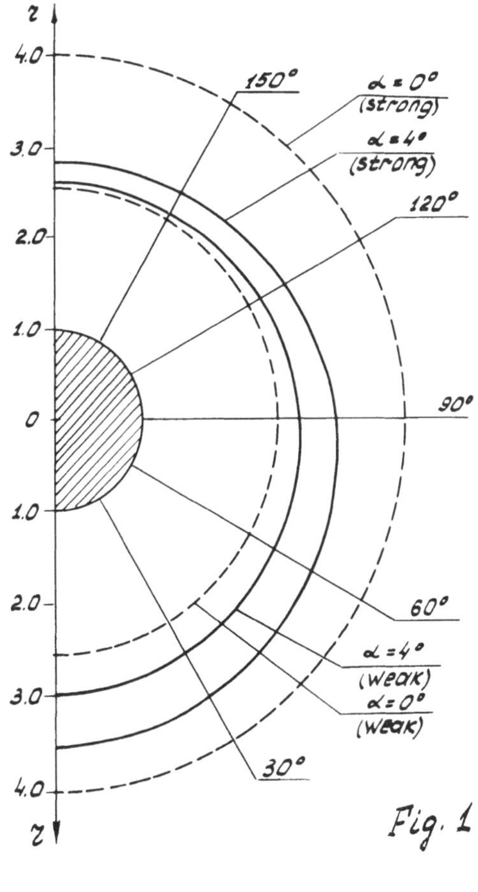

Fig. 1

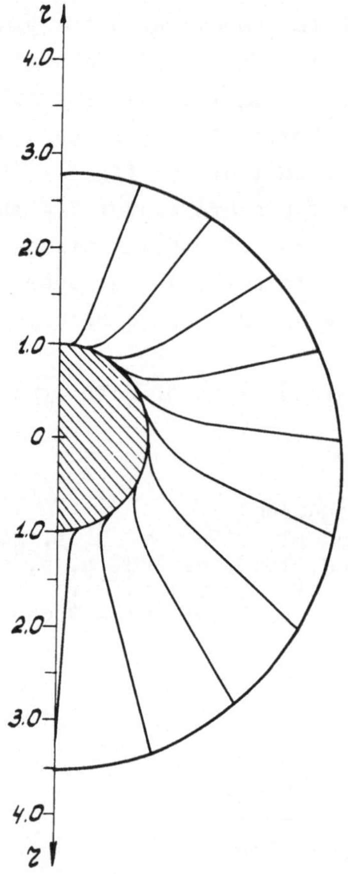

Fig. 2

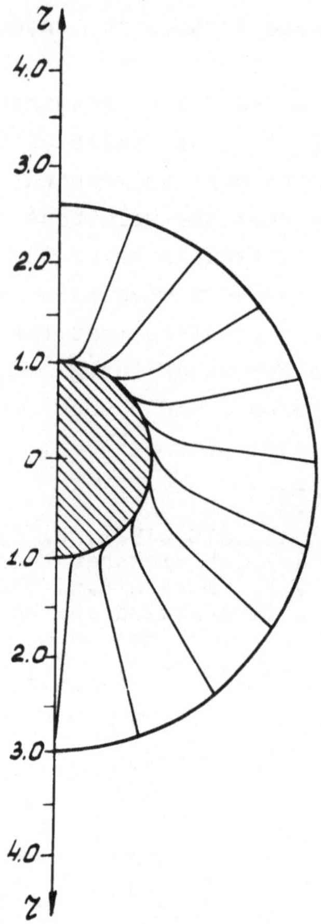

Fig. 3

HIGHER-ORDER METHOD OF LINES FOR THE NUMERICAL SIMULATION OF TURBULENCE

Nobuyuki Satofuka, Haruyoshi Nakamura and Hidetoshi Nishida

Department of Mechanical Engineering
Kyoto Technical University
Matsugasaki, Sakyo-ku
Kyoto 606, Japan

ABSTRACT

A new method is devised for the numerical simulation of turbulences. The spatial derivatives of the time-dependent incompressible Navier-Stokes equations are discretized by using the modified differential quadrature (MDQ) method. The resulting system of ordinary differential equations in time are then integrated by a class of fourth order explicit Runge-Kutta schemes. The simulations of two and three-dimensional homogeneous isotropic turbulence suggest that the present method is more efficient and versatile than the pseudospectral method.

INTRODUCTION

Recent advances in computer and numerical methods have made meaningful numerical simulation of turbulence possible. Almost all code being used for the direct simulation of turbulence are based on either spectral [1] or pseudospectral [2,3] approximation to spatial derivatives. Although the pseudospectral (P.S.) method is much simpler than the spectral method, it is still quite time consuming to compute, even with the use of fast Fourier transform (FFT) algorithm. Development of a more efficient method is needed before more complicated simulations become feasible. The present paper describes such a development.

The new method consists of the modified differential quadrature (MDQ) method [4] combined with a class of fourth order explicit Runge-Kutta time integration schemes. In the proposed method, spatial derivatives of a flow variable are approximated by a weighted sum of the values of an unknown function at the closest M grid points. The weighting coefficients are determined in the same manner as Lagrangian interpolation. It should be noted that the approximation with M=3 coincides with the usual second-order, centered difference approximation, and $M=\infty$ corresponds to the Fourier series approximation, which is of infinite order of accuracy. Thus by changing the value of M, we can easily devise a numerical approximation to the spatial derivatives with as much accuracy as desired. Substitution of the approximate relation into the Navier-Stokes equations yields the set of ordinary differential equations (ODEs) in time. The resulting system with suitable initial and boundary conditions is integrated by using appropriate time-integration scheme. The Poisson equation is solved by the standard

Fourier transform method.

The present paper is organized in the following way. The outline of numerical procedure of the present method for a nonlinear scalar model equation, that is, Burgers equation, is described briefly in Sec. 2; application to incompressible Navier-Stokes equations and the results for numerical simulation of homogeneous isotropic turbulence are discussed in Sec. 3.

OUTLINE OF THE METHOD

In this section we will describe the outline of the present method when it is applied to the following nonlinear scalar model equation, viz., Burgers equation,

$$\frac{\partial u}{\partial t} = -u \frac{\partial u}{\partial x} + \nu \frac{\partial^2 u}{\partial x^2} \quad , \tag{1}$$

with the initial condition

$$u(x,0) = f(x) \ , \tag{2}$$

and boundary conditions

$$B_j u(x,t) = 0 \ , \quad j = 1, 2 \tag{3}$$

Spatial Discretization

If the function u satisfying Eq. (1) is sufficiently smooth, we can write the approximate relation

$$\partial u_i(t)/\partial x \cong \sum_{j=1}^{N} a_{ij} u_j(t) \ , \quad i = 1,2,\ldots,N, \tag{4}$$

where we adopt the notation $u_i(t) = u(x_i,t)$. Viewing Eq. (4) as a linear transformation of u, we see that the second-order derivatives can be approximated by

$$\partial^2 u_i(t)/\partial x^2 \cong \sum_{j=1}^{N} b_{ij} u_j(t) \tag{5}$$

where $b_{ij} = \sum_{k=1}^{N} a_{ik} a_{kj}$. In this paper we have modified the approximate relations, Eqs. (4) and (5), to use the values of u at the nearest M mesh points centered around x_i, instead of using those at all mesh points in the computational domain, as is the case in the original DQ method [5]. By using these values of u, the number of arithmetic operations to be performed for every mesh point is significantly reduced; moreover, in the case of a uniform mesh, the weighting coefficients a_{ij} become independent of index i. Therefore, the approximate relations, Eqs. (4) and (5), can be rewritten as

$$\partial u_i(t)/\partial x = \sum_{j=-m}^{m} a_j u_{i+j}(t) \equiv D_M(u_i) \quad , \tag{6}$$

$$\partial^2 u_i(t)/\partial x^2 = \sum_{j=-m}^{m} b_j u_{i+j}(t) \equiv D_M^2(u_i) \quad , \tag{7}$$

where $a_j = a_{mj}$, $b_j = b_{mj}$ and $m = (M - 1)/2$.

There are many ways of determing the coefficients a_{ij}. In the DQ method, Bellman et al. determined a_{ij} explicitly, choosing x_i to be the root of shifted Legendre polynomial of degree of N, $P_N^*(x)$. In this paper we have determined a_{ij} numerically, similarly to Lagrangian interpolation. The coefficients a_j and b_j are computed once and for all at the beginning of the calculation and stored.

Time Integration

Substitution of the approximate relations, Eqs. (6) and (7), into Eq. (1) yields the set of N ODEs in time,

$$u_i'(t) = -u_i(t)D_M(u_i) + \nu\, D_M^2(u_i) \tag{8}$$

or in matrix form

$$\vec{U}' = \vec{F}(\vec{U}) \tag{9}$$

where $\vec{U} = (u_1,u_2,\ldots,u_N)^T$ and the prime denotes differentiation with respect to time. The numerical solution of such a system, Eq. (9), is a simple task, using an appropriate time-integration scheme. In this paper, we adopt the classical fourth order explicit Runge-Kutta-Gill (RKG) scheme and a low storage Runge-Kutta scheme [6]. As applied to Eq. (9), the latter scheme can be written in the following form,

$$\begin{aligned}
\vec{Q}^{(1)} &= \Delta t\ \vec{F}(\vec{U}^n)\\
\vec{U}^{(1)} &= \vec{U}^n + \vec{Q}^{(1)}\\
\vec{Q}^{(2)} &= -5\vec{Q}^{(1)} + \Delta t\ \vec{F}(\vec{U}^{(1)})\\
\vec{U}^{(2)} &= \vec{U}^{(1)} + \vec{Q}^{(2)}/8\\
\vec{Q}^{(3)} &= \vec{Q}^{(2)}/16 + \Delta t\ \vec{F}(\vec{U}^{(2)})\\
\vec{U}^{n+1} &= \vec{U}^{(2)} + 2\vec{Q}^{(3)}/3
\end{aligned} \tag{10}$$

where the superscripts n and n+1 refer to time $t = n\Delta t$ and $t = (n + 1)\Delta t$, respectively. The stability regions of two schemes are shown in Fig. 1. The low storage scheme have enhanced stability properties along the negative real axis over the conventional scheme.

SIMULATION OF HOMOGENEOUS ISOTROPIC TURBULENCE

The Navier-Stokes equations for incompressible fluid flow may be written

$$\frac{\partial \vec{u}}{\partial t} + (\vec{u}\cdot\nabla)\vec{u} = -\nabla p + \nu\nabla^2\vec{u} \quad , \tag{11}$$

$$\nabla\cdot\vec{u} = 0 \quad . \tag{12}$$

where $\vec{u} = (u,v,w)$ is the velocity field, p is pressure and ν is the kinematic viscosity. In two dimension, it is convenient to solve the Navier-Stokes equation in terms of the vorticity and stream function. With the stream function $\Psi(x,y,t)$ related to the velocity by $u = (\partial\Psi/\partial y, -\partial\Psi/\partial x)$ and to the z component of vorticity by $\omega(x,y,t) = \partial v/\partial x - \partial u/\partial y$, the dynamical equations are

$$\frac{\partial \omega}{\partial t} + \vec{u}\nabla\omega = \nu\nabla^2\omega \quad , \tag{13}$$

$$\omega = -\nabla^2\Psi \quad . \tag{14}$$

Approximately homogeneous turbulence may be most conveniently realized numerically by imposing periodic boundary conditions,

$$\vec{u}(\vec{x} + L, t) = \vec{u}(\vec{x}, t) \tag{15}$$

where L is a periodic length. In this section, we demonstrate the accuracy and efficiency of the present method comparing the computed results with those of the P.S. method.

Comparison with P.S. Method

In our comparisoin with the P.S. method, the vorticity equation, Eq. (13), is solved by using tenth order MDQ method (M = 11) on a 128x128 space grid. The initial flow field is chosen to be a realization of a statistically homogeneous, isotropic Gaussian ensemble with energy spectrum,

$$E(K) = 2/3\ K\ \exp(-2K/3) \quad , \tag{16}$$

and the same initial condition is used in all the calculations. In both methods, time integration was done by fourth order explicit RKG scheme. The solution of the Poisson equation necessary to obtain stream function and velocity components from the vorticity is accomplished by a non-iterative method utilizing fast Fourier transforms (FFT). The vorticity contours for the case with $\nu = 0.0025$ at $t = 2$ are compared in Fig. 2. Note that $t = 2$ is well into the evolution of this flow and that the vorticity field is a sensitive measure of small-scale structure. The results obtain by using the MDQ method are plotted in Fig. 2(a) and the corresponding results for the P.S. method are shown in Fig. 2(b). The difference between Fig. 2(a) and (b) are hardly noticeable. From these comparisons, we can conclude that the MDQ method with M = 11 is roughly equivalent in accuracy to the P.S. method with the same resolution in each space direction.

Table 1 gives the comparative efficiency of the MDQ method with that of the P.S. method for the case with 128x128 grid points. For each method, the table shows, the maximum time step used, the computer time required per time step on a Fujitsu FACOM M-200 computer to solve the vorticity equation, Poisson equation and the total. As the table shows, the MDQ method requires about one tenth the computer time of the P.S. method to solve the vorticity equation, whereas it allows 50% longer time step. In fact, to compute the solution at $t = 2$, the MDQ method uses only 15 minutes of CPU time on a FACOM M-200 while the P.S. method requires about 95 minutes. The efficiency of the MDQ method is remarkable.

High Reynolds Number 2-D Simulation

As an example of simulation for high Reynolds number, Fig. 3 shows the vorticity contours at $t = 3$ for $\nu = 0.0001$ obtained by using the MDQ method with M = 11 on a 512x512 space grid. The integral-scale Reynolds number R_L is about 25500 at $t = 0$. The time step used in this case is 1/400 and total computation time required to compute the solution at $t = 3$ is 30 hours on a FACOM M-200. The corresponding enstrophy dissipation spectrum for $t = 3$, $\nu = 0.0001$ is plotted in Fig. 4. With the present grid of 512x512, we may posiblely study the inertial range of two-dimensional turbulence. According to the results of Orszag with 512x512 grid point, there is a range of wave number $K \leq 50$ for which $K^4E(K)$ is roughly a linear function so $E(K)$ is roughly proportional to K^{-3}. The spectrum shown in Fig. 4 provides some support for his result.

Simulation of 3-D Turbulence

A result of direct simulation of three-dimensional homogeneous isotropic turbulence using 32x32x32 grid points is shown in Fig.5 in which computed energy dissipation rate for $\nu = 0.02$ is compared with that of the pseudospectral method [7]. Agreement of two results is quite excellent.

CONCLUSIONS

A new method has been presented for the numerical simulation of turbulence. For many application, this method is more efficient than the pseudospectral method in use today. The present method has the following features; 1) explicit, 2) fourth order accurate in time, 3) arbitrary order accurate in space, 4) simple and straightforward to program, and 5) should easily adapt to vector computers.

REFERENCES

[1] Herring, J.R., Orszag, S.A., Kraichman, R.H. and Fox, D.G, 1974, J. Fluid Mech., 66, 417-444.

[2] Fox, D.G. and Orszag, S.A., 1973, J. Comp. Phys., 11, 612-619.

[3] Fornberg, B., 1977, J. Comp. Phys., 25, 1-31.

[4] Satofuka, N. and Morinishi, K., 1982, NASA TM 81339.

[5] Bellman, R., Kashef, B.G. and Casti, J., 1972, J. Comp. Phys., 10, 40-52.

[6] Williamson, J.H., 1980, J. Comp. Phys., 35, 48-56.

[7] Orszag, S.A. and Patterson, G.S., 1971, Lecture Notes in Physics, vol. 12, 127-147.

Table 1 Comparison of Computational Time

	CPU TIME PER TIME STEP(msec)	
	M.D.Q. METHOD	P.S. METHOD
Δt	1/100	1/150
VORTICITY EQ.	1585	16437
POISSON EQ.	2784	2784
TOTAL	4369	19221

COMPUTER: FACOM M-200

MESH: 128x128

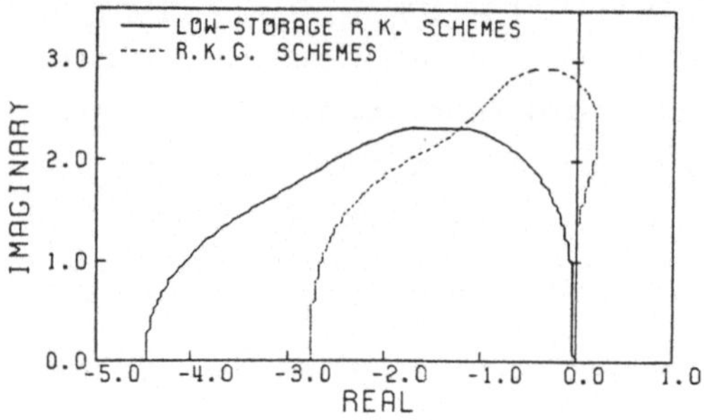

$(h\sigma)_{crit}$ = -4.520 for Low-Storage R.K. Schemes

$(h\sigma)_{crit}$ = -2.785 for R.K.G. Schemes

Fig. 1. Stability regions of time integration schemes

(a)

(b)

Fig. 2. Vorticity contours for (a) MDQ method and (b) P.S. method at t = 2 with ν = 0.0025.

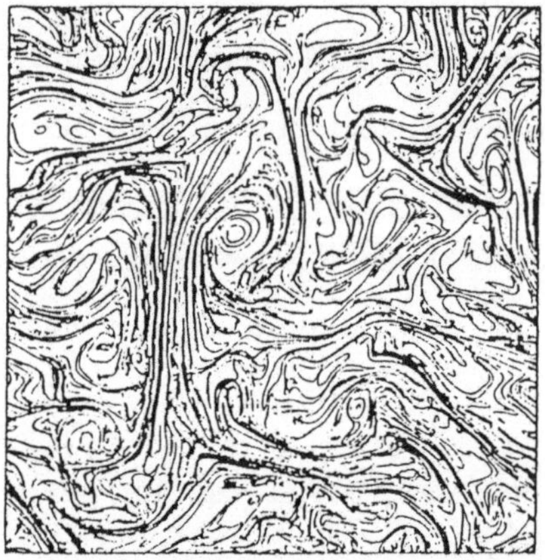

Fig. 3 Vorticity contours at t = 3 with ν = 0.0001 and R_L = 25500.

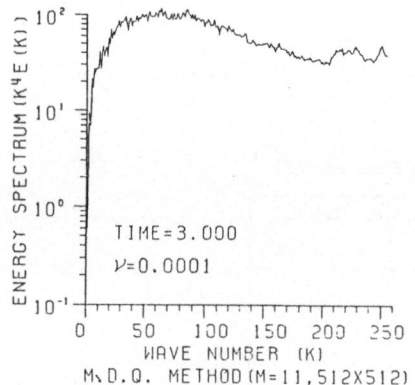

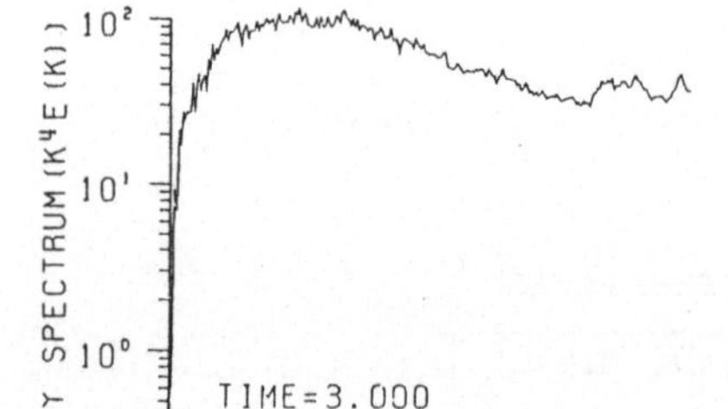

Fig. 4 Plot of $K^4E(K,t)$ vs. K at t = 3 with ν = 0.0001 and R_L = 25500.

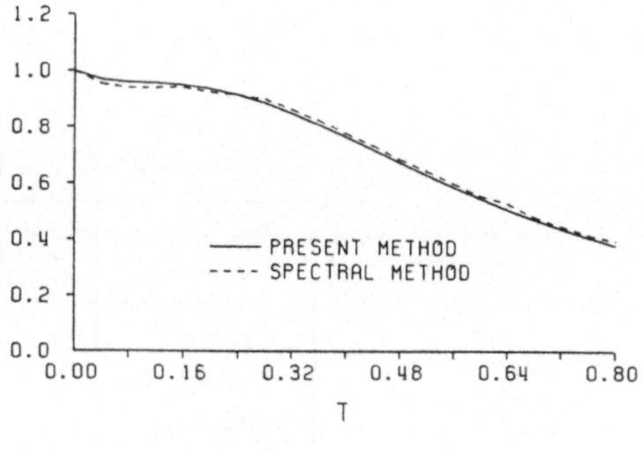

Fig. 5. Comparison of the energy dissipation rate with ν = 0.02.

A NUMERICAL STUDY OF THE FLUID DYNAMICS IN EXTRACTIONS COLUMNS

M. SCHOENAUER, Attaché de recherche, Centre de Mathématiques Appliquées, ECOLE POLYTECHNIQUE, 91128 Palaiseau, FRANCE.

W.S. YEUNG, Associate Professor, University of Lowell, Massachusetts, 01854, USA.

INTRODUCTION

Liquid-liquid extraction is a widely used method of separating the components of a solution by means of a suitably chosen solvent. (see Treybal, 1963 and Hansen,1971). The method is applied extensively in the extraction and purification of uranium. Figure 1 shows the schematic of the physical system. The solvent is injected at the bottom through a distributor, and the solution is fed at the top of the extraction column. A countercurrent flow is thus established. To maximize the extraction rates, the solvent is usually dispersed into droplets. Henceforth, we shall use the terms droplet phase and continuous phase for the solvent and solution respectively.

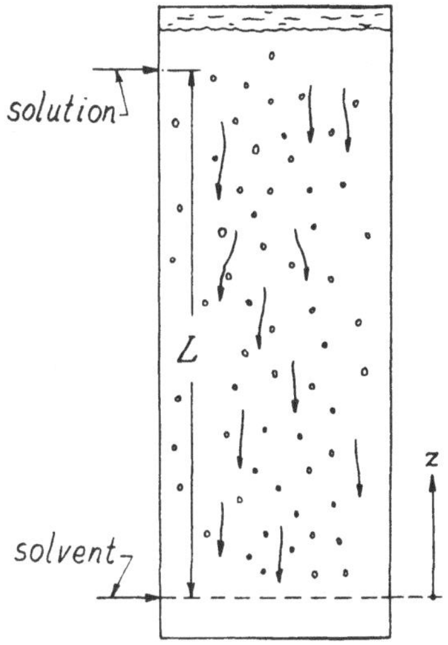

FIGURE 1. SCHEMATIC OF AN EXTRACTION COLUMN

Relatively few numerical studies have been made for the fluid dynamics of the two-phase (i,e, the droplet and the continuous phase) flow inside an extraction column. Most studies are for steady state operation and assume constant void fraction (defined as the volumetric fraction of the droplets) and constant phasic velocities throughout the entire column. However, experiments by Jiricny et al. (1979) indicated that the void fraction can vary as much as 100 % along the column. They also pointed out that interactions among droplets may be significant.

This paper discusses a transient one-dimensional model for the fluid dynamics inside an extraction column. We do not assume constant void fraction or phasic velocities. The droplet phase is taken as polydispersed with splitting and coalescence included in the model.

MATHEMATICAL MODEL

Consider countercurrent flow of a polydispersed droplet phase and a continuous phase. We shall discretize the droplet size distribution to N groups. Each size group is treated as a distinct phase. In addition, we assume that the motion of both phases is negligibly affected by the mass transfer, and the material density of each phase is constant.

One important phenomenon in extraction columns is longitudinal mixing, or backmixing, due to various processes such as entrainment in the wakes of the droplets and turbulence. To account for backmixing, we shall use a phenomenlogical approach similar to the "backflow model" (see Hansen, 1971) used sucessfully in mass transfer calculations. In this approach, a backflow velocity is assumed for each phase and is superimposed on the phasic velocity. This backflow velocity is postulated to be proportional to the void fraction gradient of the corresponding phase.

Figure 2 shows the control volumes used in the present analysis. Denote V as the phasic velocity and B as the backflow velocity. Subscripts c and i respectively refer to the continuous and the i-th size droplet phase. Our sign convention is as follows : the velocity of the continuous phase is positive downwards, while its backflow is positive upwards ; and the velocity of the droplet phase is positive upwards while its backflow is positive downwards. The independent variables are the time, t and the spatial coordinate, z, taken positively upwards. The length of the column is denoted by L.

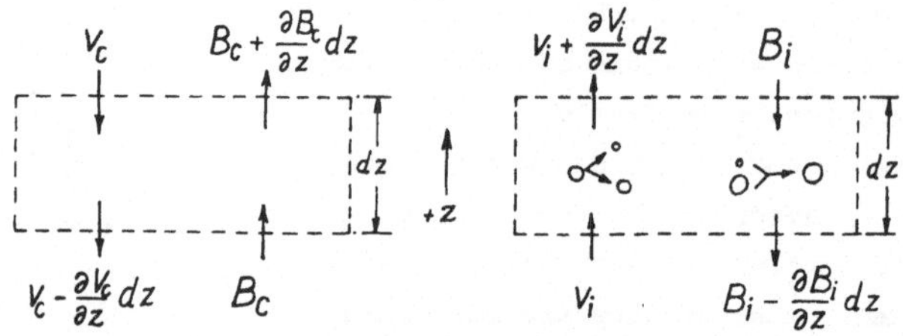

FIGURE 2. CONTROL VOLUMES

In accordance with the aformentioned backflow model, we have

$$B_c = -D_c \, \partial(1-\beta)/\partial z = D_c \, \partial\beta/\partial z \quad (1)$$

$$B_i = D_i \, \partial\beta_i/\partial z \quad (2)$$

where D_c and D_i are empirical backmixing coefficients, β_i is the void fraction of the ith size droplet, and β is the void fraction of all droplets, given by

$$\beta = \sum_{i=1}^{N} \beta_i \quad (3)$$

Performing a simple mass balance on the control volumes, we arrive at the following continuity equations :

$$\frac{\partial\beta}{\partial t} + \frac{\partial}{\partial z}\left(V_c(1-\beta) - D_c \frac{\partial\beta}{\partial z} \right) = 0 \quad (4)$$

$$\frac{\partial\beta_i}{\partial t} + \frac{\partial}{\partial z}\left(V_i\beta_i - D_i \frac{\partial\beta_i}{\partial z} \right) = \Gamma_i \quad (5)$$

Γ_i is the source term for the ith size droplet phase due to splitting and coalescence. We adopt the model proposed by Jiricny et al (1979), which gives

$$\Gamma_i = \sum_{j=1}^{N} a_{ij} \beta_j + \sum_{j,k=1}^{N} c_{ijk} \beta_j \beta_k \qquad (6)$$

where a_{ij} and c_{ijk} are probability coefficients.

The phasic velocities are related through a relative velocity correlation introduced by Wallis (1969)

$$V_C + V_i = V_i^{\infty} (1-\beta)^k \qquad (7)$$

where V_i^{∞} is the terminal velocity of a single ith size droplet, and k is an empirical constant.

For the boundary conditions, we propose the following :

$$\beta_i = \beta_i^{+}(t) \qquad \text{at } z=0 \qquad (8)$$

$$V_C(1-\beta) = Q_F(t) \qquad \text{at } z=L \qquad (9)$$

$$\partial\beta_i/\partial z = 0 \qquad \text{at } z=L \qquad (10)$$

The functions $\beta_i^{+}(t)$ are governed by the dynamics of the solvent distributor, and $Q_F(t)$ is the given volumetric flux of the continuous phase. Equation (10) is postulated based on the assumption that droplet splitting and coalescence are minimal in the vicinity of z=L. The initial conditions are given by the initial steady state or any other imposed states. Thus

$$\beta_i(0,z) = \beta_i^{0}(z) \qquad 0<z<L \qquad (11)$$

TRANSFORMATION OF THE PROBLEM

The basic formulation given by equations (3) to (7) is quite complicated for numerical computations. We have transformed the basic formulation to a parabolic system which is fully explicit in β_i and is more amenable to numerical calculation. We briefly describe the transformation : Equation (5) is summed over i, noting that $\sum \Gamma_i = 0$, and combined with equation (4) to yield :

$$V_C = \sum_{i=1}^{N} [\, V_i^{\infty}\beta_i(1-\beta)^k + (D_C - D_i)\frac{\partial\beta_i}{\partial z}\,] + \phi(t) \qquad (12)$$

where $\phi(t)$ is a function of time which can be obtained by applying (12) at z=L and using equations (9) and (10) :

$$\phi(t) = -\sum_{i=1}^{N} [\, V_i^{\infty}\beta_i(L,t)\,(1-\beta(L,t))^k\,] + \frac{Q_F(t)}{1-\beta(L,t)} \qquad (13)$$

Finally, substituting V_C into equation (4), we obtain the desired parabolic system

$$\frac{\partial \beta_i}{\partial t} - \frac{\partial}{\partial z} \left[D_i \frac{\partial \beta_i}{\partial z} + \sum_{j=1}^{N} (D_C - D_j) \beta_i \frac{\partial \beta_j}{\partial z} \right] +$$

$$\frac{\partial}{\partial z} \left[(1-\beta)^k (V_i^{\infty} - \sum_{j=1}^{N} V_j^{\infty} \beta_j) \beta_i \right] + \phi(t) \frac{\partial \beta_i}{\partial z} = \Gamma_i \qquad (14)$$

for $z \in (0,L)$, $t>0$, and $i \in [1,N]$.

The problem we shall work on is equation (14), subject to the initial condition (11) and boundary conditions (8) and (10), together with equation (13) for $\phi(t)$ and equation (6) for Γ_i.

NUMERICAL METHOD

Equation (14) represents a quasi linear parabolic system, which happens to be well-posed, although it has non-standard terms involving traces of the unknowns on the boundary (see Schoenauer,1984). Thus, we may apply a standard finite element method : the so-called linearized Crank-Nicholson-Galerkin method (see Thomee,1980).

We choose for the approximation space V_h the space of piecewise linear functions on our grid $(0,\Delta z, \ldots ,M\Delta z=L)$. For the nonlinear terms in equation (14) at time $n\Delta t$, we replace β_j by $3/2[\beta_j((n-1)\Delta t)]-1/2[\beta_j((n-2)\Delta t)]$.

We multiply the so-modified equation (14) by test functions ψ_p, $p=1,\ldots M$, (the usual base for V_h) and integrate over the column to obtain an M.N system of linear algebraic equations of centered finite-difference type, whose unknowns are the nodal values of the β_i, $i=1,\ldots,N$. We solve this system by Gauss elimination. It is useful to notice that the matrix of the system, although non-symmetric, consists of N.N blocks of tridiagonal submatrices of order M. This enables us to use Gauss elimination by blocks, or iterative methods.

RESULTS AND DISCUSSION

We present here (see Figure 3) some qualitative results we obtained using a simplified scheme (the integral of the finite elements method are approximated rather than exactly calculated). We have 5 different sizes of droplets, and 21 nodes in an 80 cm column, while the time step is 0.1 second. The volumetric flow rate of the continuous phase is 2 cm/s. The initial conditions are constant void fractions β_i throughout the column (.02, .06, .16, .06, and .02 respectively), and the boundary conditions at $z=0$ linearly decrease from β_i to $\beta_i/2$ in 1s (10 time steps), and remain constant thereafter. The following plots represent the different void fractions as functions of z at different times t.

We notice on the curves two phenomena. Firstly, far away from the bottom of the column, the void fractions tend to the equilibrium values consistent with the imposed rate of splitting and coalescence (the bold horizontal lines in the first four plots recall the initial levels of the void fractions). In our case, this is

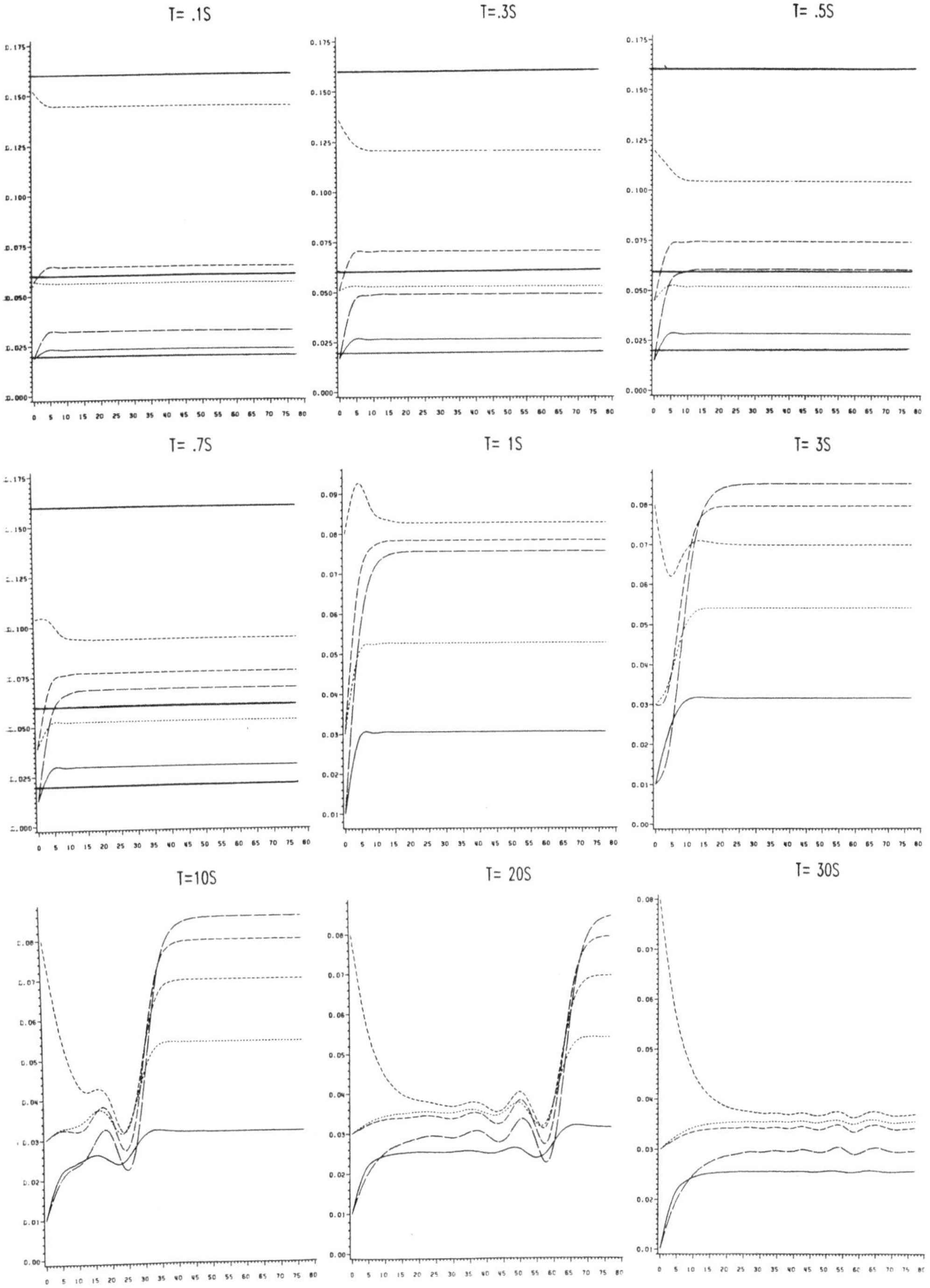

FIGURE 3 : TRANSIENT VOID FRACTIONS DEVELOPMENT.

——— β_1 ------- β_2 - - - - β_3 — — — β_4 -·—·- β_5

achieved in about 3 seconds (notice the change of scale at time t=1s). Secondly, the perturbation of the inflow rate of droplets at z=0 propagates upwards. Finally, this perturbation reaches the top of the column, which leads to a new equilibrium state corresponding to the prescribed boundary conditions. Some numerical oscillations appear, after about 100 time steps. We mention that these oscillations disappear for greater time t.

A few remarks on the influence of some parameters on the simulations :

- the splitting/coalescence have the expected effect. For instance, the same simulation without splitting/coalescence, is exactly what it is supposed to be : the constant initial values do not change downstream of the perturbation, and the steady state is constant void fractions throughout the column, equal to the ones injected at z=0.
- the longitudinal mixing coefficients affect the transient states, but not the steady state near the top of the column, which is only governed by the rate of splitting/coalescence.
- a change in the ratio of the time step and the space mesh (i.e. the Courant number) does not seem to avoid the oscillations.

CONCLUSION

We have presented here a new model for the fluid dynamics of an extraction column, on which we can apply standard schemes to obtain numerical values, for both the steady state and the time evolution of the column. The scheme we used is one of the simplest one which can be applied, and yet gives reasonable results for the first several time steps, and very good stability as the steady state is reached. Some exact finite elements method, with possible mesh refinment near z=0, or some upwind scheme might give better accuracy on the transient states and avoid the numerical oscillations.

W.S. Yeung is greatful to Dr. J.P. Boujot, of CISI at CEN Saclay, France, for support of this research project.

REFERENCES

C. HANSEN Recent advances in Liquid-Liquid extraction. Pergamon Press. 1971

V. JIRICNY, M. KRATKY, J. PROCHAZKA Counter current flow of dispersed and continuous phase. I. Chem. Eng. Sci. 34 (1979) pp 1141-1149.

M. SCHOENAUER An existence result for a parabolic system arising in liquid liquid extraction. To appear.

V. THOMEE Galerkin Finite Elements Methods for Parabolic Problems. Springer Verlag. 1984.

R.E. TREYBAL Liquid Extraction. McGraw Hill Book Co. 1963.

G.B. WALLIS One Dimensionnal Two-phase Flow. McGraw Hill Book Co. 1969.

MULTIGRID SOLUTION OF THE NAVIER-STOKES EQUATIONS FOR THE FLOW IN A RAPIDLY ROTATING CYLINDER

W. Schröder and D. Hänel
Aerodynamisches Institut. RWTH Aachen
Aachen, Germany

Summary

The steady axisymmetric flow in a gas centrifuge is calculated by a finite-difference solution of the linearized Navier-Stokes equations. For these equations a streamfunction-vorticity formulation could be derived with a structure similar to those for incompressible flow. The equations are solved with a multigrid method. The investigation covers the numerical problems of optimization of the algorithm, the implementation of the boundary conditions and the treatment of boundary-layer regions by means of subgrids. The results show that the multigrid algorithm improves the rate of convergence of the solution for the present problem substantially.

Physical Problem and Governing Equations

The steady axisymmetric flow of a compressible viscous gas between two concentric cylinders rotating with an angular velocity Ω and closed at both ends is investigated by means of a numerical solution of the Navier-Stokes equations. The calculation of the secondary motion generated by differences of temperature or angular velocity on the walls is of great interest for the construction of gas centrifuges for the enrichment of uranium isotopes [1,2].Typical for such flows are the strong radial changes of density and pressure due to solid body rotation and the secondary countercurrent flow with thin boundary layers along the endcaps (Ekman-layers) and the rotor wall (Stewartson-layers). The assumption used most often is that the perturbations about the solid body rotation are weak and as a consequence the Navier-Stokes equations can be linearized [1]. For these equations a streamfunction-vorticity formulation could be derived which reduces the computational effort. The equations for perturbated quantities are:

$$\left(\frac{1}{rE_R}\psi_r\right)_r + \left(\frac{1}{rE_R}\psi_z\right)_z - \omega = 0 \tag{1}$$

$$\left(\frac{1}{r}(rv)_r\right)_r + v_{zz} - 2uE_R/E_K = 0 \tag{2}$$

$$\frac{1}{r}(rT_r)_r + T_{zz} + 4\,Br\,r\,u\,E_R/E_K = 0 \tag{3}$$

$$\left(\frac{1}{r}(r\omega)_r\right)_r + \omega_{zz} - (rT - 2v)_z E_R/E_K = 0 \tag{4}$$

Herein $\omega = u_z - w_r$ is the vorticity component, ψ a streamfunction, defined by $\psi_z = E_R ur$ and $\psi_r = -E_R wr$. The parameters E_K and Br are the Ekman and the Brinkman number, resp., and $E_R(r)$ is the radial density distribution of the solid body rotation which changes over 7 orders of magnitude for high peripheral speed. In the derivation of equations (1) to (4) an additional assumption was made, namely that the axial pressure gradient in Eq. (4) can be neglected in comparison to T_z and V_z, which is justified in thermally and mechanically driven flows. A comparison with a solution of the full equations showed no differences for the flows considered. The boundary conditions of the problem are

$$u = w = \psi = 0$$

$$T = T_w(r,z) \qquad v = v_w(r,z) \qquad \text{(on all rigid walls)}$$

$$\omega = \left(\frac{1}{rE_R}\psi_z\right)_z \qquad \text{(on the endcaps)} \tag{5}$$

$$\omega = \left(\frac{1}{rE_R}\psi_r\right)_r \qquad \text{(on the cylinder walls)}$$

Method of Solution

To solve the elliptic set of equations (1) - (4) by a multigrid method, the full approximation storage mode of operation [3] was chosen because it is well suited for nonlinear problems and for composite grid arrangements. The SLOR method with alternating directions (ADI-SLOR) was used for the relaxation process. The interpolation of the residuals and of the variables from the fine to the coarse grid was carried out by the full residual weighting operator and the interpolation from the coarse to the fine grid by bi-linear or cubic interpolation.

In test calculations the number of grids and of relaxation sweeps and the type of relaxation schemes and of the interpolation operators were varied to minimize the computational time.

For the Poisseuille flow, Fig. 1 shows the influence of the number of relaxation sweeps on the efficiency of the algorithm. The number of work units which are a measure for the computational effort decreases by a factor of about 2 if the number of sweeps is reduced from 10 to 5.

In the course of the investigation it was found that the numerical representation of the boundary conditions for the vorticity has a very strong influence on the rate of convergence. These conditions in Eq. (5) can be discretized for z = 0, e.g., by

$$\left.\frac{1}{E_R r}\right|_i \frac{2}{\Delta z^2}(\psi_{i,2}-\psi_{i,1}) - \omega_{i,1} = 0 \tag{6}$$

In single-grid methods this condition is used as a Dirichlet condition for ω with ψ known from the previous iteration step. In the multigrid solution this approach leads to a poor rate of convergence and for that reason condition (6) was implemented in a way consistent with the multigrid algorithm. The residuals of the discretized condition (6) and the discrete operator of the boundary condition were transfered to the coarser grid in a way similar to the treatment of the interior points. This approach preserves the essential informations of the fine grid in the coarse grid. This formulation ensured convergence of all cases considered.

Results

The flow in a gas centrifuge at a high peripheral speed is characterized by strong radial gradients of pressure and density and by very thin boundary layers along the endcaps and along the side wall. As an example, the radial distribution of the axial mass flux in the midplane of the cylinder is shown in Fig. 2. This flow was generated by a linear temperature distribution along the side wall. Typically, the axial velocity inverts its direction within the side wall boundary layer and most of the mass is concentrated near the side wall. The correct resolution of such boundary layer structures is a crucial test for the multigrid method. The results have shown that the multigrid method becomes inefficient or does not even converge with increasing Reynolds numbers if uniform grids are used. One reason may be that the boundary layer thickness, which is the smallest characteristic length scale, restricts the coarsest mesh spacing. This problem could be solved by the use of composite grids with special subgrids in the boundary layer. Two arrangements of subgrids were studied. The first, sketched in Fig. 3 consists of the finest grid for the entire domain and coarser grids outside of the boundary layers. In the second arrangement, Fig. 4, the coarser grids are extended over the entire domain. In the boundary layer finer subgrids are used. For both arrangements convergence could be achieved for Reynolds numbers where uniform grids failed. Substantial differences were found in the computational effort, needed for convergence. The second arrangement leads to considerable less computational work in comparison to the first arrangement and to the single-grid solution. The efficiency of the MG algorithm with the second subgrid arrangement is demonstrated in Fig. 5a and 5b where the maximum residual of Eq. (4) is plotted against the work units which are a measure for the CPU-time. In comparison to the single-grid solution (with ADI-SLOR) the MG solution with 4 grids including 2 subgrid levels needs about 1/5 of the CPU-time for Reynolds numbers of 10^3 and 10^4.

References

[1] Villani, S. (Ed.): Uranium Enrichment. Topics in Appl. Physics, Vol. 35, Springer-Verlag, Berlin, Heidelberg, New York, 1979.

[2] Merten, A., Hänel, D.: Navier-Stokes Solution for Compressible Flow in a Rotating Cylinder. Proc. of the Fourth Conf. on Num. Meth. in Fluid Mech., Notes on Num. Fluid Mech., Vol. 5, Vieweg-Verlag, Braunschweig, 1982.

[3] Brandt, A.: Multilevel Adaptive Solutions to Boundary-Value Problems. Math. Comput., Vol. 31, pp. 333-390, 1977.

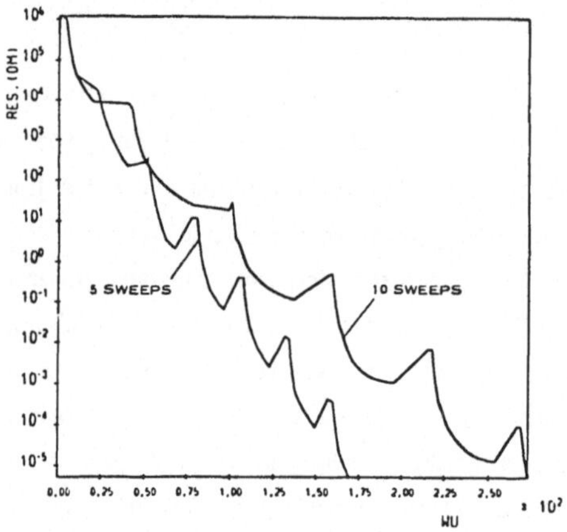

Fig. 1: Maximum residual of the vorticity transport equation as function of the work units for different numbers of relaxation sweeps

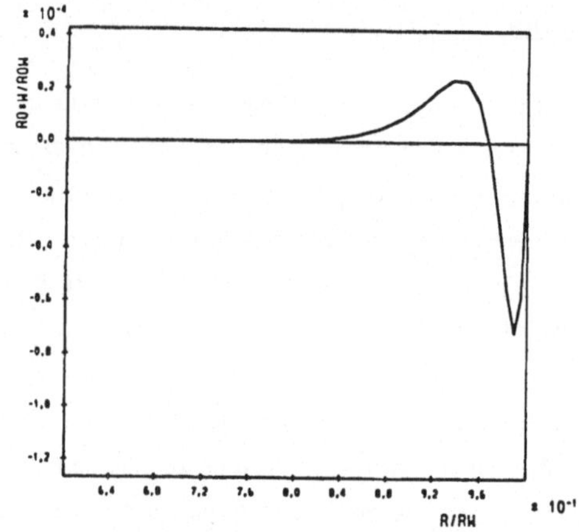

Fig. 2: Radial distribution of the axial mass flux $\rho w/\rho_0(r_0)v_0$ in the midplane of a centrifuge with v_0 = 600 m/s, Re = 10^3 in the case of a wall thermal drive

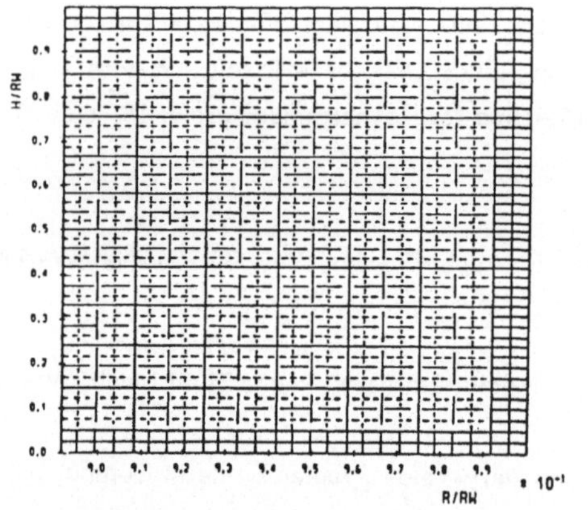

Fig. 3: Subgrid arrangement I
- global fine grid
- coarse grids outside the boundary layer

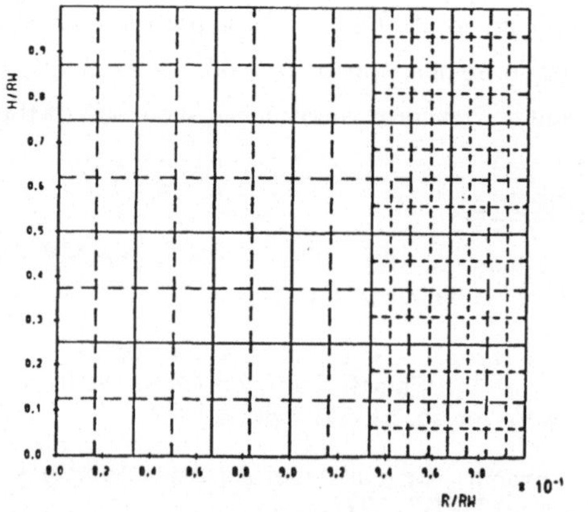

Fig. 4: Subgrid arrangement II
- fine subgrids only in the boundary layer near the side wall

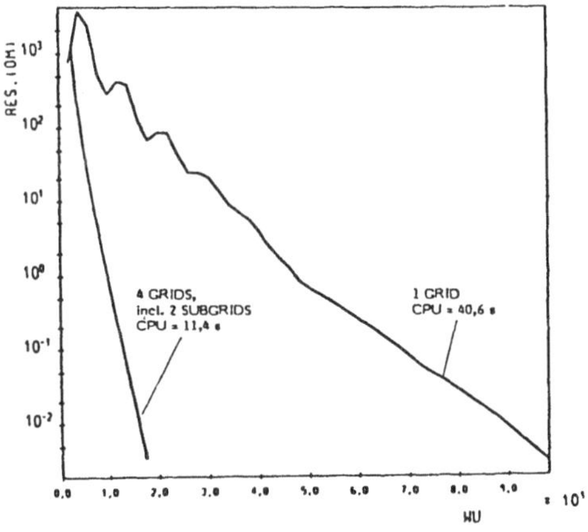

Fig. 5a: v_0 = 600 m/s, Re = 10^3

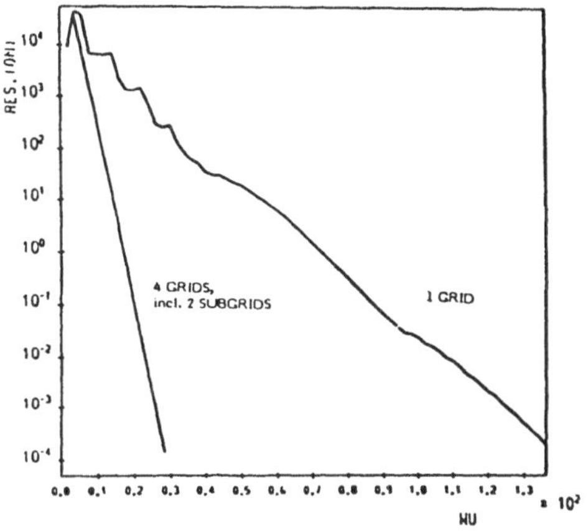

Fig. 5b: 600 m/s, Re = 10^4

Maximum residual of the vorticity transport equation as function of the work units

ALGORITHMS FOR DIRECT NUMERICAL SIMULATION OF SHEAR-PERIODIC TURBULENCE

U. Schumann
DFVLR Oberpfaffenhofen
Institute of Atmospheric Physics
D-8031 Wessling, Fed. Rep. Germany

Introduction

This paper describes algorithms for computation of pressure field and for generation of initial conditions as required for direct numerical simulation of homogeneous turbulence generated by vertical shear and buoyancy. The turbulent flow will be simulated by direct numerical integration of the three-dimensional and time-dependent Navier-Stokes equations for viscous and conductive incompressible fluid with constant material properties. Buoyancy is taken into account according to the Boussinesq approximation as induced by a mean vertical temperature gradient. The objective of this work (in progress) is to determine the structure of turbulent shear flow under stable and unstable thermal stratification and to provide data for calibration of second order turbulence closure models.

Direct simulations with e.g. 64^3 grid cells are restricted to moderate Reynolds numbers. Typical values of the Reynolds number based on Taylor-microscale and the root-mean-square velocity are 40 to 60. In the past, isotropic and axisymmetric homogeneous turbulence have been considered. Such simulations have been successful in providing data for test of turbulence models [1, 2]. In these cases, periodic boundary conditions have been applied in all three coordinate directions.

For the case of shear, periodicity boundary conditions have to be changed with respect to the shear direction. Rogallo [3,2] simulated shear with an Lagrangian approach; he used a computational domain which deforms as a function of time according to the mean shear. In the sheared coordinate system periodic boundary conditions are appropriate. This approach requires remapping of all fields from time to time onto an undeformed domain and this introduces interpolation errors.

An alternative has been proposed by Baron [4]. He uses a fixed coordinate system but boundary conditions which we call "shear-periodic": If, for example, $f_B(x,y,t)$ is a field component at the bottom (z=0) of the computational domain and $f_T(x,y,t)$ is the field at the top of the computational domain (z=H), then shear-periodicity means

$$f_T(x,y,t) = f_B(x-Ut,y,t)+(\partial f/\partial z)H$$

where $U/H=(\partial u/\partial z)$ is the mean shear of horizontal velocity in x-direction. The expression $x-Ut$ is taken modulo the horizontal length of periodicity. This type of boundary condition can be implemented easily in a finite difference scheme, see Figure 1. It does not require any interpolation if the time step Δt and the grid spacing Δx are taken such that $U\Delta t/\Delta x=1$.

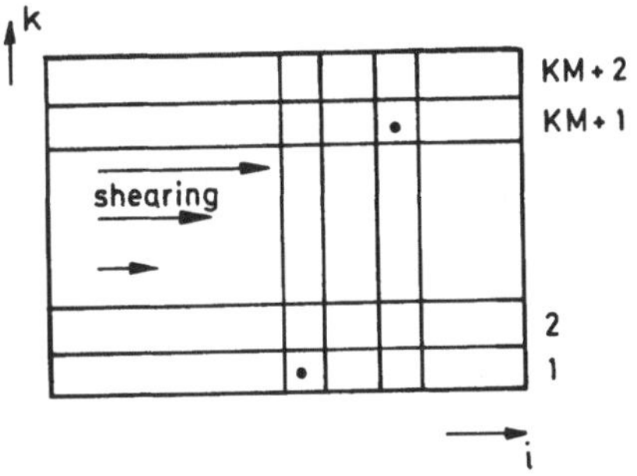

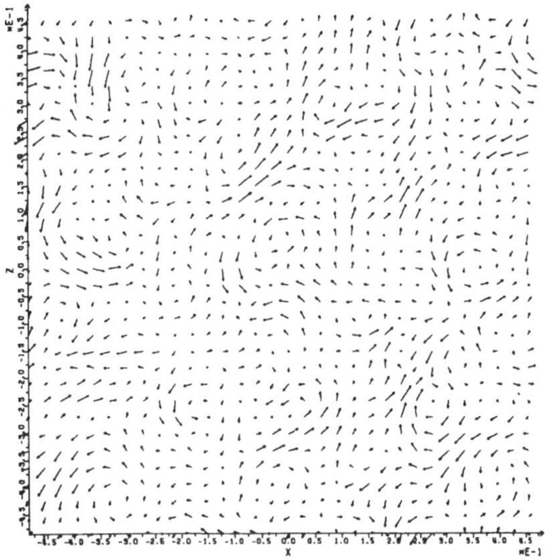

Figure 1. Left: Shear-periodic boundary condition in a finite difference grid. i and k denote the mesh indices in x and z direction; the cells marked with dots get equal values for a given value of $\mu=Ut/H$. KM $\equiv$ N.
Right: Velocity field generated for initial conditions; mean field subtracted.

On the other hand, shear periodicity is an unusual boundary condition and no fast Poisson solver - as required to compute the pressure field - is available. Baron used an iterative solution scheme which is very time-consuming. Subsequently, the new Poisson-solver and a new method to generate initial conditions is described briefly.

Direct Poisson-solver for shear-periodicity

For each time-step, we have to compute the pressure p(i,j,k) from the discrete Poisson equation

$$[p(i-1,j,k) - 2p(i,j,k) + p(i+1,j,k)]/\Delta x^2 +$$
$$[p(i,j-1,k) - 2p(i,j,k) + p(i,j+1,k)]/\Delta y^2 +$$
$$[p(i,j,k-1) - 2p(i,j,k) + p(i,j,k+1)]/\Delta z^2 = f(i,j,k), \qquad (1)$$

$i = 0,1,\ldots,N-1$; $j = 0,1,\ldots N-1$; $k= 1,2,\ldots,N$; $\Delta x=\Delta y=\Delta z=1/N$. The boundary conditions are periodic in i- and j-direction modulo N and shear-periodic in k-direction. The latter means

$$p(i,j,0) = p(i+\mu,j,N), \quad p(i,j,N+1) = p(i-\mu,j,1) \qquad (2)$$

Here μ is the integer value which corresponds to $Ut/\Delta x$. The solution method proceeds in three steps:

First, we determine the complex Fourier-modes and its conjugate pairs

$$(3)\quad f^*(m,n,k) = \frac{1}{N\,N} \sum_{i=0}^{N-1} \sum_{j=0}^{N-1} f(i,j,k)\, W_N^{-mi}\, W_N^{-nj}, \quad f^*(N-m,N-n,k)=[f^*(m,n,k)]^c,$$

$$(4)\quad W_M \equiv \exp\,[2\pi\,\sqrt{-1}/M],$$

for $m = 0,1,\ldots,N/2$; $n = 0,1,\ldots,N-1$, by means of a sequence of two-dimensional FFT's.

Second, with the ansatz

$$(5)\qquad p(i,j,k) = \sum_{m=0}^{N-1} \sum_{n=0}^{N-1} p^*(m,n,k)\, W_N^{mi}\, W_N^{nj}$$

we obtain a set of one-dimensional linear systems from eq.(1)

$$(6)\qquad [p^*(m,n,k-1) - 2p^*(m,n,k) + p^*(m,n,k+1)]/\Delta z^2 - \lambda_{m,n}\, p^*(m,n,k) = f^*(m,n,k),$$

$$(7)\qquad \lambda_{m,n} \equiv 2\,[1 - \cos(2\pi m/N)]/\Delta x^2 + 2\,[1 - \cos(2\pi n/N)]/\Delta y^2,$$

$m = 0,1,\ldots, N/2$, $n = 0,1,\ldots, N-1$, $k = 1,2,\ldots, N$, and from eq.(2) the boundary conditions

$$p^*(m,n,0) = W_N^{m\mu}\, p^*(m,n,N),$$

$$(8)\qquad p^*(m,n,N+1) = W_N^{-m\mu}\, p^*(m,n,1).$$

Subsequently we describe some aspects of the algorithm used to solve eq.(7-8) for each Fourier-mode-index (m,n). For shortness, we omit the indices m and n and define

$$\begin{aligned}
p^*(k) &\equiv p^*(m,n,k) \equiv p_R(k) + \sqrt{-1}\, p_I(k)\\
f^*(k) &\equiv f^*(m,n,k)\,\Delta z^2 \equiv f_R(k) + \sqrt{-1}\, f_I(k)\\
\alpha &\equiv -2 - \lambda_{m,n}\,\Delta z^2\\
W_N^{m\mu} &\equiv w_R + \sqrt{-1}\, w_I,\ w_R \equiv \cos(2\pi m\mu/N),\ w_I \equiv \sin(2\pi m\mu/N).
\end{aligned}$$

The complex system can be split into real and imaginary parts and ordered such that its band-width becomes minimal. This ordering is essential for effectiveness:

$$
(9)\qquad
\left[\begin{array}{cccccc|cccccc}
\alpha & 1 & & & & 1 & & & & & & \\
1 & \alpha & 1 & & & & & & & & & \\
 & \cdot & \cdot & \cdot & & & & & & & & \\
 & & 1 & \alpha & 1 & & & & & & & \\
 & & & 1 & \alpha & w_R & -w_I & & & & & \\
1 & & & & w_R & \alpha & & w_I & & & & \\
\hline
 & & & & -w_I & & \alpha & w_R & & & & 1 \\
 & & & & & w_I & w_R & \alpha & 1 & & & \\
 & & & & & & & 1 & \alpha & 1 & & \\
 & & & & & & & & \cdot & \cdot & \cdot & \\
 & & & & & & & & & 1 & \alpha & 1 \\
 & & & & & & 1 & & & & 1 & \alpha
\end{array}\right]
\left[\begin{array}{c}
p_R(2) \\ p_R(3) \\ \cdot \\ p_R(N-1) \\ p_R(N) \\ p_R(1) \\ \hline p_I(1) \\ p_I(N) \\ p_I(N-1) \\ \cdot \\ p_I(3) \\ p_I(2)
\end{array}\right]
=
\left[\begin{array}{c}
f_R(2) \\ f_R(3) \\ \cdot \\ f_R(N-1) \\ f_R(N) \\ f_R(1) \\ \hline f_I(1) \\ f_I(N) \\ f_I(N-1) \\ \cdot \\ f_I(3) \\ f_I(2)
\end{array}\right]
$$

This system is solved by Gaussian elimination. The elimination proceeds from the top line downwards and the bottom line upwards up to the equation in the centre. Hereby, the coefficients for the real and imaginary part are equal and this reduces the required computational effort.

For $N >> 1$, the number of multiplications is 5.5 and the number of divisions is 0.5 per grid point. This is the same operation-count one would have for the standard periodic case ($m\mu = 0$).

Third, from the solutions $p^*(m,n,k)$, we determine the final solution $p(i,j,k)$ according to eq.(5) again by a sequence of two-dimensional FFT's.

The algorithm is fully vectorized by interchanging loops over the indices such that the innermost loops are free of recursions. Typical computing times on a CRAY-1 are 0.36s for N=64. Details of the algorithm and a Fortran-subroutine are described in [5].

Generation of initial conditions

An important issue of all direct numerical simulations is the generation of suitable initial conditions. Here, an algorithm is described which generates random three-dimensional real initial fields for velocity (u_1,u_2,u_3), and temperatur T. The fields are generated such that the velocity field satisfies the discretized continuity equation on a staggered grid and such that the fields follow a prescribed correlation spectrum

$$
R_{ij}(k) = \sum_{k \le |\mathbf{k}| < k+1} \sum\sum f_i^*(\mathbf{k})\, f_j^*(-\mathbf{k})
$$

where $\mathbf{k}$ is the integer wave-number vector, (f_1,f_2,f_3,f_4) stands for (u_1,u_2,u_3,T), and the star indicates the complex Fourier mode. This spectrum can be selected arbitrarily except for certain realizability constraints [6].

In short, the idea is as follows: By Cholesky-decomposition we determine γ_{ij} such that $\gamma_{in}\gamma_{nj}=R''_{ij}$ and set $f^*_i(\mathbf{k}) = \gamma_{ij}\alpha_j(\mathbf{k})$, where $\alpha_j(\mathbf{k})$ are e.g. Gaussian random complex numbers with zero mean and

$$\sum\sum\sum_{k\leq|\mathbf{k}|<k+1} \alpha_i(\mathbf{k})\,\alpha_j(-\mathbf{k}) = \delta_{ij}.$$

Thereafter the Fourier modes are corrected to satisfy the Fourier-transformed continuity equation. This correction reduces the degree of anisotropy. The reduction is balanced by setting

$$(10)\qquad \begin{bmatrix} R''_{11} \\ R''_{22} \\ R''_{33} \end{bmatrix} = (1/14) \begin{bmatrix} 27 & -3 & -3 \\ -3 & 27 & -3 \\ -3 & -3 & 27 \end{bmatrix} \begin{bmatrix} R_{11} \\ R_{22} \\ R_{33} \end{bmatrix},$$

$$(R''_{12},R''_{13},R''_{23}) = (15/7)\ (R_{12},R_{13},R_{23}),$$

$$(R''_{14},R''_{24},R''_{34}) = (3/2)\ (R_{14},R_{24},R_{34}),$$

$$R''_{44} = R_{44}, \qquad\qquad R''_{ij} = R''_{ji}.$$

The anisotropy of R_{ij} is limited by the requirement that R''_{ij} has to be positive definite. By FFT the real fields are obtained. Details are given in [6].

For example, Figure 1 shows a velocity field generated with this algorithm for N=32. Shown are the u_1- and u_2-components in a x_1-x_2-plane. The velocity field has correlation spectra of the form $R_{ij}(k) \sim (k/k_p)^4\exp(-2(k/k_p)^2)$ with $k_p=6$, H=1, and $R_{11}=R_{22}=R_{33}$, $R_{12}=0.5R_{11}$, $R_{13}=R_{23}=0$. The 50% correlation between u_1 and u_2 and the energy-maximum for a wave-length of H/6 can easily be seen from the figure.

References

[1] U. Schumann & G.S.Patterson, Jr.: Numerical Study of Pressure and Velocity Fluctuations in Nearly Isotropic Turbulence. J. Fluid Mech. 88 (1978) 685-709.

[2] R.M. Kerr: Higher Order Derivative Correlation of Velocity and Temperature in Isotropic and Sheared Numerical Turbulence. 4th Symp. Turbulent Shear Flows, Sept. 12-14, 83, Karlsruhe, 14.9-12.

[3] R.S. Rogallo: Numerical experiments in homogeneous turbulence. NASA Technical Memorandum 81315 (1981).

[4] F. Baron: Macro-Simulation Tridimensionelle d'ecoulements turbulents cisailles. These de Docteur-Ingenieur, Univ. Pierre et Marie Curie, Paris 6 (1982).

[5] H. Schmidt et al.: Three dimensional direct and vectorized elliptic solvers for various boundary conditions. DFVLR-Mitteilung, to appear (1984).

[6] U. Schumann: Generation of random periodic velocity and temperature fields with prescribed correlation spectra. DFVLR-IB-553-6/84 (1984).

STEADY AND UNSTEADY NONLINEAR FLOW TREATMENT USING THE FULL POTENTIAL EQUATION

Vijaya Shankar, Kuo-Yen Szema, Joseph Gorski and Hiroshi Ide
Rockwell International Science Center
Thousand Oaks, CA 91360

Abstract

The steady and unsteady forms of the full potential equation are treated using implicit numerical methods based on the approximate factorization technique or relaxation concepts. Problems solved include supersonic flows over complex configurations with embedded subsonic regions, and flows over airfoils and spheres at all Mach numbers. The treatment involves time linearization of density, flux linearization in the marching direction, theory of characteristic signal propagation, steady and unsteady wake treatment, flux biasing concepts for artificial viscosity, and unsteady far-field based on the Riemann invariants.

Introduction

Nonlinear aerodynamics prediction methods based on the full potential equation are still very attractive for treating complex shapes because they require less computer time and memory than the Euler solvers. The full potential algorithms[1-6] are continually improved in their robustness and computational speed to enable treatment of complex mixed flows over range of Mach numbers, subsonic to supersonic. The improvements to the full potential methods largely come from concepts based on the theory of characteristic signal propagation. These include a conservative switching operator to handle supersonic flows with embedded subsonic regions, flux biasing and flux linearization procedures, better unsteady wake treatment, and far-field expressions based on the Riemann invariants.

There are two aspects that will be stressed in this paper: 1) development of approximate factorization and relaxation schemes for the unsteady full potential equation to treat steady and unsteady flows at all Mach numbers (subsonic, transonic and supersonic) within the limitations of the theory, and 2) development of a supersonic marching algorithm for the steady full potential equation with a built-in logic to detect and treat the embedded subsonic regions. Even though the unsteady algorithm, item 1 above, can handle all Mach number flows, development of a steady marching algorithm, item 2, is more efficient and a practical way to solve supersonic problems.

Methodology

The procedure for the unsteady full potential equation will be first described, followed by the marching treatment of the steady equation.

Treatment of the Unsteady Equation

The two-dimensional/axisymmetric unsteady full potential equation written in a body-fitted coordinate system represented by $\tau = t$, $\zeta = \zeta(x,y,t)$ and $\eta = \eta(x,y,t)$ takes the form

$$\left(\frac{\rho}{J}\right)_\tau + \left(\rho\frac{U}{J}\right)_\zeta + \left(\rho\frac{V}{J}\right)_\eta + S\rho\frac{v}{yJ} = 0 \tag{1}$$

where,

$S = 0$ for two dimensions, $= 1$ for axisymmetric

$\rho = \text{density} = [1 - \frac{\gamma-1}{2} M_\infty^2 (2\phi_\tau + \bar{U}\phi_\zeta + \bar{V}\phi_\eta - 1)]^{1/(\gamma-1)}$

$U = \zeta_t + a_{11}\phi_\zeta + a_{12}\phi_\eta$; $U = \bar{U} + \zeta_\tau$

$V = \eta_t + a_{12}\phi_\zeta + a_{22}\phi_\eta$; $\bar{V} = V + \eta_t$

The quantities a_{11}, a_{12} and a_{22} are functions of the transformation metrics, and J is the Jacobian.

The density ρ and the fluxes ρU and ρV are complicated nonlinear functions of ϕ, the velocity potential. Hence, to solve for ϕ from Eq. (1) will require a local linearization. The time linearization of density is given by

$$\rho^{n+1} = \rho^n + \left(\frac{\partial\rho}{\partial\phi}\right)^n \Delta\phi + .. = \rho^n - \frac{\rho^n}{a^2}\left(\frac{\partial}{\partial\tau} + U\frac{\partial}{\partial\zeta} + V\frac{\partial}{\partial\eta}\right)\Delta\phi \tag{2}$$

where $\Delta\phi = \phi^{n+1} - \phi^n$, and $\left(\frac{\partial\rho}{\partial\phi}\right)$ is a differential operator. To maintain conservation, the linearization of density is done before the time derivative is applied. The spatial derivative terms are modelled with a modified density based on flux biasing.

$$\left(\rho\frac{U}{J}\right)_\zeta^{n+1} = \frac{\overleftarrow{\delta}}{\partial\zeta}\left(\frac{\tilde{\rho}_{j,k+1/2}\, U^{n+1}_{j,k+1/2}}{J_{j,k+1/2}}\right) \tag{3}$$

where $\overleftarrow{\delta}$ stands for backward differencing. The biased density $\tilde{\rho}$ can be defined by the following three ways.

$$\text{(a)}\quad \tilde{\rho} = \rho - \max\left(0, 1 - \frac{1}{M^2}\right)\Delta\zeta\frac{\overleftarrow{\delta}\rho}{\partial\zeta}$$

$$\text{(b)}\quad \tilde{\rho} = \frac{1}{q}\left\{\rho q - \frac{\partial}{\partial\zeta}(\rho q)^-\right\} \tag{4}$$

$$\text{(c)}\quad \tilde{\rho} = \frac{1}{q}\left\{\rho q - \Delta\zeta\frac{U}{Q}\frac{\partial}{\partial\zeta}(\rho q)^- - \Delta\eta\frac{V}{Q}\frac{\partial}{\partial\eta}(\rho q)^-\right\}$$

The definition (a) above is the usual density biasing method.[7] Items (b) and (c) are termed flux biasing[8] and are more accurate and robust than the one in Item (a). Item (b) is a directional flux biasing, while Item (c) is along a streamline direction. The quantity Q is defined to be $\sqrt{U^2 + V^2}$. The term $(\rho q)^-$ is expressed as

$$
\begin{aligned}
(\rho q)^- &= \rho q - \rho^* q^* \quad \text{if} \quad q > q^* \\
&= 0 \ . \qquad\qquad \text{if} \quad q < q^*
\end{aligned}
\tag{5}
$$

The quantities ρ^* and q^* represent sonic conditions given by

$$
\begin{aligned}
(q^*)^2 &= \frac{1 + \frac{\gamma-1}{2} M_\infty^2 \, (1 - 2\phi_\tau - \zeta_t\phi_\zeta - \eta_t\phi_\eta)}{\frac{\gamma+1}{2} M_\infty^2} \\
\rho^* &= (q^* \, M_\infty)^{2/(\gamma-1)}
\end{aligned}
\tag{6}
$$

Use of sonic conditions in defining the biased value of density, Eq. (4), provides a more robust treatment of mixed flows with shocks and sonic lines.[5,6]

While solving airfoil problems with a wake behind the trailing edge, Fig. 1, it is essential to properly model the jump in ϕ across the wake by satisfying the density continuity across it. Barring details, it can be shown[6] referring to Fig. 1, the following:

$$
(\phi_\eta)_{\substack{\text{upper}\\ \text{wake}}} = \frac{\phi_3 - (\phi_{\bar{3}} + \Gamma + [\phi_{\eta\eta}]/2)}{2} \tag{7}
$$

where, the jump $[\phi_{\eta\eta}] = (\phi_{\eta\eta})_{upper} - (\phi_{\eta\eta})_{lower}$ is given by

$$
[\phi_{\eta\eta}] = -(\rho \, a_{11} \, \Gamma_\zeta)_\zeta / \rho \, a_{22} \tag{8}
$$

and the jump $[\phi] = \phi_u - \phi_\ell = \Gamma$ is obtained by solving

$$
2\Gamma_t + (U_u + U_\ell)(\phi_\zeta)_u + U_\ell \, \Gamma_\zeta = 0 \tag{9}
$$

with the initial value of Γ obtained at the trailing edge. For steady state flows, Eqs. (7-9) result in $[\phi_{\eta\eta}] = 0$, and Γ is constant along the wake.

Along the outer boundary A-B-C-D-E in Fig. 1, to avoid wave reflections, appropriate Riemann invariants are prescribed. Details can be found in Ref. 6.

Equation (1) can be solved either by approximate factorization methods[5] or by relaxation methods.[6]

<u>Treatment of the Steady Equation</u>

The steady form of the full potential equation is used to treat predominantly supersonic flows with embedded subsonic regions, using a marching algorithm.[1-4] Only a brief description here, and details can be found in Ref. 3.

The steady equation is written in the form

$$(\rho\,\frac{U}{J})_{\zeta} + (\rho\,\frac{V}{J})_{\eta} + (\rho\,\frac{W}{J})_{\xi} = 0 \qquad (10)$$

Using the characteristic theory, the direction ζ can be classified as either hyperbolic or elliptic depending on the sign of the quantity $(a_{11} - U^2/a^2)$. When this quantity is positive, the direction ζ is elliptic, and when negative ζ is hyperbolic. When ζ is elliptic, which is the case inside a subsonic bubble, relaxation methods have to be employed. Through a switching parameter θ, both the supersonic marching operator and the subsonic relaxation operator can be combined together as

$$\frac{\partial}{\partial\zeta}(\rho\,\frac{U}{J}) = \underbrace{\theta_i\,\frac{\overleftarrow{\delta}}{\partial\zeta}(\rho\,\frac{U}{J})_{i+1}}_{\text{supersonic}} + \underbrace{(1-\theta_{i+1})\,\frac{\overrightarrow{\delta}}{\partial\zeta}(\tilde{\rho}\,\frac{U}{J})_{i+1}}_{\text{marching subsonic}}$$

where $\overleftarrow{\delta}$ and $\overrightarrow{\delta}$ refer to backward and forward differencing, respectively. $\theta_i = 1$ if $(a_{11} - U^2/a^2) < 0$, and if $(a_{11} - U^2/a^2) > 0$. The forms of the supersonic and subsonic operators are given in Ref. 3.

The terms $(\rho\,\frac{V}{J})_{\eta}$ and $(\rho\,\frac{W}{J})_{\zeta}$ are the crossflow terms and are treated using a transonic density biasing scheme similar to the one described in Eq. (4).

Results

Figure 2 shows the results for the NACA0012 airfoil at $M_\infty = 0.75$, $\alpha = 2°$ performed using the unsteady method of this paper. The pressure distribution on the airfoil is obtained without having to use any user specified "constants," such as the ones described in Ref. 7, and shows no overshoot or smearing near the shock.

Figure 3 shows the schematic of a supersonic fighter with an embedded subsonic region. This type of flow can be easily handled using the marching technique applied to the steady equations. Figure 4 shows the entire surface pressure distribution for a fighter wing-body combination with vertical tails, and a comparison of the lift and drag predictions with tunnel data is shown in Table 1.

Conclusions

A complete treatment for the steady and unsteady form of the full potential equation is presented. The method employs several novel concepts, such as flux linearization, flux biasing, unsteady wake model, and nonreflecting outer boundary conditions. Extensions of the unsteady work to treat airfoils in plunge and pitch, and to three-dimensions to treat wings and wing-body combinations are currently in progress.

Acknowledgements

This work was partially funded by NASA Langley Research Center under Contract NAS1-15820.

References

1. Shankar, V., "A Conservative Full Potential Implicit, Marching Scheme for Supersonic Flows," AIAA Journal, vol. 20, No. 11, November 1982, pp. 1508-1514.
2. Shankar, V. and Osher, S., "An Efficient Full Potential Implicit Method Based on Characteristics for Analysis of Supersonic Flows," AIAA Paper No. 82-0974, June 1982; AIAA Journal, Vol. 21, No. 9, Sept. 1983.
3. Shankar, V., Szema, K. Y. and Osher, S., "A Conservative Type-Dependent Full Potential Method for the Treatment of Supersonic Flows with Embedded Subsonic Regions," AIAA Paper No. 83-1887, AIAA Journal, November 1984.
4. Szema, K. Y. and Shankar, V., "Nonlinear Computation of Wing-Body-Vertical Tail-Wake Flows at Low Supersonic Speeds," AIAA Paper No. 84-0427.
5. Shankar, V., "Implicit Treatment of the Unsteady Full Potential Equation in Conservation Form," AIAA Paper No. 84-0262.
6. Shankar, V., "Relaxation and Approximate Factorization Methods for the Unsteady Full Potential Equation, ICAS-84-1.6.2, 14th Congress of the International Council of the Aeronautical Sciences, Toulouse, France, Sept. 1984.
7. Holst, T. L., "Fast, Conservative Algorithm for Solving the Transonic Full Potential Equation," AIAA Journal, vol. 18, No. 12, December 1980, pp. 1431-1439.
8. Hafez, M., "Entropy Inequality for Transonic Flows," Transonic Unsteady Aerodynamics and Aeroelasticity Workshop, NASA Langley Research Center, June 22-23, 1983.

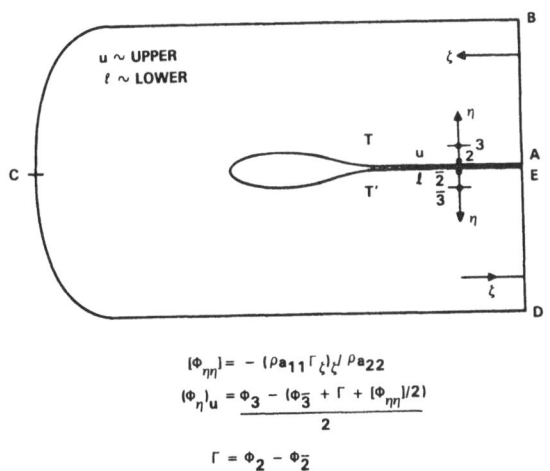

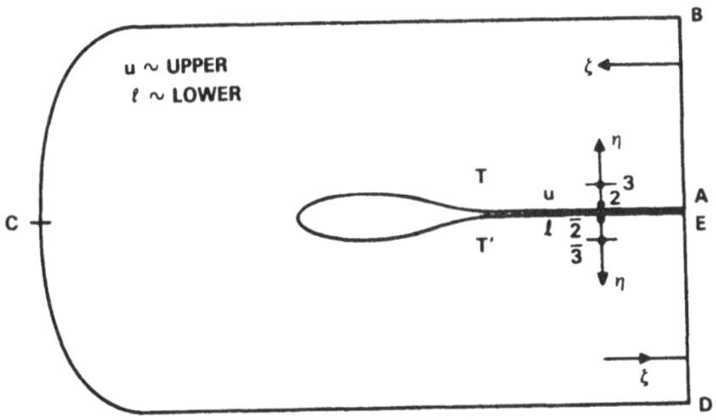

$$(\Phi_{\eta\eta}) = -(\rho a_{11}\Gamma_\zeta)_\zeta / \rho a_{22}$$

$$(\Phi_\eta)_u = \frac{\Phi_3 - (\Phi_{\bar{3}} + \Gamma + (\Phi_{\eta\eta})/2)}{2}$$

$$\Gamma = \Phi_2 - \Phi_{\bar{2}}$$

RIEMANN INVARIANTS SPECIFIED ALONG ABCDE

Fig. 1 Wake and outer boundary treatment.

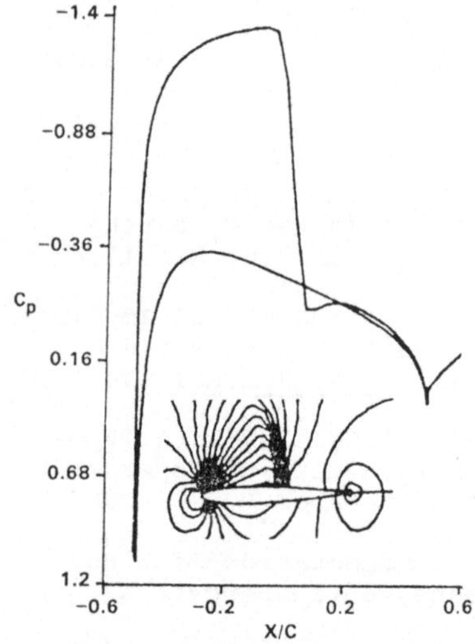

Fig. 2 Results for NACA 0012, $\alpha = 2°$, $M_\infty = 0.75$ using the unsteady code

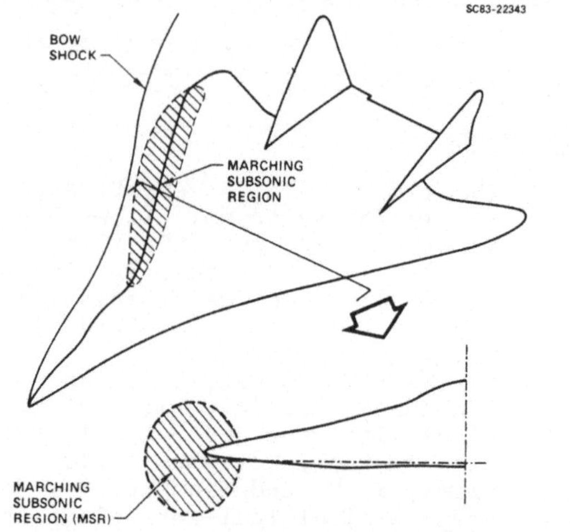

Fig. 3 Schematic of a supersonic fighter with an embedded subsonic region.

Fig. 4 Pressure distribution obtained using the marching code for a fighter-like configuration. $M_\infty = 1.6$, $\alpha = 4.46°$.

α		5°	4.5°	5°	4.5°
M_∞		1.6*	1.6†	1.4†	1.6†
Λ		48°	48°	48°	55°
C_L	CODE	0.298	0.3016	0.3561	0.29186
	DATA	0.277	0.295	0.342	0.283
C_D	CODE	0.0482	0.04916	0.04117	0.0404
	DATA	0.0457	0.0493	0.0426	0.0396

Λ= WING SWEEP ANGLE
*WITHOUT VERTICAL TAIL
†WITH VERTICAL TAIL

Table 1. Comparison of Lift and Drag for a Fighter-Like Configuration.

Vortex Method in Three-Dimensional Flow

S.Shirayama, K.Kuwahara*
Department of Aeronautics, University of Tokyo,
*The Institute of Space and Astronautical Science,
Komaba, Meguro-ku, Tokyo, JAPAN

Many vortex methods have been studied to simulate a vorticity-dominated flow of an incompressible fluid especially in two dimensions.[1)] In three dimensions, vortex stretching has prominent and complex additional effects, and usual vortex filament method fails within a short period of time because the number of elements become increased unlimitedly. Also, it is very difficult to simulate the reconnection of the vortex filaments because it is based on a continuous vortex filaments which cannot be cut. In this paper ,a new vortex method is proposed, which is based not on vortex filaments but on vortex sticks of finite lengths. It can duly treat the reconnection and the increase of the number of vortex elements is suppressed within a reasonable bound. In this direction, Chorin[2)] proposed a vortex stick method but its accuracy of integration is not good enough to simulate a complicated flow. Also Beale and Majda[3)] published a paper on a three-dimensional vortex method; where they proved its convergence to Euler equations, but its application to the computation is not easy and no computed results are available yet.

Computational Method

The vorticity field is replaced by a number of vortex segments (Fig.2). A segment Λ is defined by the following eight quantities (Fig.1): the coordinates $\mathbf{r}=(x_1,x_2,x_3)$ of the center of the stick, its vorticity $\boldsymbol{\omega}=(\omega_1,\omega_2,\omega_3)$, its length δl, and its core radius σ. The trajectory $\mathbf{r}_i$ of the i-th stick is determined as,

$$\frac{D\mathbf{r}_i}{Dt} = \mathbf{u}_i, \qquad \mathbf{u}_i = \sum_j \mathbf{u}_{ij} \; , \tag{1}$$

where $\mathbf{u}_{ij}$ is the velocity induced by the j-th stick and is determined by using Biot-Savart law:

$$\mathbf{u}_{ij} = \frac{1}{4\pi}\int \frac{\boldsymbol{\omega}_j(\mathbf{r}_j')\times(\mathbf{r}_i-\mathbf{r}_j')}{|\mathbf{r}_i-\mathbf{r}_j'|^3}\, d\mathbf{r}_j' \; . \tag{2}$$

This can be integrated, and we can get the following algebraic expression (see Fig.3 for notations):

if $h \geq \sigma_j$:

$$\mathbf{u}_{ij} = \frac{k_j}{4\pi}\left(\frac{1}{f}+\frac{1}{g}\right)\cdot\left(\frac{1}{f+g-\delta l_j}-\frac{1}{f+g+\delta l_j}\right)\cdot\frac{\boldsymbol{\omega}_j}{|\omega_j|}\times \mathbf{f}_j \ , \tag{3}$$

if $h < \sigma_j$:

$$\mathbf{u}_{ij} = \frac{k_j}{8\pi\sigma_j^2\delta l_j}\cdot\left(\frac{1}{f}+\frac{1}{g}\right)\cdot(\delta l_j - f+g)(\delta l_j + f-g)\cdot\frac{\boldsymbol{\omega}_j}{|\omega_j|}\times \mathbf{f}_j \ . \tag{4}$$

The vorticity is determined by the following vorticity equations:

$$\frac{D\boldsymbol{\omega}_i}{Dt} = \boldsymbol{\omega}_i \nabla \mathbf{u}_i = \boldsymbol{\omega}_i \sum_j \nabla \mathbf{u}_{ij} \ . \tag{5}$$

As the velocity $\mathbf{u}_{ij}$ is the function of $\mathbf{r}_i$, $\nabla \mathbf{u}_{ij}$ can be calculated analytically. The position and vorticity of the next step are determined by solving Eqs.(1) and (5). The radius σ_i is determined by Kelvin's theorem on vortex:

$$\pi(\sigma_i^n)^2|\omega_i^n| = \pi(\sigma_i^{n+1})^2|\omega_i^{n+1}| \ . \tag{6}$$

The length δl_i is determined by continuity equation:

$$\pi(\sigma_i^n)^2 \ \delta l_i^n = \pi(\sigma_i^{n+1})^2 \ \delta l_i^{n+1} \ . \tag{7}$$

In this way, the eight quantities of the stick at the next step is determined. The viscous effect is estimated by introducing the viscous vortex core:

$$\sigma = \alpha\cdot\sqrt{\nu t} \ . \tag{8}$$

Results

The interaction of vortex rings is computed by using the present method. The cross linking of two rings is shown in Figs.4,5. In the case of high Reynolds number, turbulent diffusion of vorticity propagates from closer part of two rings to all part. Fig.6 shows interaction of four rings. The vortex filaments composed of the four rings are reformed into two rings. The larger one proceeds in the same direction of the initial ones, the smaller one does in the counter direction. Game of passage of vortex ring is simulated in Fig.7. Fig.7(a) is its perspective view. After the merging, a distored ring diffuses rapidly. Fig.7(b) is the velocity distribution in the x-y plane. The decay of vortex ring is shown in Fig.8. Interaction of two rings is simulated in Fig.9(perspective view). The nearest part of two rings is affected but merging does not take place. In each case , turbulent diffusion is automatically simulated.

References

1) Leonard,A. : Vortex methods for Flow Simulation,J.Comput.Phys., **37** (1980) pp.289-335

2) Chorin,A.J. : Vortex Models and Boundary Layer Instability,SIAM J.Sci.Stat.Comput.,**1** (1980) pp.1-21

3) Beale,J.T.& Majda,A. : Vortex Methods.I:Convergence in Three Dimensions,Math.Comput.,**39** (1982) pp.1-27

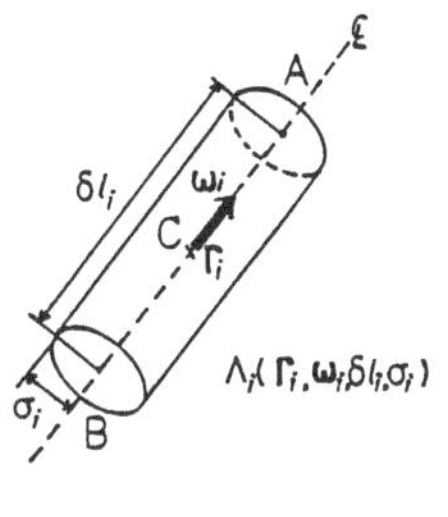

Fig.1

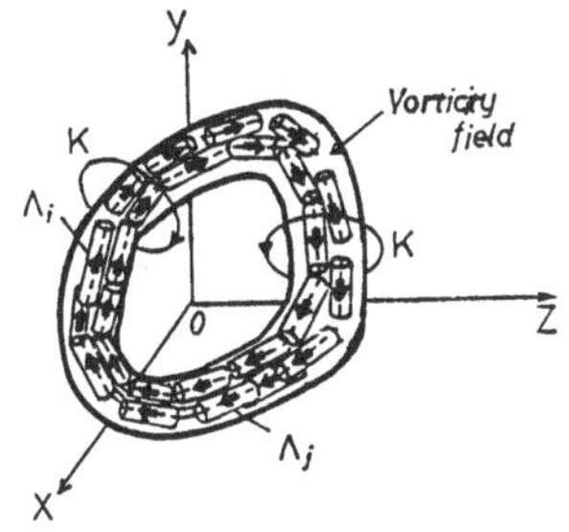

Fig.2

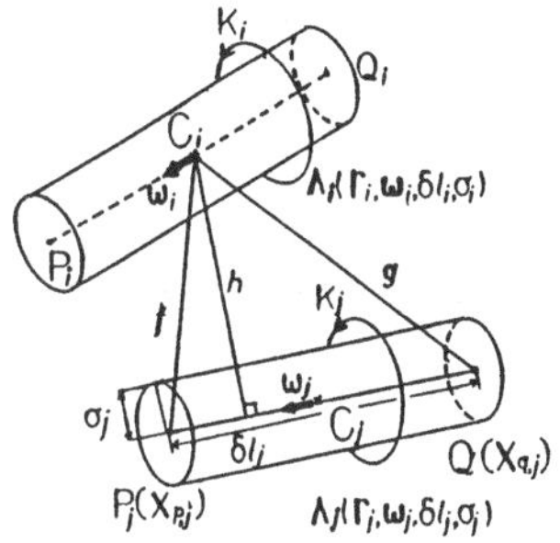

Fig.3

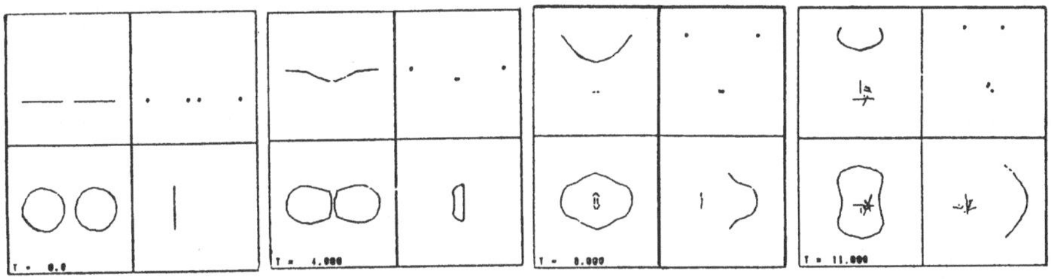

Fig.4 Interaction of two vortex rings (low Reynolds number)

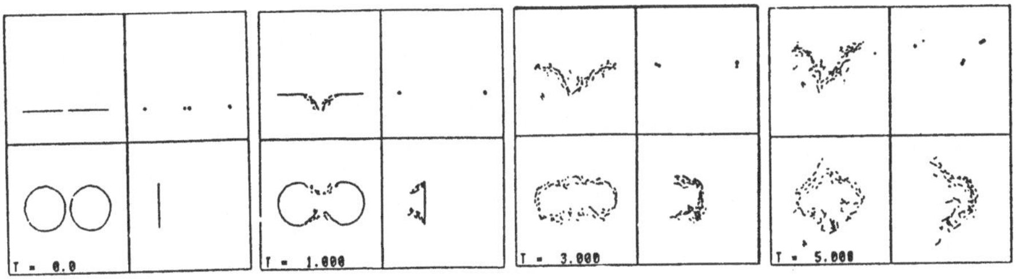

Fig.5 Interaction of two vortex rings(high Reynolds number)

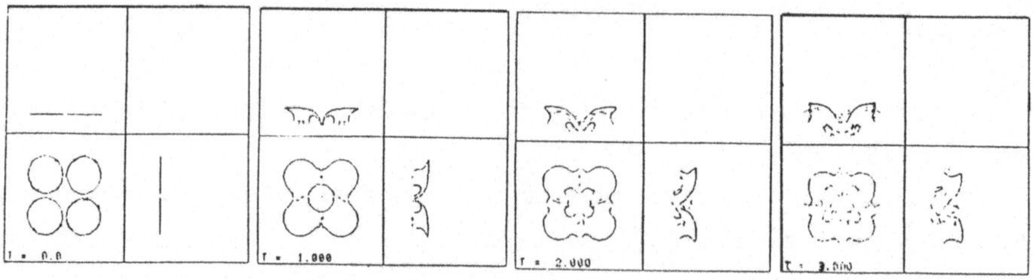

Fig.6 Interaction of four vortex rings

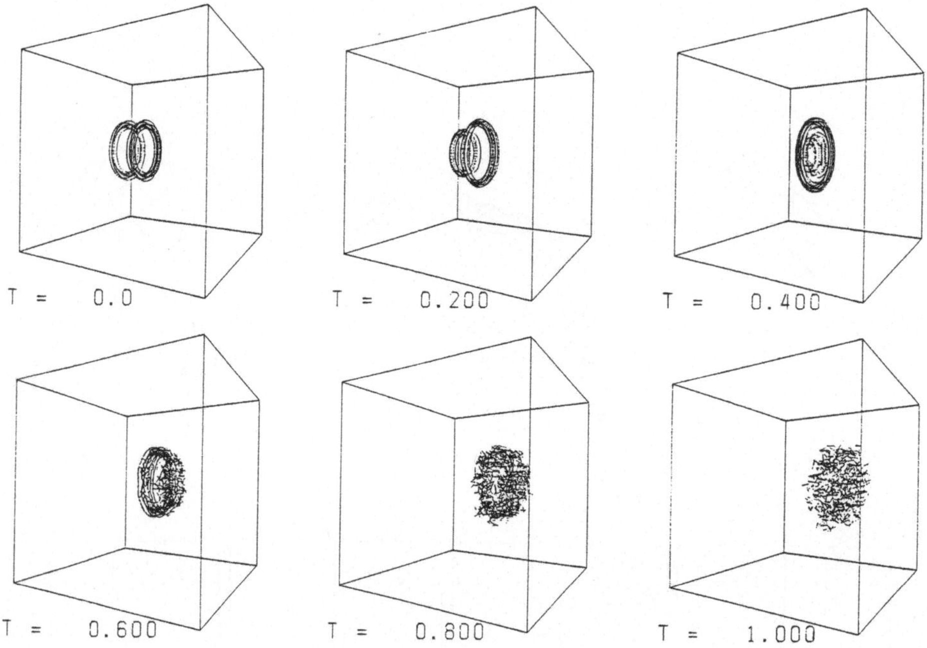

Fig.7(a) Game of passage of vortex ring

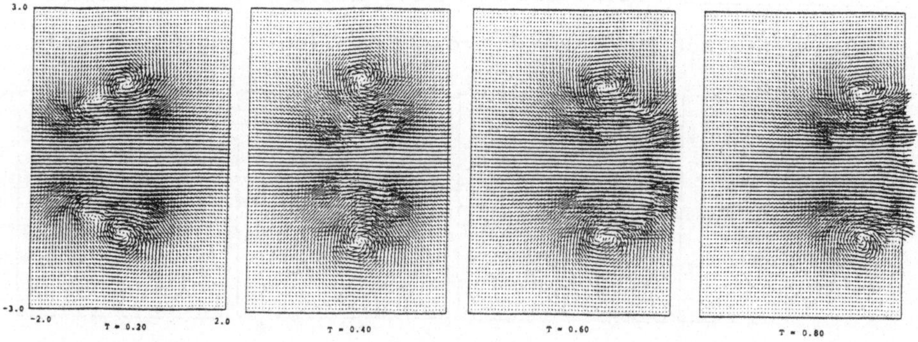

Fig.7(b) Velocity distribution (x-y plane)

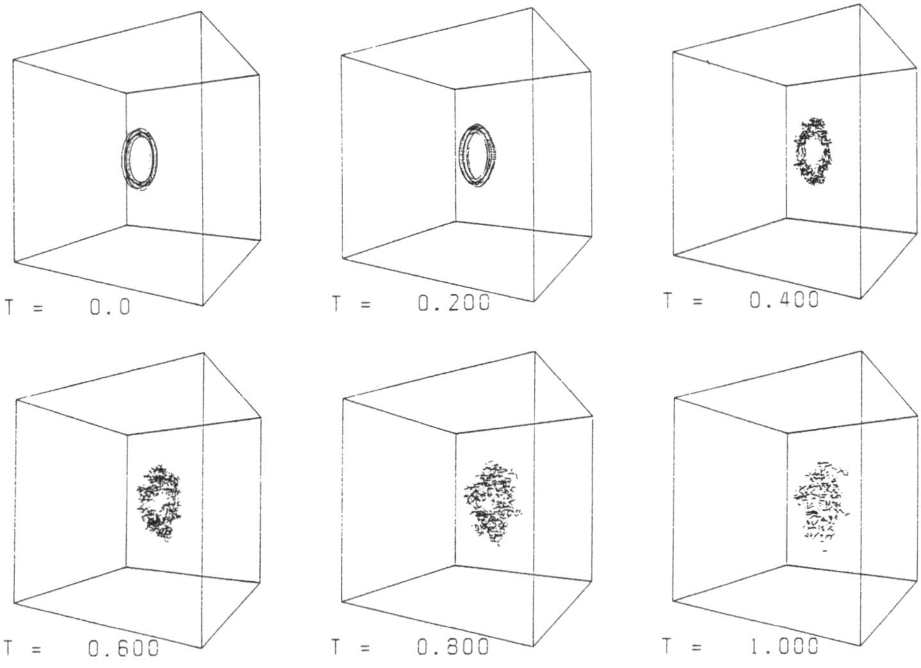

Fig.8 Decay of vortex ring

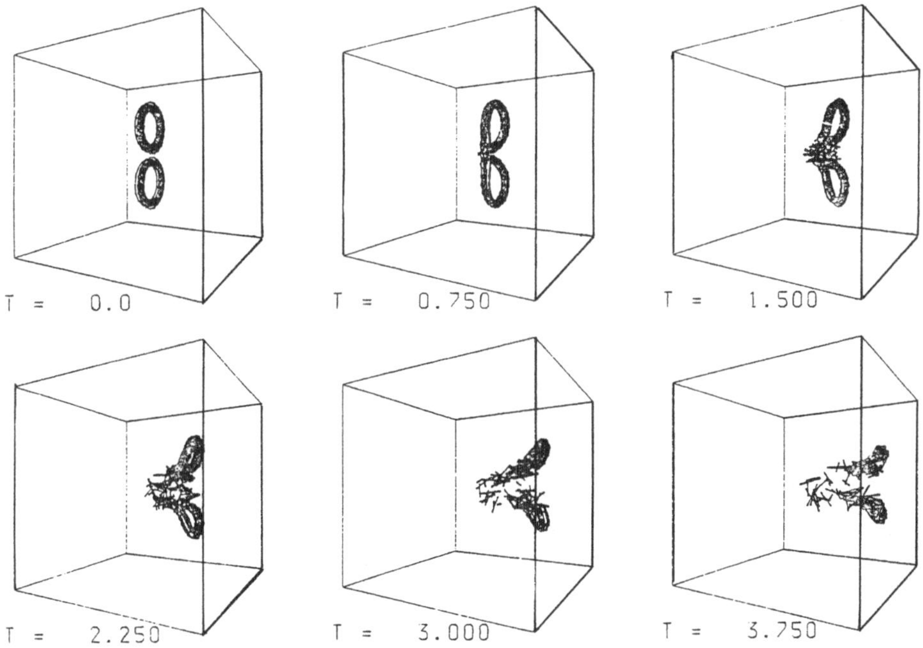

Fig.9 Interaction of two vortex rings

CALCULATION OF TRANSONIC POTENTIAL FLOW PAST WING-TAIL-FUSELAGE COMBINATIONS USING THE MULTIGRID TECHNIQUE

Arvin Shmilovich and D. A. Caughey
Sibley School of Mechanical and Aerospace Engineering
Cornell University
Ithaca, NY 14853/USA

ABSTRACT

A computer program for calculating transonic potential flow fields about three-dimensional wing-tail-fuselage combinations has been developed. The transonic potential equation is approximated and solved numerically using a finite-volume method on a boundary conforming coordinate system. The multigrid technique is utilized to accelerate convergence of the relaxation scheme and thus to allow computations about multiple-component airplanes with reasonable expenditure of computer resources. The wing-tail flow interaction is investigated and results indicate that the tail exerts a noticeable effect on the wing loading when the tail is located near the wing trailing edge.

INTRODUCTION

It is of practical interest to obtain solutions for the flow past realistic airplane configurations such as wing-tail-fuselage combinations. Such computations provide insight into the nature of the flow patterns for closely-coupled multi-wing systems (such as fighter-type aircraft) in which instances the wing and tail flowfields strongly interact and the flow analyses of the individual components in isolation (say, wing-body and tail-body combinations) cannot be expected to be adequate for realistic modeling.

Grids generated for wing-tail-body arrangements[1] have been used in the present work for the development of transonic potential flow analysis methods, within the framework of the finite-volume method. The mesh generation procedure produces a slotted computational domain and the grid is nearly orthogonal everywhere. The resulting mesh is of C-type for the wing and of H-type for the tail.

Obviously, many mesh cells are needed for modelling the flow past a complete airplane to meet the requirement of reasonable resolution in the vicinity of the individual components. Conventional relaxation schemes may exhaust computers beyond their present capabilities, rendering such calculations impractical. Therefore, the multigrid technique is used to improve convergence rates and enhance the practicality of such computations. Experience gained in two-dimensional and three-dimensional numerical calculations[2-4], from the standpoints of both programming and predictability, helped much in developing the program. Reference 5 describes in greater detail the analysis on which the program is based. The multigrid version of FLO-30 (known as FLO-30M[4]) formed the point of departure for developing the wing-tail-body code described herein.

ANALYSIS

The continuity equation for steady potential flow in an arbitrary coordinate system X,Y,Z, reads

$$(\rho hU)_X + (\rho hV)_Y + (\rho hW)_Z = 0 \tag{1}$$

where h is the determinant of the Jacobian of the transformation, ρ denotes the density (which is related to the magnitude of the velocity vector by the isentropic relation) and (U,V,W) is the contravariant velocity vector. Noting that the equation representing conservation of mass is equivalent to $\rho h/a^2$ times the potential equation in quasilinear form, the latter is utilized for formulating the iterative scheme. Here a is the local speed of sound. In order to reflect properly the limited domain of dependence, a second-order artificial viscosity is introduced in hyperbolic regions of the flowfield.[4] The solution of the algebraic equations resulting from the discretization is accomplished by an iterative scheme, formulated by embedding the steady state equation in an artificial time-dependent equation.

To be effective in conjunction with the multigrid technique the relaxation scheme must efficiently damp the high wavenumber components of the residual. Ordinarily, iterative procedures are local in nature, and therefore, quite suitable for this purpose. One such process is characterized by considering one mesh line at a time and simultaneously calculating the corrections at points that lie on it. This line-overrelaxation is successively done until the solution in the entire domain is updated. The system of equations to be solved on each line is a pentadiagonal, diagonally dominant system.

The system of equations for the XSLOR scheme, i.e., when the corrections are simultaneously calculated along X-lines (see Figure 1), is of the form

$$a_1 C_{i,j-2,k} + a_2 C_{i,j-1,k} + a_3 C_{i,j,k} + a_4 C_{i,j+1,k} + a_5 C_{i-2,j,k} + a_6 C_{i-1,j,k} + a_7 C_{i,j,k-1} = - R_{i,j,k} \tag{2}$$

for positive components of the contravariant velocity. Here $C_{i,j,k}$ is the correction applied at point (i,j,k) and $R_{i,j,k}$ is the residual calculated (by equation (1)) using values of the solution from the previous iteration. The expressions for the coefficients may be found in Reference 4.

In a similar fashion it is possible to devise a scheme in which points along Y-lines are updated simultaneously, marching in the X direction (YSLOR). Both X and Y-schemes have been implemented in the present program.

It was shown in Reference 2 that the growth factor in a local mode analysis of the XSLOR scheme should never exceed 0.78 per multigrid cycle (or 0.883 per work unit for the strategy described below) if the multigrid is effective in reducing all low wavenumber components of the error in any of the three coordinates. However, the performance suffers deterioration since errors are replenished in the course of the multigrid iteration (due to residual restriction and interpolation of corrections). Errors are most probably excited in the farfield region of the domain, in particular near the downstream boundary, where the modulus of the transformation is large and the mesh cells are highly elongated. Nevertheless, it was found effective to alternate between XSLOR and YSLOR schemes (Alternating Successive Line Overrelaxation) at each grid level within each multigrid cycle, resulting in the efficient elimination of all high wavenumber components of the error. More specifically, the discretized equations are sequentially solved on planes of constant Z (marching from the fuselage towards the lateral boundary), each of which is swept by either the XSLOR or YSLOR scheme.

A fixed strategy for changing levels has proven effective in the present calculations. The domain is swept m_1 times on each grid level except the finest until the coarsest grid is reached. Each level is swept m_2 times after coarse grid corrections are added while backing up to the finest grid. This completes a multigrid cycle. Thus, the work required for one multigrid cycle for a problem in d space dimensions is

$$1 + (1 + m_1)(1/2^d + 1/2^{2d} + \ldots) + m_2(1 + 1/2^d + 1/2^{2d} + \ldots)$$

$$< [m_1 + 2^d(1 + m_2)]/(2^d - 1) \quad \text{units} \tag{3}$$

where one work unit is the labor required for one fine grid iteration. Since it has been realized in flow calculations that the cost of computing the residuals is the better part of the overall effort for a relaxation sweep, the computational work for calculating the residuals for restriction to the next grid level was taken as equal to that required for a relaxation sweep (on that grid). The strategy using $m_1 = 4$ and $m_2 = 2$ was found to be the most effective.

Coding complications were introduced by the need to relax the solution along lines intersecting the slot which represents the images of the tail and its vortex sheet in (portions of) the computational space. Careful programming is required for correct implementation of boundary conditions on the slot (no-flux and vortex sheet conditions), for the systematic restriction of the solution and the residuals and prolongation of the correction, and the refining and coarsening of the mesh in the course of each multigrid cycle. The program was specialized for running on vector computers or array processors.

RESULTS

The geometry tested in flow computations is the ONERA wing M-6 and a tail whose cross section is identical to the wing section, mounted upon a cylindrical body. The wing is mid-mounted upon the fuselage at zero incidence so that in the absence of the tail and at zero angle of attack the flow is symmetrical about the wing plane. A perspective view of a closely coupled wing-tail arrangement is shown in Figure 1. Four grid levels were employed in the multigrid sequence, with the finest grid containing $192 \times 32 \times 24$ mesh cells. The computational labor required for the flow calculation on this grid is less than the equivalent of two CPU minutes on the CRAY I computer.

Calculations were performed to furnish information about the wing-tail interaction for a closely coupled combination. The distance between the lifting surfaces is 0.0625 of the wing root chord. The first case has a freestream Mach number of 0.88 and zero flow incidence. Figure 2 shows the effect of the tail deflection upon the wing streamwise pressure distribution at the 0.33 and 0.73 semi-span stations. (The lower symbols represent the pressure distribution on the upper surface.) In the absence of the tail the pressure distribution on the wing is symmetrical. The effect of the high tail on the wing lower surface pressure distribution is negligible, so the difference between the curves at each span location is due to the effect of the tail on the upper surface pressure distribution. The increase in the pressure in the vicinity of the leading edge of the tail reduces the suction in the region downstream of the aft shock of the wing. This, in turn, reduces the strength of the shock that terminates the supercritical region, causing it to shift upstream of its location in the absence of the tail. Figure 3 presents the tail surface pressure distributions for different tail inclinations.

To examine the range of validity of the various effects the tail has upon the wing, the wing lift is plotted as a function of stagger in Figure 4. The tail is deflected at +4°. Owing to local effects, for very narrow gaps, the tail induces negative lift. The blockage effect of the tail gradually subsides as the tail is moved further aft. At a gap of about half the wing chord, a different type of interaction takes over; at these greater distances the lifting surfaces may be considered as systems of line vortices which induce velocities upon each other. At wide enough gaps, when the local blockage effect dwindles, the numerical results indicate an increase in lift as compared to the tail-off situation. This is consistent with the upwash effect of the tail bound vortex in the upstream region where the wing is located.

The second set of results were calculated for a freestream Mach number of 0.84 and 3.06° angle of attack. Figure 5 shows the spanwise loading on the lifting surfaces for three tail deflections. Both high and low tails (positioned at the same distance from the wing plan) are considered. In contrast to the high tail, the presence of the low tail enhances the lift on the wing, especially for large positive tail angles. The flow pattern in this case is similar to that past a wing with extended Fowler flaps. The departure of the wing loading for high and low tails from the wing loading in the absence of the tail is also shown in the figure.

REFERENCES

1. Shmilovich, A. and Caughey, D. A., "Grid Generation for Wing-Tail-Fuselage Configurations", Advances in Grid Generation, Ghia, K. N. and Ghia, U., Eds., American Society of Mechanical Engineers, 1983, pp. 189-197.

2. Shmilovich, A. and Caughey, D. A., "Application of the Multigrid Method to Calculations of Transonic Potential Flow about Wing-Fuselage Combinations", J. Comp. Phys., Vol. 48, 1982, pp. 462-484.

3. Caughey, D. A. and Shmilovich, A., "Multigrid Calculation of Transonic Potential Flows", Numerical Methods in Fluids: Advances in Computational Transonics, Vol. 4, Habashi, W. W., Ed., Pineridge Press, Swansea, U.K., 1984.

4. Caughey, D. A., "Multigrid Calculation of Three-Dimensional Transonic Potential Flows", Appl. Math. and Comp., Vol. 13, 1983, pp. 241-260.

5. Shmilovich, A., "Calculation of Transonic Potential Flow past Wing-Tail-Fuselage Configurations Using the Multigrid Method", Ph.D. Thesis, Cornell University, 1983.

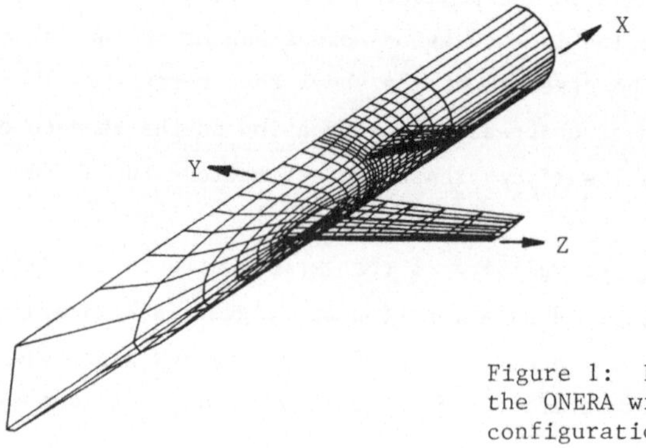

Figure 1: Perspective view of the ONERA wing-tail-fuselage configuration.

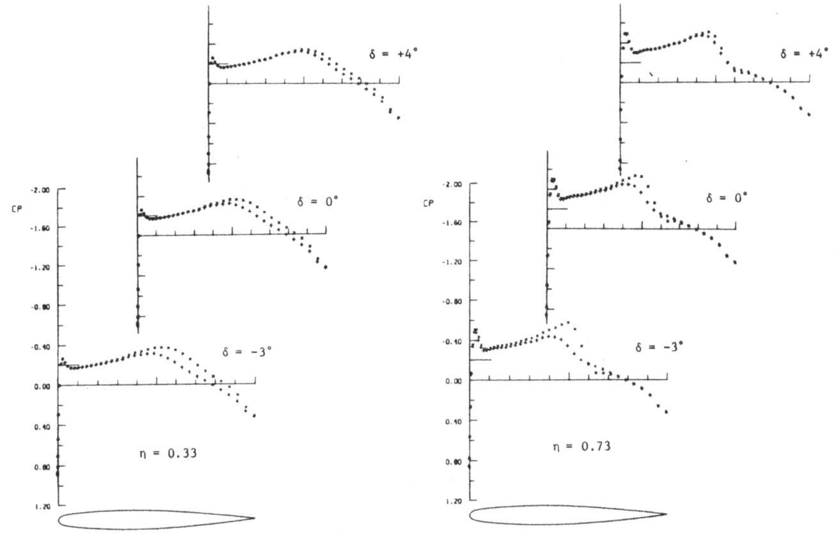

Figure 2: Effect of tail deflection upon wing pressure distribution.

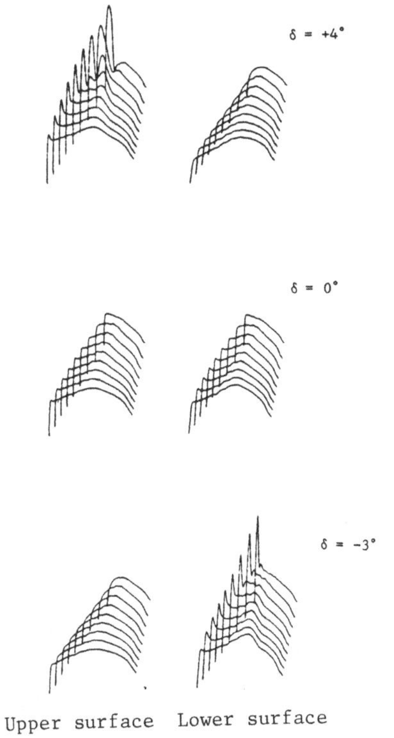

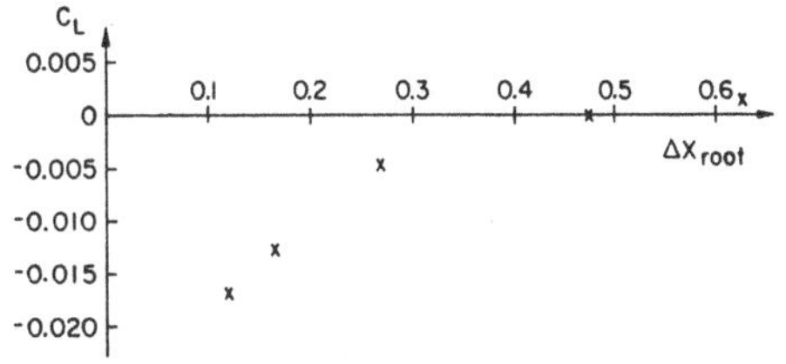

Figure 4: Variation of wing lift due to the tail as a function of stagger.

Figure 3: Tail surface pressure distributions for different deflections.

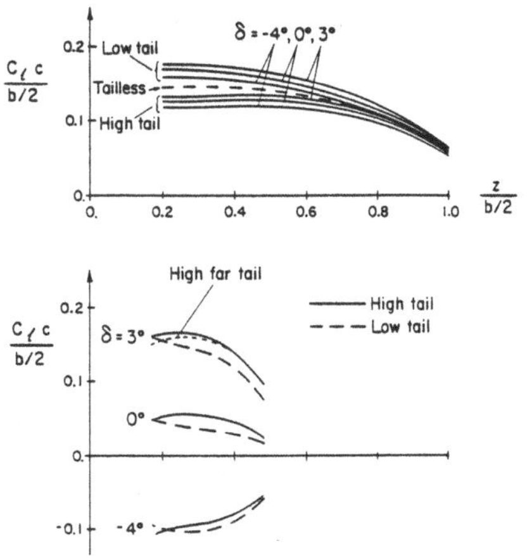

Figure 5: Spanwise loading on wing and high/low tail as affected by tail deflection.

PULSED COLUMN : TRANSIENT FLOW OF A POLYDISPERSED PHASE

S. SIEBERT and T. DUJARDIN
Commissariat à l'Energie Atomique
Centre d'Etudes Nucléaires de Grenoble
85 X - 38041 GRENOBLE CEDEX - FRANCE

ABSTRACT - The transient flow of a polydispersed phase in a pulsed column is described by a non-linear differential system with the following characteristics : high number of equations, long duration of the physical phenomena compared to the numerical time step. A suitable adaptation of a Runge Kutta method to the physical problem has permitted to reduce calculation time considerably.

DESCRIPTION OF THE PROBLEM

We study the countercurrent flow of two immiscible liquid phases in a pulsed perforated plate column represented on figure 1

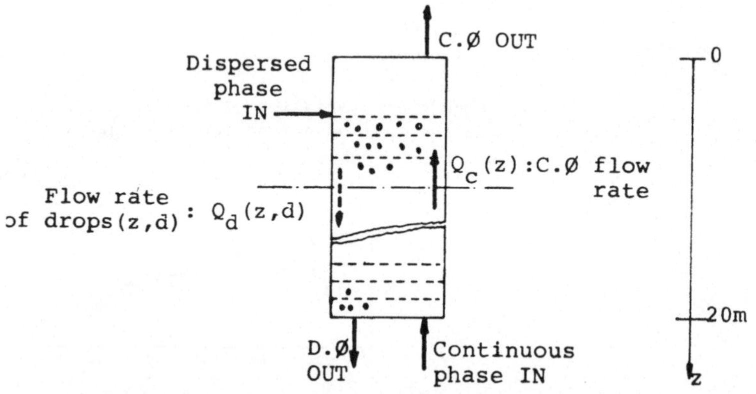

Fig. 1 : *Column diagram*

The heavy dispersed phase experiences backmixing, coalescence and breakup. It is described with the continuous, one dimensional, transport-interaction model of CASAMATTA [1]. Two problems must be solved : hydraulic steady state and response to tracer injections of the Dirac type, the representative equations of which are very similar.

The basic equation for the hydraulic problem is a local balance on drops of diameter d at height z

$$\frac{\partial}{\partial t}\beta(z,d) = -\frac{\partial}{\partial z}Q_d(z,d) + \mathcal{P}(z,d) \qquad \text{(E)}$$

- $\beta(z,d)$ is the volume fraction of drops (z,d)
- the transport term is a plug-dispersion type :

$$Q_d(z,d) = V(z,d)\ \beta(z,d) - \mathcal{D}\frac{\partial}{\partial z}\ \beta(z,d)$$

$V(z,d)$ is the velocity of drops (z,d) and the diffusion like term $\mathcal{D}$. $\frac{\partial\beta}{\partial z}$ accounts for the transient velocity variations caused by pulsation, and the subsequent backmixing effect.

- the interaction term $\mathcal{P}(z,d)$ is a sum of death and birth rates among drops (z,d) due to splitting and coalescence.

$$\mathcal{P}(z,d) = \qquad - k_{sp}(d)\ \beta(z,d)$$

(rate of death of drops (z,d) by splitting : $\mathcal{S}_d$)

$$-\ \beta(z,d)\int_0^{\infty} k_{co}(d,d_1)\ \beta(z,d_1)\ \delta d_1.\ \frac{\pi d^3}{6}$$

(rate of death of drops (z,d) by coalescence with all other drops at same height : $\mathcal{C}_d$)

$$+\ k_{sp}(d_1)\ \beta(z,d_1)$$

(rate of birth of drops (z,d) from splitting of larger drops of the appropriate volume : $\mathcal{S}_b$)

$$+\iint_{(d_1 + d_2 \rightarrow d)} k_{co}(d_1,d_2)\ \beta(z,d_1)\ \beta(z,d_2)\ \delta d_1\ \delta d_2\ .\ \frac{\pi d^3}{6}$$

(rate of birth of drops (z, d) from coalescence of smaller drops : $\mathcal{C}_b$)

The discretization with respect to diameter is based on the experimental drop size distribution observed. The space discretization is much more concerned with numerical considerations. It is very important to progressively reduce the step size near the inlet section where fast variations of β occur , while in the middle section the step size can be much larger. For a 20 m high column, NT = 11 drop sizes and NE = 80 stages are used. Then, using central finite differences for the space derivatives, we obtain for the steady state hydraulic problem a system of (80x11) non linear algebraic equations. This is solved by the direct successive substitutions method of DUHAMET [2]. Figures 2 and 3 show typical β profiles and drop size distributions along the column.

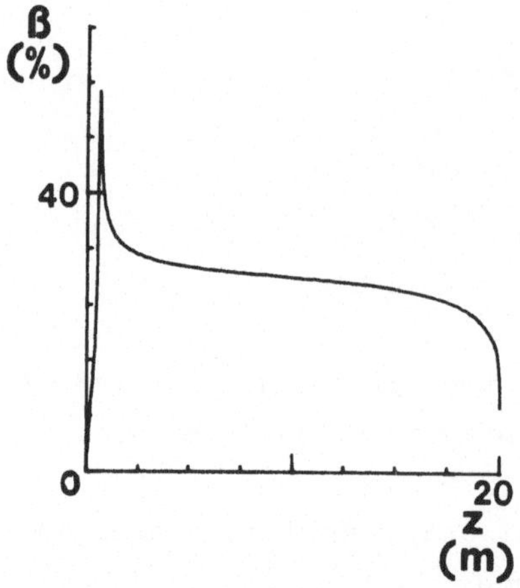

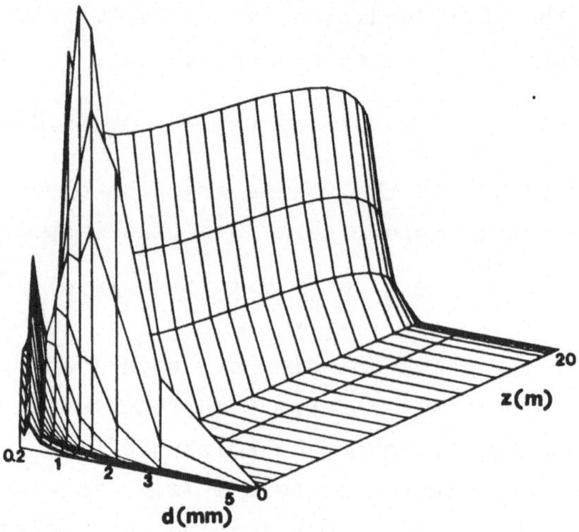

Fig. 2: *Total dispersed phase fraction* $\beta(z)$ Fig. 3: *Drop size distribution* $\beta(z,d)$

The equation describing the tracer problem is simply obtained by replacing in (E) the unknown $\beta(z,d)$ (quantity of dispersed phase) by $T(z,d)$ (quantity of tracer), with $T(z,d) = \beta(z,d)\ \overline{C}\ (z,d)$; $\beta(z,d)$ has been previously calculated (supposing that the injection of a small quantity of tracer will have no influence on the hydraulic steady state) and $\overline{C}(z,d)$ is the average concentration of tracer in drops (z,d).

METHOD OF SOLUTION

Still using the same discretization, the (i+j) th equation of the differential system (i^{th} stage, j^{th} diameter) can be written :

$$\frac{dT_{ij}}{dt} = v^+_{i-1j}\ T_{i-1j} - (v^+_{ij}+v^-_{ij})T_{ij} + v^-_{i+1j}\ T_{i+1j} + \mathcal{S}_d + \mathcal{C}_d + \mathcal{S}_b + \mathcal{C}_b$$

(the velocity components $v^{\pm}$, only determined by the hydrodynamic regime, are constant)

The tracer experiment is about 1 hour long. The simulation, using a classical Runge Kutta 4 method with adjusted time step (Merson), is very expensive :the mean step size is about 0,1 s, and the computing time is twice longer than the real phenomenon. A first attempt, consisting in a linearisation of the coalescence terms and a separate calculation of the contributions of transport and interaction, was unsuccessful. Then an improvement of the Runge Kutta method, according to the physical problem, was developped. A closer analysis showed that :

- the time consuming coalescence terms change relatively slowly
- as far as the step size is concerned, the limiting terms are the breakage ones, due to $k_{sp}(d) = d^8$, which become very large for increasing diameters. In fact, figure 3 shows that large drops exist only in a narrow region around the dispersed phase inlet. They are rapidly broken when passing through the first plates.

So we tried first to recalculate the coalescence terms only at spaced intervals. This already permits a significant time sparing without altering the result. Then we tried to divide the system into different zones, in order to isolate the "step-limiting" ones as can be seen on figure 4.

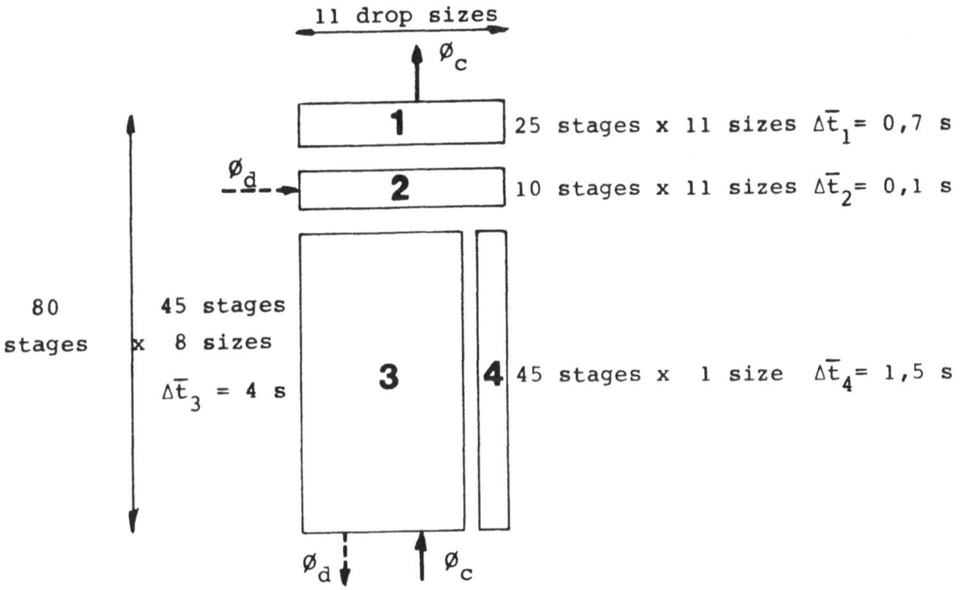

Fig. 4 : *Schematic representation of the system division*

As shown on figure 3, drops of the 2 largest sizes are in negligible quantity below the dispersed phase inlet. They are released from the system. Yet, it is still interesting to isolate the largest remaining size, forming zone n° 4. In zone n° 1, no break up occurs (there is no packing), the step size is limited by transport.

As expected the Merson algorithm shows that the individual step size, of each zone are quite different. It is possible to calculate each one separately with its own step, the largest one fixing meeting dates, so that, from time to time, the calculation of the 4 zones should be performed at the same date :

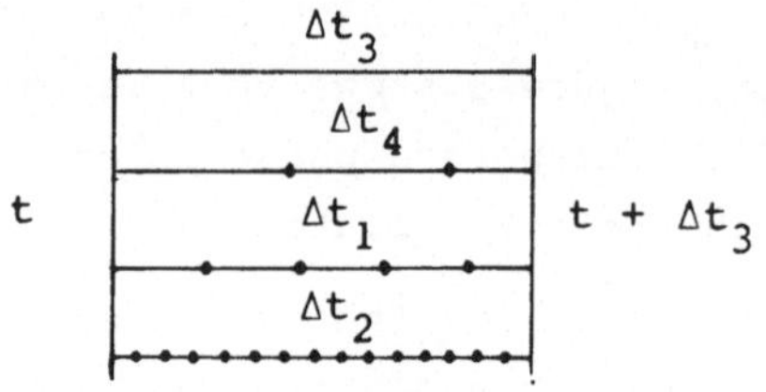

Fig. 5 : *Step sizes*

Of course, to let each zone move at maximum speed, and to avoid method error, the choice of interface junctions is the fundamental point. Rigorously, the continuity of concentrations and flux should be respected, that is :

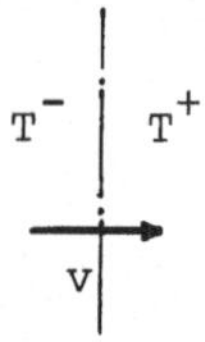

$$\begin{cases} T^- = T^+ \\ \left(vT - \mathcal{D}\,\dfrac{\partial T}{\partial z}\right)^- = \left(vT - \mathcal{D}\,\dfrac{\partial T}{\partial z}\right)^+ \end{cases}$$

thus : $$T^- = T^+ \qquad \frac{\partial T^-}{\partial z} = \frac{\partial T^+}{\partial z}$$ (v is constant)

Practically, this is not possible, and three different conditions have been tested.

1. Continuity of T^+ and $\dfrac{\partial T^-}{\partial z}$ with an intermediate point
2. Interpolation on T^+ during calculation of T^- (when $\Delta t^- < \Delta t^+$), thus refreshing the frontier point at each step Δt^-
3. β^+ simply held constant during calculation of T^-

The two first methods allow less error but they artificially reduce the step in each zone by causing frontier variations not matched with the rest of the zone. Finally the third and simplest method proved the better compromise between precision and computing cost.

It is yet careful to use the standard Runge Kutta method during the first 30 seconds of simulation, when very fast variations occurr.

RESULTS

The computing time has been reduced by a factor of 6 to 8. Compared to the "exact" result of the Runge Kutta method, the error is estimated to be less than 5% (The balance on the total quantity of tracer in the column becomes slightly erroneous). Typical responses are shown on figure 6.

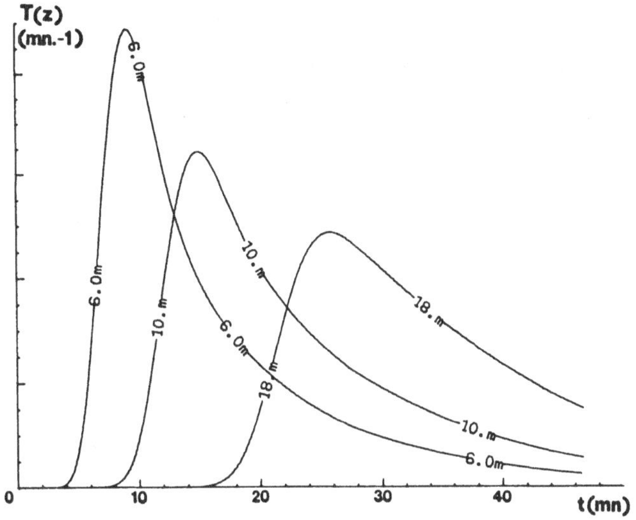

Fig. 6 : Total tracer quantity T(z) vs time for z = 6,10, 18 m

These curves show high dispersion (see the tails at the end of the response) despite of the small dispersion coefficient ($\mathcal{D}$ = 5 cm^2/s). This is to due to forward mixing caused by significant presence of fine particules. This fits well experimental results which cannot be interpreted correctly with simpler models.

REFERENCES

[1] G.CASAMATTA These de doctorat d'Etat INP Toulouse (1981)

[2] J.DUHAMET These de docteur ingénieur Ecole Centrale de Paris (à paraître)

We gratefully acknowledge Professor Destuynder of Ecole Centrale Paris for useful suggestions concerning numerical problems.

A FLAME APPROACH TO UNSTEADY COMBUSTION PHENOMENA WITH APPLICATION TO A FLAME INTERACTING WITH A COLD WALL

Gary A. Sod
Department of Mathematics
Tulane University
New Orleans, Louisiana 70118

1. Introduction

In Sod [12] a random choice method was introduced for one-dimensional laminar flame propagation. The method decomposes the flow field into normal modes which consist of elementary waves and steady-state waves (steady-state flame profiles). The steady-state profiles are sampled to provide left and right states for the Riemann problems solved in the random choice method. The diffusion and reaction terms are taken into account simultaneously in the steady-state flame profile, thereby preserving the natural balance that exists between these two processes.

In Sod [14] this method is extended to multiple space dimensions through operator splitting. However, for many reactive flows, this can be very expensive. To remedy this, the sampled values from the one-dimensional steady-state flame profiles are compiled in a dictionary. Each set of entries in this flame dictionary is the result of a detailed one-dimensional steady-state calculation including the full chemistry. For a given set of reactions, the expense associated with the establishment of the dictionary is incurred only once. The number of entries in the dictionary depends on how complex the reaction is. The method was developed for situations where solutions to problems with the same chemical properties and different hydrodynamics and geometries were sought. It is demonstrated in Sod [14] that the flame dictionary method when compared with a standard method in one space dimension is nearly four orders of magnitude faster. This becomes even faster in multiple space dimensions.

In Sod [15] the method is extended to cylindrical (or spherical) coordinate systems with axial (or radial) symmetry. The steady-state equation is modified to be translation invariant.

In this work the method is summarized and numerical results are presented for the propagation of a flame in a cylinder with cold walls.

2. Outline for the Random Choice Method in One-Space Dimension

In Sod [12] and [14] we considered the system of quasi-linear equations

$$\partial_t \underline{v} + \partial_x \underline{F}(\underline{v}) = \underline{\nu}\partial_x^2 \underline{v} + \underline{Q}(\underline{v}), \tag{2.1}$$

along with the initial condition

$$\underline{v}(x,0) = \underline{f}(x), \tag{2.2}$$

where $\underline{v} = (v_1,\ldots,v_p)$, $\underline{\nu} = \mathrm{diag}(\nu_1,\ldots,\nu_p)$ with $\nu_\ell > 0$, $\ell = 1,\ldots,p$, $\underline{Q}$ is a vector of source terms. When $\underline{\nu} \equiv \underline{0}$ and $\underline{Q} \equiv 0$, equation (2.1) reduces to the system of hyperbolic conservation laws

$$\partial_t \underline{v} + \partial_x \underline{F}(\underline{v}) = 0, \tag{2.3}$$

The method used to solve (2.1)-(2.2) is a random choice type method. The random choice method was introduced by Glimm [5] for the construction of solutions of systems of nonlinear hyperbolic conservation laws. The random choice method was developed for hydrodynamics by Chorin [1-2] and further developed by Colella [3], Glaz and Liu [4], and Sod [8-15].

The numerical method decomposes a general flow field into normal modes which consist of elementary waves for (2.3) and steady waves for (2.1). It is in the same spirit as the method developed by Liu [6] and [7] using boundary-value problems rather than initial-value problems to obtain the steady waves. In essence, the method is the random choice method using piecewise steady-state profiles rather than piecewise constant profiles.

Consider the system of equations (2.1). The homogeneous part of (2.1) is the system of hyperbolic conservation laws and in steady-state (2.1) reduces to the system of ordinary differential equations

$$\partial_x \underline{F}(\underline{v}) = \underline{\nu}\partial_x^2 \underline{v} + \underline{Q}(\underline{v}) \tag{2.4}$$

However, as discussed in Sod [14-15] we consider the steady-state equation, obtained by omitting the advection term,

$$\underline{\nu}\partial_x^2 \underline{v} + \underline{Q}(\underline{v}) = 0. \tag{2.5}$$

Consider, the case where $\nu_1 = \ldots = \nu_p \equiv \nu$. Divide time into intervals of length h. Let $\underline{u}_i^n$ approximate the solution $\underline{v}(ih,nk)$ to (2.1)-(2.2), where $i = 0,\pm1,\pm2,\ldots,$ and $n = 0,1,2,\ldots$.

Given the approximate solution $\underline{u}_i^n$ for each grid point i, consider the sequence of two point boundary value problems, one for each interval $[ih,(i+1)h]$,

$$\underline{\nu}\partial_x^2 \underline{v}^s + \underline{Q}(\underline{v}^s) = 0, \quad ih < x < (i+1)h \tag{2.6a}$$

$$\underline{v}^s(ih) = \underline{u}_i^n, \tag{2.6b}$$

$$\underline{v}^s((i+1)h) = \underline{u}^n_{i+1}, \tag{2.6c}$$

where $\underline{v}_i^s(x)$ denotes the solution to (2.6).

Consider equation (2.3) with the piecewise constant initial conditions

$$\underline{v}(x,nk) = \underline{u}^n_{i+\frac{1}{2}} , \quad ih < x < (i+1)h \tag{2.7}$$

where $\underline{u}^n_{i+\frac{1}{2}} = \underline{v}_i{}^s((i+\frac{1}{2})h)$, that is, $\underline{v}_i{}^s(x)$ is sampled at the midpoint of the interval $[ih,(i+1)h]$. This defines a sequence of Riemann problems that are centered at the actual grid points ih. If the Courant-Friedrichs-Lewy (CFL) condition is satisfied, then the waves generated by the individual Riemann problems, one for each grid point ih, will not interact. Hence, the solution to the different Riemann problems can be combined by superposition into a single exact solution, denoted by $\underline{v}^e(x,t)$, defined for $nk \leq t \leq (n+1)k$.

Let ξ_n denote an equidistributed random (or quasi-random) variable in the interval $(-\frac{1}{2}, \frac{1}{2})$. Define the approximate solution at the next time level by

$$\underline{u}^{n+1}_i = \underline{v}^e((i+\xi_n)h,(n+1)k). \tag{2.8}$$

The presence of the diffusion term in equation (2.1) places an additional requirement on the time step k. This is established in Sod [12] using the random walk solution to the diffusion equation, where the condition

$$k = h^2/8\nu \tag{2.9}$$

was obtained.

The extension to multiple space dimensions is discussed in Sod [14]. For a discussion of the case where the values of $\nu_1, \nu_2, \ldots \nu_p$ in $\underline{\nu}$ are different see Sod [12].

3. The Dictionary

The solution of the two-point boundary-value problem (2.6) can, and often is, the most expensive part of the algorithm described in Section 2. However, only the solution at the midpoint of the interval (2.7) is required.

Suppose that there are two vectors $\underline{u}_L = (u_{jL})$ and $\underline{u}_U = (u_{jU})$ such that $u_{jL} \leq u^n_{ji} \leq u_{jU}$, for $j=1,\ldots,p$ and for all i and $n \geq 0$. Choose numbers h_j, $j=1,\ldots,p$, by

$$h_j = (u_{jU} - u_{jL})/(N_j - 1),$$

where the N_j's are positive integers. Define $u^D_{j\ell} = u_{jL} + (\ell-1)h_j$, $\ell = 1,\ldots,N_j$ and $j = 1,\ldots,p$. Consider the two-point boundary value problem (2.6a) and the boundary condition

$$\underline{v}^s(ih) = (u^D_{1L_1},\ldots,u^D_{pL_p})^T ,$$

and

$$\underline{v}^s((i+1)h) = (u^D_{1R_1},\ldots,u^D_{pR_p})^T,$$

where $L_j, R_j \in \{1,2,\ldots,N_J\}$, $j = 1,\ldots,p$. The solutions of these two-point boundary-value problems is sampled at midpoint, $\underline{v}_i{}^s\ ((i+\frac{1}{2})h)$, from the basis of the dictionary, which is a 2p-dimensional table.

The details of the efficient retrieval of information from this dictionary which consists of a search and a volume weighting interpolation may be found in Sod [14-15].

4. Application to a Flame's Interaction with a Cold Wall

Consider a simple reaction given by A → B. Such a reaction is adequate to describe a homogeneous premixed combustible mixture or an overall reaction of the form FUEL + OXIDIZER → PRODUCT in which either the fuel or the oxidizer is present in very small quantities.

Under these conditions, the governing equations, in suitably chosen dimensionless variables, in an axially-symmetric geometry we zero velocity are

$$\rho\,\partial_t Y_A = \frac{1}{Le_A}\,(\Delta Y_A + \frac{1}{r}\partial_r Y_A) - K_A\Omega_A \tag{4.1a}$$

$$\rho\,\partial_t T = \Delta T + \frac{1}{r}\partial_r T + QK_A\Omega_A, \tag{4.1b}$$

$$\Omega_A = \rho Y_A e^{-N_A/T}, \tag{4.1c}$$

ρ denotes the density of the mixture, Y_A denotes the mass function of the reactants A, Le_A denotes the Lewis number (the ratio of the thermal conductivity to the mass dif-fusivity) of the reactant A, T denotes the temperature, Q_A denotes the heat released by the reaction, K_A denotes the activation energy, and $\Delta \equiv \partial_r{}^2 + \partial_z{}^2$.

In this form $0 < Y_A < 1$ and in the absence of external heat $T_0 < T < T_f$, where $T_f = 1 + T_0$.

Consider a cylinder with boundary conditions depicted in Figure 1 and with initial conditions

$$T(r,z,0) = T_0,$$

$$Y_A(r,z,0) = 1\ .$$

The premixed gas in the region is ignited at the point (0,0). This is given by the condition

$$T(0,0,t) = T_f.$$

We choose $K_A = 2.5 \times 10^5$, $N_A = 4$, $T_0 = 0.2$, and $Le_A = 1$. In this case $p = 2$ and the flame dictionary is 4-dimensional. $\underline{u}_L$ and $\underline{u}_U$ are two component vectors of the form $(T,Y_A)^T$, where $\underline{u}_L = (T_0,0)^T$ and $\underline{u}_U = (T_f+\varepsilon, 1)^T$, where ε is a small number allowing for the flame temperature to slightly overshoot T_f in the ignition stage. The values of N_1 and N_2 were both chosen to be 6 so that $h_1 = h_2 = 0.2$. The flame dictionary contains $6^4 = 1296$ entries.

Figure 1 depicts isotherms 0.3, 0.5, 0.7, 1.0, and 1.2 at different time intervals.

For further discussion of the results see Sod [14].

References

1. A. J. Chorin, J. Comp. Phys., 22, 517 (1976).
2. ________, J. Comp. Phys., 25, 253 (1977).
3. P. Colella, SIAM J. Sci. Stat. Comp., 3, 76 (1982).
4. H. M. Glaz and T. P. Liu, The Asymptotic Analysis of Wave Interactions and Numerical Calculations of Transonic Nozzle Flow, to appear.
5. J. Glimm, Comm. Pure. Appl. Math., 18, 697 (1965).
6. T. P Liu, Comm. Math. Phys., 68, 141 (1979).
7. ________, Comm. Math. Phys., 83, 243 (1982).
8. G. A. Sod, J. Fluid Mech., 83, 785 (1977).
9. ________, J. Comp. Phys., 27, 1 (1978).
10. ________, SAE Paper No. 800288 (1980).
11. ________, SAE Paper No. 820041 (1982).
12. ________, A Random Choice Method with Application to Reaction-Diffusion Systems in Combustion, Int. J. Comp. Math. with Appl., to appear Dec., 1983.
13. ________, Numerical Methods in Fluid Dynamics, Cambridge University Press, New York (1984).
14. ________, A Flame Dictionary Approach to Unsteady Combustion Phenomena, to appear.
15. ________, A Numerical Study of Oxygen Diffusion in a Spherical Cell with the Michaelis-Menten Oxygen Uptake Kinetics, J. Math. Bio., to appear.

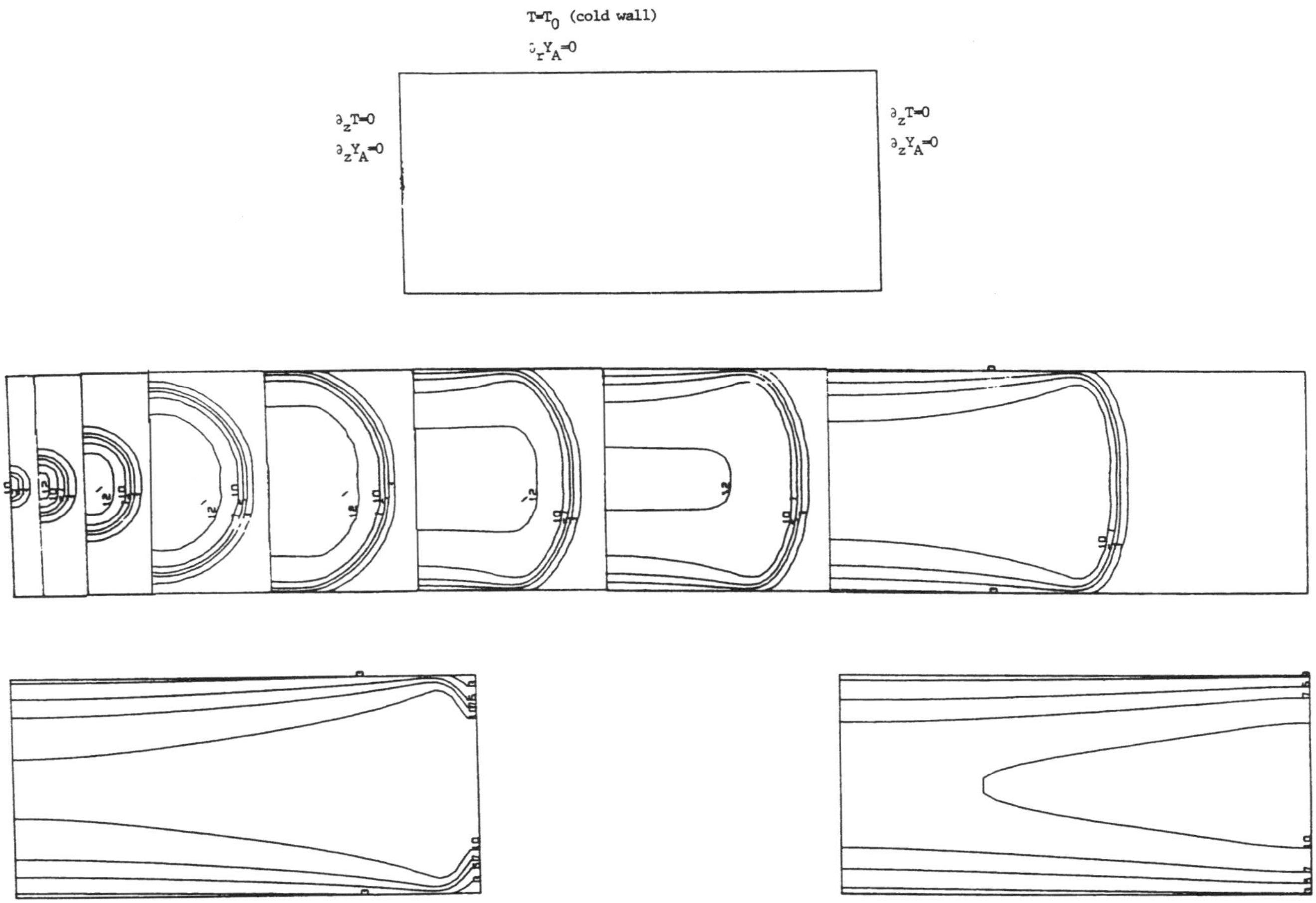

Figure 1
Isotherms at different time intervals

NUMERICAL SOLUTION FOR ENTRY FLOW IN CURVED PIPES OF ARBITRARY CURVATURE RATIO

W.Y. Soh and S.A. Berger
Department of Mechanical Engineering
University of California
Berkeley, CA 94720

I. INTRODUCTION

Fluid flow in curved pipes is of interest for both bio-fluid mechanics and engineering applications. We describe a procedure to obtain numerical solutions of the complete Navier-Stokes equations for entry flow in curved pipes of arbitrary curvature ratio.

II. MATHEMATICAL FORMULATION OF THE PROBLEM

II-1. Governing Equations

The Navier-Stokes equations can be written in conservative form for the toroidal coordinate system (see Fig. 1) as follows:

$$\frac{\partial u}{\partial t} + \frac{1}{rB}\left[\frac{\partial}{\partial r}(rBu^2) + \frac{\partial}{\partial \phi}(Buv) + \frac{\partial}{\partial \theta}(\delta ruw) - Bv^2 - \delta r\cos\phi w^2\right]$$
$$= -\frac{\partial p}{\partial r} + \frac{1}{Re}\left\{\frac{1}{rB}\left[\frac{\partial}{\partial r}(rB\frac{\partial u}{\partial r}) + \frac{\partial}{\partial \phi}(\frac{B}{r}\frac{\partial u}{\partial \phi}) + \frac{\partial}{\partial \theta}(\frac{\delta^2 r}{B}\frac{\partial u}{\partial \theta})\right]\right.$$
$$\left. - \frac{1}{r^2}(2\frac{\partial v}{\partial \phi} + u) + \frac{\delta\sin\phi v}{rB} + \frac{\delta^2\cos\phi}{B^2}(v\sin\phi - u\cos\phi - 2\frac{\partial w}{\partial \theta})\right\} \qquad (2\text{-}1\text{-}1)$$

$$\frac{\partial v}{\partial t} + \frac{1}{rB}\left[\frac{\partial}{\partial r}(rBuv) + \frac{\partial}{\partial \phi}(Bv^2) + \frac{\partial}{\partial \theta}(\delta rvw) + Buv + \delta r\sin\phi w^2\right]$$
$$= -\frac{1}{r}\frac{\partial p}{\partial \phi} + \frac{1}{Re}\left\{\frac{1}{rB}\left[\frac{\partial}{\partial r}(rB\frac{\partial v}{\partial r}) + \frac{\partial}{\partial \phi}(\frac{B}{r}\frac{\partial v}{\partial \phi}) + \frac{\partial}{\partial \theta}(\frac{\delta^2 r}{B}\frac{\partial v}{\partial \theta})\right]\right.$$
$$\left. + \frac{1}{r^2}(2\frac{\partial u}{\partial \phi} - v) - \frac{\delta\sin\phi u}{rB} - \frac{\delta^2\sin\phi}{B^2}(v\sin\phi - u\cos\phi - 2\frac{\partial w}{\partial \theta})\right\} \qquad (2\text{-}1\text{-}2)$$

$$\frac{\partial w}{\partial t} + \frac{1}{rB}\left[\frac{\partial}{\partial r}(rBuw) + \frac{\partial}{\partial \phi}(Bvw) + \frac{\partial}{\partial \theta}(\delta rw^2) + \delta rw(u\cos\phi - v\sin\phi)\right]$$
$$= -\frac{\delta}{B}\frac{\partial p}{\partial \theta} + \frac{1}{Re}\left\{\frac{1}{rB}\left[\frac{\partial}{\partial r}(rB\frac{\partial w}{\partial r}) + \frac{\partial}{\partial \phi}(\frac{B}{r}\frac{\partial w}{\partial \phi}) + \frac{\partial}{\partial \theta}(\frac{\delta^2 r}{B}\frac{\partial w}{\partial \theta})\right]\right.$$
$$\left. + \frac{2\delta^2}{B^2}(\frac{\partial u}{\partial \theta}\cos\phi - \frac{\partial v}{\partial \theta}\sin\phi - \frac{w}{2})\right\} \qquad (2\text{-}1\text{-}3)$$

$$\frac{\partial}{\partial r}(rBu) + \frac{\partial}{\partial \phi}(Bv) + \frac{\partial}{\partial \theta}(\delta rw) = 0, \qquad (2\text{-}1\text{-}4)$$

where $\delta = a/R$ and $B = 1 + \delta r\cos\phi$. Velocities, pressure, position vector, and time have been non-dimensionalized by W_0, ρW_0^2, a, and a/W_0, respectively, where a is the radius of the pipe and W_0 a characteristic velocity. The Reynolds number, Re, is defined as aW_0/ν, and R is the constant radius of curvature of the pipe bend.

II-2. Boundary Conditions

Experiments by Agrawal et al. (1978) show that the flow right after the reservoir develops into an inviscid vortex with its origin at the center of curvature; such a profile is adopted as the inlet condition here. The dimensional velocity at $\theta = 0$ is then $RW_o / (R+r\cos\phi)$, where W_o is the axial velocity at $x = 0$ at the inlet and as well the cross-section average velocity for a wide range of δ with less than 1% error. The boundary conditions are then

$$w(r,\phi,0) = \frac{1}{1+\delta r\cos\phi}, \quad u(r,\phi,0) = v(r,\phi,0) = 0 \qquad \text{(inlet)} \qquad (2\text{-}2\text{-}1)$$

$$u(1,\phi,\theta) = v(1,\phi,\theta) = w(1,\phi,\theta) = 0 \qquad \text{(no-slip condition)} \qquad (2\text{-}2\text{-}2)$$

$$\frac{\partial u}{\partial \theta} = \frac{\partial v}{\partial \theta} = \frac{\partial w}{\partial \theta} = 0 \quad \text{at some } \theta = \theta_d \qquad \text{(far downstream)} \qquad (2\text{-}2\text{-}3)$$

$$\frac{\partial u}{\partial \phi} = \frac{\partial w}{\partial \phi} = 0 \quad \text{and} \quad v = 0 \quad \text{at} \quad \phi = 0,\pi \qquad \text{(plane of symmetry)} \qquad (2\text{-}2\text{-}4)$$

III. NUMERICAL FORMULATION

III-1. Artificial Compressibility

Following Chorin (1967) we introduce the auxiliary system:

$$\frac{\partial u^*}{\partial t} + [(u^*\cdot\nabla)u^* + u^*(\nabla\cdot u^*)] = -\nabla p^* + \frac{1}{Re}\nabla^2 u^*, \quad \xi\frac{\partial p^*}{\partial t} + \nabla\cdot u^* = 0 \quad (3\text{-}1\text{-}1),(3\text{-}1\text{-}2)$$

The solution of this system, u^*, approaches u, the solution of Eqns. (2-1-1) - (2-1-4), as steady state is reached. Eqn. (3-1-2) contains an artificial sound speed $V_c = W_o/\sqrt{\xi}$ and artificial Mach number M, where $M = \sqrt{\xi}\max\sqrt{u^2+v^2+w^2} = \sqrt{\xi}\,q_{max}$. It is necessary that $M < 1$, therefore $\xi < 1/q_{max}^2$.

To be solved then are Eqns. (2-1-1) - (2-1-3) together with the modified continuity equation

$$\xi\frac{\partial p}{\partial t} + \frac{1}{rB}\left[\frac{\partial}{\partial r}(rBu) + \frac{\partial}{\partial\phi}(Bv) + \frac{\partial}{\partial\theta}(\delta rw)\right] = 0 \qquad (3\text{-}1\text{-}3)$$

with boundary conditions (2-2-1) - (2-2-3), and initial conditions

$$u = v = 0, \quad w = \frac{1}{1+\delta r\cos\phi}, \quad p = -\frac{1}{2(1+\delta r\cos\phi)^2}. \qquad (3\text{-}1\text{-}4)$$

III-2. Finite Difference Formulation

In writing the equations in finite-difference form, a non-uniform staggered grid arrangement was used (see Figs. 2 - 4) such that pressure is defined at the center of its cell and u, v, w are defined at different positions on the pressure cell boundaries. This non-uniform grid was chosen to allow accurate solutions in regions where flow conditions change rapidly.

Allen and Cheng's (1970) method for explicit differencing in time is adopted. Knowing u, v, w and p at $t = n\Delta t$, we can obtain these quantities at the (n + 1)th time step. An intermediate time step $\bar{t} = (n + 1/2)\Delta t$ is introduced, which makes it possible to achieve second order accuracy in t and allows for a larger time interval Δt. Therefore, an intermediate sweep is written as

$$\frac{\bar{\underset{\sim}{u}} - \underset{\sim}{u}^n}{1/2\Delta t} + \underset{\sim}{F}(\underset{\sim}{u}^n) = -\underset{\sim}{G}(p^n) + \underset{\sim}{H}(\underset{\sim}{u}^n; \bar{\underset{\sim}{u}}), \quad \xi\frac{\bar{p}-p^n}{1/2\Delta t} + D(\bar{\underset{\sim}{u}}) = 0 \qquad (3\text{-}2\text{-}1),(3\text{-}2\text{-}2)$$

which is followed by a regular sweep as

$$\frac{\underset{\sim}{u}^{n+1} - \underset{\sim}{u}^n}{\Delta t} + \underset{\sim}{F}(\bar{\underset{\sim}{u}}) = -\underset{\sim}{G}(\bar{p}) + \underset{\sim}{H}(\bar{\underset{\sim}{u}}; \underset{\sim}{u}^{n+1}) \qquad (3\text{-}2\text{-}3)$$

$$\xi\frac{p^{n+1} - p^n}{\Delta t} + D(\underset{\sim}{u}^{n+1}) = 0 \,. \qquad (3\text{-}2\text{-}4)$$

where $\underset{\sim}{F}(\underset{\sim}{u})$ denotes the convection of momentum and $\underset{\sim}{G}(p)$ the pressure gradient; $\underset{\sim}{H}(\underset{\sim}{u}^{\ell}; \underset{\sim}{u}^m)$ is the viscous diffusion of momentum, evaluated at the ℓ-th time step if not at the point (i,j,k), and at the m-th time step if at the point (i,j,k). $D(\underset{\sim}{u})$ denotes the divergence of the velocity.

Special treatment must be given of the cell whose center is at r = 0.

IV. RESULTS AND DISCUSSION

Calculations were carried out for Dean numbers, defined as $k = (a/R)^{1/2}\, 2a\, W_0/\nu$, in the range $108.2 \le k \le 680.3$ for $\delta = 1/7$ and 1/20. Central differences were employed for the cross flow velocities, u and v, and also for w for the first four steps in θ (up to $\theta = 10.5°$) for the case of $\delta = 1/7$, and for $w \le 0.15$ for any δ and θ. Upwind differencing in w was used for $w \ge 0.15$.

The coefficient of artificial compressibility ξ and the time step Δt were taken as 0.3 and 0.006, respectively. Non-uniform 14 X 19 intervals in the r- and ϕ-directions were used throughout, and from 17 to 30 steps in θ. Convergence usually took 9,000-14,000 regular time steps, depending on the values of δ, Re and θ_d.

Figure 6 shows for a typical case the development of the axial flow on the plane of symmetry. As the flow enters the curved bend, boundary layers begin to develop, with the boundary layer near the inner bend growing much faster than that near the outer one. In the early stages of development the potential core (the flat region in the velocity profile) appears conspicuous; this region diminishes in size as the flow develops further and far downstream the entire flow passage is occupied by the boundary layers. The location of the maximum axial velocity shifts toward the outer wall as the flow proceeds downstream.

References

Agrawal, Y., Talbot, L. and Gong, K. (1978), "Laser Anemometer Study of Flow Development in Curved Circular Pipes," J. Fluid Mech. 85, 497.

Allen, J.S. and Cheng, S.I. (1970), "Numerical Solutions of the Compressible Navier-Stokes Equations for the Laminar Near Wake," Phys. Fluids 13, 37.

Chorin, A.J. (1967), "A Numerical Method for Solving Incompressible Viscous Flow Problems", J. Comp. Physics 2, 12.

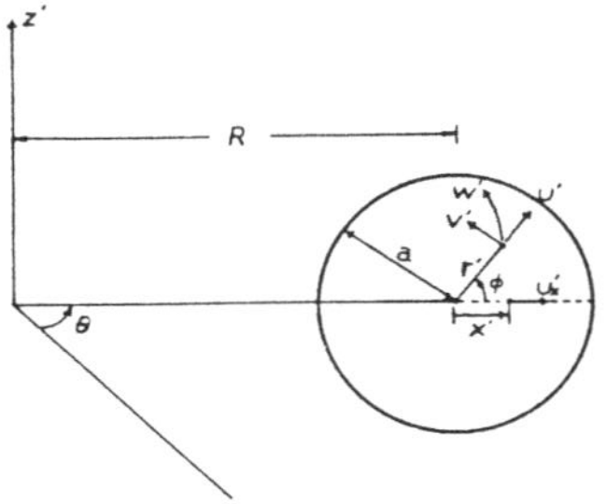

FIG 1 Toroidal coordinate system

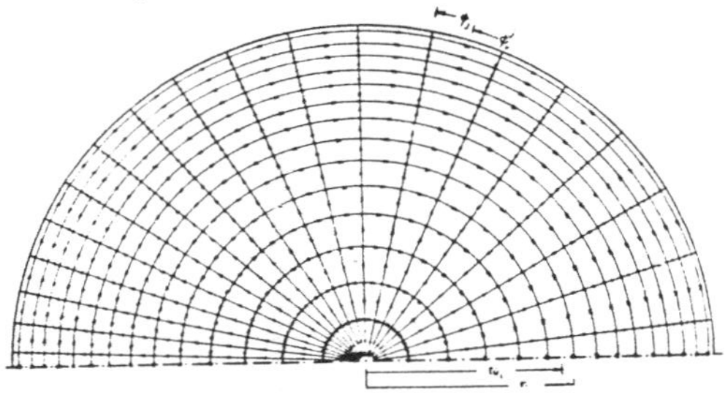

FIG 2 Grid System in the Cross-Section.
•, pressure and axial velocity; ×, circumferential velocity;
↑, radial velocity.

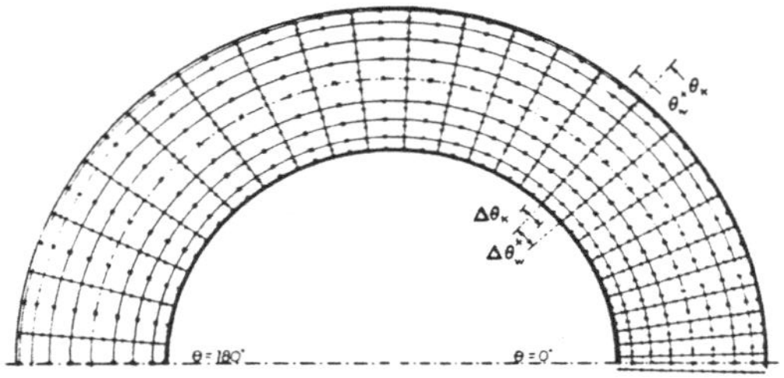

FIG 3. Grid System on the Plane of Symmetry.
•, pressure; ×, axial velocity; ↑, radial velocity.

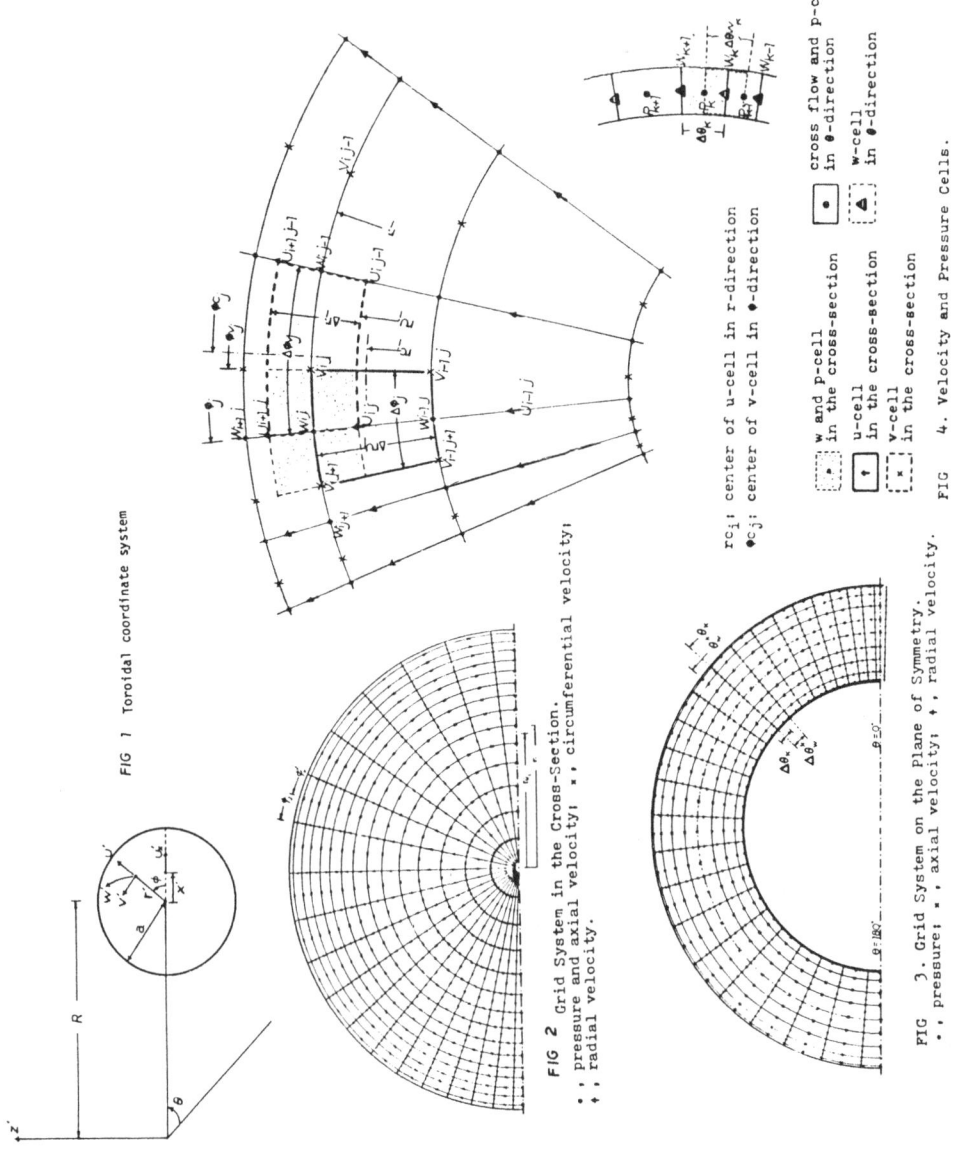

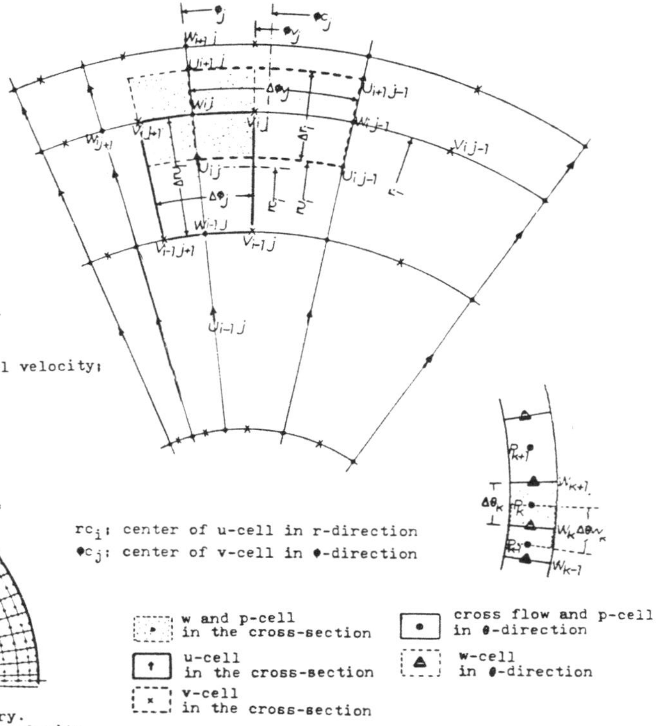

FIG 4. Velocity and Pressure Cells.

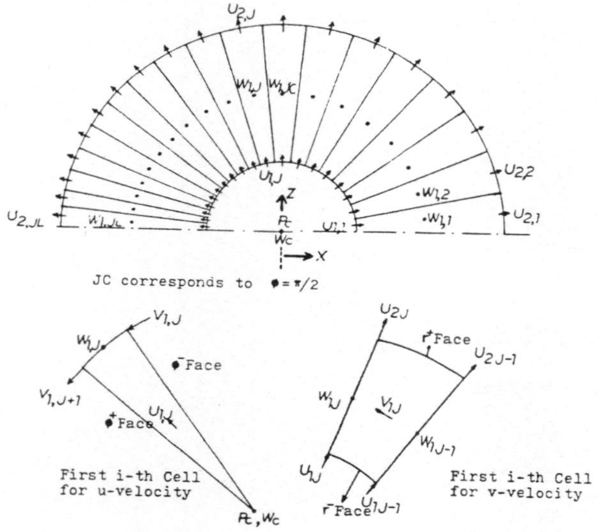

FIG 5. Center Cell for the Pressure and the Axial Velocity.

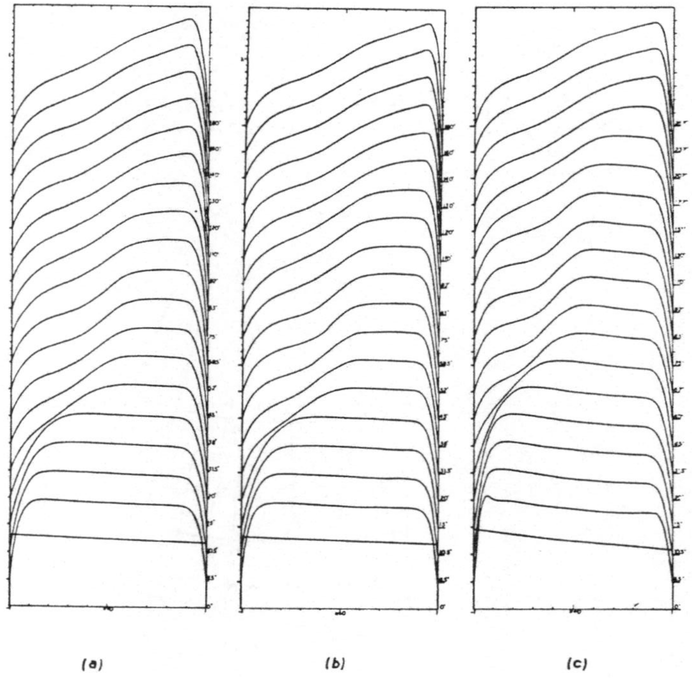

FIG 6

Development of axial velocity on the plane of symmetry for high k.
(a) k = 564.8 (Re = 1263, δ = 1/20),
(b) k = 670.8 (Re = 1500, δ = 1/20)
(c) k = 680.3 (Re = 900, δ = 1/7).

NUMERICAL SIMULATION OF BOUNDARY-LAYER TRANSITION

P. R. Spalart
NASA Ames Research Center
Moffett Field, CA 94035 USA

I. FORMULATION

The transition to turbulence in boundary layers is investigated by direct numerical solution of the nonlinear, three-dimensional, incompressible Navier-Stokes equations in the half-infinite domain over a flat plate. Periodicity is imposed in the x-direction (streamwise) and in the z-direction (spanwise); the y-coordinate extends from 0 to ∞.

The simulation is periodic in the x-direction, unlike the experiments in which the boundary-layer thickness grows in the x-direction. A forcing term is added to the x-momentum equation that approximates the convection terms associated with this spatial growth. The approximation is based on boundary-layer assumptions (applied only to the mean flow) and the self-similarity of the mean-velocity profile. With this forcing applied, the laminar velocity profile, instead of becoming an error function and thickening without bounds, is a Blasius profile. Thus, the stability characteristics are very close to the experimental characteristics. Furthermore, the equation can be written in a moving reference frame, so that the boundary-layer thickens in time while retaining a Blasius profile. The procedure of adding a forcing term allows the disturbances to extract energy from the mean flow, and is much preferable to a procedure in which the mean-velocity profile would be imposed.

The spatial representation is spectral in all directions [1]. The basis functions that are used represent divergence-free velocity fields and satisfy the boundary conditions as suggested by Leonard and Wray [2]. Leonard and Wray applied a weak formulation, which eliminates the pressure and allows an accurate and straightforward time-advance scheme. Leray's weak formulation is used here [3]. An advantage of this formulation over Leonard's is that it keeps the numerical Stokes operator real, symmetric, and negative-definite. On the other hand, Chebyshev polynomials cannot be used.

The x- and z-directions are treated by Fourier series. In the y-direction, the velocity field is first split into "irrotational" and "vortical" components in order to better accommodate two different length scales. The length scale of the vortical component is δ, the thickness of the boundary layer. The length scale of the irrotational components is Λ, the wavelength in the (x,z) plane, which is significantly larger than δ. This irrotational component can be represented by a single exponential function for each horizontal wave-vector. To represent the vortical component, an exponential mapping is applied from $[0,\infty[$ into $[0,1[$, and shifted Jacobi polynomials are used in the transformed coordinate. The vortical component is infinitely differentiable over the closed interval [0,1], so that the convergence of the polynomial method will be faster than algebraic. The cost of the transforms from real space to Jacobi space is of the order of N^2. Figure 1 is a plot of the first few basis-functions versus y. All the functions decay exponentially as $y \rightarrow \infty$ but the first function, which includes the irrotational component, decays much more slowly than the other ones.

The time-advance scheme is hybrid and second-order accurate. The convection terms are treated by a Runge-Kutta scheme which is explicit, third-order accurate, and conditionally stable; the Stokes terms are treated by the Crank-Nicolson scheme.

II. RESULTS

In order to check the convergence of the method, the Orr-Sommerfeld equation was solved for a Blasius profile and for a real wave-number. This problem is known to produce a few discrete eigenvalues and a continuous spectrum on the $C_r = 1$, $C_i < 0$ axis [4]. Figure 2 is a contour plot of the error in the principal discrete eigenvalue as a function of Yo, the length scale of mapping, and Ny, the number of points in the y-direction. The convergence as $Ny \rightarrow \infty$ with Yo fixed is very fast. It is expected to be faster than algebraic, but not as fast as exponential [5]. The plot also indicates the optimum value of Yo: about $2\delta^*$. Figure 3 shows that the numerical spectrum includes a string of eigenvalues that becomes denser and tends to the $C_r = 1$, $C_i < 0$ axis. Its convergence is much slower than that for the discrete eigenvalues; the reason is that the corresponding eigenfunctions behave like sine waves as $y \rightarrow \infty$, which makes them hard to approximate with the expansion functions in Fig. 1.

The early nonlinear stages of transition of a Blasius boundary layer, disturbed by a vibrating ribbon, were then simulated in three dimensions. A two-dimensional Tollmein-Schlichting (TS) wave of finite amplitude was introduced in the initial field, as well as three-dimensional white noise of much lower energy. The streamwise period was twice that of the TS wave; the spanwise period was chosen much longer to avoid constraining the spectrum. Spanwise lines of particles were introduced, near the critical layer, to simulate the smoke lines used in experiments.

Figure 4 summarizes the time-evolution of the flow. The energy of the fundamental TS wave and the energy carried by all the other wave-vectors are plotted separately. The TS wave grows from branch I to branch II of the TS stability diagram, then starts decaying. The energy of the other modes remains small until after the flow crosses branch II; then it grows very rapidly, and nonlinear interactions take place. The shape factor H of the boundary layer remains at the Blasius value of 2.6 until transition occurs; then it rapidly decreases. The agreement with Kachanov's experiments is excellent [6].

Simulations were conducted with the same background noise, but different values for the TS wave amplitude. Figure 5 contains top views of the particles in the boundary layer. If the maximum TS wave amplitude is less than 0.3%, transition does not occur. In Fig. 5(a), with an amplitude of 0.9%, three-dimensional breakdown occurs and is of the subharmonic or "H" type (the lambda-shaped particle lines are staggered).

In Fig. 5(b), with amplitude 5%, the lambda patterns are not staggered, indicating a Klebanoff-type breakdown. The patterns appear "broken," a result of the randomness of the initial three-dimensional disturbance. The qualitative agreement with Saric's experiments is good [7].

Figure 6 is a plot of the spectrum in an (x,z) plane at the beginning of an H-type breakdown. The fundamental TS wave still dominates the spectrum; its higher harmonic is also present. The growing three-dimensional subharmonic component is obvious; the wave number and the broadband character of the instability agree very well with Herbert's small disturbance theory [8].

III. REFERENCES

1. Gottlieb, D. and Orszag, S. A., "Numerical Analysis of Spectral Methods," NSF-CMBS Monograph No. 26, Society of Industrial and Applied Mathematics, Philadelphia, Penn., 1977.
2. Leonard A. and Wray, A., "A New Numerical Method for the Simulation of Three-Dimensional Flow in a Pipe," NASA TM-84267, 1982.
3. Temam, R., "Navier-Stokes Equations and Nonlinear Functional Analysis," NSF-CMBS Monograph No. 41, Society of Industrial and Applied Mathematics, Philadelphia, Penn., 1983.
4. Grosch, C. E. and Salwen, H., "The Continuous Spectrum of the Orr-Sommerfeld Equation. Pt. 1. The Spectrum and the Eigenfunctions," J. Fluid Mech., Vol. 87, Pt. 1, 1978, pp. 33-54.
5. Boyd, J. P., "The Optimization of Convergence for Chebyshev Polynomial Methods in an Unbounded Domain," J. Comp. Phys., Vol. 45, No. 1, 1982, pp. 43-79.
6. Kachanov, Y. S. and Levchenko, V. Y., "The Resonant Interaction of Disturbances at Laminar-Turbulent Transition in a Boundary Layer," J. Fluid Mech., Vol. 138, 1984, pp. 209-247.
7. Saric, W. S., Kozlov, V. V., and Levchenko, V. Y., "Forced and Unforced Subharmonic Resonance in Boundary-Layer Transition," AIAA Paper 84-0007, 1984.
8. Herbert, T., "Analysis of the Subharmonic Route to Transition in Boundary Layers," AIAA Paper 84-0009, 1984.

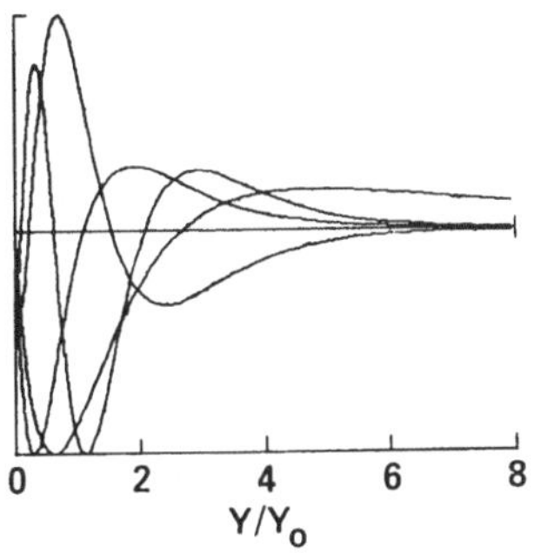

Fig. 1 First four basis-functions.

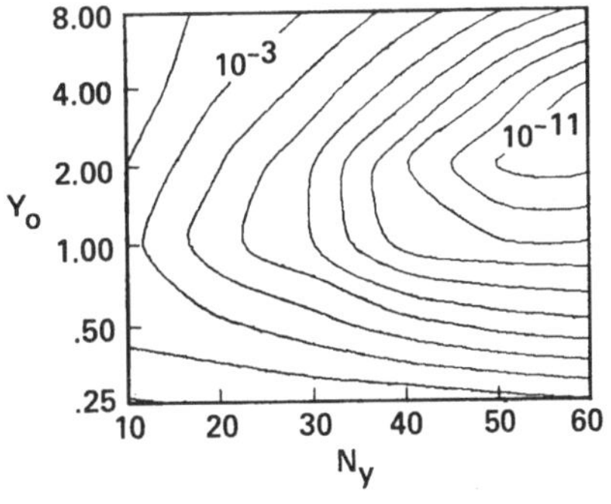

Fig. 2 Error in Orr-Sommerfeld eigenvalue.

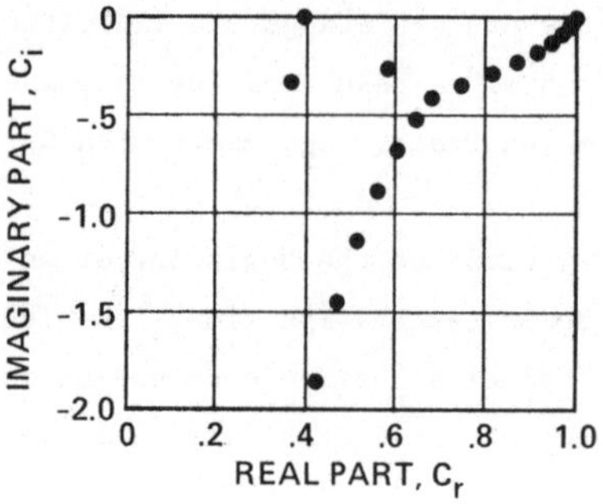

(a) Ny = 50.

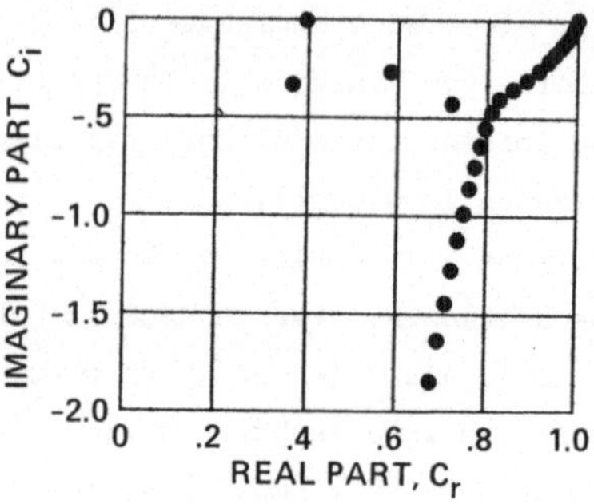

(b) Ny = 100.

Fig. 3 Numerical spectrum: Yo = 1.5.

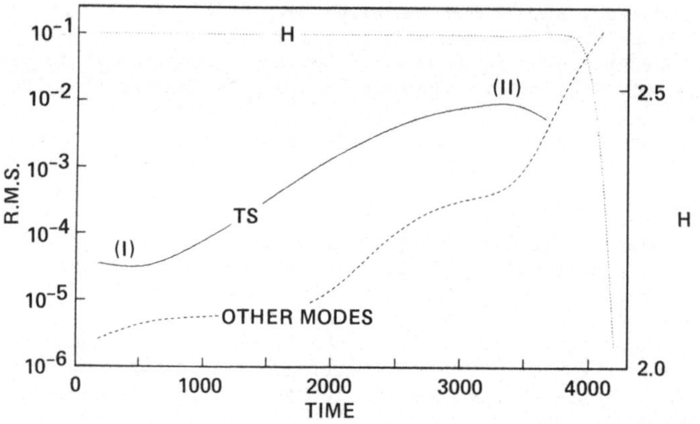

Fig. 4 Time-evolution of the disturbance energy and of the shape factor.

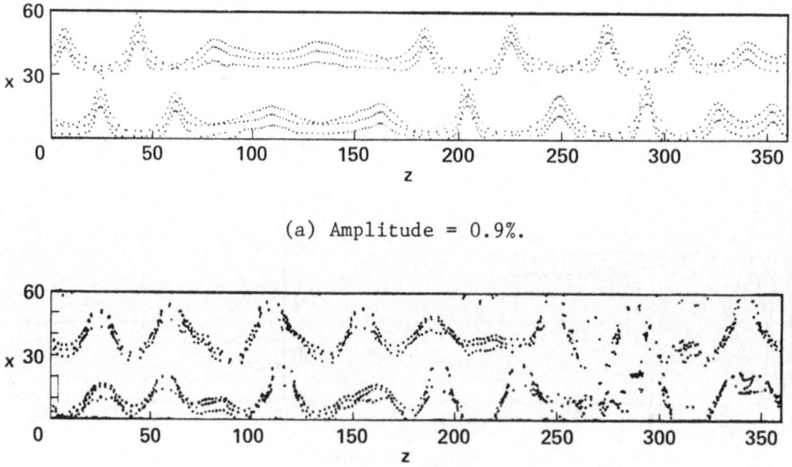

(a) Amplitude = 0.9%.

(b) Amplitude = 5%.

Fig. 5 Smoke lines in transitioning boundary layer.

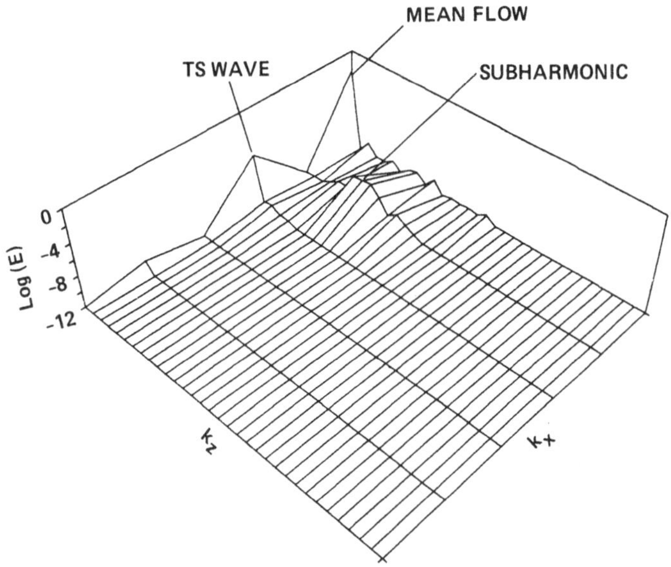

Fig. 6 Two-dimensional spectrum.

SPECTRAL METHODS FOR AERODYNAMIC PROBLEMS

C. L. Streett and P. F. Bradley
NASA Langley Research Center
Hampton, VA 23665
U.S.A.

Spectral methods approximate the solution to a differential equation by a finite series of global basis functions. The coefficients in the series expansion are determined by a projection method such as Galerkin or collocation. These methods offer certain advantages in accuracy and efficiency over finite difference techniques. Only recently have these methods been applied in aerodynamics (references 1-3).

In the present paper, a few such applications are described and results are presented showing the advantages and capabilities of the spectral collocation method. The applications include transonic potential flow past an arbitrary two-dimensional airfoil, supersonic potential flow about a conical body, and two-dimensional boundary layer flow with either external pressure gradient or displacement thickness prescribed.

The spectral discretization of the transonic full potential equation results in a set of nonlinear algebraic equations. This set of equations is solved by a relaxation scheme based on the generalized ADI method. The convergence is further accelerated by a multigrid procedure. In figures 1 and 2 are shown the surface pressure coefficient distributions from two typical solutions for flow about an NACA 0012 airfoil; figure 1 for subcritical flow and figure 2 for supercritical flow. Since the spectral solution is an expansion in terms of continuous basis functions, the solution contains more information than is apparent from just the mesh point data. The solid line in figure 1 denotes such an expanded solution, showing far more detail near the leading edge of the airfoil.

Truncation error of the spectral solution decays rapidly with mesh refinement. In figure 3 is shown the error in the lift coefficient predicted by the spectral method versus average mesh spacing in the computational plane. The solid curve indicates the relative error of a second-order method. The superior accuracy of the spectral solution is not surprising, as the solution is smooth. Shown in figure 4 are the convergence histories of the spectral multigrid solution and two state-of-the-art finite difference solutions obtained on grids yielding equivalent accuracy. As can be seen, the spectral multigrid method requires substantially less computer time to produce an equivalent solution for subcritical flow.

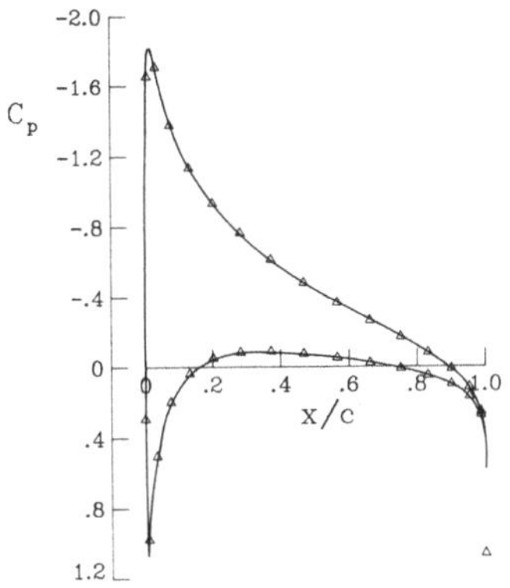

Figure 1

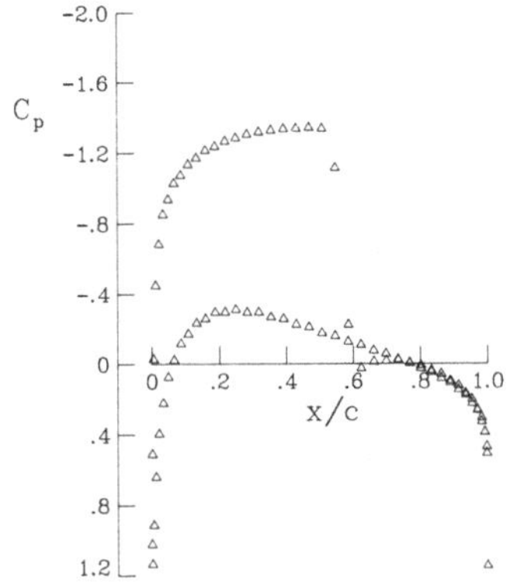

Figure 2

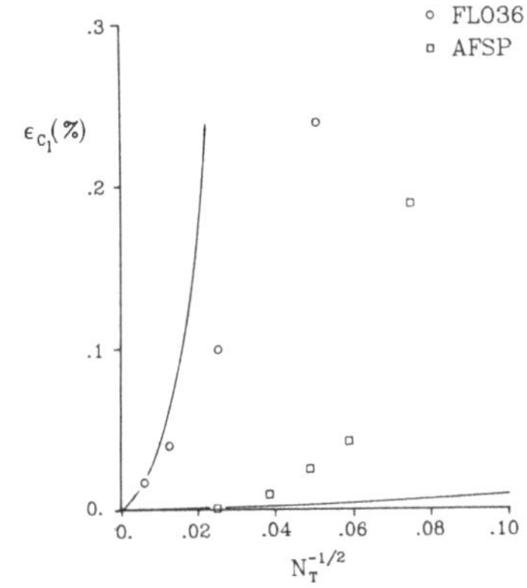

Figure 3

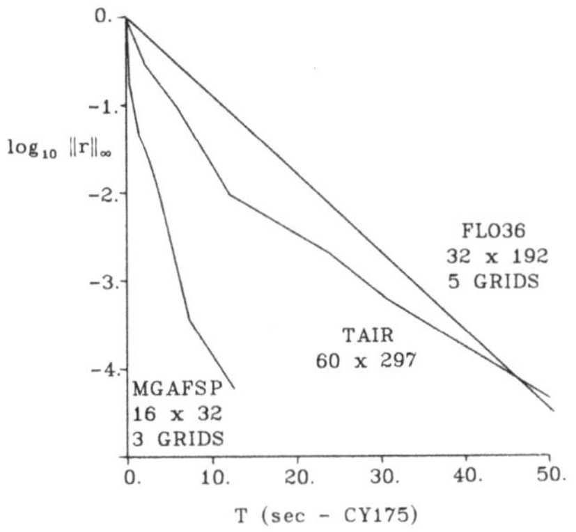

Figure 4

Somewhat surprisingly, the same rapid truncation error decay is seen for supercritical flow through the use of a formally-second-order-accurate artificial density technique. Details of the method and further results are found in reference 1.

A discretization technique similar to that used in the above was applied to the problem of supersonic conical potential flow. Shown in figure 5 are the Mach number contours in the crossplane for flow about a 10:1 elliptic cone at angle of attack. In this problem, however, since all shocks are captured and regions of supersonic flow are extensive, the low-order upwind bias/artificial viscosity disturbs the accuracy of the spectral method significantly; and the advantage of the spectral method over finite differencing is lost.

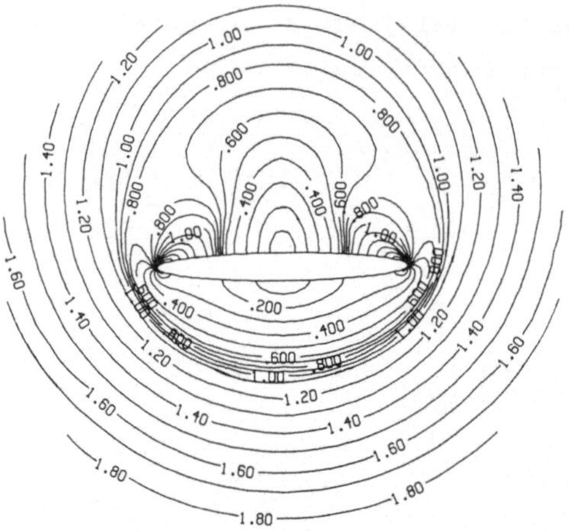

Figure 5

For the application of the spectral method to boundary layer flow, the Gortler formulation of the relevant equations is used; either Chebyshev or Legendre polynomials are used for the discretization in the direction normal to the surface. Solution is by a preconditioned Richardson iteration technique (references 6 and 3); coupling between the momentum and continuity equations is maintained in the update scheme. Figure 6 shows that the error in the displacement thickness predicted by the spectral method for the Blasius equation decays far more rapidly with increasing grid points than for a fourth-order box scheme.

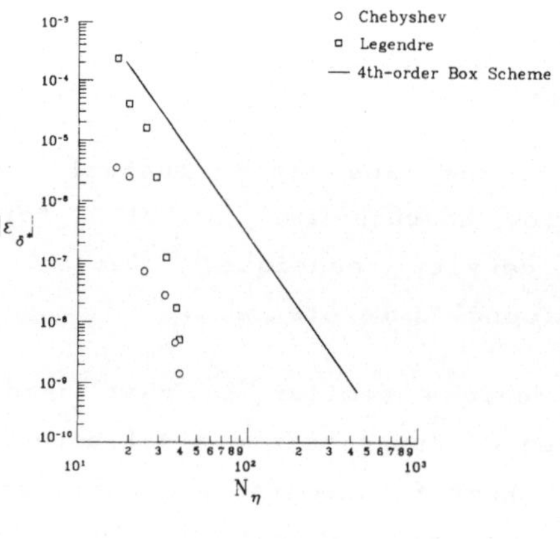

Figure 6

For the nonsimilar equations, two different discretizations are examined in the streamwise direction. The first technique involves marching in the streamwise direction, and employs a second order accurate finite difference discretization in that direction while keeping the spectral discretization in the normal direction. The second technique, called the fully spectral scheme, uses a spectral representation in both directions. Figure 7 is concerned with the inverse boundary layer formulation, where the displacement thickness is prescribed. The choice of the displacement thickness distribution shown in Figure 7a produces a fairly large region of reverse flow. The marching scheme must employ the Reyhner-Flugge Lotz approximation (reference 7) to handle the reverse flow region, whereas the fully spectral scheme needs no such approximation. For this test problem, the performance of these two methods is checked by comparing their accuracy in predicting the skin friction at the streamwise station $\xi = 1.61$ (indicated by the arrow in figure 7b). The marching scheme requires about 60 points in the streamwise direction to achieve two decimal place accuracy and about 200 points to achieve three decimal place accuracy. By contrast, only 20 and 26 points were required by the fully spectral scheme to achieve two and three decimal place accuracy, respectively.

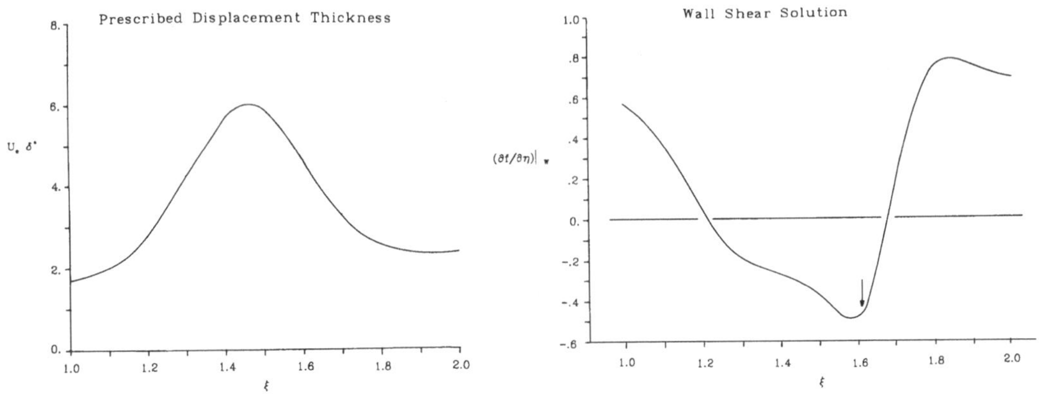

(a) (b)

Figure 7

The problems of current engineering interest considered here demonstrate the power and range of applicability of spectral methods. The efficiency advantage of these spectral techniques over finite differences for problems with smooth solutions is clearly indicated. Despite the unexpectedly good performance of the spectral discretizations used here for problems containing discontinuities (shocks), this significant efficiency advantage is essentially lost. The applications we have demonstrated here were straightforward in execution; more sophisticated techniques may improve the latter conclusion.

REFERENCES

1. Streett, C. L.: "A Spectral Method for the Solution of Transonic Potential Flow about an Arbitrary Airfoil," AIAA Paper 83-1949-CP, July 1983.

2. Streett, C. L., Zang, T. A., and Hussaini, M. Y.: "Spectral Multigrid Methods with Applications to Transonic Potential Flow," ICASE Report No. 83-11, 1983.

3. Streett, C. L., Zang, T. A., and Hussaini, M. Y.: "Spectral Methods for Solution of the Boundary-Layer Equations," AIAA Paper 84-0170-CP, January 1984.

4. Bradley, P. F., Dwoyer, D. L., South, J. C., and Keen, J. M.: "Vectorized Schemes for Conical Potential Flow using the Artifical Density Method," AIAA Paper 84-0162-CP, January 1984.

5. Wornom, S. F.: "Critical Study of Higher Order Numerical Methods for Solving the Boundary-Layer Equations," NASA TP-1302, November 1978.

6. Orszag, S. A.: "Spectral Methods for Porblems in Complex Geometries," J. Comput. Phys., vol. 37, 1980, pp. 70-92.

7. Carter, J. E.: "Inverse Boundary-Layer Theory and Comparison with Experiment," NASA TP-1208, July 1978.

TIME-DEPENDENT INVERSE SOLUTION OF THREE-DIMENSIONAL, COMPRESSIBLE, TURBULENT, INTEGRAL BOUNDARY-LAYER EQUATIONS IN NONORTHOGONAL CURVILINEAR COORDINATES*

T. W. Swafford
Sverdrup Technology, Inc./AEDC Group
Arnold Air Force Station, TN 37389 USA

INTRODUCTION

The objective of this paper is to present a time-dependent inverse boundary-layer calculation method that can be used for compressible, turbulent flows over three-dimensional adiabatic surfaces. The inverse formulation is developed from the three-dimensional time-dependent momentum and mean-flow kinetic energy integral boundary-layer equations cast in nonorthogonal curvilinear coordinates [1,2]. The streamwise and crossflow velocity profiles are those of Whitfield [3] and Johnston [4], respectively. Integral lengths resolved in the streamwise coordinate system are related to those in the nonorthogonal system using relationships as suggested by Smith [5]. Comparisons between measured and computed boundary-layer quantities for an infinite swept wing are presented.

ANALYSIS - Inverse Formulation

The x_1 and x_2 momentum, and mean-flow kinetic energy integral boundary-layer equations written in nonorthogonal curvilinear coordinates are:

$$\frac{1}{\bar{\rho}\,\bar{q}^2}\left[\frac{\partial}{\partial t}\left(\bar{\rho}\,\bar{q}\delta_i^*\right) - \bar{u}_i\frac{\partial}{\partial t}\left(\bar{\rho}\,\theta_\rho\right)\right] = \ell_i,\ i = 1 \text{ or } 2 \tag{1}$$

$$\begin{aligned}&\frac{1}{2\bar{\rho}\,\bar{q}^3}\frac{\partial}{\partial t}\left[\bar{\rho}\,\bar{q}^2\left(\theta_{11}+\theta_{22}\right)+\bar{\rho}\,\bar{q}\left(\bar{u}_1\delta_1^*+\bar{u}_2\delta_2^*\right)-\bar{\rho}\theta_\rho\left(\bar{u}_1^2+\bar{u}_2^2\right)\right]\\ &+\frac{1}{\bar{q}^2}\left[\left(\frac{\bar{u}_1}{\bar{q}}\theta_\rho-\delta_{u_1}^*\right)\frac{\partial\bar{u}_1}{\partial t}+\left(\frac{\bar{u}_2}{\bar{q}}\theta_\rho-\delta_{u_2}^*\right)\frac{\partial\bar{u}_2}{\partial t}\right] = L\end{aligned} \tag{2}$$

(Overbars denote edge condition).

where in Eq. (1), setting i = 1 and i = 2 results in the x_1 and x_2 momentum integral equations, respectively, and

δ_i^*, $\delta_{u_i}^*$, θ_{ij}, θ_ρ - integral lengths resolved in nonorthogonal coordinate system

$\bar{q}$, $\bar{u}_1$, $\bar{u}_2$ - magnitude of edge velocities, (resultant and x_1 and x_2 components, respectively)

ℓ_i, L - terms containing spatial derivatives and metric coefficients

Both direct and inverse formulations result from expanding Eqs. (1) and (2) and making appropriate choices for the dependent variables (U) and variables to be specified (S). In a direct formulation, the choice of these variables is straightforward (see [1]). However, as pointed out by Wigton, et al. [6] and Samant, et al. [7], the number of possible choices of U and S for the steady inverse three-dimensional equations is quite large. For the steady equations, Wigton [6] searched for the overall "optimum" set of variables which would allow the system of equations to remain hyperbolic in the transonic regime, and not cause the system to be ill-conditioned. For the present system of time-dependent equations, it is therefore

*The research reported herein was performed by the Arnold Engineering Development Center (AEDC), Air Force Systems Command. Work and analysis for this research were done by personnel of Sverdrup Technology, Inc., operating contractor for the AEDC propulsion test facilities. Further reproduction is authorized to satisfy needs of the U. S. Government.

appropriate to seek an analogous set of variables which will yield similar characteristics, as well as allow the coupling to an inviscid solver to be straightforward.

In order to determine this "optimum" set of variables for the time-dependent equations, Eqs. (1) and (2) are rewritten in the form of

$$A \frac{\partial U}{\partial t} + B \frac{\partial U}{\partial x_1} + C \frac{\partial U}{\partial x_2} + d = 0 \tag{3}$$

where A, B, and C are (3 x 3) matrices and U is the dependent variable vector. Obviously, the elements of A, B, and C will vary with the choice of the components of U. Therefore, the task is to find a suitable U which will cause the eigenvalues of the matrices $A^{-1}B$ and $A^{-1}C$ to be real and distinct.

Several choices of U for the inverse case have been tested with none yielding a distinct advantage over the others with respect to the system remaining hyperbolic over the range of flow variables expected. However, one choice of U and S is particularly suited for viscous/inviscid interaction calculations because of the ease of coupling, and is given by

$$\begin{aligned} U &= (\Theta_{11}, \bar{q}, A)^T \\ S &= (\Delta_1^*, \alpha)^T \end{aligned} \tag{4}$$

where Δ_1^* and Θ_{11} are streamwise displacement and momentum thicknesses, respectively, A is a parameter in the Johnston crossflow profile, and α is the angle directed from the x_1 coordinate axis to the local resultant edge velocity vector. If Eq. (4) is used in a viscous/inviscid interaction code, updating of the vector S after each inviscid pass could be accomplished by applying Carter's method [8] for Δ_1^*, and using the latest α determined from the local inviscid solution at the wall.

However, this choice of U and S dictates that certain time-dependent terms must be discarded because their values are unknown. (Note this problem does not arise when the direct formulation is used.) Guidance for this task stems from the work reported by Donegan [9] who successfully solved the corresponding two-dimensional problem. The approach taken herein was to eliminate those terms which result in the present system of equations reducing to those used in [9] for the special case of two-dimensional flow. Results presented in the following sections were obtained using equations ensuing from this rationale with all terms on the right-hand side involving time derivatives set equal to zero. Although this particular formulation reduces to the two-dimensional equations used in [9], there are possibly other formulations which will simplify to the same result.

NUMERICAL METHODS

The primary numerical method used to solve the present system of equations was a four-stage Runge-Kutta (R-K) scheme using upwind spatial differencing, local time-stepping, and CFL = 1.3 with no artificial smoothing. The use of this scheme was motivated by the success shown in [1] in solving the direct formulation of Eqs. (1) and (2). Boundary and initial conditions used herein are as described in [1]. Shown in Fig. 1 as the solid line is the convergence history using this scheme for the low speed infinite swept wing case of van den Berg, et al. [10]. Inputs used to obtain this solution were the measured Δ_1^*, and edge flow angles α which would have been present under infinite swept wing conditions (hereafter referred to as "analytic α's"). Also shown in Fig. 1 as the dashed line is the convergence obtained using as input the same Δ_1^* distribution, but the measured flow angles. Although the maximum difference in the two axial flow angle distributions is less than 2.5 deg, it can be seen that convergence using the R-K scheme is very sensitive to these relatively small differences. Numerical experiments indicate that divergence of the solution using this scheme seems to begin when eigenvalues of the $A^{-1}B$ and $A^{-1}C$ matrices at some mesh points become complex early in the solution process.

The latter set of inputs was then used in conjunction with the predictor-corrector method of MacCormack (M) [11] with "omega factor" smoothing as suggested by Kneile [12]. Although the formal accuracy of the numerical scheme is reduced, this type of artificial dissipation increases stability with little impact on efficiency. The dotted line in Fig. 1 represents the convergence obtained with this scheme ($\omega = 2$, CFL = 0.5). The obvious advantage of the M-scheme over the R-K scheme is that a solution could be obtained using the measured flow angles as input, but at the expense of additional computational time. However, as shown in the following section, some flow parameters from this solution near the trailing edge are predicted rather poorly. Furthermore, although the solution is converged, some eigenvalues of the $A^{-1}B$ and $A^{-1}C$ matrices are complex for points near the trailing edge.

RESULTS

Comparisons between computed and measured boundary-layer quantities are given in Fig. 2 [10]. Results from the present method are given for both the measured and analytical flow angle distributions. Also shown are the results obtained from the direct formulation solution reported in [1] (the direct solution was obtained using the analytically determined α's). Comparison of the inverse and direct solutions to the measured data using the analytical α's as input show a definite superiority of the inverse method. The same is true for the inverse solution using measured α's as input except near the trailing edge; whereas the solution for Θ_{11} compares well with the experimental data in this region, computed c_f and β_w are much too high. The most probable cause of these discrepancies is the crossflow representation and the resulting relationship between β_w, c_f, and A, which is not accurate in this region.

SUMMARY AND CONCLUSIONS

A method for computing three-dimensional, turbulent, compressible boundary layers in a time-dependent inverse mode has been developed and results compared with measurements. The three-dimensional inverse formulation is well suited for incorporation into an inviscid solver for viscous/inviscid interaction calculations.

It can be concluded that although the inverse formulation appears to yield more accurate solutions compared to solutions obtained from the direct formulation, the present inverse form of the equations has coefficient matrices which yield complex eigenvalues for some flow conditions. For the van den Berg infinite swept wing case, this appears to result in instabilities when solving the equations using a four-stage Runge-Kutta scheme, whereas the MacCormack scheme enabled a converged solution to be obtained. However, the latter solution magnifies the inherent inadequacies of integral methods which must rely upon empirical correlations being accurate over a very wide range of flow conditions. Accuracy of the present integral method will therefore be limited until more general correlations, particularly the crossflow velocity profile model, are incorporated into the code.

REFERENCES

1. Swafford, T. W., "Three-Dimensional, Time-Dependent, Compressible, Turbulent, Integral Boundary-Layer Equations in General Curvilinear Coordinates and Their Numerical Solution," Ph.D Dissertation, Mississippi State University, Mississippi State, MS, August 1983 (Also AEDC-TR-83-37, Arnold Air Force Station, TN, September 1983).

2. Swafford, T. W. and Whitfield, D. L., "Numerical Solutions of Three-Dimensional Time-Dependent Compressible Turbulent Integral Boundary-Layer Equations in General Curvilinear Coordinates," AIAA Paper No. 83-1674, July 1983.

3. Whitfield, D. L., "Analytical Description of the Complete Two-Dimensional Turbulent Boundary-Layer Velocity Profile," AEDC-TR-77-79, Arnold Air Force Station, TN, September 1977 (also AIAA Paper No. 78-1158, July 1978).

4. Johnston, J. P., "Three-Dimensional Turbulent Boundary Layers," M.I.T. Gas Turbine Lab, Report 39, 1957.

5. Smith, P. D., "An Integral Prediction Method for Three-Dimensional Compressible Turbulent Boundary Layers," RAE R&M No. 3739, December 1972.

6. Wigton, L. B. and Yoshihara, H., "Viscous-Inviscid Interactions with a Three-Dimensional Inverse Boundary Layer Code," Paper presented at the 2nd Symposium on Numerical and Physical Aspects of Aerodynamic Flows, January 17-20, 1983, Long Beach, CA.

7. Samant, Satish S. and Wigton, Laurence B., "Coupled Euler/Integral Boundary Layer Analysis in Transonic Flow," AIAA Paper No. 83-1806, July 1983.

8. Carter, J. E., "A New Boundary-Layer Inviscid Iteration Technique for Separated Flow," AIAA Paper No. 79-1450, July 1979.

9. Donegan, Tracy, "Unsteady Viscous-Inviscid Interaction Procedures for Transonic Airfoil Flows," M.S. Thesis, The University of Tennessee, Knoxville, TN, December 1983.

10. van den Berg, B. and Elsenaar, A., "Measurements in a Three-Dimensional Incompressible Turbulent Boundary Layer in an Adverse Pressure Gradient Under Infinite Swept Wing Conditions," NLR TR 72092 U, 1972.

11. MacCormack, R. W., "The Effect of Viscosity in Hypervelocity Impact Cratering," AIAA Paper No. 69-354, May 1969.

12. Jacocks, J. L. and Kneile, K. R., "Computation of Three-Dimensional Time-Dependent Flow Using the Euler Equations," AEDC-TR-80-49, Arnold Air Force Station, TN, October 1980.

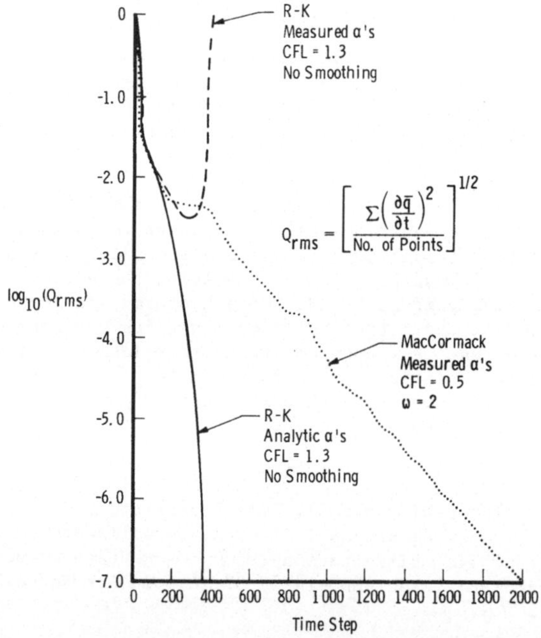

Figure 1. Convergence Histories
(Infinite Swept Wing)

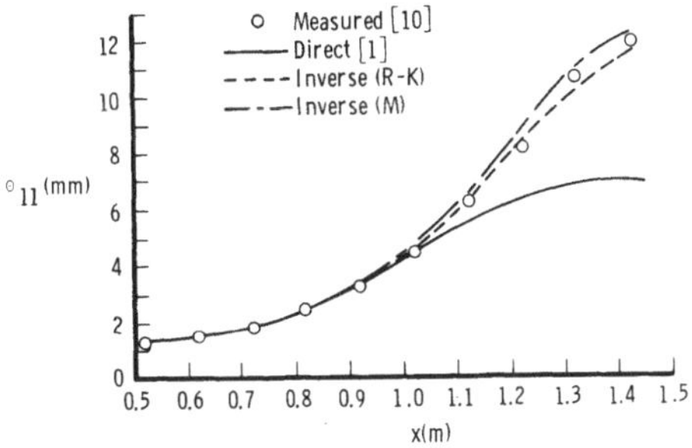

a. Streamwise Momentum Thickness

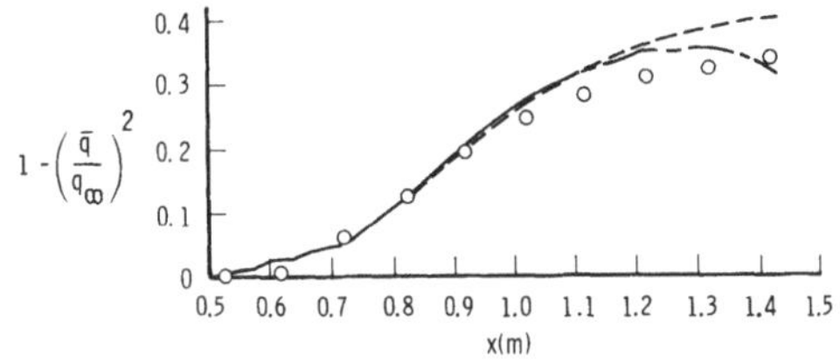

b. Edge Velocity

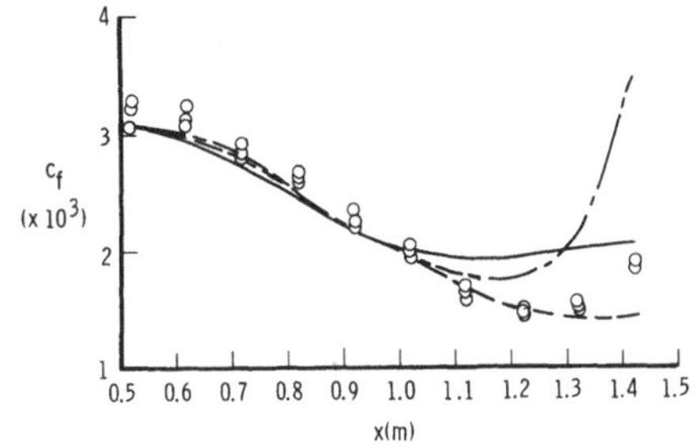

c. Skin Friction

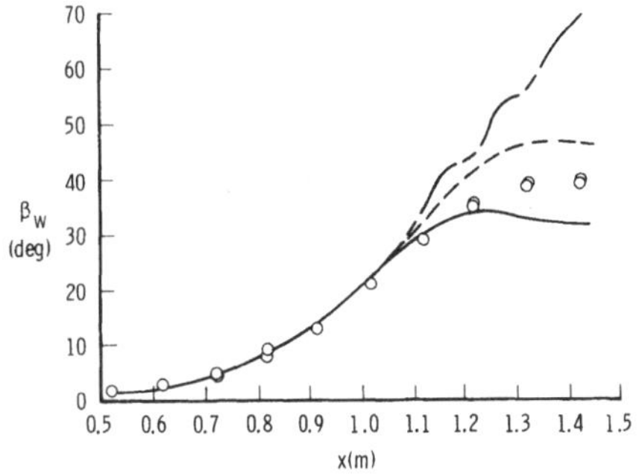

d. Wall Streamline Angle

Figure 2. Measured and Computed Boundary-Layer Quantities

A THREE-DIMENSIONAL INCOMPRESSIBLE PRIMITIVE VARIABLE NAVIER-STOKES PROCEDURE WITH NO POISSON SOLVER

T. D. Taylor
M. M. Nadworny
R. S. Hirsh

The Johns Hopkins University
Applied Physics Laboratory
Laurel, Maryland 20707 USA

INTRODUCTION

This paper outlines a new approach to solving the three-dimensional Navier-Stokes equations without introducing a Poisson equation for the pressure or stream function. The approach is utilized to study two- and three-dimensional boundary layer stability. Both the method and results are outlined.

A major problem in the solution of incompressible flows is the computation of the pressure field. The standard approaches solve either a Poisson equation for the pressure, created by combining the continuity and momentum equations, or resort to a stream function vorticity formulation and determine the pressure from the solution. The choice of either of these paths leads to the same computational difficulties, i.e., the need to solve a Poisson equation at each time step, and a means of specifying extra boundary conditions not required by the physical problem. Introducing the Poisson equation for pressure, although customary, is costly and unwise from a numerical standpoint since one replaces a first order equation (momentum) for the pressure by a second order equation. In this paper, the authors have investigated alternative procedures which avoid the above problems. There are two key points which allow the new procedures to be implemented: the first is utilization of the continuity and momentum equations in a manner different from the customary approach to incompressible Navier-Stokes calculations, and the second is the use of pseudo-spectral methods for the integration of the equations.

APPROACH

The three-dimensional incompressible equations employed in this study are:

$$\frac{\partial u}{\partial t} + u\frac{\partial u}{\partial x} + v\frac{\partial u}{\partial y} + w\frac{\partial u}{\partial z} = -\frac{\partial p}{\partial x} + \frac{1}{R}\frac{\partial^2 u}{\partial y^2} \qquad (1)$$

$$\frac{\partial v}{\partial t} + u\frac{\partial v}{\partial x} + v\frac{\partial v}{\partial y} + w\frac{\partial v}{\partial z} = -\frac{\partial p}{\partial y} + \frac{1}{R}\frac{\partial^2 v}{\partial y^2} \qquad (2)$$

$$\frac{\partial w}{\partial t} + u\frac{\partial w}{\partial x} + v\frac{\partial w}{\partial y} + w\frac{\partial w}{\partial z} = -\frac{\partial p}{\partial z} + \frac{1}{R}\frac{\partial^2 w}{\partial y^2} \qquad (3)$$

and $$\frac{\partial u}{\partial x} + \frac{\partial v}{\partial y} + \frac{\partial w}{\partial z} = 0 \qquad (4)$$

Two algorithms for solving these equations were attempted. Each differed from the standard manner in which the continuity and momentum equations are employed for Navier-Stokes solutions.

Rather than solving Eqns. (1) through (3) for the velocity components and employing a derived Poisson equation for the pressure which would replace Eqn. (4), our procedures reverse the usual roles of Eqns. (2) and (4). In Reference [1], the continuity equation replaced the momentum equation for determination of the vertical velocity, but a Poisson equation for the pressure was still retained. The present approach uses the continuity equation as in Reference [1], but uses the vertical momentum equation as a means of obtaining the pressure. Thus, in the new procedures, u and w are found by solution of Eqns. (1) and (3), but Eqn. (4) is solved to determine v, i.e.,

$$v = -\int \left(\frac{\partial u}{\partial x} + \frac{\partial w}{\partial z}\right) dy + v_o(x,z) \qquad (5)$$

where v_o is a known boundary value. In the first method, Eqn. (2) is then solved for the pressure:

$$p = p_1(x,z) + \int \left[\frac{1}{R}\frac{\partial^2}{\partial y^2} - \left(\frac{\partial v}{\partial t} + u\frac{\partial v}{\partial x} + v\frac{\partial v}{\partial y} + w\frac{\partial v}{\partial z}\right)\right] dy \qquad (6)$$

where p_1 is a pressure value at the boundary. The value of p_1 may be a function of x and z which is known or determined by integration of the x and z momentum equations along the boundaries with imposed conditions on u, v, and w. For flow past a body, where a free stream exists, p_1 can be prescribed, and no artificial pressure boundary conditions need be prescribed at the solid boundary.

The second method adds a relaxation scheme to the integration process by adjusting two separate predictions of the vertical velocity until the pressure value is correct. In addition to Eqn. (5), for what is denoted as v_c (for continuity determined), Eqn. (2) is used to get v_M (for momentum determined). The difference $v_c - v_M$ is then formed and used to iterate for the pressure at the new time step by

$$p^{n+1} = p^n + \alpha \frac{\partial}{\partial y} (v_c - v_M) \tag{7}$$

The y derivative and the negative sign in Eqn. (7) are consistent with the actual Poisson equation for the pressure obtained from Eqns. (1) through (4) and from which Eqn. (7) is derived.

The numerical procedure chosen to integrate the equations is the Chebyshev pseudo-spectral method which has proven to produce extremely accurate solutions at costs comparable to finite difference approaches, see References [2] and [3] for examples. This accuracy, based upon the global solution, was found to be essential for the integration of Eqn. (6) for the pressure. The calculations are performed in real space to avoid the complications found in time dependent problems when boundary conditions are imposed in spectral space [2]. A previously reported finite difference predictor spectral corrector method allowing time steps two orders of magnitude greater than the explicit limit for pseudo-spectral problems [4] has been used to implicitly integrate the momentum Eqns. (1) and (3) in time. For the two-dimensional x-momentum equation this appears as

$$L_{fd} u^{r+1} = L_{fd} u^r + \alpha(g - L_{sp} u^r) \tag{8}$$

where $\quad L = \frac{I}{\Delta t} - \frac{1}{R} \frac{\partial^2}{\partial y^2}$

$$g = -p_x^{n+1/2,k} + \left(\frac{I}{\Delta t} + \frac{1}{2R} \frac{\partial^2}{\partial y^2}\right) u^n - \frac{1}{2} (M_{sp} u^r + M_{sp} u^n)$$

$$Mu = u \frac{\partial u}{\partial x} + v \frac{\partial u}{\partial y}$$

The subscripts fd and sp stand for finite difference and spectral, and α is a parameter defined in Reference [4]. The integrals required in Eqns. (5) and (6) are evaluated in real space using Chebyshev expansions. The calculations employ a specified number of modes in each space direction, not limited by the power of two requirement of standard FFTs. The necessary matrix inversions are accomplished as in Reference [5] and are comparable in speed to FFTs for up to 100 modes in a direction.

RESULTS AND CONCLUSIONS

This technique has been applied to the solution of the Navier-Stokes equations for examination of boundary layer stability. The ultimate goal is to provide an understanding of compliant wall drag reduction, and as a first step the stabilization of a boundary layer by wall motion has been sought. The full equations have been solved for a flat plate flow with an upstream perturbation which causes Tollmien-Schlichting waves to propagate. No assumptions are made concerning the periodicity of the flow. Correct inflow and outflow boundary conditions are employed on the total velocities computed in Eqns. (1) through (4). That is, the total velocity, e.g., u is directly computed, not just the perturbation from some mean flow, u'. As a result, the calculations can proceed into non-linear regions without breaking down due to small perturbation assumptions, and include the effects of non parallelism and non periodic mean flow advection.

The first method, Eqn. (6) was found to be equivalent to a marching solution of a Poisson equation, and hence unstable. The second method, Eqn. (7), due to the explicit nature of its relaxation scheme, suffered from extremely slow pressure convergence. When upstream perturbations are not imposed, the method has been shown to compute the Navier-Stokes solutions for a Blasius boundary layer including the higher order Reynolds number corrections, and it maintains this solution for as many time steps as one chooses. Calculations for amplified Tollmien-Schlichting waves have been produced without a compliant surface velocity, see Figure 1, and the drag has also been computed for the same case with a surface velocity.

The two-dimensional calculations confirm the applicability of the method, and three-dimensional calculations of compliant surface flows are currently being conducted along with other means of more efficiently calculating the pressure.

ACKNOWLEDGEMENT

This work was supported by ONR through NAVSEA Contract No. N00024-83-C5301.

REFERENCES

[1] Taylor, T. D., and Murdock, J. W., Application of Spectral Methods to the Solution of Navier-Stokes Equations, in Approximations Methods for Navier-Stokes Problems, Lecture Notes in Mathematics, Vol. 771, pp. 529-537, Springer-Verlag, Berlin, 1980.

[2] Peyret, R., and Taylor, T. D., Computational Methods for Fluid Flow, Springer, New York, 1982.

[3] Gottlieb, D., and Turkel, E., Spectral Methods for Time Dependent Partial Differential Equations, NASA CR 172241, 1983.

[4] Hirsh, R. S., Taylor, T. D., and Nadworny, M. M., An Implicit Predictor-Corrector Method for Real Space Chebyshev Pseudo-Spectral Integration of Parabolic Equations, Comp. Fluids, 11, 251-254, 1983.

[5] Taylor, T. D., Hirsh, R. S., and Nadworny, M. M., Comparison of FFT, Direct Inversion and Conjugate Gradient Methods for use in Pseudo-Spectral Methods, Comp. Fluids, 12, 1-10, 1984.

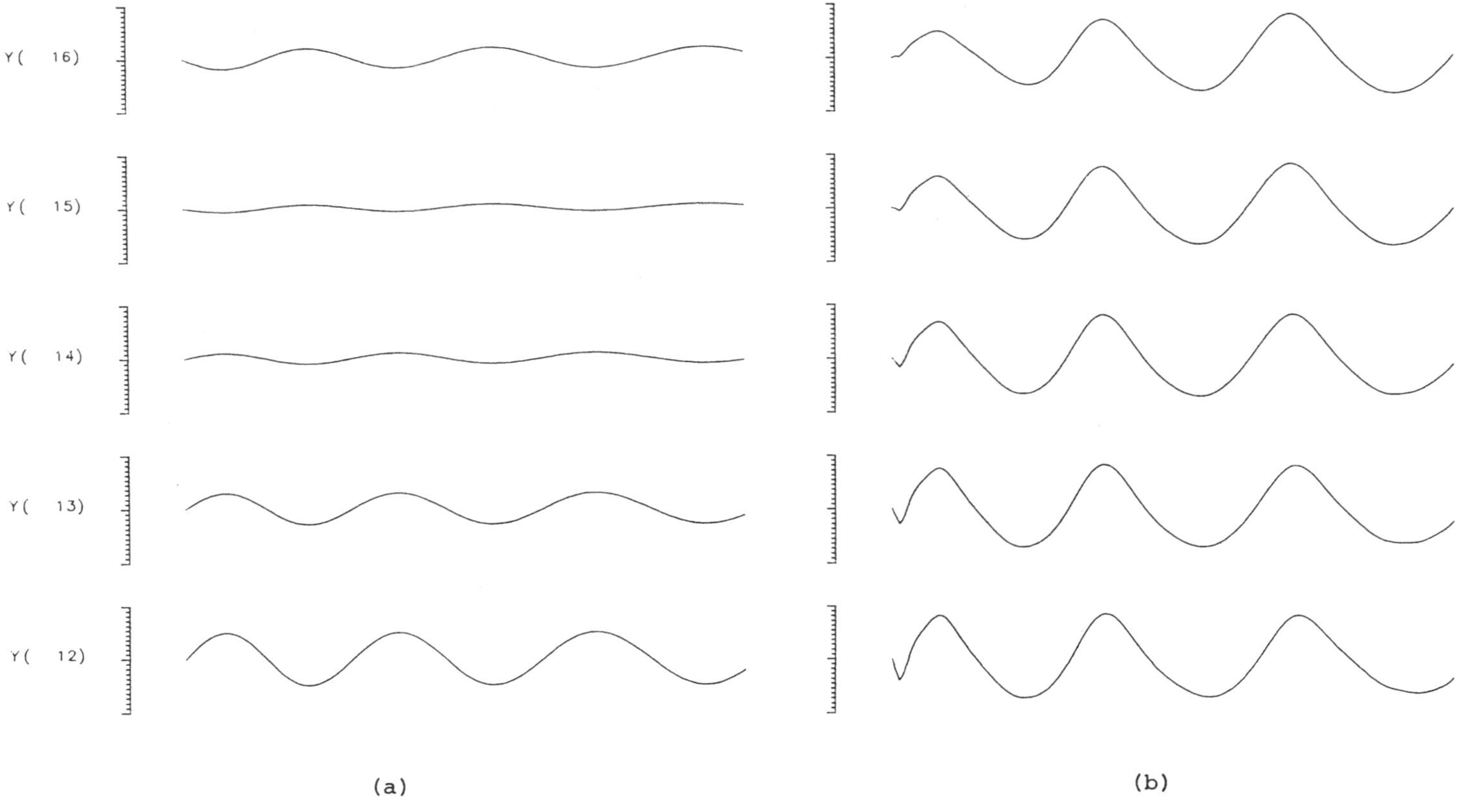

Figure 1. Amplitude of velocity perturbation along the plate at various heights within the boundary layer: (a) Initial perturbation, (b) Perturbation at two wave cycles.

FORMATION OF TAYLOR VORTICES IN SPHERICAL COUETTE FLOW

Laurette Tuckerman
CEN - Saclay
DPhG / PSRM
Orme des Merisiers
91191 Gif-sur-Yvette, France

and

Philip Marcus
Astronomy Department
Harvard University
Cambridge Mass 02138
U.S.A.

We have conducted a numerical study of spherical Couette Flow -- the flow between differentially rotating concentric spheres. When the gap ratio $\sigma \equiv (R_2 - R_1)/R_1$ (where R_1 and R_2 are the radii of the inner and outer spheres) is small, the flow near the equator resembles that between cylinders, the classic Taylor-Couette problem. As in Taylor-Couette flow, when the angular momentum gradient, measured by the Reynolds number $Re \equiv R_1{}^2 \Omega_1/\nu$ (where Ω_1 is the angular velocity of the inner sphere and ν the kinetic viscosity) exceeds a critical value, Taylor vortices form to redistribute angular momentum between radial shells.

For $\sigma = 0.18$, the gap ratio studied experimentally by Sawatzki and Zierep (1970) and Wimmer (1976), there exist three steady axisymmetric steady states, each with a different number of Taylor vortices (zero, one, or two) per hemisphere. The equilibrium attained by the flow depends on the history of its acceleration. Previous initial value codes (Bonnet & Alziary de Roquefort 1976, Yavorskaya et al. 1978, and Bartels 1982) have reproduced the steady states and some of the transitions, but have been unable to generate the one-vortex state as a transition from the basic zero-vortex flow.

We answer two questions arising from these previous studies, namely : 1) Why has generation of the one-vortex state eluded previous initial value studies ?

2) What is the mechanism by which the history of the flow determines the final steady state ?

Methods

We have written an initial value code to solve the axisymmetric incompressible time-dependent Navier-Stokes equations in a spherical geometry. We use a pseudospectral method (Gottlieb & Orszag 1977). Functions are represented as sums of basis functions -- in this case, Chebyshev polynomials in radius multiplied by sines in theta. Derivatives are taken in the spectral representation, multiplications are performed in physical space, and Fast Fourier Transforms are used to transform between the two representations. Evolution in time is accomplished by an algorithm of global accuracy $O(\Delta t)^2$: on the nonlinear terms, the

Adams-Bashforth approximation is used, while on the linear terms we employ the Crank-Nicolson approximation. We used a resolution of 16 polynomials in radius and 128 sine functions in angle θ, and 70 time steps per inner sphere revolution.

The elliptic operator resulting from the Crank-Nicolson approximation is block-upper triangular in the sine series basis. It can therefore be inverted using a sub-matrix back-solve, analogous to an ordinary back-solve for upper triangular matrices. The sub-matrix equations, one for each sine basis function, are solved using an eigenvector-eigenvalue decomposition. The sub-matrices differ only by a multiple of the identity, so the eigenvector-eigenvalue factorization need be done only once. Dirichlet and Neumann boundary conditions on the meridional stream function, which obeys a fourth-order equation, are imposed by a Greens function technique.

The algorithm requires little modification to perform linear stability analysis. To calculate the eigenvalues and eigenvectors of the Navier-Stokes equations linearized about a steady state $\vec{U}$, it suffices to replace the full nonlinear interaction $(\vec{u}.\nabla)\,\vec{u}$ by the linearized term $(\vec{u}.\nabla)\,\vec{U} + (\vec{U}.\nabla)\,\vec{u}$, and to impose homogeneous boundary conditions. Iteration in time is then equivalent to the power method, and causes an initial guess to converge to the eigenvector with the largest eigenvalue. The Rayleigh quotient is used to estimate the eigenvalue from two successive approximations to the eigenvector.

Results

We have numerically reproduced the experimentally observed states and the transitions between them for Re <1000. In particular, we find that the transition from the zero- to the one- vortex state takes place asymmetrically with respect to the equator, despite the equatorial symmetry of the initial and final states. This is why the transition was not seen in previous inital value simulations, which imposed equatorial symmetry in addition to axisymmetry. The Reynolds number Re = 652 at which the transition occurs agrees exactly with that found experimentally by Wimmer.

The reverse transition, from the one- to the zero- vortex state, when Re is decreased, takes place symmetrically with respect to the equator, as does the transition from the zero- to the two- vortex state. We observe another asymmetric transition, not in the previous published literature, from the two- to the one- vortex state. Bühler (private communication) has confirmed experimentally the qualitative form of all of these transitions.

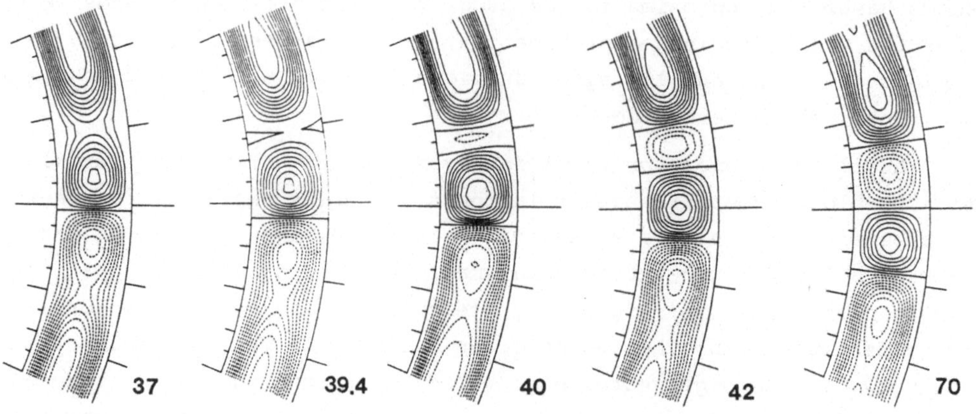

Zero- to one- vortex transition. Pictured are meridional streamlines in the equatorial region. Solid and dashed streamlines represent counter-clockwise and clockwise circulation, respectively. Time is shown in inner sphere revolutions. Note the breaking of equatorial symmetry. Re = 700.

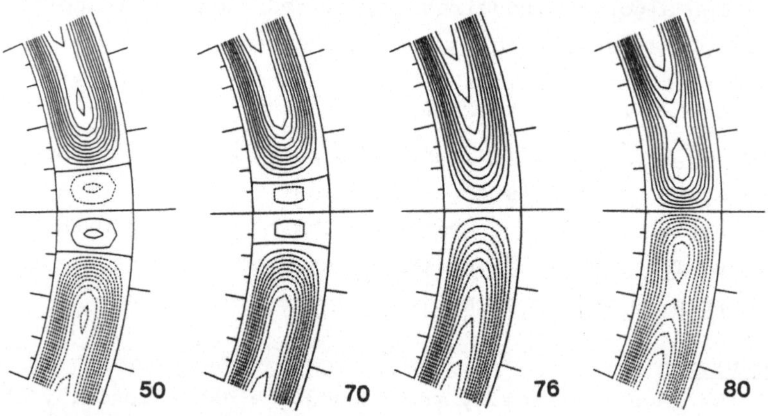

One- to zero- vortex transition at Re = 645.

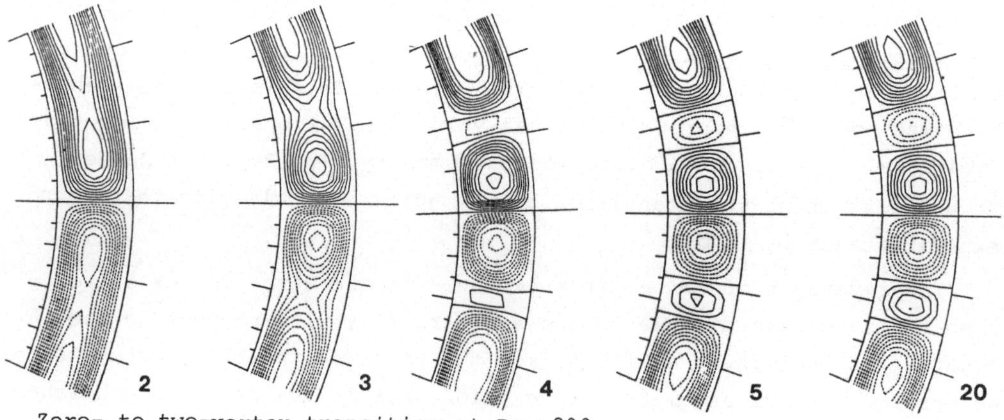

Zero- to two-vortex transition at Re = 800.

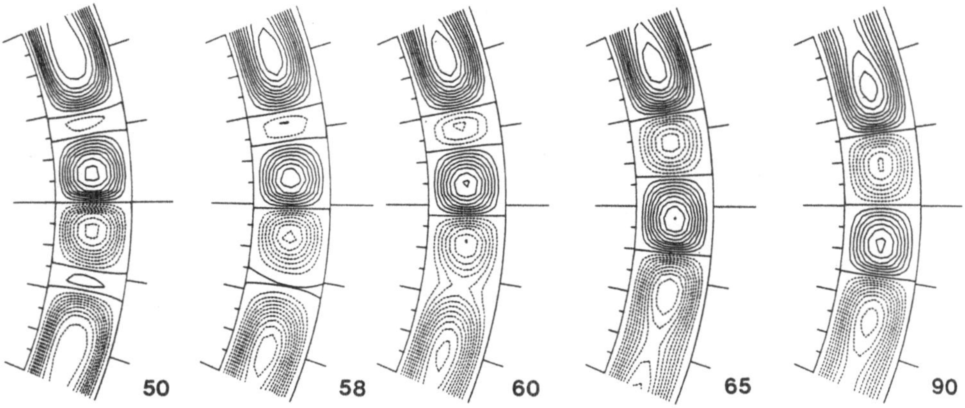

Two- to one- vortex transition at Re = 750.

Schrauf (1983)'s steady state calculation has revealed that the one-vortex states lie on a separate solution branch which never intersects the branch containing zero-vortex states. Schrauf's study and ours discovered that the zero- and the two- vortex states lie on the same solution branch, called the primary branch. That is, the zero-vortex states evolve continuously into the two-vortex states as the Reynolds number is increased, the demarkation between the two occurring at Re = 740. This branch structure is similar to that predicted by Benjamin (1978) for Taylor-Couette flow between cylinders of finite length.

By calculating eigenvectors and eigenvalues, we find that an interval of the primary branch is linearly unstable to an equatorially antisymmetric eigenvector. This instability initiates the transition to the one-vortex state ; therefore, we call the unstable interval $651 < Re < 775$ a "window" from the primary branch to the one-vortex branch. The window contains both zero- and two- vortex states, and the two-vortex states at $Re \geq 775$ are stable. This explains the non-uniqueness seen experimentally by Wimmer : in accelerating the system to its final angular velocity, a one-vortex state will be generated if the time spent in the window is sufficient for the antisymmetric instability to attain the threshold level necessary for transition. Otherwise a two-vortex state will be generated.

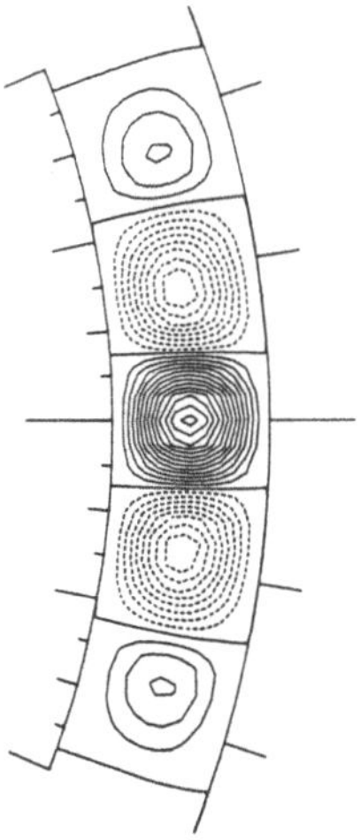

Antisymmetric eigenvector at Re = 700.

This work was supported in part by NSF Grants MEA-82-15695 and AST-82-10933. The computations were performed on the CRAY-1 at NCAR.

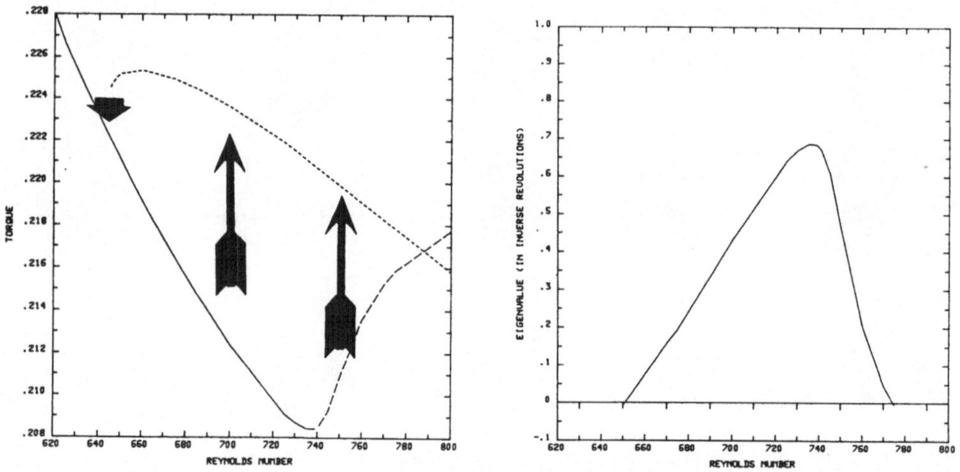

Left : Torque vs.Re of steady states in region of window. Solid, short-dashed, and long-dashed curves represent zero-, one-, and two- vortex states, respectively. Arrows show schematically transitions between states. Note that the curves representing zero- and two- vortex states join continuously, but that the one-vortex states are on an unconnected curve. ("Intersection" at Re $\simeq$ 790 is a projection effect).

Right : Growth rate vs.Re of antisymmetric eigenvector to which the primary branch is unstable.

References

Bartels, F. 1982 J. Fluid Mech. 119, 1.

Benjamin, T.B. 1978, Proc. Roy. Soc. London A359, 1.

Bonnet, J.P. and Alziary de Roquefort, T. 1976 J. de Mécanique 15,373.

Bühler, K., private communication.

Gottlieb, D. and Orszag, S.A. 1977 Numerical Analysis of Spectral Methods. SIAM Press, Philadelphia.

Sawatzki, O. and Zierep, J. 1970 Acta Mechanica 9, 19.

Schrauf, G. 1983 Ph.D Thesis, Universität Bonn.

Tuckerman, L. 1983 Ph.D Thesis, Mass. Inst. of Technology.

Tuckerman, L. and Marcus, P., to be submitted to J. Fluid Mech.

Wimmer, M. 1976 J. Fluid Mech. 78, 317.

Yavorskaya, I.M., Astaf'eva, N.M., and Vvedenskaya, N.D. 1978 Sov. Phys. Dokl. 23 (7), 461.

9th International Conference on Numerical Methods in Fluid Dynamics
(1984, June, France, Cen–Saclay)

NUMERICAL SIMULATION OF UNSTEADY FLOWFIELDS NEAR BODIES IN NONUNIFORM ONCOMING STREAM

L.I. Turchak, V.F. Kamenetsky

Comput. Centre, Acad. of Sci., USSR, Moscow

Inviscid flowfields near blunt bodies travelling at supersonic speeds in ideal gas atmosphere containing various discontinuities (shock, explosion, etc.) are discussed. The problems are solved numerically by grid – characterictic method (GC–method) [1, 2]. The results of axisymmetrical and three – dimensional difraction of plane shocks on travelling bodies of different shapes and axisym–menrical interaction of the travelling body with an explosion wave are presented. Obtained data fully reflect all physical phenomena accompanying such interactions and are shown to agree well with the available experimental evidence.

1. Introduction

Blunt – body problem in nonuniform supersonic flowfields is the subject of current intensive investigations. Such problems may be subdivided in two groups – steady and unsteady.

Steady flows of this type take place, for example, in gas jet – obstacle interactions or in the case of the body travelling in the wake. One may mention ref. [3], where a survey of the existing investigations is given in addition to the computed results.

We also computed such flows. Interactions between wake–type flows and blunt bodies for axisymmenrical case and three – dimensional one were considered. Moreover, the model of supersonic nonuniform oncoming flow was used to investigate two – dimensional separation zone of the flow in front of the step. Numerical results were obtained by GC–method with front shock fitting. The data obtained show qualitatively good agreement with experiment. Main results of this investigation are published in [4] and will not be discussed here.

Consider one of the unsteady problems mentioned above, namely, that of the body travelling in the media containing different discontinuities. Such problems were considered by many authors mainly without front shock fitting [5–10].

In many cases at fixed supersonic speed of the body before interaction the speed of oncoming flow with respect to the front shock remains supersonic during interaction. This makes possible to apply some shock fitting method. It is assumed that application of these methods to such problems allows to obtain a more accurate description of the interaction.

The results of diffraction of plane shock waves on bodies travelling with supersonic speeds, and the problem of a body travelling within explosion domain will be presented in this paper. Numerical solution was obtained by GC–method with front shock fitting.

2. Mathematical formulation

A blunt body AC (Fig. 1) moves in gas I with a supersonic speed; BD – the front shock, G – the border of discontinuity separating the gas I from a gas II. The formation of the border G is assumed to be caused by some external source. The movement of G and of gas II is considered to be known, the speed of the body AC is fixed. It's supposed that the gas velocity with respect to the shock BD remains supersonic during the interaction. In this case the solution may be determined only within shock layer III. When the problem is solved numerically discontinuity influences only on the form of the boundary conditions at the front shock. Imposing these conditions we must take into account both travelling of the border G according to a known law, and a replacement of the front shock BD, computed in each step on time.

Mathematical formulation of the problem is as follows. Within the shock layer the system to be solved is that of unsteady Euler equations for perfect gas with constant specific heat ratio κ. Boundary conditions: at the body surface – condition of the impenetrability, at the front shock – the Rankine–Hugoniot relations. Parameters of the flow of gas I about the body with Mach number $M_1 > 1$ are taken as initial data. The initial moment of time is the moment of the contact of the front shock BD with the border G.

For numerical solution the GC–method was used. The basic ideas of this method will be formulated in the next section. Here we shall only mention that in distinction from [1, 2] we used the sound velocity a and the entropy S as independent thermodynamical variables (as it was suggested in [11]).

Dimensionless Euler equations in the vector form for this case may be written as follows

$$a_t + V(\nabla a - \frac{a}{2\kappa} \nabla S) + \frac{\kappa - 1}{2} \operatorname{div} V = \frac{a}{2\kappa} S_t ,$$

$$V_t + (V\nabla)V + \frac{2a}{\kappa - 1}(\nabla a - \frac{a}{2\kappa} \nabla S) = 0$$

$$S_t + V\nabla S = 0.$$

Here the gas velocity V is scaled to $(p_1/\rho_1)^{1/2}$ time – to $L(p_1/\rho_1)^{-1/2}$, entropy – to specific heat at the constant volume. Basic parameters of the problem are: a linear size L, the pressure p and the density ρ_1 of gas I.

3. Numerical method

The GC–method is described in detail in [1, 2]. Following [2] we used first order accuracy scheme. We shall illustrate here the construction of this scheme for an one–dimensional case.

For the one–dimensional case the gas dynamic equations may be written in the next form

$$u_t + Au_x = 0 \tag{1}$$

Here u – N–component unknown vector, A – matrix of the rank $N \times N$. The matrix A obtains only real eigenvalues μ_i $(i = 1, 2, \dots, N)$. Denote by ω_i eigenvectors of the matrix A, and after multiplication of (1) by ω_i we obtain

$$\omega_i u_{ti} = \omega_i(u_t + \mu_i u_x) = 0, \tag{2}$$

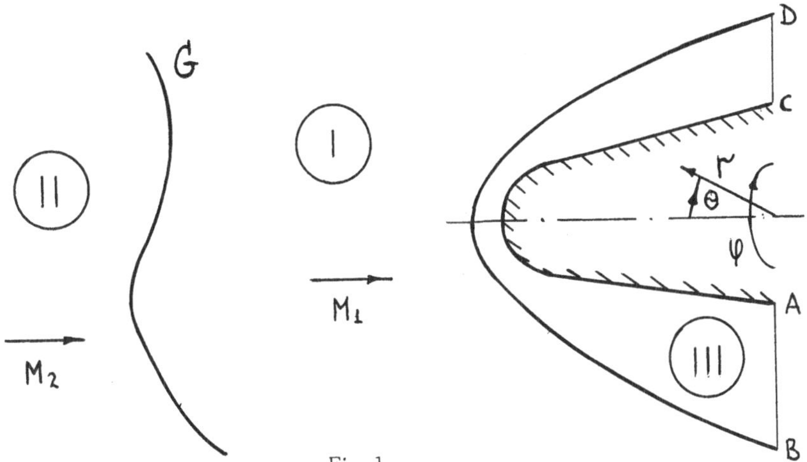

Fig. 1

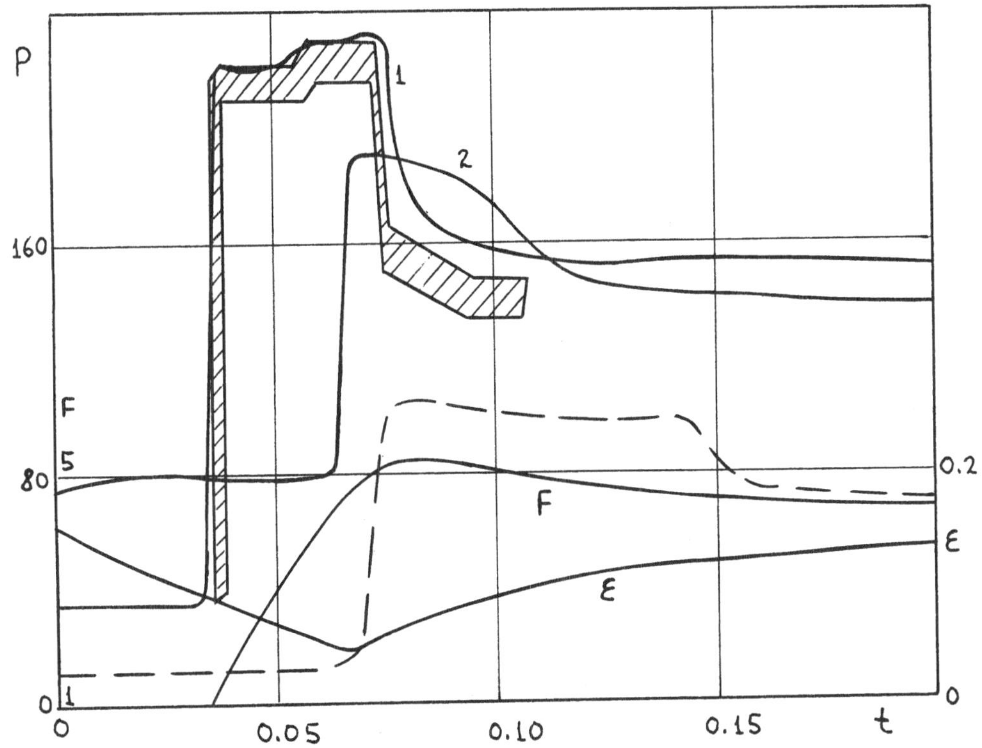

Fig. 2

where u_{t^i} is the derivative along the characterictic.

Consider the phase plane. Suppose that the solution is known at the level $t = t_n$. Determine the solution in some point k at the level $t_{n+1} = t_n + \tau$. With the first order accuracy with respect to τ one may write

$$\omega_i (u_k^{n+1} - u_i^n) = 0, \tag{3}$$

where subscript i denotes the point of the intersection of the characteristics and the line $t = t_n$ in the plane.

Introduce a fixed grid with step L along the axis x (k is the knot point). To determine values of parameters in the point i the linear interpolation is applied. Thus within the first order accuracy with respect to h one may obtain

$$\mu_i > 0, \quad u_i^n = u_k^n - \frac{\mu_i \tau}{h} (u_k^n - u_{k-1}^n); \tag{4}$$

$$\mu_i < 0, \quad u_i^n = u_k^n - \frac{\mu_i \tau}{h} (u_{k+1}^n - u_k^n). \tag{4a}$$

Denote by Ω the matrix of eigenvectors ω_i and M^+ and M^- – diagonal matrixes the main diagonals of which consist, accordingly, of positive or negative eigenvalues, or zeros. Rewriting (3) in the matrix form, using (4) and (4a) and multiplying on Ω^{-1}, one obtains

$$u_k^{n+1} - u_k^n + \frac{\tau}{h} \Omega^{-1} M \,\Omega \Delta^- u_k^n + \frac{\tau}{h} \Omega^{-1} M^{-1} \Omega \Delta^+ u_k^n = 0, \tag{5}$$

$$\Delta^+ u_k^n = u_{k+1}^n - u_k^n, \quad \Delta^- u_k^n = u_k^n - u_{k-1}^n.$$

For calculations at the border points expressions (3) and additional boundary conditions are used. The scheme (5) may be easily extended to a multidimensional case.

The solution was determined in the spherical coordinate system (r, θ, φ) connected with the body. After introducing variable $\xi = (r - R_b)/(R_s - R_b)$, where $r = R_b(\theta, \varphi)$ and $r = R_s(t, \theta, \varphi)$ are, correspondingly, equations of the body surface and the front shock, a curvilinear region of integration may be reduced to a rectangular: $0 \le \xi \le 1$, $0 \le \theta \le \theta^*$, $0 \le \varphi \le \pi$.

4. Numerical results

Initially we have solved the problem of difraction of a plane shock on sphere and bult travelling with supersonic speeds. There are two characteristic Mach numbers: M_1 – for gas I and M_G – velocity of motion of the discontinuity G scaled by sound velocity in gas I. From the point of view of an immobile observer there are two different cases: 1) the body and the shock are coming towards each other; 2) the body overtakes the shock and passes through it.

The expression for Mach number M_2 of gas II may be easily obtained. It's obvious that in the case of an overtaking interaction we have $M_2 > M_1 > 1$. The analysis of the expression of M_2 in the case of coming interaction shows that $M_2 > 1$, but the value of M_2 may differ from that of M_1. It depends on M_1 and M_G. Thus the assumption of the existence of the front shock is valid.

Shown in Fig. 2 are the computed results of coming interaction at $M_1 = 5.16$ and $M_G = 7.19$ (solid lines), experimental time dependence of stagnation point pressure, taken from [5] for this version (shaded region) and pressure in stagnation point for $M_1 = 3$ and $M_G = 5$ (dotted line). Curves 1 and 2 describe time dependence of pressure, accordingly, in stagnation point and behind the front shock at the axis of symmentry.

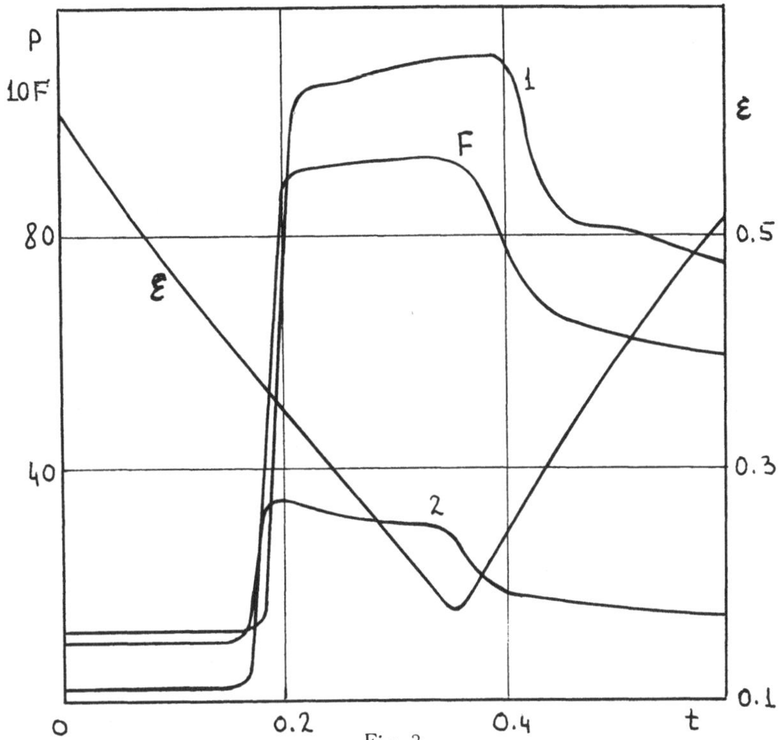

Fig. 3

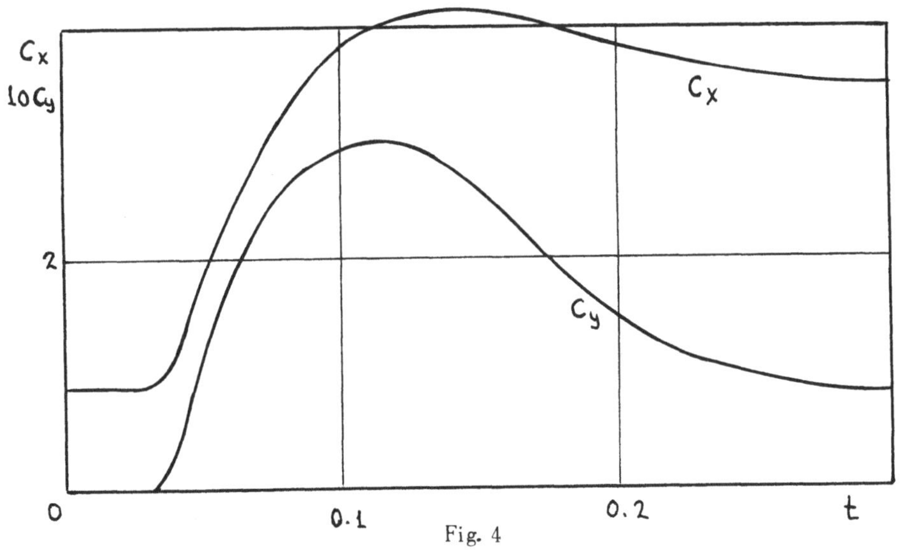

Fig. 4

Interaction occures as follows. Incident shock refracts at the bow one and penetrates within the shock layer. The pressure pas front shock increases instantaneously and becomes even more than stagnation point pressure. There is a contact discontinuity springing up between refracted and front shocks. The refracted shock approaches the body surface and reflected from it. Stagnation point pressure increases sharply. The second and rather small increase of pressure is connected with the reflection of a refracted shock from the contact discontinuity. In the case discussed the refracted shock crossing the contact discontinuity penetrates in the region of a greater density. This leads up to the reflection of a weak shock in the direction of the body. At M = 3 and M = 5 the refracted shock crossing the contact discontinuity penetrates in the region of a lower density. There is a rarefaction wave reflecting towards the body. This is the cause of the first small abatement of the dotted line. After the passing through the contact discontinuity the shock reflects from the front one as a rarefaction wave. The pressure behind the front shock increases sharply. The rarefaction wave comes to the body and induces the fast decrease of stagnation point pressure. Finally all parameters come to stationary values.

Here the curve ε is the distance between the front shock and the body surface along the axis of symmetry, and F is the dimensionless force equal to the ratio of integral of the pressure over the body surface at certain time moment to the same integral before the interaction.

Interaction with the bult occures as interaction with the sphere. The only difference is that the refracted shock begins to reflect from the body not at the stagnation point but in the corner point of the body surface. At Fig. 3 curves 1 and 2 show time dependence of pressure, accordingly, in the stagnation and corner points at $M_1 = 3$ and $M_G = 5$. Such interaction leads to the plato in the plot of the force F.

In addition to axisymmetrical problem we considered also a three-dimensional one, having the plane of symmetry about an inclined incidence of a plane shock at the sphere travelling with a supersonic speed. The problem is determined by parameters: M_1 – Mach number of the body motion in gas I; M_G – the speed of the shock in an immobile coordinate system, related to the sound velocity in gas I; α – the angle between the axis of the body movement and a normal to the chock. At Fig. 4 are shown the dependence on time of C_x and C_y for $M_1 = 6$, $M_G = 2$, $\alpha = 9^\circ$.

In overtaking interactions refracted shock penetrates within shock layer as a rarefaction wave. The subsequent flow is determined by multiple reflection of this wave from the body and the front shock. At Fig. 5 are shown the results of the axisymmetrical overtaking interaction with sphere at $M_1 = 3$ and $M_G = 2.2$. Notations are the same as at Fig. 2.

The aproach suggested above may be used for modelling of some other types of interactions. For example, we considered interaction with a plane contact discontinuity of finite thickness, and with the point explosion.

Let us consider in more detail axisymmetrical interaction with the spherical explosion. To determine the problem, besides Mach number M_1 for the body motion, the value of parameter

$$z = \frac{1}{L} \left(\frac{E_0}{\alpha_0 p_1} \right)^{1/3}$$

must be given, where: $\alpha_0 \sim 1$ – some dimensionless parameter, E_0 – the energy released by explosion. To establish boundary conditions, gas-dynamic parameters have to be known within explosion area. These parameters were determined by tables from [12]. We considered only rather bate stage of point explosion development. Initial increase of pressure within explosion area, as compared with the pressure in gas I was not greater than $0.5p_1$. In this case, if $z \gg 1$, the radius of explosion area is much greater than the characteristic size of the body. Interaction occures as follows. Initial stage is analogous to that of coming interaction with plane shock. After that the travelling of the body becomes

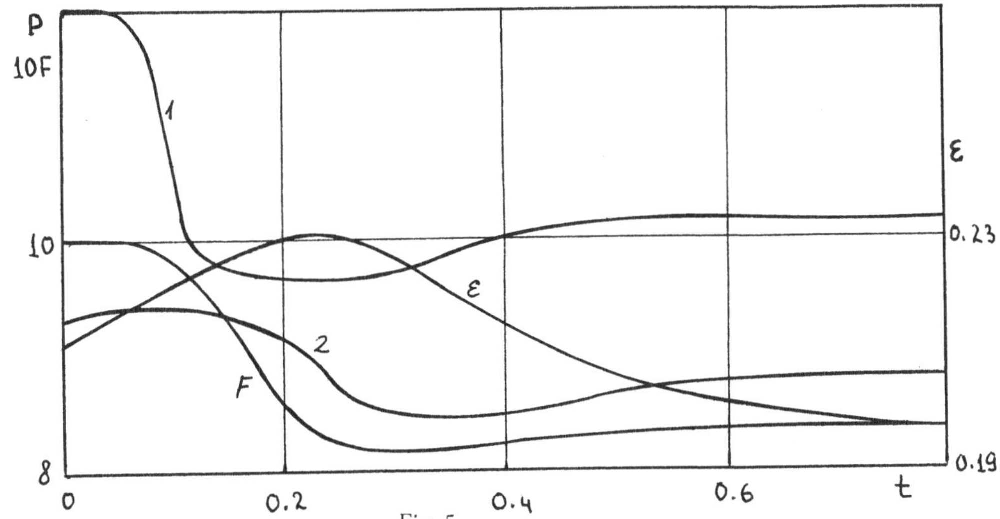

Fig. 5

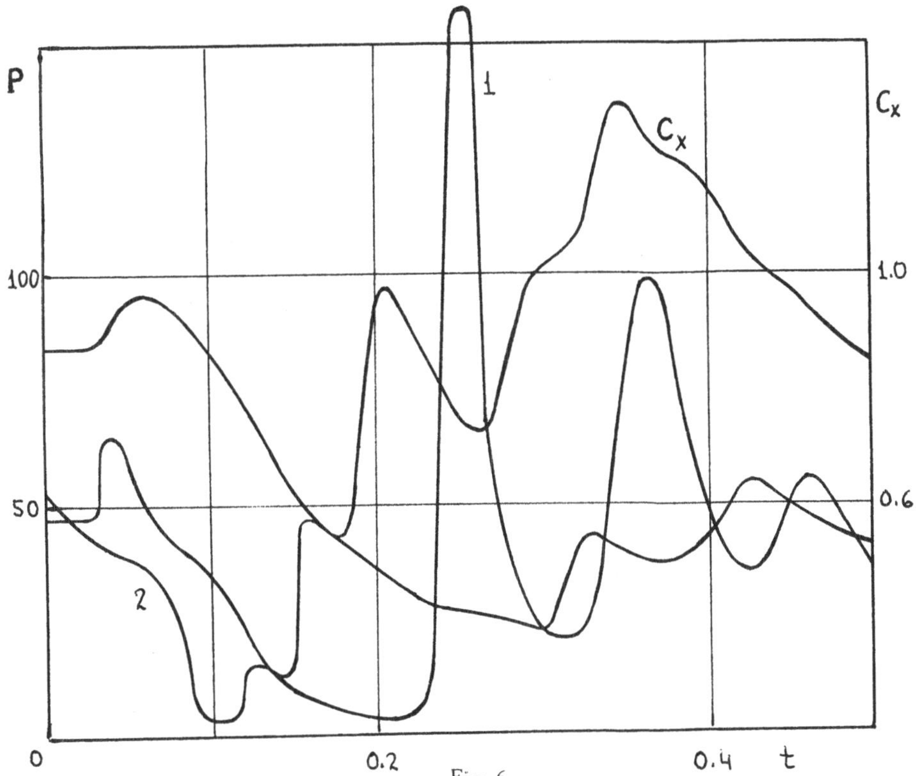

Fig. 6

quasistationary up to the moment of the approach of the body to the central region of explosion. In this part the density of gas sharply decreases and the front shock quickly moves forward. The distance between the front shock and the body surface may reach several units.

After crossing the central region the body returns to the quasistationary movement. Phenomena analogous to those of overtaking interaction with plane shock occur by the exit from the explosion area.

Shown in Fig. 6 are the time plots of the pressure in stagnation point and behind the front shock (curves 1 and 2) and C_x at $z = 1$. In this case refracted shock also reflectes from the body and moves to the front shock. Before it reaches the front shock, it penetrates in the region of sharp decrease of density. Thus the rarefaction wave is induced which travelles to reflected shock and there is no maximum in the plot of the pressure behind the front shock. The maximum appears somewhat later, after crossing the central area of explosion, when the density in front of the shock begins to increase.

The compression wave appears within the shock layer, which later transforms into the shock. It's reflection from the body leads to the second sharp increase in curve 1. The following flowfield is determined by interaction of this reflected shock, the front shock and the point explosion wave.

5. Conclusion

Application of the numerical shock fitting method to the problems of interaction of the body travelling with supersonic speed with different discontinuities in oncoming flow allows for rather accurate description of all phenomena taking place in such cases. In spite of nondivergent form of initial equations internal shocks are described rather good. For the problem of plane shock difraction numerical results show even such a minor phenomenon as secondary reflection of the shock from contact discontinuity. The appearence of the internal shock in the case of crossing the point explosion area is also rather interesting.

References

1. Магомедов К.М. Сеточно-характеристический метод для численного решения задач газовой динамики. Труды Секции по численным методам в газовой динамике II Международного коллоквиума по газодинамике взрыва и реагирующих систем (Новосибирск, 1969 г.), т. I, М., ВЦ АН СССР, 1969.

2. Численное исследование современных задач газовой динамики. Ред. Белоцерковский О.М. М., "Наука", 1974.

3. Lin T.C., Reeves B.L., Siegelman B. Blunt-body problem in nonuniform flowfields. AIAA Journal, 1977, v. 15, N 8.

4. Каменецкий В.Ф., Турчак Л.И. Сверхзвуковое обтекание тел неоднородным потоком идеального газа. М., ВЦ АН СССР, 1982.

5. Taylor T.D., Hudgins H.E. Interaction of a blast wave with a blunt body travelling at supersonic speeds. AIAA Journal, 1968, v. 6, N 2.

6. Арутюнян Г.М. К расчету давления в критической точке при падении ударной волны на тело, движущееся со сверхзвуковой скоростью. Изв. АН СССР, МЖГ, 1972, № 6.

7. Балакин В.Б., Бухманов В.В. Численное решение задачи о взаимодействии ударной волны с цилиндром в сверхзвуковом потоке. ИФЖ, 1971, № 6.

8. Тугазаков Р.Л. Дифракция ударной волны на движущемся клине. Ученые записки ЦАГИ, 1975, т. 6, № 1.

9. Липницкий Ю.М., Ляхов В.Н. Взаимодействие ударной волны с клином в сверхзвуковом потоке. Ученые записки ЦАГИ, 1976, т. 7, № 4.

10. Champney J.M., Chaussee D.S., Kutler P. Computation of blastwave-obstacle interactions. AIAA Pap., 1982, N 227.

11. Zanetti L., Moretti G. Numerical experiments on the leading-edge flowfield. AIAA Journal, 1982, v. 20, N 12.

12б Кестенбойм Х.С., Росляков Г.С., Чудов Л.А. Точечный взрыв. (Методы расчета. Таблицы). М., "Наука", 1974.

FLUX VECTOR SPLITTING AND RUNGE-KUTTA METHODS FOR THE EULER EQUATIONS

Eli Turkel - Tel-Aviv University and ICASE
Bram Van Leer - Technische Hogeschool Delft and ICASE

Introduction

We wish to solve the steady state multidimensional Euler equations with a method that is suitable for a large range of Mach numbers. At the same time we wish the method to be accurate and robust and capture shocks without excessive smearing. We also wish to reach the steady state rapidly. To achieve these goals we combine the Runge-Kutta scheme introduced in [1] with the flux vector splitting introduced in [2].

Consider the two dimensional system

$$w_t + f_x + g_y = 0 . \tag{1}$$

We advance the numerical solution in time using a N stage algorithm

$$\begin{aligned} w^{(0)} &= w^n \\ w^{(k)} &= w^{(0)} - \alpha_k \Delta t (D_x f^{(k-1)} + D_y g^{(k-1)}) \\ w^{n+1} &= w^{(N)} \end{aligned} \tag{2}$$

where $D_x f$ and $D_y g$ are difference approximations to the flux derivatives. To check the stability we freeze coefficients and Fourier transform. The amplification matrix of (2) is then

$$\begin{aligned} G &= 1 + \beta_1 z + \beta_2 z^2 + \dots + {}_N z^N \\ \beta_1 &= 1 \\ \beta_k &= \beta_{k-1} \alpha_{N-k+1} \end{aligned} \tag{3}$$

where z is the Fourier transform of $\Delta t(D_x f + D_y g)$. When central differences are used then z lies on the imaginary axis. With upwind differences z lies on some curve in the negative real half of the complex plane.

Experience has shown that one should usually choose the parameters

Research partially supported by NASA under Contract No.NAS1-17070 while the authors were in residence at ICASE, NASA Langley, Hampton, VA.

so that the time step is maximal. For central differences this implies that

$$C\Delta t/\Delta x \leqslant N - 1 \tag{4}$$

where C depends on $|z_{max}|$. With upwind schemes no general rules have been developed thus far for the optimal parameters. At present the parameters have been chosen by experimentation. One possibility is presented in the result section.

To appreciate the connection between central differences and flux vector splitting we consider a one dimensional example. Assume that we wish to compute a numerical flux at the cell interface $i + 1/2$. Quadratically interpolating yields a left side estimate of the state variables

$$w^L_{i+\frac{1}{2}} = \bar{w}_i + \frac{1}{4}(\bar{w}_{i+1}-\bar{w}_i) + \frac{1}{12}(\bar{w}_{i+1}-2\bar{w}_i+\bar{w}_{i-1}) \quad . \tag{5}$$

Interpolating the cell averages in zones i, $i+1$, $i+2$ yields the right sided estimate

$$w^R_{i+\frac{1}{2}} = \bar{w}_{i+1} - \frac{1}{4}(\bar{w}_{i+2}-\bar{w}_i) + \frac{1}{12}(\bar{w}_i-2\bar{w}_{i+1}+\bar{w}_{i+2}) \ . \tag{6}$$

The difference between these values is $O((\Delta x)^3)$. We now introduce an upwind bias in the numerical flux by using from the components of $w^L_{i+\frac{1}{2}}$ only those characteristic combinations that are convected forward and from $w^R_{i+\frac{1}{2}}$ those convected backward. In the approximation of flux splitting this becomes

$$f(w^L_{i+\frac{1}{2}}, w^R_{i+\frac{1}{2}}) = (f^+)^L_{i+\frac{1}{2}} + (f^-)^L_{i+\frac{1}{2}} \ . \tag{7}$$

This can be rewritten as

$$f(w^L_{i+\frac{1}{2}}, w^R_{i+\frac{1}{2}}) = \frac{1}{2}(f^L_{i+\frac{1}{2}}+f^R_{i+\frac{1}{2}}) - \frac{1}{2}Q_{i+\frac{1}{2}}(w^R_{i+\frac{1}{2}}-w^L_{i+\frac{1}{2}}) \tag{8}$$

where $$Q = \frac{df^+}{dw} - \frac{df^-}{dw}$$

$Q_{i+\frac{1}{2}}$ is a Roe-type [3] average of Q over the interval $(w^L_{i+\frac{1}{2}}, w^R_{i+\frac{1}{2}})$.

From (8) we see that the upward biased flux deviates from the average flux, used for central differencing, by a third order term. This leads to a fourth order viscosity with a matrix-valued coefficient. This viscosity prevents the checkerboard instability similar to the fourth order viscosity introduced in [1].

Eqs. (5) and (6) are modified before their actual use. The first order term is multiplied by a switch described in [4] while the second order term is multiplied by its square. When $\bar{w}_{i+1} - \bar{w}_i$ is large compared with $\bar{w}_i - \bar{w}_{i-1}$ and $\bar{w}_{i+2} - \bar{w}_{i+1}$, e.g. a shock profile, then the limiting yields

$$w^R_{i+\frac{1}{2}} - w^L_{i+\frac{1}{2}} \simeq \bar{w}_{i+1} - \bar{w}_i \quad . \tag{9}$$

Therefore, the viscosity term in (8) now leads locally to a second order viscosity which guarantees a monotone profile. This is similar to the second order artificial viscosity introduced in [1] and discussed in more detail in [5].

Limiting the high order terms combined with upwind differencing is a robust way of preventing numerical oscillations near discontinuities. To achieve the same effect the viscosity of [1] would have to be raised to the level of the spectral radius of Q leading to excessive smearing. In practice using the code of [1] the opposite approach is used. The coefficient of viscosity is adjusted, by trial and error, so that the shock profiles are sharp and spurious entropy production in the smooth flow is minimized. For smooth flows one can achieve viscosity levels that are smaller than that of the upwind scheme with the limiter. For violent flows the artificial viscosity is much too large. In any case the coefficient of viscosity used in [1] is very problem dependent while the present code has no adjustable parameters that need to be played with.

Flux splitting in an arbitrary multidimensional body-fitted coordinate system can be reduced to a one dimensional problem with reference direction normal to the cell face where fluxes are computed. The resultant code is a full two dimensional code and does not use time splitting to combine the coordinate directions. In all the interpolations we have ignored the geometric variation of the cells as was done in [1]. This can lead to errors for high angles of attack especially near the trailing edge where cell shape and size vary strongly.

Having discussed the spatial discretization we now introduce several techniques to accelerate the convergence to a steady state. The first technique is to use a local time step. This improves the running time by an order of magnitude due to the small cells near the airfoil. The second technique is to use residual smoothing averaging after each stage of the Runge-Kutta method. This was first introduced in [6] for the Lax-Wendroff scheme.

If one uses central differencing then the scheme is unconditionally stable when the smoothing is done after every even stage. Using an upwinded scheme the smoothing should be done after each stage of the algorithm. Even though the resultant scheme is unconditionally stable nevertheless it is not efficient to use time steps that are too large. Time steps about three times as large as those of the explicit scheme seem to be optimal. Since the residual smoothing adds only about 10% to the running time per time step the use of the residual smoothing is advantageous.

A third acceleration technique is to use a multigrid method. Jameson [7] has proposed using the Runge-Kutta scheme coupled with central differences as a smoothing algorithm for a multigrid scheme. The parameters α_k are now chosen to damp the high frequencies rather than achieving a maximal time step. We use the same technique using the upwind version of the Runge-Kutta scheme instead of central differences with an artificial viscosity. In the central difference version [7] the artificial viscosity is increased compared with the standard Runge-Kutta scheme [1]. In addition the multigrid central difference code seems to rely on the enthalpy damping in order to achieve rapid convergence to the steady state. With the upwind code there is no longer an artificial viscosity that can be tuned to damp the high modes. Furthermore, the steady state total enthalpy is not preserved by the flux vector splitting scheme [2]. Hence, the enthalpy damping introduced in [1] and [5] cannot be used. Nevertheless, the multigrid scheme does work with the upwing biased scheme. Thus, the viscosity that is implicit in the scheme seems to be sufficient to compensate for both the artificial viscosity and the enthalpy damping of the central difference scheme. However, the convergence rate of the upwind scheme is slower that that of the central difference scheme mainly because of the use of enthalpy damping.

Results

The upwind biased version of the code has been run on several different cases. The first case is a NACA0012 airfoil with $M_\infty = 0.8$, and $\alpha = 1.25°$. The mesh is an O mesh generated by a sheared parabolic transformation. We use a four stage Runge-Kutta scheme with $\alpha_1 = .17$, $\alpha_2 = .273$, $\alpha_3 = .5$ and $\alpha_4 = 1$. The residual smoothing is applied after each stage with $\beta_x = .9\lambda/8$ and $\beta_y = .6\beta_x$. λ is the local Courant number with $\lambda = 3$. Using a 64x16 mesh the residual is reduced by 4 orders of magnitude after 600 steps. The C_p curve is similar to that achieved by the central difference code [1] except that the shock profile is now sharper with one point in the middle of the shock along the airfoil. We have also run several supersonic flows about the NACA0012. The upwind version of the code converges for a larger range of Mach numbers than does the central difference version.

Becuase of the flux splitting and upwing logic entering the computation the upwind code is about two times slower per time step than the central difference version. A further slowdown is caused by the Runge-Kutta method which seems to favor the spectral distribution of central differences and which has not yet been optimized for upwind differences. Hence, the time step is about half of that for central differences. In addition the enthalpy damping described in [1] and [5] cannot be used. Hence, the present version of the upwind scheme is about 5 times slower in reaching the steady state than the central difference code of [1].

The multigrid version of the code has also been run using a four stage Runge-Kutta. The original parameters were reasonable but a better set is $\alpha_1=.15$, $\alpha_2=.3275$, $\alpha_3=.57$ and $\alpha_4=1$. We have also used a six stage formula with $\alpha_1=.073$, $\alpha_2=.138$, $\alpha_3=.22$, $\alpha_4=.334$, $\alpha_5=.5$ and $\alpha_6=1$. On a 64x16 mesh the multigrid version requires fewer iterations to converge. However, accounting for the extra work of the multigrid the two codes have about the same convergence rate per work unit. Nevertheless, if a coarser mesh is used to initialize the finer mesh, then the total number of supersonic points is predicted withing 60 iterations on the fine mesh. Moreover, it is expected that for finer grids that the multigrid will be more efficient. The central difference multigrid code is the fastest code. However, this relies heavily on the enthalpy damping. Thus, for the Navier-Stokes equations it cannot always be used. In this case the upwind multigrid scheme will be the most efficient.

References

1. A. Jameson, W. Schmidt, E. Turkel, AIAA paper 81-1259, 1982.
2. B. Van Leer, Lecture Notes in Physics 170, 507, 1982.
3. P.L. Roe, J. Computational Physics 43, 357, 1981.
4. G.D. Van Albada, B. Van Leer, W.W. Roberts Jr., Astron. Astrophy. 108, 76, 1982.
5. E. Turkel, Acceleration to a Steady State for the Euler Euqations, to appear INRIA Workshop on Numerical Methods for the Euler Equations for Compressible Flows.
6. A. Lerat, C.R. Acad. Sci. Paris, t. 288, 1979.
7. A. Jameson, J. Appl. Mech. 50, 1052, 1983.

FAST SOLUTIONS TO THE STEADY STATE COMPRESSIBLE AND INCOMPRESSIBLE FLUID DYNAMIC EQUATIONS

Eli Turkel
Tel-Aviv University and ICASE

It is well known that for low Mach flows that the use of the compressible fluid dynamic equations is inefficient. The use of an explicit scheme requires Δt to be bounded by $1/c$. However, the physical parameters change over time scales of order $1/u$ which is much larger. Hence, it is not appropriate to use explicit schemes for highly subsonic flows. Implicit schemes are hard to vectorize and frequently do not converge quickly for highly subsonic flows. We shall demonstrate that if one is only interested in the steady state then a minor change to an existing code can greatly increase the efficiency of an explicit method. Even when using an implicit method the proposed changes increase the efficiency of the scheme. We shall first consider the Euler equations for low Mach flows and then incompressible flows. We then indicate how to generalize the method to include viscous effects. We also show how to accelerate supersonic flow by essentially decoupling the equations.

Euler Equations for Subsonic Flow

We first consider low Mach flows for rotational inviscid flow. Since the flow may be rotational we consider the Euler equations rather than the potential equation. We only consider schemes in conservation form. The use of conservation form allows the same code to be used for highly subsonic, transonic and supersonic flows.

The Euler equations, in two space dimensions, can be expressed as

$$w_t + f_x + g_y = 0 \tag{1}$$

where (x,y) represent general curvilinear coordinates. Since we are only interested in the steady state we replace (1) by the system

$$M^{-1}w_t + f_x + g_y = 0 . \tag{2}$$

The requirements on M are that the matrix be nonsingular and that the original initial boundary value problem still be well posed. It is straightforward to solve (2) with an explicit scheme. With an implicit method only the diagonal portion of the matrix to be inverted is changed. Though the code solves (2) we shall only analyze the constant coefficient problem

$$M^{-1}w_t + Aw_x + Bw_y = 0 \tag{3}$$

where the matrices M, A, B are constant.
Let $w^{(0)} = Tw$, $A_0 = T\,A\,T^{-1}$, $B_0 = T\,B\,T^{-1}$, $M_0^{-1} = T\,M^{-1}T^{-1}$, where T is chosen appropriately, Then (3) can be converted to

$$M_0^{-1}w_t^{(0)} + A_0 w_x^{(0)} + B_0 w_y^{(0)} = 0 \ , \tag{4}$$

with

$$A_0 = \begin{pmatrix} q & c & 0 & 0 \\ c & q & 0 & 0 \\ 0 & 0 & q & 0 \\ 0 & 0 & 0 & q \end{pmatrix} \qquad B_0 = \begin{pmatrix} r & 0 & c & 0 \\ 0 & r & 0 & 0 \\ c & 0 & r & 0 \\ 0 & 0 & 0 & r \end{pmatrix} \tag{5}$$

$$q = Y_y u - X_y v \qquad r = X_x v - Y_x u \tag{6}$$

q and r are the contravariant components of the velocity and (X,Y) are the Cartesian coordinates.

We now consider the case of highly subsonic flows. We wish to choose M_0^{-1} so that the eigenvalues of M_0A_0 and M_0B_0 are independent of c. We also wish to choose M_0 to be positive definite. This will imply that (4) is a symmetric hyperbolic system and so well posed. One choice is

$$M_0^{-1} = \begin{pmatrix} \frac{c^2}{z^2} & 0 & 0 & 0 \\ 0 & 1 & 0 & 0 \\ 0 & 0 & 1 & 0 \\ 0 & 0 & 0 & 1 \end{pmatrix} \tag{7}$$

where $z^2 = \max(\varepsilon,\ u^2+v^2)$. ε is introduced so that the matrix M is not singular at stagnation points. Transforming back to curvilinear coordinates we find that $M = I + dQ$ with $d = (\gamma-1)\left(\frac{1}{u^2+v^2} - \frac{1}{c^2}\right)$ and

$$Q = \begin{pmatrix} s^2 & -u & -v & 1 \\ us^2 & -u^2 & -uv & u \\ vs^2 & -uv & -v^2 & v \\ hs^2 & -uh & -vh & h \end{pmatrix} \tag{8}$$

where

$$s^2 = (u^2+v^2)/2, \quad h = c^2/(\gamma-1) + s^2 \quad . \tag{9}$$

We note that given the first row of Q the following rows are derived by multiplying the first row by u, v, h respectively. Hence the product of Q times a vector requires only six multiplications.

Let $\bar{M}^2 = z^2/c^2$ then the largest eigenvalue of $D = A \sin\theta + B \sin\phi$ is given by

$$\lambda = \frac{|w|(1+\bar{M}^2) + \sqrt{w^2(1-\bar{M}^2) + 4(a^2+b^2)z^2}}{2} \tag{10}$$

where

$$w = q \sin\theta + r \sin\phi, \quad a = Y_y \sin\theta - Y_x \sin\phi, \quad b = X_x \sin\phi - X_y \sin\theta .$$

We see that at a stagnation point $M = O(\varepsilon)$, $\lambda = O(\varepsilon)$. While at $\bar{M} = 1$, $\lambda = |w| + \sqrt{a^2+b^2}\, c$. Hence, at low Mach numbers the largest eigenvalue, and hence the time step, is independent of the sound speed c. At transonic sound speeds the largest eigenvalue is comparable to the nonconditioned case. Hence, the preconditioned problem allows a larger time step for all subsonic flows.

We next consider incompressible flow. In conservation form the system is given by

$$\begin{aligned} u_x + v_y &= 0 \\ u_t + (u^2+p)_x + (uv)_y &= 0 \\ v_t + (uv)_x + (v^2+p)_y &= 0 \end{aligned} \tag{11}$$

where p is the pressure normalized by the density. We wish to integrate this system using the artificial compressibility method [1]. If p_t/c^2 is inserted in the first equation the system is hyperbolic but not well conditioned when c is small [2]. Instead we replace (11) by

$$\frac{1}{c^2} p_t + u_x + v_y = 0$$

$$\frac{u}{c^2} p_t + u_t + (u^2+p)_x + (uv)_y = 0 \quad (12)$$

$$\frac{v}{c^2} p_t + v_t + (uv)_x + (v^2+p)_y = 0$$

c is a given nonzero function. It is evident that at the steady state both systems coincide. The new system can readily be shown to be unitarily equivalent to a symmetric hyperbolic system and so is well posed and well conditioned. The extra time derivatives can be considered as a preconditioning matrix similar to that previously considered. It remains to decide how to choose the function c. For c large the coefficient of many time derivatives is small however the time step of an explicit scheme will be large. Hence, in this case c is merely a scaling of the time scale. Instead, we wish to choose c as large as possible without the need for decreasing in any substantial way the stability criterion. In one space dimension the stability criterion for (12) is given by

$$\frac{\Delta t}{\Delta x} \leqslant \frac{u + \sqrt{u^2+4c^2}}{2} \quad . \quad (13)$$

Hence, a reasonable choice for c is

$$c^2 = \max\left(\frac{u^2+v^2}{4}, \varepsilon^2\right) \quad . \quad (14)$$

Viscous Flow

We next consider the incompressible steady state Navier-Stokes euqation

$$u_x + v_y = 0$$

$$(u^2+p)_x + (uv)_y = \frac{1}{R}(u_{xx}+u_{yy}) \quad (15)$$

$$(uv)_x + (v^2+p)_y = \frac{1}{R}(v_{xx}+v_{yy}) \quad .$$

As before we shall consider a pseudo time dependent approach to the steady state. We consider two ways of extending the previous results to include viscous effects.

The first possibility is to use a leapfrog method for the inviscid part and a Dufort-Frankel scheme for the viscous portion [1]. The inviscid part bounds the time step by $\Delta x/(u+v)$ while the viscous part is unconditionally stable. The second possibility is to use a semi implicit method based on the preconditioned system. The advection terms are treated explicitly while the viscous terms are treated implicitly. As before, the explicit part restricts the time steps inversely proportional to the velocity. The implicit part only contains linear terms. Hence, the implicit part can be inverted without any need for an iterative method.

Supersonic Flow

We next consider supersonic inviscid flow. In this case the matrices A and B can be simultaneously diaganolized by a congruence transform. Hence we can find a matrix M so that (2) is similar to a diagonal system (for the linearized constant coefficient problem). Thus, in the supersonic regime we can choose a different time step for each equation. The resultant system is still hyperbolic and well posed.

Results

We consider the flow about a NACA0012 with an inflow Mach number 0.3 and zero degree angle of attack. The mesh used is a body fitted 0 mesh and the equations (2) are integrated using a Runge-Kutta method [3]. With the preconditioning the residual is reduced by five orders of magnitude within 500 steps. This is two orders better than that of the nonconditioned system.

References

1. A. Chorin, J. Comp. Phys., 2, 12-26 (1967).
2. A. Rizzi and L.E. Eriksson, Euler Workshop, INRIA (1983).
3. A. Jameson, W. Schmidt, E. Turkel, AIAA paper 81-1259 (1981).
4. Y.H. Au-Yeung, Proc. Amer. Math. Soc., 20, 545-548 (1969).

INFLUENCE MATRIX TECHNIQUE FOR THE NAVIER-STOKES PRESSURE BOUNDARY CONDITION

F.P.H.van Beckum
Twente University of Technology
Enschede, The Netherlands

In the 8th ICFD, L. Kleiser gave a suggestion, how to deal with the pressure boundary condition in implicit integration of the time-dependent Navier-Stokes system. In the present work an application of this idea is shown and a computing time comparison with explicit integration of the momentum equations is made. The model problem is a two-dimensional flow in a rectangular channel, with the outlet over the whole cross section and the inlet halfway blocked. On inlet and outlet parabolic profiles for the horizontal velocity component u are assumed, while the vertical component v is zero; no-slip conditions on the walls.

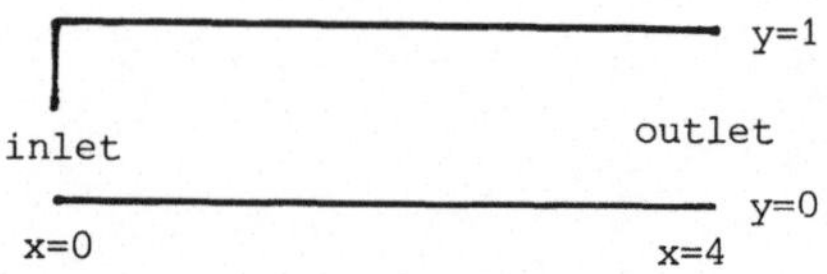

With $q \equiv p + \frac{1}{2}|\underline{u}|^2$ the time-dependent Navier-Stokes system is written:

$$\frac{\partial}{\partial t}\underline{u} + \text{grad } q = \frac{1}{Re}\Delta\underline{u} + \underline{u} \times \text{curl }\underline{u} \qquad\qquad \text{div }\underline{u} = 0$$

Explicit calculation

Every time step $\tau \equiv t_{n+1} - t_n$ the velocity is calculated from:

$$\underline{u}^{n+1} := \underline{u}^n + \tau\left(-\text{grad } q + \frac{1}{Re}\Delta\underline{u}^n + \underline{u}^n \times \text{curl }\underline{u}^n\right) \equiv \tau\left(-\text{grad } q + \underline{r}\right) \tag{1}$$

where q is the solution of a Poisson equation, derived from $\text{div }\underline{u}^{n+1} = 0$:

$$\Delta q = \text{div }\underline{r} \tag{2}$$

As $\underline{u}^{n+1}$ is given along the boundary Γ, the question how to find boundary conditions for q will not arise: integrating $\text{div }\underline{u}^{n+1} = 0$ over a grid cell adjacent to the boundary (the cell boundary being denoted by S) we find, applying (1):

$$\int_{S\cap\Gamma} \underline{u}^{n+1}\cdot\underline{n}\, ds = -\int_{S\backslash\Gamma} \underline{u}^{n+1}\cdot\underline{n}\, ds = \tau\int_{S\backslash\Gamma} (\text{grad } q - \underline{r})\cdot\underline{n}\, ds\,,$$

so $\text{grad } q \cdot \underline{n}$ along $S\cap\Gamma$ is not involved.

Semi-implicit calculation

Now we treat the diffusion term partially implicit. As the derivatives in the y-direction dominate, they are chosen for implicit treatment:

$$\begin{aligned}\underline{u}^{n+1} - \frac{\tau}{Re}\frac{\partial^2}{\partial y^2}\underline{u}^{n+1} &= \underline{u}^n + \tau\left(-\text{grad } q + \frac{1}{Re}\frac{\partial^2}{\partial x^2}\underline{u}^n + \underline{u}^n \times \text{curl }\underline{u}^n\right)\\ &\equiv \tau\left(-\text{grad } q + \underline{r}'\right)\end{aligned} \tag{3}$$

For every fixed x, this represents a tridiagonal system in y-direction, - boundary conditions $u = v = 0$ being imposed at both ends - that can easily be solved, provided

q is known. Again a Poisson equation for q is derived from the requirement $\operatorname{div} \underline{u}^{n+1} = 0$, also implying $\operatorname{div} \frac{\partial^2}{\partial y^2} \underline{u}^{n+1} = 0$, :

$$\Delta q = \operatorname{div} \underline{r}' \tag{4}$$

In this case however, the boundary treatment for q is not trivial. Proceeding as above we find:

$$\tau \int_{S\setminus\Gamma} (\operatorname{grad} q - \underline{r}')\cdot\underline{n} \, ds = \int_{S\cap\Gamma} (\underline{u}^{n+1} - \frac{\tau}{Re} \frac{\partial^2}{\partial y^2} \underline{u}^{n+1}) \, ds$$

So we need the second derivatives in y-direction, which are available at the vertical boundaries, but not at the horizontal ones. To overcome this difficulty we change over to a quite different boundary treatment (Kleiser).

Particular and homogeneous solution

The solution $(q,\underline{u}^{n+1})$ to eqs (4,3) may be thought of as the sum of a particular solution $(q_p,\underline{u}_p)$ and a homogeneous solution $(q_h,\underline{u}_h)$ to the homogeneous system:

$$\Delta q_h = 0 \tag{5}$$

$$\underline{u}_h - \frac{\tau}{Re} \frac{\partial^2}{\partial y^2} \underline{u}_h = - \tau \operatorname{grad} q_h \tag{6}$$

Boundary conditions for both $\underline{u}_p$ and $\underline{u}_h$ are zero. The question is to match the boundary conditions for q_p and q_h in such a way that the sum $\underline{u}^{n+1} = \underline{u}_p + \underline{u}_h$ satisfies $\operatorname{div} \underline{u}^{n+1} = 0$ along the horizontal boundaries H. This requirement is necessary and sufficient for $\operatorname{div} \underline{u}^{n+1}$ to vanish everywhere in the interior: as (4) holds in the interior, we find by (3) that $f \equiv \operatorname{div} \underline{u}^{n+1}$ satisfies the equation

$$f - \frac{\tau}{Re} \frac{\partial^2}{\partial y^2} f = 0 \; ;$$

together with boundary condition $f = 0$ on H, this equation has only the zero solution.

For the matching procedure we focus attention to the homogeneous system (5,6). Let n_b be the number of boundary points s_i along H in the q-problem. Assume Dirichlet conditions $q_h(s_i) = \alpha_i$, $i = 1,\ldots,n_b$. Solve (5), solve (6), and calculate $d_k' \equiv \operatorname{div} \underline{u}_h \, (s_k)$, $k = 1,\ldots,n_b$. The values d_k' are linearly dependent on α_i , which can be represented in matrix form: $d = M \alpha$; the entries m_{kj} of the influence matrix M are calculated as above: when starting with $\alpha_i = \delta_i^{\,j}$ (Kronecker delta; j fixed), the value d_k' is the matrix entry m_{kj} .

The matrix M being constructed once for all, the procedure in every time step is:

a) solve q_p from (4) with arbitrary Dirichlet conditions (e.g. from previous q_p field)

b) solve $\underline{u}_p$ from (3) and calculate $d_k = \operatorname{div} \underline{u}_p \, (s_k)$, $k = 1,\ldots,n_b$

c) solve $M \alpha = - d$

d) solve (5) with Dirichlet condition $q_h(s_i) = \alpha_i$, $i = 1,\ldots,n_b$

e) solve (6) for $\underline{u}_h$

f) add: $q = q_p + q_h$, $\underline{u}^{n+1} = \underline{u}_p + \underline{u}_h$.

Physical and numerical data for the model problem

On inlet (x=0, $0 \le y \le \frac{1}{2}$) a parabolic profile for u is assumed, with $u_{max}=1$; on outlet (x=4, $0 \le y \le 1$) the parabolic profile for u is half as high. On both inlet and outlet we take v=0. The Reynolds number is Re=100.

The grid is staggered (u and v on cell boundaries, q in cell centres) and has 32 intervals in x-direction and 16 in y-direction.

The initial velocity field is artificially chosen to be zero for $y \ge \frac{1}{2} + x/8$; for fixed x, u is parabolic between y = 0 and $y = \frac{1}{2} + x/8$, and v is adapted in accordance with the continuity equation.

The matrix for the Poisson equation in the explicit case is singular, and so is M in the semi-implicit approach; the solutions are fixed such that q is kept zero in the upper right corner.

The Poisson equations are solved with the Incomplete Cholesky Conjugate Gradient method, taking previous q-fields as starting vector. Matrix M is solved by LR-decomposition.

Results

Accuracy. Both methods are first order accurate. In the results given, the discretization error is about .0005 in q.

Stability. With respect to the stability bounds on τ the semi-implicit method is only slightly better. With τ = .1 both methods work well, with τ = .2 both methods are unstable.

Computing time. In the table results at t=2 sec are compared, reached in 20 time steps of .1 sec each. Runs were made for different accuracy eps in the ICCG-iteration. Computing times (in cpu sec on DEC 20) in the first, tenth and twentieth time step are presented, as well as the total run time for 20 steps. All implicit calculations are run with the same matrix M, constructed with eps=10^{-8} in 39 cpu sec. In most

(Notes in the table: ') not time accurate, ") tends to instability)

	eps	hbw	1^{st}step	10^{th}step	20^{th}step	total	q	divmax		
expl	10^{-6}		4.9	4.2	4.1	84	.11953	0		
s-im	10^{-6}	63	4.6	2.8	2.7	60(+39)	.11928	2.93(-8)		
		20	4.5	3.1	2.7	64(+39)	.11924	4.46(-6)	')	
		12	4.4	3.3	3.1	69(+39)	.11574	1.61(-4)	')	
expl	10^{-5}		4.5	3.8	3.8	77	.11953	0		A
s-im	10^{-5}	63	4.0	2.3	2.1	49(+39)	.11928	2.10(-7)		B
		12	3.9	2.8	2.6	56(+39)	.11574	1.61(-4)	')	
expl	10^{-4}		4.2	2.7	1.5	57	.11834	0	")	
s-im	10^{-4}	63	3.4	1.7	1.7	39(+39)	.12014	4.51(-6)		C
	10^{-3}	63	2.9	1.7	1.6	34(+39)	.11919	1.89(-5)	")	

cases the full matrix M is used (order 63), some runs are done with a restriction to banded form (hbw = half band width; hwb=63 means full matrix). To compare accuracy, the value of q in the lower left corner and the maximum continuity error in a grid cell are presented.

Remarks

Reducing M to a band matrix is not successful. The corrections α_i can work only locally, which is clearly not enough: a considerable continuity error is left (though decreasing in time) and causes more work for the next time step. On the other hand, if some inaccuracy is allowed for, it is better to maintain the full M and to raise the iteration parameter eps.

Comparison between A and B (or C) shows that the explicit method is in favour in the time range considered, but for larger time integration the semi-implicit treatment could be prefered.

Plots

As the velocity field or streamline plots are well known, we here show the pressure. Isobar patterns are given for t = 1, 2 and 6 sec respectively.

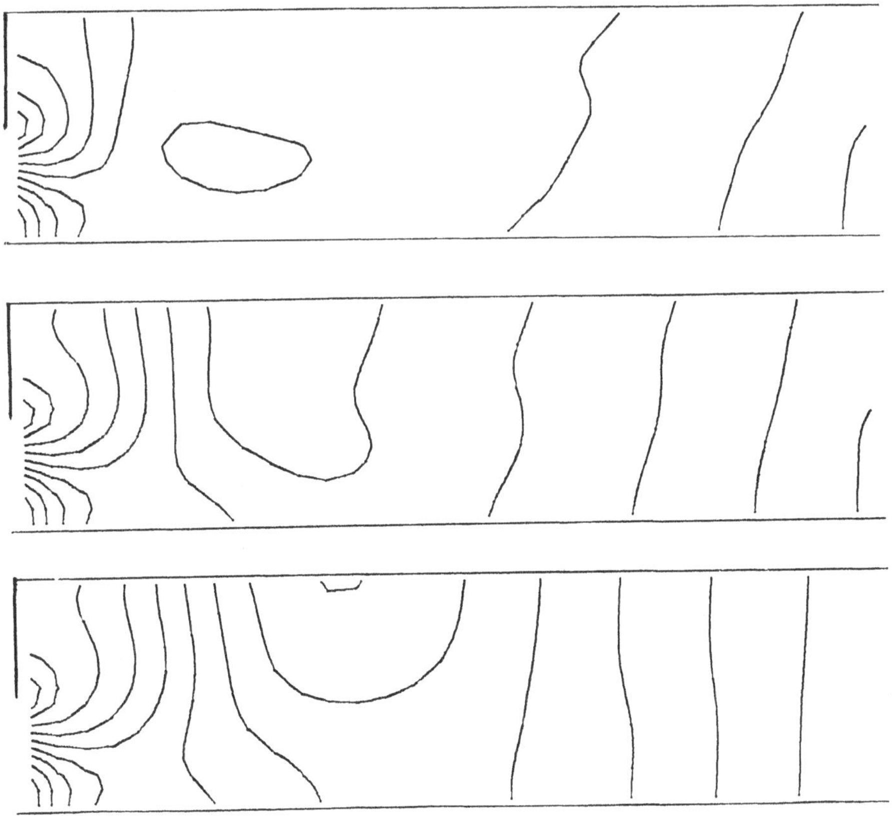

SIMULATION OF TRANSONIC SEPARATED AIRFOIL FLOW BY FINITE-DIFFERENCE VISCOUS-INVISCID INTERACTION

William R. Van Dalsem and Joseph L. Steger
Ames Research Center, NASA
Moffett Field, California

INTRODUCTION

A finite-difference viscous-inviscid interaction program has been developed for simulating the separated transonic flow about lifting airfoils, including the wake. In contrast to most interaction programs, this code combines a finite-difference boundary-layer algorithm with the inviscid program. The recently developed finite-difference boundary-layer code efficiently simulates attached and reversed compressible boundary-layer and wake flows. New viscous-inviscid interaction algorithms were also developed to couple the boundary-layer code and the inviscid transonic full-potential program (see ref. [1] for details). Transonic cases with shock-induced and trailing-edge separation are computed and compared with experimental and Navier-Stokes results.

VISCOUS ALGORITHM

The compressible boundary-layer equations for the steady, two-dimensional flow of a perfect gas are written in $\xi(x),\eta(x,y)$ coordinates:

$$\rho[u(u_\xi \xi_x + u_\eta \eta_x) + v u_\eta \eta_y] = -\beta p_\xi \xi_x + (\mu u_\eta \eta_y)_\eta \eta_y \tag{1a}$$

$$\rho c_p[u(T_\xi \xi_x + T_\eta \eta_x) + v T_\eta \eta_y] = \beta u p_\xi \xi_x + (\kappa T_\eta \eta_y)_\eta \eta_y + \mu(u_\eta \eta_y)^2 \tag{1b}$$

$$(\rho u)_\xi \xi_x + (\rho u)_\eta \eta_x + (\rho v)_\eta \eta_y = 0 \tag{1c}$$

where the equations are nondimensionalized as outlined in references [2-3]. By using a general x,y to ξ,η coordinate transformation, a complex similarity transformation is avoided and a solution-adaptive grid is easily employed. In this work, the grid height is varied as a function of the computed displacement thickness.

Near and in reversed-flow regions, the boundary-layer equations are solved in the inverse mode to avoid singular behavior at the separation point. Here the wall shear stress τ_w and the wake centerline velocity u_{wc} are used as the inverse forcing functions because they are efficiently implemented in the following numerical algorithm.

The boundary-layer equations are solved using an implicit predictor-corrector algorithm. Streamwise marching begins by predicting estimates of u, T, ρ, v, μ, and κ. This predictor step uses first-order ξ operators in the x-momentum and energy equations, but for one step produces second-order accurate values. Because the predictor step only needs to be first-order accurate, any nonlinear coefficients can be lagged in ξ. A second-order accurate corrector step is then used to calculate improved values of u, T, ρ, v, μ, and κ. During the corrector step, the nonlinear coefficients are evaluated using the most recently computed flow variables. Overall, second-order accurate solutions are obtained at the cost of two scalar bidiagonal and four scalar tridiagonal matrix inversions per streamwise station.

The algorithm is presented below in operator notation, where E is the shift operator (e.g., $E_\xi^{-1} u_j = u_{j-1}$):

$$\nabla_\xi = (1 - E_\xi^{-1})/(\Delta\xi) \qquad \Delta_\xi = (E_\xi^{+1} - 1)/(\Delta\xi)$$

$$\delta_\xi = (E_\xi^{+1} - E_\xi^{-1})/(2\Delta\xi) \qquad \bar{\delta}_\xi = (E_\xi^{+\frac{1}{2}} - E_\xi^{-\frac{1}{2}})/(\Delta\xi)$$

A predictor-step result is denoted by a tilde (e.g., $\tilde{u}$) and, unless indicated otherwise, the indices are j,k. The superscripts I, II, and III denote flow-dependent operations.

Predictor Step

Predict $\tilde{u}$ at the new station using the x-momentum equation

$$(\rho u)_{j+s'}(\xi_x \delta'_\xi \tilde{u} + \eta_x \delta_\eta \tilde{u}) + (\rho v)_{j+s'}(\eta_y \delta_\eta \tilde{u}) = -\beta \xi_x \delta_\xi p + \eta_y \bar{\delta}_\eta(\mu_{j+s'} \eta_y \bar{\delta}_\eta \tilde{u}) \quad (2a)$$

Predict $\tilde{T}$ at the new station using the energy equation

$$\rho_{j+s'} c_p [\tilde{u}(\xi_x \delta'_\xi \tilde{T} + \eta_x \delta_\eta \tilde{T}) + v_{j+s'} \eta_y \delta_\eta \tilde{T}] = \beta \tilde{u} \xi_x \delta_\xi p + \eta_y \bar{\delta}_\eta(\kappa_{j+s'} \eta_y \bar{\delta}_\eta \tilde{T}) + \mu_{j+s'}(\eta_y \delta_\eta \tilde{u})^2 \quad (2b)$$

Obtain $\tilde{\rho} = p/\tilde{T}$ and integrate the continuity equation for $\tilde{v}$

$$\nabla_\eta(\tilde{\rho}\tilde{v}) = -\frac{1 + E_\eta^{-1}}{2} \frac{\xi_x \delta'''_\xi(\tilde{\rho}\tilde{u}) + \eta_x \delta_\eta(\tilde{\rho}\tilde{u})}{\eta_y} \quad (2c)$$

The coefficients $\tilde{\mu}$ and $\tilde{\kappa}$ are then evaluated using the Cebeci turbulence model [4].

Corrector Step

Correct u at the new station using the x-momentum equation

$$\tilde{\rho}\tilde{u}(\xi_x \delta''_\xi u + \eta_x \delta_\eta u) + \tilde{\rho}\tilde{v}(\eta_y \delta_\eta u) = -\beta \xi_x \delta_\xi p + \eta_y \bar{\delta}_\eta(\tilde{\mu}\eta_y \bar{\delta}_\eta u) \quad (3a)$$

Correct T at the new station using the energy equation

$$\tilde{\rho} c_p [u(\xi_x \delta''_\xi T + \eta_x \delta_\eta T) + \tilde{v}(\eta_y \delta_\eta T)] = \beta u \xi_x \delta_\xi p + \eta_y \bar{\delta}_\eta(\tilde{\kappa}\eta_y \bar{\delta}_\eta T) + \tilde{\mu}(\eta_y \delta_\eta u)^2 \quad (3b)$$

Update ρ, v, μ, and κ as before.

Flow-dependent operators are used near reversed flow. The logic for determining the appropriate operators by specifying values of s' and s is summarized in figure 1. To incorporate the flow-dependent differencing, the operators $\delta_\xi{}'$, $\delta_\xi{}''$, and $\delta_\xi{}'''$ are defined as

$$\delta'_\xi = \begin{cases} \nabla_\xi \\ \delta_\xi \\ \Delta_\xi \end{cases} \qquad \delta''_\xi = \begin{cases} \nabla_\xi(3 - E_\xi^{-1})/2 & s = -1 \\ \delta_\xi & s = 0 \\ \Delta_\xi\left(3 - E_\xi^{+1}\right)/2 & s = 1 \end{cases} \quad (4)$$

$$\delta'''_\xi = \begin{cases} \nabla_\xi(3 - E_\xi{}^{-1})/2 & s' = -1 \\ \delta_\xi & s' = 0 \end{cases}$$

When using flow-dependent differencing values at $j+1$ and $j+2$ are obtained from a previous iterative sweep. Sweeping of the viscous flow is already required by the viscous-inviscid iterations.

The algorithm is modified to operate in the inverse mode by replacing the pressure term in the x-momentum equation with expressions containing the inverse forcing functions. These relations are obtained by applying the x-momentum equation at the wall and wake centerline. For example, at a wall ($k = 1$) the x-momentum equation yields an equality between the pressure term and the viscous term evaluated at the wall. Replacing the pressure term with a difference approximation of the viscous term in, for example, the corrector step x-momentum difference equation yields

$$\tilde{\rho}\tilde{u}(\xi_x \delta''_\xi u + \eta_x \delta_\eta u) + \tilde{\rho}\tilde{v}(\eta_y \delta_\eta u) = -2\frac{\frac{\tilde{\mu}_2 + \tilde{\mu}_1}{2}\frac{u_2}{y_2} - \tau_w}{y_2} + \eta_y \bar{\delta}_\eta(\tilde{\mu}\eta_y \bar{\delta}_\eta u) \quad (5)$$

A similar adaptation is made in the wake.

Equation (5) shows that in the inverse mode u_2 appears in the difference equation at every k index. If this term is treated implicitly, a tridiagonal-like matrix with an additional column of nonzero coefficients is obtained. This augmented scalar tridiagonal matrix system is efficiently solved using an LU decomposition algorithm (see refs. [2-3]). Because the inverse forcing functions are implemented implicitly, no iteration is required to obtain the desired τ_w or u_{wc}.

VISCOUS-INVISCID INTERFACES

If τ_w falls below a prescribed value the viscous algorithm converts from the direct to the inverse mode from that point on, including the entire wake. When operating in the inverse mode τ_w or u_{wc} must be updated so the viscous and inviscid pressures converge. A number of schemes for updating τ_w and u_{wc} were studied. The following are the fastest and most reliable of those studied and were used to obtain all the presented results:

$$\tau_w^{n+1} = \tau_w^n + \omega(p_v^n - p_i^{n+1}) \qquad \text{where:} \quad \omega = 10 \tag{6a}$$

$$u_{wc}^{n+1} = u_{wc}^n + \omega_c(p_v^n - p_i^{n+1}) \qquad \omega_c = 2 \tag{6b}$$

A transpiration velocity [5] is used to introduce the influence of the viscous region upon the inviscid flow. After δ^* has been calculated the transpiration velocity is computed and then converted to the required perturbations of the inviscid contravariant velocities used in the inviscid algorithm. This approach allows the use of inviscid grids that are not orthogonal to a body surface or wake centerline in interaction codes.

In many cases it is necessary to account for the viscous flow curvature and the pressure jump that occurs across these curved stream-tubes. Therefore, the method of accounting for the pressure variation across the shear layers developed by Lock and Firmin [6] is incorporated into the present algorithm. The reader is referred to reference [3] for details.

RESULTS

The interaction code was first tested by comparison with Navier-Stokes computations [7] of the $M_\infty = 0.720$ flow past an 18% biconvex airfoil. The agreement between the present results and the free-flight results of Levy is good in terms of both the computed pressure distributions (see fig. 2) and separation points. Also shown are experimental data [8] and the results of Levy's calculations which include the tunnel walls. It is apparent that the experimental data were affected by the tunnel walls. Experimental data and computational results are also available for this airfoil at $M_\infty = 0.754$. Experimentally, this flow can be unsteady unless a trailing-edge splitter plate is installed, which is modeled in the computations. As shown in figure 3, there is some shock-position discrepancy between the present computations and the experimental data and results of Levy; it is attributed to the wind-tunnel wall effects. Otherwise, the present results are in good agreement with the Navier-Stokes results. Both computations predict a greater trailing-edge pressure recovery than observed experimentally. The computed C_f distribution for this case (see fig. 4) indicates that shock-induced and trailing-edge separation are predicted. To indicate the convergence rate, the $C_{f_{min}}$ history for these cases is presented in figure 5. The $M_\infty = 0.720$ case has converged by the 75th iteration, whereas because of the strong shock-separated boundary-layer interaction, the $M_\infty = 0.754$ case requires approximately 160 iterations to converge.

McDevitt [9] also measured M_{peak} on the 18% biconvex (with trailing-edge splitter plate) versus M_∞. Figure 6 compares the results of the present method to this experimental data. The comparison is encouraging, especially considering the tunnel-wall effects on this data. Also, at the higher Mach numbers some of the discrepancy may be due to the isentropic inviscid flow assumption. The large difference between the inviscid and viscous peak Mach numbers is indicative of the strong viscous-inviscid interaction being modeled.

To verify the treatment of lifting airfoils, the results of computing the $M_\infty = 0.8$ flow about a NACA 0012 at $\alpha = 1°$ are compared with experimental data [10] and Navier-Stokes [11] results. Figure 7 shows good agreement between the experimental and computed C_p distributions. Both computations predict shock-induced separation from approximately 56% to 67% of chord and trailing-edge separation from 95% of chord.

As a final case, the results of computing the $M_\infty = 0.73$ flow about the RAE 2822 airfoil at $C_l = 0.803$ are presented. Figures 8-10 compare the $C_p, C_f|_e$, and δ^* distributions found experimentally [12], those computed by the present method, and those computed by Mehta [13] using a Navier-Stokes code. Mehta performed his computations at $\alpha = 2.79°$ (and computed a $C_l = 0.793$); the present computations were performed at $\alpha = 2.81°$ to match the measured $C_l = 0.803$. The present results are in good agreement with both the Navier-Stokes and experimental results. The velocity vectors near the trailing edge computed by the present method are presented in figure 11. This figure indicates the high resolution obtainable.

These simulations were obtained on 223x31 inviscid and 223x50 viscous grids. For the cases presented, the required Cray-XMP CPU time was 7 to 15 sec, and on the average, 0.0006 sec/grid point were required to obtain a converged solution. In contrast, an optimized version of the thin-layer Navier-Stokes code developed by Steger and Pulliam [11] requires about 0.06 sec/grid point to obtain a converged solution.

SUMMARY

A fast, versatile, direct-inverse finite-difference boundary-layer code has been developed and coupled to a transonic full-potential airfoil code via new viscous-inviscid interaction algorithms. The developed interaction code has been used to compute non-lifting and lifting separated transonic airfoil flows, and the results are in good agreement with experimental data and Navier-Stokes computations. Furthermore, the cost of these solutions is less than twice the cost of the inviscid solutions and close to two orders-of-magnitude less than Navier-Stokes solutions.

REFERENCES

1. Holst, T. L., **AIAA Journal**, Vol. 17, Oct. 1979, pp. 1038-1045.
2. Van Dalsem, W. R., and Steger, J. L., AIAA Paper No. 83-1689.
3. Van Dalsem W. R., Ph.D. Thesis, Stanford University, Jun. 1984.
4. Cebeci, T., AIAA Paper No. 70-741.
5. Lighthill, M. J., **Journal of Fluid Mechanics**, Vol. 4, 1958, pp. 383-392.
6. Lock, R. C., and Firmin, M. C. P., RAE Technical Memo. 1900, Apr. 1981.
7. Levy, L. L., **AIAA Journal**, Vol. 16, Jun. 1978, pp. 564-572.
8. McDevitt, J. B., Levy, L. L., and Deiwert, G. S., AIAA Paper No. 75-878.
9. McDevitt, J. B., NASA TM-78549, Jan. 1979.
10. Whitfield, J. D., et al., **Computers and Fluids**, Vol. 8, 1980, pp. 71-99.
11. Pulliam, T. H., CFD User's Workshop, Tullahoma, Tenn., Mar. 1984.
12. Cook, P. H., McDonald, M. A., and Firmin, M. C. P., AGARD AR 138, 1979.
13. Mehta, U., Second Symposium on Aerodynamic Flows, Jan. 17-20, 1983.

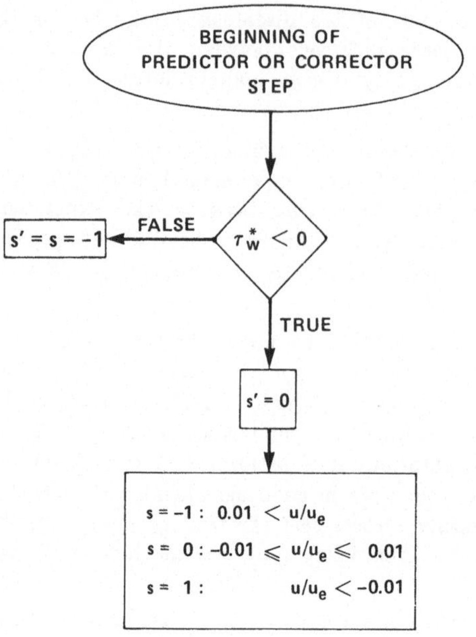

Fig. 1 Flow-dependent differencing logic.

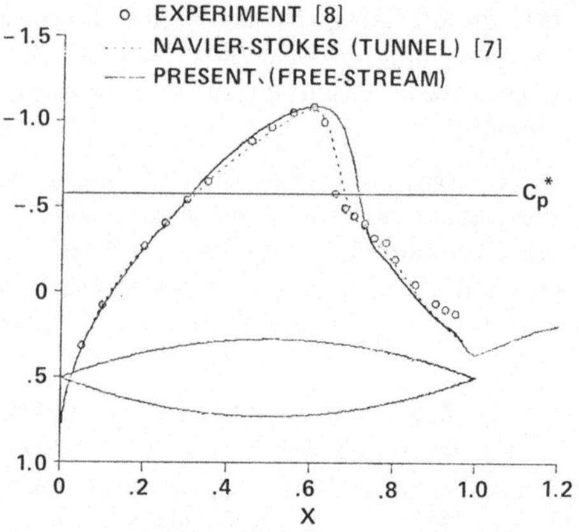

Fig. 3 C_p **distributions for an 18% biconvex at** $M_\infty = 0.754, Re_\infty = 8x10^6$.

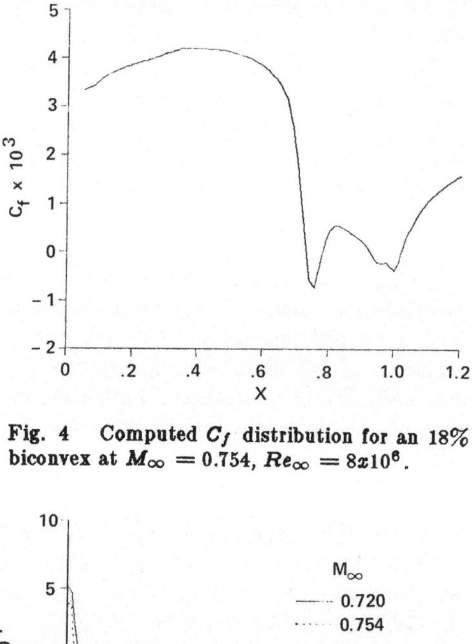

Fig. 4 Computed C_f **distribution for an 18% biconvex at** $M_\infty = 0.754$, $Re_\infty = 8x10^6$.

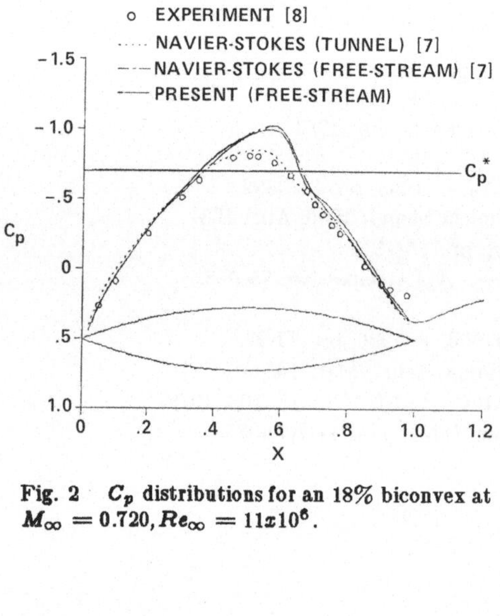

Fig. 2 C_p **distributions for an 18% biconvex at** $M_\infty = 0.720, Re_\infty = 11x10^6$.

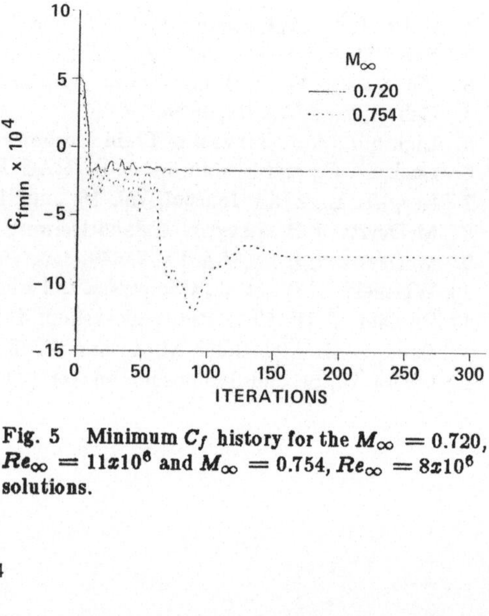

Fig. 5 Minimum C_f **history for the** $M_\infty = 0.720$, $Re_\infty = 11x10^6$ **and** $M_\infty = 0.754$, $Re_\infty = 8x10^6$ **solutions.**

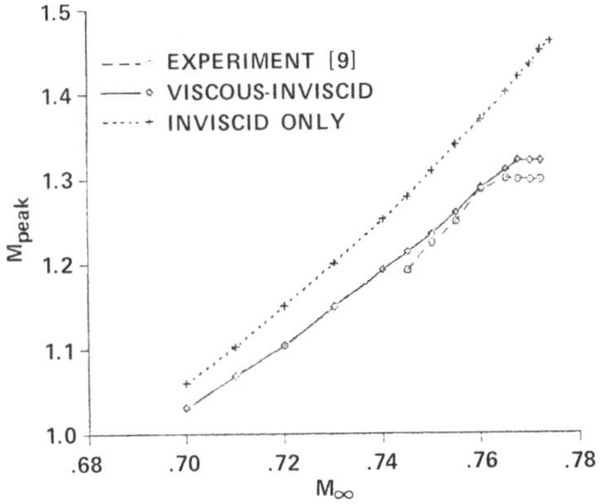

Fig. 6 M_{peak} versus M_∞ for the $Re_\infty = 8x10^6$ flow about an 18% biconvex.

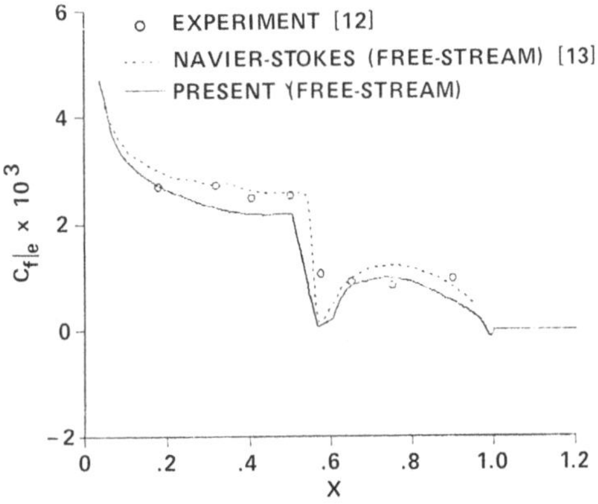

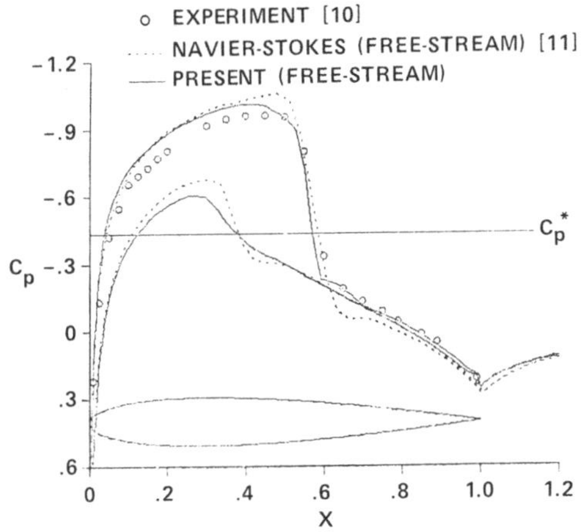

Fig. 7 C_p distributions for a NACA 0012 airfoil at $M_\infty = 0.800, Re_\infty = 2.25x10^6$, $\alpha = 1°$.

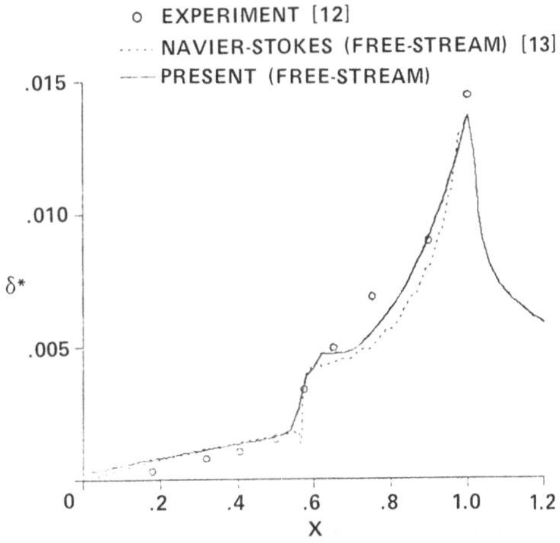

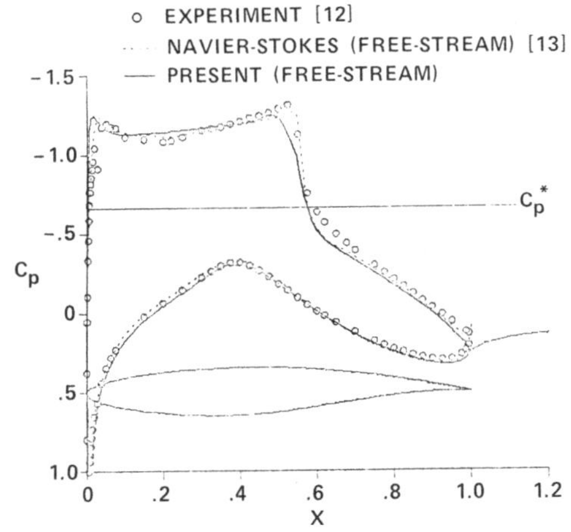

Fig. 8 C_p distributions for an RAE 2822 airfoil at $M_\infty = 0.730, Re_\infty = 6.50x10^6, C_l = 0.803$.

Fig. 9 $C_f|_e$ distributions for an RAE 2822 airfoil at $M_\infty = 0.730, Re_\infty = 6.50x10^6$, $C_l = 0.803$.

Fig. 10 δ^* distributions for an RAE 2822 airfoil at $M_\infty = 0.730, Re_\infty = 6.50x10^6, C_l = 0.803$.

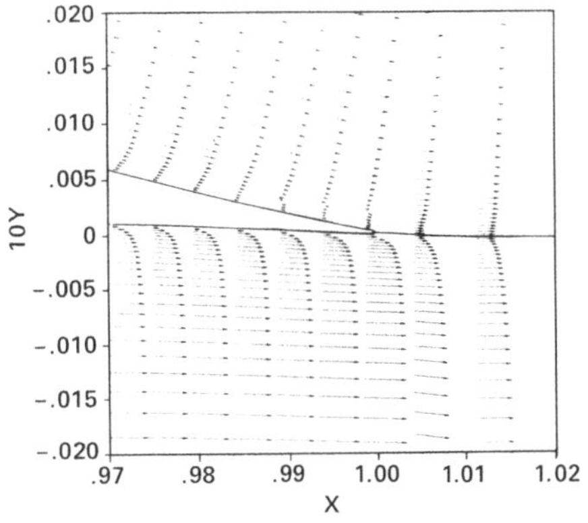

Fig. 11 Velocity vectors near the trailing edge of an RAE 2822 airfoil at $M_\infty = 0.730$, $Re_\infty = 6.50x10^6, C_l = 0.803$.

UNIVERSAL SINGLE LEVEL IMPLICIT ALGORITHM FOR GASDYNAMICS

E. Venkatapathy and C.K. Lombard
PEDA Corporation
Palo Alto, CA 94301 USA

Introduction

The Beam-Warming factored implicit algorithm[1] with the Baldwin-Lomax thin layer viscous approximation[2] has provided the basis for two similar space marching (PNS) procedures[3,4] for the compressible Navier-Stokes equations. These PNS methods which are highly efficient from the points of view of both data storage and computer time have proven effective for supersonic flows with favorable streamwise pressure gradient or with relatively small adverse pressure gradients. However, in the presence of strong adverse pressure gradient such as occurs in a wing or fin root regions, streamwise separation with flow reversal will occur. In this situation as in large elliptic regions such as occur in blunt body and base flows, the existence of eigenvalues of opposite sign render the PNS formulation ill posed and the method will diverge. Successful solution of the mixed elliptic hyperbolic problem requires global iteration with type dependent differencing. Approaches of this kind have recently been taken by Rakich[5] and by Rubin and Reddy[6] with methods that march in the predominantly streamwise direction. A new globally iterated alternating bidiagonal sweep scheme related to that which will be presented here, (but expressed in scalar equations in the Riemann variables) has been presented by Moretti[7] for the inviscid equations. Most recently, Chakravarthy[8] has presented a relaxation method based on the same diagonally dominant matrix concept advanced by Lombard, et al[9] but emphasizing unfactored effectively explicit implicit methods such as hopscotch that are less strongly coupled to boundaries than the bidiagonal approximately factored methods[10,11] of the present work.

New Universal Single Level Scheme CSCM-S

The CSCM flux difference eigenvector split upwind implicit method[12] for the inviscid terms of the compressible Navier-Stokes equations provides the natural basis for an unconditionally stable space marching technique through regions of subsonic and streamwise separated flow. In such regions the split method can be likened to stable marching of each scalar characteristic wave system in the direction of its associated eigenvalue (simple wave velocity). In supersonic flow, where all eigenvalues have the same sign, the method automatically becomes similar to the referenced PNS techniques.

Quasi 1-D Formulation

The general jth interior point difference equation for the time dependent CSCM upwind implicit method is written

$$(I + \tilde{A}^{+}\nabla + \tilde{A}^{-}\Delta)\delta q_j = -\tilde{A}^{+}\Delta q)^n_{j-1} - \tilde{A}^{-}\Delta q)^n_j \qquad (1)$$

where ∇ and Δ are backward and forward spatial difference operators. In the notation the interval averaged matrices between node points j and $j+1$ are labeled j. The right hand side of equation (1) is written for the first order method. Higher order methods in space are given with results in references 9 and 11.

Central to its accurate shock capturing capability, the CSCM conservative flux difference splitting has the "property U" put forth by Roe[13]

$$(\tilde{A}^{+} + \tilde{A}^{-})\Delta q)_j \equiv \Delta F)_j = F_{j+1} - F_j \quad . \tag{2}$$

Here q is the conservative dependent variable vector and F is the associated flux vector. The matrices $\tilde{A}^{+}$ and $\tilde{A}^{-}$ are the splittings of the CSCM interval averaged Jacobian matrix according to the signs of the averaged eigenvalues. Thus in the equation for the jth grid point, $\tilde{A}^{+}\Delta_\xi q)_{j-1}$ represents stable characteristic spatial differencing backward for positive eigenvalue contributions and $\tilde{A}^{-}\Delta_\xi q)_j$, forward for negative ones.

With $\delta q = q^{n+1} - q^n$, equation (1) defines a two level linearized coupled block tridiagonal matrix implicit scheme that can be solved by a block elimination procedure. In reference (9) a new diagonally dominant approximately factored alternating sweep bidiagonal solution procedure (DDADI) for equation (1) is presented that is shown to be very robust and is effectively explicit, i.e. requires only a decoupled sequence of local block matrix inversions rather than the solution of the coupled set.

$$D\delta q^{*}_{j} = RHS_j + \tilde{A}^{+}\delta q^{*}_{j-1} \quad , \tag{3a}$$

$$D\delta q_j = D\delta q^{*}_{j} - \tilde{A}^{-}\delta q_{j-1} \quad . \tag{3b}$$

In equations (3) $D = I + \tilde{A}^{+} - \tilde{A}^{-}$ is the central block of the unfactored tridiagonal system. For the linear problem, equation (3a) is equivalent to the single level space marching procedure

$$D\delta q^{*}_{j} = \tilde{A}^{+}q^{*}_{j-1} - \tilde{A}^{+}q^{n}_{j} - \tilde{A}^{-}\Delta q)^{n}_{j} \tag{4a}$$

Nonlinearity enters in the single level space marching form (4) in that with data updating at each step of the forward sweep the matrices $\tilde{A}^{+}$ are averaged between q^{*}_{j-1} and q^{n}_{j} rather than homogeneously at the old iteration level n. Similarly, a companion backward space marching sweep that is symmetric to equation (4a) and that is intimately related to the backward sweep of equation (3b) is

$$D\delta q_j = -\tilde{A}^{+}\Delta q^{*})_{j-1} + \tilde{A}^{-}q^{*}_{j} - \tilde{A}^{-}q^{n+1}_{j+1} \quad . \tag{4b}$$

The CSCM-S method given by equations (4) is von Neumann unconditionally stable for the scalar wave equation. Reference (9) shows heuristically the significance of retaining both sets of eigenvalues in the diagonal block D to scale the RHS contributions rendering both the forward and backward sweeps separately highly stable regardless of eigenvalue sign. Consequently as the local Courant number is taken large, the robust method becomes a very effective (symmetric Gauss-Seidel) relaxation scheme for the steady equations, a fact which substantially contributes to the extremely fast performance that will be demonstrated.

At a right computational boundary on the forward sweep we solve the characteristic boundary point approximation[9]

$$(\tilde{\tilde{A}}^{+} + \tilde{A}^{+})\delta q_N = \tilde{A}^{+} q^{*}_{N-1} - \tilde{A}^{+} q^{n}_{N} \tag{5a}$$

$q^{n+1}_{N} = q^{*}_{N}$ and at a left, on the backward sweep

$$(\tilde{\tilde{A}}^{-} - \tilde{A}^{-})\delta q_1 = \tilde{A}^{-} q^{n}_{1} - \tilde{A}^{-} q^{n+1}_{2} \quad . \tag{5b}$$

Analysis of a model system with upwind differenced scalar equations and coupled boundary conditions of the type above was related to the linearized bidiagonal scheme[9] by Oliger and Lombard[14]; the analysis also strongly supports the numerically confirmed robust stability of the present nonlinear method for gasdynamics. A final factor in the rapid convergence to steady state that the scheme affords is the use of local iteration on the solution procedure at each marching step. The iteration serves to make the eigenvectors in the coefficient matrices consistent with the advanced state and, thus, provides improved accuracy for the nonlinear system. It appears cost effective to do this inner iteration everywhere, i.e. in both subsonic and supersonic regions, as the use of two inner iterations has been found to reduce the number of global iteration steps to convergence by a factor of three to four.

Two Dimensional Formulation

For two dimensional inviscid flow the implicit block matrix equations can be approximately factored in a number of different ways[11], each leading to a scheme. A simple scheme to test in upwind implicit time dependent codes and the one for which we have obtained results is the method of lines. Then assuming a marching coordinate ξ , inviscid terms

$$(\tilde{B}^{+}\nabla_{\eta} + \tilde{B}^{-}\Delta_{\eta})\delta q \tag{6a}$$

$$-\tilde{B}^{+}\Delta_{\eta} q)_{k-1} - \tilde{B}^{-}\Delta_{\eta} q)_{k} \tag{6b}$$

are added to the left and right hand sides respectively of both the forward and backward sweep equations (4a) and (4b). For viscous flow, second centrally differenced, thin layer viscous terms are also added in the η direction as is conventionally practiced. Along each η coordinate line, one can solve the resulting coupled equations with a block tridiagonal procedure. Alternatively, a further DDADI bidiagonal approximate factorization can be employed in the η direction and solved either linearly as in reference (9) or nonlinearly as here in the ξ direction. As shown in the quasi 2-D numerical experiments of reference (9), DDADI bidiagonal approximate factorization is stable for viscous as well as inviscid terms.

References

1. Beam, R.M. and Warming, R.F., AIAA 77-645, June, 1977.
2. Baldwin, B.S. and Lomax, H., AIAA 73-257, 1973.
3. Vigneron, Y.C., Rakich, J.V. and Tannehill, J.C., NASA TM-78500, 1978.
4. Schiff, L.B. and Steger, J.L., AIAA 79-0130, Jan., 1979.
5. Rakich, J.V., AIAA 83-1955, July, 1983.
6. Rubin, S.G. and Reddy, D.R., AIAA 83-1911, July, 1983.
7. Moretti, Gino, AIAA 83-1940, 1983.
8. Chakravarthy, S.R., AIAA 84-0165, Jan., 1984.
9. Lombard, C.K., Bardina, J., Venkatapathy, E. and Oliger, J., AIAA 83-1895, 1983.
10. Lombard, C.K. and Venkatapathy, E., NASA CR 166531, Jan., 1984.
11. Lombard, C.K., Venkatapathy, E. and Bardina, J., AIAA 84-1533, June, 1984.
12. Lombard, C.K., Oliger, J. and Yang, J.Y., Lecture Notes in Physics, 170, pp. 364-370, 1982.
13. Roe, P.L., Lecture Notes in Physics, 141, pp. 354-359, 1981.
14. Oliger, Joseph and Lombard, C.K., SIAM 1983 Fall Meeting, Norfolk, Va.

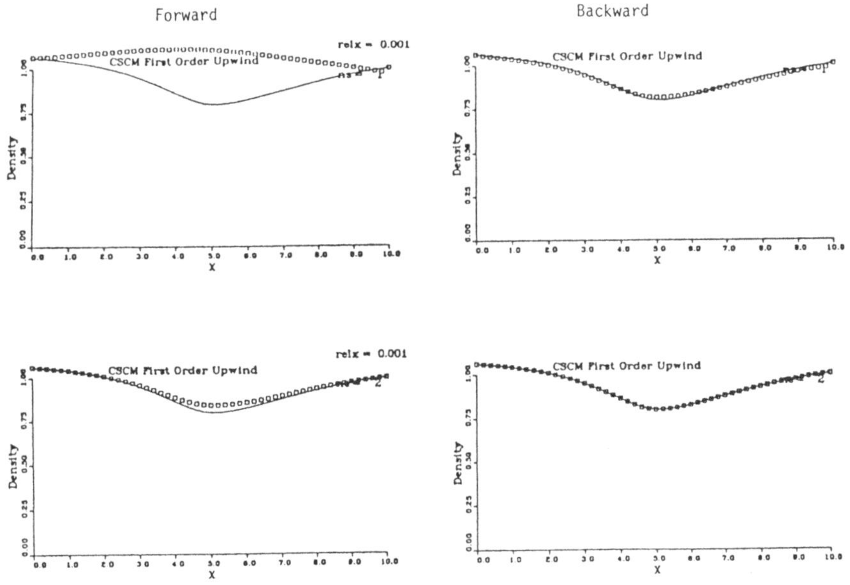

Figure 1. Subcritical solution to Blottner's converging-diverging nozzle developed with alternating sweeps in two global iteration steps. square - computed results, line - exact solution.

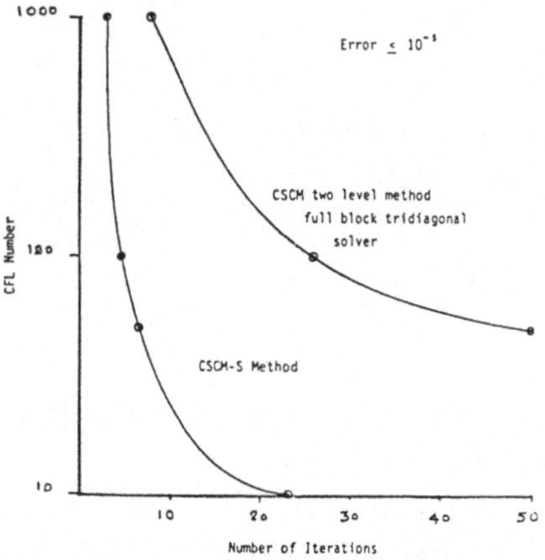

Figure 2. Convergence history comparison between the full block tridiagonal CSCM Solver and the CSCM-S method for the subcritical nozzle. The solutions have reached an RMS error less than 1 x 10^{-5}

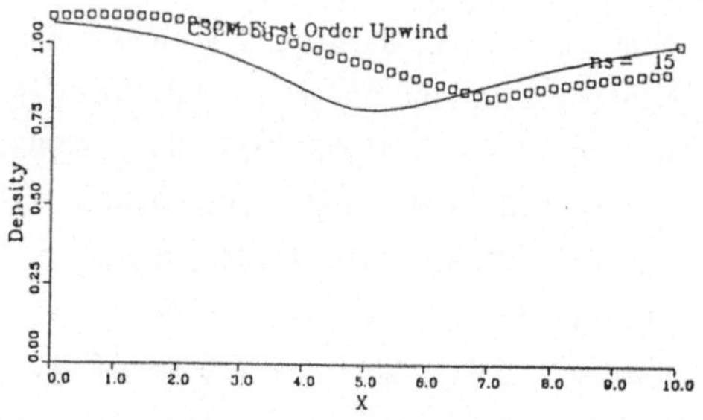

Figure 3. Solution to subcritical nozzle problem after 15 global iterations with forward (streamwise) marching sweeps only. The upstream influence has propagated only 15 mesh points showing the method without backward sweeps is explicit in the upstream waves.

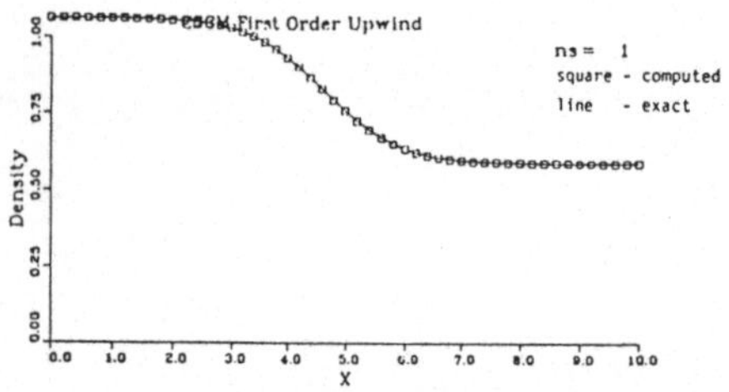

Figure 4. Shubin's diverging nozzle supersonic flow solution developed in one forward sweep from supersonic initial data.

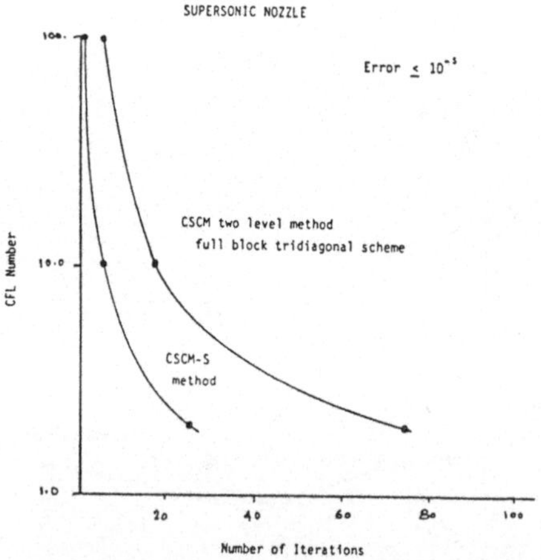

Figure 5. Convergence history comparison between the full block tridiagonal CSCM Solver and the CSCM-S method for the supersonic nozzle. The solutions have reached an RMS error less than 1 x 10^{-5}.

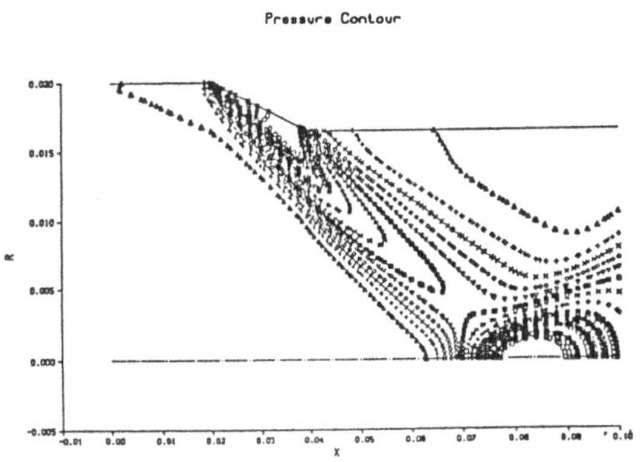

Figure 6. Pressure contours from the first order viscous solution for the inlet after 20 global sweeps with a 51 by 51 stretched mesh. Note the leading edge shock.

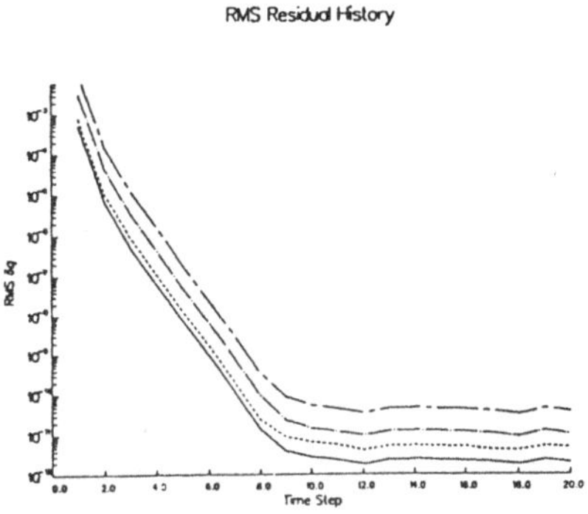

Figure 7. Convergence history of the RMS of the residuals for the inviscid first order inlet problem with 4 inner iteration per global sweep. Note at the end of the first sweep the residue has reduced and the solution has converged for all practical purposes.

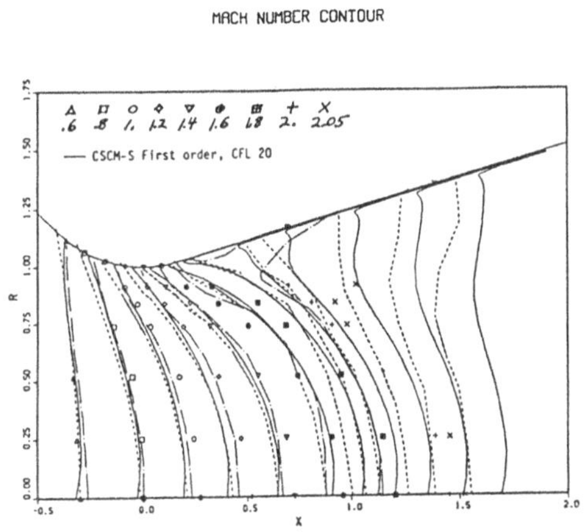

Figure 8. Mach number contours in vicinity of the throat for the axisymmetric transonic nozzle problem solved in ten global iterations. Other computed results of Cline (dashed line) and Prozan (chain dot) and experiment of Cuffel, et al (symbols) are discussed in reference 9.

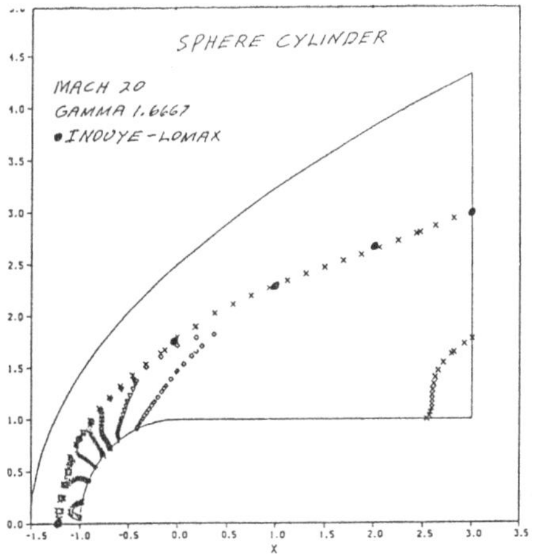

Figure 9. Pressure contours for the Mach 20 blunt body problem solved in 20 global iterations with first order CSCM-S. Inouye-Lomax data also fit experiment for shock shape.

RENORMALIZATION GROUP-BASED SUBGRID SCALE TURBULENCE CLOSURES

V. Yakhot, S. A. Orszag and R. B. Pelz
Department of Mathematics, Massachusetts Institute of Technology

Dynamic renormalization group methods are applied to the derivation of subgrid scale (SGS) closures for high Reynolds number (R) turbulence. The resulting SGS closures are attractive in that they do not require additional ad hoc treatment of wall layers and may be used to derive SGS closures for more complicated physical problems, such as flows with buoyancy effects or chemical reactions, without additional arbitrary modelling parameters. Numerical experiments with these SGS closures for turbulent channel flow at high Reynolds number show that the resulting closures give results that are in reasonable agreement with experiment.

Resolution of all space and time scales of turbulent flows is possible with present computers only at quite modest Reynolds numbers. For example, Orszag & Patera (1981) calculated turbulent channel flow at $R = 6000$ using a 64^3 grid. The calculation of flows at much higher Reynolds number requires approximations of some kind.

One way to calculate very high Reynolds number flows is by large eddy-simulation (LES) in which the eddies on scales smaller than the effective grid size are modelled, while those of grid size or larger are retained explicitly in the dynamics. Since the small scales are typically more universal than the large, the LES technique should be more faithful to the exact dynamics than Reynolds-averaging of the complete dynamical equations.

Two methods of calculating wall-bounded flows using LES ideas have been suggested. First, Deardorff (1970) and Schumann (1975) applied an LES method in the region outside the wall layers to calculate properties of turbulent shear flows. In their studies, the boundary conditions of the flow are imposed at the edge of the logarithmic layer and no attempt is made to calculate the detailed structure of the wall layers. The subgrid scale stress terms remaining after subgrid filtering are modelled using the turbulent viscosity closure due to Smagorinsky (1963)

$$\nu_{eddy} = c\Delta^2 \left| \left(\frac{\partial v_i}{\partial x_j} + \frac{\partial v_j}{\partial x_i}\right)^2 \right|^{1/2} \qquad (1)$$

where Δ is the grid size and $\vec{v}$ is the large-eddy velocity.

Present address: Princeton University, Princeton, NJ 08540

Deardorff found that c = 0.007 gave reasonable agreement with experiment. However, despite the use of wall-layer boundary conditions, Deardorff's results do not closely match wall-layer behavior at the edge of his computational domain.

More recently Moin and Kim (1982) extended LES ideas to the simulation of flows through the wall region. Their results are in quite good agreement with experiment; however, there are several ad hoc aspects to these calculations. Following Schumann, Moin and Kim included an additional term proportional to the spatially averaged stress in their SGS model. This, combined with the fact that they use horizontally averaged SGS stresses at all points in the calculation, means that their model is not very much different than a Reynolds-averaged eddy-transport model. Moin and Kim also introduced a Van Driest factor to scale the eddy-transfer coefficient near the wall, which also highlights the Reynolds-averaged eddy-transport aspects of their model. Finally, in order to avoid turbulence decay or blow-up, it was necessary to adjust their modelling constants; for example, they chose c = 0.003.

When the turbulent flow is calculated through the wall layer, it is necessary to resolve wall streaks in order to include properly the turbulent production mechanisms. Unfortunately, in order to resolve wall streaks the spatial resolution cannot be very much different from that of direct simulation (streaks require resolution of the order 500, 100 and 5 wall units in the streamwise, spanwise and normal directions respectively). More sophisticated approaches are necessary to reduce these resolution requirements.

In this paper we briefly describe a method based on the infrared dynamic renormalization group (RNG) to derive SGS models directly from the Navier-Stokes equations. The key results are:

(a) derivation of the Smagorinsky viscosity, eq (1), away from the wall layer,

(b) introduction of a random force that is large in the buffer layer and that can represent turbulent production processes (viz., the effect of random small scales on the large, resolved scales),

(c) a modified form for the eddy viscosity that accounts for the interaction of eddy and molecular viscosities in the wall layer,

(d) the resulting closures are generalizable to more complex physical flows (such as those involving bouyancy, heat transfer, and chemical reactions) without introduction of more ad hoc parameters.

We shall begin by giving a brief outline of the infrared RNG theory for homogeneous and inhomogeneous turbulence. A more detailed description of the theory may be found in Yakhot, Orszag and Pelz (1984). Then, we shall present some initial results for the application of these methods to turbulent channel flows.

RNG techniques were first applied to randomly stirred homogeneous turbulence by Forster, Nelson and Stephen (1977). Various extensions have been given by Martin and de Dominicis (1979), Yakhot (1981) and others.

The idea of the infrared RNG method is to eliminate modes from the wavenumber strip near the ultraviolet cutoff Λ. Here Λ is a wavenumber in the far dissipation range of the turbulence, so that it is justifiable to neglect dynamic degrees of freedom above Λ. Modes in the band $\Lambda e^{-\ell} < k < \Lambda$ are formally removed using diagrammatic perturbation theory, the latter being structurally similar to that introduced for homogeneous turbulence theory by Kraichnan (1961) and Wyld (1961). The system resulting from the elimination of these modes involves modified interaction coefficients, new nonlinearities, as well as modified viscosities and forces. In the RNG method, the resulting equations are then transformed to look as much as possible like the original Navier-Stokes system.

The elimination procedure just described is asymptotically exact as $\ell \to 0$. In the RNG method elimination of a finite band of modes $\Lambda e^{-\ell} < k < \Lambda$ is accomplished by iterating the above procedure of eliminating an infinitesimally narrow band of modes. The result of this iteration procedure is to generate a renormalized viscosity coefficient $\nu = \nu(\ell)$ and random force coefficient.

This infrared RNG method may be generalized to develop a SGS closure for inhomogeneous turbulent flows in finite geometries, like pipes and channels. The key assumptions in the derivation are that the renormalized scales are locally homogeneous and isotropic and belong to an inertial range characterized by the Kolmogorov $k^{-5/3}$ spectrum. This implies that the eliminated scales are much smaller than the distance z to the nearest wall, for only such scales can be isotropic.

The appropriate SGS closure is obtained by renormalizing the local rate of energy dissipation according to the procedure outlined above. The result obtained in this way is that

$$\nu_R = \nu_0[1+H(\frac{3\tilde{A}_d}{8(2\pi)^4\nu_0^3}\bar{\varepsilon}\Delta^4-C)]^{1/3} \qquad (2)$$

where

$$C = (3\tilde{A}_d/8(2\pi)^4)/\alpha^4 \approx 50,$$

$$\Delta = (\Delta_x \Delta_y \Delta_z)^{1/3}$$

$$\bar{\varepsilon} = \nu_R \left| \left(\frac{\partial v_i^<}{\partial x_j} + \frac{\partial v_j^<}{\partial x_i} \right) \frac{\partial v_i^<}{\partial x_j} \right|$$

$$H(x) = \begin{cases} 0 & x < 0 \\ x & x > 0 \end{cases}$$

and

$\tilde{A}_d$ is a function of dimension ($\approx .0118$),

α is from Kolmogorov theory (≈ 0.2),

$v^<$ is the large-eddy velocity.

Since Δ must decrease as the distance to the nearest wall (z) decreases, $\nu_R \to \nu_0$ as $z \to 0$. On the other hand, in the region far from the wall ν_R has the form of the Smagorinsky viscosity, eq. (1), with $c = .0053$.

We have written a pseudo-spectral numerical method computer code to solve the Navier-Stokes equations coupled with the RNG-SGS closure. Typical runs use 32 Fourier modes in the streamwise (x) and spanwise (y) directions and 33 Chebyshev modes in the normal (z) direction. The channel dimensions are 4π (in x), $2\pi/3$ (in y), and 2 (in z) based on channel half-width. The Reynolds number is 640 based on the wall shear velocity, channel half-width and molecular viscosity coefficient. The turbulence evolves significantly on a nondimensional time of 10. Typical runs are for a total time interval of 50.

Some results of these calculations are given in figures 1 through 3. Figure 1 shows a graph of the horizontally-averaged, renormalized viscosity as a function of z. The straight line is the molecular viscosity (1/R). The viscosity takes on the molecular value at the wall, increases and reaches a maximum for $z_+ \approx 600$, and then decreases to a minimum at mid-channel. Figure 2 shows a graph of the turbulent intensities in each direction. They compare well to experimental and other LES studies (see Moin and Kim (1982) for comparisons). The peaks in intensities of u and v in the boundary layer appear to be slightly narrower than in the experiment; this could be caused by the run time being insufficient for turbulence created at the wall to diffuse inwardly. Also the intensity of w does not attain a minimum at mid-channel but maintains nearly the same value throughout. Figure 3 shows lines of constant $u - \langle u \rangle$ ($\langle\ \rangle$ denotes horizontal average) in the plane parallel to the wall, $z_+ \approx 12.3$.

Apparent are regions of high gradient marking the sides of resolvable structures. While the resolution is not high enough to resolve the wall streaks properly, figure 3 does show the existence of strongly elongated structures.

This work was supported by the Air Force Office of Scientific Research under Contract Number F49620-83-C-0064, the Office of Naval Research under Contract Number N00014-82-C-0451, NASA Langley Research Center under Contract Number NASA-16977, and NSF Grant ATM8310210.

REFERENCES

Deardorff, J.W. 1970, J. Fluid Mech. 41, 453.
Forster, D., Nelson, D.R., and Stephen, M.J. 1977, Phys. Rev. A16, 732.
Kraichnan, R.H. 1961, J.Math Phys. 2, 124.
Martin, P.C. and de Dominicis, C. 1979, Phys. Rev. A19, 419.
Moin, P. and Kim, J. 1982, J. Fluid Mech. 118, 341.
Orszag, S.A. and Patera, A.T. 1981, Phys. Rev. Letters 47, 832.
Schumann, U. 1975, J. Comp. Phys. 18, 376.
Smagorinsky, J.S. 1963, Monthly Weather Rev. 91, 99.
Wyld, H.W. 1961, Ann Phys. (N.Y.) 14, 143.
Yakhot, V. 1981, Phys. Rev. A23, 1486.
Yakhot, V., Orszag, S.A., Pelz, R.B. 1984, submitted to J. Fluid Mech.

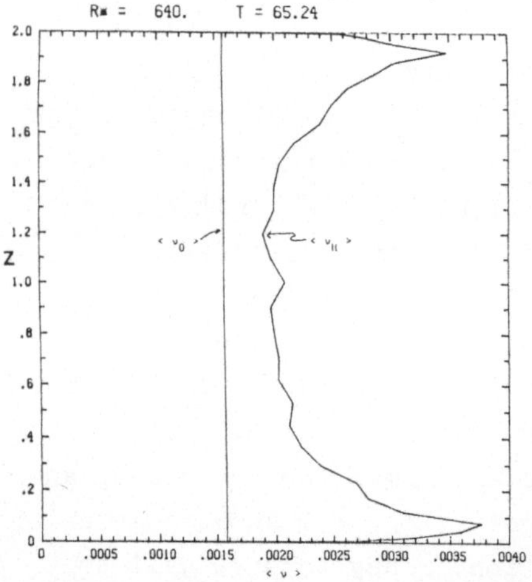

Figure 1: Horz. Avg. Viscosities

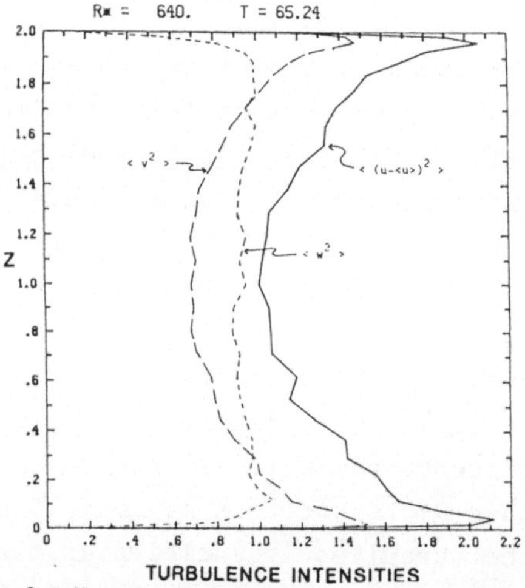

Figure 2: Horz. Avg. Turb. Intensities

Figure 3:
Contours of u - <u>
in X-Y plane at $z_+ = 12.3$
min.max velocity: -4.9/6.1
interval: 500 wall units

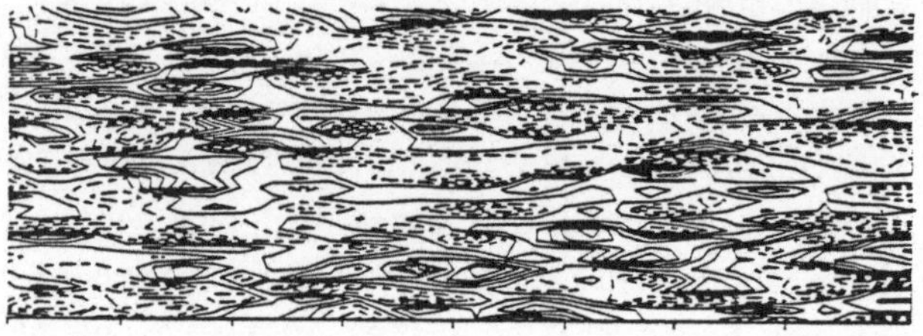

AN ITERATIVE-METHOD OF INTEGRAL RELATIONS SCHEME FOR WAKE FLOWS

W.-S. Yeung, Assistant Professor,
University of Lowell, Lowell, Massachusetts, USA
R.-J. Yang, Research Scientist,
Scientific Research Associates, Inc., Glastonbury, Connecticut, USA

INTRODUCTION

The method of Integral Relations, henceforth called MIR, was originally developed by Dorodnitsyn (1960) for laminar boundary layer flows. Since then, MIR has been applied to a wide variety of laminar flow problems with general success. An in-depth discussion on, and extensive references of, MIR is contained in Holt (1984). A crucial modification on MIR was made by Fletcher and Holt (1975) who replaced the Dorodnitsyn weighting functions by special orthonormal functions. This orthonormal MIR eliminates the complex algebra and matrix inversion associated with the original MIR. It has virtually replaced the original MIR in later numerical studies that involve MIR.

One important advantage of the orthonormal MIR is that the order of approximation can be much higher than that practically allowed by the original MIR. This enables the application of orthonormal MIR to turbulent flows as well, see Yeung and Yang (1981), and Yang and Holt (1984). However, up to now, it has been unsuccessful to apply the orthonormal MIR to wake and separated flows. Consequently, only the lowest order of approximation has been used in conjunction with the original MIR for wake and separated flows. This paper presents an efficient iterative scheme incorporating the orthonormal MIR for wake flow problems. We believe that the present study fills an important void in the application of MIR.

FORMULATION

Consider incompressible flow behind the trailing edge of a finite flat plate as shown in Figure 1. Assuming that the pressure is uniform

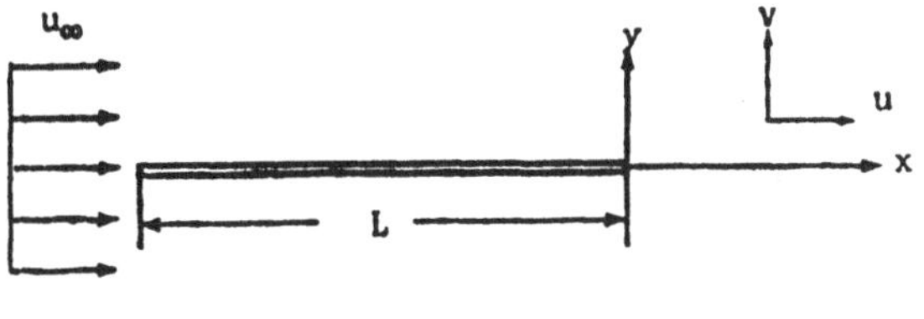

Figure 1. Notations

throughout the wake region, the governing equations are

$$\partial u/\partial x + \partial v/\partial y = 0 \qquad (1)$$

$$u\partial u/\partial x + v\partial u/\partial y = \nu\partial^2 u/\partial y^2 \qquad (2)$$

where u and v are the velocity components in the x and y directions respectively, and ν is the kinematic viscosity. The boundary conditions are

$$u = \alpha u_\infty,\ v = 0,\ \partial u/\partial y = 0 \ @\ y = 0 \qquad (3)$$

$$u = u_\infty,\ \partial u/\partial y = 0 \qquad @\ y = \infty \qquad (4)$$

where α is the nondimensional center velocity.

To understand the difficulties in applying the orthonormal MIR to wake flow, a brief account of the MIR procedure is instructive. One of the main ideas in MIR is to replace the semi-infinite y-domain by a finite u-domain, akin to the Crocco's transformation. Thus, multiply equation (1) by a weighting function $g_i(u)$ and equation (2) by $g_i'(u)$; add and integrate to yield:

$$\frac{d}{dx}\int_{\alpha u_\infty}^{u_\infty} u g_i \tilde{Z} du + [v g_i]_{\alpha u_\infty}^{u_\infty} = -\int_{\alpha u_\infty}^{u_\infty} g_i''/\tilde{Z} du \qquad (5)$$

The weighting function $g_i(u)$ is usually chosen such that $g_i(u_\infty)$ equals zero. This eliminates the variable v in equation (5). In changing from the y-domain to the u-domain, a new dependent variable, $\tilde{Z}$, arises, which is defined as

$$\tilde{Z} = (\partial u/\partial y)^{-1} \qquad (6)$$

In wake flows, $\tilde{Z}$ is represented as

$$\tilde{Z} = (\tilde{b}_0 + \sum_{j=1}^{N-2} b_j g_j)/\sqrt{u-\alpha u_\infty}\ (u_\infty - u) \qquad (7)$$

which satisfies the boundary conditions (3) and (4). N is the order of approximation. In order to yield an explicit system of ordinary differential equations represented by (5), the weighting functions are made orthonormal, in accordance with Fletcher and Holt (1975), as follows

$$\int_{\alpha u_\infty}^{u_\infty} \frac{u}{\sqrt{u-\alpha u_\infty}\ (u_\infty - u)} g_i g_j \, du = \delta_{ij} \qquad (8)$$

where δ_{ij} is the Kronecker delta. Two points can be made. Firstly, the u-domain, albeit finite, varies with x, since α varies with x*. These facts make the determination of x-independent weighting functions impossible and contribute to the failure of the orthonormal MIR in wake flows.

Thus, the rational approach is to eliminate the x-dependences in equation (8). To this end, we introduce a simple linear transformation:

$$U = (u/u_\infty - \alpha)/(1-\alpha) \qquad (9)$$

This new U-domain is now from 0 to 1, independent of x. The transformation (9) is not without problems, as we shall see.

* Secondly, the integrand in (8) depends on α.

Equations (1) and (2) are now nondimensionalized, using (9) and the following

$$V = vRe^{\frac{1}{2}}/u_\infty(1-\alpha), \quad \tilde{x} = x/L, \quad \tilde{y} = yRe^{\frac{1}{2}}/L \tag{10}$$

where L is the plate length and Re is the Reynolds number defined as

$$Re = u_\infty L/\nu \tag{11}$$

The resulting nondimensional equations are

$$\partial U/\partial\tilde{x} + \partial V/\partial\tilde{y} + (1-U)/(1-\alpha)\; d\alpha/d\tilde{x} = 0 \tag{12}$$

$$U\partial U/\partial\tilde{x} + V\partial U/\partial\tilde{y} + (\alpha/1-\alpha)\;\partial U/\partial\tilde{x}$$
$$= d\alpha/d\tilde{x}\;(1-U)/(1-\alpha)^2\cdot\;[(1-\alpha)U+\alpha] + (1/1-\alpha)\cdot\partial^2U/\partial\tilde{y}^2 \tag{13}$$

In a manner previously discussed, define a set of linearly independent weighting functions $\{f_i(U)\}$ such that

$$f_i(1) = 0, \; i = 1,2, \ldots, N \tag{14}$$

Multiply equation (12) by f_i and equation (13) by f_i'; add and integrate to yield

$$d/d\tilde{x}\int_0^1 Uf_i ZdU + \alpha d/d\tilde{x}\int_0^1 (1-U)f_i ZdU + d\alpha/d\tilde{x}\int_0^1 (1-U)\{f_i+f_i'[U+\alpha/(1-\alpha)]\}Zdu$$
$$= -\int_0^1 f_i'' dU/Z \tag{15}$$

where

$$Z = (\partial U/\partial\tilde{y})^{-1} \tag{16}$$

and is represented as

$$Z = (b_0 + \sum_1^{N-2} b_i f_i)/\sqrt{U}(1-U) \tag{17}$$

Compare equations (5) and (7) with (15) and (17) respectively. We see that equation (9) is crucial in yielding: (i) constant integration domain, and (ii) α-independent Z profile. However, it complicates the determination of f_i. Concerning the first term in equation (15), one would choose f_i to be orthonormal with respect to the weighting function $\sqrt{U}/(1-U)$; whereas concerning the second term, f_i should be orthonormal to $1/\sqrt{U}$. In general, one could not find an unique set of functions which is orthonormal to two different weighting functions.

An iterative scheme is therefore devised. Define

$$\Lambda_i = \int_0^1 (1-U)f_i ZdU \tag{18}$$

In terms of Λ_i, the second term in equation (15) can be approximated as

$$\Gamma_i = d\Lambda_i/d\tilde{x} \simeq (\Lambda_i^{\tilde{x}+\Delta\tilde{x}} - \Lambda_i^{\tilde{x}})/\Delta\tilde{x} \tag{19}$$

where $\Delta\tilde{x}$ is the integration step size. $\Lambda_i^{\tilde{x}+\Delta\tilde{x}}$ is guessed initially and iterated upon at each integration step. Thus Γ_i is known at each iteration, and equation (15) becomes

$$d/d\tilde{x}\int_0^1 Uf_i ZdU + d\alpha/d\tilde{x}\int_0^1 (1-U)\{f_i+f_i'[U+\alpha/(1-\alpha)]\}ZdU = -\int_0^1 f_i'' dU/Z - \alpha\Gamma_i \quad (20)$$

We can now define f_i as

$$\int_0^1 f_i f_j \sqrt{U}/(1-U)\,dU = \delta_{ij} \quad (21)$$

Substituting (17) into (20) and using (21), we obtain, at the kth iteration

$$a_N db_0^k/d\tilde{x} + \gamma_N^k d\alpha^k/d\tilde{x} = C_N^k - \alpha^k \Gamma_N^{k-1} \quad (22)$$

$$a_{N-1} db_0^k/d\tilde{x} + \gamma_{N-1}^k d\alpha^k/d\tilde{x} = C_{N-1}^k - \alpha^k \Gamma_{N-1}^{k-1} \quad (23)$$

$$a_i db_0^k/d\tilde{x} + \gamma_i^k d\alpha^k/d\tilde{x} + db_i^k/d\tilde{x} = C_i^k - \alpha^k \Gamma_i^{k-1},\ i=N-2,\ \ldots,\ 1 \quad (24)$$

where

$$a_i = \int_0^1 \sqrt{U}/(1-U) f_i dU \quad (25)$$

$$\gamma_i^k = \int_0^1 (1-U)\{f_i+f_i[U+\alpha^k/(1-\alpha^k)]\}Z^k dU \quad (26)$$

$$C_i^k = -\int_0^1 f_i''\ dU/Z^k \quad (27)$$

and the superscripts k and k-1 respectively denote the current and previous iteration. Equations (22) and (23) can be solved for $db_0^k/d\tilde{x}$ and $d\alpha^k/d\tilde{x}$ analytically, and then equations (24) explicitly give $db_i^k/d\tilde{x}$.

It remains to specify the weighting functions. To eliminate the singularity of equation (17) at U = 1, f_i is chosen as

$$f_i = \sum_{\ell=1}^{i} C_{i\ell}(1-U)^{\ell} \quad (28)$$

The coefficients $C_{i\ell}$ are obtained by the Gram-Schmidt process.

NUMERICAL PROCEDURES

Although some integrals in equations (25) to (27) can be integrated analytically, they are numerically evaluated by a six-point Gauss-Legendre quadrature formula for generality. The accuracy of the results obtained from the numerical integration has been found excellent when compared to those obtained from analytical integration. The system of the ordinary differential equations (22) to (24), subjected to initial conditions for α, b_0, b_1, ..., b_{N-2}, is integrated by a 4th-order Runge-Kutta method. Initial condition for α is simply taken from the centerline velocity given at an initial station. Initial conditions for b_0, b_1, ..., b_{N-2} are evaluated through $\tilde{y} = \int_0^U ZdU$ with a given initial velocity profile. For the present calculations, initial velocity profiles are taken from Rosenhead and Simpson (1936) based on Goldstein's analysis (1930) for laminar wake.

The iterative procedure is as follows. Suppose the solution at $\tilde{x}$ station is known and the integration at next station $\tilde{x}+\Delta\tilde{x}$ is to be found. The coefficients, except Γ_i, in equations (22) and (24) can be evaluated via the known solution at $\tilde{x}$. The iterative process involves the following steps: (i) Assume $(\Lambda_i^{\tilde{x}+\Delta\tilde{x}})$ and evaluate $(\Gamma_i)^{k-1}$ (ii) Integrate equation (22) to (24) with $(\Gamma_i)^{k-1}$ and obtain solutions for α^k and b_i^k's. (iii) Evaluate the new $\Lambda_i^{\tilde{x}+\Delta\tilde{x}}$ from equation (18) based on the α^k and b_i^k's. Then obtain $(\Gamma_i)^k$. (iv) If $|\Gamma_i^k-\Gamma_i^{k-1}|\leq\varepsilon$, where ε is a prescribed tolerance, the iteration is said to be convergent. Otherwise, the procedure is repeated with an improved estimate of $\Lambda_i^{\tilde{x}+\Delta\tilde{x}}$ given by

$$(\Lambda_i^{\tilde{x}+\Delta\tilde{x}})_{improved} = \omega(\Lambda_i^{\tilde{x}+\Delta\tilde{x}})_{old} + (1-\omega)(\Lambda_i^{\tilde{x}+\Delta\tilde{x}})_{new} \tag{29}$$

with ω being a relaxation constant. Our experience shows that the iteration might become divergent without using relaxation. $\omega = 0.1$ is employed at present and convergent solutions are obtained within a few iterations.

RESULTS AND DISCUSSIONS

To examine the accuracy of the present scheme, a laminar wake was first tested by using N = 2. In this case, we have analytically determined all the coefficients in equation (15) and obtained two ODE's involving α and b_0. We then combined the first two terms in equation (15). In effect, we eliminated the need for iteration. We emphasize that this approach is only practical for N = 2. Also, we can solve equations (22) to (24) as given, i.e., with iteration. Computed results obtained from these two approaches are consistent with each other. However, it is noticed that, when using the iterative approach, the accuracy of the results depends very much on the iteration tolerance, ε, and step size of integration. Figures 2 and 3 illustrate the effects. As a rule, the smaller the iteration tolerance is, the better results would be given a step size of integration.

To explore the possibility of the higher order approach of the iterative scheme, N = 3 and N = 4 were examined. Our experience showed that the iterative scheme might be divergent if we did not use relaxation during the iteration. However, this difficulty was overcome by introducing a relaxation constant mentioned previously. The number of iteration depends upon the iteration tolerance, the step size of integration and the order of approximation N. The present results were obtained within 10 iterations at each station. It took about 10 iterations at the first few integration steps and then proceeded smoothly with less than 5 iterations. Computed results for N = 3 and N = 4 were very close. Figures 4 to 6 show the comparison of numerical results obtained for N = 2 and N = 4 to Goldstein's solution.

CONCLUSION

The orthonormal MIR is extended to calculate wake flow problems. Previous difficulties are overcome by a simple linear transformation of the U-domain. An iterative scheme is devised to enable determination of unique orthonormal weighting functions. No matrix inversion is required to solve the resulting ODE's. Results are obtained for N up to 4 with relative ease. Furthermore, turbulence can easily be incorporated in the present formulation. The potential of extending the present scheme to separated flows is under study.

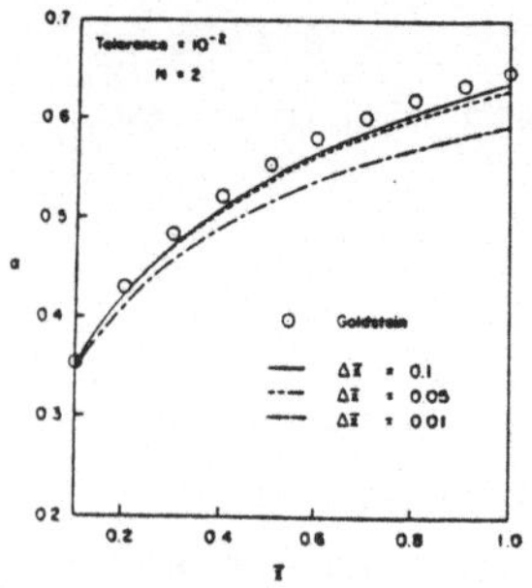

Fig. 2: Effect of step size

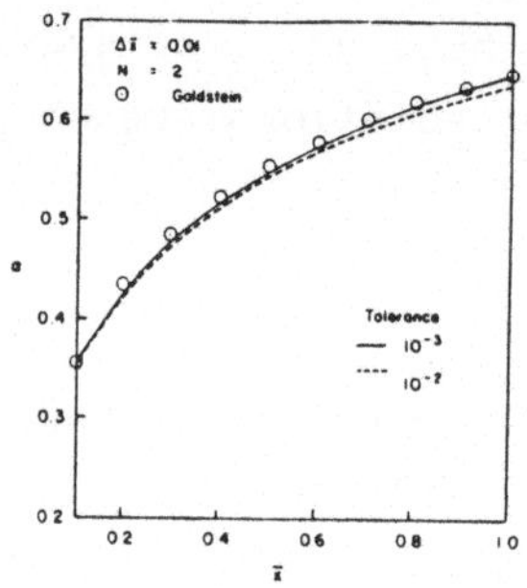

Fig. 3: Effect of ε

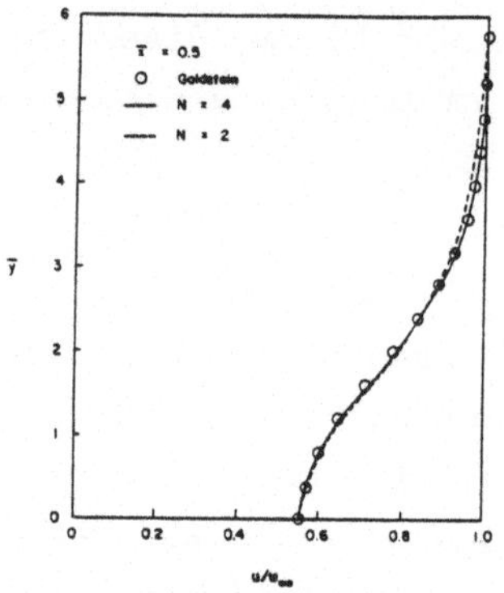

Fig. 4: Near wake profile

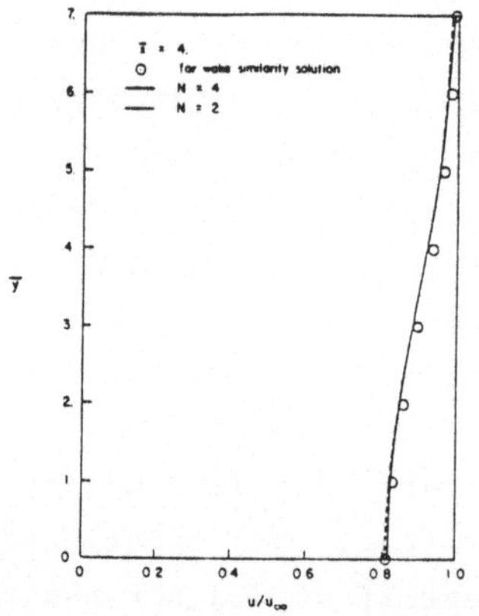

Fig. 5: Far wake profile

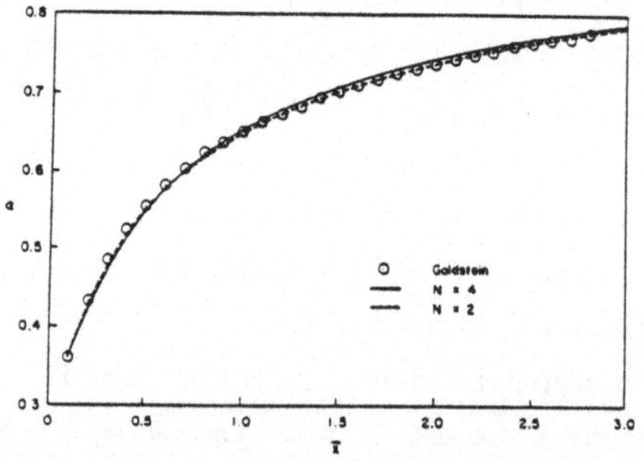

Fig. 6: Variation of center velocity

REFERENCES

Dorodnitsyn, A.A., Adv. in Aeronautical Science, 3, Pergamon Press. (1960).
Goldstein, S., Proc. Camb. Phil. Soc., 26, (1930).
Fletcher, C.A.J. and Holt, M., J. Comp. Phys. 18, (1975).
Holt, M., Numerical Methods in Fluid Dynamics, Springer-Verlag, (1984).
Rosenhead, L. and J.H. Simpson, Proc. Camb. Phil. Soc. 32, (1936).
Yang, R.-J. and M. Holt, ASME J. Appl. Mech. 51, (1984).
Yeung, W.-S. and R.-J. Yang, ASME J. App. Mech. 48, (1981).

FOURIER-LEGENDRE SPECTRAL METHODS FOR INCOMPRESSIBLE CHANNEL FLOW

Thomas A. Zang
NASA Langley Research Center
Hampton, Virginia
and
M. Yousuff Hussaini
Institute for Computer Applications in Science and Engineering
NASA Langley Research Center
Hampton, Virginia

Introduction

Fourier-Chebyshev spectral methods have been employed in several nonlinear, time-dependent investigations of transition in three-dimensional, incompressible wall-bounded shear flow ([1]-[6]). These algorithms have used a Fourier discretization in the streamwise and spanwise directions, for both of which periodic boundary conditions are presumed. The vertical direction and its no-slip boundary conditions are treated by a Chebyshev discretization. The algorithms are distinguished by the particular type of Chebyshev spectral discretization that is applied to the implicit vertical term and by how the incompressibility constraint is enforced. The Galerkin method of Moser, et al [6] incorporates the divergence-free condition into the basis functions. The tau method has been the most common approach. Incompressibility is enforced by operator splitting [1], by influence matrix techniques ([3],[4]), or by coupling the divergence-free condition with the implicit portion of the momentum equation [5]. The collocation method has been applied in connection with operator splitting [2].

In Fourier space the implicit equations for each pair of streamwise and spanwise wavenumbers are uncoupled. The Galerkin and tau methods for straight channel flow lead to tightly-banded matrices in the normal direction. If geometric curvature or coordinate stretching terms are added, the tau matrices quickly become full and the bandwidth of the Galerkin matrices rapidly increases. The collocation matrices are always full. Efficient direct inversion employs matrix diagonalization. The pre-computed transformation matrices consume a prohibitive amount of storage (and pre-processing time) unless the same transformation applies for all wavenumbers. This limits the direct collocation method to the operator splitting and influence matrix treatments of the incompressibility constraint. These methods are themselves limited to situations in which the component of the viscous term which is treated implicitly has uniform viscosity in space and time.

Here we describe an iterative collocation method which can handle implicit viscosity which varies both in the vertical direction and in time. The mean streamwise advection term may also be treated implicitly. The method is applicable to a number of wall-bounded shear flows, including channel flow, the parallel boundary layer and cylindrical Couette flow. The expansion functions in the normal direction may be any set of Jacobi polynomials, although for sufficiently fine grids Chebyshev polynomials will be more efficient because they can take advantage of the Fast Fourier Transform.

Algorithm

The basic features of the method will be outlined for a Fourier-Legendre approximation to channel flow. Let the velocities be denoted by (u,v,w) and the pressure head by p. Denote the time-step by Δt and the horizontally-averaged viscosity by $\bar{\nu}$. Let (k_x,k_y) denote the streamwise and spanwise wavenumbers and signify Fourier transformed variables by hats. The implicit momentum and incompressibility equations at each time-step have the form

$$\hat{u}^{n+1} - b\,\hat{u}_{zz}^{n+1} + ik_x\, c\,\hat{q}^{n+1} = \hat{U} \qquad (1a)$$

$$\hat{v}^{n+1} - b\,\hat{v}_{zz}^{n+1} + ik_y\, c\,\hat{q}^{n+1} = \hat{V} \qquad (1b)$$

$$\hat{w}^{n+1} - b\,\hat{w}_{zz}^{n+1} + c\,\hat{q}_z^{n+1} = \hat{W} \qquad (1c)$$

$$-ik_x\, c^*\,\hat{u}^{n+1} - ik_y\, c^*\,\hat{v}^{n+1} - c^*\,\hat{w}_z^{n+1} = 0, \qquad (1d)$$

where $b = \frac{1}{2}\bar{\nu}\Delta t$, $c = 1/\sqrt{k_x^2 + k_y^2}$, $q = (\Delta t/c)p$ and $(\hat{U},\hat{V},\hat{W})$ denotes the Fourier transforms of the explicit terms in the momentum equations. The pressure is treated by backward Euler, the horizontally-averaged vertical diffusion by Crank-Nicolson and the remaining terms by fourth-order Adams-Bashforth. The constant c is used to improve the conditioning of the implicit system of equations.

The solution to Eq. (1) is obtained by a generalization of the algorithm introduced in [7] and [8] for two-dimensional flow. The grid is staggered in the vertical direction. The velocities are defined at the Gauss-Lobatto points: $z_0 = -1$, $z_N = +1$ and z_j for $j = 1,2,\ldots,N-1$ are the roots of $P_N'(z)$--the first derivative of the last Legendre polynomial in the expansion. The pressures are defined at the usual Gauss points: $z_{j-1/2}$ for $j = 1,2,\ldots, N$ are the roots of $P_N(z)$. The momentum equations are enforced at the interior Gauss-Lobatto points, the no-slip velocity boundary conditions at $z = \pm 1$ and the incompressibility equation at the Gauss points. These conditions uniquely determine the discretized flow field at time-level n+1, except, of course, for the mean value of the pressure.

Derivatives with respect to z are evaluated by matrix multiply techniques ([9],[10]). The construction of the Legendre collocation matrices is explained in [10]. Interpolation between the velocity nodes and the pressure nodes is done spectrally, also by means of matrix multiplies. If Chebyshev polynomials are used in place of Legendre polynomials, then these matrix multiplies may be replaced with transform techniques. Although transform techniques have a smaller asymptotic operation count, matrix multiplies are still faster on a CDC CYBER 175 for N = 32, even when assembly language Chebyshev transforms are employed. Thus, Legendre polynomials are a viable alternative, especially considering that there is no particular reason to increase the number of polynomials by a factor of 2 when better resolution is desired.

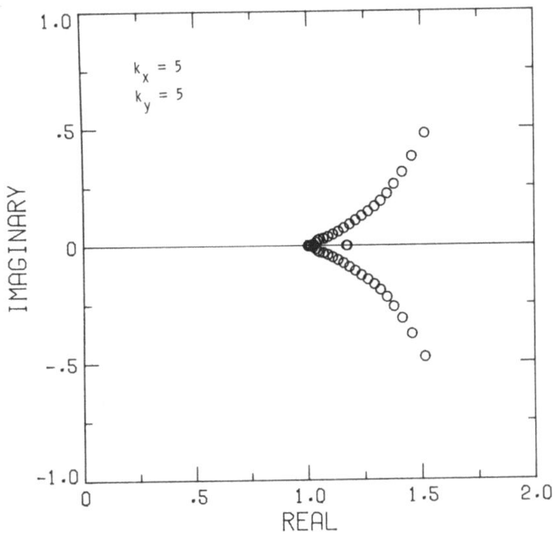

Fig. 1 - The eigenvalues of the preconditioned matrix for $k_x = k_y = 5$.

The finite-difference preconditioning described in [8] for the two-dimensional case extends to the present situation without difficulty. The eigenvalues of the preconditioned operator for a typical case are shown in figure 1. Although the eigenvalues have sizeable imaginary parts, their real parts are firmly in the right-hand plane. A Chebyshev iterative technique [11] will converge as will variational methods [12]. The former employs two parameters that depend upon the location of the eigenvalues, whereas the latter is parameter-free. The Chebyshev iterative method is insensitive to the scaling constant c, whereas for a variational method c needs to be chosen so that the Hermitian part of the preconditioned operator is positive definite. The choice of c given below Eq. (1) is based on a model problem analysis and has worked for all the problems we have done thus far. A converged solution typically requires between 5 and 10 iterations per wavenumber pair.

The major computing expense in the present Navier-Stokes algorithm is the iterative solution of the implicit equations. The finite-difference preconditioning matrix is block-tridiagonal. Our experience indicates that pivoting is not necessary if the equations are ordered properly. Thus, even this step of the algorithm is vectorizable. The relative expense of the implicit stage of this algorithm underlies our preference for an Adams-Bashforth rather than a Runge-Kutta treatment of the explicit terms. Note that the present algorithm is also capable of handling a arbitrary mean streamwise advection term implicitly. In some cases this substantially increases the maximum allowable time-step.

Results

For a demanding test of the accuracy of this algorithm we turn to low amplitude, unstable waves in plane Poiseuille flow at a Reynolds number of 7500. The initial condition for the three-dimensional Navier-Stokes code is the equilibrium parabolic streamwise velocity profile plus the growing Orr-Sommerfeld solution for a streamwise wavenumber of 1.00 and a spanwise wavenumber of 0.25. The maximum streamwise perturbation velocity is 0.01% of the mean centerline velocity. Figure 2 illustrates how rapidly the numerical results converge to the linear prediction as the vertical resolution is increased. The error in the energy ratio after 2 periods for $N_z = 32$ is less than 0.1%. These same calculations have also been performed with Cheybshev polynomials in place of Legendre polynomials. The accuracy of the Chebyshev results is comparable.

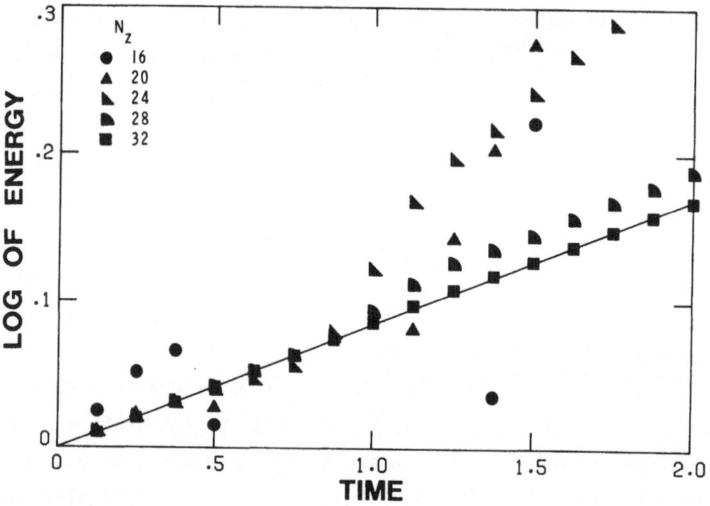

Fig. 2 - The perturbation kinetic energy relative to its initial value. The computed results are denoted by the symbols and the linear theory by the straight line. Time is measured in units of the temporal period of the low-amplitude Tollmein-Schlichting wave.

This algorithm also produces the correct nonlinear behavior of channel flow. Orszag and Patera [2] demonstrated numerically the existence of subcritical, secondary instability. Figure 3 displays the results of the present algorithm at a Reynolds number of 1500 for initial conditions consisting of the mean flow, a linearly stable two-dimensional Tollmein-Schlichting wave of 10% amplitude, as well as two linearly stable oblique three-dimensional waves of 0.1% amplitude. Both horizontal wavenumbers are 1.32. The rapid growth of the three-dimensional wave is due to nonlinear effects.

The iterative collocation algorithm can also handle boundary-layer flows under the parallel flow assumption. Our last example is an illustration of subcritical, secondary instability in a heated boundary layer at a Reynolds number of 1500. The ratio of the wall temperature to the free-stream temperature is 1.10 and both horizontal wavenumbers are 0.30. The amplitudes of the linearly stable two-dimensional and three-dimensional waves are 10% and 0.1%, respectively. The results are given in Figure 4. Note that the algorithm can also dynamically treat the temperature dependence of the viscosity.

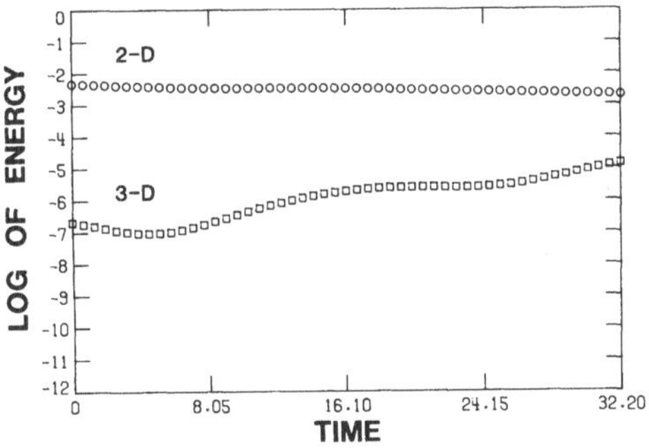

Fig. 3 - Evolution of the energy for subcritical, channel flow.

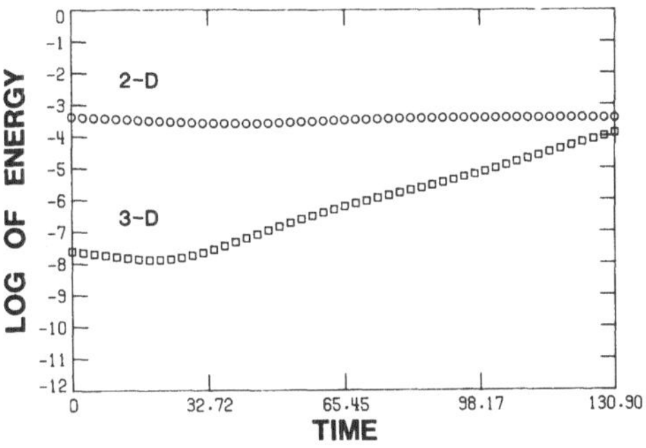

Fig. 4 - Evolution of the energy for subcritical, heated boundary layer flow.

REFERENCES

1. Orszag, S. A.; and Kells, L. C.: J. Fluid Mech., vol. 96, 1980, pp. 159-205.

2. Orszag, S. A.; and Patera, A. T.: J. Fluid Mech., vol. 128, 1983, pp. 347-385.

3. Kleiser, L.; and Schumann, U.: Proc. of 3rd GAMM Conf. on Numerical Methods in Fluid Mechanics, E. H. Hirschel, ed., Vieweg, Braunschweig, 1980, pp. 165-173.

4. Marcus, P.: Simulation of Taylor-Couette Flow--Numerical Methods and Comparison with Experiment. J. Fluid Mech., to appear.

5. Moin, P.; and Kim, J.: J. Comput. Phys., vol. 35, 1980, pp. 381-392.

6. Moser, R. D.; Moin, P.; and Leonard, A.: J. Comput. Phys., vol. 52, 1983, pp. 524-544,

7. Hussaini, M. Y.; and Zang, T. A.: Spectral Methods for Partial Differential Equations, Voigt, R. G.; Gottlieb, D.; and Hussaini, M. Y.; eds., SIAM 1984, pp. 119-140.

8. Malik, M. R.; Zang, T. A.; and Hussaini, M. Y.: A Spectral Collocation Method for the Navier-Stokes Equations, NASA Contractor Report 172365, June 1984.

9. Taylor, T. D.; Hirsh, R. S.; and Nadworny, M. M.: Comput. Fluids, vol. 12, 1984, pp. 1-9.

10. Hussaini, M. Y.; Streett, C. L.; and Zang, T. A.: Trans. 1st Army Conf. on Applied Mathematics and Computing, May 9-11, 1983, Washington, DC., pp. 883-925.

11. Manteuffel, T. A.: Numer. Math. vol. 28, 1977, pp. 307-327.

12. Eisenstat, S. C.; Elman, H. C.; and Schultz, M. H.: SIAM J. Numer. Anal., vol. 20, 1983, pp. 345-357.

ACCURATE SOLUTION OF SEVERAL COMPLICATED PROBLEMS

Y.-l. Zhu, X.-h. Wu, L.-a. Ni and Y.Wang
The Computing Center of Academia Sinica, Beijing, China

In the problems of fluid dynamics existence of discontinuities such as shocks and other singularities brings large difficulty to accurate solution of these problems. The more complicated the problems, the larger the difficulty. For example, if the equation of state is nonconvex, many schemes will not give physically relevant solutions; if the gas is combustible and there appears the transition from deflagration to detonation, it is extremely difficult to obtain an accurate result of the problem; and if a problem with a shock passing through a "strong explosion" center needs to be solved, many methods will fail because the gas temperature and the speed of shock wave tend to infinity somewhere.

We have accurately solved these problems using the singularity-separating difference method (See Zhu et al. (1980) and (1982), Wu et al. (1983) and (1984a-b)). In this paper we shall show how effective this method is through several examples.

The first example is the following initial-boundary-value problem for a single nonconvex equation:

$$\begin{cases} \dfrac{\partial u}{\partial t} + \dfrac{\partial f(u)}{\partial x} = 0 , \quad -0.001 < x < 0.001, \quad 0 < t, \\ u(x,0) = \begin{cases} 0.656 - 200(x+0.001), & -0.001 \le x < -0.0005, \\ 0.656 + 200x, & -0.0005 \le x < 0, \\ 0.014 + 170x, & 0 \le x < 0.0005, \\ 0.014 - 170(x-0.001), & 0.0005 \le x \le 0.001, \end{cases} \\ u(-0.001,t) = 0.656-4x_1(t), \\ u(0.001,t) = 0.014-3.4x_2(t). \quad 0 < t, \end{cases}$$

Here

$$f(u) = u^4/2 - 19u^3/30 + u^2/4 - 33u/1000,$$

and $x_1(t)$, $x_2(t)$ are implicitly defined by the respective relations

$$x_1(t) = tf'(0.656 - 4x_1(t)),$$
$$x_2(t) = tf'(0.014 - 3.4x_2(t)).$$

In Figure 1 the structure of solution is drawn.

In Zhu et al. (1982) several similar examples have been computed. And in Wu and Zhu (1984b) the details of the method are given. In this paper we compute the

Table I Errors of Results (t=0.066)

Schemes	$\Delta\xi$	Errors	Schemes	$\Delta\xi$	Errors
our method	1/5*	0.767×10^{-3}	Godunov Scheme	1/50	0.365×10^{-1}
				1/100	0.260×10^{-1}
	1/10*	0.19×10^{-3}		1/200	0.182×10^{-1}
				1/400	0.142×10^{-1}
	1/20*	0.48×10^{-4}		1/800	0.92×10^{-2}
E-O Scheme	1/50	0.383×10^{-1}	Lax Scheme		
	1/100	0.281×10^{-1}		1/100	0.885×10^{-1}
	1/200	0.202×10^{-1}			
	1/400	0.151×10^{-1}		1/200	0.699×10^{-1}
	1/800	0.103×10^{-1}		1/400	0.547×10^{-1}
	1/1600	0.71×10^{-2}		1/800	0.429×10^{-1}

* There are three x-subintervals.

above example using three different grids (six points, eleven points and twenty one points in each x-subinterval) in order to investigate the convergence rate of our method. In Table I we list the errors of results for every case. The data in Table I show that the convergence rate of our method is $O((\Delta t)^2)$ even for the problem with discontinuous solutions. Therefore it is quite easy to obtain very accurate results if our method is used.

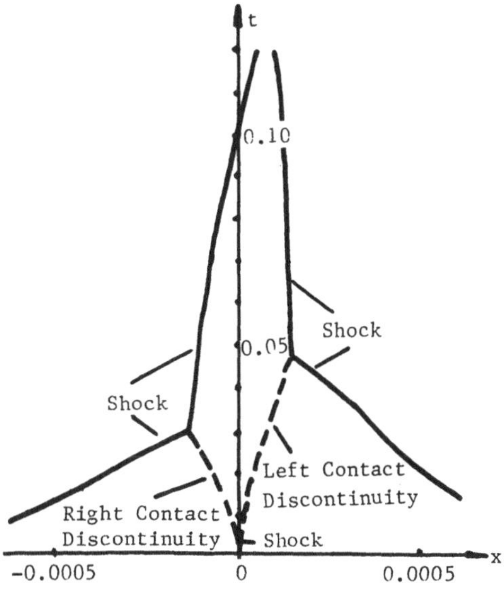

Fig. 1

By the way, we have also computed this example by using the E-O scheme, the Godunov scheme, the Lax scheme, the Murman scheme, the Courant scheme, a second-order uncentered scheme, the L-W scheme, the Richtmyer scheme and the MacCormack scheme. Unfortunately, except the E-O scheme, the Godunov scheme and the Lax scheme, all other schemes do not give physically relevant solutions. In Table I, the errors of the three schemes are listed. We know from the data in Table I that the three schemes possess nearly a convergence rate of $O(\sqrt{\Delta t})$ and that the E-O scheme and the Godunov scheme are obviously better than the Lax scheme. For more results about this example, reader is referred to Ni et al. (1984).

The second example is a combustion problem with a transition from deflagration to detonation, in which the flame is accelerated somewhere. In Figure 2 its structure of flow field is given. We know from Figure 2 that in its flow field there exist

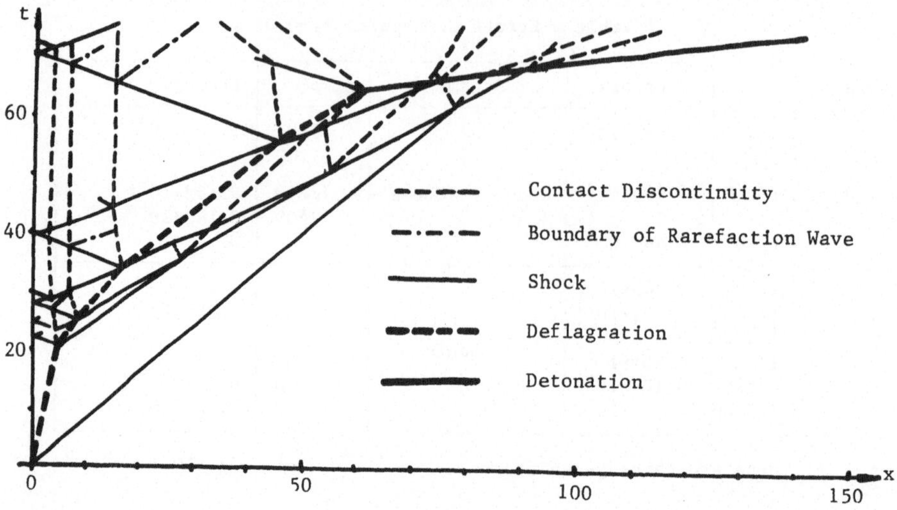

Fig.2

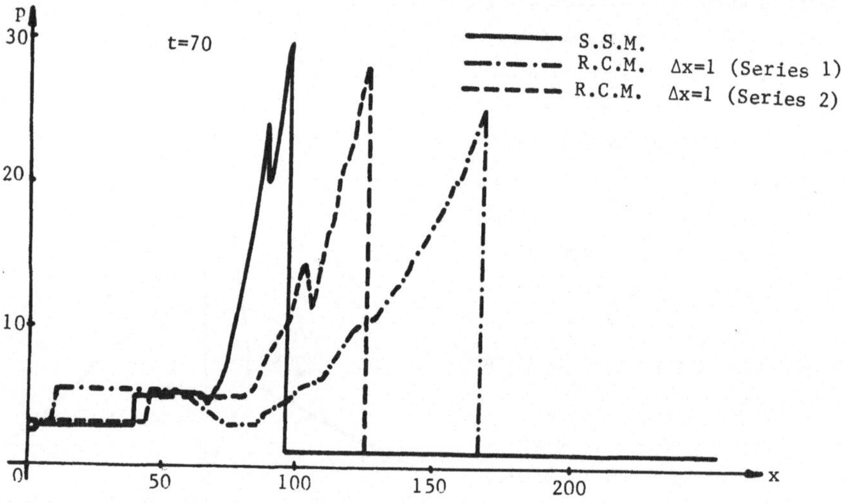

Fig.3

various discontinuities and their interactions, so its structure of flow field is very complicated and it is very difficult to obtain its accurate result. We have solved this problem using the singularity-separating method (S.S.M.). In Figure 3 we show the p-distribution obtained by S.S.M. at t=70 (Solid line), from which one can see that all the discontinuities are very sharp and there is no oscillation on our result. We have used two different grids to solve this problem. One is a grid of 4 points in each subregion and the other is a grid of 7 points in each subregion. In the case of 4 points, the maximal number of mesh points in the x-direction is 92. (At different time levels, the numbers of mesh points are different since the structures of flow field are different.) And the difference between the results obtained by using the two grids, including the difference of the locations of discontinuities,

Table II Comparison of CPU times

Methods	Numbers of mesh points or space steps	Maximal numbers of mesh points in the x direction	CPU times
R.C.M.	$\Delta x = 1$	168	11'20"
	$\Delta x=0.5$	370	60'50"
S.S.M.	4 points in each subregion	92	59"
	7 points in each subregion	161	117"

is less than 10^{-3}. That is, we have obtained very accurate results using quite a coarse grid. By the way we have compared the results and the computing time of our method with those of the random choice method (R.C.M.). In Table II the computing times are compared. In Figure 3, besides the result of S.S.M., the results of R.C.M. corresponding to different series of random numbers are given. We know from Figure 3 that both the difference between the results of S.S.M. and R.C.M. and the difference between the results corresponding to two different series of random numbers are quite large, and the former is almost as large as the latter. From this fact we think that the difference between the results of two methods is mainly caused by the error of R.C.M. These comparisons on results and computing time show that S.S.M. is quite economical and accurate. For the details of this example, reader is referred to Wu et al. (1984a).

The third example is the problem of flow with a shock wave passing through a "strong explosion" center. The concrete problem is the following. At $t=t_1 < 0, x=0.1$ there is a strong plane explosion; at t=0 the left-facing plane shock impinges upon a rigid wall and a right-facing plane reflected shock forms; then it propagates into the region disturbed by the explosion; and at $t=t_2$ the reflected wave reaches the explosion center and a new left-facing shock and an unknown singular trajectory form. In Figure 4 the structure of flow field is given. As the gas temperature on the singular trajectory and the speed of the right-facing shock at $t=t_2$ become infinity, many existing excellent methods fail. We have successfully overcame the difficulty caused by the singularities mentioned above. In our computation we introduce some new variables on the basis of analysis of the singular behavior of solution, which makes us be able to solve this problem quite accurately by using S.S.M. In Figures 5 and 6 the p-distribution and the ρ-distribution are given respectively. From them we see that the results both about shocks and singular trajectories are quite good. That is, we have obtained very good result of this problem by separating the shocks and the singular trajectory and introducing some new variables. For the details of the method, reader is referred to Wu et al. (1983).

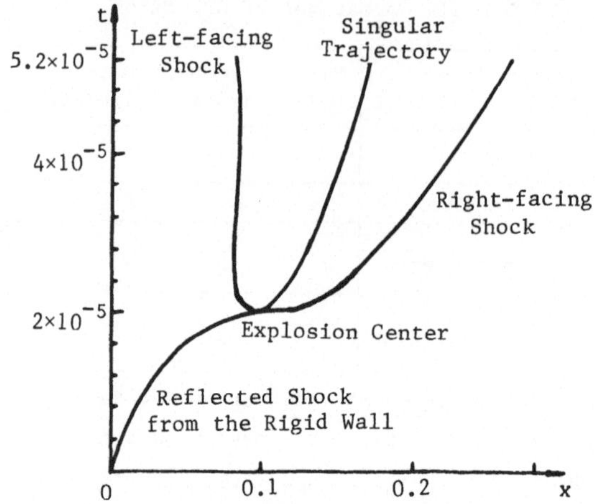

Fig.4

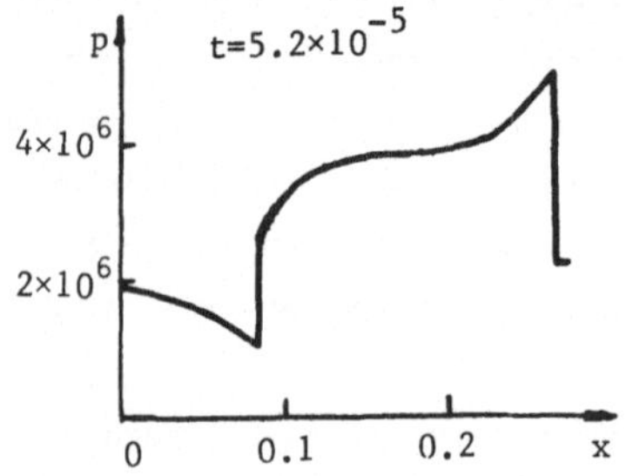

Fig. 5

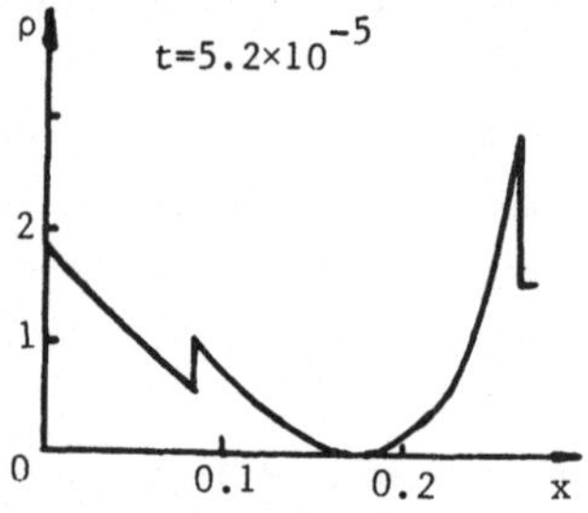

Fig.6

References

Ni, L.-a., Wu, X.-h., Wang, Y. and Zhu, Y.-l. (1984): Quantitative comparison among several difference schemes, Journal of Computational Mathematics, Vol.2, No.4 (to appear).

Wu, X.-h., Huang, D. and Zhu, Y.-l. (1983): Numerical computation of the flow with a shock wave passing through a "strong explosion" center, Journal of Computational Mathematics, Vol.1, No.3, 247-258.

Wu, X.-h., Wang, Y., Teng, Z.-h. and Zhu, Y.-l. (1984a): Numerical computation of flow field with deflagration and detonation, Journal of Computational Mathematics, Vol.2, No.3 (to appear).

Wu, X.-h., and Zhu, Y.-l. (1984b): A scheme of the singularity-separating method for the nonconvex problem, Report of the Computing Center of Academia Sinica, Beijing, China.

Zhu, Y.-l., Chen, B.-m., Wu, X.-h. and Xu, O-s. (1982): Some new developments of the singularity-separating difference method, Lecture Notes in Physics, Vol.170, Spinger-Verlag, 553-559.

Zhu, Y.-l., Zhong, X.-c., Chen, B.-m. and Zhang, Z.-m. (1980): Difference methods for initial-boundary-value problems and flow around bodies, Science Press, Beijing, China.

Selected Issues from

Lecture Notes in Mathematics